The Biology of Deer

Robert D. Brown
Editor

The Biology of Deer

With 151 Illustrations

Springer-Verlag
New York Berlin Heidelberg London Paris
Tokyo Hong Kong Barcelona Budapest

Robert D. Brown
Department of Wildlife and Fisheries
Mississippi State University
Mississippi State, MS 39762-5917 USA

Front cover photo: white-tailed deer buck, courtesy of Dr. H.A. Jacobson, Department of Wildlife and Fisheries, Mississippi State University.

Library of Congress Cataloging-in-Publication Data
The Biology of deer/Robert D. Brown, editor.
p. cm.
Proceedings of the International Symposium on the Biology of Deer, held at Mississippi State University, May 28–June 1, 1990 and hosted by the Dept. of Wildlife and Fisheries.
Includes bibliographical references and index.
ISBN 0-387-97576-4 (alk. paper)
1. Deer—Congresses. 2. Deer farming—Congresses. 3. Wildlife management—Congresses. I. Brown, Robert D. II. International Symposium on the Biology of Deer (1990 : Mississippi State University) III. Mississippi State University. Dept. of Wildlife and Fisheries.
QL737.U55B564, 1991
639.9′797357—dc20 91-9840

Printed on acid-free paper.

Production managed by Ellen Seham; manufacturing supervised by Jacqui Ashri.
Typeset by Thomson Press (India) Ltd., New Delhi, India.
Printed and bound by Edwards Brothers, Inc., Ann Arbor, MI.
Printed in the United States of America.

9 8 7 6 5 4 3 2 1

ISBN 0-387-97576-4 Springer-Verlag New York Berlin Heidelberg
ISBN 3-540-97576-4 Springer-Verlag Berlin Heidelberg New York

Preface

The first International Conference on the Biology of Deer Production was held at Dunedin, New Zealand in 1983. That meeting provided, for the first time, a forum for those with interests in either wild deer management or farmed deer production to come together. Scientists, wild deer managers, domestic deer farmers, veterinarians, venison and antler product producers, and others were able to discuss common problems and to share their knowledge and experience. The relationships formed at that meeting, and the information amassed in the resulting Proceedings, sparked new endeavors in cervid research, management, and production.

A great deal has taken place in the world of deer biology since 1983. Wild deer populations, although ever increasing in many areas of the world, face new hazards of habitat loss, environmental contamination, and overexploitation. Some species are closer to extinction than ever. Game managers often face political as well as biological challenges. Many more deer are now on farms, leading to greater concerns about disease control and increased needs for husbandry information. Researchers have accumulated considerable new information, some of it in areas such as biochemical genetics, not discussed in 1983.

This Symposium, held at Mississippi State University in 1990, easily equalled the participation and enthusiasm of the Dunedin meeting. Approximately 250 participants from 21 countries and 19 states of the United States presented 130 oral papers and posters. Ten plenary speakers gave reviews of the status of knowledge in their respective fields. Attendees participated in ten workshops, where mutual problems and needs were discussed openly. Finally, a debate on "The Commercialization of Deer" addressed the ethical and political concerns of many participating countries.

The world "Production" was dropped from the title of this Symposium and the resulting Proceedings to better reflect the comprehensive nature of the meeting. As with the Proceedings of the Dunedin conference, this Proceedings contains a wealth of up to date information which is meant to inform and to challenge deer biologists to improve management methods and to spark new research endeavors.

Robert D. Brown
Organizing Chairman

Acknowledgments

We would like to especially acknowledge the contributions of Mr. Pat Taylor of Taylor Energy Company, New Orleans, LA, and Circle Bar Ranch, Foxworth, MS, without whose support these proceedings would not have been possible.

We would also like to acknowledge the financial contributions of the following: Patrons, Safari Club International; Benefactors, The Mississippi Wildlife Federation, The North American Deer Farmers' Association, The Rocky Mountain Elk Foundation, The Samuel Roberts Noble Foundation, and The Wildlife Society; Donors, The Mississippi Chapter of the Wildlife Society; Mrs. Magalen Bryant of Tara Wildlife Management and Services, Inc.; and White-tail Institute of North America, Inc. Other support was contributed by The College of Agriculture and Home Economics, MSU; The Mississippi Agricultural and Forestry Experiment Station; The Mississippi Cooperative Extension Service; The Mississippi Department of Wildlife, Fisheries and Parks; The School of Forest Resources, MSU.

Organizing Committees for the Symposium

International Program

Robert Brown, Chairman
George Bubenik
Tim Clutton-Brock
Ken Drew
Tony English
Peter Fennessy
Alex Hawley
Peter Horn
Mike Hughes
Andrew Loudon
Mike Smith
James Suttie
Chris Wemmer
Robert White

Local Arrangements

Robert Brown, Chairman
Sheryl Brady
Bess Bragg
Marty Brunson
Will Cohen
Jeanne Hartley
Harry Jacobson
Ann Mitchell
Cindy Wasson

Manuscript Referees

Steve Albon
Robert Brown
Anthony Bubenik
George Bubenik
Ed Couvillion
Peter Dratch
Ken Drew
Anthony English
Peter Fennessy
Jerry Haigh
Alex Hawley
Eric Hellgren
Harry Jacobson
Alick Jones
David Klein
Bruce Leopold
Gerald Lincoln
Andrew Loudon
Larry Marchinton
Karl Miller
Hugh Reid
Ken Risenhoover
Don Spallanger
Mike Smith
James Suttie
Robert White

Contents

Preface v
Acknowledgments vii
Organizing Committees for the Symposium ix
Contributors xxiii

Opening Address

1 Plenary Lecture: The Status of Deer in a Changing World Environment 3
David R. Klein

Part 1: Biology of Wild Populations

2 Plenary Lecture: Cohort Variation in Reproduction and Survival: Implications for Population Demography 15
Stephen D. Albon, Timothy H. Clutton-Brock, and Rolf Langvatn

3 Habitat Relationships and Population Dynamics of Deer in the Intensively Farmed Midwestern United States 22
Charles M. Nixon and Lonnie P. Hansen

4 Current Status of Red and Fallow Deer Populations in Chile: The Need of Management 30
Claudio R. Ortiz

5 Red Deer Population Dynamics in Hungary: Management Statistics Versus Modeling 37
Sándor Csányi

6 Desert Mule Deer Mortality and the Central Arizona Project 43
Paul R. Krausman, Bruce D. Leopold, Kurt R. Rautenstrauch, John R. Morgart, and Richard C. Etchberger

7 Deer Condition Response to Changing Harvest Strategy, Davis Island, Mississippi 48
Harry A. Jacobson

8 Activity and Movement Patterns of Young White-Tailed Deer Fawns 56
Georg Schwede, Hubert Hendrichs, and Christen Wemmer

9 Fetal Growth Rate Estimation from Length and Date of Death 63
Daniel W. Soprovich

10 Neonatal Elk Habitat in Central Arizona 69
Mark C. Wallace and Paul R. Krausman

11 Recruitment and Loss in an Unhunted Subtropical White-Tailed Deer Population (Abstract) 76
Tommy R. Smith

12 Biology of a Colonizing, Unharvested Population of Elk in Western Washington (Abstract) 77
Dwayne K. Paige, Kenneth J. Raedeke, and Richard D. Taber

13 Tule Elk on Private Lands: Species Preservation, Habitat Protection, or Wildlife Commercialization? (Abstract) 78
Vernon C. Bleich and Donald B. Koch

14 The Gas Molecule Theory: Dispersal and Management of White-Tailed Deer (Abstract) 79
William F. Porter, Nancy E. Mathews, H. Brian Underwood, Richard W. Sage, and Donald F. Behrend

15 Adaptive Management of Coues White-Tailed Deer in Durango, Mexico: Is the Combined Production of Deer and Cattle Feasible? (Abstract) 80
Manuel Weber, Angeles Morales, and Carlos Galindo

16 Deer Management and Economic Development in Southern New Guinea (Abstract) 81
John Fraser Stewart

17 Stereotypic Behaviors of the Barking Deer in Nepal (Abstract) 82
Madon K. Oli and Harry A. Jacobson

18 Population Dynamics and Yield of a Dense Moose Population Without Predators (Abstract) 83
G. Cederlund and H. Sand

19 Quantification of Mule Deer Carrying Capacity in South-Central Oregon (Abstract) 84
Craig S. Bienz and Larry K. Dunsmoor

20 Spatial Component in Reproductive Performance in Desert Mule Deer (Abstract) 85
Jerry L. Cooke

21 Comparative Foraging Ecology of White-Tailed Deer and Angora Goats (Abstract) 86
John H. Roese and Ken L. Risenhoover

22 Deer/Habitat Interactions on Coronation Island, Southeast Alaska (Abstract) 87
Stephen W. Lewis and David R. Klein

23 Classification of Fallow Deer Groups According to Size Under Deer Park Conditions (Abstract) 88
Soběslav Losos and Jaroslav Madlafousek

24 Population Dynamics and Estimation of Maximum Sustained Yield (MSY) in a Black-Tailed Deer Population (Abstract) 89
Kenneth J. Raedeke

25 Resource Partitioning by Ungulates in Central Texas: A Simulation Approach (Abstract) 90
Ken L. Risenhoover and John H. Roese

Part 2: Animal Health and Disease

26 Plenary Lecture: The Biology of a Fatal Herpesvirus Infection of Deer (Malignant Catarrhal Fever) 93
Hugh W. Reid

27 Elk Disease Survey in Western Canada and Northwestern United States 101
Judit E.G. Smits

28 Vaccination to Control Brucellosis in Free-Ranging Elk on Western Wyoming Feed Grounds 107
James D. Herriges, Jr., F. Tom Thorne, and Sandra L. Anderson

29 Observations on the Relative Susceptibility to Disease of Different Species of Deer Farmed in New Zealand 113
Colin G. Mackintosh

30 The Impact of Domestication on Red Deer Immunity and Disease Resistance 120
J. Frank T. Griffin, Anthony J. Thomson, John P. Cross, Glenn S. Buchan, and Colin G. Mackintosh

31 A Review of Yersiniosis in Farmed Red Deer in New Zealand 126
Colin G. Mackintosh

32 Laboratory Assays for the Diagnosis of Tuberculosis in Farmed Red Deer (*Cervus elaphus*) 130
J. Frank T. Griffin, John P. Cross, and Glenn S. Buchan

33 Influence of Xylazine on Hematology Values in Farmed Red Deer 136
John P. Cross, J. Frank T. Griffin, and Colin G. Mackintosh

34 Monoclonal Antibodies to Leukocyte Subpopulations in Deer and Exotic Ruminants 141
Glenn Buchan, Gabrielle McCoy, Colin G. Mackintosh, and J. Frank T. Griffin

35 Social Status and Parasitic Infection in Red Deer Stags (Abstract) 149
Ludek Bartos, Pavel Rezac, Jiri Siler, Martina Buricova, and Petr Ziegrosser

36 Nonspecific Immunity in Sika Deer (Abstract) 150
E. Nivin

37 Paratuberculosis in Deer (Abstract) 151
J.F.C. Nyange, N.J.L. Gilmour, D. Buxton, and Hugh W. Reid

38 *Pasteurella*-Related Mortality in Park Fallow Deer (Abstract) 152
R. Putman, H. Davidson, and C.R. Goldspink

39 Immobilization of Norwegian and Svalbard Reindeer with Medetomidine and Medetomidine-ketamine and Reversal with Atipamezole (Abstract) 153
N.J.C. Tyler, R. Hotvedt, A.S. Blix, and D.R. Sorensen

40 Pathology of Naturally Occurring Diseases in Deer at Nandankanan Biological Park (Abstract) 154
A.T. Rao and L.N. Acharjyo

41 Causes of Mortality in Indian Deer at Nandankanan Biological Park (Abstract) 155
L.N. Acharjyo and A.T. Rao

42 Identification of 36 Strains of Mycobacteria from Deer (Abstract) 156
Wang Shuzhi, Cui Guoyin, Liu Qinghe, He Jingzhi, Wang Quarkai, and Cheng Yeguo

Part 3: Management and Behavior of Farmed Deer

43 Plenary Lecture: Requirements for Managing Farmed Deer 159
Jerry C. Haigh

44 Farming of Wapiti and Wapiti Hybrids in New Zealand 173
Anthony J. Pearse

45 Effects of Stocking Density, Feeding, and Herd Management on Mortality of Park Deer 180
Rory Putman and Jochen Langbein

46 Management Strategies for Farmed Chital Deer 189
Anthony W. English

47 Production Cycles and Characteristics of Rusa Deer in Queensland, Australia 197
Keith B. Woodford and Alison Dunning

48 Venison Production from Grazing Young Red Deer Stags: Effects of Pasture Type and Immunization Against Melatonin 203
Abu M. Ataja, Peter R. Wilson, John Hodgson, Ron M. Hoskinson, Roger W. Purchas, Hugo Varela-Alvarez, and Tom N. Barry

49 Measuring the Responses of Fallow Deer to Disturbance 211
Alick R. Jones and Sarah E. Price

50 Seasonal Pattern of Births, Female Reproductive Success and Causes of Neonate Mortality of Farmed Red Deer and Fallow Deer in Northern New Zealand (Abstract) 217
G.W. Asher and M. Langridge

51 Matching of Pasture Growth Rate, Animal Requirements, and Animal Performance to Establish Stocking Policies for Farming Red Deer Under New Zealand Grassland Farming Conditions (Abstract) 218
Peter G. Floyd

52 Profitability of Farming Red Deer in New Zealand (Abstract) 219
Peter G. Floyd

53 Responses of Farmed Red Deer to Being Handled (Abstract) 220
Sarah E. Price and Alick R. Jones

54 A Study on the Biological Characteristics of Sika Deer (Abstract) 221
Rougqian Qin and Shangren Li

55 Hungarian Red Deer and Its Potential Performance as a Farm Deer (Abstract) 222
Peter Horn and László Sugár

Part 4: Deer Products

56 Plenary Lecture: Venison and Other Deer Products 225
Ken R. Drew

57 Body Composition of Female Svalbard Reindeer: The Effects of Age and Reproductive Status 233
Nicholas J.C. Tyler

58 Evaluation of Carcass Composition Changes to Fallow Bucks Castrated Prepuberally 238
Robert C. Mulley and Anthony W. English

59 Legal Status of Exotic Deer Throughout the United States 244
R. Terry Ervin, Stephen Demarais, and David A. Osborn

60 Harvest Considerations for Free-Ranging Venison Production in Texas 253
David A. Osborn, Stephen Demarais, and David M. Hughes, Jr

Part 5: Reproduction

61 Plenary Lecture: Reproduction in Deer: Adaptations for Life in Seasonal Environments 261
Andrew S.I. Loudon and Barry R. Brinklow

62 Influence of Plane of Nutrition on Seasonal Changes in Food Intake and Reproduction in Grazing Red Deer 279
Matthew J. Heydon, Andrew S.I. Loudon, John A. Milne, and Barry R. Brinklow

63 Patterns of LH Secretion in Cycling and Seasonally Anestrous Père David's Deer Hinds 285
Bernard J. McLeod, Jonathon D. Curlewis, and Andrew S.I. Loudon

64 Reproductive Success of Female Fallow Deer in Relation to Age and Condition 293
Jochen Langbein and Rory Putman

65 Impact of Melatonin on Time of Breeding in Farmed Red Deer 300
Clare L. Adam

66 Duration of Melatonin Treatment Affects the Timing of Puberty in Young Red Deer Hinds 306
Mark W. Fisher, Lynne M. Meikle, and Peter F. Fennessy

67 Field Studies of Advancing the Breeding Season in Farmed Red Deer Using Melatonin Implants: Two Dosage Regimes and the Influence of the Stag 313
Peter R. Wilson

68 Induction of Ovulation in Père David's Deer Hinds at Two Stages of Seasonal Anestrus 320
Barry R. Brinklow, Bernard J. McLeod, Andrew S.I. Loudon, and Jonathon D. Curlewis

69 Sexual Attraction in White-Tailed Deer 327
Mark D. Whitney, Daniel L. Forster, Karl V. Miller, and R. Larry Marchinton

70 Artificial Insemination of Farmed Chital Deer 334
Gillian E. Mylrea, Anthony W. English, Robert C. Mulley, and Gareth Evans

71 Changes in Live Weight and the Reproductive Tract of Farmed Red Deer Stags from 6 to 27 Months of Age 338
James R. Webster, James M. Suttie, Barry A. Veenvliet, Ian D. Corson, and Robert E. Labes

72 Influence of Sex, Habitat, and Genotype on the Growth Patterns of White-Tailed Deer 343
Paul Leberg, Michael H. Smith, and I. Lehr Brisbin, Jr.

73 Temporal Relationship Between Estrus, the Preovulatory LH Surge, and Ovulation in Fallow Deer (Abstract) 351
G.W. Asher, M.W. Fisher, J.F. Smith, H.N. Jabbour, and C.J. Morrow

74 Reproductive Behavior of Captive Chital Deer in Australia: Seasonal Changes and a Comparison of Activity During Restricted and Unrestricted Mating Periods (Abstract) 352
Rosalie S. Chapple

75 Mating Tactics in the Male Roe Deer (Abstract) 353
Olof Liberg, Annkristin Axen, Agneta Johansson, and Kjell Wahlstrom

76 Effects of Strategic Short-term and Long-term Melatonin Treatment on Estrous Cyclicity in Fallow Deer (Abstract) 354
C.J. Morrow, G.W. Asher, and H.N. Jabbour

77 Effects of Melatonin Implants in White-Tailed Deer (Abstract) 355
Harry A. Jacobson and Stephen J. Waldhalm

78 Synchronization and Superovulation in Captive White-Tailed Deer (Abstract) 356
S.J. Magyar, C.M. Hodges, and D.C. Kraemer

79 Studies on Superovulation and Embryo Recovery in Farmed Red and Fallow Deer (Abstract) 357
H.N. Jabbour, G.W. Asher, J.G.E. Thompson, H.R. Tervit, and C.J. Morrow

80 Real-time Ultrasonography for Pregnancy Diagnosis and Fetal Age Estimation in Red Deer from Conception to 40 Days' Gestation (Abstract) 358
Brigitte Revol and Peter R. Wilson

81 Breeding Performance of Female Chinese Muntjac Deer in England (Abstract) 359
Norma G. Chapman

82 Variation in Mating Systems of Fallow Deer in Relation to Ecology (Abstract) 360
Jochen Langbein and Simon J. Thirgood

83 Variation in Conception Dates in Black-Tailed Deer in Western Washington (Abstract) 361
Kenneth J. Raedeke and Cynthia A. Kunz

84 Aspects of Placentation in White-Tailed Deer (Abstract) 362
Patrick F. Scanlon

85 Establishing Early Pregnancy by the Evaluation of the Serum Progesterone Level (Abstract) 363
Zoltan Zomborszky, László Sugár, and Tibor Feher

86 Circadian and Circannual Melatonin Secretion in Finnish Semidomesticated Reindeer (Abstract) 364
Eija Eloranta, Jouni Timisjarvi, Mauri Nieminen, Juhani Leppaluoto, and Olli Vakkuri

Part 6: Genetics

87 Plenary Lecture: Application of Biochemical Genetics to Deer Management: What the Gels Tell 367
Peter A. Dratch and Josephine M. Pemberton

88 DNA Probes for Identification of Cervidae 384
John S. Waye and Jerry C. Haigh

89 Prenatal Reproductive Losses in White-Tailed Deer 390
Olin Eugene Rhodes, Jr., Michael H. Smith, and Ronald K. Chesser

90 Genetic Variability and Secondary Productivity (Abstract) 398
Michael H. Smith and Ronald K. Chesser

91 Effect of Social Structure on Genetic Structure of White-Tailed Deer in the Central Adirondack Mountains (Abstract) 399
Nancy E. Mathews and William F. Porter

92 Establishment of Mesopotamia Fallow Deer in New Zealand and Their Hybridization with European Fallow Deer (Abstract) 400
Whitley Otway

93 Hybridization of Père David's and Red Deer (Abstract) 401
Whitley Otway

94 Genetic Variation, Parentage Testing, and Hybrid Identification in New Zealand Red Deer (Abstract) 402
Mike Tate

95 Importance of Habitat Considerations in Interpreting Electrophoretic Data for White-Tailed Deer Populations (Abstract) 403
Robert E. Vanderhoof and Raymond K. Anderson

Part 7: Nutrition and Physiology

96 Plenary Lecture: Nutrition in Relation to Season, Lactation, and Growth of North Temperate Deer 407
Robert G. White

97 Reevaluation of the Basal Metabolic Cycle in White-Tailed Deer 418
Peter J. Pekins, William W. Mautz, and John J. Kanter

98 Nutritional Strategies and Winter Survival of European Roe Deer in Norway 423
Øystein Holand and Hans Staaland

99 A Pattern of Plasma Potassium Concentrations in Adult Red Deer Stags 429
Neville B. Jopson and Peter F. Fennessy

100 Indigestible Particulate Passage in White-Tailed Deer 436
Thomas G. Barnes, Larry W. Varner, Lytle H. Blankenship, and James F. Gallagher

101 Effects of Thyroidectomy on Seasonal Patterns of Live Weight, Testicular Function, Antler Development, and Molting in Red Deer 443
Zhen D. Shi and Graham K. Barrell

102 Volatile Fatty Acids, pH, Dry Matter, and Lactic Acid of the Rumen Contents of Sika Deer 450
Shangren Li and Rongqian Qin

103 Importance of Copper Supplementation in Farmed Red Deer 453
László Sugár, Zsuzsa Lassu-Merényi, and János Kelemen

104 Effects of Dietary Protein and Energy Intake on Body Reserve Deposition and Milk Production in Caribou and Reindeer (Abstract) 456
Ann C. Allaye-Chan and Robert G. White

105 A Bioenergetics Model for Evaluating the Quality of Moose Winter Habitats (Abstract) 457
Ken L. Risenhoover and John H. Roese

106 Effects of Supplemental Feeding on White-Tailed Deer Migration in Wisconsin (Abstract) 458
Timothy L. Lewis and Orrin J. Rongstad

107 Voluntary Feed Intake and Digestion in Farmed Red Deer Stags: A Comparison with Goats and Sheep (Abstract) 459
B.M. Francoise Domingue, D.W. Dellow, P.R. Wilson, G.F. Wilson, and T.N. Barry

108 Fatty Acid Composition in the Bone Marrow Fats of Cervidae (Abstract) 460
László Sugár and Istvan Nagy

109 Urine Sampling Techniques for Semidomesticated Reindeer (Abstract) 461
Mauri Nieminen, Ulla Heiskari, and Timo Soveri

110 Annual Periodicity of Total Blood Thyroxine (T_4) in Male Fallow Deer: Problems and Experiences (Abstract) 462
Klaus Fischer

111 Seasonal Differences in Pasture Species Preferences by Red and Fallow Deer in New Zealand (Abstract) 463
W.F. Hunt and R.J.M. Hay

112 In vitro Digestibility of Fall Forages by Coexisting Deer Species in Texas (Abstract) 464
James J. Jackley, Stephen Demarais, Larry W. Varner, and Thomas G. Barnes

113 Biomechanics and Ecological Application of Escape Gaits by White-Tailed Deer, Mule Deer, and Their Hybrids (Abstract) 465
Susan Lingle

114 Intake, Digestion, and Passage Kinetics of Two Ruminants (Abstract) 466
Glen W. Stout and Ken L. Risenhoover

115 Value of "Hobble Creek" Mountain Big Sagebrush as a Winter Forage for Mule Deer (Abstract) 467
Bruce L. Welch

Part 8: Antler Physiology

116 Plenary Lecture: Recent Advances in the Physiological Control of Velvet Antler Growth 471
James M. Suttie and Peter F. Fennessy

117 Antler Growth Patterns in Young Red Deer Stags 487
Peter F. Fennessy, Ian D. Corson, James M. Suttie, and Roger P. Littlejohn

118 Antler Growth in Male and Female Reindeer Calves Occurs in the Absence of the Gonads 493
Gerald A. Lincoln and Nicholas J.C. Tyler

119 Seasonal Regulation of the Sexual Cycle and Antler Growth in Roe Deer: Evidence for an Endogenous Rhythm 499
Antoine J. Sempéré, Robert Mauget, and André Lacroix

120 Antler Growth in Roe Deer: Effects of Casting and Correlations with Growth Hormone and Insulin Like Growth Factor I 505
Dieter Schams, Dietrich Barth, Eva-Maria Heinze-Mutz, Gabriele Pflaum, and Heinrich Karg

121 Studies on Cells Derived from Growing Deer Antler 511
Colin Gray, M. Louise Taylor, Michael A. Horton, Andrew S.I. Loudon, and Timothy R. Arnett

122 Antler Cycles of a White-Tailed Deer with Congenital Anophthalmia 520
Harry A. Jacobson and Steven J. Waldhalm

123 State of Determination of the Antlerogenic Tissues with Special Reference to Double-head Formation 525
Horst Kierdorf and Uwe Kierdorf

124 Investigation of Endocrine Parameters in Chital Deer in Relation to Their Antler Cycle (Abstract) 532
George Bubenik, Robert D. Brown, and Dieter Schams

125 Antler Growth in Red Deer Stags Implanted with an Antiandrogen (Abstract) 533
Ludek Bartos

126 Artificial Production of Three-Antlered Red Deer Stags (Abstract) 534
Z. Jaczewski and R. Bartecki

127 Influence of Ecological Factors on the Development of Sika Deer Antlers (Abstract) 535
A. Kravchenko, A. Velizhanin, and E. Nivin

128 Production of 1,25(OH)2D by Deer Antler Cells (Abstract) 536
A.J. Sempéré, C. Silve, and M. Garabedian

Special Section: The Commercialization of Deer

129 Commercialization of Wildlife: A Value-Added Incentive for Conservation 539
Delwin E. Benson

130 Deer Ranching for Products and Paid Hunting: Threat to Conservation and Biodiversity by Luxury Markets 554
Valerius Geist

Closing Address

131 Biology of Seasonal Breeding in Deer 565
Gerald A. Lincoln

Workshops

Disease and Immune Systems. Chairmen: *Ed Couvillion and Richard Randall* 577

Bioenergetics and Estimation of Carrying Capacity. Chairmen: *Ken L. Risenhoover and Robert G. White* 577

Endangered Species. Chairmen: *Harry Jacobson and Anthony English* 579

Venison Marketing and International Trade. Chairmen: *K.R. Drew and T.C. Williams* 580

Chemical Immobilization. Chairmen: *Eric C. Hellgren* 581

Genetics. Chairmen: *Peter A. Dratch and Olin E. Rhodes* 582

Artificial Breeding. Chairmen: *Jerry C. Haigh and Steve J. Waldhalm* 584

Evolution of the Cervidae. Chairman: *Robert Brown* 585

Antler Development. Chairmen: *James M. Suttie and George A. Bubenik* 587

Pheromones. Chairman: *Karl V. Miller* 588

Index 591

Contributors

Clare L. Adam, Rowett Research Institute, Bucksburn, Aberdeen AB2 9SB, Scotland, UK

Stephen D. Albon, Department of Zoology, University of Cambridge, Downing Street, Cambridge CB2 3EJ, UK

Sandra L. Anderson, Wyoming Game and Fish Department Research Laboratory, Box 3312, University Station, Laramie, WY 82071 USA

Timothy R. Arnett, Department of Anatomy and Developmental Biology, University College London, Gower Street, London WC1E 6BT, UK

Abu M. Ataja, Massey University, Palmerston North, New Zealand

Thomas G. Barnes, Department of Forestry, University of Kentucky, Lexington, KY 40546 USA

Graham K. Barrell, Animal and Veterinary Sciences Group, PO Box 84, Lincoln University, Canterbury, New Zealand

Tom N. Barry, Massey University, Palmerston North, New Zealand

Dietrich Barth, MSD Research Laboratories, Sharp & Dohme, 8201 Lauterbach, Germany

Delwin E. Benson, Department of Fishery and Wildlife Biology, Colorado State University, Fort Collins, CO USA

Lytle H. Blankenship, Box 5220, Uvalde, TX 78801 USA

Barry R. Brinklow, Institute of Zoology, Zoological Society of London, Regent's Park, London NW1 4RY, UK

I. Lehr Brisbin, Jr., Savannah River Ecology Laboratory, Drawer E, Aiken, SC 29801 USA

Glenn S. Buchan, Deer Research Laboratory, Microbiology Department, University of Otago, PO Box 56, Dunedin, New Zealand

Ronald K. Chesser, Savannah River Ecology Laboratory, Drawer E, Aiken, SC 29802 USA

---

*Abstract authors do not appear in the list of contributors. Their affiliations appear as footnotes on abstract opening pages.

Timothy H. Clutton-Brock, Department of Zoology, University of Cambridge, Downing Street, Cambridge CB2 3EJ, UK

Ian D. Corson, Invermay Agricultural Centre, MAF Technology, Private Bag, Mosgiel, New Zealand

John P. Cross, Microbiology Department, University of Otago, PO Box 56, Dunedin, New Zealand

Sándor Csányi, University of Agricultural Sciences, Institute of Zoology and Game Biology, H-2103 Godollo, Hungary

Jonathon D. Curlewis, Institute of Zoology, Zoological Society of London, Regent's Park, London NW1 4RY, UK

Stephen Demarais, Department of Range and Wildlife Management, Texas Tech University, Lubbock, TX 79409 USA

Peter A. Dratch, US Fish & Wildlife Service, National Fish and Wildlife Forensics Laboratory, 1490 East Main Street, Ashland, OR 97520 USA

Ken R. Drew, MAF Technology, Ministry of Agriculture and Fisheries, Private Bag, Mosgiel, New Zealand

Alison Dunning, University of Queensland, Gatton College, Queensland 4343, Australia

Anthony W. English, Department of Animal Health, The University of Sydney, Private Bag 3, Camden, New South Wales 2570, Australia

Richard C. Etchberger, School of Renewable Natural Resources, University of Arizona, Tucson, AZ 85721 USA

R. Terry Ervin, Department of Agricultural Economics, Texas Tech University, Lubbock, TX 79409-2132 USA

Gareth Evans, Department of Animal Science, University of Sydney, New South Wales 2006, Australia

Peter F. Fennessy, MAF Technology Invermay, Private Bag, Mosgiel, New Zealand

Mark W. Fisher, Invermay Agricultural Centre, Private Bag, Mosgiel, New Zealand

Daniel L. Forster, Georgia Department of Natural Resources, Sapelo Island, GA 31327 USA

James F. Gallagher, Department of Forestry, University of California, Berkeley, CA 94720 USA

Valerius Geist, Faculty of Environmental Design, The University of Calgary, Calgary, Alberta T2N 1N4, Canada

Colin Gray, Department of Anatomy and Developmental Biology, University College London, Gower Street, London WC1E 6BT, UK

J. Frank T. Griffin, Deer Research Laboratory, Microbiology Department, University of Otago, PO Box 56, Dunedin, New Zealand

Jerry C. Haigh, Department of Herd Medicine & Theriogenology, Western College of Veterinary Medicine, University of Saskatchewan, Saskatoon, Saskatchewan, S7N 0W0, Canada

Lonnie P. Hansen, Missouri Department of Conservation, 1110 South College Avenue, Columbia, MO 65201 USA

Eva-Marie Heinze-Mutz, MSD Research Laboratories, Sharp & Dohme, 8201 Lauterbach, Germany

Hubert Hendrichs, Department of Ethology, University of Bielefeld, Morgenbreeds 45, 4800 Bielefeld 1, Germany

James D. Herriges, Jr., Wyoming Game and Fish Department Research Laboratory, Box 3312, University Station, Laramie, WY 82071 USA

Matthew J. Heydon, Institute of Zoology, Zoological Society of London, Regent's Park, London NW1 4RY, UK

John Hodgson, Department of Veterinary Clinical Sciences, Massey University, Palmerston North, New Zealand

Øystein Holand, Department of Biology and Nature Conservation, Agricultural University of Norway, N-1432 Aas-NLH, Norway

Michael A. Horton, ICRF Haemopoiesis Research Group, Department of Haematology, St. Bartholomew's Hospital, London EC1A 7BE, UK

Ron W. Hoskinson, CSIRO Division of Animal Production, Blacktown, New South Wales, Australia

David M. Hughes, Jr., Game Ranching, Inc, PO Box 530, Ingram, TX 78025 USA

Harry A. Jacobson, Department of Wildlife and Fisheries, Mississippi State University, PO Drawer LW, Mississippi State, MS 39762 USA

Alick R. Jones, Department of Pure and Applied Zoology, University of Reading, Reading RG6 2AJ, UK

Neville B. Jopson, MAF Technology Invermay, Private Bag, Mosgiel, New Zealand

John J. Kanter, Department of Forest Resources, Pettee Hall, University of New Hampshire, Durham, NH 03824 USA

Heinrich Karg, Institute of Physiology, Technical University Munich, 8050 Freising-Weihenstephan, Germany

János Kelemen, PANNON University of Agriculture, Faculty of Animal Science, H-7401 Kaposvar, PO Box 16, Hungary

Horst Kierdorf, Zoologisches Institut der Universität zu Köln, Weyetal 119, D-5000 Köln, Germany

Uwe Kierdorf, Zoologisches Institut der Universität Göttingen, Berliner Strasse 28, D-3400 Göttingen, Germany

David R. Klein, Alaska Cooperative Wildlife Research Unit, US Fish and Wildlife Service, University of Alaska, Fairbanks, AK 99775-0990 USA

Paul R. Krausman, School of Renewable Natural Resources, 325 Biological Sciences East, University of Arizona, Tucson, AZ 85721 USA

Robert E. Labes, Invermay Agricultural Centre, MAF Technology, Private Bag, Mosgiel, New Zealand

André Lacroix, CEBAS-CNRS, 79360 Villiers en Bois, France

Jochen Langbein, Biology Department, The University, Southampton, SO9 5NH, UK

Rolf Langvatn, Norwegian Institute for Nature Research, Tungasletta 2, Trondheim N-7000, Norway

Zsuzsa Lassu-Merényi, PANNON University of Agriculture, Faculty of Animal Science, H-7401 Kaposvar, PO Box 16, Hungary

Paul Leberg, Savannah River Ecology Laboratory, Drawer E, Aiken, SC 29801 USA

Bruce D. Leopold, Department of Wildlife and Fisheries, Mississippi State University, Mississippi State, MS 39762-5917 USA

Shangren Li, Department of Animal Science, Jilin Agricultural University, Changchun, Jilin, China

Gerald A. Lincoln, MRC Reproductive Biology Unit, 37 Chalmers Street, Edinburgh, EH3 9EW, UK

Roger P. Littlejohn, MAF Technology Invermay, Private Bag, Mosgiel, New Zealand

Andrew S.I. Loudon, Institute of Zoology, Zoological Society of London, Regent's Park, London NW1 4RY, UK

Colin G. Mackintosh, MAF Technology Invermay, Private Bag, Mosgiel, New Zealand

R. Larry Marchinton, School of Forest Resources, University of Georgia, Athens, GA 30602 USA

Robert Mauget, CEBAS-CNRS, 79360 Villiers en Bois, France

William W. Mautz, Department of Forest Resources, Pettee Hall, University of New Hampshire, Durham, NH 03824 USA

Gabrielle McCoy, Deer Research Laboratory, Department of Microbiology, University of Otago, Dunedin, New Zealand

Bernard J. McLeod, Invermay Agricultural Centre, Private Bag, Mosgiel, New Zealand

Lynne M. Meikle, Invermay Agricultural Centre, Private Bag, Mosgiel, New Zealand

Karl V. Miller, School of Forest Resources, University of Georgia, Athens, GA 30602 USA

John A. Milne, Macaulay Land Use Research Institute, Pentlandfiled, Roslin, Midlothian EH26 9RF, UK

John R. Morgart, Yukon Delta National Wildlife Refuge, Bethel, AK 99559 USA

Robert C. Mulley, Faculty of Agriculture and Rural Development, The University of Western Sydney—Hawkesbury, Richmond, New South Wales 2753, Australia

Gillian E. Mylrea, Private Bag, Camden, New South Wales 2570, Australia

Charles M. Nixon, Illinois Natural History Survey, 607 E. Peabody Drive, Champaign, IL 61820 USA

David A. Osborn, Department of Range and Wildlife Management, Texas Tech University, Lubbock, TX 79409-2132 USA

Claudio R. Ortiz, Asociacion Nacional de Criadores de Cervidos, Casilla 50476, Santiago, Chile

Anthony J. Pearse, MAF Technology Invermay, Private Bag, Mosgiel, New Zealand

Peter J. Pekins, NH Cooperative Extension Service, Pettee Hall, University of New Hampshire, Durham, NH 03824 USA

Josephine M. Pemberton, University of Cambridge, Department of Genetics, Downing Street, Cambridge CB2 3EH, UK

Gabriele Pflaum, Institute of Physiology, Technical University Munich, 8050 Freising-Weihenstephan, Germany

Sarah E. Price, Department of Pure and Applied Zoology, University of Reading, Reading RG6 2AJ, UK

Roger W. Purchas, Department of Veterinary Clinical Sciences, Massey University, Palmerston North, New Zealand

Rory Putman, Department of Biology, The University, Southampton S09 5NH, UK

Rongqian Qin, Department of Animal Science, Jilin Agricultural University, Changchun, Jilin, China

Kurt R. Rautenstrauch, EG&G Energy Measurements, Las Vegas, NV 89125 USA

Hugh W. Reid, Moredun Research Institute, 408 Gilmerton Road, Edinburgh EH17 7JH, Scotland, UK

Olin Eugene Rhodes, Jr., Department of Range and Wildlife Management, Texas Tech University, Lubbock, TX 79409 USA

Dieter Schams, Institute of Physiology, Technical University Munich, 8050 Freising-Weihenstephan, Germany

Georg Schwed, Institut für Haustierkunde, University of Kiel, Olshausenstr. 40, 2300 Kiel 1, Germany

Antoine J. Sempéré, CEBAS-CNRS, 79360 Villiers en Bois, France

Zhen D. Shi, Animal and Veterinary Sciences Group, Lincoln University, New Zealand

Judit E.G. Smits, Department of Herd Medicine and Theriogeneology, Western College of Veterinary Medicine, University of Saskatchewan, Saskatoon, Saskatchewan, S7N 0W0 Canada

Michael H. Smith, Savannah River Ecology Laboratory, Drawer E, Aiken, SC 29802 USA

Hans Staaland, Department of Biology and Nature Conservation, Agricultural University of Norway, N-1432 Aas-NLH, Norway

Daniel W. Soprovich, Manitoba Wildlife Branch, Box 14, 1495 St. James Street, Winnipeg, Manitoba, Canada R3H 0W9

László Sugár, PANNON University of Agriculture, Faculty of Animal Science, H-7401 Kaposvar, PO Box 16, Hungary

James M. Suttie, Invermay Agricultural Centre, MAF Technology, Private Bag, Mosgiel, New Zealand

M. Louise Taylor, Department of Anatomy and Developmental Biology, University College London, Gower Street, London WC1E 6BT, UK

Anthony J. Thomson, Deer Research Laboratory, Microbiology Department, University of Otago, PO Box 56, Dunedin, New Zealand

E. Tom Thorne, Wyoming Game and Fish Department Research Laboratory, Box 3312, University Station, Laramie, WY 82071 USA

Nicholas J.C. Tyler, Department of Arctic Biology and Institute of Medical Biology, University of Tromsø, Vreivika, 9000 Tromsø, Norway

Hugo Varela-Alvarez, Department of Veterinary Clinical Sciences, Massey University, Palmerston North, New Zealand

Larry W. Varner, Ralston Purina Company, #2 Skyvue Avenue, New Braunfels, TX 78132 USA

Barry A. Veenvliet, Invermay Agricultural Centre, MAF Technology, Private Bag, Mosgiel, New Zealand

Steven J. Waldhalm, College of Veterinary Medicine, Mississippi State University, Mississippi State, MS 39762 USA

Mark C. Wallace, School of Renewable Natural Resources, 214 Biological Sciences East Building, University of Arizona, Tucson, AZ 85721 USA

John S. Waye, Department of Pathology, McMaster University Medical Centre, Room 2N33, 1200 Main Street West, Hamilton, Ontario L8N 3Z5, Canada

James R. Webster, Invermay Agricultural Centre, MAF Technology, Private Bag, Mosgiel, New Zealand

Christen Wemmer, Conservation and Research Center, National Zoological Park, Front Royal, VA 22630 USA

Robert G. White, Institute of Arctic Biology, University of Alaska Fairbanks, Fairbanks, AK 99775-0180 USA

Mark D. Whitney, School of Forest Resources, University of Georgia, Athens, GA 30602 USA

Peter R. Wilson, Department of Veterinary Clinical Sciences, Massey University, Palmerston North, New Zealand

Keith B. Woodford, University of Queensland, Gatton College, Queensland 4343, Australia

Opening Address

1
The Status of Deer in a Changing World Environment

David R. Klein

Abstract

Changes in wildlife habitats throughout the world are accelerating as human populations continue to expand. Deer species have responded negatively to habitat loss or alteration from cutting of rain forests in Southeast Asia, Central and South America, and the Pacific Northwest of North America. Conversely, practices by the timber industry have contributed to increases in distribution and numbers of moose in Scandinavia and the Soviet Union. In China, India, and other countries of Southeast Asia, increasing pressures from agricultural expansion and livestock grazing have resulted in deer habitat loss, as well as reduced deer numbers and local extirpation of populations through increased hunting pressures, both legal and illegal. Although the production of meat from deer in captive herds continues to expand, pressures to increase agricultural crop production worldwide may have a far greater opposite effect on total deer numbers through loss of habitat. Industrial pollution, largely through its contribution to acid precipitation, is altering forests and other deer habitats in eastern North America and Eurasia with associated adverse effects on deer.

Key words: Agriculture, deer, forestry, game ranching, pollution, population status

Introduction

On Earth Day 1990, over 100 million people demonstrated in more than 120 countries throughout the world. These were not typical demonstrations protesting actions by governments or organizations, but gatherings of people who came together to share mutual concern over the continuing degradative changes that characterize our environment late in the twentieth century. In the 20 years since the first Earth Day in April of 1970, environmental degradation from increased human population pressure and expanding industrialization has accelerated. Erosive loss of natural landscapes to human developments is overwhelmingly obvious; mass media, especially television, have broadened our perspective and concern beyond our native lands to the entire world. The accelerated exploitation of resources of both land and sea is now widely recognized to be a product of the expanding drive for these resources to meet the demands generated by the affluent industrialized countries. Third World countries with populations increasing at rates that exceed their own abilities to feed and provide for themselves strive to emulate the affluence of the developed countries, further compounding the degradative pressures on the natural environment.

How have deer of the world fared in the changing global environment? A survey of the distribution and status of the 39 species of deer worldwide (Table 1.1) revealed that 14 species or subspecies are now in an endangered or threatened status. Deer are a natural and renewable resource that contributes to the sustenance and economic well-being of people throughout much of the world. Deer provide recreational opportunities, enhance esthetic returns from the landscape, and, moreover, are indispensable keystone

TABLE 1.1. Distribution and status of deer of the world (39 species in 17 genera) (Macdonald 1984; IUCN 1972)

Common name	Scientific name	Distribution Natural (introduced)	Status
CERVIDAE			
Hydropotinae			
Water-deer (Chinese water-deer)	*Hydropotes inermis*	China, Korea (England)	
Muntiacinae			
Indian muntjac	*Muntiacus muntjac*	India, Sri Lanka, Burma, SW China, SE Asia including Borneo (England)	
Reeve's muntjac (Barking deer)	*M. reevesi*	E China, Formosa (England)	
Black muntjac	*M. crinifrons*	E China	Rare
Fea's muntjac	*M. feae*	Thailand, S Burma	Endangered
Roosevelt's muntjac	*M. rooseveltorum*	N Vietnam	
Tufted deer	*Elaphodus cephalophus*	S, SE China, NE Burma	
Cervinae			
Fallow deer	*Dama dama*	Europe, Asia Minor, Iran (Australia, New Zealand)	CITES listed *D.D. mesopotamica* endangered
Chital (Axis deer)	*Axis axis*	India, Sri Lanka (Australia) (Hawaii)	
Calamian deer	*A. calamianeusis*	Calamian ls.-Philippines	Threatened
Thorold's deer	*Cervus albirostris*	Tibet	Rare
Swamp deer	*C. duvauceli*	N & C India, S Nepal	Endangered, CITES listed
Red deer	*C. elaphus*	Scandinavia, Europe, N Africa, Asia Minor, Tibet, Kashmir, Turkestan, Afghanistan (Australia, New Zealand)	CITES listed, Bukharian deer recovering in S USSR (*C..e. bactrianus*)
Wapiti	*C. canadensis*	W North America, Tien Shan Mts. to Manchuria and Mongolia, Kansu, China (New Zealand)	
Eld's deer (Thamin)	*C. eldi*	E India, Thailand, Vietnam, Hainan ls., Burma	Endangered, CITES listed
Sika deer (Japanese deer)	*C. nippon*	Japan, Vietnam, Taiwan, Manchuria, Korea, N & SE China (New Zealand, United States, British Isles)	Endangered
Rusa deer (Timor deer)	*C. timorensis*	Indonesian Archipelago (Australia, New Zealand, Fiji, New Guinea)	
Sambar	*C. unicolor*	SE Asia-Philippines to Sri Lanka (Australia, New Zealand)	
Pére David's deer	*Elaphurus davidiensis*	China (England)	Endangered
Odocoilinae			
Mule deer & black-tailed deer	*Odocoileus hemionus*	W North America, Central America (Hawaii)	*O.h. cerrosensis* endangered
White-tailed deer	*O. virginianus*	N, C & S America to Peru & Brazil (New Zealand, Finland)	*O.v. leucurus* endangered
Roe deer	*Capreolus capreolus*	Europe, Asia Minor, Siberia, N Asia, Manchuria, N China, Korea	

TABLE 1.1. (*Contd.*)

Common name	Scientific name	Distribution Natural (introduced)	Status
Moose (Elk)	*Alces alces*	N Europe, Siberia, Mongolia, Manchuria, Alaska, Canada, N USA (New Zealand)	
Reindeer/caribou	*Rangifer tarandus*	Scandinavia, Svalbard, USSR, N China, Alaska, Canada, Greenland (S Georgia, Kerguelen, Iceland, Scotland)	
Marsh deer	*Blastocerus dichotomous*	C Brazil & N Argentina	Endangered
Pampas deer	*Ozotoceros bezoarticus*	Brazil, Argentina, Paraguay, Bolivia	Endangered
Chilean huemul (Guemal)	*Hippocamelus bisulcus*	Chile, Argentina	Vulnerable
Peruvian huemaul (Guemal)	*H. antisensis*	Peru, Ecuador, Bolivia, N Argentina	Threatened, CITES listed
Red brocket	*Mazama americana*	C & S America-Mexico to Argentina	CITES listed
Brown brocket	*M. gouazoubira*	Central & South America-Mexico to Argentina	
Little red brocket	*M. rufina*	Venzuela, Ecuador, SE Brazil	
Dwarf brocket	*M. chunyi*	N Bolivia, Peru	
Southern pudu	*Pudu pudu*	Chile, Argentina	Endangered, CITES listed
Northern pudu	*P. mephistophiles*	Ecuador, Peru, Columia	Undetermined, CITES listed
MOSCHIDAE			
Siberian musk deer	*Moschus sibiricus*	C & E Siberia, N Mongolia, Korea, Sakhalin	
Southern musk deer	*M. moschiferus*	C & SW China, N India, Pakistan, Tibet	Vulnerable
Berezovski musk deer	*M. berezovskii*	Szechwan & Kansu, China	Threatened

species in complex ecological relationships in their native ranges.

There is no simple solution to conservation of deer throughout the world; nevertheless, there are lessons to be learned through a global perspective on the status of deer that may help to guide us in our relationship to the land to the benefit of both humans and deer.

Agriculture

The expanding human population has generated the need for increased agricultural production. Although much of the increased production of foods has been achieved through advances in the efficiency of crop production on existing agricultural lands, the rate of destruction of wildlife habitats in an effort to increase the extent of arable lands is alarming. This has been most pronounced in Third World countries. In Central America, 40% of the rain forest has been converted to pasture in the past 20 years to raise inexpensive beef primarily for export to the North American fast-food market (Animal Welfare Institute, Washington, D.C.). In Latin America, 7% of the people own or control 93% of the arable land. Land reforms that new governments promised often have failed to materialize. Alternatively, in many Central and South American countries, landless people have been encouraged to destroy more forest to obtain crop land for themselves to enable them to enter the expanding cash economy.

All 12 Central and South American deer species have suffered loss of habitat through deforestation in attempts to expand agricultural production. The marsh deer, pampas deer, and southern pudu are listed as endangered species and three others, the Peruvian huemul, the red brocket, and the northern pudu, are listed by CITES (Convention on the International Trade of Endangered Species of Flora and Fauna) and considered threatened.

Continuing encroachment of human settlement into habitats of South American deer has brought an additional threat from feral and uncontrolled dogs that have increased in association with human settlement, and prey heavily on the small endemic deer. These deer evolved in the absence of large canid predators and lack effective defensive behavior against dogs. Local hunters also use dogs very effectively to drive deer from dense habitats. Although hunting restrictions may exist to protect threatened or endangered species in Latin America, game law enforcement is lax.

As human populations expand throughout the Indian subcontinent and Southeast Asia, destruction of wildlife habitats through expansion and intensification of crop production is causing continuing losses of deer habitat. Grazing of domestic livestock has expanded greatly throughout this region and includes grazing within forested habitats. When natural habitats are exploited for grazing of livestock, a chain of irreversible consequences usually follows. Large predators, which may have been a deterrent to domestic livestock grazing, are hunted to local extirpation, thus permitting larger, less closely attended herds. Heavy grazing prevents regeneration of tree species and leads to eventual loss of forests. The close proximity of domestic livestock and wildlife often leads to increased parasitism and disease in both domestic and wild ungulates. Exploitation of wildlife as a subsistence food source or for sale increases as human presence expands into wildlife habitats. Several species of muntjac and axis deer have declined throughout this region as a consequence of expanded agriculture and livestock grazing. The Fea's muntjac of Thailand and Burma is now endangered. Among the *Cervus* species, the swamp deer or barasingha and the Eld's deer which previously was present throughout Eastern India, Burma, Thailand, Vietnam and Southeast China are now endangered. Thorold's deer of Tibet is rare and presumably threatened, as are populations of red deer in Kashmir, Tibet, and Afghanistan.

Throughout the Third World, political unrest stemming from burgeoning human population pressures and inequalities in land distribution has led to civil strife and extended guerilla warfare. Dependence of the insurgents on deer and other wildlife as a source of food, disregard of government regulations protecting wildlife and their habitats, and increased availability of military firearms and ammunition have had detrimental effects on deer.

In developed countries, some agricultural lands of low productivity have been abandoned or, as a result of changing local economics, have returned to natural vegetation. Deer have been quick to reoccupy these restored habitats whenever access to them from existing populations was possible. In eastern North America, white-tailed deer and moose have expanded in both numbers and distribution as a consequence of increases in habitat through abandonment of agricultural lands and protection from illegal hunting. Moose in Norway and Sweden have also benefited from abandonment of subsistence farmlands during the early successional stages of their return to coniferous forest (Ahlén 1968).

In developed countries, particularly in Europe and North America, increased crop production has been associated with large-scale agrobusiness and extreme mechanization of agriculture. This has resulted in the elimination of small fields, with associated hedgerows and windbreaks that have historically provided cover and forage for wildlife. Subsequent extensive uniform fields of single crops, interrupted by off-season fallow without either food or cover for wildlife, have been associated with reductions of roe deer in central Europe (Colic 1974) and loss of habitat for white-tailed deer in central and midwestern United States and Canada.

Ranching and Farming of Deer

Ranching of wildlife, game farming, and associated introduction of exotics in the southwestern United States have resulted in some increased hunting opportunities, occasional preservation of endangered species, and increased income to landowners, but they have also led to the demise of native deer species. In Texas and New Mexico feral populations of African or

Asian antelope, aoudad sheep, and exotic deer species have become established in natural habitats outside of fenced private lands. Native deer species often are unable to compete effectively with the exotic species, and decline or are locally extirpated as exotic species expand in number and distribution (Demarais et al. 1990). The exotic species compete for available forage and not only may they have greater resistance to disease and parasites than native deer, but they act as reservoirs for disease and parasites to which native deer are susceptible. The decline of white-tailed deer on Assateague Island in Maryland is attributed to the introduction and increase of sika deer (Keiper 1985). In Texas, experiments to determine competitive effectiveness of white-tailed deer with axis and sika deer placed together in large enclosures (39 ha), white-tailed deer were outcompeted by both exotic species (Baccus et al. 1985). Another problem has been genetic mixing when exotic deer of a closely related species become established in habitats with native species. Sika deer readily hybridize with red deer. Red deer populations in Ireland, Czechoslovakia, Great Britain, New Zealand, and southern Asia have experienced interbreeding with sika deer and wapiti, either accidentally or intentionally, which resulted in loss of typical red deer characteristics (Bartos 1990).

Interest in ranching and farming of both exotic and native wildlife is increasing worldwide as markets expand for venison and other animal products. Pressures for liberalization of existing laws governing importation, captive raising, and marketing of wildlife are increasing, particularly in Canada and the United States. The Alaska state Senate recently approved legislation to legalize farming of moose, although it failed to come to a floor vote in the House. Although some commercial production of wildlife has occurred without detriment to native deer, the potential for undesirable effects that exists requires particular scrutiny of proposals to relax existing laws governing game ranching, game farming, and the introduction of exotics.

Forest Cutting Practices

The cutting of forests for wood production has had negative and positive effects on deer throughout the world. Logging in tropical rain forests has largely been selective cutting of old-growth trees of value in the world furniture market. The effects of these practices on native deer species are not well understood, nor are they the focus of major research efforts. The direct consequences of such logging practices, however, are often irrelevant when temporary logging roads subsequently provide access for land clearing for agriculture or reforestation with monocultures of fast-growing tree species for the production of wood fiber. This pattern is characteristic in Latin America and Southeast Asia where it fosters loss or deterioration of deer habitat and increased hunting of native deer and other wildlife.

Temperature forests are habitat for many deer species in North and South America and Eurasia. Where these forests have been selectively cut or clearcut in relatively small patches on a rotational basis, deer have suffered few adverse effects and sometimes have increased their productivity. Extensive clearcutting of coniferous forests in Norway, Sweden, and Finland in the recent past has been followed by expansion of moose populations as deciduous trees, shrubs, and pine (*Pinus sylvestris*), which are important as winter moose forage, dominated the early regrowth in cutover areas (Lavsund 1987). In conjunction with intensive management of moose to maximize the annual harvest by limiting the overwintering population to primarily the breeding stock, numbers of moose killed by hunters increased many fold to over 240 000 annually by the early 1980s in these three countries. Although this is an example of enhancement of deer production through cutting of forests, the process is not as simple as it first appears. Winter snow accumulates to greater depths where trees are no longer present to intercept it, thus in regions receiving very heavy snows or in years of excessive snowfall, forage in clearcuts may be restricted in its availability. This is particularly important in northern Scandinavia where roe deer occur, or in Finland where there are introduced white-tailed deer. There are also many areas in these countries where forests have been cut extensively over a relatively short period of time. Consequently, regenerating forests are of approximately uniform age. As tree species eaten by moose grow out of their reach and the forest canopy closes, light is restricted for deciduous shrubs of the understory that are of importance to deer and moose, and productivity of forage declines profoundly. Moose populations throughout Scandinavia have begun to decline recently, apparently as a consequence of the changing age structure of the regrowing forests. Associated with this decline has been an increase in moose damage to young pine forests.

In the Soviet Union cutting of coniferous forests and protection from excessive hunting have fostered expansion of moose northward and increases in their populations since the early 1930s (L.M. Baskin, personal communication).

In the coastal temperate rain forests of northwestern North America, clearcutting of old-growth forests has resulted in deterioration of habitat for black-tailed deer. In Oregon, Washington, and southern British Columbia, initial increases in deer numbers after logging of virgin forests were associated with increased production of forage species in the flush of shrub and forb growth after clearcutting and burning of slash (Einarsen 1946). Optimism, associated with improved hunting opportunities, increased deer harvests, and easier hunter access from logging roads, waned as the forest canopy closed in the even-aged second-growth stands of young conifers, thus restricting light needed for growth of deer forage on the forest floor. In theory the staggered cutting of these forests on a sustained-yield basis should have allowed for interspersion of recently cutover areas favorable for deer with varying-age second-growth stands offering cover but little forage. In practice, however, cutting of forests has often not been done on a sustained-yield basis and when it has, the scale of the regional cutting has frequently been too large to provide associated benefits for deer and other wildlife. Furthermore, forests with highest timber values in valley bottoms were frequently cut before logging operations moved to higher elevations on mountain slopes. Deer in these areas are dependent upon seasonal movements to higher altitudes in summer to take advantage of phenological progression of growth of forage species and movements to lower altitudes in winter to avoid excessive snow depths. This practice disturbed the synchrony in availability of favorable summer and winter habitats. A compounding factor associated with the cutting of old-growth forests in the Pacific Northwest has been the elimination of arboreal lichens common on the crowns of very old trees in uneven-aged stands. The lichens, which are blown from the tops of trees during winter storms, are highly preferred by deer, and they have high digestibility and energy content, and are available on the snow surface when most other forage may be covered by snow (Stevenson and Rochelle 1984). The loss of lichens as a consequence of logging of high-altitude forests in northeastern Washington, northern Idaho, and adjacent British Columbia has been a major cause of habitat deterioration for the threatened population of mountain caribou in that area (Simpson et al. 1985).

In southeastern Alaska and on the islands of Prince William Sound where extensive clearcutting of old-growth forests has a much more recent history than in the coastal forests farther south, a controversy exists over forest cutting practices between the U.S. Forest Service and the timber industry on one side and the Alaska Department of Fish and Game supported by hunting, fishing, and recreational interests on the other side. The Sitka black-tailed deer and its habitat requirements are a major focus of this controversy (Schoen et al. 1988). Unlike areas farther south, abundant forage in recent clearcuts is unavailable to deer in most winters because of excessive snow depths. Conversely, the canopy of the uneven-aged old-growth forests captures much of the falling snow and, even though forage production may be lower in the old-growth forests than in the recent clearcuts, it remains available to deer in winter because snow depths are shallower. In the most severe winters only the old-growth stands with the largest trees remain available as a refuge for deer, and these are the stands with sufficient timber volume to be attractive to the timber industry. Because of the rugged terrain and high latitude, such high timber volume, old-growth stands make up only 4% of this temperate coastal rain forest in the Tongass National Forest in southeastern Alaska. After logging of these steady-state forests favorable to deer under the planned 100-year cutting cycle, trees in excess of 300 years of age will be lost forever. Protection of high-volume old-growth forests as critical winter habitat for deer is, therefore, in direct competition with the timber industry, whose primary interest is in cutting the forests on this same 4% of the national forest lands.

Impacts of Industrial Development Activities

Industrial developments, including mining, petroleum production, power generation, and manufacturing plants consume wildlife habitat. Cumulatively, the direct loss of deer habitat to industrial development is a significant although minor addition to losses from other human activities such as agriculture, forestry, residential sprawl, roads, and other transportation corridors.

Coal mining in the eastern United States caused large-scale deterioration of wildlife habitat when mined-out areas were left without top soil over sterile tailings of overburden, most of which were contaminated by acid drainage. The U.S. Bureau of Reclamation has restored many of these areas as wildlife habitat, and current federal regulations for mining in the United States mandate specific restoration practices after large-scale open pit mining.

Indirect effects of industrial developments often pose the greatest threats to deer populations and their habitats worldwide. Mining and oil and gas exploration and development usually increase human pressure on deer and their habitats. Construction of roads and other transportation corridors consume deer habitat, create disturbance, interrupt movements, and provide access for increased hunting or other activities that may harm deer and their habitats.

Large-scale oil development at Alaska's Prudhoe Bay, although directly consuming a relatively small portion of the habitat of caribou of the Central Arctic Herd, has resulted in a complex grid of roads, pipelines, and other oil field facilities that collectively have deflected, delayed, or obstructed movements of caribou in the area (Whitten and Cameron 1983). Experience gained during initial oil field construction now enables design and layout of new satellite oil fields that are more favorable to caribou movements. For example, separation of above-ground pipelines from roads carrying traffic has eased crossing of both pipelines and roads by caribou (Murphy and Curatolo 1987). Nevertheless, caribou still experience delays in crossing the oil fields to and from insect relief areas along the coast, and cows with young calves exhibit strong avoidance of oil field developments. The Central Arctic Herd, however, has increased during oil field development, probably from a reduction of predators and, presumably, very favorable weather.

Effects of Pollution on Deer

Industrial development, energy production, and consumption for residential and transportation use are sources of pollution that affect deer and their habitats. Many deer habitats are sufficiently removed from point sources of industrial polution, but regional or global effects of pollution, notably airborne pollution, may have subtle but long-lasting effects on deer and their habitats.

Until now, direct effects of toxic pollutants on the physiology of deer have not been widespread and concerns have focused on possible harmful effects on humans from consumption of meat with high levels of toxic substances. A particular case in point are consequences of fallout of radionuclides from the Chernobyl nuclear power plant explosion. Portions of domestic reindeer range in Sweden and both domestic and wild reindeer (caribou) ranges in Norway in the path of the explosion plume were contaminated with fallout of radioactive isotopes. While direct exposure of the reindeer to these isotopes was not sufficient for significant effects on the populations, accumulation of specific isotopes, particularly cesium^{137}, in the tissues of reindeer resulted through biological pathways from consumed forage (Pedersen et al. 1988). Lichens and fungi, which are seasonally important in the diet of reindeer, had particularly high levels of cesium^{137}. Concern for human health through consumption of the reindeer meat has reduced the marketability of reindeer with serious consequences for the reindeer industry in these countries. Public overreaction to threats to human health has persisted in spite of the fact that major reindeer production areas in northern Scandinavia were not in the path of the fallout from

Chernobyl, and nuclear isotope levels in meat from contaminated areas do not pose a significant threat to humans at moderate levels of consumption.

Heavy metals are being increasingly found in high levels in animal tissues. Forage consumed by roe deer in the Niepolomice Forest in Poland, close to a large steel mill, was found to have concentrations of the highly toxic heavy metals cadmium, lead, nickel, and chromium five to eight times greater than in unpolluted control areas (Grodzinska et al. 1983). During the 25 years since establishment of the steel mill, roe deer antlers have declined in quality, apparently independent of density-dependent relationships.

Cadmium has recently been found in high levels in the organ tisues of caribou of the George River Herd in northern Quebec and Labrador (Créte et al. 1988). Although the source of the cadmium is not known, it is thought to be from industrial pollution produced in eastern United States and Canada. Local people who depend on caribou as a major portion of their diet have been warned by government agencies to avoid eating the kidneys and livers of caribou.

Heavy metals from air pollution by a large mining and smelting complex at Norilsk in northwestern Siberia are attributed to deterioration of lichens in winter grazing areas of domestic and wild reindeer in the summer downwind pollution plume from the smelters (T. Vlasova, personal communication). Because lichens have no root system and depend on the atmosphere for their nutrients, they tend to accumulate atmospheric pollutants to levels that cause reduced growth, morphological aberrations, or death of the lichens, and, furthermore, they are a concentrated source of pollutants for herbivores that feed upon them. Because lichens make up a large portion of the winter diet of most reindeer, caribou, and musk deer and are important in the diet of other deer species, they represent a critical unit of the habitat of many deer species that is particularly vulnerable to air pollution.

The effects of acid precipitation on forests in parts of northern Europe and northeastern North America have become increasingly pronounced in recent decades. The composition of forest species changes as plant species that are sensitive to the effects of increasing acidity in the precipitation and the soil die out and are replaced by more acid-tolerant species. The changes are complex and have ramifications for all invertebrates and vertebrates in these forests. Although deer may benefit in some cases from changes in forest composition, productivity of most plants declines with increases in soil acidity and reduces growth of deer forage. Trees under the stresses of acid precipitation are more vulnerable to attacks by disease and insects. The recent irruptive spread of tent caterpillars (*Malacosoma* spp.) westward into the Appalachian mountains may be associated with reduced resistance of trees to them from acid precipitation. Defoliation by the tent caterpillar of oaks (*Quercus* spp.) generally does not kill the trees, but prevents production of acorns that are important in the fall and winter diet of white-tailed deer for accumulation of fat reserves (M. Vaughan, personal communication).

Global climate change from increase of CO_2 and other "greenhouse" gases in the atmosphere will presumably have far-reaching effects on deer worldwide through its effects on their habitat. Although there is debate at present over the extent, if any, of temperature increases from the "greenhouse effect," imminent significant climatic warming seems inevitable. Climate models generated to date have lacked the complexity to predict probable changes in temperature and precipitation by region. Consequences for deer of global warming are, therefore, difficult to anticipate. Predicted severe droughts for south temperate and tropical areas where deer forage production is already limited by low soil moisture will undoubtedly lead to declines in deer. Conversely, at the northern limits of deer distribution, warming temperatures may stimulate forage production, reduce the amount and duration of winter snow cover, and concomitantly effect increases in deer productivity. Changes, however, will be complex and both contraction and expansion in the range of distribution of deer will take place in relation to each species' adaptability. For example, the southern limits of the range of species adapted to north temperate or subarctic conditions, such as moose and caribou, may contract northward, while the northern limits of white-tailed and mule deer may be extended northward. Direct human manipulation of the environment will remain a complicating and overriding factor in the ecology of deer under a regime of global climate change.

Density-regulating mechanisms do not seem to be responsible for synchronized population increases in major arctic caribou herds in North America and Siberia in recent years. Are these population increases the product of global climatic changes that have favored caribou? In the absence of evidence to the contrary, one is tempted to draw such a conclusion, but supporting data is also missing because of the absence of detailed long-term weather records from the Arctic. Whatever the cause of increases in caribou, environmental limits to population size may already have been reached by the George River Herd in northern Quebec and Labrador. By the late 1980s, this herd of 500 000 or more animals had clearly exceeded the size of any other caribou herd in North America. Recent deterioration in body condition of animals and reduced recruitment by the herd, in combination with indications of deteriorating diet quality on the calving and postcalving ranges, suggest density-dependent food limitation (Couturier et al. 1990).

In northwestern Alaska the Western Arctic Caribou Herd has experienced a similar pattern of growth to a population of well over 300 000 without reduction in the recruitment rate (P. Valkenburg, personal communication). A previous population high in the 1960s was followed by a decline in recruitment and poor body condition of individuals, apparently a consequence of density-dependent food limitation.

Changing Attitudes Toward Deer

Public attitudes toward deer throughout the world vary widely. They are persecuted as competitors with domestic livestock or as threats to successful crop production and also revered by some religious sects and groups opposed to killing of animals. Deer are valued differentially for products derived from their harvest. When very high monetary values are placed on deer products, such as incisor teeth of some *Cervus* species for jewelry or talismen, the musk of musk deer and soft antlers for pharmaceutical use in the Orient, and trophies of rare or threatened species, illegal hunting often prevails and poses a threat to local populations. The magnitude of this threat varies dependent on development and availability of markets for such animal products, the commitment of local governments to establishment and enforcement of regulations designed to maintain wildlife populations, and the current status of the targeted species. Commercialization of deer through ranching and farming for meat, antlers, and other products, and sale of hunting opportunities and trophies, is expanding. Commercialization has the potential for substantial contributions to local economies; however, it is a threat to native deer species through competition and displacement from habitats, creation of markets for poached deer products, and deterioration of the esthetic appreciation for free-ranging deer.

Pressures by the activities of an expanding human population on deer and their habitats throughout much of the world today coincide with an increased appreciation for deer species as important components of native fauna that if managed properly provide meat and various other products as well as yielding recreational hunting and viewing opportunities. However, increased public appreciation of deer will not assure their future well-being in the face of pressures generated by the expanding demand for resources in an industrialized society and by an unchecked human population. Deer of the world are clearly indicator species. Their population dynamics and individual welfare reflect the health of the global environment. We can only provide for the protection and continued productivity of deer and their habitats into the twenty-first century if we also address the broader issues of global environmental degradation that threaten the future welfare of all of mankind.

References

Ahlén I (1968) Research on moose-vegetation interplay in Scandinavia. Proc 5th North Am Moose Workshop, Kenai, AK, pp 23–24

Baccus JT, Harmel DE, Armstrong WE (1985) Management of exotic deer in conjunction with white-tailed deer. In Beasom SL, Roberson SF (eds) Game Harvest Management. Caesar Kleberg Wildlife Research Institute, Texas A & I University, Kingsville, pp 213–226

Bartos L (1990) Sika/red deer hybridization-recognition, consequences and present status. Paper presented at Int Symp Wildlife Conservation, Tsukuba, Japan

Colic DB (1974) Utilization du sol et forêts et la population de gibier. In Kjerner I, Bjurholm P (eds) Proc 11th Int Congress Game Biologists, Stockholm, pp 369–372

Couturier S, Brunelle J, Vandal D, St-Martin G (1990) Changes in the population dynamics of the George River Caribou Herd, 1976–87. Arctic 43:9–20

Créte M, Lafebvre MA, Cooper MB, Marshall H, Benneditti JL, Carriére PE, Nault R (1988) Contaminants in caribou tissues in northern Quebec. Paper presented at 5th Int Reindeer/Caribou Symp, Arvidsjaur, Sweden

Demarais S, Osborn DA, Jackley JJ (1990) Exotic big game: a controversial resource. Rangelands 12:121–125

Einarsen AS (1946) Management of the black-tailed deer. J Wildlife Management 10:54–59

Grodzinska K, Grodzinski W, Zeveloff SI (1983) Contamination of roe deer forage in a polluted forest of southern Poland. Environ Pollution (A) 30:257–276

International Union for Conservation of Nature and Natural Resources (1972) Red data book, vol 1: Mammalia. Morges, Switzerland

Keiper RR (1985) Are sika deer responsible for the decline of white-tailed deer on Assateague Island, Maryland? Wildlife Soc Bull 13:144–146

Lavsund S (1987) Moose relationships to forestry in Finland, Norway and Sweden. Swedish Wildlife Research, Viltrevy [Suppl] 1:229–244

Macdonald D (ed) (1984) The Encyclopedia of Mammals. Facts on File Publications, New York

Murphy SM, Curatolo JA (1987) Activity budgets and movement rates of caribou encountering pipelines, roads, and traffic in northern Alaska. Can J Zool 65:2483–2490

Pedersen O, Staaland H, Hove K, Garmo T (1988) Studies on the ingestion of radiocesium from contaminated pastures in reindeer. Paper presented at the 5th Int Reindeer/Caribou Symp, Arvidsjaur, Sweden

Schoen JW, Kirchhoff MD, Hughes JH (1988) Wildlife and old-growth forests in southeastern Alaska. Natural Areas 8:138–145

Simpson K, Woods GP, Hebert KB (1985) Critical habitats of caribou (*Rangifer tarandus caribou*) in the mountains of southern British Columbia. In Meredith TC, Martell AM (eds) Proc 2nd North Am Caribou Workshop, McGill University Subarctic Res Paper 40, pp 177–191

Stevenson SK, Rochelle JA (1984) Lichen litterfall—its availability and utilization. In Meehan WR, Merrell TR, Hanley TA (eds) Proc Symp Fish and Wildlife Relationships in Old-Growth Forests, American Institute Fishery Research Biologists, pp 391–396

Whitten KR, Cameron RD (1983) Movements of collared caribou, *Rangifer tarandus*, in relation to petroleum development on the arctic slope of Alaska. Can Field Nat 97:143–146

Part 1
Biology of Wild Populations

2
Cohort Variation in Reproduction and Survival: Implications for Population Demography

Stephen D. Albon, Timothy H. Clutton-Brock, and Rolf Langvatn

Abstract

In two different species of cervid, red deer (*Cervus elaphus*) and moose (*Alces alces*) pronounced cohort differences in fitness have important demographic consequences. In particular, where the environmental conditions into which a cohort is born permanently affect reproduction and survival, even if conditions ameliorate in later life, populations may fluctuate. Although cohort variation in survival in both moose on Isle Royale, USA, and red deer on Rhum, Scotland, has been attributed to differences in nutritional status during the period of early development, there are marked contrasts in the underlying causes of the differences in the nutritional status at birth. In the moose, cohort variation in survival was related to density-dependent changes in the food availability, while in red deer density-independent, weather, factors determining the onset of spring grass growth influenced both survival and reproduction. However, where cohort differences in early development do not persist they are unlikely to have significant affects on lifetime reproductive success and population fluctuation. For example, although cohort differences in early growth and the onset of breeding in a Norwegian red deer population varied in relation to spring weather, animals were able to compensate for retarded early development, and acheive large adult size and maximum reproductive potential. The available evidence suggests that, where food supply is limited by random stochastic processes, population fluctuation is likely to be small compared with long-term changes in climate like the predicted global warming.

Key words: Climate, cohort, demography, moose, red deer, reproduction, survival

Introduction

Demographic studies of mammalian populations typically quantify mean age-specific birth and death rates, as well as the extent of immigration and emigration (Caughley 1977). Estimates of these population processes permit the prediction of the numbers of animals in the population in successive years (Begon and Mortimer 1986). The accuracy of this prediction depends on many factors, including the length of the study, whether all young born in a year (cohort) may be followed to establish cohort life-tables, and whether variation in population processes can be related to density and/or stochastic, density-independent events (Sinclair 1989). Another potential problem, that has so far largely been the subject of theoretical interest, is that populations are not homogeneous in space and time, and the use of mean values obscures important heterogeneity that may lead to very different population responses (Lomnicki 1980, 1988; Hassell and May 1985).

In this paper, we investigate the extent and causes of variation in reproductive success of cohorts and explore the consequences of temporal heterogeneity for population dynamics. Our review centers around the study of red deer on Rhum, Scotland (Clutton-Brock et al. 1982), and compares the life-long cohort differences in the low performance population on Rhum (Albon et al. 1987) with the temporary cohort variation of the high performance population of Norwegian red deer (Burkey 1989). In addition, we contrast the climate-induced cohort variation in red deer with the density-dependent cohort variation in moose (*Alces alces*) on Isle Royale, USA (Peterson et al. 1984, 1991).

Spring Weather and Life-Long Differences in Cohorts on Rhum

One of the most important factors influencing the number of offspring raised to maturity (lifetime reproductive success) by individual female red deer on Rhum has been their birth year or cohort (Clutton-Brock et al. 1988). Although the growth of the population has been regulated by density-dependent changes in fecundity and mortality (Clutton-Brock et al. 1985), each successive cohort recruited to the breeding population has not necessarily had a lower fecundity or survival than cohorts recruited in earlier years at low population density (Fig. 2.1).

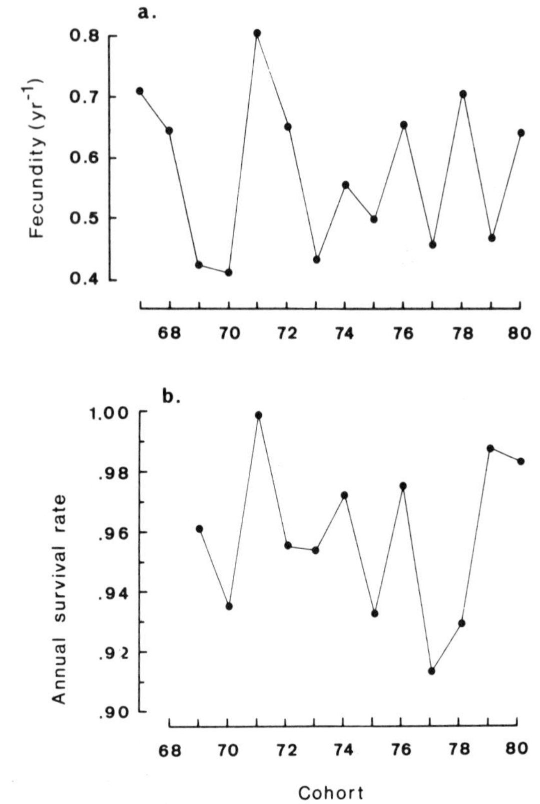

FIGURE 2.1. Cohort differences in (a) age-specific fecundity, and (b) annual survival rate (2–8 years) of female red deer on Rhum. Age-specific fecundity was standardized to 8-year-old hinds that lactated in the previous year when August-September rainfall was 450 mm and mean December-February temperature 4.5° C (Clutton-Brock et al. 1989). Annual survival rates were calculated following Caughley (1977).

Our research has investigated the stochastic factors underlying these differences in cohort reproductive success. For example, we have shown that cohort differences in offspring survival, the most important component of lifetime reproductive success, were closely related to the mean birth weight of the cohorts' offspring, which in turn was corelated to the cohorts' own birth weight (Albon et al. 1987). Since annual variation in birth weight was associated with spring (April and May) temperatures immediately prior to birth in June (Albon et al. 1983a), cohort differences in the viability of offspring were apparently due to a cross-generation lag of the environmental conditions prevailing when the cohort of mothers were in utero.

Presumably this environmental heritability reflects that in populations characterized by retarded growth and delayed maturation, like those inhabiting the impoverished moors of Scotland, cohorts are rarely able to compensate for initial growth disadvantages (Krebs and Cowan 1962). Since most fetal growth occurs in the last 2 months of gestation, differences in spring nutrition are likely to have a marked effect on birth weights (Thorne et al. 1976). Measurements of primary production on the preferred, *Agrostis-Festuca* grazings on Rhum (for methods see Iason et al. 1986) confirm that April grass growth is strongly correlated to temperature ($r^2 = 0.923$, $P < 0.05$) in the same month (Fig. 2.2a), and that birth weights are related ($r^2 = 0.919$, $P < 0.05$) to this variation in primary production (Fig. 2.2b). Furthermore, since annual differences in birth weight are likely to be a direct reflection of variation in maternal condition, we might expect differential offspring growth during the period of peak lactation to increase the initial differences in birth weight. Although we have too few adult hinds weighed to compare mean adult weight and mean birth weights for cohorts, the available data suggest that the heaviest adults come from the heaviest-born cohorts.

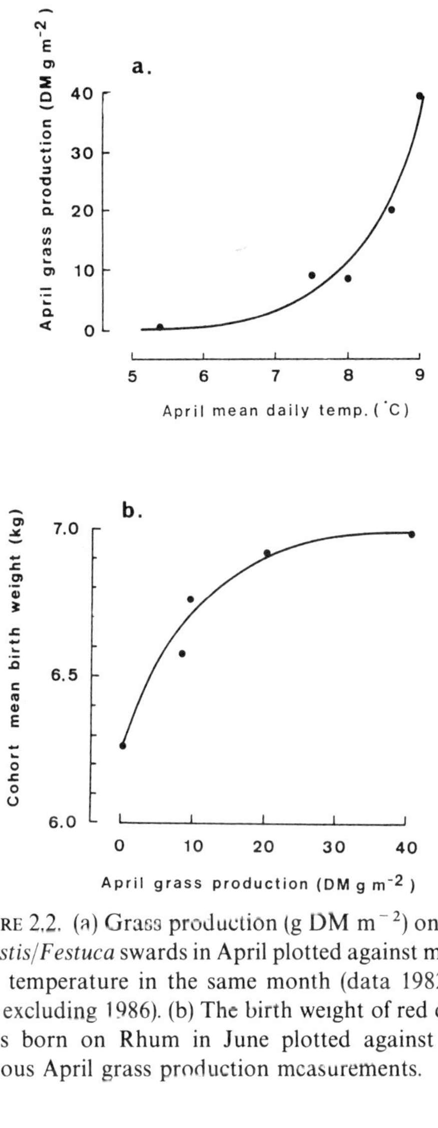

FIGURE 2.2. (a) Grass production (g DM m^{-2}) on the *Agrostis/Festuca* swards in April plotted against mean daily temperature in the same month (data 1982 to 1987 excluding 1986). (b) The birth weight of red deer calves born on Rhum in June plotted against the previous April grass production measurements.

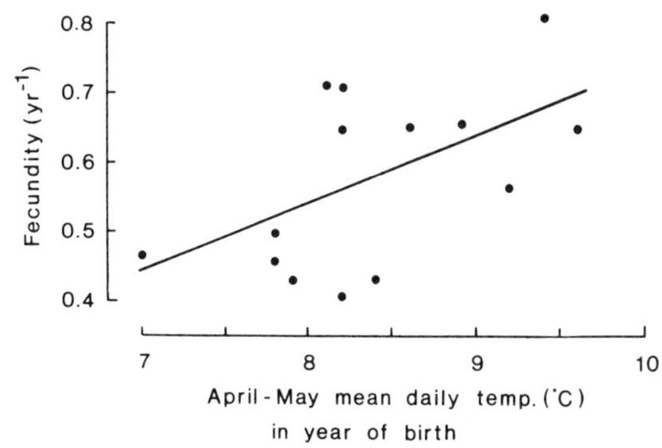

FIGURE 2.3. Cohort values of age-specific fecundity for red deer on Rhum (see Fig. 2.1) plotted against temperature in April–May of their year of birth.

If spring temperatures immediately prior to birth influence adult weight, we might expect the cohort differences in fecundity and survival, which are also weight dependent (Hamilton and Blaxter 1980; Albon et al. 1983b; Mitchell 1984), to be positively correlated to April–May temperatures. This was indeed the case. Cohort variation in age-specific fecundity (Fig. 2.1) was significantly correlated ($r^2 = 0.405$, $P < 0.02$) to April–May temperature in the year of birth (Fig. 2.3). Also, annual survival from 2 to 8 years was significantly correlated ($r^2 = 0.781$, $P < 0.001$) to spring temperature in the year of birth (Albon et al. 1987, Albon and Clutton-Brock 1988).

The Rhum study suggests that the climatic environment into which a cohort is born may shape the fate of that cohort in the future, even if conditions ameliorate (Albon and Clutton-Brock 1988). If this was a general rule, we might expect to detect cohort differences in other populations of cervids in habitats with harsh winters and variable spring weather. Unfortunately, there are currently only a limited number of studies with long-term data for individuals which can investigate stochastic environmental heritability. One such example comes from D. L. Mech's (personal communication) study of white-tailed deer (*Odocoileus virginianus*) in the National Superior Forest, Minnesota, where the weight of fawns born to cohorts of 2-year-old mothers varied with the depth of snow in the winter the mother was in utero. In Mech's study, the comparison was limited to young animals followed since the recent advent of radio-controlled dart collars permited individuals to be monitored from year to year.

A second example, from the red deer population of Norway, which is at the extreme northern limit of the species' range, provides an interesting contrast to the examples from Rhum because it shows that cohort differences are apparently not permanent.

Winter Weather and Transitional Differences in Cohorts in Norway

In Norway, data on age, weight, and reproductive status were collected from animals shot during the licensed autumn harvest (Langvatn and Albon 1986). Since individuals were aged from annuli in the cementum of the first incisor, an extremely accurate method of aging in this population (confirmed with known age animals), individuals were attributed to cohorts with confidence. Fecundity was based on whether the female was lactating and/or histological sections of ovaries. The presence of a corpus rubrum, the pigmented scar following parturition, was taken as evidence of parturition the same year (R. Langvatn, unpublished data).

Fecundity in animals aged between 4 and 19 years averaged 99% in all parts of the country in all years (Burkey 1989). Only in 2-year-olds, animals calving for the first time, and 3-year-olds was there sufficient variation in fecundity (mean 72.5% and 96.0%, respectively) to detect cohort differences. Late snowfall or late snowmelt in May, immediately prior to the birth of a cohort, depressed fecundity in females 2 and 3 years later (Burkey 1989). In general, cohorts in utero during winters characterized by low temperatures and long periods of deep snow cover showed depressed fecundity as 3-year-olds. As on Rhum, April temperature was a useful predictor of fecundity 3 years later (Burkey 1989).

Although the harsh winters and late springs might depress fecundity in the youngest females, once older than 4 years they were able to escape some initial constraint and reproduce every year. Since body weight influences puberty and subsequently fecundity, it would seem most likely that these initial cohort differences in fecundity reflect variation in early growth rates, but that asymptotic weights which are reached at about 5 years (Langvatn and Albon 1986) are always above the threshold at which reproduction is compromised. For example, cohort differences in autumn-dressed carcass weights were most pronounced as 2-year-olds varying from as little as 40 kg to 58 kg (Fig. 2.4), whereas when compared as 8-year-olds, the range in mean weights was only 4 kg (69 to 73 kg) and was not significantly different. Cohort variation in fecundity as 2-year-olds was correlated closely ($r^2 = 0.823$, $P < 0.001$) to the observed variation in their body weight (Fig. 2.4).

Cohort differences in reproduction reflected climatic conditions during the latter part of their own gestation in populations of red deer on

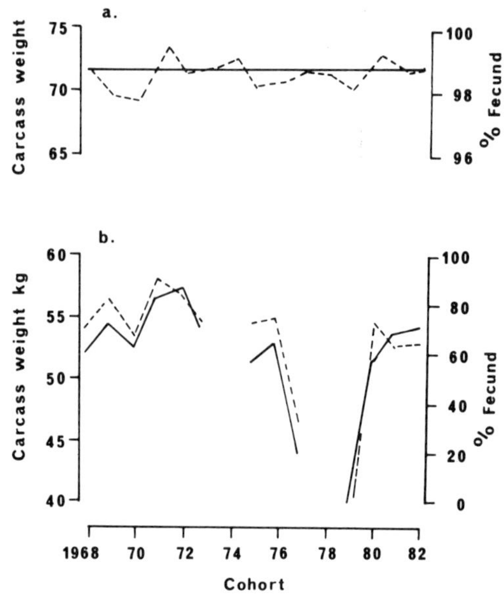

FIGURE 2.4. Mean autumn age-specific dressed carcass weight (solid lines) and fecundity (dashed lines) of cohorts of female red deer from Norway separated for samples (a) of 8-year-olds and (b) 2-year-olds. Fecundity was assessed from ovary histology.

Rhum and in Norway. In both populations, early growth and development are poor after cold springs; however, in Norway, cohorts initially exhibiting poor growth appear to catch up, and after 4 years of age calve every year, whereas on Rhum those cohorts suffering retarded early growth never appear to compensate, and in some cases, can barely calve every other year. Surprisingly, the most fecund cohort in Norway was also the most fecund cohort on Rhum (1971) while the two cohorts with lowest fecundity in Norway (1977, 1979) were among the poorest on Rhum.

Cohort Quality and Population Stability

The magnitude of cohort differences in reproduction and survival (quality) and the extent to which they are inherited across generations have important implications for the dynamics of populations (Albon et al. 1987). Using the Rhum data the potential fluctuation in numbers has been illustrated using simple Leslie matrix models. In the first model, good cohorts (those born in top third of spring temperatures tending to show highest fecundity and survival), average cohorts (middle third of spring temperatures), and poor cohorts (bottom third of spring temperatures) were recruited at random to a population where annual fecundity and survival rates were density-dependent (Albon and Clutton-Brock 1988). When the proportion of poor cohorts in the breeding population was low, numbers increased from the simple density-dependent asymptotic population size (K level). In contrast, when the proportion of poor cohorts was high, numbers fell. Thus, even though cohort quality was varied at random, the amplitude of the fluctation about the asymptotic population size was ±15% (Fig. 2.5a). Clearly, if there were long-term changes in the spring weather responsible for generating the initial cohort qualities then the amplitude of fluctuations could be much larger. This second scenario was simulated by using the actual April–May temperature data for the 100 years from 1885 to 1984 at Stornoway to generate the sequence of good, average, and poor cohorts. In the first 25 years of the period (1885–1909) 60% of springs were cold (falling in the lower third of the distribution), while from 1935 to 1959 only 16% were cold. This second model predicted a low population in the early part of this century rising to a peak population around 1960, which was approximately 50% larger than the lowest level (Fig. 2.5b).

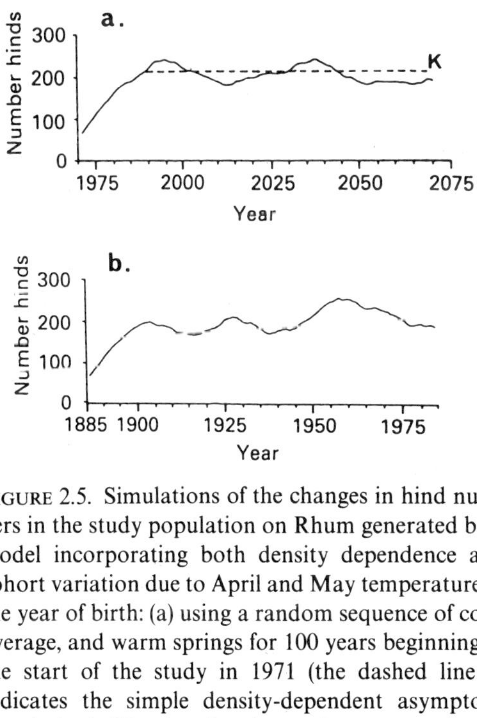

FIGURE 2.5. Simulations of the changes in hind numbers in the study population on Rhum generated by a model incorporating both density dependence and cohort variation due to April and May temperature in the year of birth: (a) using a random sequence of cold, average, and warm springs for 100 years beginning at the start of the study in 1971 (the dashed line K indicates the simple density-dependent asymptotic population); (b) using the observed sequence of cold, average, and warm springs for the 100 years from 1885 at Stornoway, Western Isles, Scotland.

In contrast to Rhum, the cohort differences in the fecundity of 2- and 3-year-old Norwegian red deer are unlikely to generate fluctuations in numbers because the presumed cross-generation environmental heritability occurs for such a small proportion of the life that there will be no life-long feedback. However, this assumes that lifespan and offspring viability do not show longer term cohort variation. Unfortunatley these data are not available.

Density-Induced Cohort Variation in Moose Survival on Isle Royale

Although so far we have dealt only with stochastic, density-independent cohort variation, differences between cohorts may arise as the result of density-dependent changes in populations and may also generate population fluctuation. For example, on Isle Royale, cohorts of moose born during a peak in population density showed poor survival, due to increased malnutrition, compared with cohorts born at lower, pre-peak population densities (Peterson et al. 1991). Although poor survival eventually prompted a decline in moose density, survival of subsequent cohorts did not improve. This lag in response may occur because of two, not necessarily mutual, reasons. First, the individuals recruited at higher densities showed lower fecundity and gave birth to less viable offspring. Second, the delay in the build-up of wolf numbers resulted in increased predation (Messier 1991). The important consequence of this lag, however it arises, is that the population will tend to exhibit perpetual waves and troughs with a periodicity measured in decades (see Peterson et al. 1984).

Conclusions

In this paper we have indicated how life-long differences in cohort reproduction and survival may lead to population fluctuation. Fluctuating numbers may have important consequences for the conservation and management of deer, especially in marginal habitats that may be particularly sensitive to long-term changes in climate like the predicted global warming.

The temporal heterogeneity we have described above ignores the within-cohort variation in fitness. Differences between subpopulations (spatial heterogeneity) and categories of animals within subpopulations may either obscure or reinforce temporal heterogeneity, and therefore may have additional consequences for population fluctuation. For example, recent investigations of genetic differences in fitness components in red deer on Rhum have shown that the depressed fecundity of cohorts born after cold springs is largely confined to a particular genotype at the polymorphic enzyme locus isocitrate dehydrogenase-2 (Pemberton et al. 1991). Selection for particular genotypes interacting with environmental change emphasizes the need for long- term, detailed studies of individuals in populations in contrasting habitats to improve our understanding of the demography of deer.

Acknowledgements

We are especially grateful to Fiona Guinness who has monitored the red deer population on Rhum for more than 20 years. In addition we would like to thank the many colleagues and volunteers who have helped during the study. In Norway, we thank the local wildlife boards for coordinating the data collection by hunters, and the Norwegian Institute for Nature Research (NINA) for financial and logistic support. The work on Rhum is financed by the NERC, SERC, and the Royal Society. The analysis of the Norwegian data was done when SDA was Visiting Scientist at NINA funded by a Royal Society Overseas (Europe) Fellowship and NINA.

References

Albon SD, Clutton-Brock TH (1988) Climate and the population dynamics of red deer in Scotland. In Usher MB, Thompson DBA (eds) Ecological Change in the Uplands. British Ecological Society Special Pub No 7, Blackwell Scientific Publications, Oxford, pp 93–107

Albon SD, Clutton-Brock TH, Guinness FE (1987) Early development and population dynamics in red deer. II. Density-independent cohort variation. J Anim Ecol 56:69–81

Albon SD, Guinness FE, Clutton-Brock TH (1983a) The influence of climatic variation on the birth weights of red deer calves. J Zool (London) 200:295–297

Albon SD, Mitchell B, Staines BW (1983b) Fertility and body weight in female red deer: a density-dependent relationship. J Anim Ecol 52:969–980

Begon M, Mortimer M (1986) Population Ecology: A Unified Study of Animals and Plants, 2nd edn. Blackwell Scientific Publications, Oxford

Burkey TV (1989) Demography of Norwegian red deer, *Cervus elaphus*. Thesis, University of Oslo, Norway

Caughley G (1977) Analysis of Vertebrate Populations. Wiley, London

Clutton-Brock TH, Albon SD, Guinness FE (1989) Fitness costs of gestation and lactation in wild mammals. Nature 337:260–262

Clutton-Brock TH, Albon SD, Guinness FE (1988) Reproductive success in male and female red deer. In Clutton-Brock TH (ed) Reproductive Success: Studies of Individual Variation in Contrasting Breeding Systems. University of Chicago Press, Chicago, pp 325–343

Clutton-Brock TH, Major M, Guinness FE (1985) Population regulation in male and female red deer. J Anim Ecol 54:831–846

Clutton-Brock TH, Guinness FE, Albon SD (1982) Red Deer: Behaviour and Ecology of Two Sexes. University of Chicago Press, Chicago

Hamilton WJ, Blaxter KL (1980) Reproduction in farmed red deer. I. Hind and stag fertility. J Agri Sci (Cambridge) 95:261–273

Hassell MP, May RM (1985) From individual behaviour to population dynamics. In Smith RM, Sibley RH (eds) Behavioural Ecology: Ecological Consequences of Adaptive Behaviour. Blackwell Scientific Publications, Oxford, pp 3–32

Iason GR, Duch CD, Clutton-Brock TH (1986) Grazing and reproductive success of red deer: the effect of local enrichment by gull colonies. J Anim Ecol 55:507–515

Krebs CJ, Cowan IMcT (1962) Growth studies of reindeer fawns. Can J Zool 40:863–869

Langvatn R, Albon SD (1986) Geographic clines in body weight of Norwegian red deer: a novel explanation of Bergmann's rule? Holarctic Ecol 9:285–293

Lomnicki A (1988) Population Ecology of Individuals. Princeton University Press, Princeton

Lomnicki A (1980) Regulation of population density due to individual differences and patchy environments. Oikos 35:185–193

Messier F (1991) The significance of limiting and regulating factors on the demography of moose and white-tailed deer. J Anim Ecol 60:377–394

Mitchell B (1984) Effects of the severe winter of 1962/3 on red deer hinds and calves in North-east Scotland. Deer 6:81–84

Pemberton JM, Albon SD, Guinness FE, Clutton-Brock TH (1991) Countervailing selection in different fitness components in female red deer. Evolution 45:93–103

Peterson RO, Page RE, Risenhoover KL (1991) Lifetime survival of moose under a regime fo wolf predation. Ecology (in press)

Peterson RO, Page RE, Dodge KM (1984) Wolves, moose and the allometry of population cycles. Science 224:1350–1352

Sinclair ARE (1989) Population regulation in animals. In Cherrett JM (ed) Ecological Concepts: The Contribution of Ecology to an Understanding of the Natural World. Blackwell Scientific Publications, Oxford, pp 197–242

Thorne ET, Dean RE, Hepworth WG (1976) Nutrition during gestation in relation to successful reproduction in elk. J Wildlife Management 40:330–335

3

Habitat Relationships and Population Dynamics of Deer in the Intensively Farmed Midwestern United States

Charles M. Nixon and Lonnie P. Hansen

Abstract

In the midwestern farm belt, where forest cover often totals < 10% of the landscape, deer (*Odocoileus virginianus*) are well fed, very fertile, and relatively short-lived. An average of 75% to 80% of fawn does and nearly 100% of older does breed each fall, and recruitment of fawns to the hunting season may exceed 90%. Available habitats may be categorized on the basis of annual deer survival: (1) larger, wooded refuges from hunting where annual survival is high (> 70%) for fawns and females and moderate (> 60%) for males; (2) large forests that are hunted each fall where survival of fawns and females is moderately high (> 50%) and that of antlered males is low (< 26%); and (3) small woodlots and strands of bottomland forest along rivers and streams where both sexes are frequently hunted to extinction. Seasonal movements of deer among these habitats are extensive and contribute to a wide dispersion of deer in summer and fall.

Key words: Habitats, midwestern United States, population dynamics, movements, survival, recruitment, white-tailed deer

Introduction

We report the demographic characteristics and seasonal movements of a sample of deer marked in east-central Illinois, a portion of the intensively farmed midwestern United States. Abundant deer populations are a relatively recent addition to the fauna of these landscapes traditionally dominated by agriculture. Restricted legal harvests of deer and carefully nurtured transplants have enabled farmland deer to recover from virtual extermination in the early years of the twentieth century. These deer have adjusted to a landscape that is nearly devoid of forest (generally < 10%), but that offers abundant foods produced on very productive farms. Deer must adapt to the annual growth and harvest of corn (*Zea maize*) which provides nearly unlimited diurnal cover that is destroyed at harvest. Although small woodlots are abundant, many are too small to provide protection for deer in winter, so deer must move to and from these small woodlots to more secure sites. This fosters extensive seasonal migrations. Hunting may exterminate deer from these woodlots, and replacement deer must locate and recolonize these "islands of cover" each spring and summer. Foods, although always abundant, are spatially variable because of annually recurring changes in crop selection.

Environment

East-central Illinois is a level, poorly drained plain of relatively young glacial drift, formerly nearly covered with prairie grasses before white settlement in the nineteenth century. Today this area is one of the world's most intensively farmed regions. Forests cover about 2% and row crops (nearly all corn or soybeans [*Glycene max*]) about 86% of the landscape. Forests are confined to areas too wet or rough to farm. Upland forests are understocked, formerly pastured mixtures of hardwoods such as elms (*Ulmus* spp.), honey locust (*Gleditsia triacanthos*), black cherry (*Prunus serotina*), sassafras (*Sassafras albidum*), oaks (*Quercus* spp.), and hickories (*Carya* spp.). Bottomland forest composition varies depending on the frequency of flooding. Stands that are frequently inundated are nearly a monotype of silver maple (*Acer saccharinum*), and better drained sites are mixtures of various hardwoods such as elms, hackberry (*Celtis occidentalis*), sycamore (*Platanus occidentalis*), bur oak (*Quercus macrocarpa*), and silver maple. Densities of shrubs and small trees are generally highest in upland stands (3630 to 13 740 stems per ha) and lowest in flooded bottoms (< 800 stems per ha).

The climate is temperate continental, with cold winters and warm summers. January, the coldest month, averages −3.1° C, and July, the warmest, averages 23.6° C. Precipitation averages about 940 mm annually and is well distributed throughout the year. Snow seldom covers the ground for an extended period, but occasional blizzards can restrict deer movement and access to ground foods for short (1 to 2 weeks) periods (Illinois State Water Survey records for Urbana, Illinois).

Methods

We used rocket nets to capture and mark 289 deer on scattered refuges in east-central Illinois. Deer were ear tagged and aged according to tooth replacement and wear as fawn (≤ 12 months), yearling (13 to 24 months), or adult (> 25 months). Females were marked with 7.5-cm-wide plastic collars bearing reflective numbers or letters ($n = 80$), with vinyl-impregnated plastic strips fastened to plastic cattle ear tags ($n = 6$) (Y-Tex, Cody, Wyo.) or with radio collars ($n = 57$) (Wildlife Materials, Carbondale, Il. or Telonics, Mesa, Ariz.). Six females were first marked with plastic collars and later with radio collars. Males were marked with the plastic ear streamers ($n = 103$) or with both ear streamers and with radio collars ($n = 37$) designed to accommodate the neck expansion associated with breeding.

Recruitment of fawns was estimated from repeated observations of marked does of known ages during the summer and early fall. Recruitment was defined as the number of live fawns at heel per marked doe of known age on 1 October, the beginning of archery hunting season in Illinois. Our estimates of fawn recruitment were compared with fetus counts from does from two other farmland populations in west-central Illinois and Iowa. Means of fawn production within age classes of does were compared using the single class ANOVA for unequal sample sizes (Sokal and Rohlf 1969).

Seasonal and annual mortalities were estimated from the fate of marked deer and were separated according to habitat as follows: (1) deer that remained within refuge boundaries after marking, (2) deer marked on large hunted forests or that migrated or dispersed to these forests after marking, and (3) deer marked within small hunted woodlots, or that migrated or dispersed to these small forest remnants after they were marked. Survival was defined as the proportion of deer surviving the prebreeding (June-September), breeding (October-December), and postbreeding (January-May) periods, as well as annually in each habitat. Habitat related survival probabilities were compared by a R × C test of independence using the G test (Sokal and Rohlf 1969).

Estimates of annual emigration and migration were based on movements of radio-marked deer and from marked deer observed by us or reported by the public and verified by us. Population trends for refuge and hunted woodlots were simulated using a population model written in Applesoft Basic. Age-, sex-, and time-specific recruitment, mortality, and seasonal movements of marked deer served as input. We were unable to develop a realistic model without including immigration to sustain known mortality and dispersal rates. Simulations were used to compare population demographics for hunted and refuge-protected deer over a 5-year period.

Results

Fawn recruitment was high in east-central Illinois based on the number of fawns observed

TABLE 3.1. Counts of fawns with marked does observed in early fall in east-central Illinois compared with fetus counts from does collected in west-central Illinois and Iowa, USA

Location	Bred as fawn: Total does	Pregnant does (%)	Fawns/fetuses per pregnant doe: N	$\bar{X}$	SE	Bred as yearling: Total does	Pregnant does (%)	Fawns/fetuses per pregnant doe: N	$\bar{X}$	SE	Bred as adult: Total does	Pregnant does (%)	Fawns/fetuses per pregnant doe: N	$\bar{X}$	SE
East-central Illinois	59	76	45	1.13	0.05	39	100	39	1.82	0.06	115	100	115	2.1	0.04
West-central Illinois[a]	30	77	19	1.11	0.07	14	93	8	2.0	0.18	50	100	49	1.92	0.05
Iowa[b]	73	84	53	1.21	0.07	37	100	27	1.96	0.06	62	98	29	2.1	0.13

[a]Source: Pederson (1982)
[b]Source: Haugen (1975)

with marked does in early fall (Table 3.1). These recruitment totals, including all mortality from conception, were not significantly different . (For fawns the values were $F = 0.52$, 2,116 df, $P > 0.90$; for yearlings values were $F = 0.92$, 2,71 df, $P > 0.90$; and for adults values were $F = 3.0$, 2,190 df, $P < 0.20$) from fetus counts taken from two other farmland populations in west-central Illinois (Pederson 1982) and Iowa (Haugen 1975). In addition we marked 17 fawns (12 males, 5 females) that were < 1 week old; all survived until the hunting season.

If fetus counts are assumed to be close to those reported for Iowa does (Table 1), fawn breeders in east-central Illinois would have lost 6%, yearlings 8%, and adults < 1% of their fawns pre- and postpartum by early fall. However, observations of marked does of known ages did not indicate that fawn and yearling mothers suffered higher fawn losses than older does (1-year-old mother = 0% loss of fawns to weaning age; 2-year-old = 7.4% loss; 3-year-old = 19.8% loss; 4-year-old = 7.9% loss). Most of these fawns were lost in late summer and fall, as summer losses were low for all age classes.

Seasonal movements among marked deer were extensive in east-central Illinois. Nearly half of male and female fawns and an average of 21% of yearling females dispersed away from the capture sites each spring (Table 3.2). Deer dispersing from these restricted (in space) capture sites settled within six watersheds in Illinois, an area of about 40 000 km^2. Six of nine spring-dispersing female yearlings with known breeding histories were either barren as fawns or lost their fawn(s) before fall. These yearlings without fawns may have reverted to the social position of a fawn before dispersing (Ozoga and Verma 1986). A few yearling males (16 to 17 months old) dispersed in the fall, but moved shorter distances than spring-dispersing fawns (Table 3.2). Marked adults of both sexes did not show dispersal behavior.

An average of 19.6% of the marked fawns and older does migrated seasonally, moving to summer ranges between April and early June and returning to marking sites between October and early January (Table 3.2). For does whose summer locations were known, yearlings migrated almost three times as far as adult does (yearling $x = 20.8$ km, adult $x = 7.3$ km). Yearling and adult males were not observed to migrate. Instead they showed seasonal shifts in home ranges within large annual ranges which were selected when males were between 10 and 17 months of age. The migratory movement of deer onto more secure sites (generally more forest or permanent cover) during winter has been documented throughout the agricultural area of the Midwest (Pietsch 1954; Sparrowe and Springer 1970; Zagata 1972; Zwank 1974).

The landscape characteristics that influence deer survival probabilities in east-central Illinois may be subdivided into three categories: (1) public and private refuges of varying size

TABLE 3.2. Extent of dispersal and migratory behavior among marked deer in east-central Illinois, 1980–1985

Seasonal movements	Age	Sex	Season	Number marked	Average of marked population (%)	Range (%)	Average distance moved (km) $\bar{X}$	SE
Dispersal	Fawn	M	Apr-Jun	113	51	38–75	40.9	5.0
		F	Apr-Jun	84	50	42–64	49.5	4.8
	Yearling	M	Oct-Nov	90	17	0–50	10.5	2.6
		F	Apr-May	43	21	0–40	20.4	3.0
	Adult	M		27	Not observed to disperse			
		F		61	Not observed to disperse			
Migration	≥ Fawn	M		141	Not observed to migrate			
		F	Mar-Jun	184	20	17–21	13.9	1.8

(< 100 ha to > 500 ha) where deer are protected from firearm hunting and public access is controlled, (2) large (> 100 ha) forests where hunting is allowed but controlled—the relatively low number of hunters and size of the forests provide deer with sanctuary during fall hunting and from inclement winter weather, and, (3) small (generally < 50 ha) woodlots and narrow strands of forest along streams and rivers that provide adequate cover from spring through fall, but do not offer adequate protection during the hunting season or in winter.

Differences in the ability of these habitats to provide seasonal protection are reflected in

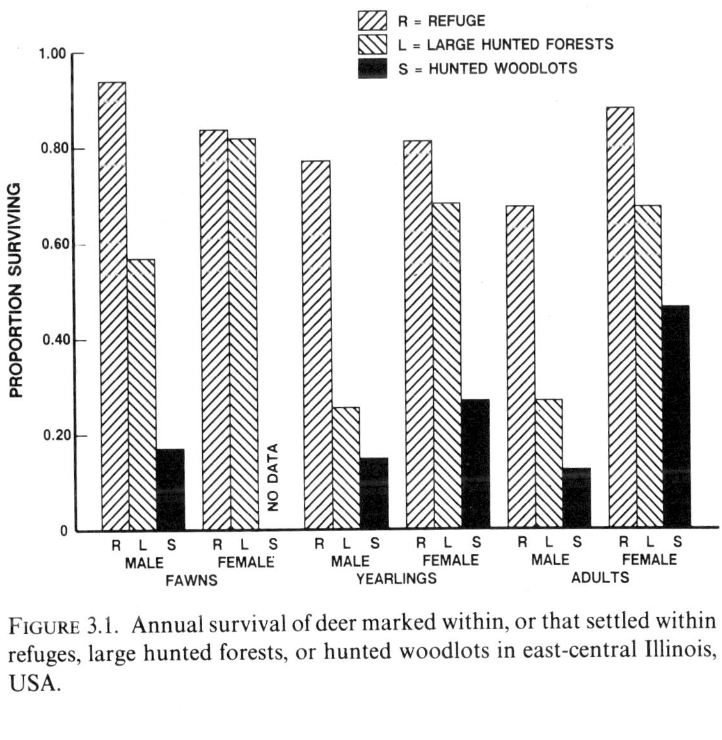

FIGURE 3.1. Annual survival of deer marked within, or that settled within refuges, large hunted forests, or hunted woodlots in east-central Illinois, USA.

TABLE 3.3. A 5-year simulation of refuge-protected and hunted deer in east-central Illinois. The initial population was 80 deer in May 1981. Survival rates were those of marked deer on refuge-protected and hunted areas in east-central Illinois[a]

	Total deer					
	Refuge			Hunted		
Period	Oct	Jan	Jun	Oct	Jan	Jun
1981–82	130	126	84	129	92	60
1982–83	144	139	87	100	73	46
1983–84	156	148	99	74	53	37
1984–85	190	166	122	61	44	43
1985–86	209	175	113	70	52	45
Finite rate of change[b]	1.12	1.09	1.08	0.543	0.565	0.750

[a]Source: Nixon et al. (1989)

[b]Calculated as = antilog$_e r$ where $r = \frac{\ln N \text{ final interval} - N \text{ initial interval}}{\text{No. of intervals}}$

survival of marked deer residing in these habitats (Fig. 1). Annual survival was significantly higher ($P < 0.01$) for both sexes on refuges compared with large or small hunted forests. These differences were the result of reduced survival for adult females during summer away from a refuge ($G = 8.54$, 2 df, $P < 0.03$) and much reduced survival of all age classes and both sexes exposed to hunting ($P < 0.001$; Fig. 1). Annual survival was also higher ($P < 0.05$) for fawn males ($G = 5.73$, 1 df, $P < 0.05$) and yearling and adult females (yearlings: $G = 14.45$, 1 df, $P < 0.001$; adults: $G = 6.52$, 1 df, $P < 0.02$) residing in large forests than for those inhabiting smaller woodlots. There was no significant difference in survival of yearling and older males in large hunted and small hunted woodlots. The larger home ranges of these males, their more frequent daily movements in the fall, and hunter preference for antlered males combined to render males vulnerable to harvest in both habitats.

A comparison of 5-year simulations (1981–1985) of deer abundance within and away from a refuge illustrates the impact of hunting on deer in east-central Illinois (Table 3.3). Simulated refuge populations showed an increase in numbers during this period while unprotected deer declined in abundance. These unprotected deer are frequently extirpated from small woodlots, which are restocked by immigrants from refuges and larger forests.

Discussion

Recent deer demographics in east-central Illinois are the result of three processes: (1) increasing deer numbers on large refuges, (2) stable to slowly increasing numbers on large hunted forests, and (3) fluctuating numbers of deer within small woodlots and bottomland forests where deer are frequently hunted to extinction (Fig. 3.2). The growth and harvest of the "corn forest" have a tremendous impact on food and cover availability each year. Deer seasonally adjust activities according to expanded cover provided by crops in summer and the reduced cover in fall and winter as these crops are harvested. The flow of deer from high-to-low density occurs in spring (both sexes) and fall (yearling males), with dispersals and migrations occurring as deer move from larger refuges to smaller woodlots and flood prone bottomland forests. In summer and fall these sites offer relatively risk-free parturition sites for females, habitat for fawns, and rich feeding areas for males building body reserves for the coming breeding season.

Postweaning survival of a fawn depends upon the fawn's location, whether on or away from a

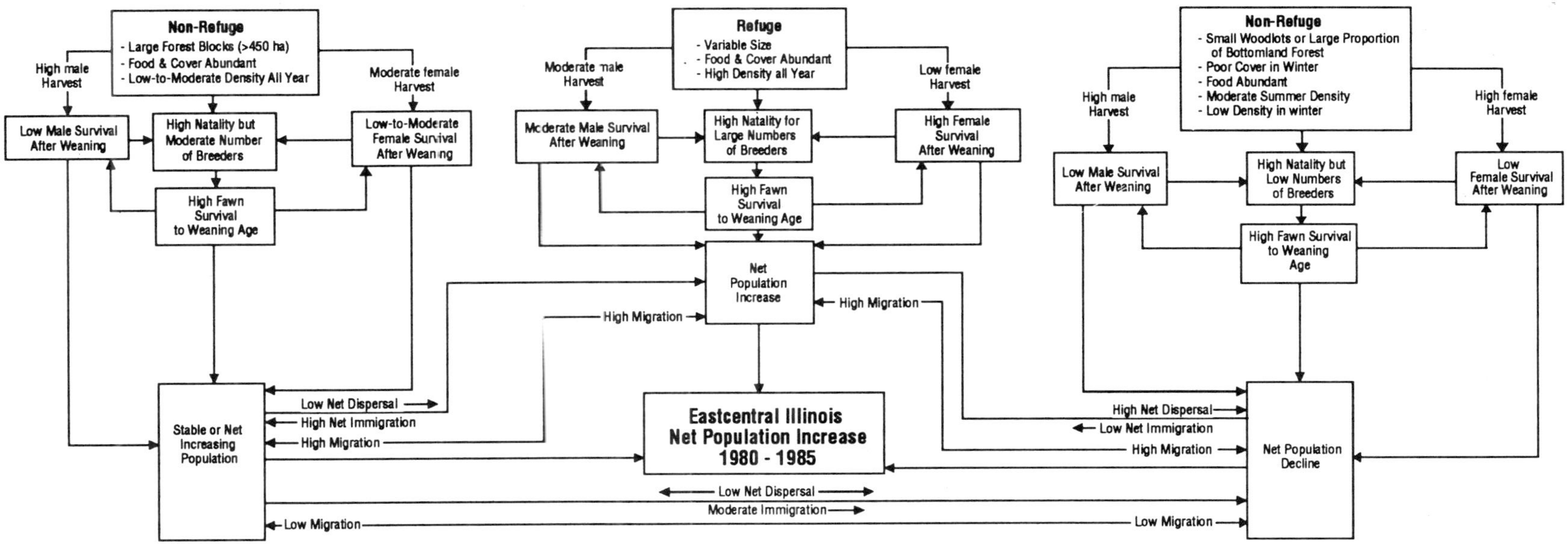

FIGURE 3.2. A conceptual summary of demographics for deer residing on, or away from, refuges in east-central Illinois, USA.

refuge, as well as the seasonal movement pattern of its mother. Seasonal movement between widely scattered summer and winter ranges described for deer in more northern ranges (Teirson et al. 1985; Nelson and Mech 1987) is not widespread in east-central Illinois. It occurred in < 20% of the females and in almost none of the males. About one-half of the male fawns alive in the spring dispersed away from their natal ranges between 1 May and 30 June at age 11 to 13 months. Another 20% dispersed in the fall at the beginning of the rut. Thus, only about 39% of the males remained close to their natal ranges after family break-up, and established a large home range surrounding their mother's ranges. About 50% of the female fawns dispersed away from their natal ranges each spring. This behavior is not typical of white-tails throughout most of their continental range. Male dispersal is considered the more normal behavior in this species (Downing et al. 1969; Nelson and Mech 1981), with females remaining close to other female relatives throughout life. Only within the greatly fragmented ranges of the midwestern United States do females show this extensive dispersal behavior (Sparrowe and Springer 1970; Zagata and Haugen 1973; Zwank 1974; Nixon et al. 1989). We attribute this behavior to the large number of fawns that survive into the spring months and the paucity of available forest cover in the spring.

Nearly 45% of our spring populations were composed to fawns, and this number may simply be too large for inclusion within existing social groups. In addition, about 75% of these primiparous females were pregnant each year, and Ozoga et al. (1982) have described the difficulties encountered by captive primiparous does in locating sites for parturition on ranges crowded with older does. These primiparous females usually dispersed later than yearling females and female migrants, as late as mid-June, after parturition for many older does. Many of these fawn does settled within woodlots and flood-prone bottomland forests that were vacant in spring because they offer poor winter shelter or have been "shot out" in the fall. Such sites are "dispersal sinks" where deer survival is low (Fig. 3.1).

The dispersion of suitable wintering sites and connecting forested corridors that promote deer dispersal and migration between these areas and existing summer ranges remains the key to continued occupancy of the intensively farmed Midwest. Retention of existing upland forest, some offering refuge protection, is particularly important because bottomland forests are seldom levied in east-central Illinois and are subject to periodic flooding. Unfortunately, for deer management, upland forests are desirable home-building sites and are likely to become more scarce in the midwestern United States as human densities increase and forests are further fragmented or destroyed.

Acknowledgements

We thank the many unpaid volunteers who helped us capture deer. P.A. Brewer and J.E. Chelsvig made invaluable contributions to the study. G.C. Sanderson and the editorial staff of the Illinois Natural History Survey reviewed the manuscript. The study was cooperatively supported by Federal Aid in Wildlife Restoration Project W-87-R, the Illinois Department of Conservation, the U.S. Fish and Wildlife Service, and the Illinois Natural History Survey.

References

Downing RL, McGinnis BS, Petcher RP, Sandt JL (1969) Seasonal changes in movements of white-tailed deer. In White-tailed Deer in the Southern Forest Habitat. US Forest Service Southern Forest Experiment Station, New Orleans, La., pp 19–24

Haugen AO (1975) Reproductive performance of white-tailed deer in Iowa. J Mammalogy 56:151–159

Nelson ME, Mech LD (1987) Demes within a northeastern Minnesota deer population. In Chepko-Sade BD, Halpin ZT (eds) Mammalian Dispersal Patterns: The Effects of Social Structure on Population Genetics, University of Chicago Press, Chicago, pp 27–40

Nelson ME, Mech LD (1981) Deer social organization and wolf predation in northeastern Minnesota. Wildlife Monogr 77

Nixon CM, Hansen LP, Brewer PA, Chelsvig JE (1989) Life history and ecology of deer in an intensively farmed region of Illinois. Illinois Department of Conservation, Federal Aid in Wildlife Restoration Performance Report Project W-87-R-10

Ozoga JJ, Verme LJ (1986) Initial and subsequent maternal success of white-tailed deer. J Wildlife Management 50:122–124

Ozoga JJ, Verme LJ, Bienz CS (1982) Parturition behavior and territoriality in white-tailed deer: impact on neonatal mortality. J Wildlife Management 46:1–11

Pederson VC (1982) Productivity of various age classes of Illinois deer. In Loomis F, Kube J, Garver J (eds) Illinois Department of Conservation, Job Completion Report W-63-R, 1981–1982, Springfield, pp 108–141

Pietsch LR (1954) White-tailed deer populations in Illinois. Illinois Natural History Survey Biological Note 34

Sparrowe RD, Springer PF (1970) Seasonal activity patterns of white-tailed deer in eastern North Dakota. J Wildlife Management 34:420–431

Sokal RR, Rohlf FJ (1969) Biometry, WH Freeman, San Francisco

Teirson WC, Mattfield GF, Sage RW Jr, Behrend DF (1985) Seasonal movements and home ranges of white-tailed deer in the Adirondacks J Wildlife Management 49:760–769

Zagata MD (1972) Range and movement of Iowa deer in relation to Pilot Knob State Park, Iowa. Ph.D. thesis, Iowa State University, Ames

Zagata MD, Haugen AO (1973) Winter movement and home range of white-tailed deer at Pilot Knob State Park, Iowa. Proc Iowa Acad Sci 79:74–78

Zwank PJ (1974) Refuges as population centers for deer in threatened habitat. M.Sc. thesis, University of Missouri, Columbia

4
Current Status of Red and Fallow Deer Populations in Chile: The Need of Management

Claudio R. Ortiz

Abstract

Red deer (*Cervus elaphus*) and fallow deer (*Dama dama*) were introduced in Chile by descendants of German immigrants. The main introductions were carried out in 1926 for fallow deer and between 1948 and 1953 for red deer. Some feral populations descending from these introductions are managed according to a system similar to that currently used in Germany. There are about 700 feral fallow deer on an estimated area of 14 000 ha. About 4200 feral red deer are spread over an estimated area of 340 000 ha. Although in many areas these populations are under management for trophies and are also heavily poached, more intense and generalized management is needed considering the impact they might have on the native fauna and flora, and the potential that they have as a natural renewable resource.

Key words: Chile, deer introduction, dispersion, fallow deer, passive migration, red deer

Introduction

Three deer species were introduced to Chile some decades ago: red deer, fallow deer, and roe deer (*Capreolus capreolus*). Of these species, red deer are of special concern due to the negative impact they may have in some ecosystems as they have in New Zealand (Clarcke 1972; Challies 1985). Research in Chile about the impact that these exotic deer species have on native ecosystems has lacked continuity. The main efforts have been oriented towards the deer's impact on vegetation (Heitzer y Ramírez 1980; Ramírez et al. 1981), social behavior and feeding habits (Eldridge 1983), habitat preferences and activity patterns (Eldridge et al. 1980), management (Schürholz 1975), and parasitology (Díaz et al. 1977; Gasic 1983; Hott 1984). Two studies have described the introduction, evaluated the distribution, and estimated the abundance of introduced deer populations (Schilling 1978; Wollenhaupt 1983). The only source of population information has been that provided by deer owners, farmers who either permanently or occasionally have deer on their lands, and estimates made by people who live in areas where these species occur. This chapter will review the current status of exotic feral deer in southern Chile based on these sources of information.

Historical Background of the Introduction of Deer

Fallow Deer Introductions

Chapman and Chapman (1980) mention an introduction of fallow deer in Chile in 1887; however, no descendants of these deer exist today. Later, in 1926 two bucks and four does were brought from a zoological park in Hamburg, Germany in 1926 (T. Grob, personal communication) and placed in a private island in Ranco Lake, X region (Fig. 4.1). In 1928, more fallow deer were imported from Germany with other game species. These were released in a fenced area within a private farm in the IX Region (J. Vidal, personal communication). Another group was imported and placed near

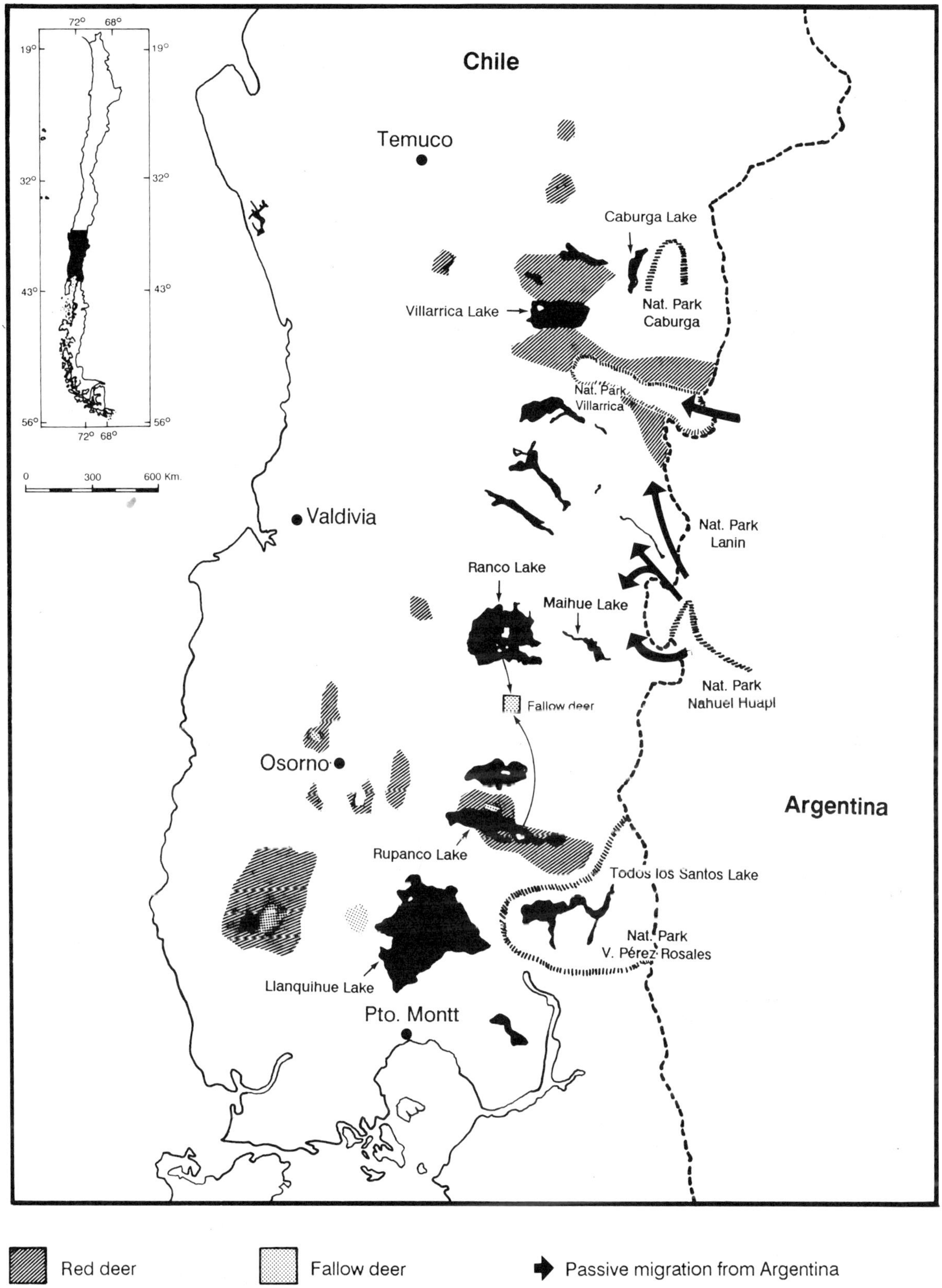

FIGURE 4.1. Current areas of fallow and red deer dispersion between the IX and X regions of Chile.

Valparaíso, in central Chile, V region. All of these imports produced stable populations, and the current feral or captive fallow deer descend from these introductions.

Red Deer Introductions

Red deer became established in Chile by active introductions and by passive migrations from Argentina. The first 11 red deer were introduced from Germany in 1928 (Wollenhaupt 1983) along with the fallow deer mentioned and some roe deer and wild boars. In 1948, another private import took place when eight deer were brought from Hacienda Collunco, Argentina, where this species was also exotic. These were placed on a private farm on the northern side of Villarrica Lake, IX region (E. Wagner, personal communication). The latest introduction occurred between 1952 and 1953 when a state-supported import of 138 calves from Argentina was undertaken (Wollenhaupt 1983). These deer were placed on four different farms and one lacustrine island of the X Region. This group became the main ancestral pool of the current populations of red deer in Chile, and also the one from which some deer were moved to different parts of the X and XI Regions. Given that red deer from Argentina provided the primary genetic basis of the current Chilean populations, it is important to note that Argentinian red deer have their origin in the old Austro-Hungarian Empire, specifically in the Hungarian Carpathian Mountains and Austrian Alps. They were introduced in Argentina in 1904 (Wollenhaupt 1983).

Passive migration from Argentina has occurred in recent years (Schürholz 1975). Currently, it takes place from (1) Lanín National Park (Argentina) through Mamuil Malal path to Villarrica National Park, and through Huahum path to private and state-owned lands (Fig. 4.1); (2) from Nahuel Huapi National Park (Argentina), colonizing areas to the east of Ranco Lake (Fig. 4.1), and (3) surroundings of Lakes Fontana and La Plata (Argentina) entering the area of Mañihuales (D. Aldridge, personal communication; Fig. 4.2).

At the beginning of the introduction these species were hunted only for sport and without regard to monetary income, but today's hunting trophies and sales of venison have become attractive to farmers involved in deer management. For those concerned about the ecological impact of exotic species, the deer are a major preocupation, especially due to the negative interactions that might take place with the native pudú deer (*Pudu Puda*) and huemul (*Hippocamelus bisulcus*) and also with the flora.

Current Distribution of Free-Ranging Populations

Chile is a country with neither a big game hunting tradition nor a big game species to be hunted. Because of this, the presence of introduced large-antlered species has caused feelings of intense curiosity but also aversion to them, especially in farmlands where feral deer are blamed for consuming pastures or damaging crops. For these reasons these deer are heavily poached, usually by poor marksmen with inadequate firearms. The damage caused by deer, however, is a common excuse for poaching a surprisingly high number of stags, in contrast to hinds which also feed on the same pastures. This poaching has strongly affected the current sex ratio, number, and distribution of these deer species.

Introduced deer are mainly located in valleys between the Andean and Coastal Cordillera of the IX and X Regions, inhabiting the so-called Valdivian Rain Forest. In general terms they are distributed on low altitude farmlands with smooth slopes, away from the large areas of native forest that are found in higher relief areas of both cordilleras (Ramirez et al. 1981). In the XI Region, red deer occur in some valleys at the foothills of the Andean Cordillera, mostly inside dense bamboo (*Chusquea*) thickets.

Fallow Deer

Ever since their release, and despite intense poaching, fallow deer have remained in almost the same surroundings as their original enclosures. This behavior has facilitated their hunting to the point that fallow deer currently occur in high numbers only where they are under permanent protection. In most other areas they are almost extinct.

The largest feral populations are located on two lacustrine islands: Pichi-Colcuma (185 ha)

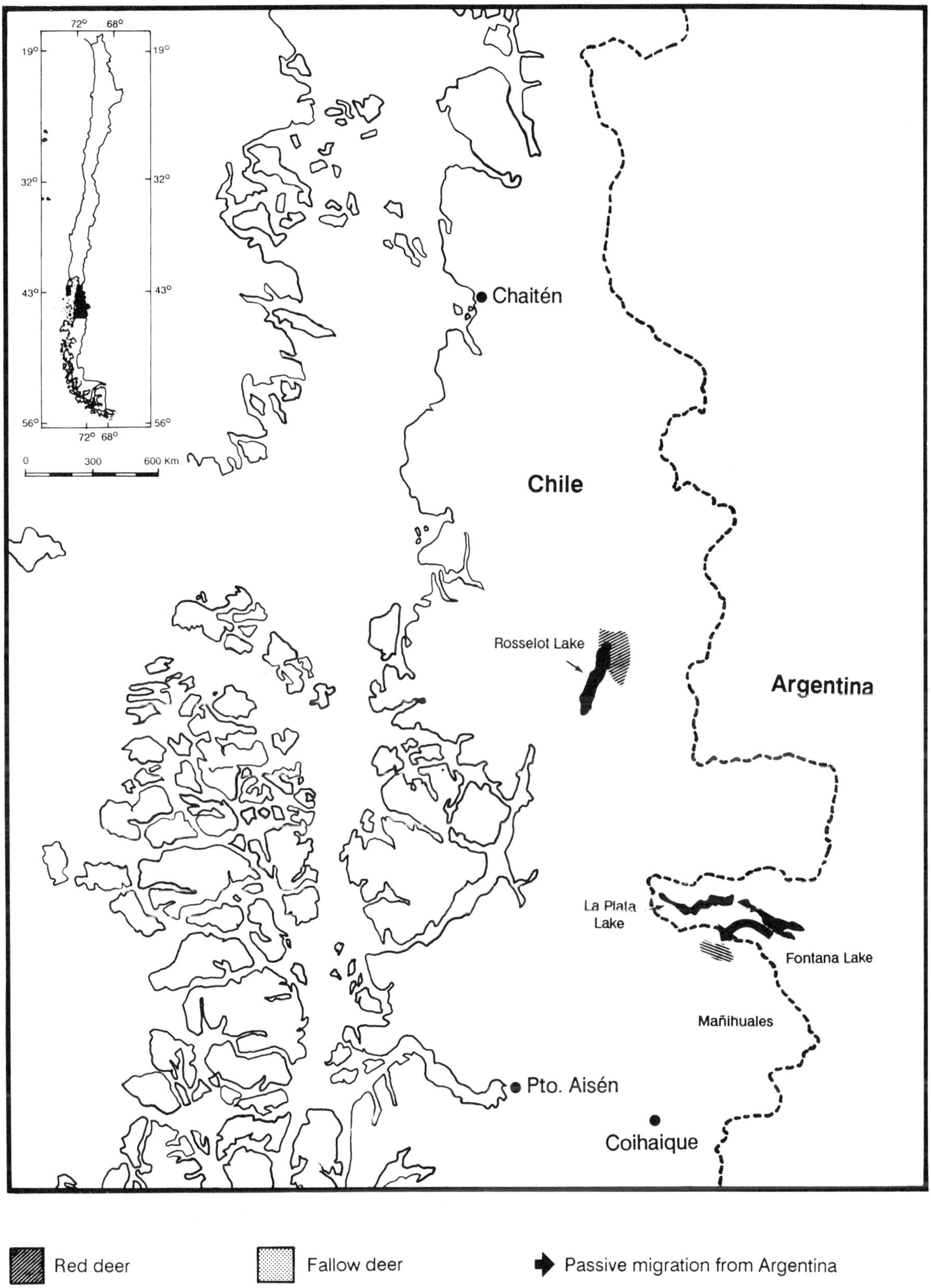

FIGURE 4.2. Current areas of red deer dispersion in the XI region of Chile.

TABLE 4.1. Estimated feral fallow and red deer populations in Chile

Region	Fallow deer	Red deer
V	—	—
VII	—	—
VIII	—	—
IX	60	1800
X	644	2050
XI	—	350
Total	704	4200

in Ranco Lake, and Altuéhuapi (135 ha) in Rupanco Lake (Fig. 4.1) with estimated populations of 250 and 130 deer, respectively (T. Grob and J. Schilling, personal communication). These populations are under permanent selective shooting, so that density is adjusted to an estimated carrying capacity. Venison and trophies are produced as by products of this management practice. The remaining free-ranging fallow deer are distributed in farmlands of the IX and X Regions, in herds seldom over 20 to 30 deer dispersed over an area estimated at about 14 000 ha (Table 4.1).

Red Deer

This species has shown a more invasive behavior, thus colonizing new areas of southern Chile. In many areas farmers have been carrying out habitat management as well as selective shooting for a number of years. This has resulted in relatively well-managed populations which produce first-rate trophies and healthy animals. However, this management has been affected by the negative impact of heavy poaching (almost always stag-oriented) and this has had a dispersing effect on red deer. As a consequence in certain areas red deer have become very shy, thus posing difficulties for selective shooting.

Red deer concentrate in specific areas of the IX and XI Regions (Fig. 4.1 and 4.2). The total estimated number of free-ranging red deer is about 4200, spread over about 340 000 ha of farmed and wild lands (Table 4.1). Although red deer are distributed only in some areas of these three regions, these regions cover more than 20 million ha.

Red Deer Population Control

Since fallow deer are almost extinct in farmlands, population control has importance only with reference to red deer.

Predation

In Chile, puma (*Felis concolor*) prey on deer calves and yearlings of both sexes (Courtin et al. 1980); these authors also mention other species which may prey mainly on newborns: chimango caracara (*Milvago chimango*), turkey vulture (*Coragyps atratus*), and gray fox (*Pseudalopex griseus*). However, the frequency of attacks from these predators is probably insufficient to control red deer populations to any significant extent. Two factors operate against the effectiveness of predation: (1) pumas and foxes are heavily hunted by man, and (2) these two carnivores usually seek easier prey such as lambs, hares, and chickens.

Diseases

All the scientific information so far reported concerns parasites. Diaz et al. (1977) diagnosed nematodes in fallow and red deer, but all the infestations were mild. Eldridge et al. (1980) pointed out that even though introduced deer in Chile are exposed to different diseases, it is very unlikely that diseases operate as a relevant natural control.

Hunting

This seems to be the logical choice for massive and selective control of feral deer. However, the Chilean hunting law includes red and fallow deer as protected animal species. This means that introduced deer are not to be hunted. Upon request of many deer owners some adjustments have been made in order to allow selective shooting provided that it takes place on grounds legally registered as hunting ranches. The authorities regard this shooting as a means to avoid migration due to overpopulation.

Venison Production

At present, venison production mostly involves harvesting wild deer. In order to better organize

TABLE 4.2. Venison processed from hunting ranches (only X Region)[a]

	Fallow deer		Red deer	
Year	Deer processed (*n*)	Venison produced (kg)	Deer processed (*n*)	Vension produced (kg)
1986	73	2100	102	7521
1987	52	1197	28	2126
1988	93	1905	89	6110
1989	122	2469	46	3554
1990[b]	67	1351	18	1817

[a]Source: Procarne, Osorno, Chile
[b]Up to 9 May 1990

the commercialization of meat derived from the deer owners' hunting, a contract operates between the National Association of Cervids Breeders (NACB) and an abbatoir and meat-processing center, Procarne. The processor receives carcasses only from members of the NACB who have registered hunting ranches. This contract provides a stable venison purchasing power only for members of the NACB, hence forcing other deer owners to join it (Table 4.2). Although small, these figures should be analyzed in light of local reality: (1) These deer represent the harvest obtained by only some members of the NADB encompassing about 7000 ha under management, (2) other meat is poached for other sources of consumption, and (3) some meat is consumed by the deer owners.

Management System Currently in Use

Selective hunting is done either on isolated hunting ranches, usually of over 500 ha and also referred as "Revier," or on integrated hunting units called "Hegeringe." They allow larger, undisturbed grounds where a higher number of deer can be sustained. This enables the production of more and better trophies, but it also implies a stronger commitment to annual culling. The management itself is based on the German system as described by Ueckerman (1987), and its basic objectives are to regulate the size of deer populations, and to qualitatively improve game through selective shooting. Although antler characteristics have been an important criterion for regulating stag selection, there is a stronger trend towards selection according to age classes as described by Bubenik (1986). However, for both sexes, body size, physical condition, and health status are also factors that determine selective shooting.

Density Regulation

Because there is no official harvest plan to be fulfilled, the deer owners themselves establish their own annual quota. Red and fallow deer are recent additions to the Chilean fauna, and since in many areas they have been hunted and poached, they have very seldom reached a density that could cause severe damage to the habitat. The existence of large territories with low human population density allow much space for deer migration, and a favorable climate and a varied, evergreen vegetation provide year-round forage availability. These factors, when added to the almost total absence of predators and competitors, the adaptive responses of deer to diverse habitats and types of food, and the legal protection which permanently forbids most hunting of deer (Ramirez 1981), may eventually cause a dangerous increase in deer density.

Final Comments

Since introduced deer in Chile are currently somewhat under control, it would not be appropriate to consider these species either as pests or as pests-to-be in the short term. Owing to the broad dispersion of red deer in Chile, their eradication would be an impossible task. Conversely, we should not pretend that they are not a potential threat to native ecosystems, should they become a pest. Therefore we may consider our current management system to be an efficient way of controlling deer populations if needed on a larger scale. Its advantages can be summarized as follows:

1. It is in the deer owner's interest to keep the deer within the limits of his property because they represent a source of additional income through sales of venison and trophies.
2. The Hege system in Chile has proven to be an efficient means for controlling population dispersal and habitat damage.

3. If exotic deer do not compete with the native fauna and can be kept in such numbers so that no detrimental effects to habitat are produced, then the utilization and management of red and fallow deer will be supported by the NACB.

It is obvious that more research concerning the interactions between introduced deer and native fauna and flora is needed. Important steps have been given through the initiative of those who are involved in deer management, although a broader participation from farmers who are receiving migrating deer is now required.

Acknowledgements

The author thanks the members of the National Association of Cervids Breeders (NACB) for their financial support and for providing information for this work. I am also indebted to F. Jaksic for his advice and for reviewing the manuscript.

References

Bubenik AB (1986) Grundlagen der Soziobiologischen Hirschwildbewirtschaftung (Mit Hinsicht auf das Problem der tragbaren Belastung der Forste). In Linn S (ed) Rotwild—Cerv rouge—Red deer. Conseil International de la Chasse et de la Conservation du Gibier GWI Druck, Munich.

Challies CN (1985) Establishment, control and commercial exploitation of wild deer in New Zealand. In Fenessy PF, Drew KR, (eds) Biology of the Deer Production. R Soc N Z Wellington, pp 23–36

Chapman NG, Chapman D (1980) The distribution of fallow deer: a worldwide review. Mammal Rev 10:61–138

Clarke CM (1972) Red deer in the northern South Island Region: their early impact. N Z J Forestry 17:37–42

Courtin SL, Pacheco NV, Eldridge, WD (1980) Observaciones de alimentación, movimientos y preferencias de hábitat del puma en el Islote Rupanco. Medio Ambiente 4:50–56

Diaz L, Rioseco H, Cubillos V (1977) Prospección y patología del parasitismo en cérvidos autóctonos y exóticos en el sur de Chile. Bol Chil de Parasitol 32:86–89

Eldridge W (1983) Impacto ambiental, alimentación y conducta social del ciervo rojo y dama en el sur de Chile. CONAF. Boletín Técnico No. 9

Eldridge WD, Pacheco NV, Courtin SL (1980) Preferencias de habitat y patrones de actividad del ciervo exótico en el sur de Chile. Medio Ambiente 4:56–74

Gasic RJ (1983) Estudio de la eliminación de ooquistes, huevos y larvas de parásitos producidos por Ciervo rojo (*Cervus elaphus* L.) en un criadero del sur de Chile a lo largo de un año. Tesis de Grado. Fac de Ciencias Veterinarias, Universidad Austral de Chile

Hott V (1984) Estudio preliminar de la tendencia en la eliminación de huevos y larvas de parásitos gastrointestinales en gamos (*Dama dama*) en el sur de Chile. Tesis de Grado. Facultad de Ciencias Veterinarias, Universidad Austral de Chile

Heitzer R, Ramirez C (1980) El impacto del ciervo rojo sobre la vegetación del islote Rupanco (Osorno, Chile). Medio Ambiente 4:75–81

Ramirez C, Godoy R, Eldridge W, Pacheco N (1981) Impacto ecológico del ciervo rojo sobre el bosque de olivillo en Osorno, Chile. Anales del Museo de Historia Natural de Valparaíso 14:197–215

Schilling J (1978) Estudio sobre la existencia de ciervos exóticos en el Sur de Chile. Mimeo. Asoc Nac de Criadores de Cérvidos

Schürholz G (1975) Proyecto sobre el ciervo común (*Cervus elaphus*) en Chile. CONAF

Ueckermann E (1987) Managing German red deer (*Cervus elaphus*) populations. In Wemmer C (ed) Biology and Management of the Cervidae. Smithsonian Institution, Washington, D.C., pp 505–516

Wollenhaupt H (1983) Die Ansiedlung, Bestandesentwicklung und Status des Rothirsches (*Cervus elaphus* L., 1758) in Chile. Doctoral Dissertation, Faculty of Forestry, Georg-August University, Göttingen

5

Red Deer Population Dynamics in Hungary: Management Statistics Versus Modeling

Sándor Csányi

Abstract

Red deer (*Cervus elaphus hippelaphus* L.) are the most important big game species in Hungary. During the last decades, their population and harvest have continuously increased, and the principles of red deer management have been strongly debated. This chapter describes a simple method for estimating red deer population size and dynamics on the basis of the available management data (spring census, harvest, recruitment, and mortality). A simulation for the period 1969 to 1989 has suggested a population size 40% to 60% greater than the annually reported spring population data. Spring population sizes are seen only as indices of the population trend. As a consequence of this underestimation, the red deer population in Hungary is underharvested, and game managers (hunters) have failed to stop population growth. The population increase and the deer-related forest damage can be reduced by increasing the cull of hinds and calves. Furthermore, for developing an appropriate and effective red deer management strategy, more reliable population estimates and monitoring of reproduction and natural mortality are essential.

Key words: Estimation, harvest, Hungary, model, population dynamics, red deer

Introduction

Red deer are the most important big game species in Hungary. For the last 20 years, both reported spring population and harvest have grown (Fig. 5.1). In the same time, the harvest rates (harvest/spring population) were also increasing and reached unreasonably high levels (> 40%), at which point the red deer population should have declined and/or become extinct (Csányi 1989). This supports the conclusion that the spring population sizes published by the Ministry of Agriculture have underestimated the actual population.

Red deer have considerable hunting value but have caused damage to forest and agricultural lands. Foresters have become increasingly intolerant of deer. The reliability of the red deer population estimates and the principles of deer management are strongly debated. In this situation, hunters are expected to stop the population increase in order to stabilize the number of deer and reduce the deer damage. These efforts have

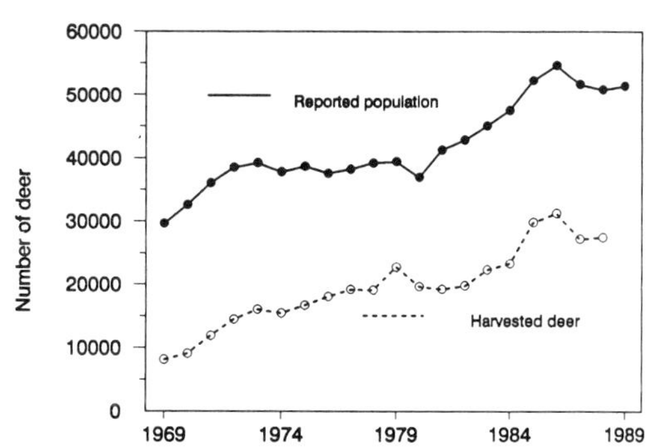

FIGURE 5.1. Reported spring red deer population (n_t) and harvest (H_t) between 1969 and 1989 in Hungary.

failed because red deer management is based on unrealistic data.

The purposes of this chapter are as follows: (1) present a simple method for estimating the spring population size from the available management data, and (2) assess red deer population dynamics using these calculated population sizes.

Materials and Methods

Management data

In Hungary, game management units (such as hunters' groups, state farms, and state forestries) are required to report the estimated February red deer population size by numbers of stags, hinds, and calves. There is no standard method of estimation. Similarly, game management units have to report the annual deer harvest for the same age and sex categories. The red deer hunting season opens in September and lasts until the end of December or mid-January.

Calculating the spring population sizes

The basic model for a change in population size is:

$$N_t = R_{t-1} * N_{t-1} \quad (1)$$

where N_{t-1} and N_t denote population at time $t-1$ and t, and R_{t-1} is the observed growth rate for the period $t-1$ to t.

Assuming reported spring population sizes (n_t) as indices of the true population sizes (N_t), the equation can be written as:

$$n_t = R_{t-1} * n_{t-1},$$

from which

$$R_{t-1} = n_t / n_{n-1} \quad (2)$$

The change in the population size can be written in a more complex form:

$$N_t = N_{t-1}*(1-d_{t-1}) + N_{t-1}*f_{t-1}*c_{t-1} - H_{t-1} \quad (3)$$

where d_{t-1} denotes the natural (nonhunting) mortality rate of the adults ($\geqq 2$ years old), f_{t-1} is the proportion of the adult females ($\geqq 2$ years old), c_{t-1} is the recruitment per female, and H_{t-1} is the harvest in year $t-1$. The recruitment rate is a product of the birth rate and the calves' survival until the hunting season (c_t = calves born per hind * calf survival).

From Eq. (2) the observed growth rate (R_{t-1}) is known, and Eq. (1) can be rewritten as:

$$N_{t-1} = N_t / R_{t-1}.$$

Replacing this expression into Eq. (3) gives:

$$N_t = (N_t/R_{t-1})*(1-d_{t-1} + f_{t-1}*c_{t-1}) - H_{t-1}$$

which can be solved for N_t as:

$$\hat{N}_t = (H_{t-1}*R_{t-1})/(1-d_{t-1} + f_{t-1}*C_{t-1} - R_{t-1}) \quad (4)$$

Using Eq. (4), the population size can be calculated from the deer management data for each year between 1970 and 1989.

On the basis of this method, the spring population sizes were determined for each year from 1970 to 1989. In these calculations n_t, f_t (number of hinds/total spring population), and H_t were taken from the annually published red deer management data (Table 5.1). R_t was calculated from the subsequent pairs of n_t (Table 5.2). For adult mortality (d_t) and recruitment rate (c_t), appropriate estimates were not available, and their values were taken from the literature or arbitrarily chosen as follows:

1. It was reasonable to assume that the adult mortality (d_t) could be low, and in the calculations its value varied randomly between 3% and 6%.
2. Values were based on embryo counts, the very high pregnancy rate of hinds (94.4%, $n = 311$), and the fact that no intrauterine mortality was found (Heltay et al. 1986). In the calculations, the birth rate was a constant of 0.95 calves born/hind. The mortality rate of the calves until the hunting season was estimated to be 10% (Páll 1985), but it seems probable that the variability of calf survival is high between years. In the calculations, the calf survival varied randomly between 75% and 99%.

To simulate the year-to-year variability, the adult mortality (d_t) and recruitment rate (c_t) were allowed to vary randomly. The calculated mean spring population size ($\hat{N}_{tm}$, where m denotes the mean), the standard error of estimates (SE $\hat{N}_{tm}$), and the coefficient of variation ($CV\ \hat{N}_{tm}$) were computed on the basis of 100 replications for

TABLE 5.1. The annual reported spring red deer population size and harvest according to age and sex categories, between 1969 and 1989 in Hungary. (Data published annually by the Ministry of Agriculture of Hungary)

	Reported spring population					Deer harvested			
Year	Stags	Hinds	Hind (%)[a]	Calves	Total	Stags	Hinds	Calves	Total
1969					29 605	2 638	3 150	2 347	8 135
1970	11 850	14 180	43.5	6 560	32 590	2 926	3 667	2 468	9 061
1971	13 285	15 116	41.9	7 638	36 039	3 490	5 241	3 138	11 869
1972	14 555	15 735	40.9	8 199	38 489	3 967	6 867	3 595	14 429
1973	14 781	16 114	41.1	8 332	39 227	4 362	7 323	4 308	15 993
1974	14 506	15 202	40.2	8 093	37 801	4 548	6 756	4 106	15 410
1975	14 723	16 016	41.4	7 926	38 665	4 886	7 340	4 416	16 642
1976	14 102	15 507	41.3	7 948	37 557	5 568	7 653	4 799	18 020
1977	14 361	15 795	41.3	8 069	38 225	5 767	8 363	5 027	19 157
1978	14 558	16 285	41.6	8 340	39 183	6 000	7 997	5 075	19 072
1979	15 097	16 019	40.6	8 300	39 416	6 966	9 259	6 485	22 710
1980	13 930	15 144	41.0	7 881	36 955	6 176	7 661	5 780	19 617
1981	15 172	17 079	41.4	9 011	41 262	6 334	7 500	5 382	19 216
1982	15 917	17 624	41.1	9 295	42 836	6 275	7 733	5 761	19 769
1983	16 503	18 304	41.6	10 225	45 032	7 034	8 967	6 274	22 275
1984	17 036	19 479	41.0	10 960	47 475	6 897	9 276	7 095	23 268
1985	18 745	21 399	41.0	12 105	52 249	7 799	12 154	9 878	29 831
1986	19 492	22 690	41.5	12 458	54 640	8 139	12 606	10 523	31 268
1987	18 532	20 778	40.3	12 247	51 557	7 736	10 360	9 090	27 186
1988	18 717	20 289	40.0	11 719	50 725	8 323	10 372	8 727	27 422
1989	18 980	20 354	39.7	11 976	51 310				

[a]Proportion of adult (≧ 2 years old) hinds

TABLE 5.2. Observed growth rates (R_{t-1}), harvest rates (H_t/n_t), mean calculated spring population sizes ($\hat{N}_{tm}$) and projected population sizes of Hungarian red deer

			Calculated spring population size					
Year	Observed growth rate (n_t/n_{t-1})	Harvest rate (H_t/n_t)	($\hat{N}_{tm}$)	(CV $\hat{N}_{tm}$)	(− 2SE)	(+ 2SE)	Bias ($\hat{N}_{tm}/n_t$)	Projected population size[a]
1969		27.48%						40 050
1970	1.1008	27.80%	42 114	14.97%	29 507	54 721	1.2922	44 651
1971	1.1058	32.93%	53 122	15.23%	36 944	69 300	1.4740	49 789
1972	1.0680	37.49%	57 431	13.58%	41 835	73 027	1.4921	53 753
1973	1.0192	40.77%	53 245	10.40%	42 167	64 323	1.3574	56 417
1974	0.9636	40.77%	47 645	9.55%	38 541	56 749	1.2604	58 365
1975	1.0229	43.04%	56 831	11.51%	43 744	69 918	1.4698	61 515
1976	0.9713	47.98%	50 291	9.34%	40 892	59 690	1.3391	64 434
1977	1.0178	50.12%	66 362	11.02%	51 737	89 987	1.7361	66 905
1978	1.0251	48.67%	72 780	11.91%	55 439	90 121	1.8574	69 023
1979	1.0059	57.62%	67 290	10.03%	53 787	80 793	1.7072	71 901
1980	0.9376	53.08%	59 852	7.14%	51 306	68 398	1.6196	72 055
1981	1.1165	46.57%	120 529	17.47%	78 421	162 637	2.9211	75 352
1982	1.0381	46.15%	78 671	11.75%	60 182	97 160	1.8366	80 098
1983	1.0513	49.46%	87 089	12.20%	65 848	108 330	1.9339	85 800
1984	1.0543	49.01%	98 394	12.37%	74 056	122 732	2.0725	90 809
1985	1.1006	57.09%	136 132	16.10%	92 293	179 971	2.6054	96 418
1986	1.0458	57.23%	123 237	11.90%	93 907	152 567	2.2554	97 248
1987	0.9436	52.73%	86 253	9.16%	70 455	102 051	1.6730	96 905
1988	0.9839	54.06%	87 778	8.86%	72 229	103 327	1.7305	100 534
1989	1.0115		102 714	10.37%	81 418	124 010	2.0018	105 082

[a] $N_t = 1.318 * N_{t-1} - H_{t-1}$

each year. The calculated mean population sizes ($\hat{N}_{tm}$) were compared with the reported population sizes (n_t) to evaluate the bias ($\hat{N}_{tm}/n_t$). Moreover, the harvest rate was recalculated for each year on the basis of the calculated mean population size ($H_t/\hat{N}_{tm}$). Statistical Analysis System (SAS Inst., 1985) was used for generating the uniform random values and for the statistical procedures.

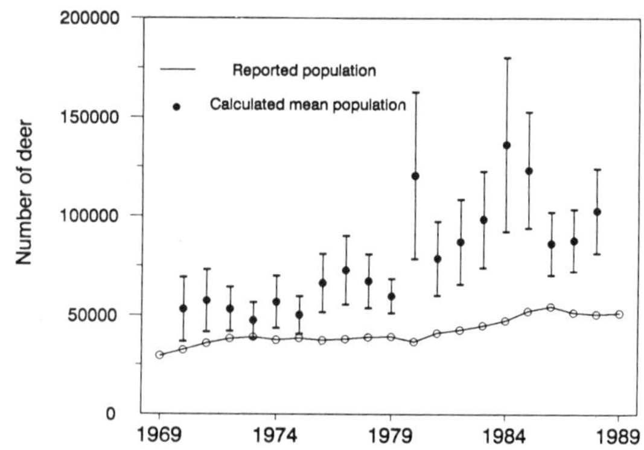

FIGURE 5.2. Calculated mean red deer population size ($\hat{N}_{tm} \pm 2$ S.E.) between 1969 and 1989 in Hungary:—, reported population; ●, calculated mean population.

Projecting the Population Trend

A simple projection model was fitted to the calculated mean population sizes for modeling the population trend:

$$N_t = \lambda * N_{t-1} - H_{t-1}$$

where λ denotes the finite population growth rate and assumes that reproduction and natural mortality can be described by a single constant. This method gives an additional estimate for the initial population size (N_{1969}) as well. The best fitting parameters (λ and N_{1969}) were achieved using a minimum sum of χ^2 procedure (Eberhardt 1987).

Results and Discussion

From 1969 to 1989, the reported spring population size (n_t) showed a 1.7-fold increase, and the 1988 red deer harvest was 3.4 times greater than in 1969. Between 1969 and 1989, the harvest rate (H_t/n_t) ranged between 27.48 and 57.62% (Table 2). The calculated mean population sizes ($\hat{N}_{tm}$) were 1.29 to 2.92 times higher than the reported population sizes (n_t). Except for 1973, the reported population size (n_t) was outside the ± 2 S.E. $\hat{N}_{tm}$ range (Fig. 2). The population size calculated for 1989 was 2.44 times greater than that of 1970. The CVs of the calculated mean population sizes ($\hat{N}_{tm}$) ranged from 7.14% to 17.47%. This level of variability seems to be acceptable for management purposes or determining the range of the real population size (Table 5.2).

Between 1970 and 1989, the ratio of the reported and the calculated population ($\hat{N}_{tm}/n_t$) generally increased. In the 1970s, the differences were smaller, and after 1980 the real population size might have been nearly two times greater than the management statistics showed (Fig. 5.2 and Table 5.2). On average, the red deer population was underestimated by 40% to 60%.

In some years the ratio of the reported and the calculated population ($\hat{N}_{tm}/n_t$) was much higher (1981, 1985, 1986), which is evidence for the failure of the assumption that the reported population sizes are indices of the true population. In these years, greater change could have been expected in the reported population size (n_t) and observed growth rate (R_{t-1}) due to the very intense harvest (Tables 5.1 and 5.2).

A serious hazard of using indices is that they will fail, and the estimates will be biased (Eberhardt and Simmons 1987). The most important assumption of the estimation method presented here is that the reported population sizes (n_t) accurately reflect actual population trends, and the relationship between the true population size (N_t) and the index (n_t) is linear. Caughley (1977) gave several examples of the nonlinear relationships and problems which occur when counts such as indices are used. In this case if a nonlinear relationship occurs (e.g., the counting efficiency declines when the population is increasing or the counting efficiency is variable), the estimate of population size can be seriously biased. A reasonable assumption is that the outermost values of $\hat{N}_{tm}$ are the results of this failure, and a further simulation study is needed to determine the bias and precision of the method as well as the behavior of the model when the linearity is violated.

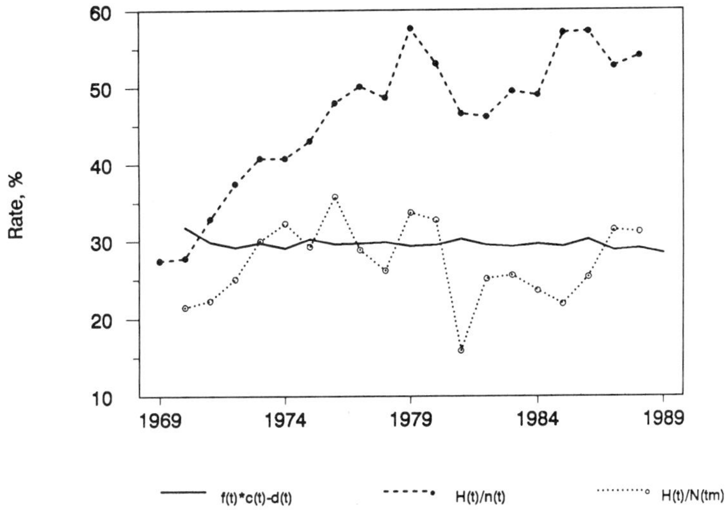

FIGURE 5.3. Red deer harvest rates calculated on the basis of the reported population size (H_t/n_t) and the calculated population size ($H_t/\hat{N}_{tm}$) as well as the mean growth rate of the population ($c_1 * f_t - d_t$), between 1969 and 1989 in Hungary.

As a consequence of the much higher calculated population sizes, the recalculated harvest rates ($H_t/\hat{N}_{tm}$) are considerably lower (Fig. 5.3), and in 12 years did not reach the level of the growth rate computed from the simulated parameters ($1 - d_t + c_t * f_t$). The recalculated harvest rates also support that underexploitation was the main source of the continued population increase.

The projection model gave the initial population size of 40 050 deer (N_{1969}) and a finite rate of increase $\lambda = 1.318$ ($\chi^2 = 38\,697$). This rate of increase is higher than the mean rate of increase ($1 - d_t + c_t * f_t$) calculated from the means of the years (1.284). The difference between the trend projected by the model and the reported population size (n_t) also increased. The projected population size in 1989 was 2.62 times higher than the initial population size (N_{1969}). The model smoothes the population trend and is unable to fit the outermost values (Table 5.2 and Fig. 5.4). The poor fit of the projection model can be related to the key assumption underlying simple models, that there exist stable and repeat able relationships, and model failure occurs when there is a persistent or periodic change in rates that are likely to arise under management (Walters 1986). In this case, rates were variable and the model averages the variability of the parameters.

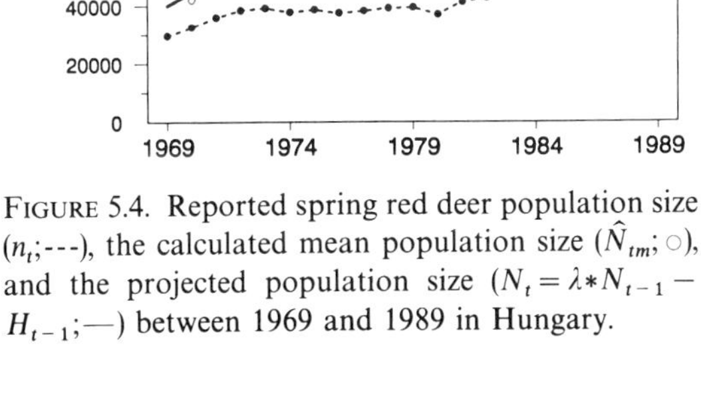

FIGURE 5.4. Reported spring red deer population size (n_t;---), the calculated mean population size ($\hat{N}_{tm}$; ○), and the projected population size ($N_t = \lambda * N_{t-1} - H_{t-1}$;—) between 1969 and 1989 in Hungary.

Conclusions

The results presented here provide a strong basis to assume that the red deer population in Hungary was always much greater than reported in the annual game management statistics. During the past decade the underestimation reached 40%

to 60%. Other studies reported similar findings that traditional estimation methods such as the counting in feeding places, drive-counts or winter track-counts considerably underestimate big game populations (Pucek et al. 1975; Bobek et al. 1987). These data can be misleading and in most cases can only be used as indices of population trends (Putman and Sharma 1987).

When the population is managed for harvesting and to limit abundance, numbers removed need to be compared with estimates of total population size, rather than with indices (Eberhardt and Simmons 1987). In the present study, country-wide data were used, and the main purpose was finding the range of population sizes which are in accordance with the harvest and other population parameters. For this purpose the calculation method seems to be acceptable.

The red deer population in Hungary was underharvested because comparing harvests to reported spring populations (indices) overestimated the harvest rates and made management plans unrealistic. By accepting the calculated population sizes (or the projected trend), the red deer harvest should be increased and kept high in order to lower and stabilize the population. In the trophy-oriented Hungarian management system, the greater harvest can be achieved by increasing the harvest of hinds and calves. In this way recruitment can be reduced and the effect on stags will be moderated.

The high reproductive potential of red deer reported in Hungary (Heltay et al. 1986) can permit the continued population growth despite increasing harvest. However, several studies indicate that the reproduction and the mortality of red deer (at least in calves) are density-dependent (Albon et al. 1983; Clutton-Brock et al. 1982; Ratcliffe 1984). The shortage of natural (or nonhunting) mortality data did not allow this aspect of the problem to be included in the study. Monitoring these parameters is also essential for developing an appropriate and effective red deer management strategy.

Acknowledgements

This work was supported by the Ministry of Agriculture of Hungary under contract AP-4.4. G.C. White reviewed earlier drafts and provided helpful comments.

References

Albon SD, Mitchell B, Staines BW (1983) Fertility and body weight in female red deer: a density-dependent relationship. J Animal Ecol 52:969–980

Bobek B, Perzanowski K, Zielinski J (1987) Red deer population census in mountains: testing of an alternative method. Acta Theriol 31:423–431

Caughley G (1977) Analysis of Vertebrate Populations. Wiley, New York

Clutton-Brock TH, Guiness FE, Albon SD (1982) Red Deer. Behavior and Ecology of Two Sexes. Edinburgh University Press, Edinburgh

Csányi S (1989) Egyszerü módszer a gimszarvas állomány minimális létszámának becslésre. Vadbiológia 89:49–55

Eberhardt LL (1987) Population projections from simple models. J Appl Ecol 24:103–118

Eberhardt LL, Simmons MA (1987) Calibrating population indices by double sampling. J Wildlife Management 51:665–675

Heltay I, Adorján Á, Sugár L (1986) A gimszarvas (*Cervus elaphus hippelaphus*) fekunditása és a magzatok fejlödése eltérö élöhelyeken. Vadbiológia 86:83–96

Páll, E (1985) A gimszarvas és vadászata. Mezögazdasági Kiadó, Budapest

Pucek Z, Bobek B, Labudzki L, Mielowski L, Morow K, Tomek A (1975) Estimates of density and number of ungulates. Polish Ecol Stud 1:121–136

Putman RJ, Sharma SK (1987) Long-term changes in New Forest deer populations and correlated environmental change. Symp Zool Soc (London) 58:167–179

Ratcliffe PR (1984) Population density and reproduction of red deer in Scottish commercial forests. Acta Zool Fennica 172:191–192

Statistical Analysis System Institute (1985) SAS language guide for personal computers, version 6 ed. SAS Institute, Cary, N.C.

Walters C (1986) Adaptive Management of Renewable Resources. MacMillan, New York

6
Desert Mule Deer Mortality and the Central Arizona Project

Paul R. Krausman, Bruce D. Leopold, Kurt R. Rautenstrauch, John R. Morgart, and Richard C. Etchberger

Abstract

The Central Arizona Project (CAP) is one of the largest water delivery systems in the western United States. The CAP is a series of open canals that divert Colorado River water east to Phoenix, then south through Tucson, Arizona. Centra Arizona Project canals dissected over 250 km of desert mule deer (*Odocoileus hemionus crooki*) habitat. The canals are an attractive nuisance to deer. We documented deer mortality during 1979 to 1989 along 179 km of the CAP (formerly called the Granite Reef Aqueduct but renamed in 1989 as the Hayden-Rhodes Aqueduct [HRA]). Deer were entrapped in the canal during the construction years (1979–1987) or drowned when water was added in 1987; however, mortality was minimized when fencing was constructed. Deer managers and biologists should be aware of the effects of creating attractive nuisances in desert mule deer habitat and of the costs associated with minimizing mortality.

Key words: Arizona, Central Arizona Project, desert mule deer, mortality

Introduction

Over 12 000 km of concrete-lined canals carry water from rivers and reservoirs to industrial, residential, and agricultural users throughout the western United States. Unfortunately, thousands of deer and other wild ungulates have drowned in > 21 canals in ten western states and Canadian provinces while attempting to cross or drink (Rautenstrauch and Krausman 1986). Fences, bridges, and alternate water sources have been built to prevent animals from entering canals. Escape structures have been developed to allow entrapped animals to escape, but few have been successful (Busch et al. 1984; Fry et al. 1984) and the effectiveness of others (e.g., fences) has not been evaluated.

Rautenstrauch and Krausman (1986) summarized ungulate losses in canals in the western United States and documented the need to establish the relevance of these losses and develop ways to minimize losses. The Mohawk Canal, southwest Arizona, has one of the highest rates of wild ungulate drowings in the United States. Desert mule deer (Furlow 1969; Guenther et al. 1979), desert bighorn sheep (*Ovis canadensis mexicana*) (Gubser 1960), and Sonoran pronghorn (*Antilocapra americana sonoriensis*) (Arizona Game and Fish Dept. 1981) have drowned in the Mohawk Canal and other Wellton-Mohawk Irrigation and Drainage District (WMIDD) canals. At least 170 desert mule deer drowned in WMIDD canals during 1977–1980 (Rautenstrauch and Krausman 1986). The U.S. Bureau of Reclamation (BR) conducted studies to reduce and minimize these losses. These studies (Rautenstrauch and Krausman 1986, 1989) examined whether entrapped mule deer could use steps and ramps with cable directors to escape from the Mohawk Canal, determined if deer would drink from alternate water sources built to decrease the number of deer drinking from the canal, and developed guidelines to prevent ungulate drownings. The studies (Rautenstrauch and Krausman 1986, 1989) indicated that escape ramps and steps with cable directors on the Mohawk Canal are well designed and spaced to assist deer to escape when entrapped during summer and probably were responsible for reducing drownings. Rautenstrauch and Krausman (1986, 1989) further suggested that freeboard levels (distance from the top of the concrete canal lining to the water level) < 30 cm or > 100 cm be maintained to decrease the number of deer falling in canals. Also, because deer appeared to approach the Mohawk Canal to drink, building alternate water sources should decrease the number of deer drinking from WMIDD canals.

Although these results (Rautenstrauch and Krausman 1986, 1989) may be applicable to the HRA and their recommendations to keep desert ungulates from drowning have been incorporated into the HRA (e.g., water developments, fences [Krausman 1985], and test escape ramps), the effectiveness of these mitigations has not been evaluated. Our objective was to document desert mule deer mortality associated with the construction and operation of the HRA and determine the effectiveness of fencing to minimize mortality related to the canal.

Study Area and Methods

The study was conducted along 179 km of the HRA that extends from the Colorado River east to Phoenix (Fig. 6.1). This section of the canal passes through fair to good desert mule deer habitat (Krausman 1985).

Topography ranges from rolling hills < 400 m to rugged cliffs extending > 1700 m. Average seasonal temperatures were 28.1°, 20.8°, 14.1°, and 10.3° C for summer (July, August, September), fall (October, November, December), winter (January, February, March), and spring (April, May, June), respectively. Rainfall averaged 6.06, 4.62, 9.06, and 1.47 cm for summer, fall, winter, and spring, respectively. The study area is described in detail by Krausman (1985).

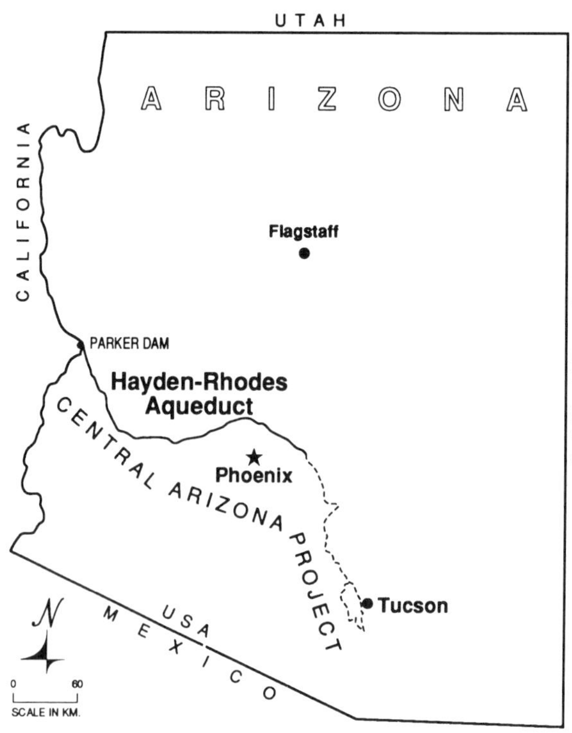

FIGURE 6.1. Hayden-Rhodes Aqueduct of the Central Arizona Project.

From 1979 through 1989 we monitored desert mule deer mortality by systematic driving along the HRA access road or flying surveys above the canal and recording deer that were actually in the canal during construction and those that were in the right-of-way (fenced canal and maintenance roads) post-construction. We recorded deer in the canal, deer rescued or guided out of the canal, and deer that died in the canal. In addition, construction personnel, Arizona Game and Fish personnel, and BR personnel provided observations of deer interactions with the canal. Only reports that could be verified are reported here. Our data represent minimum mortality of desert mule deer related to the HRA. Data collected included sex and age (Ad M, F, or <2 yr [yearling]) of deer, date, and location associated with the canal, condition of the animal if alive, and probable cause of death. When possible, we determined how deer entered the canal, especially when found in portions that had been fenced (1987–1989). Deer canal interactions were categorized into three phases: during construction, postconstruction without fences preventing entry, and postconstruction with fences.

Results

From 1979 to 1983 during construction of the HRA, ≧12 deer drowned (Krausman 1985). Other deer were entrapped in dry portions of the HRA during construction and were often chased out. During construction we initially recorded deer mortality incidental to other field activities. We decided to place more emphasis on monitoring the canal for evidence of deer interactions because water would be added in 1985, and we anticipated increased mortality until fences could be constructed. Additional deer died in the canal in 1984, but reports were incomplete until 1985 when we had adequate surveys conducted ≧1–3 times/week.

In 1985 and 1986 the HRA was complete and was being filled with water. Fence construction began but was not completed until 1987. In 1985 ≧45 deer (6M, 18F, 21 yearlings) interacted with the HRA; 27 drowned (Table 6.1). In 1986 ≧46 deer (8M, 19F, 19 yearlings) interacted with the HRA; only 10 drowned (Table 6.1).

In 1987 most of the fence was constructed and interactions decreased to ≧25 but 16 drowned (Table 6.1). After the fence was completed, interactions were reduced to one in 1988 and five in 1989. Only three deer drowned in both years.

Discussion

Krausman (1985) and Krausman et al. (unpublished data) report data from the HRA on desert mule deer movements, distribution, relationships

TABLE 6.1. Minimum number of desert mule deer interacting with the Hayden-Rhodes Aqueduct by year and month along the Central Arizona Project, Arizona

Year	Jan	Feb	Mar	Apr	May	Jun	Jul	Aug	Sep	Oct	Nov	Dec
1985												
Interactions	0	0	0	0	6	7	10	12	7	0	0	3
Drownings	0	0	0	0	3	6	4	6	6	0	0	2
1986												
Interactions	1	0	0	0	10	10	18	1	1	3	1	1
Drownings	1	0	0	0	2	3	1	0	1	2	0	0
1987												
Interactions	3	0	0	1	6	3	4	6	1	0	1	0
Drownings	2	0	0	1	4	1	3	3	1	0	1	0
1988												
Interactions	0	0	0	0	1	0	0	0	0	0	0	0
Drownings	0	0	0	0	1	0	0	0	0	0	0	0
1989												
Interactions	0	0	0	0	0	0	2	0	1	2	0	0
Drownings	0	0	0	0	0	0	2	0	0	0	0	0

to the canal, relationships to water sources, and other ungulates prior to the operation of the HRA. This section of the CAP was well suited to collect data post-HRA operation to determine how the canal has influenced desert mule deer mortality and determine if the mitigation fence for deer along the canal is adequate. The HRA bisected the home ranges of deer in western Arizona. Of the 16 males used to calculate home ranges in the mountain ranges adjacent to the HRA from 1979–1984 (Krausman 1985) most deer did not come within 1 km of the canal; three males approached to within 1 km of the canal, and four males (25%) crossed the canal. In most instances the large home ranges of males were tied to available water sources.

Krausman (1985) found that female desert mule deer used both sides of the HRA more than males. Of the 21 females monitored to calculate home ranges, only 7 (33%) did not come within 1 km of the HRA. Three (14%) had home ranges that were within 1 km of the canal and 11 (52%) crossed the canal. Female ranges usually contained $\geqq 1$ water source within the use area. Krausman (1985) documented that deer have large home ranges, and males and females crossed the dry HRA in all seasons; however, there were few locations of deer south of the HRA. Of all deer located along the HRA, only 146 of 4066 locations were south of the HRA. Of those locations south of the canal, 68 were repeated locations of a single deer. If these were removed $< 1.9\%$ of the locations would be south of the HRA.

Although habitat use by desert mule deer south of the canal was minimal, most deer monitored came within 1 km of the HRA. When water was added, the canal became an attractive nuisance and drownings increased. Before fences were completed, deer entered the canal and drowned. From 1985 through 1987, 45, 46, and 25 deer, respectively, were reported in the canal; nearly 50% of them drowned (Table 1). Most deer-canal interactions occurred from May to September (Table 1) and 84% of the interactions involved females and young. These data support the observations of Krausman and Hervert (1984) that the majority of deer was found in the canal during summer months when temperatures were high, and free-standing water was limited. That more females were found probably was due to their increased water demand during the last trimester of pregnancy (Short 1981) through fawn rearing (Hervert and Krausman 1986). We did not have a measure of deer density. However, desert mule deer populations in western Arizona are characteristically low (ca. 1/5–10 km^2) and death by drowning represented an important source of mortality.

Fencing recommended by Krausman (1985) was completed in 1988. After fences were added to keep deer from the HRA, deer in the canal and related mortalities decreased. Most deer that entered the canal gained access through open gates or holes in the fence (i.e., washouts). Nineteen reports of deer in the canal were from areas along the HRA where there was no fence. There was only one second-party observation of a deer "jumping" the wildlife fence. These reports indicate that fences can be adequate to eliminate deer-canal interactions given proper maintenance (i.e., repair holes, close gates). It is important that after mitigations are in place they should be evaluated to determine their effectiveness in reducing ungulate canal interactions. Mitigations made for wildlife should be improved if modifications are indicated.

Although fencing is an effective way to minimize drownings of desert mule deer in habitats bisected by canals, adequate fences are expensive. The 1987–1988 cost of constructing the 2.6 m high woven wire fence on both sides of the canal was $32 815 US/km. The cost versus the loss of wildlife are factors that have to be considered when natural habitats are severely altered.

Acknowledgements

We appreciate the assistance of the many University of Arizona students, Arizona Game and Fish wildlife managers, and Bureau of Reclamation biologists that assisted with data collection. This study was funded by the U.S. Bureau of Reclamation and the U.S. Fish and Wildlife Service.

References

Arizona Game and Fish Department (1981) The Sonoran Pronghorn. Ariz Game Fish Dep Spec Rep 10

Busch DE, Rorabaugh JC, Rautenstrauch KR (1984) Deer entrapment in canals of the western United States: a review of the problem and attempted solutions. In Krausman PR, Smith NS (eds) Deer in the Southwest: A Workshop. School of Renewable Natural Resources, University of Arizona, Tucson, pp 95–100

Fry ME, Wyman VL, Yoeman EH (1984) A method for evaluating the effectiveness of a deer protection system on a concrete-lined canal (with comments on the planning of these systems). In Crabtree AF (ed) Proc Third Int Symp Env Concerns in Rights-of-Way Management, Mississippi State University, pp 647–658

Furlow B (1969) The Wellton-Mohawk Canal incident. Ariz Wildlife Views 12:22–25

Gubser CE (1960) An inquiry into the effects of reservoirs and canals on big-game mitigations. Proc Annu Conf Western Assoc of Game and Fish Commissioners 40:42–52

Guenther HR, Sharpe FP, Strauss P (1979) Mule deer losses—Mohawk Canal, Arizona: a problem identified—a solution sought. In Swanson GA (ed) The Mitigation Symposium: A National Workshop on Mitigating Losses of Fish and Wildlife Habitats. For Serv General Tech Rep TM-65

Hervert JJ, Krausman PR (1986) Desert mule deer use of water developments in Arizona. J Wildlife Management 50:670–676

Krausman PR (1985) Impacts of the Central Arizona Project on desert mule deer and desert bighorn sheep. U.S. Bureau of Reclamation, Final Report, Contract 9-07-30-X069

Krausman PR, Hervert JJ (1984) Impacts of the Granite Reef Aqueduct on desert ungulates. In Crabtree AF (ed) Proc Third Inter Symp Environ Concerns in Rights-of-Way Management, Mississippi State University, pp 640–646

Rautenstrauch KR, Krausman PR (1989) Preventing mule deer drownings in the Mohawk Canal, Arizona, Wildl Soc Bull 17:280–286

Rautenstrauch KR, Krausman PR (1986) Preventing desert mule deer drownings in the Mohawk Canal, Arizona, U.S. Bureau of Reclamation, Final Report, Contract 9-07-30-X0069

Short HL (1981) Nutrition and metabolism. In Wallmo OC (ed) Mule and Black-tailed Deer of North America. University of Nebraska Press, Lincoln, Nebraska, pp 99–127

7
Deer Condition Response to Changing Harvest Strategy, Davis Island, Mississippi

Harry A. Jacobson

Abstract

A private land management program initiated in 1978 for white-tailed deer (*Odocoileus virginianus*) on 7200 ha is described. Between 1978 and 1989, an average annual harvest of 219 bucks and 242 does was achieved. Quality and productivity of deer were significantly related to intensity of past doe harvest. Within age classes, eviscerated body weight increases of about 3.5 to 5.0 kg for fawn and females and 12 to 16 kg for males were obtained. Antler measurements within age classes increased by about 1.6 cm and 14.2 cm in basal circumference and beam length, respectively. Changes from a yearly average of 1 fetus per adult doe to as high as 2.2 fetuses per adult doe were recorded. Lactation rates increased from 6% to 22% for yearling does and 46% to 85% for adult does. Prior to 1983, yearling bucks comprised most (> 75%) of the antlered harvest. When hunters voluntarily restricted harvest to bucks with antler beams of 33 cm or more, yearling buck harvest was gradually reduced to less than 15% of the antlered harvest. The number of bucks harvested greater than 2.5 years old prior to 1984 averaged 39 annually. From 1984 to 1990, this average was 144 with a high of 176 taken in the 1988–1989 season. Peak rut moved 2 to 3 weeks earlier in association with change in buck harvest strategy.

Key words: White-tailed deer, condition indices, management, populations

Introduction

Current theory for management of the white-tailed deer and other cervidae is largely based on the empirical model of McCullough and the principle of maximum sustained yield (MSY) (McCullough 1979, 1984, 1987). However, aside from McCullough's George Reserve study, published data substantiating that populations respond as McCullough predicts have been lacking.

In 1977, I began working with five contiguous hunting clubs consisting of 7200 ha on Davis Island, Mississippi. The membership of these private hunting clubs were responsive to management recommendations and agreed to participate in Mississippi's Deer Management Assistance Program (DMAP) (Gwynn et al. 1983). Their participation in DMAP made it possible to obtain data on almost all deer harvested and, therefore, evaluate condition response to changing harvest strategies. The purpose of this paper is to examine the Davis Island data in relationship to McCullough's "black box" or "stock recruitment" model (McCullough 1984).

Study Area

Davis Island is located in the alluvial floodplain of the Mississippi River, 21 km southwest of Vicksburg, Mississippi, in Warren County, Mississippi, and Tensas Parish, Louisiana. It is bordered on the east by the Mississippi River and the remaining sides by Palmyra Lake, a former river channel. About 1300 ha of the northern portion of the island is in Louisiana, and this portion was not included in the study.

The remainder of the island consisted of five separate tracts of adjoining land owned or leased by five hunting clubs: Palmyra (2024 ha), Rosedale (1215 ha), Ingamar (810 ha), Brierfield (2024 ha), and Titanic (1127 ha). The number of club members varied by club from 5 to 30, but members and their guests of all clubs would seldom exceed 150 hunters. On most days, the number of hunters would be considerably less than this.

About 84% of the island was forested, 13% agriculture (cotton, soybeans, wheat), and 3% was pasture for cattle. Dominant tree species included black willow (*Salix nigra*), sugarberry (*Celtis laevigata*), American sycamore (*Platanus occidentalis*), sweetgum (*Liquidambar styraciflua*), pecan (*Carya illinoensis*), oaks (*Quercus* spp.), cottonwood (*Populus heterophylla*), baldcypress (*Taxodium distichum*), ash (*Fraxinus pennsylvanica*), persimmon (*Diospyros virginiana*), and honeylocust (*Gleditsia triacanthos*).

There was little change in land use patterns throughout the study, except that during 1988 and 1989 some selective cut logging occurred over about 2000 ha.

Materials and Methods

Harvest data including sex, tooth wear and replacement age (Larson and Taber 1980), and eviscerated body weight were collected from most deer killed. Lactation status was recorded for does and antler measures (beam circumference and beam length) were recorded for males. In February–March of each year, a number of adult does were collected for examination of fetuses present and fetal age estimations (Hamilton et al. 1985).

Harvest records were not available for the entire study area prior to 1979; however, two clubs, Palmyra and Rosedale, had some harvest data prior to 1977. Palmyra had harvest data for the years 1968–1974 when from 33 to 54 buck deer were taken annually, or an average of 41 per year. No does were harvested by Palmyra during that period except in 1973 and 1974, when one and five does were killed, respectively. Rosedale had harvest data for the years 1969 to 1974, when from 37 to 63 bucks were killed annually with an average of 50, and no females were reported harvested.

The Davis Island deer population was not censused because of dense vegetative cover and limited resources. However, at the start of the study in 1978, the herd was at very high densities, and it was not unusual in late winter to see several hundred deer on a wheat field.

The objectives of each hunting club were determined, and after collecting an initial data base in 1978, a report was sent to each club with management recommendations for harvest. Initial harvest strategy was to reduce the deer population and obtain a more balanced sex ratio and higher productivity than the premanagement period. The clubs were not satisfied with antler qualities of bucks and in 1983 most 1.5-year bucks were protected by requesting hunters to voluntarily restrict buck harvest to bucks with antler beam length exceeding 33 cm. This measure was chosen because over 95% of the 1.5-year bucks of the Davis Island herd had less than 33 cm of beam length. In 1989, the restriction was raised to 38 cm of beam length as hunters wanted to protect more young bucks from harvest, and herd condition had improved to the point that some 1.5-year bucks with 33 to 39 cm of antler beam length were being harvested.

Results and Discussion

During the 1977–1978 and 1978–1979 hunting seasons, only the Palmyra Hunting Club collected total kill data. Thereafter, all clubs recorded total kill for each year's harvest, and body weights were collected from most deer harvested (Table 7.1). Beginning in 1979, antler characteristics, including beam circumference and beam length, were recorded for bucks (Table 7.2).

Deer 2 years of age or older were pooled for data analysis to compare changes in conditions. Aging accuracy above 2.5 years declines sharply when using wear and replacement (Jacobson and Reiner 1989). Body weights generally increased (1977–1982), decreased (1983–1985), and then increased again (1986–1989; Table 7.1). Likewise, antler characteristics followed similar trends to body weights for both yearling and 2.5-year or older deer (Table 7.2). These trends closely

TABLE 7.1 Hunter harvest and eviscerated body weights of white-tailed deer on 7200 ha of Davis Island, Mississippi, 1977–1989

Hunting[a] season	Reported[b] harvest		Eviscerated body weights (kg) by age class					
			Males			Females		
	Males (n)	Females (n)	6 months	1.5 years	2.5 + years	6 months	1.5 years	2.5 + years
1977–78	ND	71	21.4	36.4	52.3	19.1	35.5	40.0
1978–79	ND	139	22.7	40.7	56.8	22.7	37.7	41.4
1979–80	210	192	24.5	41.8	59.5	22.3	37.7	42.2
1980–81	222	232	23.6	44.0	61.4	22.7	35.5	44.1
1981–82	263	149	25.0	45.5	60.0	21.8	38.6	43.6
1982–83	215	110	23.6	46.9	68.2	23.2	40.9	42.7
1983–84	222	197	21.8	41.5	65.5	20.0	35.0	42.7
1984–85	177	333	20.5	40.4	59.5	21.8	35.5	42.7
1985–86	202	270	21.4	38.9	54.5	21.8	35.9	42.3
1986–87	234	341	23.6	44.4	57.7	20.9	36.8	42.6
1987–88	205	319	21.8	47.3	60.0	25.0	38.6	42.3
1988–89	242	279	24.5	48.4	62.7	25.0	39.5	44.1
1989–90	230	331	24.5	46.5	64.5	21.8	35.5	42.3

[a] Hunting season runs from October to January.
[b] Total harvest data not available for 1977–78 and 1978–79 seasons. Female harvest for 1977–78 and 1978–79 is based on kill figures for the only hunting club known to have harvested females in those two seasons and probably slightly underestimates actual harvest.

TABLE 7.2 Average antler characteristics of bucks harvested on 7200 ha, Davis Island, Mississippi, 1977–1990

Hunting[a] season	Age class					
	1.5 years			2.5 years		
	n	Beam circum-ference (cm)	Beam length (cm)	n	Beam circum-ference (cm)	Beam length (cm)
1979–80	81	5.3	10.4	33	7.9	28.7
1980–81	35	5.8	11.7	13	9.1	38.6
1981–82	79	6.1	14.7	14	9.4	34.5
1982–83	72	6.3	16.5	36	10.4	44.5
1983–84	103	4.3	7.1	29	9.6	39.4
1984–85	23	4.8	10.4	67	8.1	34.0
1985–86	22	5.8	15.0	115	9.1	37.6
1986–87	39	6.4	19.1	127	9.1	39.6
1987–88	19	5.6	16.7	123	9.3	39.9
1988–89	17	6.6	22.6	153	9.7	40.6
1989–90	23	6.9	24.6	144	9.9	43.4

[a] Hunting seasons extended from October through January.

matched trends in female harvest except that a 2-year lag in body weight and antler response to female harvest intensity can be seen (Fig. 7.1) Increasing or decreasing female harvest was accompanied 2 years later with a similar increase or decrease in male and female body weights and antler conditions. Eviscerated body weights changed as much as 3.5 to 5.0 kg for fawn and females and 12 to 16 kg for males. Antlers changed by as much as 2 cm in circumference and 14 to 15 cm in beam length over the period of the study.

Reproduction was seen to fluctuate in a similar manner to body weights and antler development (Fig. 7.1). Yearly averages of as low as 1.0 to as high as 2.2 fetuses per doe were recorded (Table 7.3). Lactation, which is an index of fawn survival, went from a low of 6% to a high of 22% for yearlings and a low of 46% to a high of 85% for adult does, respectively (Table 7.3).

When correlation coefficients are examined between doe harvest and condition indices 2 years later, a positive relationship was seen for all measures, with the exception of body weights of 2.5-year or older does. Significant ($P < 0.05$)

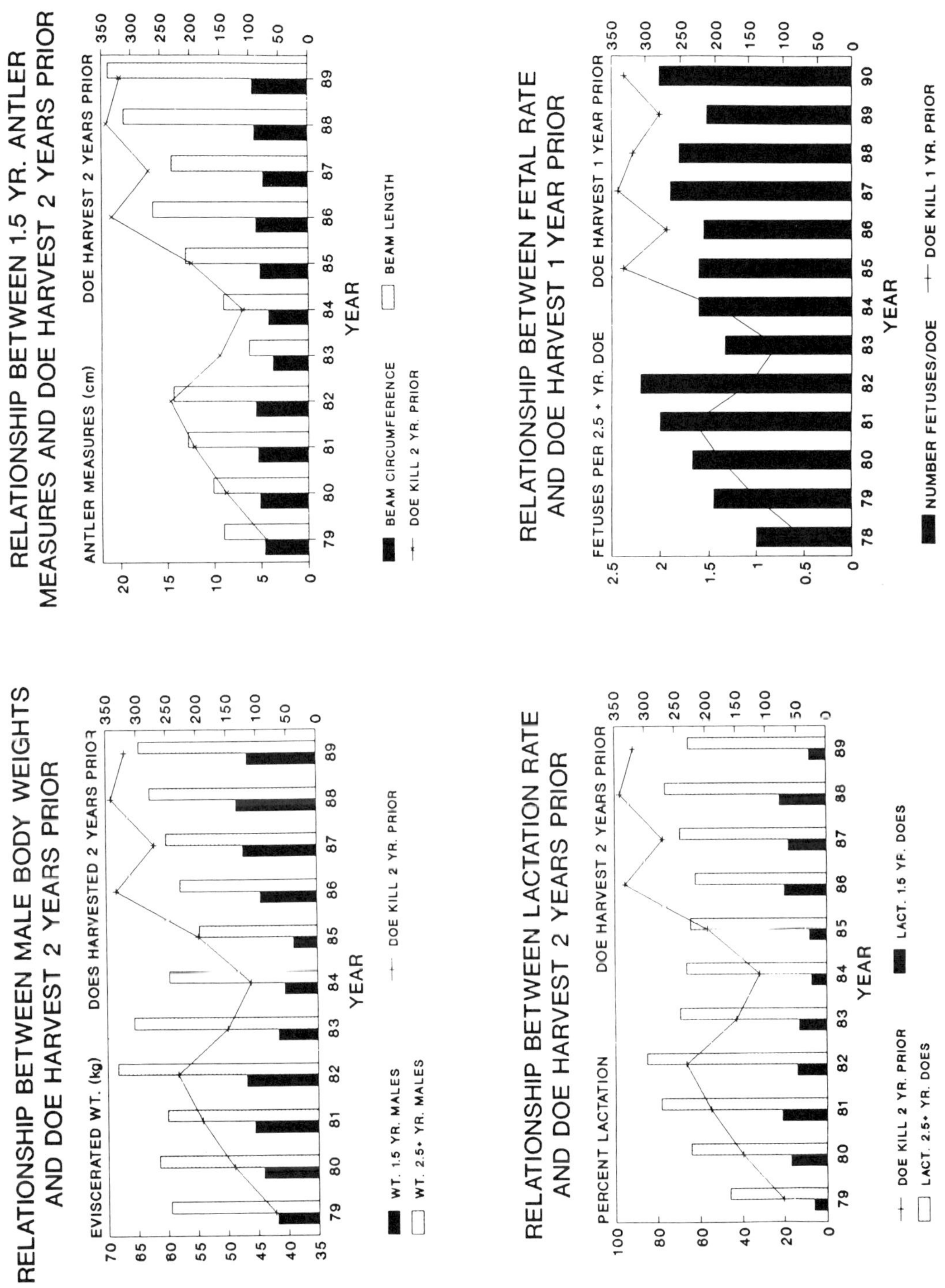

FIGURE 7.1. Relationships between selected deer condition indices and previous doe harvest history at Davis Island, Mississippi.

TABLE 7.3. Reproductive measures of does harvested on 7200 ha, Davis Island, Mississippi, 1978–1990

	Fetuses[a] per adult (2.5 + years) female		Lactation by age class[b]			
			1.5 years		2.5 + years	
Year	Adult females (*n*)	Fetuses (*n*)	Total (*n*)	Lactating (%)	Total (*n*)	Lactating (%)
1978	7	1.00	NR	NR	18	61
1979	11	1.45	35	6	99	46
1980	3	1.67	18	17	53	64
1981	4	2.00	28	21	45	78
1982	5	2.20	13	14	38	85
1983	7	1.33	23	13	71	69
1984	6	1.60	29	7	135	66
1985	10	1.60	38	8	136	64
1986	11	1.55	40	20	180	62
1987	9	1.89	51	18	160	69
1988	10	1.80	49	22	125	76
1989	16	1.52	71	8	207	65
1990	13	2.00	—	—	—	—

[a] Determined from a special research harvest conducted annually from February to March.
[b] Determined from hunter harvested deer in December and January.
NR, not recorded

correlations were derived for the doe harvest 2 years earlier and mean yearling male body weights ($r = 0.693$) and lactation prevalence of yearling ($r = 0.377$) and 2 year or older ($r = 0.403$) does. However, unlike other condition indices, fetal rate had the highest correlation with the previous year's doe harvest ($r = 0.497$). This was significant at the $\alpha = 0.05$ level and suggests fetal rate was the earliest condition indicator to respond to improved range conditions.

Highly significant ($P < 0.01$) correlations were derived for harvest of does 2 years earlier and yearling and adult antler beam circumference ($r = 0.726$ and 0.636, respectively) and yearling and adult antler beam length ($r = 0.902$ and 0.722, respectively). Does 2.5 years or older did not show a correlation between body weight and doe harvest 2 years earlier ($r = -0.025$). While this seems contrary to expected trends, it can be explained by a significant increase in fawning success following increased doe harvest. Thus, does have greater lactation demands because of increased reproductive success following population reduction periods and, as a result, have lower average body weights. This hypothesis has been supported by observations of other deer populations within the Mississippi delta physiographic region. I have observed in one situation two deer populations in close proximity to one another with differences between male body weights. The population with the highest male body weights had the lowest adult female body weights. However, the population with the lowest adult female body weights also had highest lactation rates (H.A. Jacobson, unpublished data).

Trends in body weight, antler development, and reproduction all indicated density-dependent relationships. Condition response to manipulation of female harvest was more apparent than to manipulation of total harvest. Correlation coefficients were much lower between total deer harvest and condition indicators than they were between doe harvest and condition indicators. This contrasts with McCullough's (1987) statement that "in terms of high yield management, there is no advantage to being selective from sex or age in the kill."

TABLE 7.4. Age composition of deer harvest on Davis Island, Mississippi, 1978–1990

Hunting season	Number of deer aged	Percent harvest within sex					
		Males			Females		
		6 months	1.5 years	2.5 + years	6 months	1.5 years	2.5 + years
1978–79	154	8	69	23	20	15	64
1979–80	299	9	73	18	15	34	51
1980–81	194	15	72	13	19	30	51
1981–82	337	2	85	13	19	37	45
1982–83	253	4	72	24	12	30	58
1983–84	332	6	76	19	9	20	71
1984–85	457	14	34	52	10	19	71
1985–86	408	9	23	68	11	22	67
1986–87	521	7	34	59	14	17	69
1987–88	453	10	15	75	14	24	62
1988–89	502	12	15	73	18	20	62
1989–90	471	8	19	73	9	24	67

The Davis Island deer herd can be characterized as having been a buck-harvest only, highly exploited herd before 1977. Doe harvest was initiated in 1977 and 1978, and a more balanced sex ratio was seen in the kill from 1979–1983 with 44% consisting of does (an average of 226 males and 176 does were taken annually in this period). However, the population was still heavily male-exploited as indicated by 82% of the male harvest consisting of fawn or yearling males in this period (Table 7.4). From 1984 to the present, harvest intensity shifted heavily to does, with 59% of the kill composed of females (an average of 215 males and 312 does were taken annually in this period). Male age structure of the population was also changed dramatically, and from 1984 to the present, only 33% of the male harvest consisted of fawn or yearling males.

Reproductive rates, as predicted from McCullough's model, should be directly related to density. That is, at low harvest rates, reproduction should also be low (if herd densities are at or near K) and vice versa. At low herd density, the population would have a high proportion of fawn and yearling age classes. The results indicate that initial herd reduction of the Davis Island population may have been more successful than the much higher doe harvests in later periods in stimulating reproductive response. Initially, age structure of the female segment shifted to a higher percentage of younger animals with about 50% of the doe harvest represented by fawns and yearlings from 1979–1982 (Table 7.4). The initial antlerless harvest effort also resulted in high lactation rates and interuteral fetal counts (Table 7.3). However, age structure shifted to a higher proportion of older age class does in 1983, and the relatively high 1984–1986 harvest apparently did not compensate for the increased reproductive success seen in 1981 and 1982.

Harvest figures from the Palmyra and Rosedale Hunting Clubs indicate in the "premanagement" bucks-only harvest regime of 1969–1974 an average of 91 deer per year was harvested on 3239 ha. From 1979 to the present, these two clubs have averaged 131 bucks and 123 does. Unfortunately, reliable census data are not available, and K cannot be reliably estimated. However, Davis Island does represent McCullough's "black box" management model and the use of "increasing harvest intensity and population condition" response as a means of achieving a management goal of MSY. The condition responses observed suggest population census is not necessary for achieving management goals.

At Davis Island, significant correlations were observed between doe harvest and herd condition indicators. As long as herd density is influencing range quality, these correlations would be expected. However, at low densities these correlations would not be present as herd density would then not adversely affect range conditions. Thus, when further herd reduction fails to result in improved condition responses, declining deer harvests would be expected. This threshold of herd condition response is close to MSY for highly productive and fertile areas like Davis Island. This may not be the case on areas with low soil fertility. Relationships between forage quality and quanity and deer herd density increases in areas with low site productivity may be less apparent, and lowering herd density may not result in similar condition responses to those seen at Davis Island.

That the changes in deer condition indices seen at Davis Island were in fact due to density as opposed to weather, flooding, or other variables cannot be directly substantiated. However, I have collected similar harvest data and condition data on Kings Point Island, an almost identical area in habitat to Davis Island, located 20 km north of Davis Island. Deer on Kings Point have been at much lower apparent densities than Davis Island over the period of this study. From 1979 through the 1989–1990 season, comparatively little fluctuation in body weights or antler measures data has been seen at Kings Point. During this period mean eviscerated body weights of yearling males at Kings Point have been between 47 and 52 kg (H.A. Jacobson, unpublished data) compared with 36 to 48 kg at Davis Island (Table 7.1). Similarly, other deer body weights, antler measures, and lactation rates of the Kings Point deer population generally exceeded those same measures at Davis Island over this period. The number of deer harvested per unit area has been about 1.4 times as great at Davis Island compared with Kings Point. These data support the contention that changes in deer condition at Davis Island were primarily being influenced by deer densities.

The strategy of protecting yearling males from harvest with the use of antler beam length restrictions was effective in producing desired age structure changes in the buck population. This translated to an increase in the number of older bucks harvested and production of larger numbers of deer with antler qualities desired by hunters (Fig. 7.2).

The change in male age structure in the harvest which was seen in 1984 (Table 7.4) was accompanied by an apparent shift in breeding chronology of the Davis Island herd (Table 7.5). Median breeding dates prior to 1984 ranged from 31 December to 14 January, with breeding peaks in 4 of the 5 years sampled in the second week of January (Table 7.5). From 1984 through the present, breeding peaks were from 22 December to 4 January. This 2 to 3 weeks' earlier rut was likely the result of more males available to breed

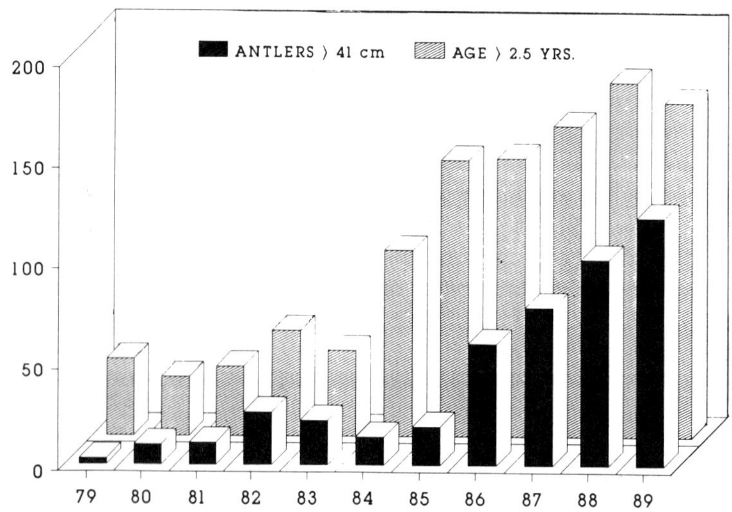

FIGURE 7.2. Harvest of 2-year-old or older bucks and bucks with greater than 41 cm of antler beam length, Davis Island, Mississippi, 1977–1989.

TABLE 7.5. Changes in white-tailed deer breeding chronology, Davis Island, Mississippi, 1978–1990

Year[a]	Does examined (*n*)	Breeding dates (range)	(Median)
1978	10	26 Dec–10 Mar	14 Jan
1979	23	16 Dec–2 Mar	11 Jan
1981	4	26 Dec–23 Feb	13 Feb
1982	9	8 Dec–31 Jan	31 Dec
1983	9	20 Dec–27 Jan	13 Jan
1984	6	9 Dec–22 Dec	27 Dec
1986	15	11 Dec–14 Feb	29 Dec
1987	10	12 Dec–24 Feb	22 Dec
1988	12	9 Dec–13 Jan	22 Dec
1989	16	12 Dec–27 Feb	26 Dec
1990	26	2 Dec–22 Jan	4 Jan

[a]Fetal age data was not collected in 1980 and 1985.

during the initial estrous cycles of the females, or sociobiological phenomena could also be involved. Verme et al. (1987) reported that phermones from adult males can stimulate earlier reproductive effort in females. Regardless, it is obvious that different sex-specific exploitation rates can affect recruitment of the white-tailed deer, and these effects should be considered in stock recruitment models.

References

Gwynn DC, Mott SF, Cotton WD, Jacobson HA (1983) Cooperative management of white-tailed deer on private lands in Mississippi. Wildl Soc Bull 11:211–214

Hamilton RJ, Tobin ML, Moore WG (1985) Aging fetal white-tailed deer. Proc Annu Conf SE Assoc Fish Wildl Agen 39:389–395

Jacobson HA, Reiner RJ (1989) Estimating age of white-tailed deer: tooth wear versus cementum annuli. Proc Annu Conf SE Assoc Fish and Wildl Agen 43:286–291

Larson JS, Taber RV (1980) Criteria of sex and age. In Schemnitz SD (ed) Wildlife Management Techniques Manual. Wildlife Society, Washington, D.C., pp 143–202

McCullough DR (1987) The theory and management of *Odocoileus* populations. In Wemmer CM (ed) Biology and Management of the Cervidae. Smithsonian Institution, Washington, D.C., pp 535–548

McCullough DR (1984) Lessons from the George Reserve, Michigan. In Hall LK (ed) White-tailed Deer Ecology and Management. Stackpole Books, Harrisburg, Pa., pp 211–242

McCullough DR (1979) The George Reserve Deer Herd. University of Michigan Press, Ann Arbor

Verme LJ, Ozoga JJ, Nellist JT (1987) Induced early estrus in penned white-tailed deer does. J Wildlife Management 51:54–56

8
Activity and Movement Patterns of Young White-Tailed Deer Fawns

Georg Schwede, Hubert Hendrichs, and Christen Wemmer

Abstract

We monitored activity and movement patterns of young white-tailed deer fawns (*Odocoileus virginianus*) at the National Zoo's Conservation and Research Center (CRC), Front Royal, Virginia, to study the adjustment of the fawn's behavior to its susceptibility to predation. Although very restricted during the first month of life, activity and movements increased continuously as fawns grew older. Activities and use of space by fawns and their mothers were completely synchronized 12 weeks after parturition. Circadian rhythms were also strongly related to age. During the first 4 weeks they were characterized by pronounced peaks of activity at sunrise and sunset and very little activity at night. Subsequently, night activity increased significantly. Males revealed higher rates of movements and activity than females after 4 weeks. Functional aspects are discussed.

Key words: Circadian activity, fawns, movements, predation, white-tailed deer

Introduction

The behavior of young ungulates has probably been shaped by strategies for avoiding predation (Ralls et al. 1986). Two behavioral patterns that have evolved as a consequence of these strategies are referred to as "following" and "hiding" (Walther 1965; Lent 1974; Ralls et al. 1986). While "following" has been viewed as a strategy for avoiding predation in open habitat, "hiding" is considered to optimize protection of young in dense habitats (Lent 1974). White-tailed deer, as most cervids, typify the hider type where young are protectively colored and spend much of their early life bedded in seclusion some distance away from their mother. Although there are reports of successful maternal defense of young against predators (Michael 1967; Litvaitis and Bartush 1980), the effectiveness of this strategy relies mainly on an optimal adjustment of the fawn's behavior to its physical development. During the first few weeks when chances to escape a predator are very limited, young hiders should choose activity patterns that represent a careful balance between necessary movements to feed and shift bedding sites and the necessary inactivity to avoid predation (Huegel 1985).

The cryptic behavior of young hiders has been the main hindrance to the collection of continuous data on individual young. Despite growing numbers of publications on neonatal behavior of young white-tailed deer due to advances in radiotelemetry (Cook et al. 1967; Jackson et al. 1972; White et al. 1972; Garner and Morrison 1977; Bartush and Lewis 1978; Ozoga et al. 1982), there is still only one study (Huegel 1985) describing activity and movement patterns of young fawns. The main objective of our investigation was to monitor the development of activity and movement patterns of fawns from birth to the time they were weaned.

Study Area

The study was conducted at the National Zoological Park's Conservation and Research Center (CRC), near Front Royal, Virginia. The main study area encompassed 390 ha and varied in elevation from 250 m to 650 m. Topography was characterized by gentle rolling hills, ridges, hollows, and steep slopes. Most (85%) of the area was covered by secondary eastern deciduous forest and the remainder by open pastures. The area supported 0.32 deer/ha. Feral dogs represented the major cause of predation on adult deer, whereas young fawns were mainly preyed on by bobcat (*Lynx rufus*). There was no legal hunting allowed.

Methods

Fawns were captured in May/June 1986–1987 by monitoring movement patterns of radio-collared females during parturition (Huegel et al. 1985). All captured fawns were aged according to physical and behavioral characteristics (Haugen and Speake 1958). Twenty-one fawns were equipped with motion-sensitive transmitters. Monthly range sizes were based on five locations per week collected in three different phases (within 1 h after dawn, within 1 h before dusk, and the remainder of the daylight period). Since long distance triangulation proved to be insufficient for precise locations in hilly and wooded terrain, fawns were carefully approached to distances < 50 m, circled, and located (Hölzenbein and Schwede 1989). Home range sizes were compared by calculating minimum convex polygons (Mohr 1947) based on all locations. Movement rates were calculated as the sum of successive straight-line distances moved during weekly 24-h locations. Thus, they express the minimum distances travelled during this time. Data on activity were obtained from changes in the pulse rate of the transmitter signal during 24-h radio locations. Signals were checked every hour from a distance for 30 s to detect changes in the pulse rate. This method proved to be accurate in 90% of all readings (Hölzenbein and Schwede 1989). Mean biweekly movement and activity rates were calculated by combining data from two successive 24-h locations. Time-of-day effects on monthly circadian activity were analyzed by pooling data from four successive 24-h locations. To adjust for possible age-related effects on activity rates, mean percentages of total activity per 2-h interval were calculated and compared between different age categories.

Differences in home range size, activity, and movements among monthly, biweekly, and time-of-day periods and between both sexes were tested using an ANOVA for repeated measures (Winer 1971). Results were considered significant at $P \leqslant 0.05$.

Results

Home Range Patterns

Monthly range sizes of fawns varied significantly from June through November ($F = 4.762$, $P = 0.001$, df = 5; Fig. 8.1). Ranges increased significantly from 3.4 ha during the first 4-week period in June ($F = 9.4228$, $P = 0.012$, df = 1) to 16.4 ha in September. Differences between adjacent 4-week periods were significant between the first and second 4-week period ($F = 27.444$, $P = 0.000$, df = 1). Following September, range sizes decreased corresponding to the ranges of mothers ($F = 7.123$, $P = 0.031$, df = 1) to 7.9 ha in November. No significant differences were found between sexes.

Activity Patterns

Mean biweekly activity of fawns showed a significant linear increase with age ($F = 41.834$, $P = 0.001$, df = 1). Activity increased from a low of 18.2% during the first 2 weeks to 54.7% at week 12 (Fig. 8.2), reaching almost the level observed in their mothers. The overall effect of sex on activity was not significant; however, males were more active than females after 4 weeks. Differences were significant at 6, 10, and 12 weeks ($F = 5.825$, $P = 0.048$, df = 1; $F = 18.856$, $P = 0.005$, df = 1; $F = 8.923$, $P = 0.024$, df = 1). Circadian activity patterns were strongly influenced by age ($F = 2.354$, $P = 0.001$, df = 22); (Fig. 8.3). During the first 4 weeks, activity patterns revealed no

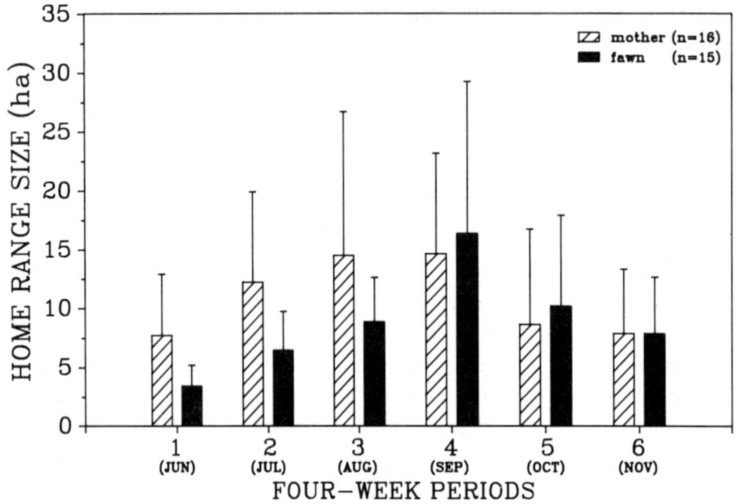

FIGURE 8.1 Mean monthly home range size of white-tailed fawns and their mothers.

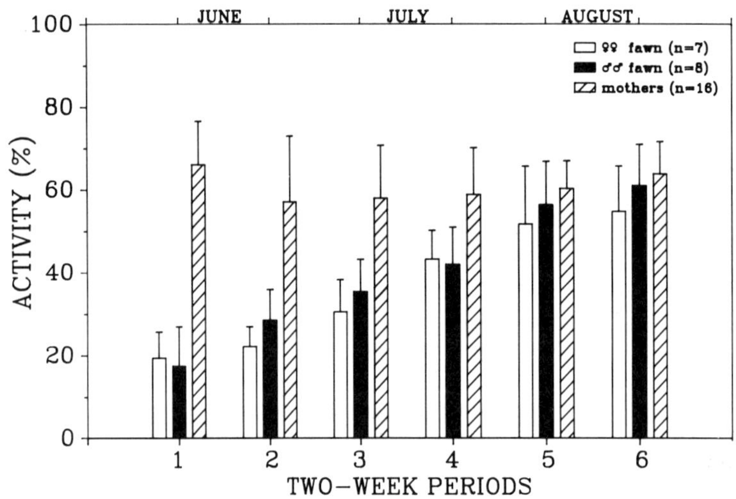

FIGURE 8.2 Mean biweekly activity rates of white-tailed fawns and their mothers.

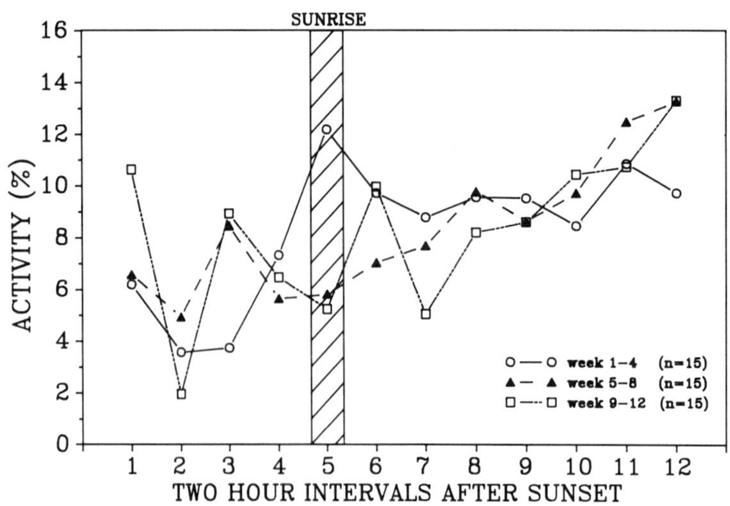

FIGURE 8.3. Mean percentage of total white-tailed fawn activity per time period.

significant variation over 24 h or any significant differences between adjacent 2-h intervals. Fawns were most active at sunrise and during hours of daylight and significantly less active during the first 6 h following sunset ($F = 5.688$, $P = 0.022$, df = 1). These patterns changed drastically after 4 weeks.

During the second and then more pronounced during the last 4-week period, activity patterns gained a more cyclic profile characterized by alternating periods of high and low activity. These alternating patterns were underlined by a highly significant variation during the second and third 4-week period ($F = 3.480$, $P = 0.000$, df = 11; $F = 8.593$, $P = 0.000$, df = 1) and by significant differences between 50% of all adjacent 2-h intervals during the last 4-week period. When comparing activity patterns of the first and last 4-week period by their corresponding 2-h intervals, differences occurred mainly at night. These differences were significant for intervals 1, 3, and 5 ($F = 4.001$, $P = 0.044$, df = 1; $F = 5.546$, $P = 0.024$, df = 1; $F = 4.206$, $P = 0.048$, df = 1). Fawn sex did not have a significant effect on circadian activity patterns at any time.

Movement Patterns

Distances moved when associated with their mother and distances moved during periods of

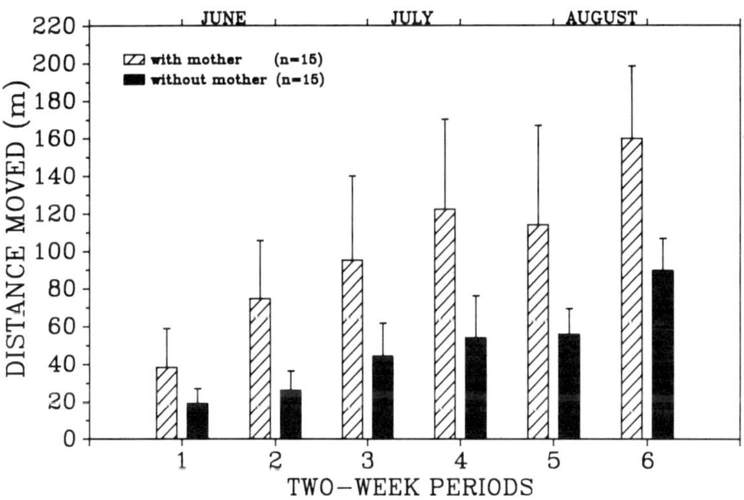

FIGURE 8.4. Mean biweekly movement rates of white-tailed fawns with and without their mothers.

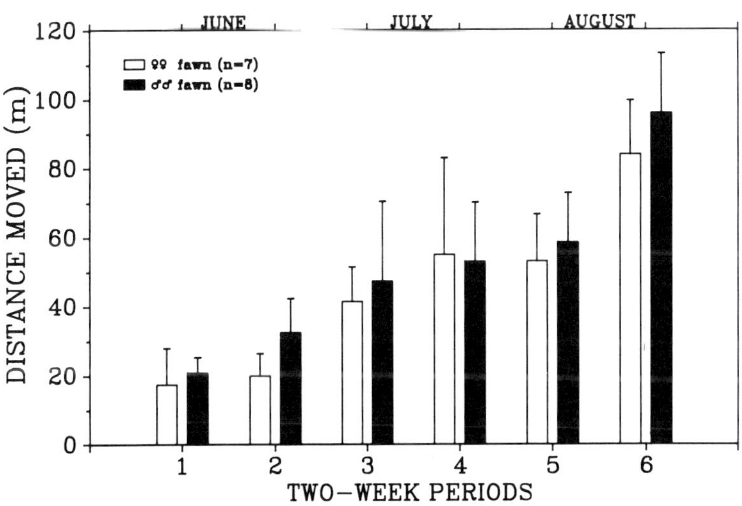

FIGURE 8.5. Mean biweekly movement rates of white-tailed fawns when not associated with their mothers.

outlaying behavior revealed a significant linear increase with age ($F = 19.060$, $P = 0.000$, df $= 5$; $F = 46.910$, $P = 0.000$, df $= 5$; Fig. 8.4). During the first 4 weeks, movements were very restricted during periods of outlaying behavior. Afterwards, movements increased from 26.2 m to 90.1 m at 12 weeks of age. Differences between adjacent periods were significant when fawns aged 4, 6, and 12 weeks ($F = 12.923$, $P = 0.007$, df $= 1$; $F = 24.297$, $P = 0.000$, df $= 1$; $F = 56.931$, $P = 0.000$, df $= 1$). Overall distances travelled during periods of outlaying behavior were not related to fawn sex. However, males travelled significantly longer distances than females when 4 weeks old ($F = 7.665$, $P = 0.024$, df $= 1$). This trend was persistent, although subsequent differences were not significant (Fig. 8.5).

During periods of reunion of mother and fawn, distances travelled showed a significant linear increase from 38.3 m during the first 2 weeks to 160.2 m when 12 weeks old ($F = 57.331$, $P = 0.000$, df $= 1$). No significant differences were found between sexes.

Discussion

Range sizes of fawns in our study were considerably smaller than those reported for low-density populations and relatively open habitat (Garner and Morrison 1977; Carrol and Brown 1977; Huegel 1985). These differences may have been caused by different deer densities (Ozoga et al. 1982) and habitat conditions (Carrol and Brown 1977). Low densities and the open habitat which is often associated with patchiness of suitable bedding site cover (Huegel 1985) may have led to larger ranges in other studies. During the first few weeks, when ranges were very small, the fawn's life was characterized by prolonged periods of outlaying behavior and very little association with its mother (Schwede and Hendrichs 1989). All activities centered around suckling and the short periods of reunion with the mother.

The seclusion and cryptic behavior shown during the first few weeks were also reflected by restricted movements and low percentages of activity. Huegel (1985) reported considerably higher rates of activity and movements for fawns in Iowa. In addition to the influence of the different densities and habitat conditions, these dissimilarities may have been also caused by the differences in data collection. Movements and activity of fawns in Iowa were determined by measuring linear distances between consecutive locations obtained by long-distance triangulation. Because this method is not as precise, rates of activity and movements may have been overestimated.

Besides the obvious benefits of conserving energy at a time of maximal growth (Verme 1963), fawn activities during the first 4 weeks may be viewed as a careful balance between necessary movements to feed and change bedding sites and the necessary inactivity to avoid predation. Together with their cryptic coloration and anogenital licking by the mother, the observed sedentary behavior represents an adaptation well suited to their undeveloped physical skills, protecting them from predators (Lent 1974). The extreme vulnerability of fawns to predation during the first few weeks was underlined by the observed cases of mortality. Five of six fawns were killed within the first 2 weeks. Studies on mortality rates in fawns have shown that in habitats with adequate cover where canids are the major predators, fawns younger than 3 weeks were relatively well protected (Nelson and Woolf 1987). In habitats where suitable resting cover was inadequate and canid predation relatively intense, mortality was highest during the first 4 weeks (Cook et al. 1971; Garner et al. 1976; and Bartush and Lewis 1978). The apparently higher vulnerability in open habitats provides further evidence that "hiding" has probably evolved in forest-dwelling animals as a successful antipredator strategy (Lent 1974).

Circadian activity patterns of young white-tailed deer fawns may have also been shaped to a major extent by predation. Similar to Huegel's findings (1985) our fawns showed pronounced peaks in activity during hours nearest sunrise and preceding sunset throughout the first few weeks. Assuming that night active predators represented the major cause of predation (all five cases of bobcat predation occurred during nighttime), fawns should have minimized their activities during night.

When fawns had reached a level of physical development, giving them a greater chance to escape predation, their activities gradually increased. This increase was predominantly related to a higher frequency and longer duration of association with their mothers (Schwede and Hendrichs 1989). At the age of about 4 weeks when adult stomach proportions were reached (Short 1964) and the process of weaning (Gauthier and Barrette 1985) started, fawns began to join their mothers on foraging trips for progressively longer periods. Thereby, they gradually adopted their mother's home range, activity, and movement patterns.

Taber and Dasmann (1954) viewed a higher mortality rate in older male black-tailed deer (*O. hemionus*) fawns to be a result of their greater curiosity, increased activity, and earlier independence of the mother. This view was supported by studies on white-tailed deer (Jackson et al. 1972) and red deer (*Cervus elaphus*) (Guinness et al. 1979). In our study, behavioral differences began to unfold at the age of 3 to 4 weeks when males moved significantly longer distances and also showed a continuously higher activity than females after 4 weeks. Together with an earlier independence from the mother, as indicated by a lower association frequency and a stronger tendency to join other deer (Schwede and Hendrichs 1989), these early traits may indeed be indications of a greater curiosity and exploration behavior in males. Both attributes could serve in males as a means to get more experience of other deer and other areas in preparation for leaving the natal range.

Acknowledgements

We thank S. Binder, B. Greenwalt, C. Sahley, and W. Horn for field assistance. Special thanks go to S. Hölzenbein for his help in organizing and conducting the fieldwork as well as for his critique of parts of this study. The research was funded by the Smithsonian Institution and Friends of the National Zoo, Washington, D.C.

References

Bartush WS, Lewis JC (1981) Mortality of white-tailed deer fawns in the Wichita Mountains. Proc Oklahoma Acad Sci 61:23–27

Bartush WS, Lewis JC (1978) Behavior of white-tailed deer does and fawns during the parturition period. Proc Annu Conf SE Assoc Fish Wildlife Agen 32:246–255

Carrol BK, Brown DL (1977) Factors affecting neonatal fawn survival in south-central Texas. J Wildlife Management 41:63–69

Cook RS, White M, Trainer DO, Glazener WC (1971) Mortality of young white-tailed deer fawns in south Texas. J Wildlife Management 35:47–56

Cook RS, White M, Trainer DO, Glazener WC (1967) Radiotelemetry for fawn survival in south-central Texas. Wild Disease Assn Bull 3:160–165

Garner GW, Morrison JA (1977) Diurnal range and movements of young white-tailed deer fawns in southwestern Oklahoma. Proc Annu Conf SE Assoc Fish Wildl Agen 31:126–133

Garner GW, Morrison JA, Lewis JC (1976) Mortality of white-tailed deer fawns in the Wichita Mountains, Oklahoma. Pro Annu Conf SE Assn Fish Wildl Agen 30:493–506

Gauthier D, Barrette C (1985) Suckling and weaning in captive white-tailed and fallow deer. Behavior 94:128–149

Guinness FE, Hall MJ, Cockerill RA (1979) Mother-offspring association in red deer (*Cervus elaphus* L.) on Rhum. Anim Behav 27:536–544

Haugen AD, Speake DW (1958) Determining age of young white-tailed deer. J Wildlife Management 22:319–321

Hölzenbein S, Schwede G (1989) Activity and movements of female white-tailed deer during the rut. J Wildlife Management 53:219–223

Huegel CN (1985) Predator-avoidance behaviors in white-tailed deer that favor fawn survival. Ph.D. thesis, Iowa State University, Ames, Iowa

Huegel CN, Dahlgren RB, Gladfelter HL (1985) Use of doe behavior to capture white-tailed deer fawns. Wild Soc Bull 13:287–289

Jackson RM, White M, Knowlton FF (1972) Activity patterns of young white-tailed deer fawns in south Texas. Ecology 53:262–270

Lent PC (1974) Mother-infant relationships in ungulates. In Geist V, Walther F (eds) The Behavior of Ungulates and Its Relation to Management, vol 1. IUCN Publications 24, Morges, Switzerland, pp 14–55

Litvaitis PA, Bartush WS (1980) Coyote-deer interactions during the fawning season in Oklahoma. Southwest Nat 25:117–118

Michael ED (1967) Behavioral interactions of deer and some other mammals. Southwest Nat 12:156–162

Mohr CD (1947) Table of equivalent populations of North American small mammals. Am Midland Nat 37:223–249

Nelson TA, Woolf A (1987) Mortality of white-tailed deer fawns in southern Illinois. J Wildlife Management 51:326–329

Ozoga JJ, Verme LJ, Bienz CS (1982) Parturition behavior and territoriality in white-tailed deer: impact on neonatal mortality. J Wildlife Management 46:1–11

Ralls K, Kranz K, Lundrigan B (1986) Mother-young relationships in captive ungulates: variability and clustering. Anim Behav 75:134–145

Schwede G, Hendrichs H (1989) Mother-infant relations in white-tailed deer (*Odocoileus virginianus*). Proc 5th Int Theriological Congr 2:834–835

Short HL (1964) Postnatal development of white-tailed deer. J Wildlife Management 28:445–448

Taber RD, Dasmann RF (1954) A sex difference in mortality in young Columbian black-tailed deer. J Wildlife Management 18:309–315

Verme LJ (1963) Effect of nutrition on growth of white-tailed deer fawns. Transcript N Am Wildl and Nat Res Conf 28:431–443

Walther F (1965) Verhaltensstudien an der Grantgazelle (Gazella Granti Brookee, 1982) im Ngorongoro-Krater. Z Tierpsychol 22:167–208

White M (1966) Population ecology of some white-tailed deer in South Texas. Ph.D. thesis, Purdue University, Lafayette, Indiana

White M, Knowlton FF, Glazener WC (1972) Effects of dam-newborn fawn behavior on capture and mortality. J Wildlife Management 36:163–164

Winer BJ (1971) Statistical Principles in Experimental Design, 2nd ed. McGraw-Hill, New York

9
Fetal Growth Rate Estimation from Length and Date of Death

Daniel W. Soprovich

Abstract

Stochastic simulation revealed that linear regression of fetal length on date of death allows accurate prediction of the fetal growth rate of white-tailed deer (*Odocoileus virginianus*). Precision improved with increased sampling intensity, wider sampling period, narrower conception distribution, and systematic over random sampling. Bias resulted from extension of sampling towards conception or parturition, and increased with wider conception distribution and for systematic sampling. To utilize this technique, attention should be paid to sample size, sampling period, and distribution of conception.

Key words: Accuracy, fetus, growth, precision, simulation, white-tailed deer

Introduction

The relationship between fetal growth (crown-rump and forehead-rump length) and age in white-tailed deer has been reported by Cheatum and Morton (1946), Verme (1963), Short (1970), and Hamilton et al. (1985). The ability to compare growth rates between studies is restricted by small sample sizes, differing ranges of fetal age examined, and other potentially different factors which may have influenced the data (e.g., condition of the doe).

Fetal growth is curvilinear to approximately 60 days (Cheatum and Morton 1946; Verme 1963) or perhaps earlier (Hamilton et al. 1985) and then linear to some indeterminate age. Cheatum and Morton (1946) provided data which suggest curvilinear growth commences between approximately 164 days and parturition; however, Jackson and Hesselton (1973) indicate Cheatum and Morton's curve had limitations for older fetuses. Specifically, fetuses were found which were much larger than the upper limit of Cheatum and Morton's growth curve. Cheatum and Morton (1946) also utilized primarily twin fetuses (4 singletons, 11 sets of twins) and more than one datum from each of 3 does. In addition, it is unclear if maternal nutrition was adequate. Short (1970) analyzed Verme's (1963) data and found a strong linear relationship for fetuses between 58 and 180 days of age; however, inspection of Verme's (1963) data suggests possible curvilinear growth for older fetuses. Verme's (1963) data set, in contrast to Cheatum and Morton (1946), consisted primarily of singletons (13 singletons, 4 sets of twins). Although Verme (1963) believed his does to be in excellent health, his data resulted from a study designed to assess the impact of doe nutrition on fetal growth, and food quality may have influenced fetal growth. In addition, Verme's (1963) data is unclear on the effect of nutrition on linear growth. Length appears related to nutrition for fetuses aged 151 to 180 days but unrelated for fetuses aged 181 to 200 days (Table 9.1; Verme 1963).

TABLE 9.1 Growth rate (B_1, mm/day) estimated from 30 simulations of length (crown-rump, forehead-rump, in mm) vs date in white-tailed deer fetuses.

Method	Sampling period/dates[a] (days)	Sample size (n): 40 $\bar{x}$	40 SD	60 $\bar{x}$	60 SD	100 $\bar{x}$	100 SD	200 $\bar{x}$	200 SD
1	184–244[b]	3.036	0.294	3.143	0.254	3.135	0.179	3.095	0.126
	194–234	3.126	0.390	3.090	0.302	3.106	0.300	3.131	0.165
	164–244	3.145	0.237	3.074	0.139	3.159	0.136	3.108	0.098
	124–244	2.984***	0.165	2.960***	0.142	2.939***	0.099	2.965***	0.072
	184–264	3.212*	0.198	3.092	0.121	3.130	0.115	3.102	0.087
	184–304	2.941***	0.186	2.980***	0.130	2.962***	0.104	2.970***	0.070
	164–264	3.094	0.162	3.115	0.140	3.116	0.109	3.108	0.067
2	164–244	3.156	0.253	3.109	0.208	3.106	0.121	3.116	0.088
	124–244	2.900***	0.125	2.927***	0.089	2.945***	0.068	2.919***	0.056
	184–264	3.120	0.197	3.135	0.161	3.167	0.127	3.112	0.098
	184–304	3.004***	0.158	2.993***	0.082	3.020***	0.113	3.026***	0.050
	164–264	3.142	0.172	3.086	0.130	3.096	0.086	3.140	0.073
2[c]	184–244	3.155	0.634	3.014	0.428	3.093	0.289	3.110	0.261
	184–304	2.784***	0.314	2.793***	0.247	2.845***	0.164	2.828***	0.121
3	184, 244	3.203**	0.138	3.138	0.132	3.100	0.105	3.124	0.087
	164, 244	3.130	0.114	3.140	0.080	3.116	0.075	3.119	0.048
	124, 244	2.740***	0.056	2.745***	0.052	2.741***	0.040	2.745***	0.025
	184, 264	3.157	0.146	3.116	0.088	3.149	0.086	3.134	0.061
	184, 304	2.855***	0.067	2.842***	0.063	2.859***	0.049	2.860***	0.029
	164, 264	3.117	0.082	3.121	0.069	3.133	0.060	3.125	0.043

Model attributes included random conception from a normal distribution ($\bar{x}$ date = 100, SD = 10 days, except where specified), individual variation, and random birth from a normal gestation distribution ($\bar{x}$ = 199.2 days, SD = 4.36 days). Methods 1 and 2 sampled randomly within the sampling period, method 3 sampled one-half of the sample at each of two dates. Method 1 deleted fetuses < 22.4 mm or > 459.9 mm in length, methods 2 and 3 utilized all fetuses > 0.4 mm in length.
[a] Sampling period or dates relate to mean of conception distribution.
[b] Methods 1 and 2 equivalent at this sampling period and SD = 10 days for the conception distribution.
[c] SD = 20 days for conception distribution.
* $P < 0.05$; ** $P < 0.01$; *** $P < 0.001$

The most comprehensive data on fetal growth are that of Hamilton et al. (1985), who examined 64 fetuses and found a strong linear relationship between 38 and approximately 191 days of age, and that fetal growth was influenced by sex and litter size. Compared with Verme (1963), Hamilton et al. (1985) maintained their does on a high protein diet (16% crude protein). Therefore, the rate of fetal growth may have been the maximum biologically attainable.

The examination of data on deer fetuses from Manitoba suggested that growth rate during the linear portion of fetal growth could be estimated by linear regression of size and date of death. Subsequently, it was found that Salwasser and Holl (1979) and Bartmann (1986) had previously utilized this method. The objective of this paper was to stochastically simulate the influence of sampling and data handling methodology on the accuracy and precision of fetal growth rate, when estimated from fetal length and date of death.

Methods

The growth function—length = 3.125 age − 115.06—from Hamilton et al. (1985), was utilized for fetuses between 44 and 184 days. Inspection of residuals indicated the youngest fetus to be in the curvilinear portion of growth, hence linear growth was set to begin at the estimated age (Figure 1 in Hamilton et al. 1985) of the next oldest fetus. Although growth was linear to an estimated 191 days of age, the estimated age of the second oldest fetus was used as the upper limit for linear growth, because 191 days falls within the gestation distribution utilized, and other data suggest that growth could be curvilinear below 191 days of age. The model allowed linear or curvilinear growth between 185 days of age and parturition. At age <44 days, a curvilinear function—(age + 1) = 0.519 + 14.109 ln (length + 1)—which reasonably approximated available data (Cheatum and Morton 1946; Hamilton et al. 1985) was used. Embryos < 0.4 mm in length were treated as too small to detect (Cheatum and Morton 1942), whereas all other embryos were assumed to be found.

Conception date was randomly selected from a normal conception distribution ($\bar{x}$ date = 100, SD = 10 days). This range of conception dates ($\bar{x} \pm 4$SD) would have included all observations from each of three areas examined in Manitoba by Ransom (1966). Date is defined as the number of days beyond an arbitrary start date (start date = 0). To examine the effect of width of conception distribution, simulations with SD = 20 days were conducted. Gestation length was randomly selected from a normal gestation distribution ($\bar{x}$ = 199.2 days, SD = 4.36) as described by Verme (1989). Gestation length was varied between ± 3SD of the mean.

Individual variation was incorporated using data from Figure 1 in Hamilton et al. (1985). Each datum was estimated with Vernier calipers, and a linear regression of the resulting data set produced a slope of 0.321 and intercept of 38.36. The slightly greater estimated intercept may have resulted from the smaller sample size of estimated data ($n = 56$) and/or measurement error. The datum for the youngest fetus was deleted, and the linear model recalculated. Inspection of residuals suggested that residual variability increased with length, which was supported by a positive linear trend between the residual squared and length ($F = 10.64$, df = 1,51). This relationship—(residual^2 = 0.050 length − 0.945) was used to approximate individual variation about lengths > 22.4 mm (44 days). For each length, a normal distribution was constructed with length as the mean and SD calculated from the above function. A new length was randomly selected from within 4 SD of the mean length. As fetal length is related to sex and litter size, this approximation would simulate variability in a data set which lumps sexes and litter sizes.

Sample size and sampling period were varied, and 30 simulations were conducted for each combination. Mean values of $\hat{B}_1$ and $\hat{B}_0$ were tested with a t test and $H_o{:}B_1 = 3.125$, $H_o{:}B_0 = 427.5625$. The test value for B_o results from the arbitrary value of 100 utilized for the mean of the conception distribution (i.e., age = 0.32 length + 36.82 can be converted to date by addition of 100, or date = 0.32 length + 136.82, which is equivalent to length = 3.125 date − 427.5625).

Several data-handling methods were examined. The first method sampled randomly within the sampling period and fetuses < 22.4 mm or > 459.9 mm in length and potentially in curvilinear growth were deleted from the sample. In the second method, all fetuses were utilized for analysis, and linear growth was assumed beyond 184 days. The third method, sampled systematically at two dates, utilized all fetuses for analysis and assumed linear growth beyond 184 days.

Results

Test Simulations

The model always returned the same values for $\hat{B}_1$ (3.125139) and $\hat{B}_0$ (−427.5921) with $n = 300$, conception distribution of 100 ± 0.01 (SD), all fetuses in linear phase of growth (i.e., sampling dates 185 to 240), random sampling, and no individual variation. The mean $\hat{B}_1$ value was 3.128 (SD = 0.104, $t = 0.13$) when 20 simulations with a conception distribution of 100 ± 10 (SD) was used. When date was regressed against size, with the same conception distribution and 20 simulations, the slope ($\hat{B}_1 = 0.233$, $t = -75.41$, $P < 0.001$) and intercept ($\hat{B}_0 = 157.35$, $t = 64.84$, $P < 0.001$) was different from expected (expected values were $B_1 = 0.32$, $B_0 = 136.82$).

Method 1

Sampling when all fetuses were in linear growth (dates 184 to 244) produced estimates which were not biased (i.e., Table 9.1; $P > 0.05$). Shortening the sampling period to dates 194 to 234 resulted in estimates which were less precise at $n = 100$ ($F = 2.81$, df $= 29{,}29$, $P < 0.01$) and exhibited the same tendency at the other sample sizes. Extending the sampling period substantially to sample early (dates 124 to 244) or late (dates 184 to 304) produced biased estimates (averaging approximately 5.1% and 5.0% respectively of the true value of B_1 at $n = 200$) as observations of fetuses in curvilinear growth, nonpregnant does, fetuses in potential curvilinear growth (i.e., age > 184 days), and fetuses which had been born were deleted. Increasing the sampling period by 20 days on either or both sides increased precision with limited detectable bias (one $\hat{B}_1$ and one $\hat{B}_0$ value out of 24 tests had $P < 0.05$).

Method 2

Extension of the sampling period substantially in one direction produced biased estimates of B_1 which averaged approximately 6.6% less for early sampling and 3.2% less for late sampling than the true value (at $n = 200$), as fetuses in curvilinear growth, nonpregnant does, and postpartum does were sampled. The precision was sometimes better (at $n = 60$, $F = 2.54$, $P < 0.01$ for dates 124 to 244 and $F = 2.51$, $P < 0.01$ for dates 184 to 304; at $n = 100$, $F = 2.12$, $P < 0.025$ for dates 124 to 244; at $n = 200$, $F = 1.96$, $P < 0.05$ for dates 184 to 304) for these biased estimates as compared with those which resulted from method 1. It was not possible to detect bias in B_1 for sampling periods which were extended by 20 days in one or both directions. For these unbiased estimates of B_1, method 1 produced a more precise estimate at dates 164 to 244 and $n = 60$ ($F = 2.24$, $P < 0.025$); however, this result was not consistent across sample size for this sampling period. With the SD of the conception distribution doubled, precision decreased substantially, and bias increased when the sampling period was extended.

Method 3

When one sampling date was substantially extended from 184 to 124 or from 244 to 304, $\hat{B}_1$ produced by systematic sampling averaged 12.2% less for an extended early sampling date and 8.5% less for an extended late sampling date than the true value. With the exception of a bias detected when fetuses were sampled at dates 184 and 244 (at $n = 40$, certainly a spurious result), no significant divergences from predicted values could be detected. Precision was substantially improved when systematic sampling was conducted on two dates as opposed to random sampling between the dates. For methods 1 to 3, results of tests of significance for $\hat{B}_0$ were almost identical to those for $\hat{B}_1$.

Discussion

The simulation model demonstrates that it is possible to accurately estimate fetal growth rate when length is regressed against date of death, as suggested by Bartmann (1986). Growth rate estimates generated from small sample sizes should be treated cautiously due to low precision. When sampling is random, as might occur when deer are collected following vehicle collision, extension of the sampling period towards conception or parturition produces relatively precise but potentially biased estimates, depending on how far sampling is extended. This bias may be insignificant (e.g., sampling period 164 to 264) or minor (i.e., bias was not severe when sampling was random and potentially close to mean conception or parturition). For the extended late sampling period, use of all observations was less biased than deletion of observations, probably due to the assumption of linear growth to parturition. Deletion of fetuses potentially in curvilinear growth produced no obvious advantage. In fact, a disadvantage of this method is that the analyst must know, a priori, what length of fetus should be deleted. An exception to this might be the deletion of an obvious outlier.

Systematic sampling increased precision, sometimes considerably, over random sampling. Extension of sampling dates towards conception or parturition produced estimates of substantially greater bias than those resulting from random sampling. Additionally, any such extension can lead to an increase in the sample size of does required to meet the sample size objectives for fetuses (e.g., 87 does, on average, were required to obtain 40 fetuses at sampling dates of 184 and 304).

The model parameters used in this study generally were selected to provide conservative estimates of precision. Fetuses were assumed to begin linear growth at 44 days of age, although limited data (Cheatum and Morton 1946; Hamilton et al. 1985) suggest that growth is likely curvilinear to approximately 60 days of age. Utilization of a curvilinear function to 60 days of age would likely have had a minimal impact on tests for bias and precision estimates because the linear model appears to produce only small errors between 44 and 60 days of age (data in Hamilton et al. 1985). If linear growth in late gestation is biologically impossible in natural environments (Cheatum and Morton 1946; Verme 1963), this study understates accuracy and precision.

The width of the conception distribution can influence precision and bias, while the form could also have an impact. Additionally, growth rate will be biased and less precise if doe condition causes fetal growth to become curvilinear during the anticipated linear period. This would not be a problem if poor doe condition produces linear growth at a lower rate throughout, as suggested by Bartmann (1986). In general, it might be expected that the precision of a set of real data would be less than reported in this study, the extent dependent upon the parameters of the population.

Salwasser and Holl (1979) used hindfoot length of mule deer (*O. hemionus*) to generate a growth function which was used to estimate mean conception dates. As demonstrated in this study, regression of date against size can lead to biased estimates, perhaps related to the number of days between the arbitrary start date and the mean conception date. The data of Salwasser and Holl (1979) contain residuals below the regression line at large hindfoot size, which violates an assumption for linear regression (Kleinbaum and Kupper 1978), and may help explain the biases found when date is regressed against size. In addition, Salwasser and Holl (1979) deleted outliers but did not define them and sampled into the gestation period which can bias results. It appears that the growth function of Salwasser and Holl (1979) is biased which might explain why it poorly predicts age when compared to the function developed by Short (1970) for known-age white-tailed deer fetuses.

Bartmann (1986) regressed forehead-rump length against date of death for mule deer fetuses. All of these fetuses appeared to be either in the linear portion of growth or near the edge of that period, such that the estimates produced would be expected to be biased minimally, if at all. Based on the results of this study, these estimates would be of low precision, given the sample sizes used; however, it appears that Bartmann's (1986) inferences, although perhaps surprising, are sound.

Management Implications

Wildlife scientists and managers typically utilize a growth function developed from a specific population to estimate conception dates from a sample of fetuses from another population (Illige 1951; Ransom 1966; Haugen 1975). To generate accurate data, fetal growth rate must be the same for both populations. Application of a wrong growth function and/or a correct growth function to a set of fetuses whose growth has been influenced by poor maternal condition will influence the variability of the estimated conception distribution, as demonstrated by Bartmann (1986). Since the above assumptions are often not met, it is likely that variance measures often are overstated for conception distributions, and mean conception dates have been incorrectly estimated.

It is important to recognize that sampling periods which lead to deletion of data or use of fetuses in curvilinear growth and/or collection of nonpregnant or postpartum does, will produce biased estimates of growth rate. As one proceeds south from Manitoba, environmental constraints on the length of time available to successfully raise a fawn allow a wider conception distribution (e.g., Jackson and Hesselton 1973; Jacobson et al. 1979) than that of northern white-tailed deer (Ransom 1966). As both precision and accuracy are sensitive to the conception distribution, caution is recommended in application of this technique to populations with wide distributions.

The advantage of the method examined in this study is that readily available data can be used to estimate fetal growth rates which may be different from those in the literature and specific to populations. Such a method can then be used to provide more accurate estimates of the conception distribution or to explore the influence of maternal condition on fetal growth. For the parameters used in this study, estimates are imprecise at small sample size. Further work will focus on methodology which may allow more efficient use of samples (e.g., deletion of outliers).

To best use this method, the sampling period or sampling dates should be extended as much as possible, while remembering that extension too far will cause bias. Knowledge of the approximate conception distribution a priori will allow obvious outliers to be deleted. Systematic sampling will increase precision. Residuals should be examined to detect obvious incomplete distributions of length at early or late dates, while recognizing that residuals will be greater for larger fetuses. If possible, data from different sexes, litter sizes, areas where mean conception dates may differ (Cheatum and Morton 1946; Jacobson et al. 1979), fawns and older deer (Cheatum and Morton 1946; Jackson and Hesselton 1973), and years should not be combined. It should also be recognized that the sample of deer, if collected fortuitously, could vary dependent upon the source (O'Gara and Harris 1988), and may not be representative of the population which could result in a biased inference.

Acknowledgements

I thank M. Mattson for reviewing a draft of this paper, R. Young for producing the tables, B. Johnson for direction on Monte Carlo simulation, and D. Stardom for his support.

References

Bartmann RM (1986) Growth rates of mule deer fetuses under different winter conditions. Great Basin Naturalist 46:245–248

Cheatum EL, Morton GH (1946) Breeding season of white-tailed deer in New York. J Wildlife Management 10:249–263

Cheatum EL, Morton GH (1942) Techniques used in determining the period of the rut among white-tailed deer of New York State. Trans North Am Wildl Conf 7:334–342

Hamilton RJ, Tobin ML, Moore WG (1985) Aging fetal white-tailed deer. Proc Annu Conf, SE Assn Fish Wildl Agen 39:389–395

Haugen AO (1975) Reproductive performance of white-tailed deer in Iowa. J Mammal 56:151–159

Illige D (1951) An analysis of the reproductive pattern of whitetail deer in south Texas. J Wildlife Management 32:411–421

Jackson LW, Hesselton WT (1973) Breeding and parturition dates of white-tailed deer in New York. N Y Fish Game J 20:40–47

Jacobson HA, Guynn DC Jr, Griffin RN, Lewis D (1979) Fecundity of white-tailed deer in Mississippi and periodicity of corpora lutea and lactation. Proc Annu Conf, SE Assn Fish Wildl Agen 33:30–35

Kleinbaum DG, Kupper LL (1978) Applied Regression Analysis and Other Multivariable Methods. Wadsworth Publishing, Belmont, Calif.

O'Gara BW, Harris RB (1988) Age and condition of deer killed by predators and automobiles. J Wildlife Management 52:316–320

Ransom AB (1966) Breeding seasons of white-tailed deer in Manitoba. Can J Zool 44:59–62

Salwasser H, Holl SA (1979) Estimating fetus age and breeding and fawning periods in the North Kings River deer herd. Calif Fish Game 65:159–165

Short C (1970) Morphological development and aging of mule deer and white-tailed deer fetuses. J Wildlife Management 34:383–388

Verme LJ (1989) Maternal investment in white-tailed deer. J Mammal 70:438–442

Verme LJ (1963) Effect of nutrition on growth of white-tailed deer fawns. Trans North Am Wildl Nat Res Conf 28:431–443

10
Neonatal Elk Habitat in Central Arizona

Mark C. Wallace and Paul R. Krausman

Abstract

In 1985 and 1986 we investigated neonatal elk (*Cervus elaphus*) habitat use in central Arizona. We captured calves and fitted them with radio collars. We documented measurements of calves, distances moved/day, and neonatal hiding habitats. Neonatal elk hid until 16 days old. Calves < 10 days old moved less than calves ≧ 10 days old. Neonatal hiding sites were characterized as midelevational ponderosa pine (*Pinus ponderosa*) summer range habitats with gentle southwest slopes. Use of elevations and associated habitats by calves changed with seasonal snowmelt and green-up. Type of cover neonatal elk used varied with habitat. Hiding cover 0.36 m to 1.7 m tall was the most important component of hiding sites.

Key words: Calf, elk, habitat, hiding cover

Introduction

Elk have evolved a hider strategy (Geist 1982) where neonates are dispersed and hide in seclusion until they have developed enough to escape predation by running. Elk calves are highly susceptible to predation during this period (Schlegel 1976), and their survival may depend upon the availability and selection of suitable hiding sites.

This important component of elk ecology has not been reported in previous studies of elk movements and habitat use that have focused on adults. Johnson (1951), Reichelt (1973), and Kuck et al. (1985) have described calf habitats in the northern Rocky Mountains. However, information on elk calving and habitat use in the Southwest is limited (Bell 1981). Our objectives were to identify the dominant characteristics of elk calving habitats in central Arizona, to determine how long calves used these habitats, and to produce a working definition of the habitat requirements of calving elk in central Arizona.

Methods

We located and captured calves by observing radio-collared females and conducting intensive foot and vehicle surveys. Nine radio-collared adult females were followed during the calving period. Changes in patterns of use (e.g., shifts from normally used areas or repeated locations in a single small area; Waldrip and Shaw 1979) were considered indicative of calving. Observations of avoidance or aggression towards other elk (Altmann 1952; Clutton-Brock and Guinness 1975; Geist 1982), increased wariness (Clutton-Brock and Guinness 1975), obvious changes in appearance, or reluctance to leave an area when approached (McCullough 1969; Clutton-Brock and Guinness 1975; Bell 1981) suggested females might have a calf in the area and prompted an immediate search.

Intensive foot surveys, using teams of volunteers hiking at 15-m intervals through suspected calving areas, searched for lone females or hiding calves. Where lone (or small groups one to three) females that appeared reluctant to flee were observed, the area was carefully searched for hiding calves. Finally, roads were driven during crepuscular hours (dawn to 0900 and 1700 to dark). Where elk that appeared reluctant to flee were observed an immediate search was made. Observers searched carefully for 10 to 40 ($\bar{x} = 20$) min in an expanding search pattern for hiding calves.

Neonatal calves were captured by surrounding or approaching them carefully from behind and grabbing them. Captured calves were measured, weighed, and aged (Johnson 1951). Expandable, vinyl coated elastic, radio collars (Telonics, Mesa, Ariz.) were fitted on captured calves. The collars stretched from a minimum circumference of 35.5 cm to a maximum of 101.6 cm.

Radio-collared calves were carefully relocated ≤ every 2 days after capture until they joined nursery groups. Relocation attempts were delayed if the cow was observed in the immediate area, and further relocations were excluded from analyses if the observer caused the calf to flee from its hiding site. Distance calves moved was calculated as the straight line distance between two consecutive locations divided by the time between locations. We recorded the location, time, and weather conditions for each calf location. After the calf vacated the site, we recorded slope, slope position, and aspect. Elevation (ELEV), distances to nearest road (DISTRD), known water (DISTH2O), and large meadows (DISTMEAD), which were the principal foraging habitat for elk in May–June (DelGuidice and Rodiek 1984), were recorded from on-site measurements and topographical maps. We determined dominant overstory and understory species by ocular estimation. We estimated overstory canopy cover (CANCO) with a spherical densiometer (Strickler 1959) and basal area (BASAL) with a wedge prism (Avery 1975). Overstory was further characterized by estimating the number and height of canopy layers (HT1 = height of dominant canopy, HT2 = height of secondary canopy), and averaging the diameter at breast height (dbh) (Avery 1975) of three trees measured for each height class (DB1 = mean dbh of HT1 trees, DB2 = mean dbh of HT2 trees).

We defined neonatal hiding cover as any obstruction that hid the prone calf from the view of an observer 25 m away. The type of and distance to hiding cover (MEANHIDE) was measured in four directions on two, randomly oriented, bisecting perpendicular 50-m transects intersecting at the calf bedsite. Density of slash and forest residues was estimated using photo guides (Maxwell and Ward 1976).

We estimated habitat availability from timber maps. Calf locations were plotted on the same maps. We compared habitat use to availability using techniques described by Byers et al. (1984). Nonparametric tests were used for most comparisons because subsets of data had small sample sizes ($n = 28.33 \pm 15.63$) or were not normally distributed.

Results

Elk calves were found using each method (Table 10.1). The vehicle survey was the most effective. The earliest neonatal calf location was 24 May and the latest 12 June; however, later calving did occur in this herd. Very young spotted calves, which were noticeably smaller and had less motor development, were seen in September and October each year. Also, an abnormally late (4–6 November) birth was documented during this study (Wallace et al. 1988).

TABLE 10.1. Elk calves located per man-hours of search time by observing radio-collared adult cows, foot surveys, and vehicle surveys in the White Mountains of Arizona, 1984, 1985, 1986

	Radio-collared cows		Foot surveys	Vehicle surveys	
	1984	1985	1985	1985	1986
Man-hours	12	92	70.3	62.5	98.5
Calves located	2[a]	1	1	9	13
Calves captured	1	0	0	5	9
Calves collared	0	0	0	4	3
Man-hours/location	12	92	70.3	6.9	7.6

[a]One calf noted here did not belong to the adult cow under observation. It was found accidentally at the beginning of a search for the adult and is not included in the calculation of man- hours/location.

Eleven of the 26 calves found eluded capture. Of these, only two escaped when more than one person was trying to catch them. The minimum collar circumference (35.5 cm) was too large or the elastic too loose to keep the collars secure. Collars put on calves captured after 3 June 1985 were secured snugly with electrician's tape. These elk showed no apparent discomfort from the collars as the tape weathered and broke in $\leqq 3$ months allowing the collars to expand as designed. Collar material weathered rapidly. No expandable collar remained on animals >2 years. Calf 480, legally shot in November 1985, was the only known radio-collared calf mortality.

Radio-collared calves were relocated from May through July between 0900 and 1600. Weather was not correlated to type of site selected for the range of conditions encountered in this study. Calves $\geqq 10$ days old moved significantly ($P < 0.0001$) greater distances/day than calves < 10 days old (Fig. 10.1). Calves 10 to 16 days old generally moved with the cows at some time during the day. Older calves joined nursery cow/calf bands and fled with adults when approached. No calf > 16 days old was observed using neonatal hiding behavior.

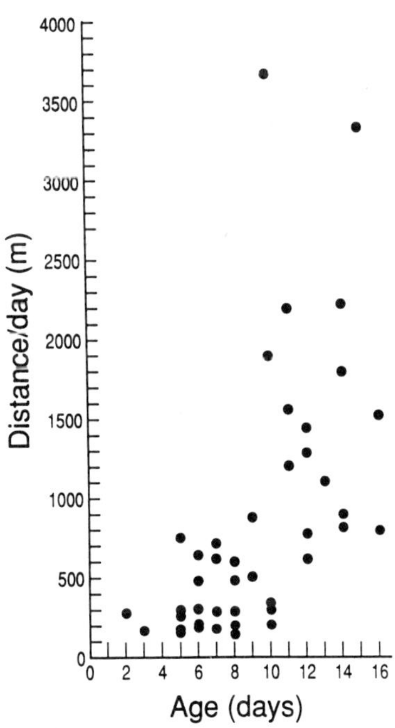

FIGURE 10.1. Distances neonatal elk moved/day in relation to age for calves captured in the White Mountains, Arizona 1985–1986.

Calf Habitats

Habitat characteristics were measured at 72 sites (26 initial capture and 46 telemetry relocations; Table 10.2). Eighty-three percent of bedsites were between 2256 and 2590 m and on $\leqq 10°$ slopes. Median aspect of 224° was significant (Rayleigh's $r = 39.72$; $P < 0.0001$). Eighty-nine percent of sites were $\leqq 1000$ m from known water and meadows; however, no sites were in open meadows.

Initial locations were closer ($U = 246.5$, $P < 0.0001$) to roads and had less BASAL ($U = 345$, $P = 0.0030$) and greater MEANHIDE ($U = 336$, $P = 0.0021$) than subsequent telemetry relocations. Males were found at lower ELEV ($U = 47$, $P = 0.0076$) than females in 1985, but there was no difference in 1986. Median ELEV of calf sites was higher ($P = 0.0043$) and MEANHIDE may have been less ($P = 0.0983$) in 1986 than in 1985.

Ponderosa pine (PIPO) habitats were used most (69.4%), followed by regenerating burns and clear cuts (FF/CC), aspen and spruce (POTR/PIEN), and mixed conifer (MIXD) habitats; 12.5%, 9.7%, and 8.3%, respectively (Table 10.3). Ponderosa and FF/CC habitats were used more, and MIXD habitats were used less than they were available. No calf bedsites were found in meadows, chaparral, or pinyon-juniper (OTHER) habitats. There were no differences in habitat use between sex or age classes. However, calves were located in nonponderosa habitats more frequently ($P = 0.0015$) in 1986 than in 1985.

Most (54.2%) sites had more than one overstory canopy layer. Single canopies were most common in FF/CC (23.6%), and PIPO (12.5%) sites. Ponderosa sites had smaller trees ($P = 0.0021$ and $P = 0.0039$; HT1 and DB1, respectively) and less CANCO ($P = 0.0006$) than MIXD sites, but had greater BASAL ($P = 0.0017$) than FF/CC sites.

South to west facing aspects (135°–293°) were used most (84.7%). Calf beds were usually slightly sloped ($\bar{x} = 4°$) on mid- (41.7%), top (33.3%), and bottom (22.2%) slopes. On steep terrain, beds

TABLE 10.2. Summary of habitat variables measured at neonatal elk bedsites in the White Mountains, Arizona. 1985–1986

Variable	Male	Female	Location	Relocation	1985	1986
N[a]	27	36	24	46	33	37
ELEV (m)	2402[b] (2316–2975)	2515 (2256–3109)	2442 (2256–2975)	2457 (2317–3109)	2408 (2317–2560)	2481 (2256–3109)
SLOPE (degree)	4 (0–22)	3 (0–45)	4 (0–44)	4 (0–45)	4 (0–18)	4 (0–45)
ASPECT (degree)	225 (90–360)	225 (22–338)	203 (22–338)	225 (90–360)	203 (90–360)	205 (22–338)
BASAL (km^2/ha)	33.07 ±14.34	28.95 ±16.77	21.85 ±17.11	33.41 ±14.02	27.97 ±13.72	29.72 ±18.25
CANCO (%)	74.37 ±15.55	66.67 ±22.19	67.68 ±23.14	68.68 ±21.07	66.64 ±18.47	69.05 ±24.40
MEANHIDE (m)	6.81 ±4.28	7.71 ±5.44	4.23 ±4.04	8.05 ±5.01	7.55 ±4.19	6.19 ±5.58
HT1 (m)	18.06 ±5.49	19.81 ±7.64	18.64 ±7.64	18.89 ±6.89	18.24 ±5.31	19.28 ±8.58
HT2 (m)	5.80 ±4.63	6.52 ±6.16	5.63 ±5.67	6.15 ±5.34	6.07 ±4.88	5.69 ±5.98
DB1 (cm)	32.42 ±13.87	35.42 ±19.14	32.33 ±18.40	33.70 ±17.03	36.20 ±16.07	30.79 ±18.61
DB2 (cm)	8.14 ±8.01	8.63 ±8.63	7.13 ±7.94	8.56 ±8.45	8.48 ±8.35	7.53 ±8.43
DISTH20 (m)	400 (12–1500)	325 (5–1200)	363 (5–1250)	425 (12–1500)	400 (5–1500)	400 (25–1200)
DISTRD (m)	150 (11–2400)	168 (0–1500)	50 (0–1500)	200 (0–2400)	175 (0–2400)	100 (0–1500)
DISTMEAD (m)	300 (0–2850)	237 (0–2500)	185 (0–1600)	288 (0–2850)	250 (0–2850)	250 (0–2500)

[a] Number of calf bedsites sampled; two bedsites sampled in 1984 are not included and seven were for calves of undetermined sex.
[b] Means are reported with ± standard deviations; medians are reported with (ranges).
CANCO = canopy coverage; MEANHIDE = mean distance to hiding cover; HT1 = height of dominant overstory; HT2 = height of subdominant overstory; DB1 = mean diameter at breast height of HT1; DB2 = mean diameter at breast height of HT2; DISTH20 = distance to known water; DISTRD = distance to road; DISTMEAD = distance to large meadow.

TABLE 10.3. Neonatal elk habitat use in the White Mountains, Arizona, 1985–1986

Habitat	Available (%)	Observed (%)	95% C.I.
PIPO	52.27	69.44	$56.06 \leqq P \leqq 82.82 >$ *
MIXD	18.38	8.33	$-0.50 \leqq P \leqq 17.16 <$ *
FF/CC	1.62	12.50	$2.78 \leqq P \leqq 22.22 >$ *
POTR/PIEN	13.91	9.72	$1.01 \leqq P \leqq 18.43$
OTHER	13.82	0.00	

PIPO = ponderosa pine; MIXD = mixed conifer; FF/CC = regenerating burn or clearcut; POTR/PIEN = aspen/spruce; OTHER = meadows, chaparral, pinyon-juniper; *indicates use significantly different from availability at $P = 0.05$; > indicates use greater than availability; < indicates use less than availability.

TABLE 10.4. Types of hiding cover used by neonatal elk in the White Mountains, Arizona, 1985–1986

Habitat	Cover type					
	LOG	SL	R/T	TR	VEG	None
PIPO	30	93	18	15	11	33
MIXD	2	6	7	0	9	0
FF/CC	0	3	2	15	10	6
POTR/PIEN	8	7	4	4	1	4

LOG = large downed trees in varying states of decay; SL = woody residues of silvicultural activities; R/T = rocks or topographic features; TREE = live standing trees; VEG = live herbaceous material; NONE = no cover within 25 m of bedsite; PIPO = ponderosa pine; MIXD = mixed conifer; FF/CC = regenerating burn or clearcut; POTR/PIEN = aspen/spruce.

were generally in a hollow or small depression in the slope.

Five hiding cover types were identified for neonatal bedsites (Table 10.4). Logs (LOG) were large fallen trees in varying stages of decay. Slash (SL), comprised of residues from silvicultural operations, was usually piled, but thinning prescriptions left high densities (> 22 tons/ha) of unconsolidated woody material in some areas. Hiding cover was classified as rocks or topography (R/T) when rocks, ground, or slope alone prevented an observer from seeing a bedded calf at ≦ 25 m. Trees (TR) and vegetation (VEG) were live standing plants that provided cover at the ground level in the form of trunks, branches, or vegetative growth. Type of cover predominant at the bedsite was related to habitat ($G = 97.44$, 15 df, $P < 0.0001$). Slash was used most (46.5%) in PIPO habitats, and VEG (37.5%) and R/T (29.2%) were used most in MIXD. Trees (41.7%) and VEG (27.8%) provided the most cover in FF/CC sites but, LOG (28.6%) and SL (25.0%) were used most in POTR/PIEN habitats. Mean height of cover (excluding TR) was 0.70 m ± 0.97 m primarily due to the height of slash piles. Cover < 0.36 m tall seldom hid the calf from an observer 25 m away. Only one (an 11-day-old female) calf site was found without hiding cover within a 25-m radius. Calves were usually found ≦ 1 m ($\bar{x} = 0.97$ m ± 2.8 m) from hiding cover in some direction.

Discussion

Techniques used to locate elk calves were successful because pregnant elk leave the herd before giving birth in seclusion (Altmann 1952; Boyd 1978; Geist 1982). Cows will remain in the vicinity (≦ 200 m) of their newborn calf and show recognizable protective behaviors (Schlegel 1976; Waldrip and Shaw 1979, DelGuidice 1982) when the calf is approached.

Most calves were found using the vehicle survey method, which also required the least man-hours of search time per calf location. Some inherent visibility biases may occur with this technique. Initial locations were closer to roads and more open than subsequent telemetry relocations. Used exclusively to identify calving areas, the vehicle survey method will underestimate use in dense habitats or those that are not accessible by road.

Our experience indicates that a minimum expandable collar circumference of 25.0 cm is necessary for neonatal elk, and that simple elastic designs provided too much opportunity for slippage or snagging on branches with calves ≦ 1 month old.

Sweeney (1975) suggested that calving elevation and associated habitat varies inversely with latitude. In central Arizona, near the southern extent of elk range, calving occurred on summer ranges dominated by ponderosa pine and lacking significant understory brush cover.

Calves most frequently hid on south-to-west facing gentle (≦ 10°) slopes in regenerating PIPO habitats 50 to 1000 m from meadows and water. Use of southerly exposures (Altmann 1952; Reichelt 1973; Sweeney 1975; Waldrip and Shaw 1979) and gentle slopes (Johnson 1951; Reichelt 1973) has been reported for other calving areas. Free water in May and June calving areas in this study was most associated with large meadows. Skovlin (1982) suggested that free water was normally within 400 m of calving areas. Elk use during the calving season in Arizona was greatest within 840 m of free water (DelGuidice and Rodiek 1984). These habitat components were common on the midelevational rim country where southwest aspects and gently sloped PIPO habitats interspersed with large, wet meadows predominated. However, calves were found on steeper slopes in 1986 when locations were more frequent in the higher elevation and steeper MIXD habitats. Between-year differences reflect the earlier snowmelt and green-up in 1986 similar to that reported by Sweeney (1975).

We were unable to test whether individual cow elk returned to traditional calving areas. We found calves in the same areas each year. However, locations of radio-collared cows that calved during the peak calving periods indicated that mean distance between locations through May and June of 1985 and 1986 was 4.17 km $\pm$ 3.35 km ($n = 10$ cows). Three cows had locations that were each clustered within 2 km^2 areas between years during calving, but most showed no similar site fidelity.

Calves used neonatal hiding habitats until 16 days old. Altmann (1952) and Knight (1970) reported that the hiding period normally lasts 10 days to 3 weeks. Calves in this study began to travel part-time with their mothers by 10 days old but did not abandon neonatal hiding behavior until 16 days old.

Where calving occurs on transitional spring ranges (Reichelt 1973) shrub species most commonly provide neonatal hiding cover. In mixed-open and forested ranges (Johnson 1951; Roberts 1974; Sweeney 1975) newborn calves are often found a short distance into the forest. We found calves bedded almost exclusively (94.4%) in forested sites. Calf hiding cover was the most important feature of neonatal bedsites. We found some form of visual obstruction served as neonatal hiding cover at all but one calf bedsite. Slash (Skovlin 1982) and rocks or woody vegetation (Waldrip and Shaw 1979) have been listed as important for neonatal elk cover. Calves in PIPO areas were in midsuccessional stands under thinning prescriptions that left piled or unconsolidated residues that provided the most ready source of hiding cover for calves. In FF/CC sites, young regenerating trees or herbaceous growth provided cover while calves used what was available in MIXD and POTR/PIEN sites.

The time that elk require neonatal hiding habitats can readily exceed 2 months, even if no calf hides when > 16 days old, due to variation in parturition times (Wallace et al. 1988). Our evidence suggests some flexibility in habitats used for calving, but neonatal hiding cover is essential. Neonatal hiding cover should be $\geqq 0.36$ m tall, 50 to 1000 m from open foraging areas and water, on gentle ($\leqq 10°$) southwesterly slopes. It should be available along normal elk migration routes at elevations commonly reached by May and June. Cover should be dispersed to avoid attracting predators to localized hiding sites (Schlegel 1976). In forest types lacking vegetative cover or downed woody materials, neonatal hiding cover can be readily provided by modifying prescriptions to leave some residues in situ.

Acknowledgements

This study was supported by the White Mountains Fort Apache Game and Fish Department and the University of Arizona. Wildlife and Fisheries Science students from the University of Arizona assisted with elk calf captures in 1985–1986.

References

Altmann M (1952) Social behavior of elk, *Cervus canadensis nelsoni*, in the Jackson Hole area of Wyoming. Behavior 4:116–143

Avery TE (1975) Natural resources measurements, 2nd ed. McGraw-Hill, New York

Bell M (1981) Elk research. Fed Aid Wildl Restor, New Mexico Department of Game and Fish

Boyd RJ (1978) American elk. In Schmidt JL, Gilbert DL (eds) Big Game of North America: Ecology and Management. Stackpole Books, Harrisburg, Pa., pp 11–29

Byers CR, Steinhorst RK, Krausman PR (1984) Clarification of a technique for analysis of utilization-availability data. J Wildlife Management 48:1050–1053

Clutton-Brock TH, Guinness FE (1975) Behavior of red deer (*Cervus elaphus* L.) at calving time. Behavior 55:287–300

DelGuidice GD (1982) A habitat analysis of spring-summer elk range on the Apache-Sitgreaves National Forest, Arizona, M.Sc. thesis, University of Arizona

DelGuidice GD, Rodiek JE (1984) Do elk need free water in Arizona? Wildl Soc Bull 12:142–146

Geist V (1982) Adaptive behavioral strategies. In Thomas JW, Toweill DE (eds) Elk of North America: Ecology and Management. Stackpole Books, Harrisburg, Pa., pp 219–277

Johnson DE (1951) Biology of the elk calf, *Cervus elaphus nelsoni*. J Wildlife Management 15:396–410

Knight RR (1970) The Sun River elk herd. Wildl Monogr 23:1–66

Kuck L, Hompland GL, Merrill EH (1985) Elk calf response to simulated mine disturbance in southeast Idaho. J Wildlife Management 49:751–757

Maxwell WG, Ward FR (1976) Photo series for quantifying forest residues in the: ponderosa pine type, ponderosa pine and associated species type, lodgepole pine type. USFS, Pac Northwest For and Range Exp Sta, Gen Tech Rep PNW-52

McCullough DR (1969) The Tule Elk: Its History, Behavior, and Ecology. University of California, Berkeley

Reichelt LR (1973) Characteristics of elk calving sites along the west fork of the Madison River, Montana. M.Sc. thesis, Montana State University

Roberts HB (1974) Effects of logging on elk calving habitat. Moyer Creek, Salmon National Forest, Idaho

Schlegel M (1976) Factors affecting calf elk survival in north-central Idaho—a progress report. Proc West Assoc State Game Fish Comm 56:342–355

Skovlin JM (1982) Habitat requirements and evaluations. In Thomas JW, Toweill DE (eds) Elk of North America: Ecology and Management. Stackpole Books, Harrisburg, Pa. pp 369–413

Strickler GS (1959) Use of the densiometer to estimate density of forest canopy on permanent sample plots. USFS, Pac Northwest For and Range Exp Sta Res Note PNW-180

Sweeney JM (1975) Elk movements and calving as related to snow cover. Ph.D. dissertation, Colorado State University

Waldrip GP, Shaw JH (1979) Movements and habitat use by cow and calf elk at the Wichita Mountains National Wildlife Refuge. In Boyce MS, Hayden-Wing LD (eds) North American Elk: Ecology, Behavior, and Management. University of Wyoming, Laramie, pp 177–184

Wallace MC, Jojolla J, Krausman PR (1988) Unusual reproduction in an Arizona elk. Southwest Nat 33:249

11
Recruitment and Loss in an Unhunted Subtropical White-Tailed Deer Population

Tommy R. Smith*

Reproduction and mortality of white-tailed deer (*Odocoileus virginianus seminolus*) in eastern Everglades National Park, Florida, have been studied by live-capture and radiotelemetry since March 1987. As of 1 September 1989, 106 captures of 68 individuals have been made, and 58 deer have been radio-monitored for 81 deer-years. Estimated minimum pregnancy rates of adults and yearlings have averaged 82 and 33%, respectively. There is no evidence of breeding among female fawns. Only singleton fawns have been observed. Most (57%, $n = 49$) experienced ≧ 3-year-old mothers have recruited young (i.e., raised fawns to yearlings) compared with 1 of 9 (11%) inexperienced 2-year-old mothers. Estimates of annual fawn mortality have varied from 18% to 58% ($\bar{x} = 34\%$). Estimated annual mortality of yearlings (37%) approximates that of fawns. Radio-tagged deer ≧ 2 years old have been lost at an annual rate of 17%; there is some indication that the mortality rate of 2-year-olds is higher than that of older adults. Predation by Florida panthers (*Felis concolor coryi*) has accounted for most (67%, $n = 15$) deaths of deer ≧ 1 year old. Assuming a field-observed sex ratio of 1:2 for yearlings and adults and given preliminary estimates of age-specific rates of natality and mortality, annual net recruitment to the population is estimated at 7%. This may be offset by the loss of dispersing young males. The Everglades deer population appears to be highly k-selected and relatively stable, indicative of a system with consistently low primary productivity.

*Florida Museum of Natural History, University of Florida, Gainesville, Florida 32610, USA

12
Biology of a Colonizing, Unharvested Population of Elk in Western Washington

Dwayne K. Paige[1], Kenneth J. Raedeke[1], and Richard D. Taber[2]

The indigenous Roosevelt elk (*Cervus elaphus roosevelti*) of the northwest coast of North America was largely extirpated from the west slopes of the Cascade Mountains of Washington State in the late 1800s. In the early 1900s Rocky Mountain elk (*C.e. nelsoni*) were introduced in the Cascades. By the late 1960s these elk, having interbred with the surviving *roosevelti*, had spread to the Cedar River Watershed, a protected water source for the city of Seattle. The biology of this natural, unharvested elk population was intensively studied from 1970 to 1988, through its establishment, increase, and stabilization, by means of radiotelemetry and direct observation of individually marked elk. Three major hypotheses were tested, and supported by the data. These were: (1) The rate of population growth exhibited by elk colonizing an unoccupied area will decrease as resources become limited through reduction in reproduction and recruitment rates. (2) Pregnant cow elk in late gestation will select foods that are phenologically early in their annual growth, primarily sedges and grasses. (3) Reproductive females occupy the most productive habitats while other sex and age classes of elk forage elsewhere. Additional information collected in the course of the study includes periodic estimates of total population, population structure, age-specific survival rates, food habits, forage production and diet quality, and seasonal habitat selection and movement patterns.

[1]College of Forest Resources AR-10, University of Washington, Seattle, Washington 98195, and [2]625 Continental Way, Missoula, Montana 59803, USA

13
Tule Elk on Private Lands: Species Preservation, Habitat Protection, or Wildlife Commercialization?

Vernon C. Bleich[1] and Donald B. Koch[2]

Tule elk (*Cervus elaphus nannodes*), the only North American elk endemic to California, once occurred throughout much of that state. Historically, the population numbered approximately 500 000, but that was reduced to critically low levels in the late nineteenth century as a result of habitat loss and commercial hunting. As recently as 1975, only 500 tule elk existed; recent efforts to reestablish this subspecies on historical ranges have resulted in a current population of at least 2700. Much of the previously suitable tule elk habitat has been altered by commercial development, or for agricultural uses. However, substantial tracts of suitable habitat remain in private ownership. In several instances, landowners have accepted translocated tule elk in cooperative ventures with the California Department of Fish and Game. Others have indicated a willingness to cooperate in this program, pending assurance that they will be able to recoup losses associated with elk damage to fences and agricultural endeavors. The California Private Lands Wildlife Management Program provides incentive for private landowners to preserve and improve suitable remaining tule elk habitat in exchange for providing hunting opportunities to those parties willing to pay appropriate fees. This results in direct benefits by expanding the total number of subpopulations, preserving critical habitats, providing income for landowners, and helping to ensure the continued existence of this unique species.

[1]California Department of Fish and Game, 407 W. Line St., Bishop, California 93514 and [1]Institute of Arctic Biology, University of Alaska, Fairbanks, Alaska 99775, and [2]California Department of Fish and Game, 1416 Ninth Street, Sacramento, California 95814

14
The Gas Molecule Theory: Dispersal and Management of White-Tailed Deer

William F. Porter[1], Nancy E. Mathews[1], H. Brian Underwood[1], Richard W. Sage[1], and Donald F. Behrend[2]

While the notion is pervasive, that vertebrate populations behave like systems of gas molecules, we contend it is an inaccurate metaphor for how deer populations expand. The mechanism by which white-tailed deer (*Odocoileus virginianus*) populations spread to fill available habitat is hypothesized to be rapid, density-related dispersal. We have tested this hypothesis using data from seven radiotelemetry studies. No trend is evident between the proportion of marked individuals dispersing and population density in populations ranging from $1/km^2$ to $70/km^2$; the probability of a female dispersing is 2% to 5%. Data from our own studies in the central Adirondack Mountains of New York, involving >200 marked females, suggest an alternative metaphor. Analyses show that a population expands slowly as a series of overlapping home ranges because female offspring remain near their dams for life. Each female added to a social group increases the cumulative summer home range of the group by $3.6 \pm 0.6\ km^2$. This social behavior of deer precludes rapid influx of populations to areas of low density and allows population voids to develop as a result of harvest. A model of population growth combined with geographic expansion suggests "surgical" removal of deer by family units could eliminate deer from areas of 500 to 1000 ha for 10 to 15 years. Because of increasing conflicts between deer populations and human activity in suburban areas and parks, this has important implications for a creative management approach.

[1]Adirondack Ecological Center, State University of New York, College of Environmental Science and Forestry, Newcomb, NY 12852, and [2]University of Alaska, Anchorage, Alaska 99501 USA

15 Adaptive Management of Coues White-Tailed Deer in Durango, Mexico: Is the Combined Production of Deer and Cattle Feasible?

Manuel Weber[1], Angeles Morales[1], and Carlos Galindo[2]

Due to the possible depletion in the deer population, a research program concerning Coues white-tailed deer (*Odocoileus virginianus couesi*) was initiated in 1987 in La Michilia Biosphere Reserve, Durango, Mexico (420 km^2). In order to determine the possible causes of the depletion, intensive population density estimates were conducted by means of pellet group counts. The counts were then correlated with the presence/absence index of cattle feces in the oak/pine forest of the reserve, following an altitudinal and land use gradient for the model of random sampling used. Results indicate a strong negative correlation between the presence of cattle and deer population densities. However, forage competition was not indicated as origin of this depletion and further studies are needed to better understand the local deer/cattle interactions. Toward this aim, a 3 ha enclosure was constructed in the buffer zone of the reserve, within which six tame Coues deer (three bucks and three does) were penned as an initial experimental group. Cattle grazing constitutes the principal economic activity of the local inhabitants within the buffer zone, providing a unique opportunity for research into the study of other aspects of interactions between cattle and deer, such as diseases, spatial competition, and behavioral interactions. Another advantage is that poaching is controlled through local vigilance. Our studies support the hypothesis that livestock grazing and hunting may be more detrimental for Coues deer than for other subspecies of white-tailed deer.

[1]Instituto de Ecologia, Apartado Postal 632 Durango, Dgo, Mexico, and [2]Ecology Group, Department of Zoology, University of British Columbia, 2204 Main Mall, Vancouver, B.C. Canada V6T 1W5

16

Deer Management and Economic Development in Southern New Guinea

John Fraser Stewart*

A wild, introduced population of Rusa deer (*Cervus timorensis*) inhabits most of the suitable lowland habitat in both the Merauke district in the southeastern portion of the Indonesian province of Irian Jaya and much of the southwest of neighboring Papua New Guinea. The total population size is of the order of 500 000. Perhaps 60% occur in the Indonesian portion of their range. Both the social/economic conditions and the strategy for deer management which currently exists in the two neighboring sovereign states are very different. In Papua New Guinea a system of social control of exploitation of wildlife is based on the concept that the subsistence people most closely associated with the habitat benefit from approved forms of utilization. This system has resulted in the preservation of a large wild deer resource. Private investment aimed at establishing an export-oriented industry based upon the sustainable management of the Bensbach deer population constitutes the area's principal economic development. Conversely, the hunting and commercial utilization of deer in Indonesia is still prohibited by a law which predates the establishment of the large introduced deer population in Irian Jaya. This legislation is now so inappropriate that it is ignored by local residents. Consequently large scale, illegal, and, therefore, uncontrolled commercial exploitation has developed producing some undesirable side effects. A proposal to establish a deer management program which emulates developments in Papua New Guinea has been made.

*Box 480, Port Moresby, NCD, Papua, New Guinea

17
Stereotypic Behaviors of the Barking Deer in Nepal

Madon K. Oli and Harry A. Jacobson*

Observations on barking deer (*Muntiacus muntjac*) in Nepal suggest this solitary forest dweller may be the most communicative of all deer. A total of 147 h of visual observations on 9 captive and 110 wild barking deer were recorded. Thirteen stereotypic vocalizations are described including the alarm bark, distress call, displacement whine, moan, alarm wheeze, castanet, submissive squeak, scream, grooming whine, bleat, nursing whine, maternal squeak, and pursuit call. Natives report another call, the rut call, is also given. Visual and olfactory stimuli are likewise very prominent. Antagonistic behaviors include side-saddling, foot-stomping, head bobbing, antler pointing, canine exposure, flaring of preorbital glands, scent marking with frontal glands, chasing, butting, and biting. Submissive behavior include grazing and crouching with head lowered and chin resting on the ground. Head bobbing was also used by females to call fawns. Conspecific and sexual interactions included allogrooming, stretching postures, and flehmen responses. Scent marking and use of dung piles were important social behaviors.

*King Mahendra Trust for Nature Conservation, Kathmandu, Nepal, and Department of Wildlife and Fisheries, Mississippi State University, Mississippi 39762, USA

18
Population Dynamics and Yield of a Dense Moose Population Without Predators

G. Cederlund and H. Sand*

The moose (*Alces alces*) population in Fennoscandia is thoroughly regulated through hunting. The annual harvest has in conjunction with the population density increased dramatically during the past two decades. In 1989, the total harvest was approximately 200 000 moose. The major reasons for the increase are considered to be regulated hunting, favorable food conditions, and low nonhunting mortality. This study describes the population dynamics and potential yield of a dense moose population with no predators in south central Sweden. Data were collected in Grimsö Research Area and adjacent areas (approximately 37 000 ha), which are situated in the flat, southern part of the taiga region. Forestry practices are intensive, and a considerable part of the forested area provides moose with large amounts of food in the young pine stands. Climate is typical to inland but with moderate snow accumulation. The study was accomplished during 1983 through 1988. Data on sex, age, carcass weight, and fecundity were obtained by material collected from moose killed during the regular hunting season in October to December. Analyses were done on approximately 500 adult males, 500 adult females, and 600 calves. Moose density was obtained by aerial counts and was rather stable throughout the study (approximately 1.3 moose/km^2). Data on dispersal and faithfulness to the previous year's home ranges were available from studies on 30 radio-marked moose. Nonhunting mortality was calculated by means of number of found carcasses and death rates among marked moose. A high and rather stable age-specific fecundity, low death rate, and low dispersal rate give a high rate of increase in the moose population at Grimsö (approximately 50% annually).

*Grimsö Wildlife Research Station, 770 31 Riddarhyttan, Sweden

19
Quantification of Mule Deer Carrying Capacity in South-Central Oregon

Craig S. Bienz and Larry K. Dunsmoor*

The relationship between mule deer (*Odocoileus hemionus*) populations and summer habitat was evaluated in south-central Oregon from 1986 to 1989. Habitat effectiveness was calculated for more than 200 000 ha of diverse habitats 1250 to 2500 m in elevation. Populations were monitored throughout the year to determine animal density, annual production, and physiological condition for specific habitat conditions. The study area was partitioned by township for analysis of the relationships between habitat effectiveness, annual production, animal density, and condition. Townships with the same carrying capacity (habitat effectiveness) were grouped and the variables represented as means. Multiple regression analysis of carrying capacity to annual production and animal condition was significant ($r^2 = 0.94$, $P < 0.001$). The relationship of carrying capacity to animal density and annual production was also significant ($r^2 = 0.76$, $P = 0.01$). Carrying capacity regression models were useful in predicting population responses to changes in habitat and significant reductions in populations due to severe winters and drought.

*Klamath Tribe, Chiloquin, Oregon 97624, USA

20
Spatial Component in Reproductive Performance in Desert Mule Deer

Jerry L. Cooke*

The output of a theoretical spatial model of desert mule deer (*Odocuileus hemionus*) movement was compared to reproductive predictions based on regressions from field collections. Regressions were prepared that relate doe age and body condition (kidney-pelvic fat categories) to fetal count and conception date for 207 desert mule deer does collected in Texas. These regressions implied that reproductive performance (in utero fetal counts) and conception dates could be inferred from these values. Fetal counts have been shown to be related to body condition in previous studies. However, a relationship between body condition and the timing of conception has not been described in the literature. I predicted fetal counts and conception dates within the deer movement model incidental to animal foraging. In this model, breeding occurred when two individuals met during breeding season and this season was arbitrarily selected by photoperiod. Fetal counts and conception dates for the modeled population matched those predicted from the regressions that used age and body condition of the modeled individuals as independent variables. It is submitted that fetal production, conception chronology, and body condition in desert mule deer are all functions of geographic location of the animal rather than being independent characteristics related to each other.

*Department of Wildlife and Fisheries Sciences, Texas A&M University, College Station, Texas 77843, USA

21
Comparative Foraging Ecology of White-Tailed Deer and Angora Goats

John H. Roese and Ken L. Risenhoover*

Seasonal foraging strategies of white-tailed deer (*Odocoileus virginianus*) and angora goats were compared in three separate pastures under different grazing management schemes at the Texas Agricultural Experiment Station near Sonora, Texas. Seasonal changes in diet composition and quality, foraging behavior, and nutrient ingestion rates of deer and goats were compared to identify the relative importance of factors influencing animal foraging efficiency and performance. Comparisons of deer and goat foraging efficiency and weight dynamics in different pastures were used to estimate the nutritional carrying capacity of the three study pastures investigated and to assess the potential impacts range management practices are having on the quality of deer and goat habitats.

*Department of Wildlife and Fisheries Sciences, Texas A&M University, College Station, Texas, USA

22
Deer/Habitat Interactions on Coronation Island, Southeast Alaska

Stephen W. Lewis and David R. Klein*

This study assesses the impact of forage quality and availability on deer body condition for a recovering population of Sitka black-tailed deer (*Odocoileus hemionus sithensis*) on Coronation Island. Wolves were introduced to Coronation in 1961 with a subsequent reduction in deer numbers. Since then, wolves have died out and deer numbers are rebounding. Comparisons are made with data from Klein's 1959–1961 study when the population of deer was at a moderately high and sustained level in relation to food resources. We compare the abundance of plant species and diversity at 30 locations in four habitat types located as closely as possible to Klein's original transects. A sample of 20 male deer was collected during the summer of 1988. Age, body size, condition, and growth rate parameters were measured and compared with results from Klein's earlier study. Data collected from 149 deer remains located over a summer and winter field season were used to augment sample size whenever possible.

*Alaska Cooperative Wildlife Research Unit, University of Alaska Fairbanks, Fairbanks, Alaska 99775, USA

23
Classification of Fallow Deer Groups According to Size Under Deer Park Conditions

Soběslav Losos and Jaroslav Madlafousek*

Until now, two methods have been used to study the various problems associated with the size of fallow deer (*Dama dama*) groups. The first compares arithmetic means of groups. The second works with frequencies in class intervals defined by the observer. In the present study, the graphical form of the "survivor function" was used. The function is based on a construction of a diagram in which the values on the y-axis refer to the number of the observed groups, whereas those on the x-axis refer to the sizes of groups ordered from the smallest to the largest. The plotted points are connected by a curve. If the groups of different sizes are formed at random, only one function will pass along the curve (negative exponential). However, if the size of the groups is influenced by one or more factors, several partial curves with different courses will be obtained. Their mutual intersections will then define the size category of the groups on the x-axis. The three methods were tested using data obtained in a fallow deer park during the period April to August (when the animals live separately according to sex) in two consecutive seasons.

*Group of Ethology, Research Institute of Animal Production, CS-104 00 Phaha 10—Uhříněves, Czechoslovakia

24
Population Dynamics and Estimation of Maximum Sustained Yield (MSY) in a Black-Tailed Deer Population

Kenneth J. Raedeke*

The population characteristics of a free-ranging deer (*Odocoileus hemionus columbianus*) population in western Washington were studied from 1979 to 1989 to determine the impacts of sport hunting on the dynamics of the deer herd. Annual measurements of relative population density, age-specific reproductive rates, sex and age structure, harvest rates, and hunting effort have been used to understand the response of the deer populations to different harvest rates and resultant changes in deer density. A sustain yield (SY) curve was empirically derived from the 10-year data set, and compared with a theoretical curve, based on the logistic model. The SY curve derived from the data set indicated a higher maximum sustained yield (MSY) than the theoretically derived MSY value. The latter MSY was taken from a population at approximately 50% of carrying capacity, while the former was taken from a population at approximately 75% of carrying capacity. These observations are likely the result of the highly nonlinear relationship of reproductive rates and population density, which was also derived from the data set. The population dynamics model is now being applied to other deer herds for extensive field testing.

*College of Forest Resources Ar-10, University of Washington, Seattle, Washington 98195, USA

25
Resource Partitioning by Ungulates in Central Texas: A Simulation Approach

Ken L. Risenhoover and John H. Roese*

A simulation model was developed to evaluate potential patterns of resource partitioning among sympatric wild and domestic ruminants in central Texas. Empirical evidence relative to body size, physiological and morphological adaptations, and nutritional requirements were combined with information of environmental conditions to predict patterns of foraging behavior and animal performance. Predictions of nutrient ingestion rates and weight dynamics were used to identify feeding niches and habitat conditions which resulted in optimal performance for each herbivore species. Spatial overlap in habitat use and patterns of resource depletion were used to evaluate the potential for forage competition among herbivore species. Model predictions were used to generate hypotheses regarding factors limiting animal performance during different seasons.

*Department of Wildlife and Fisheries Sciences, Texas A&M University, College Station, Texas, USA

Part 2
Animal Health and Disease

26
The Biology of a Fatal Herpesvirus Infection of Deer (Malignant Catarrhal Fever)

Hugh W. Reid

Abstract

Malignant catarrhal fever (MCF) affects many species of ruminant and is one of the most important infectious diseases of farmed deer. It is almost invariably fatal causing widespread lymphoproliferation and tissue destruction. Despite the dramatic pathological changes which have the characteristics of a virulent virus infection, no etiological agent has been identified. Research that will be described has sought both an explanation for the pathogenesis and the nature of the causative virus. In contrast to the disease in cattle, MCF can readily be transmitted by inoculating cells from affected deer into other deer, laboratory rabbits, and hamsters. Studies of these laboratory models have suggested an explanation for the disease mechanism which satisfies many of the observed features of the infection. It is proposed that when deer become infected with the virus of MCF, specific subpopulations of lymphocytes are targeted. Virus transcription is only partial and results in (1) a very limited production of infective virus, or viral proteins, and (2) deregulation of host immune processes, giving rise first to a nonspecific T-lymphocyte expansion followed by a fatal autoimmune cell-mediated destruction of normal tissue. The cloning of part of the genome of the presumptive cause, a herpesvirus, has been achieved from tissues of MCF-affected animals. The nature of the causative virus can now be predicted with confidence providing the opportunity for its isolation and characterization and should be designated Ovine herpesvirus-2. It would appear that the virus is a normal commensal of domestic sheep in which species infection is prevalent and causes no clinical signs. Transmission of this virus to deer and other ruminants, however, results in fatal disease.

Key words: Malignant catarrhal fever, deer, lymphocyte, gamma-herpesvirus, antigen, nucleic acid

Introduction

Malignant catarrhal fever (MCF) has been recognized as a disease of cattle for many years (Gotze 1930). Despite the dramatic clinical and pathological nature of the disease, throughout much of its geographical distribution the etiology is obscure, and the pathogenesis is difficult to explain. In recent times its devasting effect on deer has become apparent (Huck et al. 1961; Reid et al. 1979), and it has been due largely to the realization that MCF is one of the major obstacles to the development of deer farming that efforts to elucidate this highly enigmatic disease have been renewed.

The incident involving Pere David's deer (*Elaphurus davidianus*) at Whipsnade Park, England, was the first indication that MCF could be a serious problem in deer (Huck et al. 1961), although it is intriguing to speculate that a report by Lupke (1906) of extensive mortality in the German Kaiser's Axis deer (*Axis axis*), diagnosed as *periarteritis nodosa,*[1] may have been MCF. Although MCF was subsequently recognized to

[1] *Periarteritis nodosa* is a pathological condition no longer recognised in deer but which histologically resembles the changes characteristic of malignant catarrhal fever.

occur sporadically in zoological collections (Straver and Beekum 1979; Heuschele et al. 1988), the true magnitude of the problem was not appreciated until the establishment of deer as agricultural animals, when it rapidly became apparent in the U.K. (Reid et al. 1979), New Zealand (McAllum et al. 1982), and Australia (Denholm and Westbury 1982) that MCF was a major threat to the development of deer farming.

The Disease

The disease is a fatal lymphoproliferative condition affecting ruminants, including 13 species of deer, which may be present as peracute to chronic forms (Table 26.1). In deer, the disease tends to be peracute or acute, with animals succumbing before the more florid lesions, characteristic of protracted cases, develop (Reid and Buxton 1989).

Typically affected animals are first seen as inappetant, standing away from the herd, with rectal temperatures above 40° C, and may go on to develop dysentery and die within 12 h (Reid and Buxton 1984). In other cases the disease runs a longer course: peripheral lymph nodes are enlarged, while ocular and nasal secretions initially serous become mucopurulent, and bilateral corneal opacity characteristically affects the periphery initially before spreading centripetally. Erosion of the buccal cavity, particularly of the tips of the papillae and ulceration of the skin, especially the perineum, may be present. In all cases death is the eventual outcome. Even in those few cases where recovery apparently occurred, animals succumbed after a period of remission (Wilson et al. 1983; Heuschele et al. 1985).

At postmortem examination the lesions reflect the nature of the clinical response. In the more acute cases there is severe hemorrhagic gastroenteritis in the absence of other lesions, while where the disease has run a longer course, lesions affecting most systems can be found. The catarrhal lesions associated with erosion of the turbinates may extend to involve the larynx, trachea, and bronchi with emphysema and inflammatory lesions in the lungs. Erosion and hemorrhage may affect most areas of the intestine and are particularly severe in the regions of gut-associated lymphoid tissue. The urinary tract often shows characteristic lesions. The cortex of the kidney is affected by multiple, white, raised areas 1 to 2 mm in diameter while there are ecchymotic hemorrhages in the urinary bladder. All lymph nodes, but in particular the mesenteric, submandibular, and retropharyngeal are often

TABLE 26.1. Species of deer in which malignant catarrhal fever has been described

Species	Source of infection	Reference
Alces alces	Z	Williams et al. 1984
Capreolus capreolus	E/S	HW Reid, unpublished data
Cervus axis	Z	Clark et al. 1970
Cervus duvauceli	Z	Hatkin 1980
Cervus elaphus	F	Reid et al. 1979
Cervus eldi	Z	Heuschele 1982
Cervus marianus	Z	Heuschele 1982
Cervus nippon	Z/F	Sanford et al. 1977 Wilson et al. 1983
Cervus timorensis	F	Denholm and Westbury 1982
Elaphurus davidianus	Z/F	Huck et al. 1961 Reid et al. 1987
Odocoileus hemionus	F	Pierson et al. 1974
Odocoileus virginianus	Z/F	Wobeser et al. 1973 Wyand et al. 1971
Rangifer tarandus	Z	Heuschele et al. 1988 Altmann 1973

Z, zoological collection; E/S, experimental transmission with sheep-associated form of disease; F, farmed, sheep probably the source of infection grossly enlarged, edematous, and sometimes hemorrhagic and necrotic. However, the only way of reaching an unequivocal diagnosis is by histological examination of a number of tissues for evidence of the pathognomonic changes (Reid and Buxton 1989). These consist of hyperplasia of lymphoid organs and accumulation of lymphoid cells in many nonlymphoid tissues, epithelial and lymphoid degeneration, and vasculitis. The finding of these typical lesions in a variety of tissues form the basis for the diagnosis of MCF. The organs that show the most specific changes are kidney, brain, cornea, and epithelial surfaces such as buccal cavity, urinary bladder, and turbinates.

Etiology

It is one of the extraordinary features of MCF that despite these spectacular clinical and pathological changes, the identification of its cause has proved to be a most frustrating microbiological problem, though resolution of this question can now be predicted with confidence.

It was apparent to the first investigators that MCF occurred in at least two circumstances (Table 26.2). In Europe, MCF was recognized as a disease that generally occurred following contact with sheep (Gotze and Leiss 1930), and this form of the disease became recognized as sheep-associated (SA)-MCF. In Africa the disease was reported to occur following contact with wildebeest (*Connochaetes spp.*), and Mettam (1924) was able to show that the disease could be transmitted to cattle with blood collected from normal healthy wildebeest. Both forms of the disease produce similar clinical and pathological changes and cannot be distinguished on these grounds.

SA-MCF could not be transmitted from its putative reservoir, and experimental transmission studies from affected cattle generally proved fruitless (Magnusson 1940; Duncan and Pearson 1956; Roderick 1959; Selman et al. 1978; Pierson et al. 1979; Flanagan and Hoffman 1988). In contrast, not only was MCF associated with wildebeest readily transmitted, but by 1960 a viral etiology was established, and the virus was isolated (Plowright et al. 1960) and partially characterized (Plowright et al. 1965). Thus, it was confirmed that the natural, wildebeest host of the virus of MCF experienced no clinical reaction following infection, and it was only when the virus transferred across a species barrier that fatal disease developed. The ability to propagate the wildebeest virus in vitro has also provided the essential reagents for further investigation, in particular comparative studies of the other form of the disease.

Epidemiological studies soon established that the wildebeest virus (*Alcelaphine herpesvirus*-1 [AHV-1]) was an extremely efficient parasite infecting all free-living wildebeest by about 6 months of life (Plowright 1967). Probably a few calves are infected in utero (Plowright 1965), and thereafter there is a period when spread occurs by contagion, each calf experiencing a short inapparent phase of infection when productive virus replication occurs with excretion in ocular and nasal secretions (Mushi et al. 1980). Thereafter, lifelong latent virus infection is established, though infectious virus may again be shed during periods of stress (Rweyemamy et al. 1974; Plowright 1986).

Infection in the wildebeest herds is thus efficiently ensured, and only a few isolated populations in zoological collections are free of infection (Heuschele et al. 1984). Disease due to AHV-1 is thus restricted to areas of Africa where wildebeest are present and to zoological collections worldwide; however, in Africa disease

TABLE 26.2. Principal features of the two causes of malignant catarrhal fever

	Alcelaphine herpesvirus 1	Sheep associated agent
Natural host	Wildebeest (*Connochaetes spp.*)	Domestic sheep
Distribution	Africa, zoological collections	Worldwide
Disease	Deer, other ruminants	Deer, other ruminants
Experimental transmission	Readily from all species	Readily only from deer

occurs only in domestic cattle, whereas experimentally and in zoological collections a wide variety of other ruminants, including other antelope native to Africa, have been reported to be susceptible (Heuschele et al. 1988).

The identification of AHV-1 not only provided insights into the ecology of this virus in wildebeest populations but was the source of antigenic substrate for a number of serological tests which have been used for epidemiological studies in other species (Reid et al. 1975; Heuschele et al. 1988); however, as serological tests for herpesviruses are liable to be cross-reactive (Heuschele et al. 1984; Rossiter et al. 1988), evaluation of results must be approached with caution. It should also be noted that herpesviruses are characteristically highly species-specific, establishing lifelong latent infections in most of the potential natural hosts while they are still juvenile. From the numerous studies of sera collected from a variety of ruminant species it would now appear certain that there is a large family of at least 25 herpesviruses related to AHV-1 present in four subfamilies of bovidae (Tables 26.3 and 26.4). The viruses recovered from the three species of wildebeest have been tentatively classified as a single virus AHV-1, while the viruses isolated from Hartebeest (*Alcelaphus buselaphus*), Jimela topi (*Damaliscus lunatus*) (Seal et al. 1989) and Roan antelope (*Hippotragus equinus*) (Reid and Bridgen 1991) appear to be distinct. An as yet unexplained feature of these viruses is that while they are antigenically and biologically very similar to AHV-1, and two of the three other viruses which have been isolated from African antelope can cause MCF experimentally, only AHV-1 and the as yet unidentified virus of sheep have proved to be a cause of disease under natural circumstances.

TABLE 26.3. Subfamilies of Bovidae for which there is serological evidence that infection with gamma-herpesviruses related to Alcelaphine herpesvirus is prevalent

Subfamily	Species with antibody	
	Neutralizing (*n*)	Immunofluorescent (*n*)
Alcelaphinae	7	7
Hippotraginae	7	7
Caprinae	0	12
Ovibovinae	0	1

It is also necessary to reconcile the observation that no causal virus has been formally identified in the other species, the domestic sheep, implicated as a source of MCF. The word formally is chosen judiciously as there is convincing evidence that this virus has in fact been propagated in vitro and partially characterized.

Characterization of the Sheep-Associated Agent

Three lines of evidence support the proposal that the sheep-associated agent is a herpesvirus biologically, antigenically, and genomically closely related to AHV-1.

TABLE 26.4. Proposed nomenclature of the bovid gamma-herpesviruses which have been studied

Host	Proposed designation of virus	Authority
Connochaetes spp.[a]	Alcelaphine herpesvirus-1	Plowright et al. 1960 Seal et al. 1989
Alcelaphus buselaphus	Alcelaphine herpesvirus-2	Anonymous 1973
Damaliscus korrigum	Alcelaphine herpesvirus-3	Mushi et al. 1981
Hippotragus equinus	Hippotragine herpesvirus-1	Reid and Bridgen 1991
Domestic sheep	Ovine herpesvirus-2[b]	Bridgen and Reid 1991

[a] Isolates from all three species of wildebeest (*C. taurinus taurinus*, *C.t. albojubatus*, *C. gnu*) appear to be very similar and are assumed to be the same.

[b] Originally proposed as a caprine herpesvirus, revised ICTV terminology suggest Ovine herpesvirus is appropriate.

Biological Relationship

The disease MCF caused by either agent is fundamentally similar, and a common pathogenesis arising through a profound T-lymphocyte deregulation has been proposed (Reid and Buxton 1989). In addition, seroepidemiological studies of both agents indicate that they spread efficiently in their respective natural hosts causing no appreciable disease and infecting all individuals at a young age (Rossiter 1981; Herring et al. 1989; Reid and Bridgen 1991).

Antigenic Relationship

Antibody has been found consistently in sheep sera by employing AHV-1 as antigen in indirect immunofluorescent (IIF) antibody tests (Rossiter 1981; H.W. Reid, unpublished data). The cross-reactivity between herpesviruses in this test has made interpretation of this result problematic, but confidence that sheep were infected with a virus which shared many of the major structural proteins of AHV-1 was provided by immunoblotting studies (Herring et al. 1989). However, this information did not establish a relationship between the presence of this putative virus in sheep and the cause of SA-MCF. Examination of sera collected from Syrian hamsters reacting with experimentally induced SA-MCF did establish the connection (Reid et al. 1989a). In that study it was found that hamsters with SA-MCF transmitted from red deer (*Cervus elaphus*), and Pere David's deer, as well as from cattle, developed specific antibody that reacted with AHV-1. Thus, a relationship between the development of SA-MCF and the production of antibody reacting with AHV-1 was established proving that the two causes of MCF were antigenically similar.

Genomic Relationship

Although attempts to isolate the SA virus have consistently proven unsuccessful, it has been found possible to generate lymphoblastoid cell lines from deer and cattle affected with MCF (Reid et al. 1983, 1989b). These cell lines have the surface phenotype of T cells (Burrells and Reid 1991) and at least those derived from deer will transmit the disease when inoculated into deer experimentally. Thus, it has been proposed that T lymphocytes or a specific T cell subset represent important targets for the SA virus and that their immortalization in vitro is a virus-induced phenomenon. By definition at least part of the viral genome must be retained in these cells.

Concurrent studies of the molecular biology of AHV-1 have involved the preparation of cloned fragments of the viral DNA (Bridgen 1991). These fragments were used in hybridization experiments with total DNA extracted from the lymphoblastoid cell lines (Bridgen and Reid 1991). In this way it was possible to demonstrate that there were sequences in the cell lines very similar to the AHV-1 genome and that these sequences were distributed throughout much of the unique region of the viral genome (Fig. 26.1).

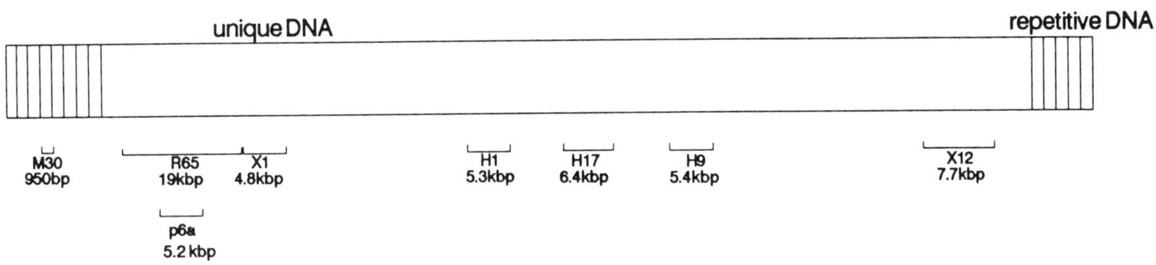

FIGURE 26.1. Diagrammatic representation of the structure of the genome of the attenuated WC11 isolate of AHV-1, which comprises approximately 130 kbp unique DNA plus a total of 25–30 kbp repeat DNA. The figure shows the genomic location of the AHV-1 clones used in the hybridization analyses with DNA extracted from lymphoblastoid cell lines derived from animals with SA-MCF (Bridgen and Reid 1991). All the clones hybridized to SA-MCF DNA except M30 and X1. Clone p6a hybridized to the DNA clone of the SA-MCF virus.

Furthermore, the conditions of hybridization indicated that the level of homology was relatively high, and only two of the eight AHV-1 clones tested showed no homology. From this it is concluded that all or part of the SA-virus genome is contained in the cell lines and that the virus is a gamma-herpesvirus closely related to, though distinct from, AHV-1. Furthermore, as cell lines derived from red deer, cattle, and experimentally infected rabbits (Table 26.5) all contained similar sequences, it would appear that within the U.K. there is one cause of SA-MCF which affects both cattle and deer.

In addition, from a genomic library of one of the cell lines, 5 kilobase (kb) pairs representing approximately 4% of the viral genome has been cloned. Thus, at least part of the SA virus genome has now been characterized and will provide the essential reagents for finally identifying the virus in sheep, defining the epidemiology of SA-MCF, and developing rational control strategies.

From the foregoing it is evident that the SA virus is a gamma-herpesvirus similar to AHV-1 and there is substantial evidence to formally nominate the virus as Caprine herpesvirus-3. (Caprinae is the subfamily to which domestic sheep belong, and as there are two previously identified herpesviruses of Caprinae, the SA virus is the third. ICTV proposals on terminology have now changed and the virus would be more appropriately designated Ovine herpesvirus-2.)

TABLE 26.5. Source of lymphoblastoid cell lines from which viral sequences have been identified with cloned Alcelaphine herpesvirus-1 fragments

Reference no.	Source	Period in culture	Agent
MF/629	Bovine (CSF)[a]	5 years	SA-A
BJ/576	Bovine (cornea)	1 year	SA-A
MF/816	Red deer (lymph node)	3 years	SA-A
MF/824	Red deer (lymph node)	5 months	SA-A
BJ/615	Red deer (lymph node)	9 months	SA-A
BJ/393	Rabbit (lymph node)	7 months	SA-A
BJ/610	Rabbit (lymph node)	9 months	AHV-1
BJ/465	Rat (lymph node)	1 year	AHV-1

[a] All animals were affected with malignant catarrhal fever when sampled.
SA-A, sheep-associated agent.

References

Anonymous (1973) Malignant catarrhal fever (MCF). Record of Research Annual Report (1972) East African Vet Res Organisation, pp 9–13

Altmann D, Kronberger H, Schuppel KF (1973) Bösartiges Katarrholfieber (*Coryza gangraenosa*) bei zwei Elchen, zwei Rentieren und eine Hausziege im Thüringer Zoopark Effurt. Proc 15th Int Symp on Diseases of Zoo Animals, Kolmarden, Berlin, Akademie-Verlag, pp 41–49

Bridgen A (1991) The derivation of the restriction endonuclease map for Alcelaphine-herpesvirus-1 DNA. Arch Virol 117:183–192

Bridgen A, Reid HW (1991) The derivation of a DNA clone corresponding to the viral agent of sheep-associated malignant catarrhal fever. Res Vet Sci 50:38–44

Burrells C, Reid HW (1991) Phenotypic analysis of lymphoblastoid cell lines derived from cattle and deer affected with "sheep-associated" malignant catarrhal fever. Vet Immunol Immunopathol (in press)

Clark KA, Robinson RM, Morbung RG, Jones LP, Orchard JH (1970) Malignant catarrhal fever in Texas cervids. J Wildl Dis 6:376–383

Denholm LJ, Westbury HA (1982) Malignant catarrhal fever. Vet Immunol Immunopathol (in *sis*). 1. Clinico-pathological observations. Australian Vet J 58:81–87

Duncan DW, Pearson IG (1956) Bovine malignant catarrh. Australian Vet J 32:156–161

Flanagan M, Hoffman D (1988) Malignant catarrhal fever research in Queensland. In Daniels PW (ed) Malignant Catarrhal Fever in Asian Livestock. Sudarisman and Ronohadjo, P. ACIAR, Canberra, pp 118–123

Gotze R (1930) Untersuchungen über das bösartige Katarrhalfieber des Rindes III. Deutsche tierarztliche Wochenschrift 38:487–491

Gotze R, Liess J (1930) Untersuchungen über das bösartigen Katarrhalfieber des Rindes. II Schärfe als Übertrager. Deutsche tierarztliche Wochenschrift, 38:194–200

Hatkin J (1980) Endemic malignant catarrhal fever at the San Diego wild animal park. J Wildl Dis 16:439–444

Herring A, Reid H, Inglis N, Pow I (1989). Immunoblotting analysis of the reaction of wildebeest sheep and cattle sera with the structural antigens of Alcelaphine herpesvirus-1 (malignant catarrhal fever virus). Vet Microbiol 19:205–215

Heuschele WP (1982) Malignant catarrhal fever in wild ruminants—a review and current status report. Proc 86th Annu Meeting US Animal Health Assoc 86:552–570

Heuschele WP, Seal BS, Klieforth RB (1988) Malignant catarrhal fever in wild ruminants. Sonderdruck aus Verhandlungsbericht des 30 Internationalen Symposiums über die Erkränkungen der Zoo- und Wildetiere, Sofia 1988. Akademie-Verlag, Berlin, pp 45–55

Heuschele WP, Nielsen NO, Oosterhuis JE, Castro AE (1985) Dexamethasone-induced recrudescence of malignant catarrhal fever and associated lymphosarcoma and granulomatous disease in a Formosan sika deer (*Cervus nippon taiouanus*). Am J Vet Res 46:1578–1583

Heuschele WP, Fletcher HR, Oosterhuis J, Janssen D, Robinson PT (1984) Epidemiologic aspects of malignant catarrhal fever in the USA. Proc 88th Annu Meeting US Animal Health Assoc 88:640–651

Huck RA, Shand A, Allsop PJ, Paterson AB (1961) Malignant catarrh of deer. Vet Rec 73:457–465

Lupke F (1906) Über Periarteritis nodosa bei Axishirscher Tierhandlungen. Deutsch Ges Pathol 10:149–157

McAllum JJF, Mavor NM, Hemmingsen P (1982) A malignant catarrhal fever-like disease in red deer (*Cervus elaphus*) in New Zealand. N Zealand Vet J 30:99–101

Magnusson H (1940) Om elakartad katarralfeber Nagra forsok att experimentellt overfora sjukdomen. Beretn 5, nord. Vet-mode, Copenhagen, 1939, pp 369–409

Mettam RWM (1924) Snotsiekte in cattle. South African Agricultural 9th and 10th Reports of the Director of Veterinary Research for 1923, pp 395–432

Mushi EZ, Rossiter PB, Jesset DM, Karstad L (1981) Isolation and characterisation of a herpesvirus from topi (*Damaliscus korrigum*). J Comp Pathol 91:63–68

Mushi EZ, Karstad L, Jesset DM (1980) Isolation of bovine malignant catarrhal fever virus from ocular and nasal secretions of wildebeest calves. Res Vet Sci 29:168–171

Pierson RE, Hamdy FM, Dardiri AH, Ferris DH, Schloer GM (1979) Comparison of African and American forms of malignant catarrhal fever. Transmission and clinical signs. Am J Vet Res 40:1091–1095

Pierson RE, Storz J, McChesney AE, Thake D (1974) Experimental transmission of malignant catarrhal fever. Am J Vet Res 35:523–525

Plowright W (1986) Malignant catarrhal fever. Rev Sci Techn Office Int Epizooties 5:897–918

Plowright W (1967) Malignant catarrhal fever in East Africa. III. Neutralising antibody in free-living wildebeest. Res Vet Sci 8:129–136

Plowright W (1965) Malignant catarrhal fever in East Africa. 1. Behaviour of the virus in free-living populations of blue wildebeest. Res Vet Sci 6:56–68

Plowright W, Macadam RF, Armstrong JA (1965) Growth and characterisation of the virus of bovine malignant catarrhal fever in East Africa. J Gen Microbiol 39:253–266

Plowright W, Ferris RD, Scott GR (1960) Blue wildebeest and the aetiological agent of bovine malignant catarrhal fever. Nature (London) 188:1167–1169

Reid HW, Bridgen, A (1991) Recovery of a herpesvirus from a roan antelope (*Hippotragus equinus*). Vet. Micro (In press)

Reid HW, Buxton D (1989) Malignant catarrhal fever. In Wittman G (ed) Developments in Veterinary Virology. Herpesvirus Diseases of Cattle, Horses and Pigs. Kluwer Academic, Boston, Dordrecht/London, pp 116–143

Reid HW, Buxton D (1984) Malignant catarrhal fever in deer. Symposium on Deer. Proc R Soc Edinburgh 82B:261–293

Reid HW, Pow I, Buxton D (1989a) Antibody to Alcelaphine herpesvirus-1 (AHV-1) in hamsters experimentally infected with AHV-1 and the "sheep-associated" agent of malignant catarrhal fever. Res Vet Sci 47:383–386

Reid HW, Buxton D, Pow I, Finlayson J (1989b) Isolation and characterisation of lymphoblastoid cells from cattle and deer affected with "sheep-associated" malignant catarrhal fever. Res Vet Sci 47:90–96

Reid HW, Buxton D, McKelvey WAC, Milne JA, Appleyard WT (1987) Malignant catarrhal fever in Pere David's deer. Vet Rec 121:276–277

Reid HW, Buxton D, Pow I, Finlayson J, Berrie E (1983) A cytotoxic T-lymphocyte line propagated from a rabbit infected with sheep associated malignant catarrhal fever. Res Vet Sci 34:109–113

Reid HW, Buxton D, Corrigall W, Hunter AR, McMartin DA, Ruston R (1979) An outbreak of malignant catarrhal fever in red deer (*Cervus elaphus*). Vet Rec 104:120–123

Reid HW, Plowright W, Rowe LW (1975) Neutralising antibody to herpesviruses derived from wildebeest and hartebeest in wild animals in East Africa. Res Vet Sci 18:269–273

Roderick LM (1959) Malignant catarrhal fever. Vet Med 54:509–532

Rossiter PB (1981) Antibodies to malignant catarrhal fever virus in sheep sera. J Comp Pathol 91:303–311

Rossiter PB, Gumm ID, Mirangi PK (1988) Immunological relationships between malignant catarrhal fever virus (Alcelaphine herpesvirus-1) and bovine cytomegalovirus (Bovine herpesvirus-3). Vet Microbiol 16:211–218

Rweyemamu MM, Karstad L, Mushi EZ, Otema JC, Jessett DM (1974) Malignant catarrhal fever virus in nasal secretions of wildebeest: a probable mechanism for virus transmission. J Wildl Dis 10:478–487

Sanford SE, Little PB, Rapley WA (1977) The gross and histologic lesions of malignant catarrhal fever in three captive Sika deer (*Cervus nippon*) in Southern Ontario. J Wildl Dis 13:29–32

Seal BS, Klieforth RB, Welch WH, Heuschele WP (1989) Alcelaphine herpesvirus 1 and 2 SDS-PAGE analysis of virion polypeptides restriction endonuclease analysis of genomic DNA and virus replication restriction in different cell types. Arch Virol 106:301–32

Selman IE, Wiseman A, Wright NG, Murray M (1978) Transmission studies with bovine malignant catarrhal fever. Vet Rec 102:252–257

Straver PJ, Beekum van JG (1979) Isolation of malignant catarrhal fever virus from a European bison (*Bos bonasus*) in a zoological garden. Res Vet Sci 26: 165–171

Williams ES, Thorne ET, Dawson HA (1984) Malignant catarrhal fever in a Shira's Moose (*Alces alces shirasi Nelson*). J Wildl Dis 20:230–232

Wilson PR, Alley MR, Irving AC (1983) Chronic malignant catarrhal fever: a case in a sika deer (*Cervus nippon*). N Zealand Vet J 31:7–9

Wobeser G, Majka JA, Mills JHL (1973) A disease resembling malignant catarrhal fever in captive white-tailed deer in Saskatchewan. Can Vet J 14:106–109

Wyand DS, Helmboldt CF, Nielsen SW (1971) Malignant catarrhal fever in white-tailed deer. JAVMA 159:605–610

27
Elk Disease Survey in Western Canada and Northwestern United States

Judit E.G. Smits

Abstract

This 10-year study of elk (*Cervus elaphus canadensis*) mortality in western Canada and United States is based upon 231 necropsy reports from seven pathology labs. Twenty-five percent of fetuses had confirmed *B. abortus* infections, and all these came from the National Elk Refuge in Wyoming. Among neonates, dehydration and emaciation accounted for 66% of deaths, followed by enteritis at 25%. Coronavirus or *E. coli* was isolated from more than half the enteritides. *E. coli* was cultured from seven of ten cases with septicemia. The leading causes of calf morality (26% each) were emaciation, and *E. coli* or *Cl. perfringens* enteritis. Pneumonia, septicemia and trauma each accounted for 17% of calf deaths. Trauma, the major cause of death in mature elk, was seen in 40% of bulls and 26% of cows. Emaciation in mature elk (23%) was seen in conjunction with pneumonia, footrot, trauma, arthritis, and nephritis. Pneumonia (18% of adults) commonly yielded *A. pyogenes* and *P. multocida*.

Key words: Elk, disease, wapiti, western Canada, northwestern United States

Introduction

There is a strong interest in many parts of the world in the captive raising of traditional game animals for meat production. The Canadian game farming industry, yet in its infancy, is developing and expanding rapidly. Elk is the most popular cervid species for farming in western Canada in provinces which allow the raising of native species on game farms. There is a paucity of information about the types and prevalence of diseases which affect this species. The intent of this study was to identify major disease problems that have caused death in captive or semi-captive elk in the northern part of the western Great Plains and Rocky Mountain regions of Canada and United States.

Materials and Methods

A retrospective study of elk mortality was conducted based upon necropsy reports from several provincial, state and veterinary college pathology laboratories. The loboratories were chosen in an attempt to survey elk populations which exist in environments similar to those found in game farming areas of Saskatchewan and Alberta. Ten years of data were required in order to get a meaningful number of cases. Not all laboratories had 10 years of accessible data.

The necropsy records came from the Western College of Veterinary Medicine (WCVM) at the University of Saskatchewan (Sk.), the veterinary diagnostic offices at South Dakota (S.D.) and North Dakota (N.D.) State Universities, the Alberta Agriculture Pathology Laboratories in Edmonton and Airdrie, Alberta, the Sk. Agriculture Pathology laboratory in Regina, Sk., and the Wyoming Game and Fish Department in Laramie, Wyoming.

The basis for selection of cases included the following: (1) cases had to come from a captive (zoo, game farm, research facility) or a semicaptive management system (game park or refuge) and (2) cases used were only from animals which were found dead and did not include hunter-killed animals.

All animals were categorized into one of five possible age groups: fetus, neonate (under 3 weeks of age), calf (over 3 weeks but under 1 year of age), mature (any animal at least 1 year of age) and unknown. The data were sorted into categories which included 22 specific diseases, plus "other" and "no diagnosis" categories. The sex of the animals was often not known. Therefore, only when there were clear differences among mature animals of known sex were data broken down by gender. Mature elk, the largest group, had "major" diseases which were diagnosed in 18% to 26% of cases and "minor" diseases which were seen in 5% to 10% of the cases studied.

Results

There were 231 necropsy case reports represented in this study. Certain diseases occurred primarily in specific age categories. Only diseases that occurred in four or more calves and five or more other individuals were included in the summary (Table 27.1).

Fetal Elk

The major disease diagnosed in fetal elk was brucellosis due to *Brucella abortus*. All eight of the confirmed cases of brucellosis came from the Wyoming Game and Fish laboratory which was also where 30 of the 32 fetuses were examined. Etiology was confirmed in 9 of the 32 cases. There was one aborted fetus from which a heavy growth of *Ps. aeruginosa* was cultured, one fetus with goiter and two anomalous fetuses.

Neonatal Elk

In the neonatal elk group, dehydration was the major cause of death in animals under 2 to 3 days of age. Between 3 days and 3 weeks of age, emaciation was the main reason for death. These diseases accounted for 66% of neonatal deaths. At 25%, enteritis was the third most common diagnosis in neonatal elk. Five of 11 cases of neonatal enteritis were due to coronavirus infections whereas two had coliform enteritis.

TABLE 27.1. Results of disease survey of elk in western Canada and northwestern United States

Disease	Fetus ($n = 32$)		Neonate ($n = 44$)		Calf ($n = 23$)		Adult ($n = 114$)	
	n	%	n	%	n	%	n	%
Major diagnoses								
Brucellosis	8	25						
Dehydration	—	—	17	39	—	—	—	—
Emaciation	—	—	12	27	6	26	27	23
Enteritits	—	—	11	25	6	26	—	—
Pneumonia	1	3	6	14	4	17	20	18
Septicemia	1	3	10	23	4	17	—	—
Trauma	—	—	5	11	4	17	30	26
Unknown	19	59	6	14	1	4	7	6
Other	3	9						
Minor diagnoses								
Capture myopathy							11	10
Nephritis							11	10
Footrot							10	9
Peritonitis							9	8
Arthritis							8	7
Myocarditis							6	5

The fourth most frequently diagnosed disease was septicemia (25%). *Escherischia coli* was isolated from the majority of septicemic neonates. Other organisms cultured from the remaining cases of septicemia included *Klebsiella pneumonia*, *Pseudomonas aeruginosa*, and *Proteus* sp. Meningitis was commonly seen concurrently with coliform septicemia.

Pneumonia was another important disease of neonatal elk. In only one of the six cases was an infectious agent (mycotic) isolated from the lung. Another neonate with pneumonia also had coronaviral enteritis although the association is not clear.

One hand-reared neonate developed a perforated colonic ulcer, which caused a fatal diffuse peritonitis.

Elk Calves

The two most important diseases seen in the calf group were emaciation and enteritis at six cases each, representing over 50% of the calf diagnoses. Among the calves dying of emaciation, one had concurrent *Pasterurella* sp. pneumonia, and another had moderate numbers of fecal trichuris eggs and coccidian oocysts. Several infectious agents were identified from the calves with enteritis. These included rotavirus (1 case), *E. coli* (1 case), and *Clostridium perfringens* (2 cases). There were four calves each diagnosed with pneumonia, septicemia, and trauma. In every case with pneumonia there were other concurrent diseases. These were omphalitis, emaciation, and traumatic reticuloperitonitis.

Infectious agents associated with the diseases seen were identified in 44% of the calf necropsies.

Adult Elk

The most frequently diagnosed problem was trauma, which accounted for 26% (30 elk including those of unknown sex) of mature elk deaths. This was considerably higher than the overall population rate of 18% trauma-related deaths. Among the mature males examined postmortem, 40% were diagnosed as dying from trauma, while 26% of the mature females shared this diagnosis.

Emaciation, observed in 27 elk, was the second most common diagnosis. Approximately the same percentage of adult males and females were emaciated (19% and 18% respectively). Many of these animals had other pathological conditions as well. Seven elk had pneumonia, five had footroot, five were traumatized, while others suffered from concurrent (in order of frequency) arthritis, nephritis, peritonitis, myocarditis, enteritis, capture myopathy, dental attrition, or gastrointestinal parasitism. In only three cases was emaciation the sole diagnosis.

There were 20 adult elk with pneumonia, which was the third most commonly diagnosed disease. Forty-five percent of these cases had etiological diagnoses. *Actinomyces (corynebacterium) pyogenes* and *Pasteurella multocida* were the most frequently isolated agents. Mycoses were seen as both primary and secondary infections. Other infectious bacteria among the pneumonia cases seen were *E. coli* with *Streptococcus* sp. plus one *Mycobacterium bovis*. Animals with pneumonia tended to have other diseases as well. Besides emaciation, footrot, trauma, and lungworm were also seen.

Diseases seen with less frequency, but which represented 5% to 10% of the mature elk in this survey, were categorized as "minor" diseases. Nephritis and capture myopathy were both seen in 11 (10%) of the adult elk. Nephritis was seen much more often in cows (73%) than bulls (9%). Positive bacterial cultures in two of these infections yielded *E. coli*, and *Pseudomonas* sp. with *Enterobacter* sp.

Deaths due to capture myopathy (CM) were reported most frequently from captive environments (92%). Suspected anesthetic overdose or postimmobilization complications were associated with 67% of the cases, while the rest were related to transport or misadventure.

Nine percent of the adult elk had footrot. These animals came from all three habitats with almost equal frequency, but no footrot was seen by the Sk., N.D., or S.D. laboratories. Some of the cases yielded *E. coli*, *A. pyogenes*, and *Streptococcus* sp.

Peritonitis, diagnosed in nine elk, was seen with a variety of conditions. Two cases of peritonitis were secondary to GI tract impaction with dried feed material. One case was due to a mycotic infection which caused localized intestinal thickening and induced intestinal obstruction. One elk had a postcalving intestinal torsion, and another had hepatic abscesses and myocarditis. *P. multocida*, *E. coli*, and *Streptococcus* sp. were agents isolated from cases of peritonitis.

Arthritis, 7% of submissions, was most often seen in farmed elk. It was seen together with emaciation, trauma, and footrot.

In the Wyoming research herd, five elk deaths were confirmed to be due to spongiform encephalopathy or "chronic wasting disease," a disease with an unidentified etiology, suspected to be a "slow virus" similar to scrapie in sheep. This is the only occurrence reported in elk, although it has been described in captive mule deer (*Odocoileus hemionus hemionus*), black-tailed deer (*O. hemionus columbianus*), and a mule deer X white-tailed deer (*O. virginianus*) hybrid in wildlife facilities in Wyoming and Colorado (Williams and Young 1982).

Unknown Age Group

Among animals of unknown ages (which were therefore not included in Table 27.1), there were several cases of peritonitis and arthritis as well as death due to unknown causes. There were two cases of pneumonia, two cases of encephalitis, and one each of blackleg, enteritis, fungal dermatitis, and nephritis.

Discussion

Abortion due to *B. abortus* is a well-documented entity in western Wyoming. A recognized focus of brucellosis has existed in the National Elk Refuge, Jackson, Wyoming since the 1930s (Thorne et al. 1978a, b). Serological surveys for brucellosis across the United States have shown that outside the National Elk Refuge the prevalence of infection is less than 1% (Adrian and Keiss 1977; Thorne et al. 1979a, b). The most probable factor responsible for maintaining this focus of infection is the concentration of animals at the winter feeding grounds (Thorne et al. 1978a).

Greater than 75% of the fetuses in this study were from the Game Research Facility of the Wyoming Game and Fish Department. Because of this, the information on fetal deaths is undoubtedly biased and should not be considered a balanced overview of problems associated with pregnancy wastage in elk.

The major factors leading to death of neonatal elk were problems common to most domestic neonates. Both dehydration and emaciation result from various causes. According to people managing captive deer herds mismothering in which the neonate is abandoned, has been observed among first-time mothers. Post-natal rejection of calves has been described in severely stressed red deer hinds (*Cervus elaphus*) (Blaxter and Hamilton 1980).

Neonates born under a certain "critical weight" (11.5 kg for elk) are subviable and have a lower than 50% chance of survival (Thorne et al. 1976). Birth weight of the calf is directly affected by the nutritional status of the dam both prerut and in late gestation (Thorne et al. 1976; Moore et al. 1988). Low birth weight in this study was a factor in viral or coliform enteritides. Both hand-rearing and low body weight were associated with septicemia in calves. There were multiple deaths due to septicemia on some premises suggesting a problem with environmental conditions.

There were three cases of mycotic pneumonia. Munro et al. (1984) and Munro (1986) considered fungi to be opportunists causing disease secondary to social or dietary stress or heavy environmental fungal contamination. Trauma accounted for 11% of neonatal death. Sources of trauma range from misadventure to savaging by other mature elk.

The majority of calves died before or shortly after weaning. Emaciation and enteritis were the major diseases of this group. Presumably, the main reason leading to uncomplicated emaciation in calves was inadequate lactation of the dam. Clostridial enterotoxemia was the most common enteropathy in the calves of this study. Enterotoxemia is usually associated with sudden change from a low to a high energy diet. Neither the dietary history nor the vaccination status (vs clostridial diseases) of these calves was available.

Pneumonia, septicemia, and trauma were reported with equal frequency among elk calves. *P. multocida* was the most common isolate from pneumonic lungs of calves. Omphalitis or umbilical abscesses causing generalized infections, yielded *A. pyogenes* and *F. necrophorum* from abscesses and affected lymph nodes. The umbilical route of infection suggests environmental contamination since *A. pyogenes* and *F. necrophorum* are both ubiquitous organisms (Rosen 1981).

Trauma was the number one cause of mortality in mature elk. In intensive, high density game farm systems, herd management is the single most important factor in the prevention of trauma (English 1984). Emaciation was a common finding among adult elk and accompanied a variety of other diseases. Malnutrition can predispose elk to other diseases, but conversely, pneumonia, footrot, arthritis, and nephritis will cause emaciation. Uncomplicated deficiency malnutrition occurs when the highly seasonal nutritional requirements of mature elk are not met (Kay 1979; Hudson et al. 1985).

Pneumonia was a major problem of mature elk. Bronchopneumonias and pulmonary abscesses most commonly yielded *A. pyogenes* and *P. multocida.* These lung infections had several possible origins. *P. multocida*, a resident of the upper respiratory tract, may become pathogenic under stressful conditions (Rosen 1981). Pneumonic lungs from two animals yielded fungi. Possible sources of fungal infection are mouldy feeds, and poorly ventilated conditions (Munro et al. 1984; Munro 1986).

Almost all of the nephritis cases seen were in females. Of the few cases from which bacteria were cultured, ubiquitous opportunistic organisms were found (e.g., *E. coli*, *Pseudomonas*, and *Enterobacter*). The routes of infection are likely to be similar to those in other domestic species.

The majority of CM deaths described in this report occurred following chemical immobilization. This diagnosis is difficult to prove. It is based on circumstantial evidence, history, typical postmortem findings, and a lack of any other explanation. Some CM deaths were due to drug overdoses; some were due to post-immobilization complications. Hyperthermia and acidosis are key factors in the development of capture myopathy (Chalmers and Barrett 1982).

Footrot occurred with equal frequency from all habitats (farm, park, and research herd), but no cases were reported from Saskatchewan or North or South Dakota. These are semiarid climatic zones. It is possible that wetter environments predispose hoof stock to foot lesions or create conditions that allow more persistent environmental contamination.

Conclusions

Due in part to the nonrandom sampling procedures, the results in this survey suffer from a degree of bias. They cannot be considered representative of the entire population of captive and semicaptive elk in western Canada and the northwest United States. Because the animals in this study came from relatively confined environments, there may have been cluster effects with infectious diseases. Numerous animals in one group could share and communicate infectious diseases. Brucellosis in western Wyoming illustrates such a cluster effect.

References

Adrian WJ, Keiss RE (1977) Survey of Colorado's wild ruminants for serological titres to brucellosis and leptospirosis. J Wildl Dis 13:429–431

Blaxter KL, Hamilton WJ (1980) Reproduction in farmed red deer. Calf growth and mortality. J Agric Sci 95:275–284

Chalmers GA, Barrett MW (1982) Capture myopathy. In Hoff GL, Davis JW (eds) Noninfectious Diseases of Wildlife. Iowa State University Press, Ames, Iowa, pp84–94

English AW (1984) Veterinary aspects of deer farming in New South Wales—an update. Post Graduate Committee in Veterinary Science, University of Sydney, Deer Refresher Course, Proceedings 72:427–459

Hudson RJ, Watkins WG, Pauls RW (1985) Seasonal Bioenergetics of Wapiti in Western Canada. Biology of Deer Production. R. Soc N Zealand, Bull 22, Wellington, pp447–452

Kay RNB (1979) Seasonal changes of appetite in deer and sheep. Agric Res Council Res Rev 5:13–15

Moore GH, Littlejohn RP, Cowie GM (1988) Liveweights, growth rates and mortality of farmed red deer at Invermay. N Zealand J Agric Res 31:293–300

Munro R (1986) Respiratory system—fungal infections. In Alexander TL (ed) Management and Diseases of Deer, 1st ed. Deer Society London, W1M OAT, pp69–70

Munro R, Hunter AR, Bonniwell M, Corrigal W (1984) Systemic mycosis in Scottish red deer (*Cervus elaphus*). J Comp Pathol 95:281–289

Rosen MN (1981) Necrobacillosis. In Davis JW,

Karstad LH, Trainer DO (eds) Infectious Diseases of Wild Mammals, 2nd ed. Iowa State University Press, Ames, Iowa, pp332–338
Rosen MN (1981) Pasteurellosis. In Davis JW, Karstad LH, Trainer DO (eds) Infectious Diseases of Wild Animals, 2nd ed. Iowa State University Press, Ames, Iowa, pp244–252
Thorne ET, Dean RE, Hepworth WG (1976) Nutrition during gestation in relation to successful reproduction in elk. J Wildlife Management 40:330–335
Thorne ET, Morton JK, Blunt FM, Dawson HA (1978a) Brucellosis in elk II. Clinical effects and means of transmission as determined through artificial infections. J Wildl Dis 14:280–291
Thorne ET, Morton JK, Thomas GM (1978b) Brucellosis in elk I. Serologic and bacteriologic survey in Wyoming. J Wildl Dis 14:74–81
Williams ES, Young S (1982) Spongiform encephalopathy of Rocky Mountain elk. J Wildl Dis 18:465–471

28
Vaccination to Control Brucellosis in Free-Ranging Elk on Western Wyoming Feed Grounds

James D. Herriges, Jr., E. Tom Thorne, and Sandra L. Anderson

Abstract

Approximately 22 500 elk (*Cervus elaphus nelsoni*) are fed during winter months at 23 feed grounds in Wyoming to reduce winter mortality and depredation on agricultural crops. Brucellosis was documented on 18 feed grounds and seroprevalence among adult females averaged 37% (range 16% to 50%). Nonfed elk populations were generally free of brucellosis. Strain 19 *Brucella abortus* vaccine was shown to reduce infection and abortion in elk. A vaccination program was initiated to increase calf production and decrease the potential of transmission of the disease to livestock. An airgun was used to implant a methylcellulose pellet containing lyophilized vaccine in muscle tissue. Vaccinated elk were marked with a gelatin-encased, oil-base paint ball. Elk were first acclimated to human activity and then to the report of the airgun through repeated dry-firing at gradually increased volumes. Cow and calf elk were marked and vaccinated from feed sleds as they fed on hay. Over 13 000 doses were administered during 1985–1990 on nine feed grounds. An estimated 4700 elk were vaccinated during winter 1990, including an average of 95% of calves on eight feed grounds and 72% of cows on four feed grounds.

Key words: Elk, brucellosis, Brucella abortus, *vaccination, disease, Wyoming*

Introduction

Brucellosis, a disease caused by the bacteria *Brucella abortus*, is most commonly associated with cattle. It was first discovered in wildlife in 1917 when bison (*Bison bison*) of Yellowstone National Park were found to be infected (Mohler 1917). Brucellosis was not known to occur in elk until 1930 when three of nine elk on the National Elk Refuge (NER), near Jackson, Wyoming, were found to be seropositive (Murie 1951). The NER was established by Congress, anbd feeding began in 1912 to reduce winter elk mortality and depredation of elk on stored hay. The Wyoming Game and Fish Department subsequently established 22 other feed grounds in western Wyoming. Approximately 22 500 elk are now supplementally fed, and brucellosis has been shown to occur in elk at most of these feed grounds. Except for bison, and a low percentage of elk in Yellowstone National Park, other wildlife and elk in the area are essentially free of brucellosis (Thorne et al. 1979; E.T. Thorne, unpublished data). The sources of brucellosis in Wyoming's wildlife are unknown, but probably originated from domestic livestock around the turn of the century. The most common biovars of *B. abortus* affecting cattle (1 and 4) have been isolated from elk at the Greys River feed ground. This suggests that elk were infected on at least two occasions.

The effects of brucellosis are similar in elk and cattle. Between 50% and 70% of cow elk that became infected either aborted their next fetus or the calf was weak and died shortly after birth (Thorne et al. 1978, 1981; Herriges et al. 1989). Most subsequent calves were born healthy. In some chronically infected elk, brucellosis caused synovitis, carpal bursitis, and lameness (Thorne et al. 1978). Brucellosis is acquired through exposure to reproductive products; elk may consume contaminated placentae, lick aborted fetuses, or eat contaminated hay. The disease was shown to readily transmit from elk to elk (Thorne et al. 1978) and from elk to cattle when confined together during the latter half of pregnancy of infected elk (Thorne et al. 1979).

Brucellosis is the target of a cooperative eradication program begun in 1940 and involving the U.S. Department of Agriculture, state departments of agriculture, and livestock products. The program has been successful in eliminating brucellosis from livestock in 27 states. Wyoming and Montana were classified brucellosis free in 1985, and Idaho is expected to receive certification soon (R. Hillman, personal communication). As the prevalence of brucellosis in livestock decreases, wildlife reservoirs will come under increasing scrutiny; continued presence of the disease in elk may limit future elk management options.

The Wyoming Game and Fish Department began studies with captive elk in 1970 to determine the effects of brucellosis and subsequently the efficacy of strain 19 vaccine in elk. The latter studies showed strain 19 *B. abortus* vaccine to be effective in decreasing infection and abortion in elk following challenge with virulent *B. abortus* (Thorne et al. 1981; Herriges et al. 1989). In early studies, vaccine was administered by hand-held syringe. This required capture of free-ranging elk, a difficult and time-consuming method. A remote-delivery system was developed by BallistiVet (White Bear Lake, MN) by which a vaccine-containing implant, or biobullet, was fired from an airgun. With a relatively efficient method of vaccine delivery, a vaccination program was initiated in 1984 with the goal of controlling brucellosis in elk, thereby reducing abortions and the potential for transmission to livestock.

Materials and Methods

Elk feed grounds are located in western Wyoming as far south of Yellowstone National Park as Big Piney and Big Sandy (approximately 220 km).

Feedground elk were trapped in wood corral traps and blood-tested to determine prevaccination prevalence of brucellosis and aid in setting program priorities. Blood was collected by jugular venipuncture with the elk free-standing in a chute. Standard procedures of the U.S. Department of Agriculture, Animal and Plant Health Inspection Service (National Animal Disease Laboratory Diagnostic Reagents Manual, 65E and F) were used for conducting the standard plate agglutination test (SPT), buffered *Brucella* antigen rapid card test (BBA), and rivanol test (Riv). The complement fixation test (CFT) was patterned after the microtiter technique for the diagnosis of bovine anaplasmosis (Martin and Ritchie 1973). Serum producing any degree of reaction to two or more tests was considered positive. If a sample reacted to only one of the four tests, standards established for nonvaccinated cattle were used for the SPT ($\geqq +100$), BBA (+), and Riv ($\geqq +25$). Complete fixation at the 1:40 dilution or higher was considered positive on the CFT. Test results reported here do not include vaccinated elk.

Biobullets were 6.5 mm in diameter and 15 mm or 21 mm in length; the longer bullets were used in 1990 for improved accuracy. Composed of hydroxypropyl methylcellulose, biobullets had a hollow cavity in the rear into which a lactose pellet containing lyophilized strain 19 vaccine was loaded. Biobullets were loaded at National Veterinary Services Laboratory, Ames, Iowa (Angus 1989). Ten biobullets were loaded in a plastic clip, sealed with cellophane, and placed in a plastic/foil pouch in which air was replaced with nitrogen. Biobullets used during 1986 and 1987 contained approximately 20×10^9 colony forming units (cfu) *B. abortus* strain 19; in 1988–1990 the dose was approximately 7×10^9 cfu. Loaded biobullets were stored at $\leqq 0°$ C until immediately before use. Biobullets were propelled at 213–282 m/s by compressed air (62–84 kg/cm^2) contained in a diving tank and implanted in the large muscle mass of the rear leg. Gelatin-encased, oil-base paint balls, 17.3 mm in diameter (Nelson Paint, Iron Mountain, MI) were used to mark vaccinated elk. During 1984–1986 one double-barrelled gun which shot paint balls as well as biobullets was used at one feed ground; single-barrelled biobullet airguns and separate CO_2-powered paint ball pistols (Nelson Paint, and Pursuit Marketing, Northbrook, IL) were used at other feed grounds and at all feed grounds during 1987. In 1988, paint ball pistols were attached to biobullet rifles by means of a bracket and were converted from CO_2 to compressed air power.

Elk were marked with a paint ball on the shoulder or chest, and a biobullet was immediately ballistically implanted. All elk were consistently marked on the same side to facilitate future identification. Vaccination was conducted from a horse-drawn feed sled or tractor-drawn feed trailer. Feed was distributed around a stationary feed sled and elk vaccinated as they approached, or feed was distributed in lines and elk vaccinated as the tractor or sled drove back along the line of feeding elk. Calves of both sexes and cows were vaccinated. Vaccination was conducted during January through early April. Elk feeders generally vaccinated elk for 1–1.5 h, 3 days per week over a 2 to 6-week period.

Elk on some feed grounds did not normally come onto the feed ground in the presence of the feeder. To encourage them to do so, the feeder distributed hay late in the afternoon and waited at a distance. Once elk were accustomed to feeding in the presence of the feeder, feed was distributed only near the stationary sled. When elk routinely approached the feed sled or allowed the feed sled to closely approach them, they were gradually accustomed to the sound of the airgun by repeated dry-firing at increasing volumes.

Results and Discussion

Brucellosis has been documented in elk at 18 of 23 feed grounds through serology or isolation of field strain *B. abortus* from aborted fetuses. Elk on untested feed grounds are assumed to be infected due to the proximity of those feed grounds to tested feed grounds. The seroprevalence among adult females at six individual feed grounds for which > 70 samples were available, varied from 16% at Soda Lake feed ground near Pinedale to 50% at Greys River feed ground near Alpine. Weighting the prevalence estimate for each of these feed grounds by the average population size, yielded an overall prevalence estimate for adult female elk on feed grounds of 36.8%. True differences in prevalence between feed grounds probably relate to differences in elk behavior and distribution and resultant opportunities for transmission. For example, the low prevalence at Soda Lake feed ground may have been related to the habitual movement of elk to tree cover > 1 km from the feeding site. A relatively low prevalence at Forest Park feed ground (16%) was probably due to the recent establishment of this feed ground and limited time for spread of the disease. In addition, prevalence may be somewhat cyclic and thus affect estimates based on sampling during any one winter.

Extensive long-term prevalence data exist only for elk at the NER, a herd of approximately 7500. Prevalence estimates varied considerably from year to year; however, sample sizes were not sufficient to assure that these accurately reflected true prevalence. Prevalence of brucellosis in adult females during 1971 through 1990 appeared to have a stable trend, varying around the 39% average ($n = 1059$). Among other age and sex classes, prevalence averaged 37% in 29 adult males, 19% in 102 yearling females, 11% in 76 yearling males, and 4% in 457 male and female calves. The increase in prevalence with age probably reflected increased opportunities for exposure with time.

Initial field trials of biobullet vaccination were conducted on the Greys River feed ground during winter 1984; 122 elk were implanted with placebo biobullets. In 1985, strain 19 loaded biobullets were used, and an estimated 70% of cow and calf elk were vaccinated (Table 28.1).

Delays in start of vaccination until late winter or early spring, when elk were less dependent on supplemental feed, limited the number of elk which could be vaccinated in all years except 1989. Numbers of elk vaccinated in 1989, a winter of near-normal snowfall, were probably most representative of future vaccination efficiency. An estimated 71% to 91% of adult and yearling females and 88% to 100% of calves were vaccinated on each feed ground. During 1990, the number of elk vaccinated (2653 cows and 2038 calves) was limited by the number of loaded biobullets received.

Vaccination at three feed grounds was limited to calves only in 1990 (Table 28.1). On these feed

TABLE 28.1. Proportion of elk vaccinated with strain 19 *Brucella abortus* on several western Wyoming feed grounds

		Adult females		Calves	
Feed ground	Year	Counted (*n*)	Vaccinated[a] (%)	Counted (*n*)	Vaccinated[a] (%)
Greys River	1985	523	70 (estimated)	172	70 (estimated)
	1986	616	47.4	223	87.0
	1987	520	61.9 (57.0)	142	103.5 (95.2)
	1988	613	76.5 (67.5)	280	113.2 (100.0)
	1989			279	133.7 (100.0)
	1990			191	94.8 (89.5)
Green River Lakes	1986	386	37.0	118	77.1
	1987	436	73.2 (62.9)	135	101.5 (87.3)
	1988	394	54.8 (51.5)	156	101.3 (95.5)
	1989	405	81.0 (71.3)	169	99.4 (87.5)
	1990			130	83.8 (79.2)
Forest Park	1988	357	91.9 (86.3)	154	100.6 (94.8)
	1989	458	84.5 (78.6)	229	105.7 (98.4)
	1990			211	102.4 (99.4)
Black Butte	1989	333	95.8 (92.5)	187	104.8 (100.0)
	1990	346	94.5 (89.0)	131	127.5 (100.0)
Horse Creek,	1989	1560	76.8 (68.3)	531	126.2 (100.0)
Camp Creek	1990	1501	67.5 (60.8)	406	111.1 (100.0)
National Elk	1989			1970	28.8 (27.9)
Refuge	1990	5312	10.2 (9.6)	1234	45.8 (43.5)
Dog Creek	1990	674	72.7 (67.5)	258	112.0 (100.0)
South Park	1990	641	81.3 (71.3)	221	113.1 (100.0)

[a]The number of doses administered to calves often exceeded the number of calves counted. Therefore, revaccination of elk that had lost paint marks was assumed to have occurred, and estimates of the percentages of both cow and calf elk vaccinated were adjusted accordingly (in parentheses).

grounds, all adult females that were likely to come within range of the airgun were assumed to have been vaccinated in at least 1 of the previous 2 to 4 years. Calves were easier to vaccinate because they were more likely to approach within airgun range. This factor, combined with lower numbers of calves than cows in the population, made vaccination of at least 273 calves on Greys River feed ground possible during seven sessions. Once whole-herd vaccination status is attained through vaccination of cows and calves, it should be maintained by annual vaccination of calves.

The number of doses administered to calves on a feed ground often exceeded the number of calves counted there, indicating that some calves lost paint marks and were vaccinated twice. Vaccination occurred over a 6-week period at Greys River feed ground in 1988, and revaccination of calves was estimated at 12%. A weekly revaccination rate of 2% was assumed in estimating the proportion of cows and calves vaccinated at all feed grounds (Table 28.1). In some cases, the differences between number of doses administered and number of calves on a feed ground may have partially resulted from arrival or departure of elk in the period between counts and completion of vaccination, or from misidentification of yearling cows and calves.

Vaccination of elk at the NER was more difficult than at most other feed grounds. In addition to problems inherent in trying to deal with such a large number of elk, the mechanized method of feeding pelleted alfalfa meant that elk were in association with humans during only a brief time. Also, pelleted alfalfa was much more rapidly consumed than baled hay; within 30 min of feed dispensing, elk were often satiated and no longer willing to remain near the feed truck.

Only one mortality due to biobullet-induced injury was documented—a calf was hit in the lungs in 1986. Seven calf mortalities suspected to be due to biobullet injuries occurred in 1989; four showed wounds that were possibly due to biobullet injury and the other three were observations of paint-marked calves found dead with no obvious cause of death, but no necropsies were performed. These were difficult to evaluate because some mortality of calves normally occurs, and most calves were marked within a few weeks of the initiation of vaccination. If all seven mortalities were due to biobullet injury, less than 0.2% of the estimated number of elk vaccinated in 1989 were killed. Vaccine-induced abortion was not observed.

Elk at two feed grounds which were originally considered too timid for biobullet vaccination were successfully acclimated to closer contact with elk feeders during a single season. At one of these, Forest Park, elk were also acclimated to the report of the airgun, and a high proportion of cow and calf elk vaccinated (Table 28.1).

Success of the program will be measured in the short term by the number and proportion of elk vaccinated. Controlled studies demonstrated that strain 19 vaccine (0.4–76 × 10^8 cfu) reduced infection and abortion in elk following challenge with 4.5–7.5 × 10^6 cfu *B. abortus* strain 2308 (Herriges et al. 1989). Calving success was 23 (62%) of 37 among vaccinates and 3 (33%) of 9 among nonvaccinates. At necropsy, 19 (45%) of 42 vaccinates and 9 (69%) of 13 controls were infected. The number of tissues infected and numbers of bacteria isolated were also generally lower among vaccinated than nonvaccinated elk. Earlier studies conducted with a higher vaccine dose and larger numbers of elk, especially in control groups, showed similar rates of calving success and infection within vaccinated and nonvaccinated control groups (Thorne et al. 1981). These protection rates approached those obtained in vaccine trials with domestic cattle (Manthei 1959; Woodard and Jasman 1983; Deyoe et al. 1979). However, precise comparison is difficult due to varying vaccine and challenge doses, fetal ages at challenge, and resultant abortion and infection rates in nonvaccinated animals.

Reports of few, if any, large-scale programs of vaccination to control a bacterial disease in wildlife are available for use in predicting the effectiveness of this program. Brucellosis was eliminated from bison of the National Bison Range, Moiese, Montana, apparently through vaccination of calves (J. Malcolm, personal communication). In addition, two studies on the effects of vaccination in control of brucellosis in domestic cattle support Nicoletti's (1983) statement that vaccine efficacy is much greater on a population basis than individual basis, as is examined in controlled challenge studies. Safford (1959), in Montana, and Stuart et al. (1959), in California, reported that brucellosis reactor rates among beef cattle herds, which were 76% to 100% vaccinated, averaged 27% and 10%, respectively, of the reactor rate in nonvaccinated herds. Similarly, among partially vaccinated herds in the same studies in Montana and California, the reactor rate of vaccinated animals was lower and averaged 35% and 0.4% to 15%, respectively, that of nonvaccinated animals. In California, brucellosis was apparently eliminated from 3 large dairy herds through vaccination alone (Stuart et al. 1959).

Ultimately, the prevalence of infection as measured by serology will be used to gauge success of the program in controlling brucellosis. In the short term, seroprevalence rates may be confused by vaccine-induced positive reactions in noninfected animals. Persistent vaccinal titers are more common among elk vaccinated as adults, so eventual cessation of adulthood vaccination will improve the accuracy of currently used serologic tests. Promising new serologic tests capable of distinguishing strain 19 vaccinates from infected animals (Gorrell et al. 1983; Wright et al. 1990) would also aid in this endeavor.

Acknowledgements

We sincerely appreciate the assistance of the Wyoming Game and Fish Game Division District 1 and Habitat and Technical Services Division personnel. We thank J. Kopec, D. Woody, and G. Frye, APHIS, Veterinary Services; S. and L. Kilpatrick; F. Paul, T. McDermott, and M. Chevalier of BallistiVet; National Elk Refuge personnel; D. Angus and others at NVSL, APHIS; and E. Nelson, Veterinary Services, Wyoming State Veterinary Laboratory. Funding for this project was provided by Veterinary Services, APHIS, USDA; U.S. Fish and Wildlife Service, USDI; and Wyoming Game and Fish Department under Cooperative Agreement No. 12-16-028.

References

Angus RD (1989) Preparation, dosage delivery, and stability of a *Brucella abortus* strain 19 vaccine ballistic implant. Proc U.S. Animal Health Assoc 93:656–659

Deyoe BL, Dorsey TA, Meredith KB, Garrett L (1979) Effect of reduced dosages of *Brucella abortus* strain 19 in cattle vaccinated as yearlings. Proc U.S. Animal Health Assoc 83:656–659

Gorrell MD, Milliken GL, Anderson BJ, Pucci A (1983) An enzyme immunoassay for bovine brucellosis using a monoclonal antibody specific for field strains of *Brucella abortus*. In Valette L, Hennessen W (eds) Developments in Biological Standardization vol. 56, IIIrd Int Symp on Brucellosis. S. Karger, Basel, pp 491–494

Herriges JD Jr, Thorne ET, Anderson SL, Dawson HA (1989) Vaccination of elk in Wyoming with reduced dose strain 19 *Brucella*: controlled studies and ballistic implant field trials. Proc U.S. Animal Health Assoc 93:640–653

Manthei CA (1959) Summary of controlled research with strain 19. Proc U.S. Animal Health Assoc 63:91–97

Martin WH, Ritchie WH (1973) The microtiter technique for the diagnosis of bovine anaplasmosis. Proc 6th Natl Anaplasmosis Conf, Las Vegas, Nevada, pp 141–142

Mohler JR (1917) Annual report of the United States Bureau of Animal Industry, Pathological Division. Ann Repts of the Dept of Agriculture, p 39

Murie OJ (1951) The elk of North America. Stackpole, Harrisburg, Pa. and Wildlife Management Institute, Washington, O.C.

Nicoletti P (1983) The epidemiology of brucellosis in animals. In Valette L, Hennessen W (eds) Developments in Biological Standardization, vol. 56, Third Int Symp on Brucellosis. S. Karger, Basel, pp 623–628

Safford JW (1959) Montana beef cattle brucellosis vaccination survey—1952–58. Proc U.S. Livestock Sanitary Assoc 63:98–103

Stuart JE, Bills CB, DeMattei JD, Mace DL (1959) The results of eleven years' vaccinating with strain 19. Proc U.S. Livestock Sanitary Assoc 63:83–90

Thorne ET, Morton JK, Blunt FM, Dawson HA (1978) Brucellosis in elk II. Clinical effects and means of transmission as determined through artificial infection. J Wildl Dis 14:280–291

Thorne ET, Morton JK, Ray WC (1979) Brucellosis, its effect and impact on elk in western Wyoming. In Boyce MS, Hayden-Wing LO (eds) North American Elk: Ecology, Behavior and Management. University of Wyoming, Laramie, pp 212–220

Thorne ET, Walthall TJ, Dawson HA (1981) Vaccination of elk with strain 19 *Brucella abortus*. Proc U.S. Animal Health Assoc 85:359–374

Woodard LF, Jasman RL (1983) Comparative efficacy of an experimental S45/20 bacterin and a reduced dose of strain 19 vaccine against bovine brucellosis. Am J Vet Res 44:907–910

Wright PF, Nielsen KH, Kelley WA (1990) Primary binding techniques for the serodiagnosis of bovine brucellosis—enzyme immunoassay. In Nielsen K, Ducan JR (eds) Animal brucellosis. CRC Press, Boca Raton, FL, 199–236.

29

Observations on the Relative Susceptibility to Disease of Different Species of Deer Farmed in New Zealand

Colin G. Mackintosh

Abstract

The four major types of deer farmed in New Zealand are European red deer (*Cervus elaphus*), North American wapiti (*C. elaphus spp.*), wapiti X red hybrids, and fallow deer (*Dama dama*). There are also small numbers of sika (*C. nippon*), rusa (*C. timorensis*), sambar (*C. unicolor*), and Père David's deer (*Elaphurus davidianus*). Field observations over the past 20 years have shown a number of species differences in susceptibility to various diseases of deer farmed in this country. Malignant catarrhal fever has never been reported in fallow deer, while wapiti and red deer seem relatively resistant, and sika, rusa and Père David's deer are very susceptible. Fallow deer appear relatively resistant to yersiniosis which commonly affects young red deer. Fallow deer are more susceptible than red deer to sporidesmin, a fungal toxin produced by (*Pithomyces chartarum*) which causes liver damage and "facial eczema." Wapiti and wapiti X red deer appear more susceptible than red deer to a number of conditions including copper deficiency and enzootic ataxia, the fungal endophyte neurotoxicity which causes "ryegrass staggers," internal parasitism, "dietary scours," and "the fading elk syndrome."

Key words: Diseases, farmed deer, relative susceptibility

Introduction

There are approximately 1 000 000 deer on nearly 5000 farms in New Zealand, and according to the 1988 Department of Statistics survey, the majority are red deer (81%), with fallow (9%), wapiti and wapiti X red hybrids (7%) and other species comprising sika, rusa, sambar and Père David's (3%) making up the rest. Most of the farmed deer originate from animals captured from wild populations established throughout New Zealand in the last 130 years. All the red, fallow, and sika deer were of British origin, although some came via Australia. The sambar came from Sri Lanka, the rusa came from New Caledonia, whereas the wapiti and white-tailed deer (*Odocoileus virginianus*) came from North America (Challies 1985). The long sea voyages of the late 1800s and early 1900s probably served as a natural quarantine period, and recent importations of deer from the United Kingdom, Europe, and North America have required extensive quarantine residence and testing. Consequently the wild deer in New Zealand are remarkably free of serious diseases, although most populations carry low burdens of helminth parasites. No blood-borne parasitic diseases are recognized, probably due to the lack of suitable insect vectors. The most significant disease to emerge in wild deer populations in certain areas in the past 20 years is tuberculosis (Tb), caused by *Mycobacterium bovis*; it has been isolated from wild red, fallow, and sika deer (de Lisle and Havill 1985). Apart from Tb, none of the important diseases of farmed deer dealt with in this paper have been reported to cause clinical disease in wild deer.

Since deer farming in New Zealand started

TABLE 29.1 Summary of relative disease susceptibility of various deer species farmed in New Zealand

Disease	Very Susceptible	Susceptible	Resistant
Malignant catarrhal fever	WTD, Ru, PD, S	R, W	F
Yersiniosis	—	R, W	F
Facial eczema (sporidesmin)	F	R, W	—
Ryegrass staggers	W	R	F
Enzootic ataxia	W	R	—
Lungworm	R(w), W(w), F(w)	W(a)	R(a), F(a)
Gastrointestinal nematodes	W(w)	W(a), F(w)	R(a), F(a)
Preputial ulceration and prolapse	W	—	R, F
"Dietary scour"	W	—	R, F
"Fading elk syndrome"	W	—	R, F

Species: WTD, white-tailed deer; R, red; Ru, rusa; S, sika; P, Père David's; W, wapiti; F, fallow. Age: w, weaners; a, adults

over 20 years ago people have tried to farm all of the above species, but some, especially white-tailed deer, have not adapted well to being farmed, whereas red deer, wapiti, and fallow deer have thrived. One of the factors that has influenced this adaptation is their susceptibility to a combination of stress and the diseases that they encounter in the farming environment. This chapter reviews the evidence for species differences in susceptibility to a number of diseases of deer in New Zealand based largely on field observations (Table 29.1).

Malignant Catarrhal Fever

In New Zealand, malignant catarrhal fever (MCF) of deer and cattle are both caused by a sheep-associated gamma-herpesvirus which appears to be closely related to the wildebeest-associated virus which causes a similar disease in cattle in Africa (Reid et al. 1989) and in deer in zoos in North America (Saio-Kun Wan et al. 1988). The prevalence of infection in deer is not known, but once disease develops it is invariably fatal, although the course varies from hyperacute (< 24 h) to acute (2 to 3 days) to chronic (1 to 4 weeks) (Fennessy 1988). Deer are literally a "dead-end" host, as there appears to be no natural deer-to-deer transmission. It is believed that all infections result from direct or indirect contact with sheep-excreted virus. The majority of cases (70%) in farmed deer occurs between June and September with a peak in July. This is 3 to 4 months earlier than in cattle, but it coincides with the peak stress on deer, especially stags which enter the winter with poor fat reserves (<5% body weight) following the rut in late autumn.

Throughout New Zealand, approximately 1% of red deer die annually from MCF with the incidence slightly higher in the southern South Island and slightly lower in the northern North Island (Fennessy 1988). This contrasts strongly with the incidence of MCF in other species of deer. White-tailed deer and Père David's deer kept on farms appear highly susceptible. Early attempts to farm white-tailed deer in the lower South Island foundered when all 50 live-captured animals succumbed to MCF within a year (R. Brooks, personal communication). Only 11 of the original 70 or so Père David's deer originally imported remain alive 5 years later, the majority having died in the first 2 years from MCF (Orr and Mackintosh 1988). A similar experience occurred with Père David's deer farmed in Scotland (Reid et al. 1987). Hybrids between Père David's and red deer appear to be more resistant, with only one of seven F1 hybrids and none of about 20 3/4 red X 1/4 PD hybrids dying of MCF so far.

Rusa and sika deer are also more susceptible to MCF than red deer. In Northland, where MCF occurs only rarely in red deer, a Moluccan rusa deer herd of 20 loses two or three animals annually (Thorley 1990). Similarly, sika deer farms have experienced high mortality due to

MCF, especially in the early days of farming, with losses of up to 29% (Anonymous 1980; Hunter 1981). Under Australian conditions sika, chital (*Axis axis*), rusa and white-tailed deer appear to be the most susceptible.

At the other end of the scale it appears that fallow deer are highly resistant to MCF, and despite there being 80 000 to 90 000 fallow deer farmed, there have not been any confirmed cases reported in New Zealand. Wapiti contract MCF but have a lower incidence rate than red deer (C. Mackintosh, unpublished data) and appear to be less susceptible to experimental infection than red deer (R. Oliver, personal communication).

Yersiniosis

Enteric yersiniosis is characterised by a watery green, often blood-stained, diarrhea, and it affects young deer in their first winter. It is caused by the bacterium *Yersinia pseudotuberculosis* and is predisposed to by stressors, especially transport, underfeeding, and bad weather (Mackintosh and Henderson 1984). It appears that almost all young deer are exposed to infection, but with good management they develop only subclinical infection. However, under certain conditions, outbreaks occur and can affect 5% to 40% of a group. Overall around 1% of young red, wapiti and wapiti X red hybrid deer die annually (Mackintosh 1988). Yersiniosis is very uncommon in fallow deer and is rarely reported in other species, probably due to the small numbers farmed.

Sporidesmin Toxicity ("Facial Eczema")

Sporidesmin toxicity is a disease of ruminants which occurs mainly in the warmer northern half of the North Island. It is caused by a toxin, sporidesmin, produced by a fungus *Pithomyces chartarum* which grows on dead plant litter in the pasture especially in warm, humid periods in late summer and autumn (Mortimer 1984). The ingestion of sporidesmin results in severe liver damage often leading to the build-up of the chlorophyll breakdown product, phylloerythrin, in the blood. This is a photodynamic compound and may cause photosensitivity—thus the name "facial eczema," because the lesions are most noticeable on thinly haired areas, especially the face. However, many animals suffer liver damage without any sign of photosensitivity. Elevation of the serum level of the liver enzyme gamma-glutyltransferase (GGT) is a good indicator of the level of damage sustained by the liver. Dosing trials with sporidesmin have shown that red deer are more resistant, but fallow are more susceptible than sheep (Mortimer and Smith 1981; Mortimer 1984). A few isolated cases have occurred naturally in red deer, but serious disease is not uncommon in fallow deer, and the mortality in the natural disease can be high.

"Ryegrass Staggers"

The majority of deer in New Zealand are grazed on perennial ryegrass (*Lolium perens*) and clover pastures. Many of the ryegrass cultivars are infected with an endophyte (*Acremonium* sp.) which produces a number of toxins (Lolitrems) especially in the seed head and the base of the plant, and these reach their highest concentrations during hot, dry summers (Mortimer et al. 1984). Ingestion of these toxins leads to "ryegrass staggers" which is characterized by head tremor, muscle tremors, and unsteady gait. The signs become worse if the animal is made to move quickly, and severely affected animals may fall over on their sides and paddle. All domestic livestock, including deer, can be affected and it is much more common in wapiti than in red deer (Mackintosh et al. 1982; Brooks and Cahill 1985; Bringans 1986; Brooks 1986). At Invermay Agricultural Centre, on a number of occasions wapiti have developed ryegrass staggers when grazing alongside red deer that have been apparently unaffected, suggesting that wapiti are more susceptible to the toxins or that their grazing habits may result in the ingestion of larger amounts of toxins. There are no published reports of ryegrass staggers in fallow, rusa, or sika deer.

Copper Deficiency and Enzootic Ataxia

Wapiti and wapiti X red hybrids appear to be more prone to copper deficiency and enzootic ataxia than red deer. Signs of copper deficiency in deer include loss of condition and ill-thrift, dry dull coats, enlarged limb joints, and brittle bones. Enzootic ataxia, due to demyelination of the spinal cord, occurs in a proportion of copper-deficient deer. This was first reported in New Zealand in a red deer (Wilson et al. 1979), but subsequently wapiti have been shown to be more likely to be affected (Mackintosh et al. 1986a). In a group of 55 2-year-old wapiti, red, and wapiti X red hybrids grazing together as a group at Invermay, five of the wapiti animals developed enzootic ataxia and none of the others were clinically affected. Samples showed red deer had the highest and wapiti the lowest liver copper levels. Copper deficiency problems have not been reported in other deer species farmed in New Zealand.

Liver copper concentrations of less than 100 μmol/kg (wet weight) indicate copper deficiency in red deer, and deer with levels below 60 μmol/kg are likely to develop enzootic ataxia (Mackintosh et al. 1986b).

Nematode Parasites

There is a dearth of published information on the relative susceptibilities of the various breeds of deer to parasites. However, the lungworm (*Dictyocaulus viviparus*) is generally regarded as the most important nematode parasite of young red deer in New Zealand and Britain (Charleston 1980; Fletcher 1982; Watson and Charleston 1985). Although young fallow deer can sometimes develop heavy lungworm burdens under warm wet autumn conditions, they appear to be more resistant than young red deer (G. Asher, personal communication). In Denmark, problems are experienced with lungworm in red deer calves in the autumn and gastrointestinal parasites in fallow deer calves in the late winter (Jorgensen and Vigh-Larsen 1988).

In our experience at Invermay, Canadian wapiti also appear more susceptible to lungworm and take longer to develop age or exposure-related resistance. Watson (1986) reported that wapiti X red hybrids captured from Fiordland, New Zealand, were more susceptible to lungworm than red deer. These hybrids continued to shed fecal larvae throughout the year, and one 5-year-old animal had a heavy lungworm burden (> 1200) at necropsy. In wild populations of Rocky Mountain elk (*C. elaphus nelsoni*), the lungworm is a serious pathogen whereas gastrointestinal parasites only rarely reach pathogenic levels (Worley 1979).

Gastrointestinal parasites are regarded as of little importance in farmed red deer except for recently captured, aged, or debilitated animals. However, there have been a number of reports of abomasal parasites causing clinical disease in adult Canadian wapiti and wapiti X red hybrids which is often associated with elevated pepsinogen levels (Mason 1984; Bringans 1986; Pearse 1988; C.G. Mackintosh, unpublished data).

Preputial Ulceration and Prolapse

Three cases of severe preputial prolapse and swelling, which necessitated surgical removal, occurred in a group of ten Canadian wapiti bulls at Invermay over a 2-year period. All three cases occurred just prior to or early in the rut (Mackintosh 1990). At least two other cases have occurred in wapiti on other farms (N.S. Beatson and L. Hodgkinson, personal communication). There have been no reported cases of this condition in other species of deer in New Zealand. The prolapses were preceded by chronic ulceration and eversion of the preputial lining. In cattle and sheep, preputial ulceration is associated with high protein diets and is caused by ammonia which is produced by urease-forming micro-organisms from urine. The etiology of this condition in wapiti is probably similar and may also relate to physiological changes occurring in males at that time of year. Why it should occur so frequently in wapiti and not in red deer grazing the same pastures is not known.

Dietary Scour

Experience in New Zealand and Canada has shown that Canadian wapiti are prone to a "dietary scour" when grazed on lush pastures in spring and autumn. There is some suggestion that wapiti require more roughage than red deer, and the provision of lucerne (or alfalfa) hay at these times of year can alleviate this problem (Bringans 1986, 1987).

Preliminary investigations at Invermay also suggest that some food-related factor, possibly a fungal mycotoxin, may also be involved in this "dietary scour" because elevated serum enzyme levels indicative of liver damage often accompany this condition. If animals are removed from pasture and fed on hay as soon as signs of diarrhea appear, they usually recover quickly. Red deer grazed side-by-side with wapiti under these conditions have been unaffected (C. Mackintosh, unpublished data).

"Fading Elk Syndrome"

This syndrome is peculiar to Canadian wapiti (elk) and wapiti hybrids, and unless treated early it is usually fatal. It is characterized by progressive weight loss, low serum albumin levels, and may be with or without scour (Orr et al. 1990). It has been seen in New Zealand (Bringans 1987; Fennessy et al. 1988) and Canada (M. Bringans, personal communication), and it is also called "chronic ill-thrift in elk" and, erroneously, "wasting elk disease." It is definitely *not* the same etiology as the chronic wasting disease described in North American elk which is a type of spongiform encephalopathy and is caused by a scrapie-like agent (Williams and Young 1982). The etiology is complex and may involve "dietary scour," copper deficiency and parasitism. It is also not associated with Johnes disease.

Discussion

Disease is the end result of an interaction between an infectious agent (or agents), the environment, and the host animal (Mackintosh and Beatson 1985). In the farming environment, stress plays an especially important part in the equation (Griffin 1989). Deer are still in the process of domestication, and stressors such as handling, increased stocking density, increased social interaction, reduced vegetative cover, contact with other domestic livestock, and exposure to various environmental influences contribute to increasing their disease susceptibility. Some breeds cope with these stressors better than others. White-tailed and Père David's deer appear more "highly strung" than red deer, which may contribute to their increased susceptibility to MCF; however, fallow deer are also very flightly, but are highly resistant to this disease. Therefore, other factors, such as their exposure to disease agents during their evolution and selection for resistance, have probably been important. For example, Père David's, white-tailed, rusa, and sika deer, evolved in parts of China, North America, southeast Asia, and Japan, respectively, where they would not have had exposure to domestic sheep or antelope and, therefore, have had no selection pressure for resistance to MCF. It is likely that fallow and red deer have had more exposure to sheep during their evolution in the Middle East and Europe and consequently have developed some resistance to MCF.

Pasture-based farms provide a relatively restricted range of environmental conditions. Animals that naturally depend heavily on browse are probably more likely to develop certain deficiencies or have increased susceptibility to diseases associated with a pasture-based diet. For example, the susceptibility of wapiti to ryegrass staggers, copper deficiency, and the "fading-elk syndrome" probably relates to their requirements for higher levels of roughage or browse in their diet (Hofman 1985). A naturally higher level of grazing in the wild may have helped red deer to become more resistant to fungal endophytes and parasites. Hybridization between red deer and Père David's, wapiti, and sika deer may produce animals that adapt better to a farming environment than purebred deer of these latter species.

In the past, veterinarians and farmers in New Zealand have tended to assume that all farmed deer, irrespective of species, have similar susceptibilities to various diseases. The diagnostic laboratories have also tended to collect data on deer as a whole rather than by species. However, the realization that farmed deer of different species and subspecies respond differently to various diseases will help improve the diagnosis, treatment, and prevention of these conditions in deer. This knowledge may also assist new farmers to choose the best type of deer to suit their particular farming environment.

References

Anonymous (1980) Malignant catarrhal fever outbreaks in sika deer. Surveillance 7:13

Bringans M (1986) Post-quarantine experiences with imported elk. Proc Deer Branch N Z Vet Assoc Course 3, pp 183–189

Bringans M (1987) Wapiti health. Deer Farmer 40:43–44

Brooks HV (1986) Ryegrass staggers in Canadian wapiti. Surveillance 13:16

Brooks HV, Cahill JJ (1985) The susceptibility of Canadian wapiti to ryegrass staggers. N Z Vet J 33:126

Challies CN (1985) Establishment, control and commercial exploitation of wild deer in New Zealand. In Fennessy PF, Drew KR (eds) Biology of Deer Production. R Soc N Z, Wellington, Bull 22:21–36

Charleston WAG (1980) Lungworm and lice of the red deer (*Cervus elaphus*) and the fallow deer (*Dama dama*), a review. N Z Vet J 28:150–153

de Lisle GW, Havill PF (1985) Mycobacteria isolated from deer in New Zealand from 1970–1983. N Z Vet J 33:138–140

Fennessy PF (1988) Malignant catarrhal fever: workshop report. Proc Deer Branch N Z Vet Assoc Course 5, pp 186–195

Fennessy PF, Beatson NS, Bringans MJ (1988) Reproduction workshops. Proc Deer Branch N Z Vet Assoc Course 5, pp 216–219

Fletcher TJ (1982) Management problems and disease in farmed deer. Vet Rec 111:219–223

Griffin JFT (1989) Stress and immunity: a unifying concept. Vet Immunol Immunopathol 20: 263–312

Hofman RR (1985) Digestive physiology of the deer—their morphophysiological specialisation and adaptation. In Fennessy PF, Drew KR (eds) Biology of Deer Production. R Soc N Z, Wellington, Bull 22:393–407

Hunter JW (1981) Malignant catarrhal fever—Taupo area. Proc of a Deer Seminar for Veterinarians. Deer Advisory Panel, N Z Vet Assoc, Queenstown, pp 121–124

Jorgensen RJ, Vigh-Larsen F (1988) Epidemiology and control of parasitic diseases in Danish deer farms. In Reid HW (ed) The Management and Health of Farmed Deer. Kluwer Academic, Dordrecht, pp 23–25

Mackintosh CG (1988) Yersiniosis: workshop report. Proc Deer Branch N Z Vet Assoc Course 5, pp 178–185

Mackintosh CG (1990) Preputial prolapse in North American wapiti farmed in New Zealand. Second Int Wild Ranching Symp, Edmonton, June 1990

Mackintosh CG, Beatson NS (1985) Relationships between diseases of deer and those of other animals. In Fennessy PF, Drew KR (eds) Biology of Deer Production. R Soc N Z, Wellington, Bull 22:77–82

Mackintosh CG, Henderson TG (1984) The epidemiology of yersiniosis in deer. Proc Deer Branch N Z Vet Assoc Course 1, pp 34–42

Mackintosh CG, Orr MB, Gallagher RT, Harvey IC (1982) Ryegrass staggers in Canadian wapiti deer. N Z Vet J 30;106–107

Mackintosh CG, Orr MB, Turner K (1986a) Enzootic ataxia in wapiti. Proc Deer Branch N Z Vet Assoc Course 3, pp 165–169

Mackintosh CG, Wilson PR, Beatson NS, Turner K, Johnstone P (1986b) Preliminary report of the liver: serum copper relationship in red deer. Proc Deer Branch N Z Vet Assoc Course 3, pp 156–164

Mason PC (1984) Gastrointestinal nematodes in farmed deer. Surveillance 11:24–25

Mortimer PH (1984) Facial eczema in red and fallow deer. Proc Deer Branch N Z Vet Assoc Course 1, pp 59–63

Mortimer PH, Smith BL (1981) Facial eczema in deer. Proc Ruakura Farmers Conf 1981, pp 109–112

Mortimer PH, Young PW, di Menna ME (1984) Perennial ryegrass staggers research—an overview. Proc N Z Soc Animal Prod 44:181–184

Orr MB, Mackintosh CG (1988) An outbreak of malignant catarrhal fever in Pére David's deer (*Elaphurus davidianus*). N Z Vet J 36:19–21

Orr MB, Mackintosh CG, Walker G (1990) Chronic ill thrift in elk—a review of 1988 Invermay Animal Health laboratory cases. Surveillance 17:11–12

Pearse AJ (1988) Wapiti and hybrids—special management needs. Proc Deer Branch N Z Vet Assoc Course 5, pp 164–177

Reid HW, Buxton D, McKelvey WAC, Milne JA, Appleyard WT (1987) Malignant catarrhal fever in Père David's deer. Vet Rec 121:276–277

Reid HW, Buxton D, Pow I, Finlayson J (1989) Isolation and characterisation of lymphoblastoid cells from cattle and deer affected with "sheep-associated" malignant catarrhal fever. Res Vet Sci 47:90–96

Saio-Kun W, Castra AE, Heuschele WP, Ramsay EC (1988) Enzyme-linked immunosorbent assay for the detection of antibodies to the alcelaphine herpesvirus of malignant catarrhal fever in exotic ruminants. Am J Vet Res 49:164–168

Thorley A (1990) The Moluccan rusa dilemma. Deer Farmer 66:31–34

Watson TG (1986) Efficacy of drenching red deer and wapiti with particular reference to (*Elaphostrongylus cervi*) and (*Dictyocaulus viviparus*). Proc Deer Branch N Z Vet Assoc Course 3, pp 170–182

Watson TG, Charleston WAG (1985) The significance of parasites in farmed deer. In Fennessy PF, Drew KR (eds) Biology of Deer Production. R Soc N Z, Wellington, Bull 22:105–117

Williams ES, Young S (1982) Spongiform encephalopathy of Rocky Mountain elk. J Wild Dis 18:465–471

Wilson PR, Orr MB, Key EL (1979) Enzootic ataxia in red deer. N Z Vet J 27:252–254

Worley DE (1979) Parasites and parasitic diseases of elk in the Northern Rocky Mountain Region: a review. In Boyce MS, Haden-Wing LD (eds) North American Elk: Economics, Behaviour and Management, Univ Wyoming, Laramie, WY USA pp 206–211

30
The Impact of Domestication on Red Deer Immunity and Disease Resistance

J. Frank T. Griffin, Anthony J. Thomson, John P. Cross, Glenn S. Buchan and Colin G. Mackintosh

Abstract

Domestication of red deer (*Cervus elaphus*) in New Zealand has caused changes in patterns of infectious diseases such as tuberculosis (*M. bovis*), malignant catarrhal fever (MCF), and yersiniosis (*Y. pseudotuberculosis*). Among the factors known to influence disease susceptibility and immunity are: capture from the wild and adaptation to farm conditions, transport, handling, and weaning. A series of experiments has examined the influence of some of the above stressors on immune function in farmed deer. Hematologic parameters, plasma proteins, nonspecific lymphocyte mitogenesis (using concanavalin A), and specific humoral (using ELISA) and cellular immune lymphocyte transformation have been measured in stressed immunized deer. Transport, handling, and nutritional deprivation have an acute but transient impact on immune function, which may be expressed as enhancement or suppression of immunity. Transport and fasting increase the susceptibility of deer to experimental infection with *Y. pseudotuberculosis*. Behavioral responses may not give an indication of the negative impact of management stress on immunocompetence. The present data show that aggressive animals, which appear to be adversely affected by restraint, have normal or enhanced levels of immune reactivity.

Key words: Domestication, disease resistance, farmed deer, immunity, red deer

Introduction

For centuries it has been recognized that nature resists change. Claude Bernard (1878) identified the physiologic changes that occur in animals exposed to a changing environment. He identified a steady state within the animals' interior (milieu interieur), characterized by its constancy, and the external environment typified by its variability. Cannon (1929) described the "flight-fight" reflex of animals exposed to extreme changes in the environment (stress) and identified pathology associated with extreme levels of stress. Selye (1946) described the "general adaptation syndrome" (GAS) as the response of animals exposed to stress. These early workers considered the central physiological response to environmental change was directed through adrenal function with increased levels of corticosteroids causing altered physiologic responses within the host. More recent studies (Mason 1968; Dantzer and Mormede 1985) have demonstrated that more complex endocrine responses are produced following exposure to environmental stress, which may vary depending on the type of stress or species of host affected. Selye (1973) recognized that all forms of stress need not produce pathology within the host. Stress which had an adverse affect was defined as distress, while it caused eustress when the animal was not adversely affected.

Domestication of wildlife represents a unique and extreme example of environmental stress. During domestication, animals are exposed to a rapidly changing environment which imposes well-recognized stressors on the animal. Physical stressors include capture, restraint, transport, exposure to novel sounds, odors, and climatic extremes. Psychological stressors may involve hierarchical dislocation, overcrowding or isolation, and exposure to unfamiliar surroundings (Ewbank 1985; Griffin 1989).

Materials and Methods

In the present study, two stress paradigms have been used to evaluate the impact of management stress on immunologic and inflammatory parameters and disease susceptibility patterns in deer exposed to typical stressors found in intensified management of farmed deer. In the first study, young deer (3 to 5 months) were fasted for 24 h, transported over 30 km, and relocated in a new farm setting. Animals were then challenged orally with 3×10^{10} live *Yersinia pseudotuberculosis* to evaluate resistance to experimental infection. Animals were maintained on a limited nutritional plane for 24h following experimental challenge. All animals were monitored closely for clinical evidence of yersiniosis as seen by scouring or changed behavior. When evidence of clinical disease was present, individual animals were treated with antibiotics and electrolytes. Immunologic function in this group of animals was evaluated by immunizing the animals with a novel antigen, keyhole limpet hemocyanin (KLH), either 28 days prior to the imposition of stress or 1 day following completion of the stress regime. Blood samples were taken from the animals at weekly intervals from the outset. Specific antibody (Ab) titers were measured using an ELISA assay (Hibma and Griffin 1988) and cellular immune function was measured using nonspecific mitogens (concanavalin (Con) A) by in vitro culture. In all, 30 animals were stressed, and 30 nonstressed control animals were challenged experimentally with infection. Specific immune function and Con A reactivity was measured using two groups of ten animals either immunized 28 days prior to stress or challenged 1 day following stress.

In a second study, groups of four animals were studied to monitor the effect of restraint on inflammatory cells and imunologic function. The animals were restrained in a standard crush, or exposed to differing doses of ACTH, to evoke levels of cortisol as a probe for steroid-related changes in hematological and immunologic function. Two control groups were used; one of which were well-adapted "placid" animals, the other of which were "aggressive" (needle shy) animals which had developed a fractious response to earlier handling. Circulating neutrophil levels in peripheral blood were assessed at 30 min, 1, 4 and 24 h following restraint or injection of ACTH. These were measured using automated (HC6000-Technicon) counting of leukocytes using techniques adapted for studies in deer (Cross et al. 1988). In addition, immune function, as measured by lymphocyte transformation using specific antigen (KLH), was evaluated over 6 weeks from the time of initial restraint or injection with ACTH. All animals were immunized with 500 μg of KLH in Freund's complete adjuvant (FCA) by subcutaneous injection.

Results

Deer showed increased susceptibility to experimental infection following exposure to transport and nutritional stress (Table 30.1). Control animals showed a 30% incidence of clinical disease and no fatalaties following challenge with these high doses of live *Y. pseudotuberculosis*. By contrast, 60% of stressed animals developed clinical disease, and 13% died.

Animals immunized 4 weeks prior to stress showed a classic antibody response, with significant and optimal antibody (Ab) titers developing 14 and 21 days following immunization (Fig. 30.1). These titers had waned by 35 days following

TABLE 30.1. Increased susceptibility to experimental infection[a] in deer exposed to transport and nutritional stress

Animal	Status Clinical disease	Death
Stressed	18/30 (60%)	4/30 (13%)
Normal control	9/30 (30%)	0/30 (0%)

[a]Animals were challenged orally with 3×10^{10} live Y. pseudotuberculosis organisms following transport, nutritional deprivation, and relocation.

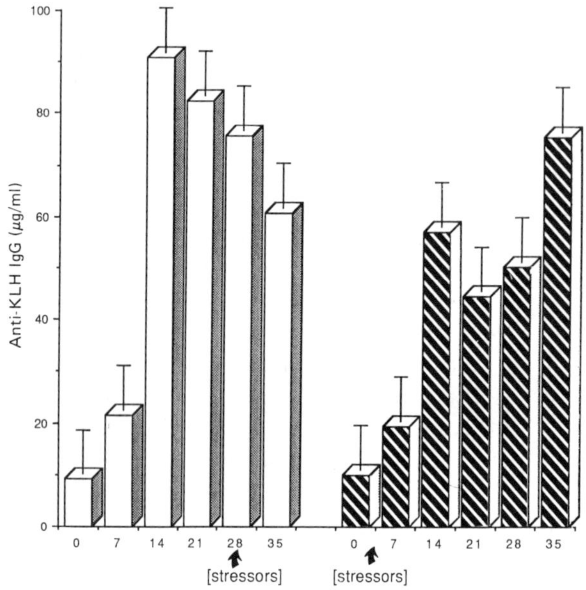

FIGURE 30.1. Levels of Immunoglobulin G specific for KLH in groups of red deer exposed to transport, nutritional, and climatic stress, immunized 28 days prior to or 1 day following exposure to stress. Results are expressed as mean + SE.

immunization. By contrast animals immunized 1 day after the imposition of transport and nutritional stress showed marked alteration in immunologic activity. Responses at 14 to 21 days following immunization were significantly lower than in the control group, and a bimodal response was seen with reduced antibody peaks evident at day 14 and day 35 following immunization.

Nonspecific cellular immunity in animals sampled at 7 days and 1 day prior to stress and at 1, 7, and 14 days following stress was evaluated. The results show a significant fall in lymphocyte reactivity 1 day following stress which decreased even further in the week following stress; there was only 30% of prestress reactivity at 7 days following stress, with recovery in the ensuing week (day 14; Fig. 30.2).

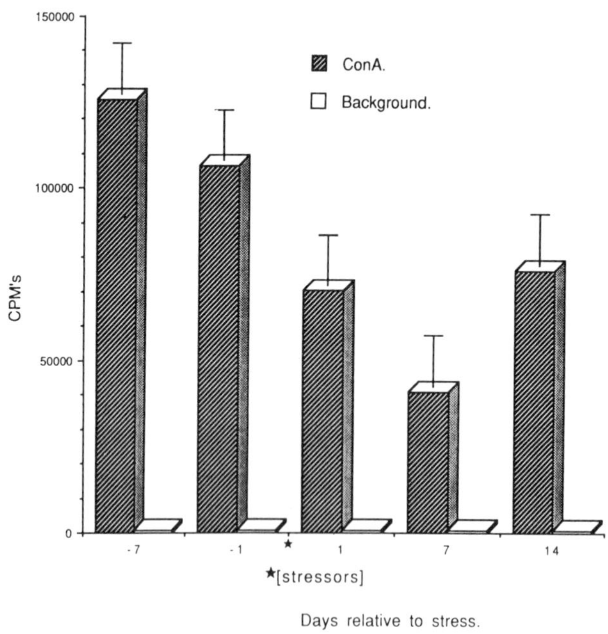

FIGURE 30.2. Concanavalin- (Con-) A induced lymphocyte transformation in a group of 50 red deer sampled prior to and following exposure to transport, nutritional, and climatic stress. Unstimulated background control values are also given. Results are expressed as mean + SE.

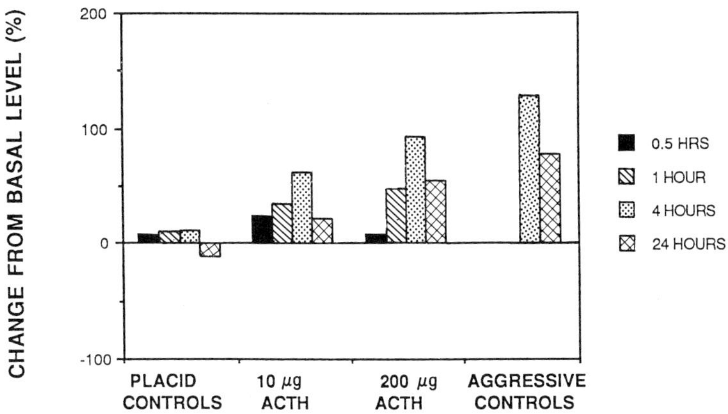

FIGURE 30.3. Neutrophil levels measured at 0.5 h, 1 h, 4 h, and 24 h after exposure to ACTH injection.

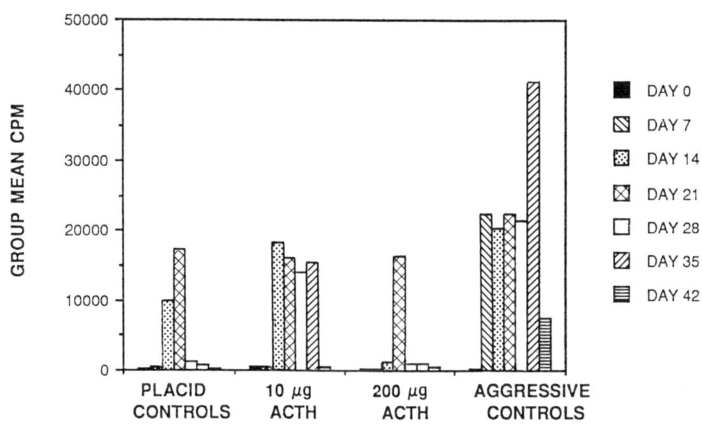

FIGURE 30.4. Lymphocyte transformation to KLH in animals exposed to ACTH simultaneously to immunization with KLH. Nonimmunized controls were segregated into "placid" or "aggressive" groups depending on their response to handling and restraint.

The results in Fig. 3 show that ACTH, used to induce steroid levels similar to those found in animals exposed to acute stress, had a significant impact on neutrophil levels compared with control animals subjected to physical restraint. "Placid" control animals had no change in neutrophil levels over a 24-h period following restraint, while "aggressive" animals showed a significant increase in neutrophil levels at 4 and 24 h following restraint. Animals treated with a low dose (10 µg) or a high dose (200 µg) of ACTH showed increasing levels of neutrophil leukocytosis with increasing dosage of ACTH. Animals treated with 200 µg of ACTH had steroid levels (50 to 120 ng/ml) similar to those found in the "aggressive" controls. Animals treated with 10 µg ACTH had steroid levels ranging from 20 to 60 ng/ml.

Antigen-specific lymphocyte transformation to KLH in the different groups shows that animals treated with 200 µg of ACTH had the lowest response to KLH, with moderate levels of cellular reactivity being evident only at 14 days following immunization. The "placid" control group showed moderate levels of activity at day 7 and day 14 following immunization. Animals treated with 10 µg ACTH had moderate levels of activity from 14 to 35 days following immunization. The response in the "aggressive" control group contrasted markedly with other groups studied. These animals showed significant specific cell reactions within 7 days of immunization, which remained unchanged for 21 days and peaked at 33 days, at very high levels.

Conclusions

The data presented here showed that management stress involving transport, fasting, or physical restraint can each impact on immunologic function or inflammatory activity in farmed deer. Transport and nutritional deprivation caused a significant increase in susceptibility to clinical disease and death following experimental challenge with *Y. pseudotuberculosis*. Specific antibody production was also influenced by transport and nutrition where animals exposed to stress immediately prior to immunization showed reduced patterns of antibody production, by comparison with animals immunized 28 days prior to stress. The overall recovery of antibody production at 35 days following immunization would suggest that these stressors do not produce blanket impairment of antibody production but rather altered patterns of reactivity. Nonspecific mitogenic activity with Con A was also significantly influenced by transport and nutritional stress with a significant reduction in reactivity evident in the week following stress.

The response in the restraint experiment was interesting in that it demonstrated a reduction in immunologic activity in animals exposed to high doses of ACTH, sufficient to induce high levels of plasma steroids. The pattern of reactivity in these animals was marginally lower than that found in normal "placid" controls. By contrast low dose ACTH, which induced intermediate levels of plasma steroids, appeared to potentiate the immune response with significant reactivity being evident from 7 to 35 days following immunization and injection with low dose (10 μg) ACTH. These data would infer that moderate levels of steroids do not inhibit immunologic reactivity and may cause some level of enhancement. The striking observation from this experiment was the extremely high levels of immunologic reactivity in "aggressive" animals subjected to restraint prior to immunization. Such animals showed significantly elevated immune cell reactivity from 7 to 42 days following immunization. The time required to develop immune reactivity was shorter and the overall levels of immune reactivity were significantly higher in this group than in "placid" controls. These results suggest that the physiologic response in "aggressive" animals under restraint, though producing high levels of steroids and neutrophil leukocytosis, had no inhibitory effect on immunologic function, and may in fact have caused enhancement. This argues against steroids per se being the key factor in immunosuppression when animals are subjected to acute stress.

The results infer that acute stress may have little impact on immunologic function and even cause enhancement of function in animals adapted to cope with the acute stress. By contrast, transport and nutritional stress applied over a 48-h period cause a significant reduction in immunocompetence. This form of severe acute stress also increased susceptibility of animals to infection following oral challenge with live *Y. pseudotuberculosis*.

These findings suggest that care must be taken in the design of experiments to elucidate the relationship between stress and behavior with immunocompetence or disease susceptibility. One could argue that in fact the only normal response being produced was the response of the "aggressive" animal exposed to restraint, while 'placid" animals, though appearing behaviorly normal, may, in fact, have produced a response to restraint which impaired their immunocompetence. It further highlights the need for design of sampling and experimental protocols to ensure that true control values can be obtained in studies involving deer. Data from our laboratory (Cross et al. 1988) demonstrate that animal handling can cause acute and precipitous change in the inflammatory profile and produce inaccurate values in studies designed to establish reference data for wild animals or normal farmed animals. Splenic contraction and production of steroids which affect margination and redistribution of circulating leukocytes dramatically alter the hematologic profile of animals exposed to adversive stimuli such as transport or restraint.

Independent studies have evaluated the impact of capture, weaning and nutritional deprivation on hematological values and immunity in red deer hinds and weaners (Griffin et al. 1990). Capture causes a significant depression in Con A induced lymphocyte reactivity, which is greatest at 7 days postcapture and which recovers gradually to normal levels over a 6-week period. Simultaneously, there is a significant neutrophil leukocytosis, which is resolved by 1 month postcapture.

Weaning appears to have little direct impact on hematological or immunological reactivity.

However, a gradual decrease in Con A mitogenesis and lymphocyte transformation with antigen (tuberculin) was seen in animals subjected to nutritional deprivation in the month following weaning. This was paralleled by a leukopenia and anemia in nutritionally deprived weaned animals (Griffin et al. 1990).

From these studies it is evident that transport, restraint, and nutritional stress all can impact directly on immunocompetence, while weaning appears to have little impact on immunity or hematology.

References

Bernard C (1878) Les phenomenes de la vie. Bailliere, Paris

Cannon WB (1929) Bodily Changes in Pain, Hunger, Fear and Rage: An account into the Function of Emotional Excitement. New York, Appleton

Cross JP, Mackintosh CG, Griffin JFT (1988) The effect of physical restraint and xylazine sedation on hematological values in red deer (*Cervus elaphus*). Res Vet Sci 45:281–286

Dantzer R, Mormede P (1985) Stress in domestic animals: a psychoneuroendocrine approach. In Moberg GP (ed) Animal Stress. Am Physiol Soc, Waverley Press, Bethesda, Md., pp 81–95

Ewbank R (1985) Behavioral responses to stress in farm animals. In Moberg GP (ed) Animal Stress. Am Physiol Soc, Waverley Press, Bethesda, Md., pp 71–80

Franzmann AW, Flynn A, Arneson PD (1975) Serum corticoid levels relative to handling stress in Alaskan moose. Can J Zool, 53:1424–1426

Griffin JFT (1987) Stress and disease in farmed deer. Vet Deer Soc 2:3–9

Griffin JFT (1989) Stress and immunity: a unifying concept. Vet Immunol Immunopathol 20:263–312

Griffin JFT, Mackintosh CG, Cross JP, Buchan GS (1990) Influence of management stress on the immune profile and disease resistance of farmed deer. Proc Second Int Wildlife Symp, Edmonton, Canada

Gwazdauskas FC, Gross WB, Bibb TL, McGillard ML (1978) Antibody titers and plasma glucocorticoid concentrations near weaning in steers and heifer calves. Can Vet J 19:150–154

Hibma M, Griffin JFT (1988) Optimization of antibody production and immunization schedules in farmed deer. Proc Deer Branch NZ Vet Assoc 5, pp 105–110

Mason JW (1968) "Over-all" hormonal balance as a key to endocrine function. Psychosom Med 30: 791–808

Selye H (1946) The general adaptive syndrome and diseases of adaptation. J Clin Endocrinol 6:117–230

Selye H (1973) The evolution of the stress concept. Am Scientist 61:692–699

31

A Review of Yersiniosis in Farmed Red Deer in New Zealand

Colin G. Mackintosh

Abstract

Yersiniosis, caused by *Yersinia pseudotuberculosis*, commonly affects young farmed red deer (*Cervus elaphus*) in New Zealand. It is characterised by diarrhea, dehydration and sudden death. Lesions include severe focal enteritis with patchy ulceration and edema of the intestinal mucosa and submucosa and mesenteric lymphadenitis. Treatment with tetracyclines and fluid is usually successful if given early in the disease. Most young farmed deer experience subclinical infection in their first winter; however, concurrent stress may precipitate clinical disease in up to 40% of a group. Stressors may include inclement weather, under-feeding, and transportation. The organism is carried by a wide range of wildlife and domestic animals, and it persists in the environment, especially in cold wet conditions. Good husbandry will help prevent the disease. Natural immunity develops after initial exposure. A vaccine is being developed.

Key words: Epidemiology Yersinia pseudotuberculosis, *New Zealand, red deer, yersiniosis*

Introduction

The first recorded outbreak of yersiniosis in farmed red deer occurred in 1978 (Beatson and Hutton 1981). Since then yersiniosis has become one of the most common causes of death in young farmed red deer in New Zealand (Beatson 1984; Mackintosh and Henderson 1984a, b; Wilson 1984; Mackintosh 1990). The disease is caused by the bacterium *Yersinia pseudotuberculosis* (*Y. pstb*) serotypes I, II, and III. A related organism, *Y. enterocolitica* may occasionally cause this disease, but it is usually part of the normal intestinal flora (Henderson 1983). Clinically yersiniosis is characterized by a green or brown watery diarrhea which often contains blood, and staining may be seen around the anus, or on the tail, perineum, or hocks, but is often not very obvious. Affected animals have elevated temperatures (> 40° C) and they soon become dehydrated. The first cases are often found dead, and close inspection of the in-contact animals reveals animals with diarrhea. Outbreaks of the disease can affect up to 40% of a group, although usually 5% to 20% are affected.

Yersiniosis is primarily a disease of young red deer calves in their first winter when they are 5 to 9 months old, with a peak of clinical cases in June, July, and August (Mackintosh and Henderson 1984a). Sporadic cases occur throughout the year, but these usually involve recently captured or debilitated animals.

Subclinical Infection

It is believed that virtually all red deer are exposed to *Y. pstb* in the farming environment, and if they are fed and managed well the majority experience a subclinical infection. At Invermay, near Dunedin in the South Island, New Zealand, a group of 38 weaned, red deer calves were monitored from autumn to late spring with fecal and blood samples taken at 3 to 4 weekly intervals. It was found that 11% to 20% of calves yielded *Y. pstb* isolates on each sampling occasion, between May and August. A total of 20 calves (53%) had positive fecal cultures on at least one occasion and 37 (97%) seroconverted during this period. None of the animals developed clinical signs of disease. A survey of fecal samples from 527 healthy deer calves in the North Island of New Zealand in late June and July yielded a 9% isolation rate (Hodges et al. 1984b), which is similar to the periodic point prevalences found at Invermay (11% to 20%). This contrasts with the 1% isolation rate obtained from sampling adult stags at slaughter (Mackintosh and Henderson 1985). These findings, together with the very low incidence of disease in adult red deer, suggest that young animals develop resistance to infection subsequent to exposure in their first winter.

Pathogenesis

The development of clinical disease in young deer is precipitated by various stressors. These may include underfeeding, transportation, exposure to adverse weather, lack of shelter, rapid changes in diet, high stocking densities, social stresses, and capture from the wild. It is believed that these stressors may affect the localization and multiplication of *Y. pstb* organisms in the gut and may suppress humoral and cell-mediated immunity (Mackintosh and Henderson 1984a; Griffin 1989).

Potential Reservoirs of Infection

Infected animals shed *Y. pstb* in their feces, and transmission appears to be via contaminated food and water. *Y. pstb* infection is endemic in some species of animals which act as long-term reservoirs and excretors, whereas others are accidental hosts and short-term excretors. In the northern hemisphere, rodents, lagomorphs, and various birds are regarded as the principal reservoirs of infection (Mair 1968, 1973, 1975; Hubbert 1972; Obwolo 1976; Stovell 1979). Investigations of New Zealand wildlife showed isolation rates of 27.8% in feral cats (*Felis domesticus*), 8.6% in Norway rats (*Rattus norvegicus*), 5.5% in mice (*Mus musculus*), 3.8% in hares (*Lepus europeaus*), 1.9% in rabbits (*Oryctolagus cuniculus*), 5.3% in ducks (*Anas platyrhynchos*), 2.3% in sparrows (*Passer domesticus*), 2.3% in seagulls (*Larus dominicanus*), and 1.7% in starlings (*Sturnus vulgaris*) (Mackintosh and Henderson 1984b). It is not known to what extent domestic animals act as reservoirs of *Y. pstb*, but in New Zealand it has also been isolated on many occasions from domestic livestock which have died or shown signs of disease, especially diarrhea (Hodges et al. 1984a). These included isolates from 56 cattle, 8 sheep, 13 goats, 7 pigs, 6 rabbits, 5 guinea pigs, and 22 aviary species of birds, in addition to 117 deer. *Y. pstb* has also been isolated from apparently healthy domestic animals including cattle (Hodges et al. 1984b) and pigs (C.G. Mackintosh, unpublished data).

Y. pstb organisms can survive for long periods in the environment in cold, wet winter conditions, whereas they do not persist in hot, dry environments (Borg 1968). Environmental surveys in France (Barre et al. 1979) and New Zealand (C.G. Mackintosh and T.G. Henderson, unpublished data) have yielded *Y. pstb* isolates from pasture and water samples in winter months. This property of cold tolerance is utilized in the laboratory to enhance the recovery of organisms from samples that are subcultured into media and kept at 4° C for up to 3 weeks, during which time *Y. pstb* multiplies, and most other bacteria remain dormant or die.

Pathology

Enteric yersiniosis causes profound damage to the gastrointestinal tract in deer. Usually the ileum, cecum, and colon are most severely affected, and occasionally the changes extend to the upper small intestine and abomasum. On necropsy, there is usually severe reddening of the intestinal mucosa with blood-stained contents in the lumen, and sometimes focal ulceration with pseudomembranes. Mesenteric lymph nodes are often swollen and edematous. The abomasal wall may be reddened, and there may be petechial hemorrhages. Histopathological changes in the intestine can include acute enteritis with patchy ulceration and edema of the mucosa and submucosa (Beatson and Hutton 1981). This may be accompanied by mesenteric lymphadenitis with focal necrosis. This profound damage to the mucosa causes dehydration and protein loss and allows *Y. pstb* organisms to invade the body via the lymphatic system.

Treatment

Treatment with parenteral antibiotics, proprietary scour medicines, and fluid therapy for 3 or 4 days is usually successful if instituted early in the disease (Mackintosh 1984). The antibiotics of choice are oxytetracycline or trimethoprim/sulphonamide combinations to which the majority of isolates are sensitive (Hodges et al. 1980). To curtail an outbreak, in-contact deer may be treated en masse with long-acting tetracyclines or neomycin or they may be fed concentrate pellets medicated with tetracyclines.

Prevention

Good management practices, adequate feeding, provision of shelter from bad weather, and minimizing stress will help to prevent yersiniosis. The stress of weaning can be minimized by weaning prior to the rut when feed is more plentiful, environmental temperatures are higher and severe storms less likely. Once weaned, the calves can be brought into the yards regularly for anthelmintic dosing in autumn and given preferential grazing over the winter. Alternatively, indoor wintering of weaners is proving popular in the colder areas in the south of New Zealand and can reduce the likelihood of yersiniosis if the level of feeding is adequate and if the accommodation is warm and dry.

Immunity and Vaccination

Investigation of the immune response of deer to *Y. pstb* infection and killed bacterins has been carried out at Invermay for a number of years. A previous paper (Mackintosh et al. 1986) reviewed the literature and presented the results of field studies, experimental infections, and vaccination trials. In essence, these studies showed that preliminary trials using killed *Y. pstb* organisms with aluminum hydroxide adjuvant gave promising results, but future vaccines should have a higher antigenic load, especially of the "virulence" antigens and have a better adjuvant to ensure a higher and more prolonged immunological response. Subsequent work has led to the development of an ELISA test for *Y. pstb* in deer which can measure the humoral response of deer to both somatic and virulence antigens (S. Hook, unpublished data) based on methods developed by Hibma and Griffin (1978) and modified specifically for deer (Hibma and Griffin 1988). Studies have also shown that there is a cell-mediated immune (CMI) component in the response of deer to infection and immunization with *Y. pstb* (Mackintosh et al. 1986). Recent vaccination studies have confirmed that deer produce a good humoral and CMI response to an oil adjuvanted killed *Y. pstb* vaccine which gave significant protection against a live virulent *Y. pstb* experimental challenge compared with unvaccinated controls (C. Mackintosh, unpublished data). Interestingly, this experimental challenge also showed that some sires had a significantly higher proportion of their calves affected with yersiniosis than others, suggesting that there is a genetic component to susceptibility to this disease.

The vaccination studies are continuing, and it is hoped that a commercial vaccine will be available in the near future. This, along with good management practices, should minimize losses due to yersiniosis in young weaned deer.

References

Barre N, Bercovier H, Treignier M, Brault J (1979) Summary of an epidemiological study of *Yersinia* infections in a field and forest ecosystem of the Paris region. I. Research on *Yersinia* on the soil, earthworms and vegetation. Med Malades Infect 9:34–39

Beatson NS (1984) Yersiniosis—clinical aspects. Proc N Z Vet Assoc Deer Branch Course 1, pp 43–45

Beatson NS, Hutton JM (1981) An outbreak of yersiniosis in farmed red deer. Proc Deer Advisory Panel, New Z Vet Assoc Deer Seminar, Queenstown, pp 136–139

Borg K (1968) Sylvatic pseudotuberculosis in Scandinavia. Int Symp on Pseudotuberculosis, Paris, 1967. Symp Series Immunobiol, Standard 9:129–132

Griffin JFT (1989) Stress and immunity: a unifying concept. Vet Immunol and Immunopathol 20:263–312

Henderson TG (1983) Yersiniosis in deer from the Otago-Southland region of New Zealand. N Z Vet J 31:221–4

Henderson TG, Hemmingsen P (1983) Faecal survey of deer for *Yersinia pseudotuberculosis* and *Salmonella sp.* N Z Vet J 31:225–226

Hibma M, Griffin JFT (1987) Altered humoral immunity during pregnancy in the guinea-pig. J Reprod Immunol 10:299–307

Hibma M, Griffin JFT (1988) Optimisation of antibody production and immunisation schedules in farmed deer. N Z Vet Assoc Deer Branch Course 5, pp 105–110

Hodges RT, Carman MG, Holland JTS (1980) In vitro antimicrobial sensitivity of isolates of *Yersinia pseudotuberculosis* from deer. N Z Vet J 28:191–192

Hodges RT, Carman MG, Mortimer WJ (1984a) Serotypes of *Yersinia pseudotuberculosis* recovered from domestic livestock. N Z Vet J 32:11–13

Hodges RT, Carman MG, Woods EP (1984b) *Yersinia pseudotuberculosis* recovered from the faeces of clinically healthy deer. N Z Vet J 32:79

Hubbert WT (1972) Yersiniosis in mammals and birds in the United States. Case reports and review. Am J Trop Med Hyg 21:458–463

Mackintosh, CG (1984) Workshop—handling yersiniosis outbreaks. N Z Vet Assoc Deer Branch Course 1, pp 138–139

Mackintosh CG (1990) Diseases of farmed deer in New Zealand. Vet Annu 30:59–63

Mackintosh CG, Henderson TG (1984a) The epidemiology of yersiniosis in deer. N Z Vet Assoc Deer Branch Course 1, pp 34–42

Mackintosh CG, Henderson TG (1984b) Potential wildlife sources of *Yersinia pseudotuberculosis* for farmed red deer (*Cervus elaphus*). N Z Vet J 32:208–210

Mackintosh CG, Henderson TG (1985) Survey of red deer stags for yersiniosis at slaughter. In Fennessy PF, Drew KR (eds) Biology of Deer Production. R Soc N Z, Wellington, Bull 22:159–162

Mackintosh CG, Henderson TG, Griffin JFT (1986) *Yersinia* vaccination trials in red deer (*Cervus elaphus*). N Z Vet Assoc Deer Branch Course 3, pp 138–145

Mair NS (1968) Pseudotuberculosis in free-lining wild animals. Symp Zool Soc Lond 24:107–117

Mair NS (1973) Yersiniosis in wildlife and its public health implications. J Wildl Dis 9:64–71

Mair NS (1975) Yersiniosis (Infections due to *Yersiniosis pseudotuberculosis* and *Y. enterocolitica*). In Hubbert WT, McCulloch WF, Schnurrenberger PR (eds) Diseases Transmitted from Animals to Man. Charles C Thomas. Springfield, Ill. pp 174–185

Obwolo MJ (1976) A review of yersiniosis (*Yersinia pseudotuberculosis*) infection. Vet Bull 46:167–171

Stovell PL (1979) Pseudotubercular yersiniosis. In Stoener H, Torten M and Kaplan W (eds) CRC Handbook Series in Zoonoses. Section A: Bacterial, Rickettsial and Mycotic Diseases. CRC Press, Boca Raton, FLla., pp 209–256

Wilson PR (1984) Diseases of farmed deer. Deer Refresher Course, Post Grad Comm Vet Sci, Sydney University, Proc 72, pp 505–530

32

Laboratory Assays for the Diagnosis of Tuberculosis in Farmed Red Deer (*Cervus elaphus*)

J. Frank T. Griffin, John P. Cross, and Glenn S. Buchan

Abstract

An assay for tuberculosis (Tb) has been developed that measures relative levels of lymphocyte transformation (LT) of deer mononuclear cells cultured with avian or bovine tuberculin (PPD). The LT assay has been shown to have a sensitivity of 94% for disease diagnosis. The specificity in excluding disease varies with the incidence of Tb within an affected herd with levels ranging from 78% to 99%. An ELISA assay has been developed which measures antibody specific for mycobacterial antigens. This assay has been shown to be of special value in the identification of false (−) skin test reactors harboring tuberculosis. Whereas the ELISA assay has an overall sensitivity of 87% in detecting Tb, it is extremely sensitive (> 95%) in diagnosing animals with serious Tb disease. The ELISA has a specificity of 92%. Soluble inflammatory proteins are also used as a marker of disease, and both plasma viscosity and haptoglobin have a high predictive value in identifying animals with proliferative tuberculosis. The results are combined to produce the blood test for Tb which is expressed as a diagnostic radial plot that can identify nondiseased, diseased, and immune deer.

Key words: ELISA, farmed deer, lymphocyte transformation, red deer, tuberculosis

Introduction

The cellular immune reaction to *Mycobacterium bovis* has classically been measured by skin testing animals or humans. Intradermal skin testing in deer is more technically demanding than in cattle (Carter et al. 1984). Limitations in skin test sensitively in deer in some herds has made it impossible to eradicate disease from infected herds using skin testing as the sole diagnostic method (Beatson et al. 1984). The caudal fold skin test, used widely for tuberculosis (Tb) testing in cattle is unsuitable for deer, where cervical testing is far superior (N.S. Beatson and P.G. Livingstone, personal communication). Aside from sensitivity limitations, Tb skin tests (ST) using *M. bovis* purified protein derivative (PPD) lacks specificity to distinguish between Tb reactivity and sensitization with other mycobacterial species, producing false positive test reactions (Griffin and Cross 1989b). The use of a single intradermal skin test (ST) in New Zealand deer herds has produced an unacceptably large percentage (60% to 70%) of false positive reactions. Because Tb is present in less than 1% of deer herds, most reactions found with bovine tuberculin are nonspecific and unrelated to Tb exposure. The unnecessary wastage due to false positive reactions prompted Carter et al. (1985, 1986) to produce a comparative cervical skin test (CCT) using tuberculin from *M. bovis* and *M. avium* inoculated at different sites.

The undue wastage of animals due to false positive, nonspecific reactions to the ST and the perceived risk of false negative, diseased animals remaining undetected by the CCT led to our laboratory being commissioned to look at alternative laboratory tests for tuberculosis in farmed deer.

A Laboratory Test for Tb Diagnosis in Deer

Results from initial studies carried out in our laboratory developed a blood test for Tb (BTB) which appeared to have significant potential for diagnosis of Tb in ruminants (Griffin and Cross 1986, 1987, 1989a; Griffin and Batchelor 1988; Griffin et al. 1988a, 1988b). To date these studies have examined more than 34 000 deer blood samples from which more than 3000 deer have been subjected to necropsy and detailed laboratory examination to validate diagnostic parameters for the identification or exclusion of tuberculosis in farmed deer. This test is registered by MAF-NZ for use as an ancillary test in the national control program for deer Tb in New Zealand.

Material and Methods

Multiple blood samples were obtained from individual animals to establish optimal conditions for sampling, transport, and processing of deer blood for laboratory testing. Samples were obtained in a single draw using an 18-gauge needle and transported to the laboratory for processing within 48 h. Complications in mustering or restraint of animals and the use of sedatives were identified so that changes associated with stress or sedation could be taken into account by the laboratory (Cross et al. 1988a).

Three separate parameters were evaluated to produce the composite BTB assay. These involved lymphocyte transformation (LT) with tuberculin, measurement of antibody (Ab) levels to tuberculin using an ELISA test, and the estimation of inflammatory cell and protein levels.

Mononuclear leukocytes were isolated from deer blood using buoyancy density gradients, after which the cells were washed, concentrated to a standard level (2×10^5 cells), and cultured with bovine tuberculin, avian tuberculin, nonspecific mitogen (Concanavalin A), or unstimulated. Following 5 days of incubation at 37° C in a CO_2-rich atmosphere, cells were pulsed with a tritium label and harvested 18 h post labeling using an automated cell harvester (PHD Cell Harvester). Radiolabel uptake was estimated for triplicate sets using a beta scintillation counter (LKB). Relative reactivities in bovine or avian stimulated cultures were compared to establish the specificity of lymphocyte transformation. Minimal reactivity in the positive and negative control cultures was necessary to establish a valid culture set.

Using a modified ELISA (Engvall and Ruoslahti 1979), our laboratory developed a sensitive and specific ELISA assay for detection of Ab to Tb in deer and other ruminants (Griffin and Buchan 1989). An indirect ELISA system, using a horse radish peroxidase (HRP) labeled antibody to deer immunoglobulin, was used to detect bound deer antibody in immunoplates (NUNC) coated with avian or bovine tuberculin or specific peptide extracted from M. bovis (MPB70). The ability to produce a precise ELISA assay for Tb is dependent on a critical standardization of all reagents used in the assay system so that limiting dilutions (1/100) of deer serum can be used in the primary detection system for Ab analysis.

Differential WBC cell counts expressing neutrophil number (NEUT), plasma viscosity (PV), and fibrinogen (Fib) were carried out according to the method of Cross (1987). Haptoglobin (HP) levels were estimated according to the method of Cross et al. (1988b). Data obtained from these three individual assay systems was used individually or in combination to produce diagnostic values which allowed for specific and sensitive laboratory diagnosis for Tb in deer, or by extrapolation, in other ruminants or humans.

Results

In order to evaluate the performance of two ancillary tests for Tb diagnosis we compared the sensitivity and specificity obtained using lymphocyte transformation (LT) and CCT in a herd which underwent complete depopulation, and harbored Tb in over 40% of the stock. Lymphocyte transformation was significantly superior to CCT in both sensitivity (94% (33/35) vs 65% (18/28)) and specificity (91% (49/54) vs 65% (36/56)). The LT also identified infected animals up to 3 months earlier than was found using

CCT. An additional advantage was that the LT identified all seriously diseased animals, with multiple lesions or generalized tuberculosis, whereas the CCT failed to identify over 25% of animals harboring multiple Tb lesions.

Whereas the specificity of LT appeared satisfactory in the herd tested, it was noted that in other herds which had a significant incidence of tuberculosis (> 70%), it appeared that the specificity of LT was reduced. Whereas sensitivity remained high in all infected herds (94%; 148/157), the specificity was significantly lower in seriously infected Tb herds (78%; 207/265) than in herds known to be free of disease (98.6%; 534/542). It is debatable whether animals with LT positive reactivity from known infected herds were in fact false positives or represented immune reactivity to M. bovis following development of protective immunity and clearance of the infectious agent. Ultimately the test of specificity is its ability to exclude Tb where it is known to be absent, so the high level of specificity of LT in disease-free herds suggests it to be totally satisfactory for use in Tb exclusion.

While the sensitivity and specificity of LT appeared adequate for diagnosis or exclusion of Tb in deer herds, data were also obtained using the ELISA assay alone for Tb diagnosis. The ELISA assay, using differential response to avian and bovine tuberculin to identify specific Ab to M. bovis, was highly sensitive and specific for diagnosis and exclusion of Tb. The test had a sensitivity of 87% (77/89) for diagnosing diseased animals and 92% (132/144) specificity to exclude Tb in nondiseased animals. The ELISA was especially sensitive in detecting animals with multiple lesions, the group most likely to produce false negative skin test reactions.

The combination of results obtained from LT and ELISA were used in the composite BTB assay. The combined tests not only improved the precision of laboratory diagnosis for Tb, but they also identified the possibility of identifying "immune" animals in Tb-infected herds. A persistent problem in all Tb diagnostic tests has been to distinguish between immune reactivity which characterizes disease and lesions rather than the immunity produced when protection develops following exposure of an animal to infection. It is widely recognized that herd immunity develops in a number of animals exposed to infection but which do not develop active disease following the initial encounter with the infectious agent. When LT and ELISA are combined, it is possible to identify and rank animals which have been exposed to M. bovis as to their likely disease status at necropsy (Table 32.1). Animals which demonstrated high grade LT and which were ELISA positive invariably harbor active disease with typical lesions at necropsy. By contrast, animals which showed lower levels of cellular sensitization to M. bovis in LT and had no ELISA antibody were often disease-free and may have developed protective immunity rather than disease. Whereas the program did not attempt to salvage such animals, the identification of animals that have developed protective immunity would be valuable because these may represent a genetically elite group with superior ability to develop protective immunity following exposure to M. bovis.

TABLE 32.1. Laboratory parameters for disease and immunity in skin test positive deer

Animal no.	LT[a] Bovine PPD	ELISA (units) Bovine PPD	TB status	Disease severity[b]
1	7	203	Positive	4
2	99	40	Positive	4
3	73	160	Positive	3
4	69	180	Positive	3
5	31	110	Positive	2
6	46	110	Positive	2
7	7	45	Positive	2
8	38	11	Negative	0
9	23	0	Negative	0
10	24	0	Negative	0
11	31	0	Negative	0
12	17	30	Negative	0
13	6	0	Negative	0
14	16	0	Negative	0
15	23	0	Early Tb?	2

[a]Counts/min × 1000.
[b]Disease ranking 0, no lesions → 5, generalized Tb.

A computer plot involving a combination of lymphocyte transformation, ELISA, and inflammatory markers shows the pattern of reactivity for disease diagnosis or exclusion (Fig. 32.1). Figure 32.1a shows typical patterns of reactivity

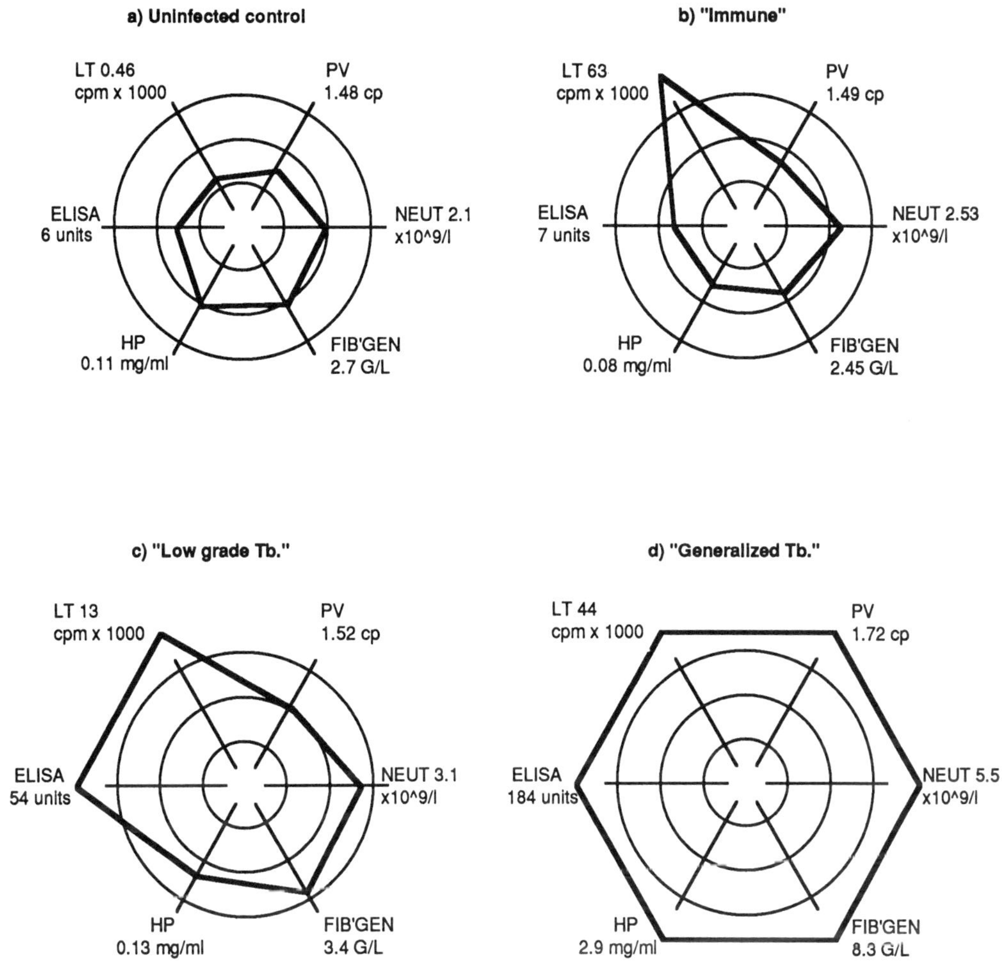

FIGURE 32.1. Radial plots of immune and inflammatory parameters in nondiseased, immune, and Tb-infected deer. Center circle, mean; inner circle and outer circles, SD limits. Abbreviations and reference ranges for healthy deer: LT (lymphocyte transformation to bovine Tb antigen), $0.2–2 \times 10^3$ cpm; PV (plasma viscosity), 1.44–1.59 centipoise, NEUT (peripheral blood neutrophils), $0.92–3.62 \times 10^9$/l; FIB'GEN (plasma fibrinogen), 1.81–3.65 g/l; HP (serum haptoglobin), 0.05–0.17 mg/ml Hb binding capacity; ELISA (serum antibody to bovine Tb antigen by enzyme-linked immunosorbent assay), 0–20 units.

in a normal animal without any specific evidence of exposure to M. bovis. Figure 32.1 b shows the pattern of reactivity in an animal with specific LT reactivity but no ELISA Ab or inflammation; this type of animal is considered likely to be "immune" to Tb. Animals harboring limited lesions (Fig. 32.1 c) usually have reactivity in the LT and ELISA, without any evidence of active inflammation. Figure 32.1 d shows a typical pattern of reactivity in animals harboring multiple Tb lesions and actively spreading disease. In such cases there is significant cellular immune reactivity as measured by lymphocyte transformation, Ab reactivity as measured by ELISA, and increased inflammatory reactivity as an indicator of active proliferative disease.

Conclusions

Laboratory assays such as the BTB offer considerable advantages over CCT as ancillary tests in a Tb control program for farmed deer. Earlier studies in humans (Schreck 1963; Verma et al. 1974) and animals (Charapas et al. 1970) have identified that lymphocyte transformation tests can detect serious Tb infection in individuals that are anergic to skin test. Similar findings have been observed in the present studies where LT has been shown to be significantly superior to skin test in identifying seriously infected deer. In addition, the LT assay can identify a higher incidence of infected individuals than skin testing. A further advantage of the LT is that the test can be used in relatively close proximity to skin testing. Because of skin test suppression, an interval of 90 days is required between the application of successive skin tests (Carter et al. 1986). The LT can be applied accurately within 10 days of skin testing (Griffin and Cross 1987). This means that valuable time can be gained in the implementation of ancillary tests to establish the specificity of skin test reactors using the single intradermal skin test (ST). Apart from its sensitivity, the LT also offers a high degree of specificity for the exclusion of nonspecific reactors sensitized by M. avium or other nonspecific or saprophytic mycobacteria (Muscoplat et al. 1975, 1979).

Recent recommendations (Wilson 1986) are that the CCT should not be used in deer herds known to harbor M. bovis infection because of the reduced sensitivity of CCT to detect infected deer. Similar limitations in the CCT have been identified by de Jong and Ekdahl (1969) who found unacceptably low levels of sensitivity of CCT for Tb diagnosis in New Zealand cattle. While the CCT has an acceptable level of specificity to exclude M. bovis, its use is contraindicated in herds with an unknown disease status or those known to harbor M. bovis. This is an important limitation in the use of CCT in countries where M. bovis is endemic or the overall national herd status is ill defined. In our experience, the LT has superior immunologic specificity to the CCT, as we find less than 50% of CCT-positive animals have reactivity to bovine tuberculin when tested in the LT. Such spurious reactions may be due to cross reactivity between M. bovis antigens and those found in other mycobacterial saprophytes (Hattikudar and Kamat 1985).

The composite results using LT, ELISA, and inflammation markers (Fig. 32.1) show that the BTB can be used as a composite assay to identify animals with no prior exposure to M. bovis, animals that may have developed immunity to M. bovis, or animals harboring M. bovis lesions. The ability to exclude naive animals using a BTB test that also can identify animals with specific immunity to Tb is of considerable importance. It has already been recognized that skin test reactivity, while correlating reasonably with infection, does not measure protective immunity (Orme and Collins 1984; Shapiro et al. 1974). The inability of the skin test to identify immune animals means that animals that have developed protective immunity remain undetected by skin test screening. Because genetic factors appear to play an important role in disease resistance to Tb (Stach et al. 1984), it is important to use Tb testing regimes that can discriminate between protective immunity and disease due to M. bovis. Use of the BTB may ultimately allow for the retention of genetically elite animals from within infected herds, with unique genetic resistance to Tb. In addition, this technology may allow for the elucidation of unique pathways of immune reactivity of special relevence for immune protection.

Under field conditions in New Zealand, the BTB is used as an ancillary test to clarify the status of ST reactors found during whole herd testing. BTB tests are carried out 10 days following skin testing to clear animals as nonspecific reactors or to identify M. bovis reactors for slaughter. This approach allows for animals free of disease to be salvaged while Tb-infected animals can be identified with a high degree of sensitivity. In herds with significant levels of Tb disease and suspected false (−) skin reactions, the ELISA assay alone can be used as a second line screening test to monitor skin test negative animals and identify "anergic" animals. Because the ELISA is technically simple, it can be used inexpensively to screen animals to identify seriously infected Tb carriers following skin testing. This dual screening approach can significantly improve Tb diagnosis and speed up a disease control program in herds with significant levels (> 5%) of infection.

Acknowledgements

The authors acknowledge financial assistance and animal resources from Haldon Station, the New Zealand Deer Farmers Association, and support for field studies and laboratory assays by MAF-Qual NZ. The unique contribution made by the management of Haldon Station in the provision of sampling and pursuance of necropsy studies is acknowledged. The BTB is registered under New Zealand patent number 214400 and U.S. patent 935896 with patent applications pending in Europe (EEC) and Australia.

References

Beatson NS, Hutton JB, de Lisle GW (1984) Tuberculosis—test and slaughter. Proc Deer Branch NZ Vet Assoc 1:18–27

Carter CE, Corrin KC, de Lisle GW, Kissling RC (1986) An evaluation of the comparative cervical test in deer. Proc Deer Branch N Z Vet Assoc 3:65–70

Carter CE, Corrin KC, de Lisle GW, Kissling RC (1985) Comparative cervical test in deer. Proc Deer Branch N Z Vet Assoc 2:80–87

Carter CE, Corrin KC, de Lisle GW (1984) Evaluation of the tuberculin test in deer. Proc Deer Branch N Z Vet Assoc 1:1–8

Charapas SD, Hedrick SR, Clark RG, Garman R (1970) Comparison of the lymphocyte transformation test with the tuberculin test in rhesus monkeys and chimpanzees. Am J Vet Res 31:1427–1441

Cross JP (1987) Haematology-based prediction of lesion status in bovine tuberculosis of farmed red deer. Proc Deer Branch N Z Vet Assoc 4:147–154

Cross JP, Mackintosh CG, Griffin JFT (1988a) The effect of physical restraint and xylazine sedation on haematological values in red deer (*Cervus elaphus*). Res Vet Sci 45:281–286

Cross JP, Tate M, Griffin JFT (1988b) Haptoglobin and plasma viscosity as markers of acute phase reactions in red deer (*Cervus elaphus*.) Proc Deer Branch N Z Vet Assoc 5:111–115

de Jong E, Ekdahl MO (1969) Evaluation of a number of ancillary tuberculin tests in cattle. N Z Vet J 17:213–226

Engvall E, Ruoslahti E (1979) Immunoassays in the clinical laboratory. AR Liss. New York, pp 89–97

Griffin JFT (1988) The aetiology of tuberculosis and mycobacterial disease in farmed deer. Irish Vet J 42:23–26

Griffin JFT, Batchelor NJ (1988) Case reports: blood test (BTB) for diagnosis and management of Tb. Proc Deer Branch N Z Vet Assoc 5:66–73

Griffin JFT, Buchan GS (1989) The ELISA technique for diagnosis of severe tuberculosis in deer and exotic ruminants. Proc Deer Branch N Z Vet Assoc 6;78–86

Griffin JFT, Cross JP (1989a) The acute inflammatory response in farmed red deer (*Cervus elaphus*). Proc Deer Branch N Z Vet Assoc 6:151–157

Griffin JFT, Cross JP (1989b) Diagnosis of tuberculosis in New Zealand farmed deer: an evaluation of intradermal skin testing and laboratory techniques. Irish Vet J 42:101–107

Griffin JFT, Cross JP (1987) Cellular immunity in the aetiology and diagnosis of tuberculosis in farmed deer. Proc Deer Branch N Z Vet Assoc 4:137–146

Griffin JFT, Cross JP (1986) In vitro tests for tuberculosis in farmed deer. Proc Deer Branch N Z Vet Assoc 3:71–77

Griffin JFT, Chinn ND, Buchan GS (1988a) Do deer develop immunity to tuberculosis? Proc Deer Branch N Z Vet Assoc 5:74–83

Griffin JFT, Rodgers C, Cross JP (1988b) Blood testing for diagnosis of Tb within a national control programme. Proc Deer Branch N Z Vet Assoc 5:55–65

Hattikudur S, Kamat RS (1985) Polymorphisim of mycobacterial antigens participating in cell mediated immunity. J Med Microbiol 19:69–75

Muscoplat CC, Thoen CO, Chen AW, Johnsone DW (1979) Development of specific in vitro lymphocyte responses in cattle infected with Mycobacterium bovis and with Mycobacterium avium. Am J of Vet Res 36:395–398

Muscoplat CC, Thoen CO, Chen AW, Rakish PM, Johnson DW (1975) Development of specific lymphocyte immunostimulation and tuberculin skin reactivity in swine infected with *M. bovis* and *M. avium*. Am J Vet Res 36:1167–1177

Orme I, Collins FM (1984) Adoptive protection of *M. tuberculosis* infected lung. Dissociation between cells that passively transfer protective immunity and delayed type hypersensitivity to tuberculin. Cell Immunol 84:113–120

Schreck R (1963) Cell transformation and mitosis produced in vitro by tuberculin purified protein derivative in human blood cells. Am Rev Resp Dis 87:734–738

Shapiro CDK, Harding GE, Smith DW (1974) Relationship of DTH and acquired cellular resistance in experimental airborne tuberculosis. J Infect Dis 130:8–15

Stach JL, Gros P, Forget A, Skamene E (1984) Phenotypic expression of genetically controlled resistance to *M. bovis* BCG. J Immunol 132:888–892

Verma IC, Gupta MI, Ghai OP (1974) In vitro lymphoblastoid transformation with PPD as a diagnostic test in tuberculosis. Indian J Med Res 62:615–620.

Wilson PR (1986) Tuberculosis in deer: non-specificity and the comparative cervical test. Proc Deer Branch N Z Vet Assoc 3:49–64

33
Influence of Xylazine on Hematology Values in Farmed Red Deer

John P. Cross, J. Frank T. Griffin, and Colin G. Mackintosh

Abstract

Reductions in circulating basophils and lymphocytes following xylazine administration to farmed red deer (*Cervus elaphus*) previously observed using automated technology have been confirmed using standard manual procedures. The kinetics of the changes were followed and observed to be maximal at 2–5 min after intramuscular injection, whereas red cell and platelet changes follow a different time course, commencing and peaking later. There is clearly a need to be aware of such changes when interpreting hematological results. The changes seen in basophils and lymphocytes were not associated with splenic sequestration, unlike red cells and, to some extent, platelets. Manual and automated techniques indicated that disruption of basophils appeared to take place on in vitro exposure of blood to xylazine. Although all cell types were affected in vitro, the basophil was much more sensitive, with disruption occurring in under 30 s, compared with loss of other cell types taking place between 5 and 10 min. If basophil disruption occurs in vivo this may play a part in occasional stag deaths due to a hypersensitivity-like reaction.

Key words: Basophil, hematology, red deer, spleen, xylazine

Introduction

Our experience suggests that about 12% of blood specimens from farmed red deer in New Zealand are collected with the aid of sedation. It is well-established that the red cell mass falls during xylazine sedation in several species (Hawkey et al. 1980). Biochemical and some hematological changes take place when sedation is used in white-tailed deer (*Odocoileus virginianus*) and antelope (*Antilocapra americana*) (Seal and Bush 1987). The data for sedated red deer are much more sparse, involving few animals and limited information with regard to formed elements of the blood other than red cells (Drescher-Kaden and Hoppe 1972, 1973; McAllum 1978). Chapman (1977) concludes from the Drescher-Kaden and Hoppe papers that red cell count, hemoglobin, and packed cell volume (PCV) showed small decreases following sedation, with no change in mean red cell volume, mean cell hemoglobin concentration, mean cell hemoglobin, and white cell count.

We have reported on 24 farmed red deer in which the animals showed a substantial fall (22%–30%) in circulating red cell mass in all cases, with reductions in circulating platelets (25%) and lymphocytes (36%) (Cross et al. 1988). Preliminary reports have described a fall in basophils following xylazine administration, and a possible link between the loss of basophils from the circulation and occasional deaths of stags following velvet removal conducted under xylazine sedation (Cross et al. 1989; Mackintosh and Cross 1989). The aim of this study was to assess the short-term kinetics of leukocyte changes during xylazine sedation in vivo, and the effect of the drug on blood cells in vitro.

Materials and Methods

Physical restraint was achieved using a pneumatic crush designed for use with deer. Short-term kinetic studies were carried out on four 18-month-old female red deer. All animals were farmed at the Invermay Ministry of Agriculture and Fisheries (MAF) research station. All four were sampled using physical restraint, then two were given xylazine (2-(2,6-xylidino)-5,6-dihydro-4H-1,3-thiazine, "Rompun," Bayer) intramuscularly at a dose of 1 mg/kg, and all four had blood samples taken at 1, 2, 5, 10, 15, 30, and 60 min. Manual differential counts were performed on ten yearling male red deer. In vitro studies were carried out on samples from four mixed-age adult hinds.

Dipotassium ethylenediaminetetraacetic acid (EDTA) was used as the anticoagulant except where stated otherwise. In the in vitro studies where the cell morphology changes were assessed visually, specimens were obtained from five deer, and xylazine was used at a final concentration of 2, 5, and 10 mg/ml, the reaction being stopped by the addition of 10 volumes of ice-cold phosphate buffered saline. Cytocentrifuge smears were then stained using the Jenner-Giemsa method. This work was done on three occasions, two using EDTA anticoagulated blood (three deer) and one using heparinized blood (two deer). Hematological changes were also assessed by means of the Technicon H6000/C, and the final concentrations of xylazine were 0, 0.25, 0.5, 1, and 2 mg/ml. In this latter case specimens from eight animals were presented to the machine in a sequence to ensure each had approximately 6 min contact with the drug before counting was completed. The Technicon H6000/C was also used to obtain counts on blood specimens used for short-term kinetic studies in vivo.

Manual differential counts on cytocentrifuge smears were carried out by selecting an area where cell morphology was clear, differentiating all the cells in the field with the aid of a 10 × 10 eyepiece micrometer grid, and traversing in a straight line to a new field and again identifying all the cells present. This was repeated until 500 cells had been identified. The 95% probability bars were obtained using a computer program (Cross 1985) adapted for 500-cell counts.

Manual differential counts on whole blood were carried out on well-made blood smears prepared using the spreader-slide technique from freshly collected blood in di-potassium EDTA within 10 min of venepuncture. After staining with Jenner-Giemsa, the smears were scanned using a search pattern commencing near the tail of the film and traversing in a straight line towards the origin until white cell morphology began to appear indistinct. Immediately, two microscope fields were traversed at right angles before tracking back towards the tail. This process was repeated until 200 consecutive leukocytes had been identified.

Time for red cell lysis was found by adding 150 μl of 20 mg/ml xylazine to 150 μl of whole deer blood and stopping the reaction by adding 2.7 ml of saline to each tube after 0.5, 1, 2, 3, 4, 5, 6, 7, 8, 9, and 10 min. After centrifuging the tubes, the optical density of the supernatant was measured at a wavelength of 540 nm, within 30 min of commencing the experiment.

Scanning electron microscopy (SEM) was carried out after treating freshly collected blood with xylazine at a final concentration of 10 mg/ml and adding 1 drop to 1% buffered glutaraldehyde after 1, 5, and 10 min. The control consisted of an aliquot of the same cells treated with saline for 10 min.

Results

Intramuscular injection of xylazine was associated with reductions in platelet, lymphocyte, and basophil numbers, and red cell mass. The reductions in red cell mass and platelet count were evident at 5 min and peaked around 30 min, whereas the basophils and lymphocyte reductions were evident at 1 min and peaked around 5 min. The basophil count had returned to initial levels at 60 min, whereas the lymphocyte count was still substantially reduced after 60 min (Fig. 33.1).

The results of manual differentials (Table 33.1) confirmed those found using the automated system, with a moderate increase in neutrophils, significant decreases in lymphocytes and basophils, and no change in eosinophils. No change was seen in the monocytes.

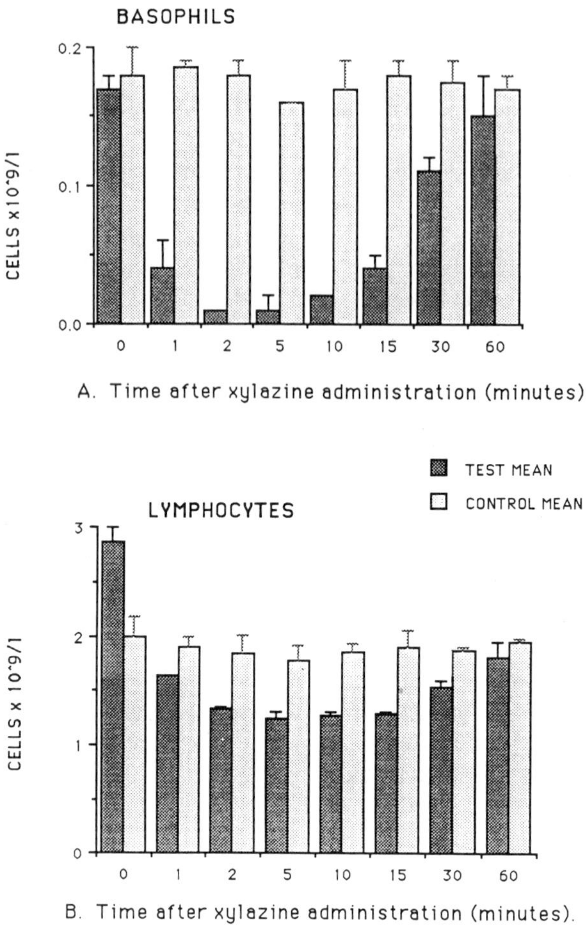

FIGURE 33.1. In vivo effects of xylazine on circulating red deer leukocytes measured using an automated technique (Technicon H6000/C hematology analyser).

In vitro exposure to xylazine brought about a very rapid reduction in basophils, within 25 s, whereas the reduction in lymphocytes was not seen until 5 to 10 min had elapsed (Fig. 33.2). Neutrophils, eosinophils, and monocytes showed similar patterns to lymphocytes. At the same time, an increase in "smear cells" was seen, and occasionally one could see disrupted basophils. Morphological changes were also noted in the nuclear chromatin of the lymphocytes, which developed a more open appearance between 5 and 10 min, suggestive of imminent disruption.

Release of hemoglobin from red cells exposed in vitro to xylazine showed a similar time scale to the loss of all leukocytes except basophils, with lysis commencing at 8 to 9 min, and complete between 10 and 15 min. The lytic process for red cells was shown by SEM studies to involve a spherocyte transformation prior to lysis. The drug preparation used for these experiments is isotonic according to the suppliers (Ippen, personal communication), and freezing point depression measurements confirmed this, giving an osmolality of 270 mosmol.

TABLE 33.1. Circulating leukocyte levels (10^9/l) before and 30 min after intramuscular injection of xylazine in farmed red deer using standard manual techniques ($n = 10$)

Cell type	Mean post-xylazine	Mean post-xylazine	Wilcoxon P value
Neutrophils	2.05	2.78	0.0125
Lymphocytes	2.45	1.44	0.0051
Monocytes	0.26	0.23	0.5073
Eosinophils	0.35	0.34	0.8785
Basophils	0.16	0.06	0.0217

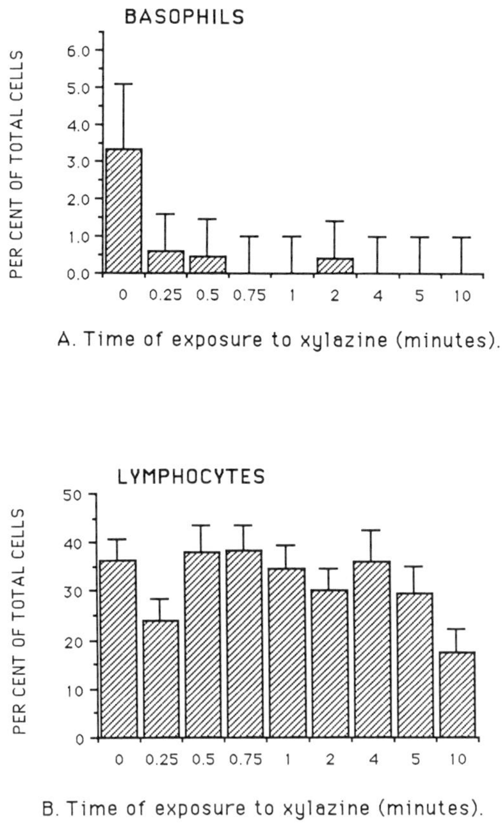

FIGURE 33.2. In vitro effects of xylazine on red deer blood cells measured using manual techniques.

In vitro effects detected using the H6000/C indicated that basophils and lymphocytes were significantly reduced in number at doses of xylazine as low as 0.5 mg/ml, whereas there was no evidence that neutrophils or platelets were reduced under these conditions.

Conclusions

The use of precise technology for leukocyte differential counts (Technicon H6000/C) has been an important aid to the study of leukocyte changes with xylazine. We considered it important that these results be confirmed by standard manual techniques, and the results shown in Table 33.1 clearly show this to be the case.

There are two generally stated hypotheses to account for the red cell mass changes. The first is that splenic contraction prior to sedation leads to a high circulating red cell mass which decreases as red cells return to the spleen once sedation occurs. The second hypothesis is that tissue fluid enters the blood stream due to the blood pressure fall that occurs during sedation with xylazine (Presidente et al. 1973). We have established that the red cell mass changes are greatly reduced in splenectomized red deer, but that the lymphocyte and basophil changes are unaffected (Cross et al. 1989). Therefore, the main reason for the change in circulating red cell mass is return of red cells to splenic stores, and the observed change in platelets can be partly explained in this way.

Since the basophil and lymphocyte falls in vivo precede the red cell and platelet falls, this is a further indication that the mechanisms may differ. The in vitro observations suggest that the basophils are particularly prone to disruption when xylazine is present, and that this is not an IgE-mediated phenomenon. A further indication that it is not due to specific immune hypersensitivity is that the effect is regularly seen in deer not previously exposed to xylazine. It is also evident that a moderate amount of dilution of the drug can take place before the in vitro disruption of basophils is nullified. Experiments are now under way to discover whether in vivo basophil disruption occurs. It is possible that the effects noted in vitro are due to different mechanisms to those seen in vivo. For example, the in vitro loss of lymphocytes could be the result of the action of enzymes released from the basophils.

A wide range of factors can bring about basophil and mast cell degranulation directly, including drugs and chemicals (Fadal 1985). We have drawn attention to the possible role this could play in the occurrence of occasional deaths following xylazine administration (Mackintosh and Cross 1989). Several authors have pointed out that reference values obtained using physical restraint may not be applicable to those obtained using sedation. It has been suggested that "anesthetised animals yield more nearly normal values" (Seal and Bush 1987). While this seems likely to be true in the case of red cell parameters, there must be some doubt that it is always the case for xylazine-treated ones as far as lymphocytes and basophils are concerned.

Acknowledgements

This work was partly funded by a grant from the Deer Branch of the New Zealand Veterinary Association.

References

Chapman DI (1977) Haematology of the deer. In Archer RK, Jeffcott LB (eds) Comparative Clinical Haematology. Blackwell Scientific Publications, Oxford, pp 345–364

Cross JP (1985) Program for confidence limits for leukocyte differentials. Lab Med 17:296

Cross JP, Mackintosh CG, Griffin JFT (1988) Effect of physical restraint and xylazine sedation on haematological values in red deer (*Cervus elaphus*). Res Vet Sci 45:281–286

Cross JP, Mackintosh CG, Griffin JFT (1989) Further observations on xylazine and haematological parameters in red deer (*Cervus elaphus*): the effect on reference values, and on splenectomised animals. Proc N Z Vet Assoc Deer Branch 6:146–150

Dacie J, Lewis SM (1975) Practical Haematology, 5th ed. Churchill Livingstone, London

Drescher-Kaden UV, Hoppe P (1972) Vergleichende hämatologische Untersuchungen an wildlebenden Wiederkauern (Rehe, Rotwild, Gemsen, Rentiere) 2 Mitteilung: Erythrozentahl, Erythrozytendurchmesser, Hämoglobingehälte und Hämatokrit. Zeit Jadgdwiss 18:121–132

Drescher-Kaden UV, Hoppe P (1973) Vergleichende hämatologische Untersuchungen an wildkebenden Wiederskauern (Rehe, Rotwild, Gemsen, Rentiere). 2. Mitteilung: Das weibe Blutbidle (Zahl der gesamten und der eosinophilen Leucozyten, Differentialiblutbild). Zeit Jadgdwiss 19:65–75

Fadal RJ (1985) The immunobiology and immunopharmacology of the allergic response. Otolaryngol Clin N Am 18:649–676

Hawkey C, Frankel T, Jones D, Jones A, Nevill G, Hart M, Alderson C (1980) Preliminary report of a study of changes in red blood cells of zoo animals during sedation. In Montali RJ, Migaki G (eds) The Comparative Pathology of Zoo Animals Smithsonian Institution, Washington D.C., pp 625–632

Mackintosh CG, Cross JP (1989) Xylazine study report. Proc N Z Vet Assoc Deer Branch 6:136–143

McAllum HJF (1978) Post capture myopathy syndrome in red deer (*Cervus elaphus*): M.V.Sc. thesis, Massey University, New Zealand, pp 36–39

Presidente PJA, Lumsden JH, Presnell KR, Rapley WA, McCraw BM (1973) Combination of etorphine and xylazine in captive white-tailed deer: II. Effects on haematologic, serum biochemical and blood gas values. J Wildl Dis 9:342–348

Seal US, Bush M (1987) Capture and chemical immobilization of cervids. In Wemmer CM (Ed) Biology and Management of the Cervidae. Smithsonian Institution, Washington D.C., pp 480–504

34
Monoclonal Antibodies to Leukocyte Subpopulations in Deer and Exotic Ruminants

Glenn Buchan, Gabrielle McCoy, Colin G. Mackintosh, and J. Frank T. Griffin

Abstract

A significant proportion of monoclonal antibodies developed against specific ovine (54%) and bovine (16.2%) leukocyte subpopulations cross-react with populations of cells from deer. These monoclonals have been of use to identify subpopulations of leukocytes within the cervine immune system and to characterize homologous monoclonal antibodies that have been developed against cervine peripheral blood leukocytes. Antibodies to ovine and bovine cells have also been found to cross-react with a variety of exotic ruminants including Père David's deer (*Elaphurus davidianus*). Poor levels of cross-reactivity were seen between human or mouse specific monoclonal antibodies and the ruminants tested in this study. The fact that some leukocyte markers are conserved between ruminants offers researchers the chance to use reagents produced against domesticated ruminants to characterize exotic species.

Key words: Deer, exotic ruminants, lymphocyte subpopulations, monoclonal antibodies

Introduction

The use of monoclonal antibodies to examine subpopulations of leukocytes in the peripheral blood and tissues of animals has allowed rapid advances to be made in our understanding of immune responsiveness and disease susceptibility. Only recently has this technology been taken out of the laboratory and applied to a broad range of economically or esthetically important animal species.

The intensive farming of newly domesticated species such as deer has given rise to new problems in disease diagnosis and control (Griffin 1989). The future economic viability of the deer industry demands an ability to understand, analyze, and ultimately manipulate the immune response of animals such that productivity/unit can be maximized. There is also a growing awareness that management of any wild animal in captivity is associated with an increased susceptibility to disease. The immune system of many of these species has been poorly characterized. The production and characterization of mononuclear antibodies to even the basic leukocyte subpopulations of these species will take time and are beyond the scope of most small laboratories. To expedite our ability to analyze the immune response of several exotic species, and red deer (*Cervus elaphus*), in particular, we tested a large number of well-characterized monoclonal antibodies directed against leukocyte subpopulations from better characterized species such as human, mouse, ovine, and bovine in order to determine the usefulness of heterologous antibodies in identifying cervine leukocyte subpopulations.

Materials and Methods

Cross-reactive Monoclonal Antibodies

We studied the following antibodies: ovine specific antibodies that included SBU I(41-17) (Gogolin-Ewens et al. 1985), SBU II(Puri et al. 1985), SBU T19 (MacKay et al. 1986), ST4 (C. MacKay, unpublished data), ST8 (Ezaki et al. 1987) and 197 (M.F. Beya, unpublished data); bovine-specific antibodies that included IL-A24 (Ellis et al. 1988), IL-A29 (Baldwin et al. 1988), 9-9F4 (own unpublished data), CC15 (unpublished data), H4 (Lewin et al. 1985), anti-bovine interleukin-2 receptor monoclonal (Dobbelaere, unpublished data); mouse-specific antibodies-Pgp-1 (Trowbridge 1982); and porcine-specific antibodies-MSA-1 (Hammerberg and Schurig 1986).

Lymphocyte Isolation

Samples of venous blood (9 ml) were drawn into heparinized vacutainers from healthy red deer. Samples were diluted 1:2 with tissue culture medium (RPMI 1640, Gibco), layered onto Ficoll-conray (1.08 g/ml), and centrifuged for 30 min at 400 g. The mononuclear fraction was harvested and washed three time in tissue culture medium. When antibodies specific for macrophage/polymorphonuclear leukocytes were assayed, blood samples were sedimented using 3 ml of 5% Dextran B (150 000–200 000 MW, BDH). Erythrocytes were allowed to sediment for 30 min at 37° C. Residual erythrocytes were lysed for 3 min in 0.85% buffered NH_4Cl before the leukocytes were removed and washed twice in RPMI.

Leukocytes were isolated from lymph node tissue within minutes of animal slaughter and necropsy. Nodes were minced finely with scissors into heparin (10 units/ml) containing RPMI. A single cell suspension was obtained by filtering this material through a fine cotton gauze after which the cells were washed twice, counted, and stained for FACS analysis.

Staining and Analysis of Leukocytes

Using an appropriate dilution (previously determined) of primary antibody 1–2 × 10^6 leukocytes were incubated for 20 min on ice. The reaction was terminated by washing three times in PBS, 0.1% BSA, and 0.1% NaN_3 (buffer). Cells were then treated with 1:50 dilution of FITC labeled goat anti-mouse IgG (Sigma, FO257) followed by three further washes in buffer. The last wash included propidium iodide (5 ml/ml of stock solution, 2 mg/ml) to stain dead cells.

A FACSCAN flow cytometer (Becton Dickinson) was used to detect positive cells. Dead cells were gated-out using the red channel (Fl 2), and percent positive cells were defined as those live cells giving readings above those obtained on the same cells using a primary antibody (SBU T1, anti-ovine CD5) shown not to cross-react with deer leukocytes. Ten thousand gated cells were counted in each sample.

Forward vs side scatter was used to determine the purity of cell preparations, and where erythrocyte or platelet contamination occurred these were gated out before statistical analysis.

Leukocytes from the peripheral blood or lymph nodes of at least three animals (except in Fig. 34.2, where cells from only one mature and fetal thymus were obtained, and Table 34.2 where leukocytes from only one ibex, (*Capra ibex*), muscat (*Gazella gazella cora*), and oryx (*Oryx leucoryx*) were tested against each monoclonal antibody to verify cross-reactivity.

In Vitro Cell Culture

Cells isolated as above were resuspended at 2 × 10^6/ml in RPMI 1640 supplemented with 10% fetal calf serum (Gibco). In 24-well plastic macrotitre plates (Falcon) 1-ml aliquots were cultured for 5 days with 15 mg/ml of concanavalin (Con) A, bovine purified protein derivative (PPD), or avian PPD. Recombinant bovine IL-2 (Immunex) was then added to 10 ng/ml. Every 2 to 3 days, cells were counted, split to 1 × 10^6/ml, and fed with fresh media containing rIL-2. After 12 days, cells were removed from the wells and stained for FACS analysis.

Results

We assayed 73 leukocyte-specific monoclonal antibodies. Of these, 15 have been shown to cross-react strongly with deer leukocytes. The

TABLE 34.1. Heterologous monoclonal antibodies cross-reactive with cervine leukocyte subpopulations

Monoclonal antibody specificity	Species specificity (% Reactive)		% Cervine PBMs stained positive	
Leukocytes			Mean	SE
9.9F4	Bovine	?	94.7	1.6
MHC				
SBU I (class I)	Ovine	> 99	99	0.9
SBU II (class II)	Ovine	?	18.9	7.2
B Lymphocytes				
H4	Bovine	33	31	7.9
T Lymphocytes				
ST-4 (CD4)	Ovine	26.6 ± 9.9	24.7	9.2
ST-8 (CD4)	Ovine	15 ± 2.5	17.3	5.8
IL-A29 (T19)	Bovine	7 ± 5.6	15.9	5
CC15 (T19)	Bovine	7.6 ± 5	28	11.2
SBU T19	Ovine	10–30	17.7	6.3
197 (T19)	Ovine	12.8 ± 4.6	17.9	1.6
SBU T6 (CD1)	Ovine	20–30	5.4	1.6
IL-2 Receptor (Tac)	Bovine	?	18.1	5.9
Monocyte/PMN				
IL-A24	Bovine	8.3 ± 2.1	6.8	1.9
MSA-1	Porcine	26.3 ± 3.8	27.5	8.3
Maturation/activation				
Pgp-1 (Ly 24)	Murine (bone marrow)	25–80	87.1	5.9

proportions of cells staining with each of the cross-reactive monoclonals and their target subpopulation in their homologous species are shown in Table 34.1. Of the monoclonals shown to cross-react, none (0/10) were human-specific, and only 1 (1/10) was mouse-specific (Pgp-1). Six bovine-specific antibodies (6/37, 16.2%) were shown to detect cell-surface markers on cervine leukocytes. One of these showed similarities to a leukocyte-common antigen (9-9F4), and IL-A24 detects a subpopulation of monocytes and polymorphonuclear leukocytes (PMNs). The anti-bovine B-cell antibody H4 was also cross-reactive and detected a nonimmunoglobulin marker. Its specificity for cervine B cells was determined by comparing the proportions of H4-positive and surface immunoglobulin-positive staining cells using anti-cervine monoclonal antibodies developed in this laboratory (Hibma and J.F.T. Griffin, unpublished data). A similar proportion of positive lymphocytes was found in density gradient purified mononuclear cells. H4 appears to be more specific for B cells than anti-immunoglobulin, as in total leukocyte preparations the anti-immunoglobulin monoclonal bound to some myeloid cells (data not shown).

Two bovine-specific monoclonals cross-reacted with a subpopulation of T lymphocytes similar to ovine T19 (MacKay et al. 1986), although CC15 appears to stain less specifically (28% vs 15%–16% for other T19 markers). The T-cell activation marker Tac, a chain of the IL-2 receptor recently found in cattle, also cross-reacted strongly in the cervine system.

Of 13 anti-ovine monoclonals a high proportion (7/13, 54%) was found to cross-react with deer cells (Table 1). Antibodies to the immunologically important CD4, CD8, and major histocompatibility antigens cross-react as did the anti-thymocyte monoclonal T6 (CD1). One of three anti-porcine leukocyte antibodies tested had cross- reactivity with a marker on a subpopulation of cervine monocytes and/or polymorphonuclear leukocytes.

These monoclonal antibodies were used to determine relative proportions of lymphocyte subpopulations in peripheral blood and lymph nodes of two species of deer, red and Père David's (Fig. 34.1). There was little difference between the two species in the proportions of lymphocytes in each subpopulation, and there was no significant difference between peripheral blood and lymph nodes except in the T19 subpopulation. This population was virtually absent from samples of mesenteric lymph nodes.

When lymphocytes from the thymus of a fetal red deer were compared with those from an adult, significant differences were seen (Fig. 34.2). Once again the T19 subpopulation was absent in both. The fetal thymus contained a large proportion

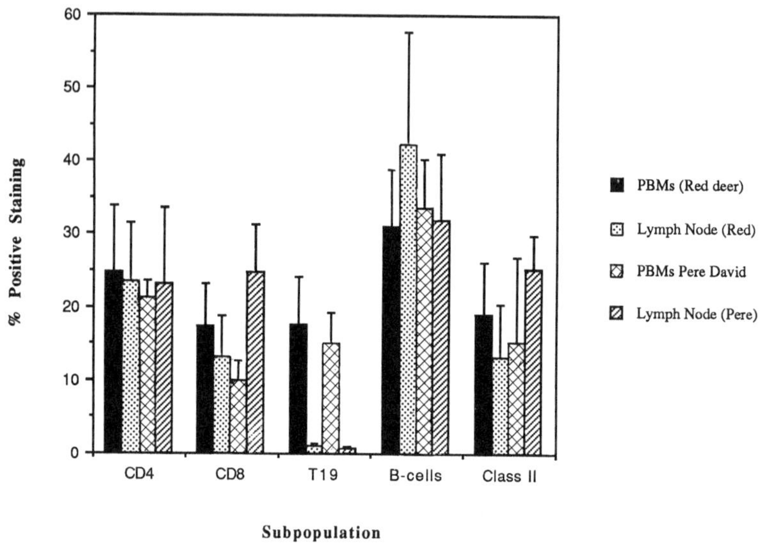

FIGURE 34.1. The relative proportions of B lymphocytes, MHC class II+ leukocytes, CD4+, CD8+ and T19+ T lymphocytes found in peripheral blood mononuclear cells (PBMs) and mesenteric lymph-node preparations from red deer ($n = 3$) and Père David's deer ($n = 3$) identified by staining with heterologous monoclonal antibodies and FACS analysis.

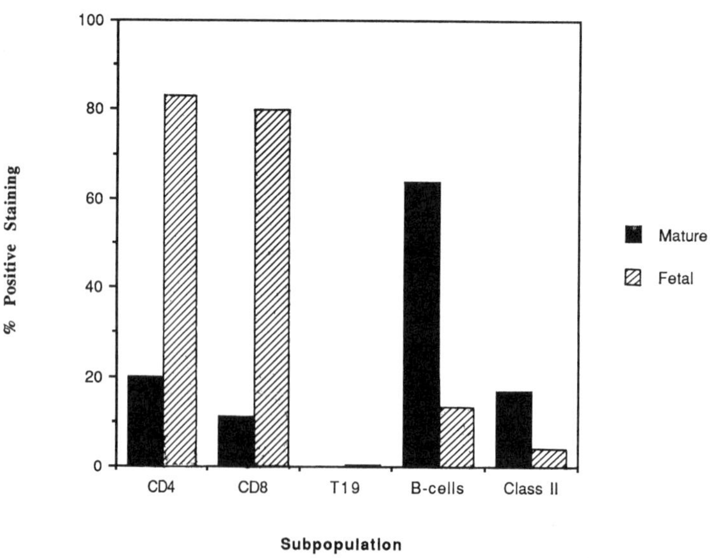

FIGURE 34.2. The relative distribution of lymphocytes among the CD4+, CD8+, B-cell, T19+, and MHC class II subpopulations in the thymus of an adult and fetal red deer using heterologous monoclonal antibodies and FACS analysis.

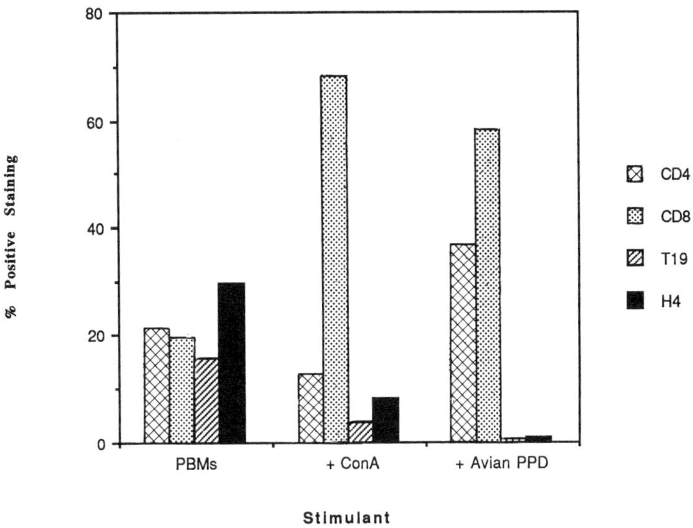

FIGURE 34.3. Changes in the proportions of red deer peripheral blood lymphocytes subpopulations upon in vitro activation with mitogen (concanavalin A) or avian purified protein derivative (PPD). Lymphocytes from an animal with high lymphocyte reactivity to avian PPD were cultured for 5 days with stimulant, then expanded for 9 days with recombinant bovine IL-2.

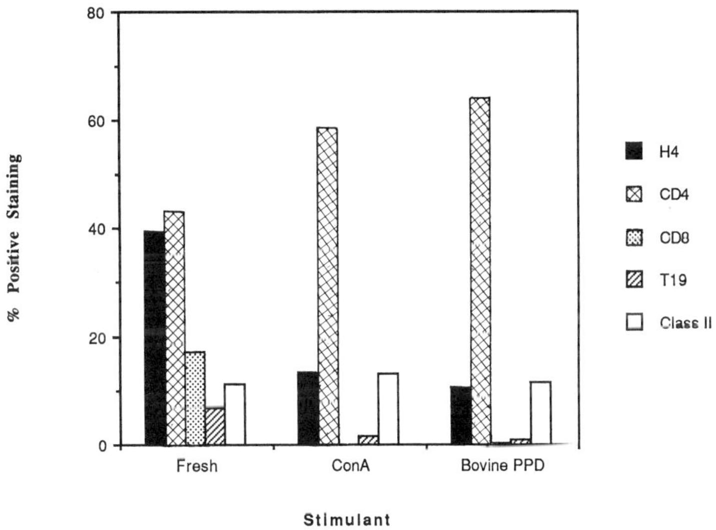

FIGURE 34.4. Proportions of tonsil lymphocyte subpopulations before and after in vitro stimulation with mitogen (concanavalin A) or the specific antigen bovine purified protein derivative (PPD). Lymphocytes from the tonsil of a tuberculosis red deer were incubated for 5 days with stimulant followed by 9 days with recombinant bovine IL-2

($\sim$80%) of T lymphocytes which were CD4+ and/or CD8+, few B lymphocytes and few activated (class II+) T lymphocytes. The adult thymus had a large proportion of B cells (63.8%) and relatively normal proportions of CD4 (19.7%), CD8 (10.9%), and class II (16.8%) positive lymphocytes.

The use of these cross-reactive monoclonals to identify cells which proliferate in vitro to specific antigen or mitogen was also investigated. Figure 34.3 shows that when grown in culture for 12 days in the presence of rIL-2, the majority of cells exposed to mitogen had the CD8 phenotype, whereas when exposed to specific antigen, avian PPD, both CD4 and CD8 cells proliferated. The numbers of B cells and T19-positive cells decreased with time.

A tonsil preparation from a tuberculous red deer was analyzed for changes in lymphocyte subpopulations during in vitro culture with bovine PPD antigen (Fig. 34.4). The phenotypic compositions of cells from the tonsil were different from the peripheral blood, and upon culture CD4+ lymphocytes grew out to the exclusion of all other phenotypes.

Discussion

A variety of species possess functionally similar subpopulations of leukocytes which express common surface determinants. The class II reactive (Th) population is identified by the CD4 antigen in man and the L3T4 antigen in mice. A class I reactive population is identified in man by the CD8 antigen and in mice by the Ly2 antigen. Whether antigens such as these are sufficiently conserved to allow for the use of heterologous monoclonal antibodies, raised against one species but used to detect a similar cell type in another species, has not been determined.

While immunological cross-reactivity has been described between major histocompatibility complexes of various species (Lunny et al. 1979; Aasted et al. 1988), the incidence of cross-reactivity in cell phenotyping appears to be much more restricted (Aasted et al. 1988). Our data show that the incidence of cross-reactivity is dependent on the species being studied. None of the human monoclonals used were found to cross-react in deer. Murine- specific antibodies were also found to give a low incidence of cross-reactivity. The solitary success was Pgp-1 which is directed against an antigen that is highly conserved among various species. These data argue against random association between markers on the cells and monoclonal antibodies tested as reasons for the observed cross-reactivity. It appears more likely that the subpopulations detected by cross-reactive antibodies vary between species in the proportion of total leukocytes they contain.

By choosing an appropriate heterologous donor species, in this case ovine or bovine, a much higher incidence of cross-reactivity can be obtained. This may reflect a closer evolutionary relationship between these species or alternatively that ruminants may have evolved similar immune systems to deal with a common antigenic environment. Four of the cross-reactive monoclonals (IL-A29, 197 SBU-T19, and CC15) appear to detect a subpopulation that is similar, if not the same, as that reported in the sheep by MacKay et al. (1986) and seems to be strongly conserved between species.

As pointed out by Aasted et al. (1988) a positive reaction with phenotyping antibodies does not necessarily mean that the cross-reactive monoclonal is binding to the same subpopulation of leukocytes as in the species to which they were developed. Functionally, we have evidence that SBU I and II are seeing similar molecules on cervine and ovine cells (Buchan and Griffin 1990).

The correlation between numbers of lymphocytes staining with H4 and anti-cervine immunoglobulin suggests that H4 does in fact stain cervine B cells. This observation is compelling because it shows that monoclonal antibodies to a surface antigen in one species do identify a functionally similar population, identified by a homologous marker (anti-immunoglobulin) in the host species. Double labeling with anti-cervine immunoglobulin monoclonals will verify this. In addition, the similarities in both reaction patterns and proportions of positive cells between ovine T19, bovine 197, and IL-A29 suggest that the these antibodies also identify a similar marker on cervine leukocytes.

We have begun to use these heterologous monoclonals to probe the immune response of deer, and red deer in particular, in order to better understand the immunological basis of disease problems associated with intensive farm management. Preliminary studies have shown that some subpopulations differ significantly in thier tissue distribution.

Lymphocytes with the T19 phenotype are virtually absent from lymphoid tissues such as mesenteric lymph nodes and thymus, despite being well represented in the peripheral blood and tonsil. These cells are reported to be null cells probably carrying the $\gamma\delta$-antigen receptor. In other species it has been suggested that lymphocytes with these characteristics are localized in specific tissues, usually near epithelial surfaces (Bluestone and Matis 1989).

Although limited to two animals, the phenotypic profiles found in the adult and fetal thymus are intriguing. First it would appear that instead of involuting with age, the cervine thymus develops to take on a role more akin to a lymph node, although a 21-year-old hind was found to have very little thymic tissue when autopsied. The high proportion of both CD4+ and CD8+ thymocytes in the fetal thymus suggests that many lymphocytes in the thymus are double positive (CD4+/CD8+), immature T lymphocytes, as has been reported in other species (Hokland et al. 1987). In adults this profile has altered to one in which the B-cell predominates while CD4+ and CD8+ numbers are similar to those found in the blood and mesenteric lymph node.

The ability to determine the phenotype of the immunoreactive cells responding to an antigen is a basic requirement when studying the immune response of an animal. Using heterologous monoclonals we have been able to detect differences in the type of lymphocyte that responds to a nonspecific mitogen such as concanavalin A, compared with a specific antigen such as avain or bovine PPD in both peripheral blood and lymphoid tissues. This will allow us to identify the phenotype of cloned T cells and determine which cell phenotypes are important in immune protection to cervine tuberculosis.

Until homologous, leukocyte, subpopulation-specific antibodies become available in newly domesticated species, valuable information on the immunology of these species can still be gained by using the huge number of monoclonal antibodies available as markers of leukocyte subpopulations in better characterized species. By searching for cross-reactive antibodies and then verifying that they identify a functionally similar cell type in the cross-reactive species, we will be able to more rapidly advance the study of immunology in the more exotic species of ruminants.

Acknowledgements

We would like to thank N. Hogg, M. Brandon, C. Mackay, G. Splitter, J. Elis, I. Morrison, C. Howard, Terasaki, D. Dobbelaere, R. Hyman, C. Hammerberg, Dakopatts (Denmark) and Becton and Dickinson for their kind gifts of monoclonal antibodies.

References

Aasted B, Blixenkrone-Moller M, Larsen E, Bielefeldt Ohmann H, Simesen R, Uttenthal A (1988 Reactivity of eleven antihuman leukocyte monoclonal antibodies with lymphocytes from several domestic animals. Vet Immunol Immunopathol 19:31–38

Baldwin C, Black S, Brown W, Conrad P, Goddeeris B, Kinuthia S, Lalor P, MacHugh N, Morrison W, Morzaria S, Naessens J, Newson J (1988) Bovine T-cells, B-cells and null cells are transformed by the protozoan parasite *Theileria parva*. Infect Immunol 56:462–467

Bluestone JA, Matis LA (1989) TCRgd cells—minor redundant T cell subset or specialized immune system component? J Immunol 142:1785–1788

Buchan G, Griffin JFT (1990) Identification of heterologous monoclonal antibodies that cros-react with cervine leukocyte subpopulations. Vet Immunol Immunopathol 245:247–257

Ellis J, Davis W, MacHugh N, Emery D, Kaushal A, Morrison W (1988) Differentiation antigens on bovine mononuclear phagocytes identified by monoclonal antibodies. Vet Immunol Immunopathol 19:325–340

Ezaki T, Miyasaka M, Beya MF, Dudler L, Trnka Z (1987) A murine anti-sheep T8 monoclonal antibody, ST8, that defines the cytotoxic T-lymphocyte population. Int Ann Allergy Appl Immunol 82:168–177

Gogolin-Ewens K, MacKay C, Mercer W, Brandon M (1985) Sheep lymphocyte antigen (OLA). Immunology 56:717–733

Griffin JFT (1989) Stress and immunity: a unifying concept. Vet Immunol Immunopathol 20:263–312

Hammerberg C, Schurig GG (1986) Characterization of monoclonal antibodies against swine leukocytes. Vet Immunol Immunopathol 11:107–121

Hokland P, Hokland M, Daley J, Ritz J (1987) Identification and cloning of a prethymic precursor T-lymphocyte from a population of common acute lymphoblastic leukemia antigen (CALLA)-positive fetal bone marrow cells. J Exp Med 165:1749–1754

Isacke C, Sauvage C, Hyman R, Lasley J, Schulte R, Trowbridge I (1986) Identification and characterization of human Pgp-1 glycoprotein. Immunogenetics 23:326–332

Lewin H, Calvert C (1985) Cross reactivity of a monoclonal antibody with bovine, equine, ovine and porcine peripheral blood B-lymphocytes. Am J Vet Res 46:785–788

Lewin H, Davis W, Bernoco D (1985) Monoclonal antibodies that distinguish bovine lymphocytes-T and lymphocytes-B. Vet Immunol Immunopathol 9:87–102

Lunney JK, Mann DL, Sachs DH (1979) Sharing of la antigens between species III la specificities shared between mice and human beings. Scand J Immunol 10:403–413

MacKay C, Maddox J, Gogolin-Ewens K, Brandon M (1985) Characterization of two sheep lymphocyte differentiation antigens SBU-T1 and SBU-T6. Immunology 55:729–748

MacKay C, Maddox J, Brandon M (1986) Three distinct subpopulations of sheep T- lymphocytes. Eur J Immunol 16:19–25

Puri N, MacKay C, Brandon M (1985) Sheep lymphocyte antigens (OLA) 2. Major histocompatibility complex class II molecules. Immunology 56:725–733

Trowbridge I, Schulte L, Hyman R, Trotter J (1982) Biochemical characterization and cellular distribution of a polymorphic, murine cell-surface glycoprotein expressed on lymphoid tissues. Immunogenetics 15:299–312

35 Social Status and Parasitic Infection in Red Deer Stags

Ludek Bartos[1], Pavel Rezac[2], Jiri Siler[1], Martina Buricova[3], and Petr Ziegrosser[4]

The number of parasites in a host population is often dispersed unequally so that a few hosts harbor most of the parasites. Social factors may be involved. In this pilot study, droppings of 12 stags of a "white" red deer (*Cervus elaphus*) bachelor group with known social hierarchy were collected. The samples were investigated for the presence of the first stage larvae of *Elaphostrongylus cervi*, *Varestrongylus sagittatus*, and *Dictyocaulus viviparus*. Only larvae of *Elaphostrongylus cervi* were detected in significant numbers. No signs of *Varestrongylus sagittatus* infection were found. A low density of *Dictyocaulus viviparus* was found in only 5 samples of the 12; hence, this parasite was omitted from further analysis. The deer were divided into two groups according to rank position of a stag: "high ranking" (the first to the sixth position within the hierarchy) and "low ranking" (all others). High-ranking stags had lower mean densities of the *Elaphostrongylus cervi* larvae per sample than the low-ranking stags (the first quartile, median, the third quartile: high-ranking 0.00, 0.50, 23.00; low-ranking 14.00, 32.50, 54.00; U-test, $U = 6.50$, $P = 0.03$). The results suggest that a relationship between social status of a stag and infection with some parasites may exist. Whether it is the parasite that influences the rank position of a stag or vice versa remains to be answered.

[1]Group of Ethology, Research Institute of Animal Production, CS-104 00 Praha 10-Uhrineves, [2]Institute of Parasitology, Czechoslovak Academy of Sciences, CS-370 05 Ceske Budejovice, [3]Faculty of Natural Sciences, Charles University, CS-128 00 Praha 2, and [4]State Forestry Enterprises, CS-538 42 Ronov nad Doubarovou, Czechoslovakia

36
Nonspecific Immunity in Sika Deer

E. Nivin*

We investigated phagocytosis, complement, lysozyme, and immunoglobulins in domesticated sika deer (*Cervus nippon*) in the south of the Primorye Territory. We determined the physiological status of cellular and humoral links of nonspecific immunity depending on season, age, and sex. We ascertained considerable fluctuation of activity of the immunologic indexes according to seasons, which illustrates physiological adaptation of the sika deer to changing environmental conditions. Most of the nonspecific immunity indexes were maximum in autumn, the period when mortality is minimal (10.6% of loss of all animals a year). In addition to seasonal changes of immunobiological indexes of sika deer, differences due to age and sex were also apparent. So cellulose factors of the immune defense were more active at a mature age; and indexes of humoral barriers increased during aging, yet greater activity of resources ensuring an oxidative reparative process was observed when deer were young. This demonstrates that physiological systems function intensively at different stages of development of the deer. Male sika deer have greater nonspecific immunity parameters (intensity of phagocytosis, concentration of lysozyme, titre of hemolysin, and glutinins) than does. These are probably connected with the level of metabolic processes.

*Primorksy Agricultural Institute, Ussuriisk, USSR

37
Paratuberculosis in Deer

J.F.C. Nyange, N.J.L. Gilmour, D. Buxton, and Hugh W. Reid*

Paratuberculosis (Johne disease) a chronic infectious disease of ruminants caused by *Mycobacterium paratuberculosis*, is potentially an important problem in deer farming. Primarily a disease of cattle, sheep, and goats, paratuberculosis is characterized by progressive wasting and diarrhea and granulomatous inflammation of the intestines and mesenteric lymph nodes. Infection usually occurs early in life, but the disease manifests itself after a long period of incubation. In deer, paratuberculosis affects a range of animals from yearlings to mature breeding stock. The clinical disease is characterized by a sudden onset, rapid loss of condition and weight and in many but not all cases, diarrhea; and it runs a more acute course than in other ruminants. The typical gross pathological changes of *M. paratuberculosis* infection in deer are often variable and mild, but the microscopic changes are often quite extensive. In severe cases, lymph node changes are sometimes associated with necrosis, mineralization, and reactive fibrosis. Characteristics of isolates of *M. paratuberculosis* from deer by culture, pathogenicity tests, SDS-PAGE, and Western blots did not show significant differences from the organism isolated from other ruminants. From experimental infection, the pathogenesis of paratuberculosis in deer was found to be basically similar to that of the disease in other ruminants.

The control of paratuberculosis in deer as in other ruminants is hampered by lack of sufficiently sensitive and specific tests for detection of infection in individual animals, particularly the clinically normal carriers. As an adjunct to improvement of general hygiene and pasture management, vaccination of calves has shown promising results in reducing the prevalence of infection in an affected herd.

*Moredun Research Institute, 408 Gilmerton Road, Edinburgh, Scotland

38

Pasteurella-Related Mortality in Park Fallow Deer

R. Putman, H. Davidson, and C.R. Goldspink*

In one enclosed population of fallow deer (*Dama dama*) in the north of England, major mortality events have been recorded in a number of years where up to 20% of the herd has been found dead or dying in midsummer. Deaths were concentrated among male deer, and particularly those of middle age. Postmortems showed extensive infection in each case by *Pasteurella septica*, but the actual cause of the epidemics was unclear. Why should deer in this population succumb when deer in other parks equally exposed to *Pasteurella* are apparently less susceptible? In an attempt to assess what factors may be causing some nutritional or other physiological stress among susceptible animals predisposing them to infection, we initiated in March 1989 a research program investigating differences in the behavior, ecology, and use of space by different age and sex classes of animals within the population, and also assessing differences in behavior, use of habitat, and diet of animals in this sensitive population by comparison with other local deer herds. Parasitic burdens are also being investigated and blood samples are being screened for various condition markers. In addition, the possibility that the animals are being stressed by purely human disturbance is considered.

*Department of Biology, University, Southampton, and Department of Biological Sciences, Manchester Polytechnic, Manchester, UK

39
Immobilization of Norwegian and Svalbard Reindeer with Medetomidine and Medetomidine-ketamine and Reversal with Atipamezole

N.J.C. Tyler, R. Hotvedt, A.S. Blix, and D.R. Sorensen*

The sedative action of medetomidine (-ketamine), a novel, selective and specific alpha$_2$-adrenoreceptor agonist, was studied in 12 captive Norwegian semidomisticated reindeer (*Rangifer tarandus tarandus*, NR), including four newborn calves, and in seven free-living Svalbard reindeer (*R.t. platyrhynchus*, SR). Medetomidine, with or without ketamine, caused effective, reliable immobilization in NR. Doses of 50 to 200 g/kg body weight medetomidine alone or 30 to 125 μg/kg medetomidine combined with $\geqq$ 300 g/kg ketamine induced complete immobilization good muscle relaxation, and persistent, deep sedation in NR; SR required higher doses. Atipamezole, a potent, highly selective alpha$_2$-adrenoreceptor antagonist, successfully antagonized medetomidine (-ketamine), resulting in smooth, rapid, and persistent reversal of immobilization in all cases (NR and SR). Both medetomidine and atipamezole had wide safety margins and no conspicuous lasting side effects after reversal. The combination of characteristics in medetomidine (-ketamine) and atipamezole of high potency, rapid action, good sedation, good reversibility, and safety was satisfactory both for capturing reindeer by immobilization and for minor surgery.

*Department of Arctic Biology, Institute of Medical Biology and University Hospital, University of Thomsø, Tromsø, Norway

40

Pathology of Naturally Occurring Diseases in Deer at Nandankanan Biological Park

A.T. Rao[1] and L.N. Acharjyo[2]

Gross and histological examinations were carried out on 88 Indian deer which succumbed at Nandankanan Biological Park during 1967–1988. Tissues were collected from 36 spotted (*Axis axis*), 20 barking (*Muntiacus muntjak*), 12 hog (*A. porcinus*), and 20 sambar deer (*Cervus unicolor*). Typical lesions of foot-and-mouth disease caused by "O" type virus were seen in four spotted deer and six sambar. Bluetongue-like disease was encountered in three barking deer along with a similar disease in other ruminants. Pulmonary tuberculosis was seen in all species. One case of pulmonary nocardiosis and one of rhinopneumonic zygomycosis was found in hog-deer. Fascioliasis was found in ten spotted deer and amphistomiasis in five sambar. Intestinal moneziasis was found in one barking deer, sarcocystosis in the heart of a sambar, and pulmonary hydatidosis in one spotted deer. Nodular hyperplasia in liver, perhaps caused by aflatoxins, was the most common lesion in hog-deer (ten cases). Among systemic diseases, nephrosis followed by pneumonia and hepatitis of undetermined etiology were common in all species. We encountered a case of rarifying osteitis affecting the nasal chamber of a sambar and a congential anomaly affecting the skeleton of a newly born spotted deer fawn. A trichobezoar and a trichobezoar-cum-phytobezoar were recovered from the rumen of one spotted deer. This study indicated that captive deer suffer from a wide range of diseases either alone or in combination.

[1]Department of Pathology, Orissa Veterinary College, OUAT, Bhubaneswar-751 003, India and
[2]Nandankanan Biological Park, Orissa, India

41
Causes of Mortality in Indian Deer at Nandankanan Biological Park

L.N. Acharjyo and A.T. Rao*

We completed necropsy wherever appropriate with histopathological studies on 351 Indian deer [140 Chital (*Axis axis*), 29 hogdeer (*Axis porcinus*), 83 sambar (*Cervus unicolor*), 94 barking deer (*Muntiacus muntjak*), and 5 thamin or brow-antlered deer (*Cervus eldi eldi*) at Nandankanan Biological Park from 1967 to 1989. The major causes of mortality were traumatic injuries (21.94%), pneumonia (15.10%), tuberculosis (9.40%), debility (7.69%), still-birth (6.27%), stress (5.70%), hepatitis (5.13%), drowning (4.27%), and predation (2.56%). Sporadic deaths also occurred due to toxemia, septicemia, colibacillosis, gastroenteritis, peritonitis, fascioliasis, maternal rejection, foot-and-mouth disease, bluetongue-like disease, pulmonary nocardiosis, zygomycosis, intussusception of intestines, and other miscellaneous causes. Furthermore, 61.25% of deaths in all species occurred in adults (2 to 10 years of age), whereas 38.75% were during the first year of life.

*Nandankanan Biological Park, Orissa, and Department of Pathology, Orissa Veterinary College, Bhubaneswar-751 003, India

42

Identification of 36 Strains of Mycobacteria from Deer

Wang Shuzhi, Cui Guoyin, Liu Qinghe, He Jingzhi, Wang Quarkai, and Cheng Yeguo*

During the years 1986–1988, tuberculosis was detected in farmed deer by tuberculin eye testing. Seventy-eight deer from ten deer farms in Jilin and Beijing were killed because of being OT positive. Ninety-six lungs and lymph nodes, which had gross lesions at necropsy, were collected from 48 of these deer. Thirty-six strains of mycobacteria were isolated from the lungs and lymph nodes, with an isolation rate of 75.0% (36/48). Seven were identified as *M. bovis* using identification media. Of 29 atypical mycobacteria strains, 4 were identified as Runyon's group 1, 9 Runyon's group 2, 12 Runyon's group 3, and 4 were Runyon's group 4. By further biochemical tests, 8 strains were identified as *M. avium–M. intracellulare* Complex (MAIC), 4 strains were *M. fortuitum*, 4 strains were *M. Kansasii*, 4 strains were *M. gordonae*, 2 strains were *M. szulgai*, 2 strains were *M. nonchromogenicum*, 1 strain was *M. scrofulaceum*, and 4 strains were without definite results. It would be better to first treat the deer specimens with 4% sulfuric acid solution. Sodium pyruvate medium is suited to isolate and culture *M. bovis*, while Lowenstein-Jensen (L-J) medium adapts to isolating and culturing all mycobacteria in deer. The growth of bacteria was facilated by adding 5 mg/ml cytochrome C and ubiquinone (CoQ) respectively to L-J media. It was demonstrated that staining techniques used in the detection of mycobacteria are Ziehl-Neilsen methods. Morphology, size, and cultural character of mycobacteria in deer are the same as those of other animals.

*Jilian Agricultural University, Changchun, People's Republic of China

Part 3
Management and Behavior of Farmed Deer

43
Requirements for Managing Farmed Deer

Jerry C. Haigh

Abstract

The successful management of farmed deer has several basic requirements. Among them are a knowledge of the industry; an understanding of the availability of the animals and a grasp of the biology of the species involved; an ability to handle the animals and a thorough understanding of the nutritional needs of the stock; and an awareness of the types of disease to which the animals are susceptible. The industry may involve anything from the operation of hunting preserves to the production of venison and velvet antlers as well as several ancillary products. An operation may be involved in one or more of these activities, and the manager must be aware of the developments within a given sector in order to ensure maximum profitability. The biology of the species must be grasped in order to allow proper management of reproductive programs, as well as variations in nutritional needs and behavioral aspects. Handling systems may vary from locale to locale and from species to species. The evolution of such systems has allowed a great deal of detailed husbandry to be practiced. A good handling system is an absolute must. Once a handling system has been developed, the single most important factor in any commercial deer operation is almost certainly the nutrition of the stock. The requirements may vary from season to season and among different cohorts.

Key words: Biology, confinement, deer, domestication, handling, management systems, nutrition, reproduction

Introduction

Any management system for farmed deer has several basic components. The first is that production animals should have appropriate characteristics to allow them to be domesticated, at least to some degree. Other basic components of a deer farm include confinement and handling, nutrition, reproduction, and prevention of disease. For the system to be successful, these components must be integrated and at the same time managed with an underlying grasp of the biology, including the behavior, of the animal in question.

I will concentrate my discussion on the farming of deer rather than on their ranching, herding, or hunting. Hudson (1989a) has developed a model that elegantly describes the degree of confinement and management intensity required in each of these activities, but it is important to note that his model has no rigid divisions and considers the entire concept in terms of a continuum. While subsistence hunting certainly lies in a totally different category than an intensive grazing system, hunting cannot be entirely dismissed as a farming practice when one considers the hunting ranches that have been established in many countries. These are sometimes run in conjunction with operations that involve many intensive farming practices. Furthermore intensive farming operations that are physically remote from hunting ranches may target the production of animals specifically for the latter, although the potential for growing public resentment of such a practice may curtail it.

Farming itself involves a variety of husbandry techniques that cannot be rigidly separated. For instance, large-scale farms of several thousand hectares may, within their own confines, have areas in which a specific group of stock may be managed in an intensive fashion. A range of techniques, again lacking rigid boundaries, would involve decreasing area from an extensive operation through semiconfinement, to an intensive system and finally to a feedlot. There seem to be two basic and opposite strategies for farming deer. On the one hand, an approach that involves as little direct management as possible is perceived as being desirable; on the other hand, a more intensive system with frequent handling is espoused. The decision on which method to adopt will largely depend upon available resources and capital costs, but the species chosen for the farm will also govern the method by which the animals are managed. In the United States of America (USA) and Argentina there are many deer farms on which minimal handling is practiced. In New Zealand intensive grazing systems are common, while in China virtually all deer are managed in feedlot situations.

The development of a business plan is the first step to establishing a deer farm. Without this step, the entire enterprise is fraught with risk. The establishment of objectives is a primary concern. These may be governed by legislation in some cases. For instance in the USA, some states specifically ban deer farming. In the United Kingdom (UK) and some states in the USA, velvet antler production is proscribed (Fletcher 1987). In Canada the rules governing species that may be farmed vary from province to province.

Objectives in deer farming include the production of meat (mainly venison), the production of medicinal compounds (mainly from velvet antler), the production of breeding stock, the production of trophy deer for hunting, the development of a tourist facility, the preservation of genetic diversity, and finally the production of musk. By-products exist in many systems. Within each of the objectives there may be differences of detail, both in management and market. For instance, requirements for the velvet antler trade differ from place to place (Pinsan 1982; Renecker 1987; Sim 1987; Fennessy 1989a; Wallis 1989). Currently, the market for antler as a medicinal product lies predominantly in the Orient and in Russia, but at one time the medicinal properties of deer parts, including antlers, were a subject of interest in Europe (Turbevile 1576a). Even befoe Turbevile, Pliny, writing in the first century A.D. ascribed medicinal properties to deer parts (Bostock and Riley 1855).

In many instances, production objectives can overlap, but they can also be diametrically opposed. The choice of which of them will be targeted is governed by a variety of factors. The most important will always be the marketplace. For instance, in China, velvet antler is the most important component, and even the by-products of slaughter of surplus animals are more valuable than the meat. Indeed the meat may be processed into components of traditional medicines (Pinsan 1982). Musk deer (*Moschus* spp.) are predominantly farmed in China, where the value of the musk lies both in its role in traditional medicine and in the production of material for scent (Macartney 1983; Qin and Liang 1983; Zhang 1983; Green 1989). On the other hand, in Mauritius, the production of venison is the prime objective (Lalouette 1985, Drew et al. 1989). In Canada, no wapiti (*Cervus elaphus*)-based venison industry has been developed. This is in part due to the small numbers of animals on farms, but this segment of the industry could be further governed by both high velvet prices and the fact that mature males in hard antler can be sold to hunting ranches. However, there is a growing North American market for venison from other species.

Deer venison should not be confused with meat products of other species that have also been called venison. Most marketable venison has been produced from juvenile (16- or 27-month-old) males. The rigid criteria of some associations that govern the naming and selling of the product as venison are designed to ensure that the consumer is offered a consistently high quality product. The objective for the farmer must be to get the product to market weight as quickly as possible, at the most economical price. There is even debate on whether the industry should be "market led" or "product driven." Disciples of the former tend to see a need for manipulation of the animals in order to provide a market at all times. The latter system requires that the natural cycle of the deer be used to its fullest advantage when it is available.

Obviously velvet antler production requires the retention of numerous young male animals, and it is the group that also provides the venison. Commercial quantities of velvet are seldom available from red deer (*Cervus elaphus*) stags under 3 years of age. In New Zealand at the moment, there is a tendency to retain this cohort because of the high prices for velvet antler. This inevitably creates a shortfall in venison supplies. If velvet antler prices decrease, there could well be a large number of slaughter animals that will not conform to the rigid age and grade requirements of a high quality venison trade.

The scale of breeding stock within any given jurisdiction is often governed by the stage of development of the industry. In North America, the demand for farmed stock of cervine deer currently far exceeds the supply, even to the extent that importation of deer from New Zealand and Europe has taken place. This mimics the scene in New Zealand as little as 6 or 7 years ago. In such situations almost anything may be acceptable to the buyer. More discriminating buyers, or those shopping where there is an adequate number of animals, will be more interested in quality than quantity (Yerex and Spiers 1987; Haigh and Moore 1990). Records will become an essential component of any farm management system when breeding stock are being offered to a mature industry. There is a small white-tailed deer (*Odocoileus virginianus*) farming (as opposed to ranching) industry that currently supplies superior bucks for use as breeding animals for prices ranging from \$15 000 to \$50 000 (Amass, personal communication).

Almost any species of deer, and indeed often a mix, can be successfully used in a tourist situation. Sometimes the tourist option is used as an extra way to generate farm income, and has the added benefit, if so structured, that it can be used to educate the public about other aspects of the industry.

Domestication

A critical component for the successful establishment of any new domestic animal is availability (Fletcher 1986a). There can be little doubt that the success of the New Zealand industry over the past 20 years, should be ascribed to the ready availability of stock, especially red deer (Challies 1985). Indeed the rapid growth seen there cannot be matched in most other parts of the world because of a lack of an abundant, readily available, free-ranging source. Other countries in which deer farming has been or is becoming important have either had an established industry for many years (China), or have growth rates governed by obtainable stock (UK, Canada, USA, and others).

Galton (1865) listed six requirements for successful domestication of species: The animals should (1) be hardy, (2) have an inborn liking for man, (3) be comfort-loving, (4) be found useful, (5) breed freely, and (6) be easy to tend.

Clutton-Brock's (1987) interpretation of criteria 2, 3, and 6 respectively suggests that the animals must be social, with behavioral patterns based upon dominance heirarchy; not highly adapted for instant flight; and gregarious so that they will keep together and can easily be controlled by a herdsman. The need for hardiness, usefulness, and reproductive success is self-evident. To the need for availability, Fletcher (1986a) added the need for variability (by which he meant an ability to hybridize), marketability (almost the same as useful), and a preference for grazing rather than browsing eating habits. He also suggested that the animals should preferably be polygamous in order to reduce the number of males required in a herd.

There is a real need to distinguish between the two main methods of domestication, imprinting and habituation (Parker 1983). There is no doubt that deer species can be habituated and will become exceedingly quiet and tolerant of human presence. Imprinting of deer on to humans is a technique that has been described since ancient times and was advocated by the Roman agricultural author Columella. He suggested that the introduction of one or two hand-raised animals could help train the rest of the herd (Forster and Heffner 1968). It is almost certain that he would have been talking about female deer, as it is well known that imprinted entire male deer become exceedingly dangerous when in hard antler.

By and large, the Galton criteria can be well fitted to the various species of deer that are being successfully farmed in many parts of the world.

The members of the genus *Cervus* appear to fit the pattern most closely, and this genus is the most common cervid on farms (Fletcher 1986a). Of the six criteria, the one that fits least well is the tendency to instant flight. It takes weaned calves a considerable time to settle, even if they are trained through handling systems. Selection for temperament, and by implication increased domesticity and reduced flight response, is considered an integral part of deer farming programs (Beatson 1986; Milne 1987; Yerex and Spiers 1987). A successful domestication program for fallow deer (*Dama dama*) has been reported (Hemmer 1987). Reports from China (Pinney 1981) tell both of a remarkable translocation of a herd of deer by herding along public roadways over a distance of 100 km, and of techniques for herding deer using methods more often associated with domestic sheep and cattle. Fletcher (1987) recounted reports of red deer movements by herding between towns in medieval England.

Two other species that are farmed with some success are the fallow deer and the chital or axis deer (*Axis axis*). Again the Galton criterion which they fit the least well is related to flight, which appears to be more highly developed than in *Cervus*. However, there are farms that have settled their animals by training and taming to such a degree that handling problems associated with flight are overcome. There are records from the nineteenth century of fallow deer being herded from one location to another (Fletcher 1987).

An example of a species which has failed to become widely established as a farm species is the moose (*Alces alces*). It does appear that moose can be readily domesticated and historical records in both Russia and Canada indicate that they have been used as draft animals for at least 300 years (Galton 1865). However, they are more of a solitary than a herd animal and their high productivity is balanced by high mortality rates (Renecker 1989; Syroechkovsky et al. 1989). They can even be hand-milked (Yazan and Knorre 1964; Renecker 1989), but efforts in both Russia and Canada to develop large-scale moose operations have not been successful. Hemmer (1987) rates the moose as one of the least likely species, from a behavioral viewpoint, to succeed as a farm animal.

One of the many species translocated to New Zealand in the nineteenth century was the white-tailed deer (Challies 1985), but this species has not become established under New Zealand farming conditions. Part of the reason may be that they are not a herd animal, even though they can be seen in large aggregations at certain times of year. They are more excitable than most other species and even fawns will jump a 2.13 m fence, especially if put under herding pressure (Amass, personal communication).

Biological Considerations

Successful management systems for farming deer depend upon a sound understanding of several basic needs of any farming operation. It would be naive to avoid applying many of the basic tenets of traditional livestock practice to these species. Because deer are rediscovered species, as far as their farming is concerned, there is perhaps a greater need to understand and apply information relating to their biology than would appear to be the case in the farming of more common domestic species.

The student of biology will gain enormous benefit when it comes to the husbanding of the various deer species. The biological considerations that must be taken into consideration cover ethology, reproduction, and nutrition and to some extent disease. Underlying the reproduction and nutrition fields must be the recognition that tropical and temperate species differ from one another. Even the discussion of these differences must be tempered by the fact that a plastic species such as the white-tailed deer has a rigid breeding season in temperate zones, but in the tropics conforms to no apparent calender (Verme and Ullrey 1984).

Within the deer family, there is another division that appears to be reflected in the degree of success with which various species are being farmed. This is the life history strategy of "r" and "K" selection which Harrington (1985) used to differentiate deer species. A tendency to early maturity and multiple births makes eminent sense if projected on a spreadsheet. Based purely on these two criteria, the "r" selected moose, white-tailed and mule deer (*Odocoileus hemionus*) would be obvious choices as farm animals.

Harrington (1985) states that they tend to be less disease resistant than the "K" selected cervine deer (*Cervus*, *Dama*, and *Axis*). This division is certainly borne out by examination of 15 years of data comparing medical records of wapiti, sika deer (*Cervus nippon*), fallow deer, moose, mule deer, and white-tailed deer at one facility (J.C. Haigh, unpublished data). Disease susceptibility has been cited as a reason for discontinuing studies on farmed moose (Renecker 1989; Syroechkovsky et al. 1989). An "r" selected species that continues to be farmed, despite problems typical of this group, is the musk deer. In this case the great value of the product creates a situation in which the usefulness supersedes the disadvantages (Macartney 1983; Qin and Liang 1983; Zhang 1983; Green 1989).

Confinement and Handling

Deer in Greek, Roman, and European medieval times were enclosed in a variety of boundary fences and were farmed in substantial numbers using many of the techniques employed today. These included predator control, supplementary feeding, wild source restocking, and venison production (Hudson 1989a). The decline in farmed deer numbers in Europe has been linked to the devastation caused by the black death and the disintegration of the feudal system (Fletcher 1989). Modern deer farming has evolved in concert with the development of custom-designed fencing that has allowed the retention of animals inside specific enclosures (Hudson 1989a).

For any farming enterprise to really succeed, some sort of handling will periodically be required. This may be simply for an annual removal of velvet or hard antler, more frequent administration of anthelmintics, or specialized interventions such as are required for the manipulation of estrus cycles and artificial breeding. In some situations, immobilization by remote injection is considered a standard tool (Fletcher 1989). This particularly applies where the removal of antler in velvet is not permitted (such as in the UK). In these cases, aggressive behavior would create severe problems if deer were yarded for antler removal after the onset of rutting behavior. Virtually everywhere that deer are farmed, systems of handling complexes, most often in a shed, are linked to an alley and chute. There are numerous designs which make use of reduced light, movable gates, circle or half circle tubs, and finally chutes that may have drop floors, belly bands for lifting deer, or consist of darkened box-like structures with access panels.

One of the reasons for managing the breeding seasons of tropical deer is that yarding of mixed age and sex mobs is likely to lead to complications and trauma. Chital have traditionally been considered a difficult animal to handle (Thomson 1986), but the segregation of mobs has helped improve this component of their management (English 1987).

Nutrition

Nutritional requirements of deer vary according to reproductive status and, in temperate species, season of the year. Different species of deer have different requirements, which have been broadly categorized by Hofman (1985). Farmed deer tend to fall into the intermediate or mixed group. Indeed the concentrate-selecting deer species pose problems for the farmer in that suitable food may be difficult to obtain on a regular basis. Metabolizable energy requirements have been published for the most commonly farmed species (red deer and fallow deer) (Fennessy 1982; Adam 1986; Suttie et al. 1987; Mulley 1989a). However, it is important to recognize that these requirements are guidelines. For instance, the requirements of red deer may increase by as much as 50% if they are exposed to outdoor conditions, instead of being indoors (Adam 1986; Suttie 1987). It is also important to recognize that temperate species, relying on the cue of photoperiod, have evolved an efficient reproductive system that allows for calf production and lactation when natural forage is at its highest nutritional value. Requirements vary markedly through the year, being much reduced in winter, and it is this variation that allows the deer farmer to reduce costs, both in labor and feed bills. Local conditions and stocking rates will dictate the degree to which supplementary feeding is required, as will the type of feed that is locally available at the most economical price.

The feed requirements of farmed tropical

species of deer that have evolved without the benefit of a photoperiodic cue must be primarily geared to meet the needs of lactation and breeding in females, and of the rut in males. It has been reported that birth season peaks occur in response to plant phenology (Dinerstein 1982; Mishra and Wemmer 1982). As parturition in nature may occur almost at any time, nutritional needs may not really be related to pasture condition (Denholm 1984). Therefore management of these species on farms in temperate zones may require timed matings in order to match the two.

Temperate species of deer show a distinct voluntary appetite cycle (Adam 1988). While thermal conditions undoubtedly govern nutritional needs, it is the adaptations to available metabolizable forage energy that are of greatest significance to the animal's productivity (Moen 1985). Studies have shown that while overfeeding during times of reduced need (such as winter) may not be harmful for pregnant wapiti, they are certainly wasteful (Kozac 1988). Animals that are fed to excess in winter will not gain as much weight in summer as will properly fed individuals. Summer grazing will balance the situation; the lighter animals will gain more weight (Hudson et al. 1985). Good nutritional managers must take into account not only the given needs, in terms of metabolizable energy, but also such factors as shelter, shade, and season of the year. In the United Kingdom there is an increasing tendency toward winter housing of deer (Milne 1987; Fletcher 1989). This not only reduces feed requirements (Adam 1986), but has the obvious potential to spare pastures from damage when conditions may be wet and muddy and grass is not growing.

Supplementary feeding is used to increase stocking rates, balance rations, particularly trace minerals, and to assist in habituation (Hudson 1989b; Yerex and Lindeman 1982). The highest stocking rates are probably seen in China, where deer are fed a variety of nutrients in feedlot situations (Pinney 1981; Qin and Liang 1983). In one case, it is reported that 800 deer were kept in 15 enclosures in an area of 40 000 square meters (Pinsan 1982).

Feeding management also provides an opportunity for segregation of the various classes of stock on a deer farm (Denholm 1984). In China, where deer are maintained in the equivalent of a feedlot, males, females, and weaned calves are kept separately and fed different rations at different times of year according to production needs (Pinsan 1982). Stratification within the industry exists just as in other livestock operations (Hudson 1989b). Milne (1987) has described how stock are moved from Scottish farms where breeding is carried out to finishing operations where high quality forage is available.

Depending upon local conditions, deer may be hand-fed at any season of the year. In temperate climates, this is almost always during winter months, and it is at these times that deer, particularly female deer, have their lowest requirements. Indeed the requirements of lactating red deer, wapiti, and fellow deer in summer are virtually double those of the winter period (Fennessy 1982; Adam 1986; Hudson 1989b; Mulley 1989a). This allows the farmer the opportunity to take advantage of natural grass growth to supply a maximum amount of feed when it is required. Where deer have been translocated, maximum grass growth may not always be in concert with reproductive and lactational needs. This will mean that supplementary feeding may be needed at times other than winter. The manipulation of reproductive cycles and the changing of parturition dates have been one of the tools used in New Zealand and Australia to either take advantage of early spring pasture growth or avoid drought or winter conditions. The latter particularly applies to the tropical species of deer that may have year-round calving if left unmanaged.

Different members of a deer herd may have different nutritional needs at a given time of year (Moore 1988b). For instance, male deer require supplementation after the end of the rut. This is particularly so for tropical deer that are farmed in subtropical or temperate zones to which they have been translocated (Denholm 1984). During the intense period of the rut, mature male deer may hardly eat at all and can lose as much as 20% of their body weights (Denholm 1984). Supplementary feeding is essential to restore some of this lost condition. The consequences of an inadequate supply of feed at this time include an increased likelihood of death during the winter (from so-called winter death syndrome) (Denholm 1984). An increased production of antler velvet has been associated with increased prerut weight in the preceding autumn, as well as winter nutrition (Muir and Sykes 1988).

It has become standard practice to avoid overfeeding pregnant females in late gestation, as overfat, lazy wapiti and red deer hinds are inclined to dystocia. This is particularly so when cross breeding between widely disparate sized animals is practiced, for instance, wapiti and red deer (an increasingly common practice in New Zealand) (Pearse 1988), or even Xinjiang red deer (*C.e. xanthopygus*) and spotted or sika deer (*Cervus nippon*) as practiced in China (Drew et al. 1989).

Maximal weaning weights are important for production. They are linked to birth weights, but are particularly relevant in the juvenile female cohort. The proportion of females that will conceive at the age of 16 months is directly linked to their weaning weights at 3 months (Asher 1988). The importance of nutrition on overall calving percentages and therefore economic return has been demonstrated in red deer. Not only were weaning weights important, but subsequent nutrition targeting increased weight at 16 months had a marked effect on calving rates (Hamilton 1982, 1987).

The subject of weaning remains an area of contention in deer farming circles. The debate centers not only on the correct time to wean (before or after the rut), but even in some cases, whether to wean at all. Studies on free-ranging deer, most notably on the island of Rhum, have shown that hinds rearing calves came into estrus later than dry hinds. Lactating hinds in poor nutritional condition may not conceive at all (Guinness et al. 1978; Clutton-Brock et al. 1982). Loudon et al. (1983) showed that hinds on good quality pasture came into estrus earlier than those that were on a low plane of nutrition on hill pasture.

Lactational anestrus does not appear to occur in farmed fallow deer (Asher 1988). Studies indicate that the onset of estrus and the degree of synchronization of wapiti hinds are not improved by weaning. There is even an indication that, unless carefully managed, hind calves weaned before the rut may be less likely to conceive as yearlings because they are not as well grown as those weaned later or not at all (R.J. Hudson, personal communication).

Even if there is little or no effect of weaning upon the females, there may be advantages to the practice when it comes to management of the calves or fawns. Prerut weaning may be an advantage when reproductive manipulation or single sire mating is planned. Weaned juveniles can be preferentially fed and also trained to handling systems (English 1987; Asher 1988). Furthermore, selection of hinds in a breeding program can be based upon calf performance, weaned fawns can be yarded for treatments without the problems that would occur if the entire mob, including a stag, had to be handled, and early weaning may be better than late weaning because the young are less likely to be subjected to cold stress at the same time as weaning stress (Yerex and Spiers 1987). However, it is certain that weaning is the single most stressful event in the first year of a young deer's life, and deaths associated with this stress must be prevented by appropriate management.

Reproduction

There are profound differences in the reproductive strategies of temperate and tropical species of deer. Temperate species, responding to photoperiodic cues, have a well-defined rut, during which the bulk of breeding occurs. Tropical species, lacking such a cue, can breed almost at any time, although in some circumstances there may be a tendency towards some degree of seasonality (Turbevile 1576b; Dinerstein 1982; Mishra and Wemmer 1982; Lincoln 1985; English 1987; Fletcher 1986b). For some species of temperate deer, the rut is well defined and will occur at more or less the same time independent of the latitude at which the animals are maintained (Fletcher 1974). This is not always the case; the most plastic species of all is the white-tailed deer, which is highly seasonal in the northern part of its range, and less so in central America and northern South America (Verme and Ullrey 1984).

It is absolutely necessary to grasp that the rut of temperate deer is not equivalent to the entire potential breeding season. Both males and females are fertile well past the end of the rut (Asher 1988; Haigh et al. 1985). Serious consequences are inevitable if farm practice does not restrict the breeding season. Most reports indicate that a period of 6 or at most 8 weeks is adequate. There is evidence that a high proportion of females above a threshold minimum weight will conceive at the first estrus (Asher 1988; Wilson 1984; Hudson et al. 1991), so it is not the adult, well-nourished hinds or does that are of concern in an unmanaged system. Problems arise when yearlings that are underweight are run with males. These females may reach pubertal weights late in the breeding season. The consequences of allowing unrestricted mating are that their young will be born so late in the year that not only is survival unlikely, but the dams may not even breed in the following year. The normal calving season for wapiti in the northern hemisphere is in late May and most of June. The latest calving of which I have a record is 23 November. This implies a breeding in mid-March.

For similar reasons, it is common practice to restrict breeding seasons of tropical deer when they are farmed in temperate or subtropical climates. This ensures that young are born at a time of year when feed availability is optimal and calf survival most likely (English 1987; Hart 1986). As chital are capable of producing viable semen at any stage of the antler cycle, and may remain in hard antler for variable periods, occasionally exceeding 12 months, reproductive management and segregation of sexes are essential (English 1987).

Puberty in some "r" selected species of deer often occurs in the first year of life, and it is not uncommon for white-tailed deer and mule deer to achieve pubertal weights and be fertile at 6 months of age (Verme and Ullrey 1984). Musk deer are also capable of breeding in their first year of life (Qin and Liang 1983; Green 1989). On the other hand, puberty in "K" selectors occurs in the second year of life. The onset of estrus is governed not so much by age as by weight and occurs at about 70% of adult weight. There is a direct relationship between weights during the rut and percentage conceptions of wapiti yearlings (Hamilton 1987; Hudson et al. 1990).

There is an increasing interest in the manipulation of breeding seasons of temperate deer. This subject has been extensively reviewed by Wilson (1989) and Fennessy and Fisher (1988). The reasons are mostly related to attempts to either produce a calf crop at a time when natural forage is at its most nutritious, or to bring forward the calving dates and therefore allow a longer period for growth to slaughter weight to meet market demands (Yerex and Spiers 1987). The methods have included attempts at hybridization with Père David's deer (*Elaphurus davidianus*), because of their earlier breeding season (Yerex and Spiers 1987; Asher et al. 1988), nutritional management and more recently manipulation with a variety of hormonal regimes (Wilson 1989).

A management strategy that has attracted a good deal of interest in several deer farming regions is the use of large strains of stags over smaller hinds, either in natural mating situations (Qin and Liang 1983; Adams 1986; Milne 1987; Fennessy 1989b) or by artificial insemination (Pearse 1988; Haigh and Bowen 1991). This allows the maintenance of females of relatively small body size, which can be more densely stocked than larger animals. The objective is to produce rapidly growing offspring that either achieve market weight sooner than smaller young, or that grow to a larger size in a given period of time (Fennessy 1989b; Yerex and Spiers 1987). Artificial insemination, principally in the interests of increased velvet antler production, is also practiced in New Zealand and China (Pearse 1988; Qin and Liang 1983).

Disease Prevention

As greater numbers of deer are confined and farmed, and more research is conducted, it has become evident that deer species have some specific disease problems and share some diseases with other classes of livestock. Writing in 1969, at what appears to have been the first deer farming convention of modern times, at least in the West, McDiarmid concluded that of 31 well-recognized diseases of domesticated stock, only 7 occurred in deer, as compared with much higher numbers in sheep and cattle (McDiarmid 1969). Much research since then has elucidated many of these problems, but there is little doubt that the most important causes of loss on deer farms are related to mismanagement rather than infectious diseases. Trauma, malnutrition, and out-of season parturition are all largely preventable. It is an old adage that prevention is better than sure, and in any successful livestock management scheme attention to this subject alone will have a great effect upon the economics of the operation (Morris 1988). It should be remembered that once proper handling facilities have been constructed, more can be achieved in the area of disease prevention by good nutritional management than by any other procedure. Vaccination and anthelmintic treatment may be needed in some circumstances, but these vary according to locale and the species being farmed. There are circumstances everywhere in which preventive management, by screening and/or quarantine, is essential.

Techniques

Production systems for each of the listed objectives in deer farming may differ in detail. Selection of suitable stock at start-up and their continued evaluation year-by-year are particularly important for venison, velvet, and breeding stock operations. Genetic selection for superior stock is evolving from the use of phenotypic characteristics to the use of progeny testing methods (Beatson 1986; Fennessy 1989b; McCall and Parrat 1988; McManus 1989). While velvet and venison are to some extent mutually exclusive, in terms of marketable product, the recognized links between growth rates, body weights, and velvet antler weights allow the farmer to use similar criteria in selection (Moore et al. 1988a; Muir and Sykes 1988). A small percentage of the top animals in each category can be retained as potential breeding stock. As most deer farms rely on growth rates to slaughter weight and velvet antler production, buyers will take particular note of these parameters in selecting breeding animals (Yerex and Spiers 1987).

Many management systems rely entirely on venison production. The majority of deer farmers aim to produce a marketable carcass at 14–17 or 26–28 months of age (Fletcher 1989). Recent developments with cross breeding may allow younger animals to be used. By taking advantage of winter inappetance, and growth spurts the following summer, maximum use of available forage can be made (Drew 1985). The effects of castration on both red and fallow deer growth have been studied. Little advantage was found to this treatment for red deer (Blaxter et al. 1974), but in fallow deer the technique has gained acceptance in some quarters. Although carcass weights of castrates were reduced when compared with intact males, the proportion of the high priced cuts on the carcass was increased (Mulley 1989b). This makes the economics attractive not only because there is little difference to the producer, but the handling of castrates is much easier than intact males, and fighting in close confinement is not a factor. This has a major impact upon the degree of carcass bruising and trim (English 1987; Mulley 1989b).

Another technique that has gained some attention is the polling of young males before antler development occurs (English 1987; Milne 1987). This requires surgical removal of the specialized preantler periosteum from the frontal bone. It has the advantage that at no time does the stag have any head weaponry with which to inflict damage on cohorts or handlers, but it may be more of a management tool for easy handling rather than a component of venison production, as the cost of the procedure may overshadow the economic return. Therefore it is predominantly used for potential herd sires.

It is only members of the genus *Cervus* and *Rangifer* that produce velvet in marketable quantities, although small amounts of fallow velvet do find their way onto the market. Velvet is a mainstay of some segments of the industry. Most velvet from farmed deer is derived from wapiti, red deer, or sika deer, although large quantities of velvet are also produced from reindeer, especially in Russia (Fennessy 1989a; Wallis 1989). Improving quality, both in terms of weight and configuration, is an important goal for deer farmers, and as has been shown, such improvements can be achieved through selective breeding within a subspecies, or by cross breeding, especially with the largest member of the genus *Cervus* (Qin and Liang 1983; Beatson 1986; Fennessy 1989b; Zhu and Wu 1985).

It is seldom that top velveting stags will be sold

into the hunt ranch market, certainly not at least while the prices for velvet remain high. The balance must be struck between velvet quality, usefulness as a sire, and the likely survival of the animal over winter. The development of artificial insemination allows the storage of large amounts of genetic material that can be used in future years, so that such an animal is not necessarily lost to the farm population (in the genetic sense) when he leaves the property. The hunt ranch operator requires males in hard antler, and these antlers must usually be of superior quality, judged by the standards of the species and the day. Such animals may be used either for genetic improvement of the ranched herd, or directly as potential trophies for clients (Morrill 1988). The management of these animals after they shed their velvet, and their transportation to the ranch will require quite different techniques than those that lack antlers. They should not be offered the opportunity to compete across fences, they should not be confined with antlerless males, and they must be transported in custom-designed vehicles that allow sufficient headroom and segregation of each individual.

The development of a tourist facility is a specialized component of deer farming that may be conducted hand-in-hand with other aspects of the industry. An attractive environment, and the use of extremely quiet animals, selected above all for temperament, are important in such an enterprise. In particular the use of supplementary feed and the training of animals to come at a given signal may be components of the operation.

Species

The "r" selected species of deer are those that are also most commonly farmed worldwide. Their hardiness and longevity have established them in many locales. The most widely distributed genus of deer in nature is *Cervus*. It is probably no accident that they are also the most widely farmed. Above all other genera they have the potential to be cross-bred or hybridized (the choice of word depends upon which taxonomic camp the reader chooses to join). This provides the farmer the opportunity to develop an ideal animal to meet the needs of the marketplace (Fletcher 1986a, b). They will thrive under a variety of nutritional regimes, particularly grass, but they are farmed in systems that range all the way from feedlots to extensive "hands off" operations. Apart from meat they produce a variety of saleable products of considerable value.

The next most commonly farmed deer is the fallow deer, and these too have a global distribution on farms, having proved themselves able to thrive at widely differing latitudes. Management practices are very similar to those for red deer, with the exception that handling facilities and techniques have evolved in a slightly different direction to meet the temperament of the animals.

There are large numbers of axis deer being ranched in Texas (Harwell et al. 1987), but from a farming point of view most of the development has been in Australia. It is evident that most of the basic principles of livestock farming also apply to this species. The most important specific requirements call for reproductive management to ensure calvings timed to meet forage availability, and handling systems that allow control of the animal's characteristics.

Père David's deer are farmed in a few locales, and their management appears to differ little from that for red deer or wapiti, apart from details relating to reproductive seasonality and antler growth (Gray, personal communication). However, they require very careful handling, and are particularly susceptible to malignant catarrhal fever.

The "K" selected deer species are much less widely farmed. Their dietary requirements, solitary nature, disease susceptibility, and apart from *Alces*, their flightiness have made them difficult to manage.

One of the most valued animal products in the world is musk, derived from male musk deer. The production of musk is a specialized deer farming system conducted predominantly in China. It has lately developed considerably due to the fact that wild stocks are being depleted by overhunting (Macartney 1983; Green 1989). Published reports indicate that dietary requirements, solitary nature, and disease susceptibility make this a difficult species to farm (Qin and Liang 1983; Zhang 1983; Green 1989).

The white-tailed deer is the most abundant species in North America, but they are not widely farmed. Ranches, particularly in Texas, rely upon them to generate considerable income from hunting leases. There are a few white-tailed deer farms becoming established in North America. In general the best management system for them on commercial farms is considered to be a "hands-off" approach, and there appears to be a direct relationship between the amount of handling and the incidence of disease and mortality (K. Amass, personal communication). Fences should be at least 253 cm high and have small mesh near the bottom. Handling systems, where developed, work well if walls in approach alleys are high (300 cm) and principles similar to those for fallow deer (English 1984) are followed (Schmitt et al. 1983). In farmed situations, white-tailed deer do not appear to be as hardy as cervine deer, and regular parasite control programs are important (K. Amass, personal communication). They do well in adequate-sized pens on good diets with minimal handling, but in close confinement with frequent handling losses can be high-up to 33% over a 2-year period. The management of antlered deer appears to differ from place to place. If antlers are left in place, males of 3 years of age and older that are kept in wooded lots must be segregated in order to avoid fighting losses (K. Amass, personal communication). On the other hand, as many as six antlered bucks to a 1.5 ha. pen have been satisfactorily kept for several years. The difference appears to be that the animals can see one another at all times, and once they establish a dominance heirarchy among themselves fighting seems to be minimal. Breeding bucks are de-antlered before being placed with females (J.D. Nellis, personal communication).

A 30-year study of this species in a research facility housing as many as 140 deer has seen the development of systems that allow regular handling, a reproductive rate of about 1.8 offspring per year, and limited mortality (D.E. Ullrey, personal communication). Whether the high capital cost of numerous small pens, including individual fawning pens, small paddocks, and a feeding regime predominantly based upon supplementation, can be justified on a commercial farming basis is open to question. On the other hand, the potential for sale of high-priced bucks for use as breeding sires can quickly justify such expenditures.

Another common species in some parts of North America is the mule deer, and farming efforts are just beginning with this species. They are less individualistic than the white-tailed deer and can be handled in a fashion similar to the fallow deer, but there is little or no published information on them as a farm animal. They have been successfully managed in research institutions where individual penning was practiced (Wood et al. 1961), but this would hardly be appropriate as a farming system. A study of free-ranging black-tailed deer (*O. columbianus*) has shown that they are susceptible to at least 25 species of parasite, and the author considered that some of these played a significant role in mortality (Cowan 1946). There have been numerous recent attempts at raising these animals in Alberta, but problems associated with diet and parasites have been noted (R.J. Hudson, personal communication). Several deaths have occurred and some farmers have discontinued efforts to farm them. My own initial observations with a group of 12 animals maintained on 1 ha. are that they will move together as a group when herded, and can thrive if properly fed. Mother-raised young appear to habituate to human presence within 6 months of birth and when approached by familiar persons at feeding time their flight distance does not exceed 5–10 m. They will probably require solid barrier fences at least 250 cm high when herded under pressure, as they will certainly jump a 2.1-m fence in these circumstances.

Conclusion

Modern deer farming has primarily evolved in concert with available stock, the presence (or absence) of societal and legislative restrictions, and the suitability of species. The evolution of modern fencing techniques and materials has played a prominent role in this development. Handling systems have undergone changes as farmers have learned more about their animals and will no doubt continue to change as farming practices alter. Successful farm managers have taken into consideration the biological events governing their stock and have learned to offer feeds appropriate to the species, the cohort, and the time of year. Reproductive management is as important in deer farming as in any livestock industry. For deer, calving is usually targeted to occur when natural fodder is readily available. In many circumstances this only requires the use of normal breeding seasons, but recent developments at manipulation of reproductive cycles have allowed refinement. Good management, especially good farm design and proper handling, plays a crucial role in disease prevention.

References

Adam CL (1986) Nutritional requirements. In Alexander TL (ed) Management and Diseases of Deer. Veterinary Deer Society, London, pp 181–187

Adam JL (1988) Growth. In Allen PL, Asher GW (eds) Progressive Fallow Farming. Proc of a Course on Fallow Deer Farming. MAF Ruakura Agricultural Centre, pp 47–55

Adams K (1986) Javan Rusa/hybrids. In Owen P (ed) Deer Farming into the Nineties. Owen Art and Publishing, Brisbane, pp 45–48

Asher GW (1988) Reproduction. In Allen PL, Asher GW (eds) Progressive Fallow Farming. Proc of a Course on Fallow Deer Farming. MAF Ruakura Agricultural Centre, pp 33–38

Asher GW, Adam JL, Otway W, Bowmar P, van Reenen G, Mackintosh CG, Dratch P (1988) Hybridization of Père David's deer (*Elaphurus davidianus*) and red deer (*Cervus elaphus*) by artificial insemination. J Zool London 215:197–203

Blaxter KL, Kay RNB, Sharman GAM, Cunningham JWM, Hamilton, WJ (1974) Farming the Red Deer. Dept of Agric and Fish for Scotland. Her Majesty's Stationery Office

Beatson N (1986) Selecting a breeding herd. In Owen P (ed) Deer Farming into the Nineties. Owen Art and Publishing, Brisbane, pp 149–151

Bostock J, Riley HT (1855) The Natural History of Pilny, vol 2, Book 7, pp 299–302

Challies CN (1985) Establishment, control and commercial exploitation of wild deer in New Zealand. In Fennessy PF, Drew KR (eds) Biology of Deer Production. R Soc N Z Wellington Bull 22:23–36

Clutton-Brock J (1987) A Natural History of Domesticated Animals. University of Texas Press, Austin

Clutton-Brock TH, Guinness FE, Albon SD (1982) Red Deer: Behaviour and Ecology of Two Sexes. University of Chicago Press, Chicago

Cowan I McT (1946) Parasites, diseases, injuries, and anomalies of the Columbian black-tailed deer, *Odocoileus hemionus columbianus* (Richardson), in British Columbia. Can J Res 24 (sec D): 71–103

Denholm LJ (1984) The nutrition of farmed deer. Deer Refresher Course. Proc No. 72, University of Sydney, pp 661–707

Dinerstein E (1982) Deer, plant phenology and succession in the lowland forests of Nepal. In Wemmer C (ed) Biology and Management of the Cervidae. Smithsonian Institution Press, Washington, DC, pp 272–288

Drew KR (1985) Meat production from farmed deer. In Fennessy PF, Drew R (eds) Biology of Deer Production. Roy Soc N Z Wellington Bull 22:285–290

Drew KR, Bai Q, Fadeev EV (1989) Deer farming in Asia. In Hudson RJ, Drew KR, Baskin LM (eds) Wildlife Production Systems. Cambridge University Press, Cambridge, pp 334–345

English AW (1987) The annual management cycle of farmed deer in Australia. Proc First Int Conf on Zool Avian Med, Hawaii, pp 425–434

English AW (1984) Fallow deer—their biology, behaviour and management in Australia. Deer Refresher Course. Proc No. 72, University of Sydney, pp 285–294

Fennessy PF (1989a) Pharmacology of velvet. Proc of a Deer Course for Vets. Deer Branch Course NO. 6, N Z Vet Assoc, pp 96–103

Fennessy PF (1989b) Stag selection, progeny testing and recording. Proc of a Deer Course for Vets. Deer Branch Course No. 6 N Z Vet Assoc, pp 118–128

Fennessy PF (1982) Growth and nutrition. In Yerex D (ed) The Farming of Deer: World Trends and Modern Techniques. Agricultural Promotion Associates, Wellington, New Zealand, pp 105–114

Fennessy PF, Fisher MW (1988) Advancing the calving season in Red Deer. Proc of a Deer Course for Vets. Deer Branch Course No. 5. N Z Vet Assoc, pp 17–27

Fletcher TJ (1989) Deer farming in Europe. In Hudson RJ, Drew KR, Baskin LM (eds) Wildlife Production Systems. Cambridge University Press, Cambridge, pp 323–333

Fletcher TJ (1987) The transport of deer. In Reid HW (ed) The Management and Health of Farmed Deer. Kluwer Academic Publishers, Boston, pp 181–190

Fletcher TJ (1986a) Red deer as the most suitable species for venison production. Technical papers. N Z Deer Farmers Assoc, 11th Annual Conf, pp 44–45

Fletcher TJ (1986b) Reproduction. In Alexander TL (ed) Management and Diseases of Deer. Vet Deer Soc London, pp 17–18 and 173–180

Fletcher TJ (1974) The timing of reproduction in red deer (*Cervus elaphus*) in relation to latitude. J Zool 172:363–367

Forster ES, Heffner EH (transl) (1968) Lucius Junius Moderatus Columella on Agriculture, vol 2. Loeb Classical Library 407. Heinemann, London

Galton F (1865) The first steps towards the domestication of animals. Trans Ethnolog Soc Lond N.S. 3:122–138. Reprinted in *Inquiries into Human Faculty*. Macmillan, New York

Green MJB (1989) Musk production from musk deer. In Hudson RJ, Drew KR, Baskin LM (eds) Wildlife Production Systems. Cambridge University Press, Cambridge, pp 401–409

Guinness FE, Gibson RM, Clutton-Brock TH (1978) Calving times of red deer (*Cervus elaphus* L.) on Rhum. J Zool 185:105–114

Haigh JC, Bowen G (1991) Artificial insemination of Red Deer with frozen/thawed Wapiti semen. Second Int Wildlife Ranching Symp (in press)

Haigh JC, Moore N (1990) Selling stock. Game Farm Practice: Notes for the Game Farming Industry. University of Saskatchewan, Saskatoon, Saskatchewan

Haigh JC, Barth AD, Cates WF, Glover GJ (1985) Electro-ejaculation and semen evaluation of wapiti. In Fennessy PF, Drew KR (eds) Biology of Deer Production. R Soc N Z Wellington Bull 22:197–203

Hamilton WJ (1987) Early nutrition, growth and reproductive performance in young Scottish red deer (*Cervus elaphus*) hinds, and their economic significance in commercial herds. In Reid HW (ed) The Management and Health of Farmed Deer. Kluwer Academic Publishing, Boston, pp 191–198

Hamilton WJ (1982) Farming red deer (*Cervus elaphus*)—some aspects of research and development in Scotland and their implications for the management of the deer and the pastoral resources. In Wemmer C (ed) Biology and Management of the Cervidae. Smithsonian Institution Press, Washington, DC, pp 461–466

Harrington R (1985) Evolution and distribution of the Cervidae. In Fennessy PF, Drew KR (eds) Biology of Deer Production. R Soc N Z Wellington Bull 22:3–11

Hart D (1986) Moluccan Rusa. In Owen P (ed) Deer Farming into the Nineties. Owen Art and Publishing, Brisbane, pp 53–55

Harwell G, Robinson RM, Davis DS, Hughes DM (1987) Texas game ranching: overview of a new industry. Proc, First Int Conf on Zool Avian Med, Hawaii, pp 435–444

Hemmer H (1987) Ethological aspects of deer farming. In Reid HW (ed) The Management and Health of Farmed Deer. Kluwer Academic Publishers, Boston, pp 129–137

Hofman RR (1985) Digestive physiology of the deer—their morphophysiological specialisation and adaptation. In Fennessy PF, Drew KR (eds) Biology of Deer Production. R Soc N Z Wellington Bull 22:393–407

Hudson RJ (1989a) History and Technology. In Hudson RJ, Drew KR, Baskin LM (eds) Wildlife Production Systems, Cambridge University Press, Cambridge, pp 11–27

Hudson RJ (1989b) Nutrition of farm game. Proc Western Nutr Conf 10:273–287

Hudson RJ, Kozac HM, Adamczewski JZ, Olsen CD (1991) Reproductive performance of farmed wapiti (*Cervus elaphus nelsoni*). Small Rumin Res (in press)

Hudson RJ, Watkins WG, Pauls RW (1985) Seasonal bioenergetics of Wapiti in Western Canada. In Fennessy PF, Drew KR (eds) Biology of Deer Production. R Soc NZ Wellington Bull 22:447–452

Kozac H (1988) Effect of winter feeding programs on weight changes, calving and lactation of wapiti on game farms. MSc thesis, University of Alberta

Lalouette JA (1985) Development of deer farming in Mauritius. In Fennessy PF, Drew KR (eds) Biology of Deer Production. R Soc N Z Wellington Bull No. 22:379–380

Lincoln GA (1985) Seasonal breeding in deer. In Fennessy PF, Drew KR (eds) Biology of Deer Production. R Soc N Z Wellington Bull 22: 165–179

Loudon AS, McNielly AS, Milne JA (1983) Nutritional and lactational control of fertility in red deer. Nature 302:145–147

Macartney C (1983) The deer with the fatal scent. Unasylva 35:18–19

McCall J, Parrat AC (1988) Genetics working for fallow deer farming. In Allen PL, Asher GW (eds) Progressive Fallow Farming. Proc of a Course on Fallow Deer Farming. MAF Ruakura Agricultural Centre, pp 59–63

McManus C (1989) The genetic improvement of farmed deer. Deer Farming 24:17–20

McDiarmid A (1969) Diseases of red deer and their possible significance in deer farming projects, veterinary preventative medicine and public health. In Bannerman MM, Blaxter KL (eds) The Husbanding of Red Deer. The Highlands and Islands Development Board and the Rowett Research Institute, Edinburgh, pp 50–53

Milne JA (1987) Management and nutrition of farmed deer. In Reid HW (ed) The Management and Health of Farmed Deer. Kluwer Academic Publishing, Boston, pp 1–9

Mishra HJ, Wemmer C (1982) The comparative breeding ecology of four cervids in Royal Chitwan National Park. In Wemmer C (ed) Biology and

Management of the Cervidae. Smithsonian Institution Press, Washington, DC, pp 259–271
Moen AN (1985) Energy metabolism of deer in relation to environmental variables. In Fennessy PF, Drew KR (eds) Biology of Deer Production. R Soc N Z Wellington Bull 22:439–446
Moore GW, Littlejohn RP, Cowie GM (1988a) Live weights, growth rate and antler measurements of farmed red deer stags and their usefulness as predictors of performance. N Z J Agric Res 31:285–292
Moore GW, Littlejohn RP, Cowie GM (1988b) Liveweights, growth rate and mortality of farmed red deer at Invermay. N Z J Agric Res 31:293–300
Morrill WI (1988) Overview of wildlife ranching in the USA. In Valdez R (ed) Proc First Intl Wildl Ranching Symp, Las Cruces, NM, pp 30–36
Morris RS (1988) The effects of disease on productivity and profitability of livestock: how should it be assessed? Proc N Z Soc Anim Prod 48:117–125
Muir PD, Sykes AR (1988) Effect of winter nutrition on antler development in red deer (*Cervus elaphus*): a field study. N Z J Agric Res 31:145–150
Mulley RC (1989a) Feed and nutrition. Deer Farmers J N South Wales Bull 63
Mulley RC (1989b) Fallow deer carcass research. Fed Deer Breeder 8:4–8
Parker I (1983) Oh Quagga! Ian Parker Publishers, pp 99–120
Pearse AJ (1988) Wapiti and hybrids—special management needs. Proc of a Deer Course for Vets Deer Branch Course No. 5, N Z Vet Assoc, pp 164–177
Pinney B (1981) Delegation to China. Deer Farmer, Spring 1981, pp 22–35
Pinsan Z (1982) World trends: China. In Yerex D (ed) The Farming of Deer: World Trends and Modern Techniques. Agricultural Promotion Associates, Wellington, pp 11–14
Qin RQ, Liang FX (1983) The Science of Deer Farming. Jilin Agricultural University, Jilin, Peoples Republic of China
Renecker LA (1989) Overview of game farming in Canada. In Valdez R (ed) Proc First Int Wildl Ranching Symp, pp 47–64
Renecker LA (1987) Velvet production and the market goal. In Renecker (ed) Focus on a New Industry. Proc of the Alberta Game Growers Assoc Conf, pp 59–61
Sim JS (1987) Uses of traditional medicines in Korea—deer antlers. In Renecker (ed) Focus on a New Industry. Proc of the Alberta Game Growers Assoc Conf, pp 68–70
Schmitt SM, Cooley TM, Schrader LD, Bradley MA (1983) A squeeze chute to restrain captive deer. Wildl Soc Bull 11:387–389
Suttie JM (1987) Deer growth and nutrition. Proc of a Deer Course for Vets. Deer Branch of the N Z Vet Assoc, 4:94–108
Suttie JM, Fennessy PF, Van Vleit BA, Littlejohn RP, Fisher MW, Corson ID, Labes RE (1987) Energy nutrition of young red deer (*Cervus elaphus*) herds and a comparison with young stags. Proc N Z Soc Anim Prod 47:111–114
Syroechkovsky EE, Rogacheva EV, Renecker LA (1989) Moose husbandary. In Hudson RJ, Drew KR, Baskin LM (eds) Wildlife Production Systems. Cambridge University Press, Cambridge, pp 369–386
Thomson R (1986) Chital deer. In Owen P (ed) Deer Farming into the Nineties. Owen Art and Publishing, Brisbane, pp 49–55
Turbevile G (1576a) Of the vertue and properties of the harte. In The Noble Arte of Venerie or Hunting. Clarendon Press [1908], pp 39–40
Turbevile G (1576b) Of the rut and vaulte of hartes. In The Noble Arte of Venerie or Hunting. Clarendon Press [1908], pp 44–46
Verme LJ, Ullrey DE (1984) Physiology and nutrition. In Halls LK (ed) White-Tailed Deer: Ecology and Management. Stackpole Books, Harrisburg, PA, pp 91–118
Wallis T (1989) Velvet—the world scene. Proc of a Deer Course for Vets. Deer Branch Course No. N Z Vet Assoc, pp 91–95
Wilson P (1989) Advanced calving in deer: practical aspects. Proc of a Deer Course for Vets. Deer Branch Course No. 6, pp 54–68
Wilson P (1984) Reproduction in deer. Deer Refresher Course, Proc No. 72. University of Sydney, Sydney, pp 375–389
Wood AJ, Nordan HC, Cowan I McT (1961) The care and management of wild ungulates for experimental purposes. J Wildl Manage 25:295–302
Yazan Y, Knorre V (1964) Domesticating elk in a Russian national park. Oryx 7:301–304
Yerex D, Lindeman B (1982). Behaviour. In Yerex D (ed) The Farming of Deer: World Trends and Modern Techniques. Agricultural Promotion Associates, Wellington, N Z, pp 167–170
Yerex D, Spiers I (1987). Modern deer farm management. Ampersand Publishing Associates, Carterton, NZ
Zhang B (1983) Musk deer, their capture, domestication and care according to Chinese experience and method. Unasylva 35:16–24
Zhu S, Wu S (1985) Velvet antler production of Sika deer. In Fennessy PF, Drew KR (eds) Biology of Deer Production. R Soc N Z Wellington Bull 22:282

44 Farming of Wapiti and Wapiti Hybrids in New Zealand

Anthony J. Pearse

Abstract

The New Zealand deer farming industry has now reached the stage where production of young lean animals for venison over an extended period of the year is becoming the cornerstone of deer farming profitability. There are worthwhile incentives for producers who can finish stags to market weights of 55 to 65 kg at 10 to 13 months of age. In addition, some exporters offer premiums for lean heavyweight carcasses which are targeted at well-grown but lean 1.5- to 2.5-year-old red deer (*Cervus elaphus*) or the younger wapiti (*C. canadensis*) x red hybrids. Consequently, with declining emphasis on live sales of breeding females and increasing emphasis on venison production from male and female deer, many farmers are now actively pursuing the strategic use of larger strains of deer as sires. It is in these situations that the wapiti and its hybrids are finding a niche. There are also considerable potential advantages for the wapiti in terms of velvet antler production. In contrast to red deer, wapiti do have some special management needs. These are particularly evident in terms of their animal health requirements and in the management of red hinds mated to these larger sires.

Key words: Deer farming, management, production, red deer, velvet, venison, wapiti

Introduction

Deer farming in New Zealand (NZ) is based predominantly on red deer (*Cervus elaphus*) and in particular on the production of venison at carcass weights of 55–60 kg, which are reached by young stags at 12–18 months of age. Because of the strong seasonality of growth patterns in stags and to a lesser extent hinds (Suttie 1987), optimum carcass weights for efficient slaughter systems are not easily achieved in red deer before 12 months of age. Venison exporters offer attractive schedule prices based on carcass quality, weight, and leanness (as expressed by the GR, or tissue depth measurement), and these vary according to the season. Markets for farm-raised venison are dependent largely on September–December demand in the northern hemisphere, although an all-year-round supply of chilled product into other markets has become desirable. This requirement is somewhat out of phase with the growth characteristics of young red stags in the southern hemisphere. Typical seasonal financial incentives offered by exporters to attract product are contrasted with a normal red deer growth curve in Fig. 44.1 (Wilson 1989). The growth curve of a wapiti × red hybrid is shown for comparison. The mismatch between the growth pattern of young red deer and the demand reflected in price per kg of carcass is clearly evident.

Research work and on-farm production systems have been developed to improve the early growth rates of red deer. A number of systems are being developed to increase live weights of red deer; these include early calving, systems involv-

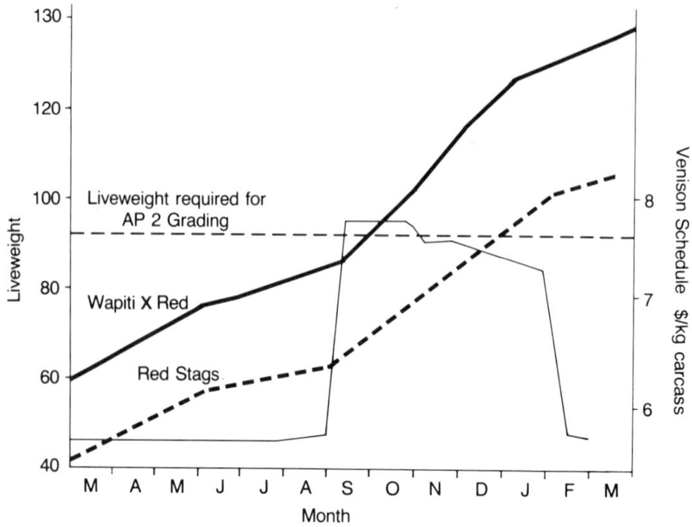

FIGURE 44.1. Venison schedule fluctuations (1989–90) and a typical weight gain curve for red deer stags and F1 stags (wapiti × red) to 15 months of age.

ing wintering in sheltered or indoor pens, developing new pasture species which might provide a better match between feed supply and demand, and selection of genetically superior animals in terms of growth rate and early maturity. Hybridization of red deer females with males from various wapiti or wapiti hybrid strains in order to improve performance and the efficiency of production have also proven highly successful for venison production. The genus *Cervus* shows an incredible diversity in body size, and many species are able to interbreed and produce fertile hybrids offering commercial advantages in live weight gain and velvet antler production (Fennessy and Dratch 1986). Consequently, NZ researchers and farmers have been evaluating the role of the wapiti and their hybrids with red deer.

Hybridization

Venison Production

In productive terms, wapiti or their hybrids are larger, faster growing, and leaner at the same age than their red counterparts (Drew and Hogg 1990). In terms of relative antler yield per unit live weight, wapiti are comparatively superior antler producers (based on velvet antler weight and grading quality) at an earlier age and maintain that advantage in maturity. This versatility offers a number of options for the farmer. The potential advantages in growth rate due to the larger mature size can lead to advantages in terms of the body weight of offspring and hence age at slaughter or carcass weight at slaughter. The potential profitability can be estimated by using concepts of biological efficiency (Fennessy and Thompson 1989) when biological efficiency is defined as the output (kg of carcass produced) relative to input (MJ of feed eaten). Changes in economic efficiency can be expected to follow changes in biological efficiency. The most efficient approach involves mating a genetically large male to a genetically small female as long as there is little effect on calving rate and the survival of the progeny.

New Zealand red hinds mated to large wapiti males have successfully reared their hybrid calves (Moore and Littlejohn 1989). At Invermay the larger red hinds (100–110 kg) are generally used to produce the hybrid. Such red hinds have been routinely mated by natural mating or artificial insemination to the largest Canadian wapiti bulls available. Table 44.1 compares the carcass production data from yearling Canadian wapiti × red crosses (F_1) and 2-year-old red deer and

TABLE 44.1. Comparison of slaughter data for 2-year-old red deer stags with data for 11-month-old wapiti x red hybrids slaughtered in October (from Drew and Hogg 1990)

	Red deer	Hybrids
Number	53	8
Live weight (kg)	110	116
Hot carcass weight (kg)	63	68
Dressing (%)	57	59
Carcass compositon (% cold carcass weight)		
Saddle	15	18
Hindquarters	39	40
Forequarters	19	20
Neck	16	14
Ribs	11	8
Fatness (GR in mm)	10	4.7

demonstrates the substantial advantage in carcass growth gained by hybridization (Drew and Hogg 1990; Drew Chapter 56, this volume). As indicated, calving rate is important in considering the total system. Over 3 years, Moore and Littlejohn (1989) reported a 72% calving rate for 106 red hinds mated to wapiti, with 89% of these hinds rearing calves to weaning. The overall weaning rate is lower than that achieved with red hinds mated to red stags, due mainly to the lower proportion of hinds mated to wapiti actually producing calves.

The wapiti × red hybrid, however, is considered to be at the extremes of practical management. Hinds must be managed to shed surplus fat during winter and kept fit in late pregnancy to avoid dystocia problems. Growth rates of around 550 g/day have been recorded for Canadian wapiti × red calves during lactation in contrast to 380–400 g per day in a typical red × red mating (Moore and Littlejohn 1989). This nutritional load can cause loss of condition during lactation and may result in a lowered conception rate in the following mating season.

Natural mating of red hinds with wapiti bulls also presents management problems. Bulls appear to require previous mating experience and be 5 or 6 years of age. Young bulls are timid, especially if in audial or visual competition with more aggressive red deer stags, even in single sire mating groups. Mating ratios of 1:30 are the maximum recommended (Moore 1988). In commercial deer farming where cost-effective management and profitable venison production are the objectives, it is the wapiti × red (F_1) stag that is finding an increased demand as a sire. Special mating and calving management requirements are reduced with apparently little or no effect on calving percentage. This quarter-bred hybrid has produced carcass weights of 56–58 kg in October/November, or 66–68 kg at 15 months of age (i.e., about 25% heavier than red deer of the same age) and satisfies all the existing market requirements for weight, leanness, and quality.

Velvet Production

There is a strong relationship between antler size and live weight (Fennessy 1989) with a 10 kg increase in body weight associated with a 0.1–0.2 kg increase in velvet antler weight in red deer (Moore et al. 1988; P. Fennessy, unpublished data). Analysis of the work of Huxley (1931) shows that hard antler weight increases at a relatively faster rate than body size in European red deer. Velvet antler production from mature wapiti a known in NZ is in the range of 5.6–7.8 kg with some markedly higher weights (NZ Wapiti Society Velvet Pool data) and supports this live weight/antler weight relationship. If increased velvet antler production is the primary objective, then the two alternative strategies of selection within a strain, and hybridizing across strains can be considered to improve performance. Fennessy (1989) showed comparisons of the gains likely to be achieved through these selection procedures (Table 44.2), assuming a heritability coefficient for velvet antler weight of 0.4 (Zhou and Wu 1979). Data for velvet antler production of red deer, wapiti, and their hybrids are presented in Table 44.3.

Management

The first wapiti-type animals used in the NZ farmed situation were themselves a hybrid type sourced from the original liberation of *C. elaphus nelsoni* in the Fiordland region of southern New Zealand. On blood type (based on protein polymorphisms) the average wapiti component is estimated at 56% (Tate and Rummel 1989). These animals have readily adapted to a pasture graz-

TABLE 44.2. Comparison of various breeding options to improve velvet antler production (comparisons as 4-year-olds; from Fennessy 1989)

Breed option	Velvet antler production (kg) as 4-year- olds Breeding animals Stags × red hinds		Genetic progress per generation	Yield of progeny
Selection within strain				
Average stags	2.5	2.5[c]	0.0	2.50
Top 3% of stags	3.6[b]	2.5	0.22	2.72
Hybridization between strains[a]				
A) Average stags	3.6	2.5	0.55	3.05
Top 3% of stags	5.2	2.5	0.87	3.37
B) Average stags	5.2	2.5	1.35	3.85
Top 3% of stags	7.5	2.5	1.81	4.31

[a]Calculations assume no hybrid vigor (i.e., progeny of average superior strain and average red hinds are midway between the parents).

[b]Assuming S.D. = 0.5 kg, C.V. = 20%, and a heritability of 0.40.

[c]Hinds are average red hinds that in their male progeny from mating with average red stags would be expected to produce 2.5 kg of velvet antler.

TABLE 44.3. Typical velvet antler yields (kg; days of growth) NZR, CW × NZR hybrids, and CW at Invermay, and the expected velvet antler weight for CW[a] assuming no hybrid vigor and the ratio of actual to expected velvet antler weight for CW (Fennessy and Pearse 1990)

Age (years)	Actual velvet antler weights NZR	CW × NZR	CW	Expected velvet antler weight of CW	Actual/expected velvet weight
2	1.01 (55)	2.05 (64)	1.99 (68)	3.09	0.64
3	1.60 (58)	2.63 (64)	2.50 (71)	3.66	0.68

[a]Expected velvet antler weight of CW = 2(CW × NZR) − NZR

ing regime, although they usually require roughage supplementation in autumn and spring to avoid a diarrhea of unknown origin in times of lush feed growth. Wapiti have the reputation of quietness and docility in the paddock situation, but often find yarding and intensive handling confrontational.

Overcrowding in yards and prolonged periods of confinement are the major stress factors. Wapiti must be drafted into small groups as soon as possible to minimize aggression and handling problems. Unlike red deer, wapiti will kick savagely with both hind and fore legs if provoked, or else become stubborn and difficult to move. Handlers often provoke aggression in the animals through impatience or frustration. Handling systems that utilize raceways exploiting the natural desire to run are effective particularly if the handler is working outside the raceway and is not threatening. Open boarded yards that give good visibility from pen to pen have proved successful in settling groups of animals. Above all, ample time and space allow the largest wapiti to be handled easily and with confidence. Prolonged and unnecessary yarding only increases handling difficulty.

Poor performance of wapiti and their hybrids has been reported in many situations in NZ (Mackintosh et al. 1982, 1986), with the problems being particularly apparent in the purebred wapiti. Data for the comparative weight performance of wapiti, red deer, and their hybrids at

TABLE 44.4. Typical live weights (kg)[a] by age for New Zealand red deer (NZR, *Cervus elaphus scoticus*), Canadian wapiti (CW, *C.e. manitobensis*) × NZR hybrids and CW at Invermay with the expected live weights[b] for CW (assuming, A, no hybrid vigor and, B, 10% hybrid vigor in CW × NZR progeny) and the ratio of actual to expected live weights for CW (Fennessy and Pearse 1990)

Age	Actual live weights NZR	CW × NZR	CW	Expected live weight to CW[b] A	B	Actual/expected live weights of CW A	B
Males							
Birth	9.6	14.1	18.1	18.6	16.0	0.97	1.13
3 months	48	74	73	100	86	0.73	0.85
1.3 years	110	155	155	200	172	0.78	0.90
2.2 years	150	210	210	270	232	0.78	0.91
3.2 years	188	264	265	340	292	0.78	0.91
4.2 years	210	(295)[c]	305	380	326	0.80	0.94
5.2 years	220	(310)[c]	330	400	344	0.83	0.96
Females							
Birth	8.8	13.3	17.2	17.8	15.3	0.97	1.12
3 months	43	71	70	99	85	0.71	0.82
1.3 years	85	129	140	173	149	0.81	0.94
2.3 years	96	158	180	220	189	0.82	0.95
3.2 years	105	167	215	229	197	0.94	1.09
4.2 years	110	170	240	230	198	1.04	1.21

[a] Moore 1982, 1983, 1984; Moore et al. 1988a, b; Moore and Littlejohn 1989; A.J. Pearse, unpublished data

[b] Expected live weight of CW: A-2(CW × NZR)-NZR, and B = 0.909(CW × NZR)-NZR.

[c] Actual live weights for (CW × NZR) males at 4.2 and 5.2 y are based on limited data.

Invermay are presented in Table 44.4. The weights of the hybrids are clearly superior to those expected for an intermediate animal, and point to considerable hybrid vigor and/or severe underperformance of the purebred wapiti (Fennessy and Pearse 1990).

As many farmers are upgrading the NZ wapiti-base animal with imported strains to increase the proportion of wapiti genes in their animals, these productive losses are significant. Anecdotal evidence throughout NZ suggests the problem is widespread and possibly has some genetic basis in terms of the comparative susceptibility to disease of wapiti compared with farmed red deer (Mackintosh 1990).

Animal Health

Canadian wapiti are highly susceptible to ryegrass staggers (toxicity due to the presence of an endophyte in the ryegrass). The condition is particularly marked in late summer and autumn under tight grazing conditions or drought (Orr and Mackintosh 1985). Staggers exhibit as a slight head tremor, lethargy, ataxia, and weight loss in mild cases, to an excited uncontrolled gait, collapse, and convulsion in advanced cases, particularly under herding pressure. Avoidance of such pasture is the only cure, and recovery occurs by supplementary feeding with safe feeds such as lucerne hay or silage or concentrates. Many wapiti farmers have introduced low endophyte ryegrasses or alternative pasture and herb species to overcome the problem. Hybrids are less susceptible, and the condition is rarely seen in red deer (Mackintosh Chapter 29, this volume).

Wapiti and hybrids are also more prone to the effects of copper-deficiency-induced enzootic ataxia than red deer and may have a greater absolute requirement for copper, although recommended blood serum levels have not been well defined (Mackintosh et al. 1986). Seasonal fluctuations are recorded with deficiencies most commonly observed in spring in 2- and 3-year-old males during velvet growth. Ill-thrift, loss of condition, hind leg incoordination, and ataxia are common physical effects. Treatment is by prevention through oral dosing of copper needles (4–12 g cupric oxide wire in capsules) or injectable copper. Farm fertilizer history, including molybdenum and sulfur application and high soil pH through liming can affect normal copper uptake in the diet and should be assessed in diagnosis and treatment.

Purebred wapiti (like the hybrid wapiti mentioned previously) may develop diarrhea particularly during spring, but also in autumn. A high quality meadow hay or lucerne should be available ad lib during this period. Improved health status has been reported when wapiti have access to unimproved pastures including browse, and are run at low stocking densities.

Many wapiti farmers have experience of pure wapiti and hybrids developing a wasting condition, usually in spring or autumn and spring. Poor calving performance, infertility, massive weight loss, and a lingering death may result. The animals, often without scouring, lose weight rapidly and become depressed and lethargic. Secondary contributing symptoms are common, including parasitic worm build-up (particularly Dictyocaulus and Ostertagia), inflamed abomasal mucosa, liver damage, and on occasions low serum levels of copper and cobalt. Some evidence suggests unidentified fungal mycotoxins present in rank and dead pasture may initiate the condition which also may have a genetic basis in susceptibility (Pearse 1988). Preventative measures include offering supplementary feed such as high quality lucerne hay at these times of the year. Control of internal parasites is critical, as wapiti appear to show a lesser degree of acquired immunity to lungworm than red deer (C.G. Mackintosh, personal communication).

In general terms, wapiti respond to low stocking rates and less restriction in paddock size than red deer. Red, hybrid, and wapiti stags do not mix well as young animals where there is competition for supplementary feeds and should be yarded and handled in separate groups to reduce behavioral stress with intensive handling, particularly during velveting. Wapiti farmers have developed many different yarding and management systems for wapiti. However, successful farmers generally exhibit a high level of stockmanship.

Conclusion

If the primary role of the wapiti is to enhance the venison industry through their use in hybridization across the base herd of red deer hinds, then producers can limit the specialist management required to a small number of highly selected sire animals without additional yard alterations or wide-ranging feeding and animal health programs. With both male and female hybrids being suitable for the current venison market at an early age, the dilemma of what to breed hybrids to in the next generation is removed. The wapiti, if used as a terminal sire and a premium velvet producer, will continue to make a major contribution to the profitable development of the New Zealand deer industry.

References

Drew KR, Hogg BW (1990) Comparative carcass production from red, wapiti and fallow deer. Proc Australian Assoc Animal Breeding Genetics 8:491–494

Fennessy PF (1989) Stag selection, progeny testing and recording. Proc of a Deer Course for Veterinarians. Deer Branch, N Z Vet Assoc 6:118–128

Fennessy PF (1987) The farming of deer towards 2000: an agricultural scientists' perspective. Proc 12th Annu Conf, N Z Deer Farmers Assoc, pp 26–29

Fennessy PF, Dratch PA (1986) Red deer and wapiti: diversity and hybridisation. Proc 11th Annu Conf, N Z Deer Farmers Assoc, pp 22–24

Fennessy PF, Pearse AJ (1990) The relative performance of Canadian wapiti and their hybrids. Proc Australian Assoc Animal Breeding Genetics 8:497–500

Fennessy PF, Thompson JM (1989) Biological efficiency for venison production in red deer. Proc N Z Soc Anim Prod 49:5–10

Huxley JS (1931) The relative size of the antlers of red deer (*Cervus elaphus*). Proc Zool Soc London, pp 819–864

Mackintosh CG, Orr MB, Turner KT (1986) Enzootic ataxia in deer. Proc of a Deer Course for Veterinarians. Deer Branch N Z Vet Assoc 3:165–169

Mackintosh CG, Orr MB, Gallagher RT, Harvey IC (1982) Ryegrass staggers in Canadian wapiti deer. N Z Vet J 30:106–107

Moore GH (1988) Crossbreeding Canadian wapiti bulls with red hinds. Deer Farmer 43:33–35

Moore GH, Brown G (1987) Crossbreeding with N Z wapiti-type bulls. Deer Farmer 40:23–27

Moore GH, Littlejohn RP (1989) Hybridisation of farmed wapiti (*Cervus elaphus manitobensis*) and red deer (*Cervus elaphus*). N Z J Zool 16:191–198

Moore GH, Littlejohn RP, Cowie GM (1988) Liveweights, growth rates and antler measurements of farmed red stags and their usefulness as predictors of performance. N Z J of Agr Res 31: 145–150

Orr MB, Mackintosh CG (1985) Ryegrass staggers in deer. Proc of a Deer Course for Veterinarians. Deer Branch N Z Vet Assoc 2:39–43

Pearse AJ (1988) Wapiti and hybrids: special management needs. Proc of a Deer Course for Veterinarians. Deer Branch N Z Vet Assoc 5:118–127

Suttie JM (1987) Deer growth and nutrition. Proc of a Deer Course for Veterinarians. Deer Branch N Z Vet Assoc 6:94–108

Tate ML, Rummel RT (1989) Bloodtyping for herd improvement. Proc of a Deer Course for Veterinarians. Deer Branch N Z Vet Assoc 6:187–194

Wilson PR (1989) Orchestrated manoeuvres with the dark. Deer Farmer 63:75–82

Zhou S, Wu S (1979) A preliminary study of quantitative and character inheritance of antlers. Acta Genetica Sinica 6:434–440

45
Effects of Stocking Density, Feeding, and Herd Management on Mortality of Park Deer

Rory Putman and Jochen Langbein

Abstract

Patterns of mortality were studied in 15 park populations of red (*Cervus elaphus*) and fallow deer (*Dama dama*) in England in an attempt to determine relationships between levels of overwinter mortality and various characteristics of population structure or management. Our data show that high levels of mortality may be expected among animals of poor condition (low body weight) at the end of cold winters (especially in years where late-winter temperatures in January and/or February are particularly low). Lower mortalities are experienced where stocking densities remain at or below two stock units per hectare of good grazing, where feeding of supplementary fodder overwinter commences in November rather than later in the winter, and when that winter supplement provides the equivalent of 12 MJ or more ME per stock unit per day.

Key words: Fallow deer, management, mortality, red deer, stocking density

Introduction

This chapter describes an analysis of mortality among populations of red and fallow deer in an attempt to determine which of various population and management factors may influence levels of mortality sustained. Our work has focused on deer in English deer parks; the populations are enclosed by a perimeter fence or wall, but are otherwise free-ranging within parkland areas of between 5.7 and 820 ha.

Prompted by a major "die-off" among the deer at Richmond Park in London in 1985–86, when 59 red deer and 47 fallow (11.3% total population) were found dead within the park towards the end of winter, the research has been directed to a study of the factors associated with differing levels of mortality in this and other parks, in order to discover where changes in management practice might help reduce unnecessary losses in the future. In 1989 we presented a preliminary evaluation of these questions based on an analysis of data researched from past records maintained by 21 individual estates. Results of that initial analysis (Putman and Langbein 1989) suggested. (1) The factors implicated in major "die-offs" are no different from those controlling "normal" variation in levels of natural mortality. (2) The factors influencing levels of natural mortality were essentially: stocking density in relation to available grazing area and productivity of natural forage, percentage of a park providing a high degree of cover from adverse weather conditions (winter and summer), and the regime of supplementary feeding practiced overwinter (quantity, quality, and timing of supplementation). Degree of public access to a park for recreation was found to affect mortality through disturbance, only indirectly acting in combination with some limitation of one or more of these other predisposing factors (Putman and Langbein 1989), except where directly involved in causing death by accident.

Such analysis was however based on incomplete and perhaps unreliable data. We were relying heavily on records maintained by the estates themselves; furthermore, not all parks considered were equally represented in our analyses. Since 1985 we have been monitoring performance ourselves, within 15 separate populations. We present here an analysis of this new material to reassess the relationship between mortality, population attributes, and management.

Methods

Data Recording

All analyses presented here were based on data collected from October 1985 to April 1989, from 15 separate park deer populations. Parks studied were widely dispersed throughout England (50° 30′ to 53°30′) and ranged from 20 to 400 m above sea level; total area varied from 5.7 to 820 ha. Stocking densities (as stock units: fallow = sheep = 1; red = 2; see Putman and Langbein 1989) were calculated as the mean daily number of animal units supported per hectare between November and March and ranged between 0.5

TABLE 45.1. Summary of the population and management characteristics of deer populations in 15 English deer parks

	Minimum	Maximum	Mean
Total park area in ha	5.7	820.0	217.0
Deer species (red only, fallow only, red + fallow)			
*Presence/absence other stock (cattle/sheep)			
Winter stocking density (mean stock units per ha)	0.5	8.2	3.3
Mean winter density over previous 3 years	0.6	7.6	3.3
*Proportion of park providing tree cover	0	0.5	0.2
Proportion of park as good pasture	0.4	1.0	0.6
*Grazing density (mean stock units per ha pasture)	0.8	11.5	5.3
Mean grazing density over previous 3 years	0.8	9.6	5.2
*Date winter feeding commenced (before/after end November)			
*Daily ration of supplementary food over winter			
As ME (MJ per animal)	0.1	26.7	5.8
As crude protein (g per animal)	1.2	394.4	66.1
Average winter supplement per animal, previous year			
*Mean weight (within a park) of			
Adult fallow doe (live weight, kg)	34.7	48.9	42.1
Yearling fallow buck (live weight, kg)	35.3	55.9	44.5
Adult red hind (carcass weight, kg)	40.4	58.8	51.8
Yearling red stag (carcass weight, kg)	36.3	64.1	47.1
*Mean monthly temperature (°C, Nov–March)	1.7	6.9	4.9
*Mean monthly temperature (°C, Jan, Feb)	0.8	6.5	
*Total winter precipitation (mm, Nov–March)	167	747	360
*Number of winter days snow on ground at 0900	0	60	11.7
Number of winter days of ground frost at 0900	60	110	85.5
*Mean monthly temperature (°C, May, June)	9.7	17.5	12.8
*Total precipitation (mm, May, June)	70	306	130
Percentage of park area open to public	0	100	47.2
*Level of public use (scale 0–9)	1	9	4.4

*Variables carried forward in multivariate analyses

and 8.2. Nine parks contained only fallow deer, one had only red, and five had mixed populations of red and fallow.

Population and management parameters recorded are listed in Table 45.1, with range and mean value across study parks; data are available for all parks in all years. Methods used in collection of particular data sets are briefly summarized below; fuller details of the methodology are presented by Putman and Langbein (1990) and Langbein (1990).

Population data derive from direct censuses carried out in each park in September and March each year. Mean weights of animals were based, for red deer, on dressed carcass weight of animals culled between November and January (total weight less viscera, head and lower limbs); for fallow on whole carcass weights (total weight less bleedable blood) or live weights obtained during live-capture operations in the different parks between November and February. Weights of adult females and yearling males were chosen as indicators of population condition in general as offering high sample size and lowest variance. Mortality figures included should be considered minimum estimates, based on the number of animals found dead overwinter up until the end of March in each year. Finally, meteorological data were based on records from the nearest weather station to each park, published monthly by the Meterological Office (HMSO).

Analysis

Relationships between the level of mortality sustained within a park (separately as red deer only, fallow only, and total mortality) and the different population or management parameters were investigated individually by correlation analysis (or, appropriately *t*-tests) and in multiple regression. Analyses were undertaken at three distinct levels: (1) Data sets were explored using each year's records separately for all parks (offering a total of 60 park-years). Mortality levels vary markedly within a park between years; analysis at this level of all park-years allows discrimination of those factors characteristic of a particular park in a particular year in relation to a given level of mortality. No bias is introduced by overrepresentation of one park or 1 year, since data are presented for all 15 parks considered in the same 4 years. (2) Analysis at the level of park-years (above) may overemphasize, by replication in successive years, those features of a park or population that are unchanged between years, or not-independent of previous values. Analysis was thus repeated across all 15 parks for which data were available, but using (a) data for single years and (b) mean values over the period 1985–89 in each park. (3) In six of the parks included in our study, winter mortality in one or more year(s) exceeded 7% of the total population. Wilcoxon's matched pairs tests were used to compare population and management characteristics in the year of the heavy mortality with average values in the same park over 3 years of low mortality.

Results

Levels of overall mortality varied over the 60 park-years considered from 0% to 11.3% of deer populations, with six parks among those considered sustaining mortality in excess of 7% in at least 1 year (note that in all our analyses, figures for mortality exclude cull deaths). Variation in mortality between years within a park was substantial, with Richmond Park in London, for example, recording mortality of 11.3% in 1985–86 and 0.5% in this last winter (1988–89). Variation between populations was also marked with a 4-year average ranging from 0.5% to 6.2% in different parks.

Winter losses showed significant individual correlations ($P < 0.01$) with the index of population body condition used (adult female weight), length of the growing season of the previous year, average winter temperature (over all months, November to March) and specifically, temperatures in January/February. Weak relationships were apparent for both red and fallow deer between mortality and levels of supplementary feed offered over winter (as ME or crude protein per animal) and time feeding commenced (early or mid-November vs December) (Table 45.2). Most of these associations were significant both in analyses considering all park-years separately and those using park averages; where relationships missed

TABLE 45.2. Correlations between levels of mortality in English park deer and associated population or management variables[a]

	All park-years		Average park-year	
Fallow deer mortality	*r*		*r*	
Winter temperature	−0.36	**		NS
Days of snow	+0.41	***		NS
Jan/Feb temperature	−0.33	***		NS
Length of growing season	−0.50	**	−0.75	*
Adult body weight	−0.41	*	−0.45	*
Red deer mortality				
Winter temperature	−0.40	*		NS
Days of snow	+0.53	**		NS
Jan/Feb temperature	−0.38	*		NS
Adult body weight	−0.75	*		NS
Grazing pressure	−0.34	*	−0.64	*
Supplementary ME	−0.48	**		NS
Total natural mortality				
Winter temperature	−0.40	***		NS
Days of snow	+0.46	***	+0.46	*
Jan/Feb temperature	−0.41	***	−0.41	**
Length of growing season	−0.55	***	−0.81	*
Doe weight	−0.40	**	−0.42	*
Hind weight	−0.81	*		NS
Total mortality including accidents				
Days of snow	+0.31	**		NS
Jan/Feb temperature	−0.29	**		NS
Length of growing season	−0.39	*		NS
Public access	+0.32	***	+0.43	*

[a] Results shown from analyses using: (1) all individual park-years; (2) park means over 4 years (1985–89)
* $P < 0.05$; ** $P < 0.01$; *** $P < 0.001$ (one-tailed)

significance in the latter analysis, trends were still in the same direction.

Much of the analysis presented in Table 45.2 relates primarily to natural mortality; however, analyses were repeated, using information for total winter mortality—that is natural mortality and accidental death (dog chases, road traffic accidents, etc.). In such analyses, relationships are apparent with both measures used to assess public disturbance. Thus, public access increased the level of total mortality within a park, but was not related to natural mortality.

Results from matched pairs analyses comparing characteristics of those years within any park where high mortality levels were recorded, and an average of those years within the same park of low or zero mortality, offer independent confirmation of a significant relationship between high levels of mortality and mean winter temperature, food supplements (as ME and CP per animal per day), winter precipitation ($P < 0.005$); days of ground frost, temperature during January/February ($P < 0.025$); winter grazing density ($P = 0.03$; one-tailed tests throughout).

Considerable intercorrelation was apparent among the different measures of mortality (red deer with fallow deer mortality within a park, and each species independently with overall natural mortality; $P < 0.0001$, $r > 0.9$), although relationships between measures of natural mortality and those including accidental death were less significant. A high degree of intercorrelation was also apparent among some of the independent variables assessed. Within parks, body weights of adult females and yearling males within a species were highly correlated (fallow, $P = 0.0001$; red, $P = 0.05$), and weights of yearlings also correlated between the two species ($P = 0.006$), validating the use of these measures as indicators of population condition.

The proportion of a park's area open to the public was highly correlated to level of public use ($r_s = +0.85$; $P < 0.001$); winter grazing density (stock units per hectare of good grazing within a park) correlated with absolute density (stock units per hectare overall), and perhaps not surprisingly, with densities of previous years ($P < 0.001$; $r > 0.8$ in all cases). Finally, a number of climatic variables were clearly interrelated ($P < 0.001$, $r > 0.7$ unless otherwise indicated). Mean winter temperature was significantly negatively related to number of days of ground frost, days of snow and temperature in January/February; winter temperature was also significantly associated with winter precipitation ($r = -0.40$). Number of days with snow lying was highly correlated with January/February temperatures ($P < 0.0001$; $r = -0.80$). Length of growing season in any park year (defined as number of days a year when mean temperature was $> 5°$ C) was predictable from mean winter temperature and temperature in May/June. In view of such clear interactions within the variable sets considered, subsequent multivariate analyses were carried out using a restricted variable list (asterisked in Table 45.1).

Multiple regression analyses were carried out at the level of all park-years and (except for mortality of red deer where sample sizes were too small) individual years or 4-year park averages. Analyses were repeated with and without the inclusion of body weight parameters—since while poor body weight may indeed be a factor contributing directly to overwinter mortality, it may also merely reflect, as an indicator, pasture productivity or summer grazing pressures. Mortality of fallow deer was related significantly to winter temperature or number of days of snow. Days of snow as a single variable accounted for 21.4% of the variance in mortality across all park-years; inclusion of adult female body weight in the model increased adjusted R^2 to 25.7%. If climatic data were not included in the analysis, adult doe body weight and level of supplementary feeding (ME) combined to offer an adjusted R^2 of 25.9%. In individual years, mortality related to doe weight in 1986 and 1987 ($R^2 = 54\%$, 20.4% respectively), but more strongly to January/February temperature in 1985 ($R^2 = 12.6\%$).

The number of red deer populations included in our analyses was too small to permit year by year analysis in multiple regression (providing insufficient degrees of freedom). In analysis based on all park-years, regression of mortality on days

TABLE 45.3. Multiple regression analyses of mortality[a]

Dependent variable	Independent variables selected	R^{2b}	P
Fallow deer mortality			
	Adult body weight + supplementary ME per animal	25.9	<0.001
	OR Adult body weight + days of snow	25.7	<0.001
	Days of snow, alone	21.4	<0.001
Red deer mortality			
	Adult body weight	49.9	<0.03
	OR Days of snow + supplementary ME per animal	31.4	<0.01
	Days of snow, alone	24.8	<0.008
Total deer mortality			
	Adult body weight + supplementary ME per animal	24.4	<0.001
	OR Days of snow	21.2	<0.001
Total mortality, including accidental deaths			
	Disturbance/public access	15.8	<0.01

[a] For simplicity, results are shown for analyses over all park-years
[b] Adjusted; %

of snow ($R^2 = 24.8\%$; $P < 0.001$) was improved by addition of values for supplementary feeding (total adj. $R^2 = 31.4\%$, $P < 0.05$), but this could not explain more of the variance in mortality levels than could adult hind body weight alone ($R^2 = 49.9\%$, $P < 0.05$).

Total natural mortality (of red and fallow combined in any park) was significantly related to days of snow, whether all park-years were considered together ($R^2 = 21.2\%$), or in analysis of park averages, and in the individual years 1985 and 1987 ($R^2 = 45.7\%$; 27.9%). In 1986 and 1987, mortality was highly correlated with fallow doe body weight, selected as an indicator of body weights in general since all are so closely correlated, with adjusted $R^2 = 55.3\%$, 17.2% respectively (Table 45.3). When mortality measures were extended to include all noncull mortality, thus including accidental deaths, significant relationships were recorded with both measures of public access. The index of public disturbance ($P = 0.002$) or proportion of park accessible to the public ($P = 0.015$) each independently offered adjusted R^2 of 15.8% and were the only variables selected in the model.

Throughout, whether in analysis of the relationships between mortality and individual population or management characteristics, or in multivariate analyses, few significant relationships were detected between mortality and stocking density; however, all such analyses seek linear correlation. Examination of the actual form of the relationships between mortality and density suggest instead, critical thresholds. Mortality levels in excess of 2.5% of the population were recorded in single species or mixed herds only where densities exceeded two stock units per hectare of available grazing. Similar thresholds were also apparent in relationships between mortality and levels of supplementary feeding; mortality within any population did not exceed 2% when 12 MJ or more of ME were provided daily per stock unit.

The importance of stocking levels, supplementary feeding overwinter, and late winter weather conditions are well illustrated here by comparison of mortality levels over an 8-year period in two deer parks in London, both containing mixed populations of red and fallow and situated within 5 km of each other (Table 45.4). Mortality levels were very high only in 1 year where late winter temperatures were particularly low—and only in the park where little or no supplementary feed was provided. Note, however, the park where little supplementary feeding was carried out (Richmond) only sustained heavy mortalities following this single severe winter.

Discussion

Among most wild populations of red and fallow deer or any other temperate species, overwinter

TABLE 45.4. Mortality levels and associated characteristics of two deer parks in London, England, located less than 5 km apart[a]

Bushy Park					Richmond Park		
Total mortality (%)	Supp feed (MJ/SU)	Grazing pressure (SU/ha)	Year	Jan/Feb temperature (°C)	Grazing pressure (SU/ha)	Supp feed (MJ/SU)	Total mortality (%)
1.5	5.4	2.6	1981–82	4.6	2.6	0.1	1.7
0.2	5.2	2.7	1982–83	4.9	2.7	0.1	1.3
0.5	6.4	2.2	1983–84	4.7	2.4	0.2	1.8
1.6	6.7	2.1	1984–85	2.4	2.6	0.1	1.0
0.5	5.9	2.4	1985–86	2.0	2.8	0.1	11.3
0.8	5.7	2.5	1986–87	3.0	2.2	5.4	3.9
0.6	6.2	2.2	1987–88	5.8	2.2	1.9	1.0
1.4	7.5	1.8	1988–89	6.3	2.1	3.9	0.5

[a]Both parks are very similar in character, have mixed populations of red and fallow deer at approximately the same densities, and are subjected to similar levels of public access; because of their geographical proximity, they are also exposed to similar climate.

losses contribute the major proportion of annual mortality, and losses are increased following particularly severe winters. Thus, in the intensively studied red deer population of the Isle of Rhum, mortality of hinds (and particularly that of yearlings) increased in those years with particularly cold winters (Clutton-Brock et al. 1985). Skogland (1985) showed increased mortality among Norwegian reindeer populations (*Rangifer tarandus*) following severe winters and suggested an interaction with density such that density-independent factors of this kind had a greater effect among populations at higher density. Mortality of mule deer (*Odocoileus hemionus*) and white-tailed deer (*Odocoileus virginianus*) in Piceance Basin was shown (Bartmann and Bowden 1984) to be significantly correlated over a 7-year period with mean winter temperature and snow depth. Snow depth and snow compaction are themselves strongly correlated with mortality in mule deer and white-tailed deer populations within the U.S. (Severinghaus 1947; Verme and Ozoga; 1971, Mech et al. 1987); winter snow depths in England are rarely sufficient to prevent animals feeding.

What is perhaps surprising is that these same factors still seem to have an effect upon levels of mortality even among enclosed park populations subject to more intensive management. In our analyses, mortality levels in both deer species were significantly negatively correlated with mean winter temperature (November to March), January/February temperatures in that year, length of previous growing season, and animal body weights (doe, hind, and yearlings). With lower levels of significance, winter losses were also associated (positively) with winter precipitation and winter grazing pressure (stock units per hectare of available grazing), and negatively with the amount of supplementary food provided per animal overwinter; significant differences in mortality also accompanied differences in date of commencement of such winter feeding. These same variables emerge consistently at all levels of analysis: whether we include all years separately in all parks or restrict analysis to particular years or park averages (and in matched pairs analyses of high vs low mortality years within parks), offering confidence in the relationships detected.

Many of these independent variables were, however, themselves intercorrelated. The ME and crude protein content of food supplements were closely associated; thus, wherever one measure showed correlation with levels of mortality, either or both might be implicated. In most analyses, levels of ME were used as the more direct measure of food supplementation. Relationships among the different climatic variables are more difficult to resolve. Cold winter temperatures and high rainfall may be associated with increased levels of mortality directly by depleting energy reserves more rapidly in response to windchill, or they may act indirectly through their relationship to the length of the previous growing season. In the event, both effects would seem indicated. Mortality was more highly correlated with January/February temperatures than with winter temperature overall, strongly suggesting direct influence of temperature. Equally, mortality was strongly negatively associated with all measures of animal body weight (a measure of the bodily reserves available overwinter to compensate for energy shortfall), while autumn body weights themselves are strongly associated with length of previous growing season (Langbein 1990).

Multiple regression analysis resolves the issue further: highly intercorrelated variables which independently contribute no additional explanation to the relationships observed are not selected. Such analysis also permits investigation of interactive or summative effects of various combinations of independent variables in influencing mortality. Mortality among fallow populations related primarily to number of days of snow over winter and adult female body weight. Most significantly, in years of low or "normal" mortality, mortality levels were best predicted on female body weight; in years where high losses occurred in a number of populations, mortality was best predicted by late winter temperature and snowfall. Red deer mortality related most strongly to adult female body weight. When this variable was excluded from analysis, days of snow and supplementary feeding (ME) could together explain 31.4% of the variance. Total noncull mortality (including accidental deaths as well as natural mortality) was highly correlated with winter grazing pressure and levels of public access within a park (indeed, most nonnatural mortality recorded was due to road traffic).

Earlier analyses of factors associated with mortality in park deer (Putman and Langbein 1989) also suggested linear relationships between winter mortality levels and both density (stock units per hectare; stock units per hectare of good grazing) and autumn body weight. Multiple regression analyses of that admittedly weaker data set explained only 12% of the variance in total natural mortality; with the two species considered separately, 57% of the variance in fallow deer and 41% of the variance in red deer could be explained by winter densities, adult female body weights, and the proportion of a park providing some form of tree cover. Mortality was also significantly higher in parks where the deer shared the grazing with other stock and in regression analyses, both measures of potential disturbance (percentage of park open to the public and our index of disturbance level) were implicated in natural mortality as well as total noncull mortality. In our present analysis, presence of other livestock within the park showed no relationship with mortality beyond that occasioned through overall increases in grazing pressure; differences in mortality levels of deer in parks with or without sheep grazing were never significant. Furthermore, levels of public access or potential disturbance were not related to mortality other than that including accidental losses.

It seems clear that high levels of mortality may be expected among animals of poor condition (low body weight) at the end of cold winters (especially in years where late-winter temperatures in January and/or February are particularly low, reflected here in a high number of days with snow recorded). Lower mortalities are experienced where stocking densities remain at or below two stock units per hectare of good grazing, where winter feeding commences in November rather than later in the winter, and when that winter supplement provides the equivalent of 12 MJ or more ME per stock unit per day. (It is perhaps relevant to note in this context that this level of 12 MJ per day equates precisely to the resting energetic requirements estimated for an adult female fallow deer (11.5 MJ per day) or red hind (22 MJ per day) in winter (fallow deer, from Ueckermann 1971, Ueckermann and Hansen 1983; red deer, from Suttie 1983).

Animals in populations at higher densities, fed a lower level of winter supplements, or entering winter with low body weights, may survive equally well in relatively mild winters; thus major "die-offs" among such susceptible populations appear only sporadically. It seems that it is the added energy demands of a protracted cold spell at the end of winter in January and/or February that tip the balance. Late winter temperatures cannot easily be predicted in advance! However, a good indicator of populations at risk may be taken from body weight of adult females or yearlings in autumn at the time of the normal annual cull. Estates with average weights for adult fallow does or yearling males at <41 kg (live weight; dressed carcass weight <24 kg) or adult red hinds/yearling stags with dressed carcass weight <46 kg and 50 kg respectively should ensure increased levels of supplementary winter feeding commence in November.

Acknowledgements

We would like to thank all the keepers and managers of the various deer parks represented in these analyses for their support and help in supplying management information. The study was financed under contact from the Royal Parks Office of the Department of the Environment.

References

Bartmann RM, Bowden DC (1984) Predicting mule deer mortality from weather data in Colorado. Wildl Soc Bull 12:246–248

Clutton-Brock TH, Albon SD (1982) Winter mortality in red deer (*Cervus elaphus*). J Zool 198:515–520

Clutton-Brock TH, Major M, Guinness FE (1985) Population regulation in male and female red deer. J Anim Ecol 54:831–846

HMSO (1985–89) Monthly Weather Report of the Meteorological Office. Her Majesty's Stationery Office

Langbein J (1990) Effects of density and age on body condition, reproduction, survival and behaviour of fallow deer (*Dama dama*). PhD thesis, University of Southampton, UK

Mech LD, McRoberts RE, Peterson RO, Page RE (1987) Relationship of deer and moose populations to previous winters' snow. J Anim Ecol 56:615–627

Putman RJ, Langbein J (1990) Factors affecting performance of deer in parks. Report PECD 7/2/65, Department of Environment

Putman RJ, Langbein J (1989) Factors affecting mortality in deer parks. In Goldspink CR, Baxter Brown M (eds) Management, Conservation and Interpretation of Park Deer. British Deer Society, pp 38–54

Severinghaus CW (1947) Relationships of weather to winter mortality and population levels among deer in the Adirondack region of New York. Trans North Wildl Conf 12:212–223

Skogland T (1985) The effects of density-dependent resource limitations on the demography of wild reindeer. J Anim Ecol 54:359–374

Suttie JM (1983) Deer energy requirements. Br Deer Farmers Assoc Newsl 12:7–8

Ueckermann E (1971) Die Futterung des Schalenwildes, 2nd ed. Paul Parey, Hamburg and Berlin

Ueckermann E, Hansen P (1983) Das Damwild, 2nd ed. Paul Parey, Hamburg and Berlin

Verme LJ, Ozoga JJ (1971) Influence of winter weather on white-tailed deer in upper Michigan. In Haugen AD (ed) Proceedings of a Symposium on Snow and Ice in relation to Wildlife and Recreation. Iowa State University Press, Ames, Iowa

46
Management Strategies for Farmed Chital Deer

Anthony W. English

Abstract

Chital deer (*Axis axis*) are one of a number of species of deer being farmed in Australia. After early problems with restraint and handling, and an often high perinatal mortality rate, the perception developed that chital could not be farmed as readily as was the case with red deer (*Cervus elaphus*) and fallow deer (*Dama dama*). In order to establish whether this was true a study was commenced at Camden, New South Wales in 1986, to examine those aspects of the behavior and biology of the species which would assist in the development of management strategies on farms. The information which has been derived from this study has enabled the development and evaluation of practices which have been extremely effective. In most cases it has been possible to compare and contrast chital deer with fallow deer, and the former is not found wanting.

Key words: Behavior, carcass, growth, health, reproduction, rusa deer

Introduction

The Indian spotted or chital deer is one of four species of deer being farmed in Australia. A number of deer species were introduced into Australia, largely during the nineteenth century, and of these only six now survive in the wild—fallow deer, red deer, rusa deer (*Cervus timorensis*), sambar deer (*C. unicolor*), hog deer (*Axis porcinus*), and chital deer. After several failed earlier attempts, chital deer were successfully introduced into Queensland in 1863 (Bentley 1978). All the original breeding stock for deer farms came from this population, which is reputed to have developed from only four animals.

There was early interest in the potential of chital deer as a farm species, with small numbers becoming available in the late 1970s as a result of trapping in Queensland. They acquired an early reputation as excellent venison producers, and for their potentially high reproductive rate (Anderson 1978). It was not long before the chital also developed a reputation for being an extremely nervous animal, and it became apparent that it could be a most difficult animal to manage. If the species was to be farmed in Australia, both economically and humanely, there was an obvious need for a reappraisal of management strategies. In many cases, the owners of chital deer farmed other deer species as well—often fallow deer. It became a matter of comparing and contrasting the behavior and performance of the two species, which were about the same size, to determine where modifications might be necessary to those practices which were proving to be successful with fallow deer.

Tropical Vs Temperate Deer

The major difference between fallow and chital deer is the temperate origins of the former, with strict seasonality in reproduction imposed by photoperiodic variations. Chital deer are a tropical species, and perhaps their greatest attribute as a farm animal was thought to be the ability of hinds to produce three offspring in 27 months, compared with the single young produced annually by fallow and red deer. The chital was believed to be virtually aseasonal, and certainly it was known that calves were being born in all months of the year on farms in Australia. The only substantial information available on the species was the result of observations on wild populations in India (Schaller 1967), Nepal (Mishra 1982), Hawaii (Graf and Nichols 1967), Texas (Ables 1977), and California (Wehausen and Eliot 1982).

It was apparent that there were major problems being encountered whenever an attempt was made to bring chital deer into handling systems. It seemed inevitable that deer would be killed or injured in these attempts, and many of the original owners of chital deer either sold their animals or allowed them to breed without any form of control or management.

The losses from trauma and postcapture myopathy (PCM) were obviously a serious problem, but so was the high perinatal mortality which occurred over much of the southern part of Australia. There were chital herds in regions which frequently suffered very inclement winter weather, and most calves born in the winter months were being lost within 1 day or so of birth. In almost no case was the species achieving the high reproductive rate which was one of its alleged major attributes.

These very real problems of the species were such that many Australian deer farmers came to the conclusion that chital deer would never be farmed successfully. There were a small number who continued to maintain that this was not the case, and with their support an intensive study was commenced in 1986 at the Deer Research Unit at Camden, 60 km from Sydney. The aim of the study was to seek sufficient information on the biology and behavior of chital deer under the conditions prevailing on farms, to enable the development and evaluation of management strategies (Chapple 1989).

Behavior of Farmed Chital Deer

Chital deer are by nature a timorous species, and if not accustomed to the close proximity of man, will readily panic. In particular, in herds which have been allowed to mate in an unrestricted way, heavily pregnant and recently calved females, newborn and very young calves, and antlered stags will be present. Such groups of deer simply cannot be safely yarded for even the most routine procedures, and indeed they often cannot be moved from one paddock to another without the mismothering of very young calves.

This simple principle resulted in a decision to impose a restricted mating period on the chital herd, to ensure that the reproductive status of each female would be similar. Groups of pregnant hinds could then be allowed to calve without any attempt at interference, and their calves could be allowed to go through their first 3 months of life without the potential disaster of attempts being made to yard the herd.

The other major benefit of a restricted mating policy with chital deer was the opportunity which this provided to determine when calves are going to be born—simply by introducing and withdrawing the stags at specific times. This potential to ensure that calves would not be born during severe winter weather provided a very good basis for pursuing chital deer as a farm species.

Restraint and Handling

The difficulties encountered by almost all owners of chital deer in the restraint and handling of their animals became the most significant obstacle to the further progress of the species as a farm animal in Australia. Given the fact that the mixed groups of all ages were a major part of the problem, it was first necessary to segregate chital deer in the same way as was done with fallow deer—a female herd, groups of growing juveniles, and stags. Once this was done, it was found that chital deer could be handled very well in yards designed for fallow deer, including the same size cradle or crush: adult chital stags weighing 75–80 kg, with hinds weighing 48–55 kg.

The second requirement was good stockmanship, perhaps at a higher level than required for fallow deer or red deer. The species is unforgiving when something goes wrong, and handlers must have a very good knowledge of the behavior of their animals in a variety of situations in order to react accordingly. Patience and common sense are essential; chital deer cannot be driven by pressure but must be coerced and trained to a particular facility. Once the animals become familiar with a farm layout, and with a routine for yarding, they will very readily move as required.

The third requirement was perhaps the most essential: the "taming and training" of groups of animals (Harthoorn 1981), so that each episode of restraint does not become an exercise in wild animal capture. This was achieved by a system of regular and gradually increasing exposure to people in the paddock, by the routine use of supplementary feeds in conjunction with this, and by using feed trails to coerce and train the animals to a farm layout.

It has been found that chital deer can be yarded as readily as other species of deer, given the adoption of these policies. Serial blood sampling of animals which were handled regularly over long periods showed that a physiological "training" could be demonstrated (Chapple et al. 1991). The behavior of chital deer in yards is such that all routine procedures can be carried out, as long as a cradle or crush is available. The species simply cannot be restrained manually from about 3 months of age.

Reproduction

Due to their tropical origin, where there is an approximately equal chance of calf survival throughout the year, chital deer have not been noted to have a well-defined rutting or breeding season, with calves being born throughout the year. However, birth peaks have been recorded in the wild, as have seasonal differences in the intensity of sexual behavior, for instance, Graf and Nichols (1967) referred to a period of heightened sexual activity of chital stags as the "true rut."

Chital Stags

Commonly used biological indicators of the reproductive cycle include the endocrine profile and physical changes in the males, the secondary sexual characteristics. Spillett (1966) recorded physical indicators of wild chital stags in India which included a thickening of the neck, a possible darkening of the coat color, and fluid secretion from the infraorbital gland. Until the present study commenced, the characteristics of the chital reproductive cycle had been derived only from observations in wild populations, although Loudon and Curlewis (1988) reported findings on the male cycle from captive stags in the United Kingdom.

Observations at Camden showed that chital stags were not growing and casting antlers at random, as had been suggested by earlier anecdotal accounts from farmers. There was a high level of synchrony, with almost all stags casting between August and October. Thus, almost all stags in a group were in velvet from August to December, with the majority in hard antler in the first half of the year. This cycle was not greatly different from that of fallow bucks, with the exception that an occasional chital stag failed to cast with the others and carried hard antlers for 22 to 24 months.

Given the fact that some chital calves were born in every month before mating was restricted, there was confirmation that chital stags were capable of achieving conception while in velvet, a situation that would never arise with temperate deer. This observation had been made on wild chital deer (Goss and Rosen 1973; Graf and Nichols 1967), but Loudon and Curlewis (1988) found that the antler cycle of chital stags was associated with changes in testicular activity.

To further clarify the extent of male seasonality at Camden, a group of stags was physically restrained and examined weekly for 2 years. Body weight, neck girth, and coat color changes were recorded, blood testosterone was measured, as were the antler cycle and changes in behavior—particularly in the level of aggression. Testicular histology was also examined in other stags after slaughter at different stages of the antler cycle.

Chital stags were confirmed to be fertile year-round, but a major rutting period was evident during the months when the majority were in hard antler. A second period of rutting activity was observed later, coincident with the winter birth peak in a group where mating was not restricted. Serum testosterone levels closely reflected the antler cycle, and despite basal levels of testosterone during antler growth, spermatogenesis was maintained. More recent studies have demonstrated that there is a marked fluctuation in semen quality, with chital stags producing poor quality semen during their velvet antler phase, compared with that collected during the hard antler phase (G.E. Mylrea and A.W. English, unpublished data) There was no annual fall-off in the liveweight of chital stags, except for a slight decrease (5%–6%) in the dominant stag in the group during antler growth. It was apparent that there was not the same cycle of fat deposition and utilization as had been observed in fallow bucks at the same location (Mulley 1989). The implications of these findings for the management of chital stags were considered to be as follows.

Management Implications

Mating Management

The major reason for adopting restricted mating was to control the timing of calving, not only to avoid the coldest winter months, but also to ensure that calves were born when pasture conditions were optimal. This reproductive flexibility was considered to be a very major attribute of the chital deer. Given that the desired period would be the spring months of September and October, it was known that mating would be necessary in February and March, given a gestation length which had been calculated at 234.5 days (SD 3.0). There was no difficulty in using this period for mating, since the stags had been shown to be fertile at that time.

The ability of chital stags to adapt to a restricted mating period of about 6 to 8 weeks was confirmed, with a stag able to modify his behavior to successfully mate with a number of hinds coming into estrus together. This is not the case during unrestricted mating, when there would be only a small number of hinds in estrus at any one time.

Antler Removal

There is little debate that there is no place for full hard antlers on farmed deer, and chital stags are no exception. The options for dealing with antlers in fallow deer have been reviewed by English (1991). However, the situation is complicated somewhat by the vigor and strength of chital stags; it is not possible to manually restrain even young stags for the cutting of hard spikes, let alone older animals.

It is not wise to wait until the antlers of adult stags have hardened before removing them. Chital stags are capable of extreme aggression within groups when in hard antler, and it is impossible to gather such animals in pens or small yards without the risk of injuries occurring. Chemical restraint is required to achieve the removal of the antlers, with most adult stags unlikely to move easily through the normal cradle or crush due to the length of their antlers. If the stags have become agitated and stressed during the yarding, it becomes likely that mortalities will occur with chemical restraint. Young stags destined for slaughter can have their hard antlers removed with large pruning shears while restrained in a cradle.

The removal of antlers while in velvet is an option, in the same way as is routine for fallow bucks. However, unless the stags are very well trained to the handling process, there is a substantial risk of antler damage. Not only does this have economic implications, if it is intended to seek a market for the antlers, but there is an animal welfare aspect as well. Owners of chital stags who adopt this procedure as the routine method for removing antlers from their breeding animals must ensure that the deer are well trained and accustomed to being handled. Very little velvet antler from chital stags has been sold in Australia, and it is certain that there is no prospect of large numbers being retained for this purpose. Chital deer are farmed for their capacity to produce venison, and velvet antler plays little or no part in their economics.

Perhaps the preferred solution with chital stags is to prevent antler growth by the use of a surgical polling procedure. This must be carried out at 5 to 7 months of age, with the developing antler pedicle being removed under general anesthesia. The procedure is a once-only solution, and can be carried out on animals selected from a group of weaned stags. The only drawback is the removal of the most obvious indicator of their reproductive status, given that stags in velvet are certain to be less fertile. However, there are other physical changes, such as neck girth and coat color, which are sufficient to identify rutting stags. Behavioral changes, including aggression and rubbing of trees provide further evidence.

The castration of fallow bucks at about 5 to 7 months of age is attracting increasing interest in Australia, in order to avoid the aggression and carcass bruising which occurs when entire bucks are transported for slaughter during the breeding season (English 1991). So far there has been no use of castration with chital deer, perhaps because there is not the same level of aggression with these young stags as is the case with fallow bucks.

Chital Hinds

The ability of chital hinds at Camden to undergo a continuous series of estrous cycles throughout the year has been verified (G.E. Mylrea and A.W. English, unpublished data). In the initial study, serum progesterone profiles and the use of a vasectomized stag fitted with a marking harness revealed that the length of the estrous cycle was 19.3 days (SD 1.3), with a high level of synchrony in the group. In the unrestricted mating group there was a winter birth peak which obviously related to the major rutting period in the early months of the year.

Perinatal Mortality

The problem of perinatal mortality in calves born in winter was confirmed. There was a 48% death rate in calves born to 45 chital hinds over 4 years at Camden, with the majority of the deaths in the winter months. This is an unacceptable rate, and compares unfavorably with the rate of 18% in fallow deer fawns at the same location (Mulley 1989).

The most frequent cause of calf death was found to be a complex of mismothering, predation, and exposure/starvation. It became apparent that the timorous nature of chital hinds leads to their being particularly sensitive to human disturbances when there are very young calves in the herd. It would seem preferable not to attempt to catch newborn calves for weighing and eartagging. Chital calves can be eartagged at weaning, at 3 months of age.

The rarity of difficult births in chital hinds reduces the need to disturb the herd by close observation during calving, which in a group which has been mated over 6 to 8 weeks will be known quite precisely. Parturient hinds must be given every opportunity to seek isolation from the herd during birth and the immediate postpartum period. Harassment of hinds by stags immediately after parturition is common, and all adult males should be removed from the females at the end of the mating period. This behavior presents a distinct risk of mismothering, or of trauma to a newborn calf.

There was significant predation of young calves by foxes (*Vulpes vulpes*) at the study site, particularly during late winter and early spring. It appeared that chital hinds protected their calves less effectively from predation than other deer species do, although the fact that fox predation of fallow deer fawns at the same location is rare may be related to when the female foxes are lactating themselves, and are especially hungry. This was probably confirmed by the much lower level of predation in groups of chital calves born later in the year, under a restricted mating policy. It was apparent that the normal hiding behavior of calves was not protecting them from foxes, and that some form of predator control, or fox-proof fencing is necessary if calving occurs when foxes are active.

The incidence of misadventure would be reduced by ensuring that calving enclosures have fences through which chital calves cannot pass, and thus become separated from their mothers. Chital calves are somewhat smaller at birth than are fallow deer fawns, mean birth weights at Camden were found to be 3.4 kg (SD 0.5) for females and 3.6 kg (SD 0.4) for males, and can often move through gaps which will stop the latter.

Restricting the birth season to those times of the year when weather conditions are favorable for calf survival and when abundant high quality pasture is available for lactating hinds has been shown to greatly reduce the problem of perinatal mortality in chital deer.

Mating Management

Stags can be joined with hinds about 3 weeks after the last hind in a group has calved, to achieve the earliest possible conceptions, and thus maximum reproductive rate, as long as calves will be born at a desirable time of year. If stags are joined 3 to 6 weeks after calving, they may then be removed from the hinds when calves are weaned at about 12 weeks of age.

The interval from parturition to conception ranged from 18 to 118 days, with an average of 48.1 days (SD 27.8). The longest intervals occurred after a severe winter in August 1986, when all neonates perished. Conception within 33 days postpartum occurred for 39% of hinds, and it can be assumed that these females conceived at their first postpartum estrus. Given that young chital stags reach puberty at 14 to 16 months, it is necessary to remove any such deer to ensure that no unplanned matings occur in recently calved females. Similarly, chital hinds reach puberty at about 22 to 25 kg, which is 50% of their adult weight, and this may be as young as 10 months. Early segregation of male and female calves will ensure that very young hinds do not conceive at a time of year which is not desired.

Weaning

The weaning of fallow deer fawns before the rut has become common practice in Australia, with perceived benefits for both dams and offspring (English 1991). In particular, it permits the early use of supplementary feeding and close contact with man to quieten and train the young deer.

The same process has been adopted with great success for chital deer, with groups of calves born under a restricted mating policy being weaned at 12 to 18 weeks. A minimum body weight of 13 kg is used before a calf is weaned, with most at this age being in the range 15 to 20 kg. The animals are best confined with some exposure to other farm activity, and hand fed daily with grain or concentrate, even if pasture is adequate. This begins the training and taming of the deer at an early age. Furthermore, the growing deer can be preferentially fed throughout the first year, to ensure maximum growth rates for both replacement breeding stock and those destined for venison production.

Venison Production

Given that chital deer are likely to be farmed solely for venison production, a knowledge of growth rates and annual live weight fluctuations is essential in assessing the most appropriate time to send males to slaughter. As with fallow bucks, some live weight gain will be sacrificed if young stags are slaughtered prior to antler hardening; however, given the fact that there is much less annual fluctuation in body weight with chital stags than occurs with the temperate deer, there is the opportunity for more flexibility in the age of slaughter. The present study verified that there is much less seasonal variation in the carcass quality of chital stags than is the case with temperate species. The fat content of chital carcasses was least during the early months of the year, but prior to this period, fat deposition was negligible when compared with that in temperate males. There was an average 5% variation in the carcass fat content (2.2% to 7.2%) of chital stags throughout the year, compared with 22% variation in red deer stags (Wallace and Davies 1985) and a maximum carcass fat content in fallow bucks of 33% (Gregson and Purchas 1985). At the time of year that red and fallow deer are commonly slaughtered, their carcass fat content is 8% to 12% (Drew 1985).

It is clear that the quality of chital deer carcasses is more consistent, with a lower fat content than red and fallow deer. Adult chital stags over 70 kg had a higher fat content than those under 24 months, and this fact would suggest that no attempt should be made to market prime venison from older males.

The live weights achieved by growing chital stags are very similar to those of fallow bucks at the same age. Chital stags at 13 to 18 months, with a mean live weight of 42.8 kg (SD 0.7), had a dressing out proportion of 60%, compared with a figure of 61% for fallow bucks at 17 months (Mulley et al. 1991). Chital stag live weights at 24 to 32 months averaged 54.9 kg (SD 1.8). Of greater interest was the finding that the high priced hindquarter cuts comprised a greater proportion of the chital carcass than was the case for red deer or fallow deer. Thus, by any criteria related to meat production and carcass quality, chital stags compare more than favorably with fallow deer.

This preliminary study has not addressed the economic consideration of the relative meat production per hectare of fallow and chital deer. Nonetheless, there is sufficient information to indicate that some of the early perceptions of the venison potential of chital deer were not misplaced. In particular, the species will add to Australian deer farming a measure of flexibility in venison production which cannot be achieved when only temperate species are farmed, even with the adoption of measures to manipulate breeding seasons.

Conclusion

The chital deer is considered by many to be the most attractive cervid of all. Whether that be the case or not, the challenge has been to come to grips with the management of the species on deer farms, to determine their place in the mix of deer species being farmed. The study at Camden has provided good evidence that chital deer can be farmed, both in economic and humane terms. They are not a deer for beginners, with a high level of stockmanship being required to manage them. There are obviously geographic limits to their ability to cope with inclement winter weather. The least surprising fact to emerge from the study has been that the principles which apply to the farming of the temperate species of deer apply equally well to the chital. Their place in the Australian deer farming industry seems assured.

Acknowledgements

The study at Camden was supported by funds from the Australian Special Rural Research Committee. Animals were generously donated by John Murphy, without whom the work could not have commenced.

References

Ables ED (1977) The Axis Deer in Texas. Kleberg Studies in Natural Resources, Texas A and M University, College Station Texas

Anderson R (1978) Gold on Four Feet. Ronald Anderson and Associates, Collingwood, Victoria, Australia

Bentley A (1978) An Introduction to the Deer of Australia. Koetong Trust Service Fund, Forests Commission, Victoria, Australia

Chapple RS (1989) The biology and behaviour of chital deer (*Axis axis*) in captivity. PhD Thesis, University of Sydney, Australia

Chapple RS, English AW, Mulley RC, Lepherd EE (1991) Hematology and serum biochemistry of captive chital deer (*Axis axis*) in Australia. J Wild Dis (in press)

Drew KR (1985) Meat production from farmed deer. In Fennessy PF, Drew KR (Eds) Biology of Deer Production. R Soc N Z Wellington, Bull 22:285–290

English AW (1991) Management strategies and health programs for farmed fallow deer. In Rennecker LA, Brown RJ (Eds) Conservation and Sustainable Development. University of Alaska, Fairbanks pp. 16–23

Goss RJ, Rosen JK (1973) The effect of latitude and photoperiod on the growth of antlers. J Reprod Fert Suppl 19:111–118

Graf W, Nichols L Jr (1967) The Axis Deer in Hawaii. Bombay Nat Hist Soc 63:629–734

Gregson JE, Purchas RW (1985) The carcass composition of male fallow deer. In Fennessy PF, Drew KR (Eds) Biology of Deer Production. R Soc N Z Wellington, Bull 22:295–298

Harthoorn AM (1981) Mechanism and prevention of capture myopathy. In Fowler ME (ed) Wildlife Diseases of the Pacific Basin and Other Countries. Proc 4th Int Conf Wild Dis Assoc, University of California, pp 248–257

Loudon ASI, Curlewis JD (1988) Cycles of antler and testicular growth in an aseasonal tropical deer (*Axis axis*). J Reprod Fert 83:729–738

Mishra HR (1982) The ecology and behaviour of chital deer (*Axis axis*) in the Royal Chitwan National Park, Nepal (with comparative studies of hog deer (*Axis porcinus*), sambar (*Cervus unicolor*) and barking deer (*Muntiacus muntjac*). PhD thesis, University of Edinburgh, Scotland

Mulley RC (1989) Reproduction and performance of farmed fallow deer (*Dama dama*). PhD thesis, University of Sydney, Australia

Mulley RC, English AW, Butterfield RM, Thompson JM (1991) Changes in growth rate and body composition of entire and pre-puberally castrated fallow bucks (*Dama dama*) treated with zeranol. 1.

Liveweight, body organs and carcass components. Anim Prod (in Press)
Schaller GB (1967) The Deer and the Tiger—a Study of Wildlife in India. University of Chicago Press, Chicago
Spillett JJ (1966) A report on the wildlife surveys in North India and Southern Nepal. Bombay Nat Hist Soc 63:492–628
Wallace V, Davies AS (1985) Pre- and post-rut body composition of red deer stags. In Fennessy PF, Drew KR (eds) Biology of Deer Production. R Soc N Z Wellington Bull 22:291–293
Wehausen JD, Eliott HW III (1982) Range relationships and demography of fallow and axis deer on Point Reyes National Seashore. Calif Fish Game 68:132–145

47
Production Cycles and Characteristics of Rusa Deer in Queensland, Australia

Keith B. Woodford and Alison Dunning

Abstract

A herd of rusa deer (*Cervus timorensis*) was monitored for 4 years at University of Queensland, Gatton College. Reported characteristics include reproductive cycles and performance, antler cycles, growth rates, live weights, carcass parameters, animal health, and behavior. Additional results obtained from private herds in Queensland are presented. Breeding stags and hinds at Gatton College were run together throughout the year. However, of mature hinds 84% calved in autumn between 4 March and 15 May. Most primiparous hinds gave birth between June and October at 15 to 18 months of age. Intercalving intervals following autumn births were about 365 days but were shorter following winter and spring births. Mean gestation was 252.6 days. Pregnancy rates in mature hinds were 97% per annum. Weaning rates were 84% per annum. Mean birth weights were 5.1 kg in males and 4.7 kg in females. Highest growth rates occurred prior to 6 months of age. Most stags were slaughtered for venison at about 14 months and produced a 47 to 50 kg carcass. Live weight changes in males greater than 14 months were associated with the antler cycle and stags older than 2 years lost about 15% of their live weight during the breeding rut.

Key words: Behavior, carcass, growth, health, reproduction, rusa deer

Introduction

Two subspecies of rusa deer known as Javan and Moluccan rusa are farmed in Queensland. The Australian history of these subspecies has been described by Bentley (1978). Farming of rusa in Queensland commenced in 1980 and has been described by Mackenzie (1985). Unpublished Gatton College survey results indicate 3685 rusa deer on 31 Queensland properties as of 30 June 1989.

Study Area and Methods

Monitoring of the Gatton College herd (lat. 27° 36′S) commenced December 1985 at the time of herd establishment with 18 females and 2 males. The herd was mainly Javan, but one Moluccan stag was introduced in January 1987 and replaced by two other Moluccan stags in January 1989. The rusa deer were run on the same farm as a herd of red deer (*Cervus elaphus*). On occasions the two species were run together in the same paddock.

The herd was run initially on 7 ha of lightly timbered elevated yellow podzolic soils derived from sandstone. The main pasture species was green couch (*Cynodon dactylon*) with lesser quantities of Pioneer rhodes grass (*Chloris gayana*) and green panic (*Panicum maximum* var. *trichoglume*). Spray irrigation was installed and kikuyu (*Pennisetum clandestinum*) progressively introduced. Legume content was minimal. Pastures were fertilized with up to 230 kg N/ha per

TABLE 47.1. Climatic data, 1986–1989, Gatton College, Queensland, Australia

January daily maximum (°C)	32.1
July daily maximum (°C)	20.6
July daily minimum (°C)	7.1
Days/annum > 35° C	17
Frosts/annum	11
Daily wind, June to August (km/day)	111
Rainfall, mm/annum (range)	838 (541–1079)

annum. By the end of 1989, 120 red deer and 110 rusa deer were running on an expanded area of 10 ha subdivided into six paddocks. Shortfalls of feed (mainly from mid-June to late September) were met by feeding barley, sorghum, and hay. The rusa received about 20% of total feed intake as supplement. The climate was characteristically hot in summer and temperate in winter (Table 47.1).

Normal practice was to inspect daily the Gatton herd by vehicle. On occasions field observations were constrained either by the risk of running over young calves hiding in the long tropical grass, or by wet conditions that prevented vehicle access. In 1988 and 1989, almost half the calves were weighed and eartagged in the field within 24 h of birth.

The deer were usually mustered every 4 weeks for weighing and any husbandry operations; however, the breeding herd was not mustered during the main calving season. Deer were mustered via a laneway into covered yards with concrete floor and solid walls 2.1 m high. For weighing they were directed into an enclosed crate placed on load cells where they stood without restraint. All live weights were nonfasted. Animals were restrained for husbandry operations in a V-shape crush with a drop floor. Venison stags were slaughtered in a commercial abattoir on the cattle chain.

Additional information presented in this chapter comes from monitoring of private herds in South East Queensland plus postal surveys of all Queensland deer farms.

Reproductive Cycles

Females

Information provided to us by Queensland farmers indicates that most Javan hinds calve between mid-March and mid-May, with Moluccan calvings being more evenly spread throughout the year. However, several farmers with Moluccan herds have succeeded through seasonal mating policies to concentrate the calving into the spring period from late August to October. It appears that once in this pattern the concentrated calving is maintained even if the stags remain with the hinds, but this requires confirmation.

The breeding stags of the Gatton College herd of Javan rusa were run continuously with the hinds from herd establishment until October 1989. The autumn calving of this herd commenced each year between 4 March and 20 March, although one animal in 1989 calved on 20 February. All hinds calving prior to 20 March ($n = 13$) were either first calvers purchased as yearlings ($n = 10$) or hinds that failed to rear a calf the year before ($n = 3$). Approximately 84% of mature hinds calved by mid- May, but there were occasional births through to early December. Most female progeny at Gatton College had their first offspring at 15 to 18 months of age, and these were born between June and October. No hind gave birth between 8 December and 20 February; however, births in Moluccan rusa on private farms have been recorded during this period.

Intercalving intervals were recorded for 56 Javan rusa (Table 47.2) at Gatton College. These results illustrate the ability of the herd to endogenously concentrate the calving into one period of the year. The shortest intercalving interval was 281 days where the first calf was reared and 271 days where the first calf was not reared. An intercalving interval of 266 days was recorded on a private farm for a Moluccan hind that had failed to rear the first calf.

We estimated the gestation period from 13 observed matings and births as 252.6 days (SD = 3.7; range, 248–258). Four additional observations exceeding 268 days were excluded as being pregnancies from later estrus cycles. This estimate compares with 249 days (SD = 13.4; range, 217–277, $n = 19$) recorded by van Mourik (1986).

Observations of gestation for Moluccan rusa are restricted to one recording from each of two farms, these being 238 and 249 days.

Observations of 24 Javan hinds born between March 1986 and April 1988 showed that 18 had

TABLE 47.2. Intercalving interval of Javan Rusa at Gatton College, Queensland, Australia

Previous calving (month)	Intercalving interval (days)				
	< 280	280–320	321–360	361–400	> 400
February	—	—	—	—	1
March	—	—	—	26	4
April–May	—	—	6	2	—
June–October	2	13	—	—	1
December	—	1	—	—	—

calved by 19 months of age with the youngest confirmed age at parturition of 476 days. Based on a gestation estimate of 253 days, then hinds became pregnant as young as 223 days with most pregnant before 270 days of age. This indicates minimum mating live weights of about 40 kg with most hinds pregnant before reaching 50 kg.

Males

We observed pedicle initiation in both Javan and Moluccan rusa by 30 kg live weight and in all animals prior to 35 kg. Depending on the level of nutrition this ranged from less than 5 to as late as 9 months of age. Antler growth appeared to commence about 2 months after pedicle initiation, but this was delayed in some Moluccan calves under poor nutrition. Spiker antlers grew for about 90 days and antler branching occurred in some animals. Spiker antlers in the Gatton herd were cast at ages ranging from 14 to 24 months of age, and there was an apparent relationship between casting and the absence of estrus females. Each year most of the spikers in the herd cast within a period of 1 month irrespective of age, but the month varied from January to March.

Antler cycles in adult stags were generally circannual, but there were considerable variations. Antler cycles at Gatton College ranged from 259 to 393 days. Stags in this herd cast in all months except July and August. These were the 2 months when most matings occurred. Some farmers reported antler cycles of up to 2 years, but we were unable to document this. It seems that this may have occurred where hinds were coming into estrus throughout the year and where there was no competition from other breeding stags. Short antler cycles appeared to be associated with situations where there were either few or no hinds in estrus, and the stags responded by casting.

Reproductive Performance

The calving rate for hinds 2 years and greater ($n = 123$) was 97%. This calving rate was measured as calves born divided by hinds potentially mated. The weaning rate from these hinds, measured as calves weaned divided by hinds potentially mated, was 84%.

There were also 18 births from 24 yearling hinds, of which 12 calves were successfully raised. These hinds were all less than 19 months at time of parturition. Typically these hinds gave birth to their second calf at 26 to 28 months of age and their third calf at about 36 months of age.

Growth Rates and Live Weights

Average live weights at birth for Javan rusa were estimated from pooled 1988 to 1990 Gatton College data. Data were not obtained for all animals. Males averaged 5.1 kg ($SD = 0.7, n = 24$) and females averaged 4.7 kg ($SD = 0.6$, $n = 38$). The difference in mean values was statistically significant at $P = 0.03$, measured using the two-sample t test for independent samples with unequal variances (Satterthwaite's method, see Snedecor and Cochran 1989).

Growth rates from birth to 420 days (Table 47.3) were estimated from those autumn born 1988 progeny for which birth dates and birth weights were available. All of these animals were born during March, and both sexes were

TABLE 47.3. Daily growth rate (g/day) in Javan Rusa from birth to 420 days (mean ± SE)

Days	Females ($n = 7$)		Males ($n = 6$)		Female:male
0–60	169	8	184	9	0.92
60–120	158	9	172	8	0.92
120–180	175	9	214	8	0.82**
180–240	121	6	178	4	0.68**
240–300	121	3	150	7	0.81**
300–360	98	4	96	7	1.03
360–420	121	14	173	16	0.70*

*,** indicate differences between females and males significant at $P < 0.05$, $P < 0.01$, respectively

TABLE 47.4. Seasonal growth rates (g/day) from 2 to 26 months of autumn-born Javan Rusa Males (mean ± SE)[a]

Age group (months)	Initial live weight on 1 June (kg)	Seasonal growth rates			
		Winter	Spring	Summer	Autumn
2–14 ($n = 11$)	18 1	212 4	171 5	157 10	163 5
14–26 ($n = 5$)	69 2	45 18	91 8	111 17	229 34

[a]Seasons are defined as 3-month periods with winter commencing 1 June

run throughout as one group. Highest growth rates occurred prior to 6 months of age. For females the increased live weight gain between 360 and 420 days corresponded to the second half of pregnancy. In males the increased growth rates during the equivalent stage corresponded with the period just before the rut. Overall female growth rates were 83% of male growth rates.

Seasonal growth rates were compared for two groups of males aged 2 to 14 months and 14 to 26 months (Table 47.4). These stags were born in the autumn, between 15 March and 15 May of 1987 and 1986 respectively. The two groups of stags were run separately up to 18 August, but thereafter all animals were run in the same paddock. The greater growth rates in the younger animals were significantly greater (separate variance t test; see above) in winter ($P < 0.001$), spring ($P < 0.001$) and summer ($P = 0.048$). The apparent difference in autumn growth rates between the two groups was not significant ($P = 0.13$). Overall annual live weight gain averaged 64 kg in the younger animals and 43 kg in the older animals. Of the annual live weight gain in the older animals 72% was achieved in the second half of the year, and 48% in the final 3 months.

The live weight changes in stags aged 14 to 26 months may not always follow the patterns shown in Table 45.4. This is because stags born outside the main season tended to cast their spiker antler at a younger age, giving a reduced period of low growth. Further clarification of these relationships is required.

Javan stags older than 2 years followed a circannual pattern of weight loss and weight gain. Average weight loss during the rut was 15% (SD = 3.7, $n = 7$) of prerut weight. The heaviest weight recorded at Gatton College was 152 kg in a 4-year-old stag.

The average weight of Gatton College hinds aged 2 years or more was 83 kg (SD = 8, $n = 42$) in January 1989, at which stage most hinds were between 5 and 6 months pregnant. Subsequent July mating live weights for these hinds averaged 74 kg (SD = 6.3, $n = 41$). These compare with average live weights of 66 kg recorded by van Mourik (1986) for the parent herd of Javan rusa in Victoria.

Comparisons of growth rates and body size between Javan and Moluccan rusa were confounded by the differences in nutrition and management between properties; however, it would appear that Moluccans are 70% to 80% of the size of Javans.

Carcass Parameters

Males from the Gatton College herd were slaughtered each year in late May or ealy June prior to the rut (Table 47.5). Average dressing percentages of 60% to 62% in stags aged 13 to 15 months compare with average figures of about 57% for red deer of the same age from the same farm. Average tissue depths of about 11 mm in these rusa compared with about 8 mm for red deer at the same age (Woodford et al. 1990). The rusa carcasses had higher visible fat content than those of the red deer, but no dissections to determine carcass composition were undertaken.

Data supplied by a commercial farmer showed Moluccan carcasses from animals aged about 23 months averaging 42.1 kg (SD = 3.4, $n = 28$) with an average dressing percentage of 58.8% (SD = 2.3, $n = 28$). Because of the different farm-

TABLE 47.5. Slaughter live weights and carcass measurements of Javan Rusa

Birth cohort	Age at slaughter (months)	Live weight (kg)	(SD)	Carcass weight[a] (kg)	(SD)	Tissue depth[b] (mm)	(SD)	Percentage (%)	(SD)
1986 ($n = 5$)[d]	23–26	107	9	67	5	18	2	62	1
1987 ($n = 10$)[e]	13–14	81	6	50	4	11	2	62	1
1988 ($n = 14$)[f]	13–15	80	4	47	2	10	3	60	1

[a]Measured before cooling after removal of head, feet, skin, and all internal organs including kidneys
[b]Total tissue including fat and muscle measured 16 cm down from the midline over the 12th rib
[c]Calculated as bone in hot carcass weights divided by non-fasted live weight
[d]Excludes one 119-kg nonslaughtered animal
[e]Excludes one 89-kg nonslaughtered animal
[f]Excludes two nonslaughtered animals weighing 88 kg and 77 kg, respectively

ing environment, these figures cannot be compared directly with the results for the Javan rusa.

Antler Weights

Only limited data were collected on antler weights. Javan spikers averaged 149 g of hard antler (SD = 57, $n = 16$). Six 2-year-old stags averaged 1.05 kg (SD = 0.25). Mature stags produced 1.0 to 2.5 kg of hard antler. It is probable that some older stags would produce heavier weights. Moluccan stags produce small spiker antlers of about 50 g, but some mature stags can produce at least 2 kg of antler.

Animal Health

Aggregate survey results from postal surveys of Queensland rusa deer farmers for the 3 years ending 30 June 1987, 1988, and 1989 showed 129 deaths from 7437 animals aged more than 3 months. The implied annual death rate was 1.7%.

The major causes of death as reported by farmers were dog attacks (21% of the total), poisoning (20%, all of which occurred on one farm), injuries from fighting and yarding (15%), and scrub tick, *Ixodes holocyclus* (9%); 6% were considered due to bacterial and viral infections and 8% were of unknown cause.

Farmers also reported 230 perinatal, neonatal, and preweaning losses in calves, representing 10.6% of the known births. The major causes as classified by farmers were predators (16%), cold (14%), stress (15%), accidents (11%), and stillbirths (6%). Twenty-one percent were of unknown cause. It is likely that not all stillbirths were identified as hinds tend to calve in long grass, and stillborn calves decompose rapidly in the Queensland climate.

At Gatton College there were two deaths in hinds aged more than 14 months, representing an average annual death rate of 1.2%. There were also two deaths in animals between weaning and 14 months, representing an annual death rate of 1.8%. Perinatal losses and deaths up to weaning totalled 22, or 15.7% of total births. Most losses were stillbirths, but others resulted from mismothering, hypothermia within 24 h of birth, and a calf with a deformed leg which the mother chewed off. There was no obvious common factor, such as low birth weight or dystocia as a cause of death, and post mortems of stillborn calves were inconclusive.

The Gatton College herd was regularly tested for internal parasites by monitoring fecal egg counts. Despite the high stocking rate on irrigated pastures there was no evidence of clinical burdens. In 1988, young stock were dosed with oral anthelminthic as a preventive measure. Three weaners were successfully treated for bacterial scouring.

Behavior and Management

Field experience at Gatton College indicated that rusa deer are large herd animals. Groups of more than about 20 animals tended to ignore the presence of a vehicle and would sometimes remain lying down when the vehicle passed within 3 m; however, smaller groups invariably moved away and remained unsettled. Also, whereas large groups could be mustered and yarded with ease, small groups of less than ten were on occasion impossible to muster.

In comparison with red deer, the rusa deer spent much less time and energy moving and showed no tendency to run along fences. The rusa deer were also much less selective grazers than red deer and in these subtropical environments showed greater ability to thrive on low quality feeds. They appeared better able than red deer to deal with varying levels of feed supply by storing body fat, and accordingly less susceptible to reproductive failure resulting from nutritional stress.

Animals showed no sign of aggression when in the paddock except for newly calved hinds which would often vigorously defend their own and sometimes other calves from human interference. Yarded rusa were readily drafted and weighed, but were much more likely than red deer to jump in an attempt to escape. Unlike red deer, it was almost impossible to perform routine operations such as eartagging and vaccinations without restraining the animals individually in a crush.

Behavioral differences between Javan and Moluccan rusa were observed, with the Moluccan rusa tending to move continually when in the yards, and more likely to break from the herd when being mustered; however, their smaller size made them less daunting in the yards.

Breeding stags were able to co-exist within the one herd and were observed to fight less than red stags. However, a definite hierarchy of stag dominance was established. At Gatton College, where group size was up to 50 females, the currently dominant stag appeared to do all the mating. Stags growing velvet antler did not compete with stags in hard antler for access to estrus females and showed no sexual interest in these females. Farmers advised us of calves resulting from matings by spiker stags as young as 8 months.

Management Implications

It is apparent that the behavioral and production characteristics of rusa deer make them a suitable species for farming. Although rusa deer have circannual cycles of reproduction and antler growth, there can be considerable variation in the length of these cycles. There is no evidence that the cycles are entrained by changing day length, but there is circumstantial evidence for pheromonal or other social signals within the herd which can result in concentration of calving and antler casting. There is an apparent relationship between reproductive cycles, antler cycles, and growth rates. Accordingly, improved understanding of these cycles is likely to have considerable practical implications.

Acknowledgements

We acknowledge the assistance of J. Winch and L. Lyne in data collection, and I. Bruce for statistical advice. We are grateful to the Queensland deer farmers who supplied information, particularly L. Townsend and D. Hart of "Oakwood" and R. Peachey and D. Naumann of Mt. Tamborine. Much of the information was collected as part of the Deer Farming Systems Project funded jointly by the Australian Special Rural Research Council and the Queensland Deer Breeder's Association.

References

Bentley A (1978) An introduction to the deer of Australia. Koetong Trust Service Fund, Forests Commission, Victoria, Australia

Mackenzie AR (1985) Reproduction of farmed rusa deer (*Cervus timorensis*) in south-east Queensland, Australia. In Fennessy PF, Drew KR (eds) Biology of Deer Production. R Soc N Z Wellington Bull 22:213–215

Snedecor GW, Cochran WG (1989) Statistical Methods, 8th ed, Iowa State University Press, Ames, Iowa

van Mourik S (1986) Reproductive performance and maternal behaviour in farmed rusa deer (*Cervus* (*rusa*) *timorensis*). Appl Anim Behav Sci 15:147–159

Woodford KB, Dunning A, Winch JB (1990) Production of red deer in a subtropical environment. Proc Aust Soc Anim Prod 18:436–439

48

Venison Production from Grazing Young Red Deer Stags: Effects of Pasture Type and Immunization Against Melatonin

Abu M. Ataja, Peter R. Wilson, John Hodgson, Ron M. Hoskinson, Roger W. Purchas, Hugo Varela-Alvarez, and Tom N. Barry

Abstract

Three grazing experiments were conducted to devise systems for growing weaner red deer (*Cervus elaphus*) stags through winter and spring to attain a slaughter weight of 95 kg by the end of spring, at 12 months of age. Grazing was conducted on perennial ryegrass/white clover pastures, and the same direct-drilled with Moata annual ryegrass. A group of animals in each experiment was immunized against melatonin in an attempt to increase winter live weight gain (LWG). Grazing the young deer a high pasture surface height (10 to 8 cm) and herbage mass (2100 to 1600 kg DM/ha) on conventional perennial ryegrass/white clover pasture resulted in 75% to 41% attaining the target slaughter weight, of which the critical factor was a consistent winter growth of 140 g/day. Introduction of annual ryegrass increased animals attaining target slaughter weight by an average of 19%, due to small but consistent increases in LWG in both winter and spring. When commenced at birth, immunization against melatonin resulted in large antibody titres during winter and spring and higher circulating concentrations of prolactin. Use of Freund's adjuvant did not produce any increase in LWG, but there were indications that use of DEAE dextran adjuvant may have increased growth.

Key words: Melatonin immunization, pasture nutritive value, red deer, venison production

Introduction

In order to meet the requirements of northern hemisphere markets, it is necessary to produce venison in New Zealand (NZ) of > 50 kg carcass weight (95 kg live weight) in the spring (August to November). In the three experiments described here this has been attempted by 1 year of age, using red deer stags.

Ruminant animal production in NZ is achieved from grazing intensively managed forages throughout the entire 12-month production cycle; there is no cereal concentrate feeding, except during short summer droughts, and no indoor housing. The forage commonly used over all of NZ consists of mixed perennial ryegrass (*Lolium perenne*): white clover (*Trifolium repens*) pasture in the approximate ratio 80:20. To achieve the above-set objectives in a venison production system, it is necessary to achieve high animal growth rates at all times. This is especially true over winter, when both rates of pasture production (Korte et al. 1987) and deer growth rates (Kay 1985) are at their low point in the annual cycle, and when both the concentration of metabolizable energy (ME) in pasture and its efficiency of utilization are lower than in spring (Waghorn and Barry 1987).

Relative to perennial ryegrass, annual ryegrasses have improved winter DM production (Armstrong 1981) and have shown higher nutritive value in experiments with sheep (Ulyatt 1981); therefore, three experiments were conducted to study the effects of introducing "Grasslands Moata" annual ryegrass (*Lolium multiflorum*) into perennial ryegrass/white clover pastures upon the growth of weaner red deer stags during winter and spring.

Barry et al. (1991) concluded that plasma melatonin concentration was the signal that synchronized rhythms (including growth) in red deer to changes in photoperiod, while the work of Domingue et al. (1991) suggested that a period of high plasma melatonin concentration is important, which normally occurs in late autumn/early winter. A second objective was to evaluate active immunization against melatonin as a potential means of increasing the growth of young red deer over winter.

Materials and Methods

Weaner red deer stags, aged approximately 4 months, were used in all three experiments. To suppress gut parasites and lungworm, they were drenched at 4-week intervals with the anthelmintic "Ivermectin" (Merck, Sharpe, and Dohme, NZ). In late November/early December of each year the stags were slaughtered at a commercial deer slaughter plant (DSP).

Direct-drilling with Moata annual ryegrass was done in late autumn each year using the herbicide Roundup (Glyphosate 360; Monsanto, NZ) as a spray to suppress existing vegetation. This was done as band spraying in Experiments 1 and 2, using drilling in one direction only in Experiment 1 (15 kg Moata seed/ha) and in two directions in Experiment 2 (20 kg Moata seed/ha). In Experiment 3, blanket spraying was used, followed by direct drilling in two directions (24 kg Moata seed/ha). This was followed by applications of urea (30 kg N/ha) in both autumn and winter, to both conventional pastures (henceforth referred to as pasture) and to direct-drilled Moata areas (henceforth referred to as Moata). Both types of pasture were managed to keep the plants in the short vegetative state in all times and, hence, of high nutritive value. This was done by measuring pasture height twice per week and relating height to pasture mass using calibration curves. When necessary, nonexperimental deer were brought in for short periods to keep the pastures at the desired height, and during spring the rotations were shortened (Experiments 1 and 3) by deleting paddocks. Throughout all three experiments, samples were regularly taken of the feed on offer (i.e., cut to ground level).

Experiment 1

In a 2 × 2 × 3 factorial experiment, 48 weaner stags weighing 55.7 ± 0.58 kg (SE) were assigned to two levels of pasture DM allowance (6.3 (high) and 4.5 (medium) kg DM/animal per day), two types of pasture (perennial ryegrass/white clover and the same pasture direct-drilled with Moata), and three types of immunization (nonadjuvant, adjuvant, and adjuvant + antigen) on 14 May 1987. Rotational grazing was used, with weekly breaks, using electric fences, with a 5-week rotation during winter and a 3-week rotation in spring. Booster immunizations were given on 16 July and 6 August 1987.

Experiment 2

In a 2 × 2 × 2 factorial experiment, 52 weaner stags weighing 60.1 ± 0.62 kg were continuously stocked on one of two levels of pasture surface height (10 cm vs 5 cm), two types of pasture (perennial ryegrass/white clover, and the same pasture direct-drilled with Moata) and two types of immunization (nonadjuvant vs adjuvant + antigen) on 30 May 1988. Because of the slow antibody response encountered in Experiment 1, the immunization procedures in Experiment 2 commenced at weaning in late February. Booster immunizations were given 9 June and 7 July 1988.

Experiment 3

In a 2 × 3 factorial experiment, 36 weaner stags weighing 52.9 ± 0.70 kg were allocated to their groups on 15 May 1989. These comprised two pasture types, perennial ryegrass/white clover and the same direct-drilled with Moata after blanket spraying in an attempt to increase the proportion of Moata in the herbage (18 animals/group). Three types of immunization against melatonin were used, these being no immunization (control group), Freund's adjuvant, and DEAE dextran as the adjuvant, using 12 animals/group. Because of the slow antibody response observed in Experiments 1 and 2, the immunization procedure was commenced at birth in Experiment 3, with booster immunizations being given at weaning (28 February) and in autumn (15 May).

Pastures were managed using rotational grazing, with a 5-week rotation during winter and a 3-week rotation during spring. Herbage mass was set at 2100 kg DM/ha when the animals were introduced to a break, and they were removed when this was reduced to 1600 kg DM/ha. The pastures were maintained in this state by taking frequent height measurements and using height vs mass calibration curves.

The antigen in all experiments was 5-methoxytryptamine hemisuccinamide conjugated to human serum albumin, which was given at 1 mg per immunization as 2 × 1 ml s.c. injections on either side of the neck. Freund's complete adjuvant (FCA) was used to give the primary immunization in Experiments 1 and 2, and to one of the two vaccinated groups in Experiment 3. Booster immunizations were given using FCA in Experiment 1 and Freund's incomplete adjuvant (FIC) in Experiments 2 and 3. In Experiment 3, a group of calves was also immunized using DEAE dextran as the adjuvant, for both the primary and booster immunizations.

Total N content of the feed was determined using the Kjeldahl method, while digestibility was predicted using the in vitro procedure of Roughan and Holland (1977). Melatonin antibody titers were determined using a method based on that of Abraham (1974), while prolactin was determined using a modified form of the method of van Landeghem and van der Weil (1978).

Data were analyzed using analysis of variance procedures for factorial experiments. In the case of live weight gain (LWG), initial weight was used as a covariate. This was especially important in Experiment 3, where the adjuvants used caused differences in weaning weight.

Results

Pasture Composition and Nutritive Value

In Experiment 1, Moata annual ryegrass comprised approximately 20% of the total DM in both the medium and high pastures during winter. During spring, this increased to 40%–50% in the Moata high group and to 30% in the Moata medium group. Due to direct-drilling in both directions, the Moata content of paddocks was initially much higher in Experiment 2, comprising 40%–50% total DM. However, its proportion declined during a very wet winter under continuous stocking, and values during spring were approximately 15% and 25% respectively for the Moata high and medium areas. Following blanket spraying in Experiment 3 to remove all competition, the proportions of Moata were much higher, being 82% of total DM during winter and 65% during spring. White clover contents of the control perennial ryegrass-based pastures were low during these experiments, being typically < 5% total DM during winter and < 10% DM during spring. Lower proportions of white clover were present in the direct-drilled Moata paddocks. Perennial ryegrass formed the other principal component of the grazed pastures in these experiments.

Nutritive value was consistently kept at a high level (Tables 48.1–48.3) due to frequent monitoring of the pastures and effective grazing control. Total N content of the herbage was very high during winter, perhaps partly due to urea application. During winter, pastures direct-drilled with Moata tended to have higher contents of total N and to have higher organic matter digestibility, with the latter attaining significance ($P < 0.05$) in Experiments 1 and 3. There was little difference in nutritive value between the two pasture types during spring.

Animal Growth

Experiment 1

Live weight gain during winter (Table 48.1) was approximately 100 g/day and was not significantly affected by either pasture allowance or the introduction of Moata; LWG increased in spring, and tended to be higher for animals grazing Moata high pastures than those grazing pasture high or Moata medium pastures ($P < 0.1$; Table 48.1). Because of temperament problems, the deer were very difficult to handle in spring and in particular to keep behind electric fences. For these reasons, data collected from the

TABLE 48.1. Chemical composition of the feed on offer, live weight gain, and percentage of stags ($n = 12$/group) attaining slaughter live weight (92–95 kg) by early December for stags offered high (6.3 kg/DM/head/per day) and medium (4.5 kg/DM/head/per day) allowances of pasture and Moata in 1987 (Experiment 1)

	Allowances (kg/DM/hd per day)				
	6.3		4.5		
	Pasture	Moata	Pasture	Moata	SE
Total N (% DM)					
Winter	4.66	4.84	4.40	4.60	0.16
Spring	4.06	3.25	4.06	3.54	0.33
Organic matter digestibility (%)					
Winter	82.1	83.3	82.3	84.6	0.65
Spring	84.3	83.1	84.1	83.8	1.35
Live weight gain (g/day)					
Winter	103	110	91	99	18.7
Spring	184	222	186[a]	176	18.8
Stags to slaughter (% total)	0	25	0	17	

[a]Not true result because of problems in keeping animals behind electric fences

TABLE 48.2. Chemical composition of the feed on offer, live weight gain, and percentage of stags attaining slaughter live weight (92–95 kg) by early December for stags offered pasture or Moata at heights of 10 or 5 cm in 1988 (Experiment 2)

	Height (cm)				
	10		5		
	Pasture	Moata	Pasture	Moata	SE
Total N (% DM)					
Winter	4.41	4.53	3.93	4.28	0.20
Spring	2.68	2.68	2.93	3.09	0.30
Organic matter digestibility (%)					
Winter	81.4	81.6	81.9	82.5	0.52
Spring	76.4	77.8	79.1	79.0	1.23
Number of stags	12	14	12	14	
Live weight gain (g/day)					
Winter	153	131	74	79	9.0
Spring	234	209	147	211	10.0
Stags to slaughter (% total)	75	79	8	36	

TABLE 48.3. Chemical composition of the feed on offer, live weight gain, and percent of stags attaining slaughter weight (92–95 kg) by end of November 1989 (Experiment 3)

	Pasture	Moata	SE
Total N (% DM)			
Winter	4.00	4.40	0.27
Spring	2.83	2.70	0.49
Organic matter digestibility (%)			
Winter	80.3	86.1	0.95
Spring	78.9	80.4	2.42
No of animals/group	17	15	
Live weight gain (g/day)			
Winter	139	164	6.5
Spring	226	235	5.4
Stags to slaughter (% total)	41	60	

pasture medium group over this period are not considered true values. Antibody titers against melatonin were slow to develop, being absent during winter and slowly increasing during spring to attain a mean value of 1:1571 ± 583 by early November. Animals given the melatonin antigen grew at the same rate as the control and adjuvant groups over both winter and spring.

Experiment 2

During winter, LWG was much greater in animals grazing at high (10 cm) than at medium (5 cm) pasture height ($P < 0.001$; Table 48.2), with the introduction of Moata having no effect. During spring there was an interaction ($P < 0.001$) between pasture height and the presence of Moata, with LWG being high on 10 cm pastures and not affected by pasture type. LWG was lower in animals grazing perennial ryegrass/white clover pasture at the 5-cm height and was increased by the presence of Moata ($P < 0.001$), with LWG of the Moata medium group being similar to that of the high groups. As a result of high LWG in both winter and spring, 75% to 79% of animals grazing at the greater pasture height reached the target live weight by early December (92 to 95 kg live weight). Antibody titers against melatonin were again slow to develop, attaining values of approximately 1:300 over winter, rising in spring to 1:600 by November. The immunization procedure produced no responses in LWG in either winter or spring.

Experiment 3

Grazing of stags on the direct-drilled pastures increased the proportion of animals attaining the target slaughter weight by the end of spring (30 November) from 41% to 60% (Table 48.3), due to increasing overall LWG, with the increase during winter attaining significance ($P < 0.05$). Commencing the immunization sequence at birth with Freund's adjuvant produced high titers of approximately 1:6000 during winter, rising rapidly in spring to peak at 1:15 000 in October (Fig. 48.1). Use of the DEAE dextran adjuvant produced titers of approximately 1:2000 in late autumn (May), which declined to zero by the end of winter (August). These were associated with plasma prolactin concentrations being higher in immunized than in control animals during winter

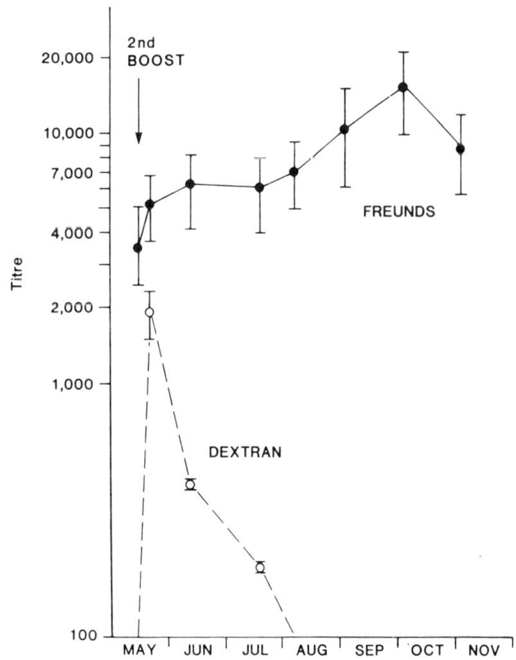

FIGURE 48.1. Development of antibody titers in young red deer immunized against melatonin in Experiment 3. ●, Freund's adjuvant; ○, DEAE dextran adjuvant; bar is 1 SEM.

and increasing in spring at an earlier time, with the responses generally being greater in animals immunized using Freund's adjuvant (Fig. 48.2).

Immunization of calves at birth with Freund's adjuvant resulted in lower weaning weights (47.7 kg) compared with control animals (51.6 kg) or those given dextran adjuvant (52.5 kg: $P < 0.05$), and this depression in growth continued to the end of autumn (Fig. 48.3). Thereafter, calves immunized using Freund's adjuvant grew at the same rate as control animals during winter and spring. There was some evidence that calves immunized using dextran grew faster in spring (Fig. 48.3), but the difference did not attain significance ($P > 0.05$).

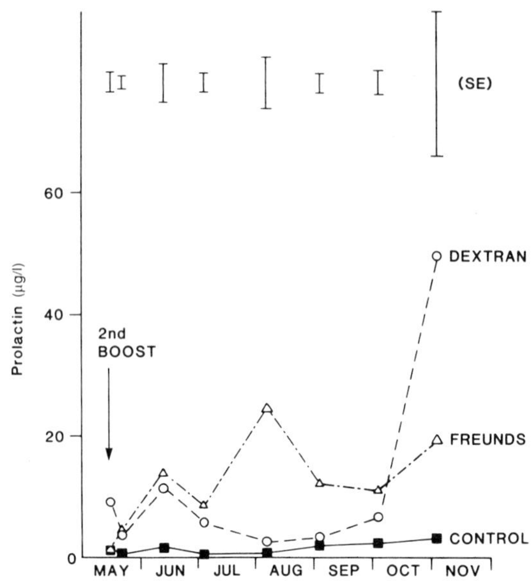

FIGURE 48.2. Plasma prolactin concentrations in young red deer in Experiment 3. ■, control deer; △, melatonin immunized using Freund's adjuvant; ○, melatonin immunized using DEAE dextran adjuvant; bar is 1 SEM.

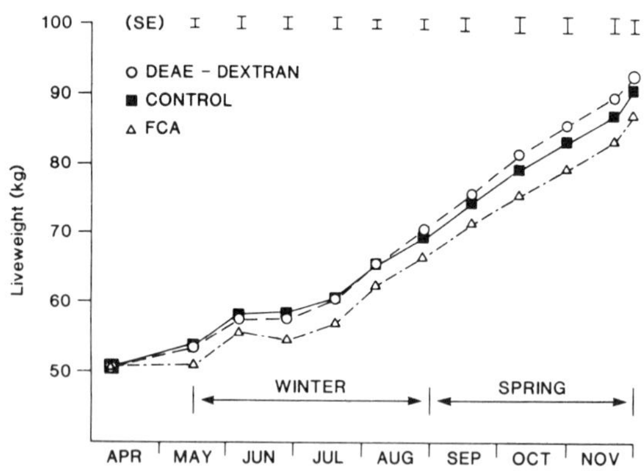

FIGURE 48.3. Effect of immunization against melatonin upon the body weight of red deer stags in Experiment 3. The data are corrected by analysis of covariance to a common postweaning weight of 50.7 kg in early April. ■, control deer; △, melatonin immunized using Freund's adjuvant; ○, melatonin immunized using DEAE dextran adjuvant; bar is 1 SEM.

Discussion

Body growth of young red deer stags in the present study followed the normal seasonal trends (Kay 1985; Barry et al. 1991), with growth rates during winter being much lower than during spring. Despite the seasonal effect, the present study has shown that winter growth rates of 140–150 g/day can consistently be obtained from a diet of perennial ryegrass/white clover pasture. Depending on weaner weight at the end of autumn (60 kg Experiment 2; 53 kg Experiment 3), this resulted in 75% and 41% of the animals attaining the target slaughter weight by the end of spring (30 November) at 12 months of age. These results are dramatically higher than the NZ average, with only 15% of the national venison kill occurring during August to November. Clearly, there is room for the application of these results in the NZ commercial deer industry.

Inclusion of Moata annual ryegrass in the pastures generally produced small but consistent increases in animal growth. Averaged over all experiments, inclusion of Moata increased the proportion of animals attaining their target slaughter weight by 19%. The effects of Moata generally became more apparent as the deer grazed pastures with a high proportion of this species, such as in Experiments 1 and 3.

To attain these high levels of deer performance, it was necessary to graze them on pastures of at least 10 cm initial height and 2100 kg DM/ha mass, and leaving residues in the range 1400–1600 kg DM/ha. Grazing under more restricted conditions (5 cm; Experiment 2) markedly reduced performance.

The present results have demonstrated systems for producing satisfactory LWG of young stags during winter, and further considerations are now the development of pastures for increasing weaning weight and autumn growth of weaners. In species selection studies, Hunt and Hay (1989) showed that red deer had a strong preference for red clover (*Trifolium pratense*). Initial studies at Massey University have showed that compared with perennial ryegrass/white clover pasture, grazing pure red clover during lactation increased weaning weight from 42 to 50 kg (Niezen et al. 1991), while postweaning it increased LWG during autumn from 197 to 263 g/day (Semiadi, unpublished data). Thus, there is considerable potential to increase deer production through the use of alternative pasture species.

Concerning the effects of active immunization against melatonin, the present studies have shown that it is necessary to commence the vaccination sequence at birth in deer in order to achieve a high antibody titer by antumn/winter. One of the effects of successful immunization appears to be higher plasma prolactin concentrations during winter, and an earlier occurrence of the spring rise in prolactin. As injections of prolactin to young reinder (*Rangifer taranadus*) (Ryg and Jacobsen 1982) and red deer (Suttie, personal communication) increased both voluntary feed intake and LWG during winter, it might be expected that the antimelatonin immunizations used here should increase winter LWG. Such effects produced using Freund's adjuvant may have been obscured by adverse effects of the adjuvant itself upon growth during the first 6 months of life (i.e., Experiment 3). Although it was shown that Freund's adjuvant itself (i.e., vehicle treatment) did not influence growth in Experiment 1, the immunizations were not commenced until the deer calves were 6 months old.

Duckworth and Barrell (1989) reported a melatonin immunization procedure, commenced at birth, which increased the weight of young stags between 9–11 and 16–20 months of age (i.e., during spring) by 7% to 10%. Immunization against melatonin in their studies reduced the time to attain 95 kg live weight by approximately 4 weeks. There is some indication (Fig. 48.3) that the deer immunized using dextran adjuvant in the present study may have been showing a reduced form of this response. Further work is needed using a longer acting form of the dextran adjuvant.

These studies have shown that efficient venison production systems can be developed using a sole diet of grazed forages. Further improvements in this efficiency may be possible by developing specific pastures for deer production, and by developing improved immunization procedures to minimize winter depressions in growth and to suppress the rut in yearling stags.

References

Abraham GE (1974) Radioimmunoassay of steroids in biological materials. Acta Endocrinol 183:7–42

Armstrong LS (1981) "Grasslands Moata" tetraploid Italian ryegrass (*Lolium multiflorum* Lam.). N Z J Expt Agr 9:337–341

Barry TN, Suttie JM, Milne JA, Kay RNB (1991) Control of food intake in domesticated deer. In Tsuda T, Sasaki Y, Kawashima R (eds) Physiological Aspects of Digestion and Metabolism in Ruminants. Proc VII Int Symp Ruminant Physiol, Academic Press, San Diego, pp 385–401

Domingue BMF, Wilson PR, Dellow DW, Wilson GF, Barry TN (1991) Effects of subcutaneous melatonin implants during long daylength on voluntary feed intake, rumen capacity and heart rate of deer fed a medium quality forage. Br J Nutr (in press)

Duckworth JA, Barrell GK (1989) Effect of melatonin immunization on the liveweight gain of red deer. Proc N Z Soc Anim Prod 49:29–34

Hunt WF, Hay RJM (1989) Alternative pasture species for deer production in the Waikato. In Asher G, Squire J (eds) Proc Ruakura Deer Industry Conf, Hamilton, N Z, MAF-Technology, pp 31–33

Kay RNB (1985) Body size, patterns of growth, and efficiency of production in red deer. In Fennessy PF, Drew KR (eds) Biology of Deer Production. R Soc N Z Wellington Bull 22:411–421

Korte CJ, Chu ACP, Field TRO (1987) Pasture production. In Nicol AM (ed) Livestock Feeding on Pasture. Occasional Publ 10, N Z Soc Anim Prod, pp 7–20

Niezen JH, Wilson PR, Holmes CW, Hodgson J, Barry TN (1991) Lactation performance of red deer hinds grazing red clover and perennial ryegrass/white clover pastures. Proc N Z Soc Anim Prod (in press)

Roughan PG, Holland R (1977) Predicting in vitro digestibilities of herbages by exhaustive enzymic hydrolysis of cell walls. J Sci Food Agr 28:1057–1064

Ryg M, Jacobsen E (1982) Effects of thyroid hormones

and prolactin on food intake and weight changes in young male reindeer (*Rangifer tarandus tarandus*). Can J Zool 60:1562–1567

Ulyatt MJ (1981) The feeding value of herbage: Can it be improved? N Z J Agr Sci 15:200–205

van Landeghem AAJ, van der Weil DFM (1978) Radioimmunoassay for porcine prolactin: plasma levels during lactation, suckling and weaning, and after TRH administration. Acta Endocrinol 88:653–667

Waghorn GC, Barry TN (1987) Pasture as a nutrient source. In Nicol AM (ed) Livestock Feeding on Pasture. Occasional Publ 10. N Z Soc Anim Prod, pp 21–37

49
Measuring the Responses of Fallow Deer to Disturbance

Alick R. Jones and Sarah E. Price

Abstract

A multitechnique approach was developed to measure effects of disturbance in farmed and park deer. The use of lactate dehydrogenase (LDH) and its skeletal muscle derived isozyme lactate dehydrogenase (LDH)-5 as indicators of disturbance and stress was evaluated. Fallow deer (*Dama dama*) at catch-ups showed levels of plasma LDH-5 proportional to the length of time from the beginning of the drive. Increases in LDH-5 compared with increases in total LDH indicated the former was of skeletal muscle origin. Plasma cortisol levels did not correlate with LDH-5 probably because of very different time-courses for release. Plasma LDH-5 is a useful stress marker for acute and possibly chronic stress lasting for more than 1 h.

Key words: Cortisol, disturbance, fallow deer, LDH, stress, welfare

Introduction

Stress and disturbance affect ungulates in a variety of ways. Although the topic is complex with debate even about the basic terminology (Griffin et al. 1988), the responses of the animals can be categorized. Relatively short-term responses usually involving the neuromedullary axis (Cannon 1929) are typified by changes in heart rate (Baldock et al. 1988). Responses involving or resulting from the release of glucocorticoids following pituitary secretion of adenocorticotropin (ACTH) are often measured by the determination of circulating cortisol (Smith and Dobson 1990). Longer-term acute stress may lead to muscular damage as well as a range of other rather variable responses (body temperature rise, panting, shivering, postural changes, etc.). In some animals, including deer, fatal myopathy may result (McAllum 1985). Such muscle changes are of interest to the meat industry as they frequently lead to reduced carcass quality and are often measured from values of circulating muscle-specific enzymes (Warris et al. 1984; Jones and Price 1990). In long-term chronic stress the outcome may be reduced productivity and lowered immunity (Griffin et al. 1988). All of these physiological and biochemical events are usually accompanied by behavioral changes which may be more or less specific (Humphries et al. 1989).

Previous studies have related behavioral observations of deer to disturbance (Sibly and Humphries 1989, Humphries et al. 1989) and also, in sheep, with heart rate measurement including using radiotelemetry to detect pulse (Baldock et al. 1987, 1988).

We have been interested in the use of the isozyme of lactate dehydrogenase (LDH) found in largest quantities in skeletal muscle, namely LDH-5. Cardiac muscle, which contains a preponderance of LDH-1, leaks LDH into the blood stream when damaged and high levels of circulating LDH-1 have long been used as a diagnostic aid in suspected cases of cardiac infarction (Latner and Skillen 1968). Additionally, when animals are subjected to stressful conditions, leakage from damaged skeletal muscles leads to high proportions of plasma LDH-5. Preliminary work (Jones and Price 1990) confirmed our prediction in deer supporting early work in pigs (Sybesma and Hessel-deHeer 1967) which suggested that LDH-5 might have value as both a stress indicator and a carcass predictor. The objective of the current study was to expand these earlier findings and compare results for LDH-5 with those for cortisol.

Methods

The blood samples were taken from fallow deer maintained in parks in southern England. Owners have requested anonimity for their parks which are therefore designated by letter. (Readers requiring more information should communicate with AR Jones.) The animals are usually handled once per year during round-ups. Deer were driven by a line of people into a relatively small area where they were enmeshed in nets hung in lines. The animals were then transferred into a trawl net with their legs restrained but able to move their heads and necks. They were blindfolded immediately after capture. Animals were weighed, measured, and blood samples were taken (approximately 10 ml by jugular puncture with a Vacutainer, Becton Dickinson). The catch-up procedure was no doubt stressful, usually involving 1 h or more of driving and handling.

Blood samples were stored on ice and spun as soon as possible. The plasma was stored at −10° C or lower. LDH activity was assayed using the spectrophotometric method of Bergmeyer (1974). LDH isozymes were determined by electrophoretic separation at pH 8.5 for 12 min at 300 V on "Helena" cellulose acetate plates followed by densitometry in a "Helena Quick-Scan." Results were expressed for each isozyme as a percentage. Given total LDH activity and percentage distribution for the isozymes, the activity of each isozyme could be calculated. Cortisol was measured by radioimmuneassay (Cunnah et al. 1987) after extraction in dichloromethane. Cross-reactivity was 19% for prednisolone, 0.45% for 1,1,-deoxycortisol, and less than 0.05% for all other steroids and steroid conjugates. The interassay precision at 100, 300, and 1000 nM/l was 13%, 7.4%, and 7% respectively. Statistical analysis of the various results was carried in "Minitab."

Results

The distribution of LDH isozymes in fallow deer (Table 49.1) was similar to those of many other mammals and, most importantly, skeletal muscle was rich in LDH-5 which is only present at low levels in heart muscle. In liver LDH-5 was also only present in relatively low levels.

Table 49.2 shows that in general there is a highly significant correlation between the time from the beginning of the catch-up and percentage LDH-5, LDH activity and LDH-5 activity. However, the relationship is most consistent with percentage LDH-5, which is technically the simplest to measure.

Figure 49.1 shows the values for percentage LDH-5 plotted against time for four parks. The data for park A indicates a tendency to plateau

TABLE 49.1. Distribution of LDH isozymes by percentage for various fallow deer tissues

	Tissues							
Isozyme	Kidney		Heart		Liver		Skeletal	
	(%)	(SEM)	(%)	(SEM)	(%)	(SEM)	(%)	(SEM)
1	39.2	0.88	46.3	4.00	34.3	3.29	6.3	1.04
2	29.2	0.95	31.6	1.82	29.9	1.04	8.3	0.90
3	14.2	0.59	11.1	1.72	18.4	2.18	10.5	0.69
4	4.8	0.54	4.7	1.26	8.0	1.02	18.5	1.32
5	12.5	2.72	6.4	2.14	9.3	1.47	56.3	0.84

TABLE 49.2. Values of the slope calculated by regression analysis of the relationships between time after start of catch-up and percentage LDH-5, LDH activity, and LDH-5 activity, in fallow deer. All values are $P < 0.01$

	% LDH-5		LDH activity		LDH-5 activity	
Park	Slope	SD	Slope	SD	Slope	SD
A	0.097	0.011	5.22	0.57	2.69	0.29
B	0.156	0.036	0.05	1.09	1.01	0.33
C	0.131	0.026	3.14	0.68	2.04	0.38
D	0.144	0.024	4.11	0.79	2.36	0.39

for the later observations, and it is this which explains the relatively low value for the slope (Table 49.2).

The hypothesis tested was that, due to stress, LDH was added to the blood stream from skeletal muscle. If this was the case then only a proportion of the added LDH would be LDH-5; in effect about 56% LDH-5 as determined from tissue analysis. Figure 49.2 shows the data for total LDH plotted against LDH-5 activity. There is a good linear relationship. Clearly if all the added LDH was in the form of LDH-5, the slope of the line would be 1.0; likewise if 50% of it was LDH-5, the slope would be 2.0. In fact the slope is 1.84, which indicates that 54.4% of the added LDH was LDH-5. The agreement with the tissue analysis is extremely good. Table 49.3 gives values for the slope and calculated percentage of LDH-5 for this data set and those from four other parks.

Figure 49.3 shows a similar graph to those shown in Fig. 49.1 (park A) except that cortisol values for individual animals are shown. In Fig. 49.4 these same cortisol values are plotted against the percentage LDH-5 values for each animal.

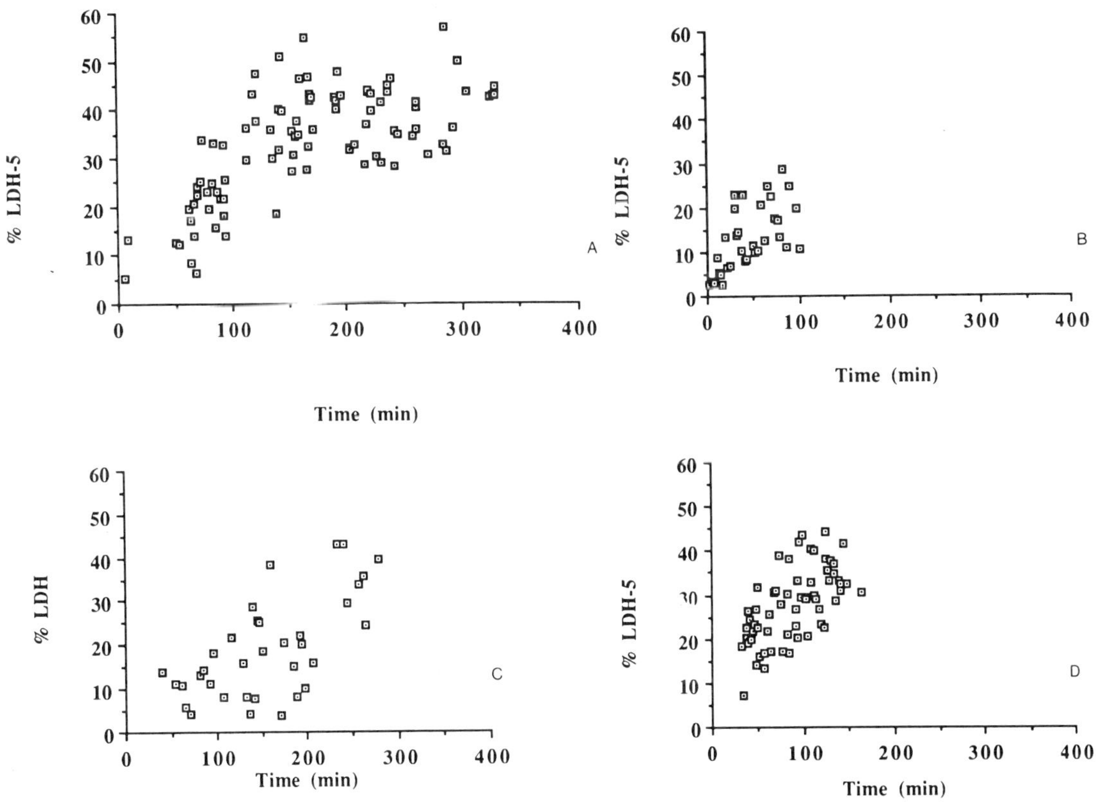

FIGURE 49.1. Scatter plots of LDH-5 (as a percentage of total LDH) plotted against the time after capture for fallow deer from four different parks (indicated by letter). Points represent mean of two determinations for an individual animal.

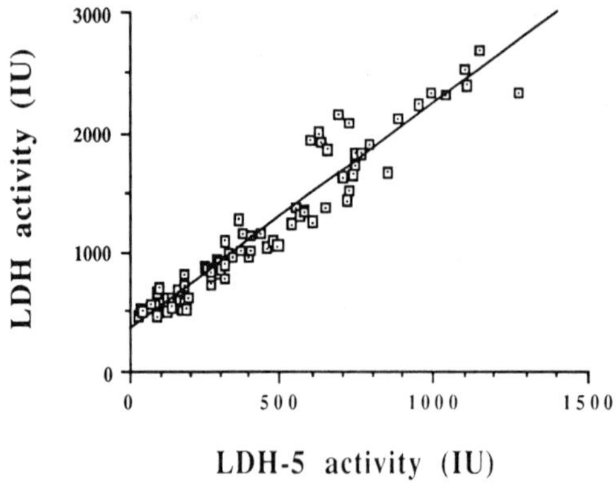

FIGURE 49.2. Values for fallow deer from park A for LDH activity plotted against LDH-5 activity.

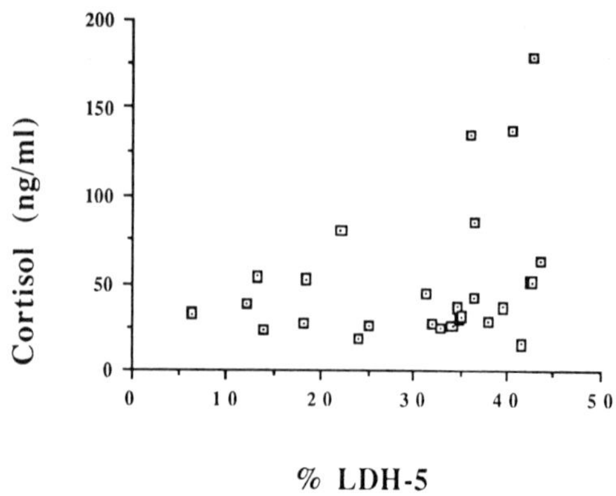

FIGURE 49.4. Values for fallow deer from park A for cortisol plotted against values for percentage LDH-5 from the same individuals.

TABLE 49.3. Calculated slopes for plots of LDH with LDH-5 activity for fallow deer from five parks with the calculated percentage of LDH accountable as LDH-5

Park	Slope	SD	% of added LDH accountable as LDH-5
A	1.84	0.07	54.4
B	1.84	0.41	54.4
C	1.47	0.14	68.0
D	1.81	0.09	55.3
E	1.97	0.04	50.8

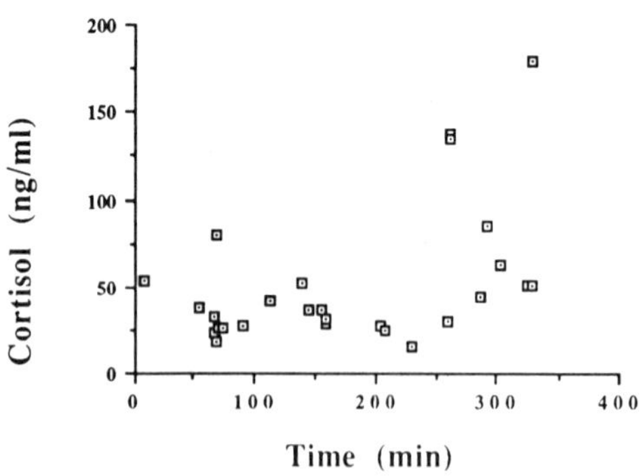

FIGURE 49.3. Values for fallow deer from park A for cortisol plotted against time after capture.

Discussion

The use of LDH-5 as an indicator of stress is not new. Sybesma and Hesse-deHeer (1967) reported significantly higher levels of LDH-5 in the blood of stress-sensitive Pietrain pigs as compared with the stress-resistant Large Whites. These levels were further raised when the animals were subjected to a period of transport to the slaughterhouse. Reddy et al. (1971) also examined the isozymes of LDH in pigs and came to similar but qualitative conclusions. Kallweit et al. (1981) recorded differences in LDH-5 in steers before and after transport. However, more usually total LDH or creatine phospholinase (CPK) assays have been preferred (Sybesma and Eikelenboom 1978) despite the lack of specificity of the former and the possible complication of cardiac CPK in the latter. In the light of the evidence presented here, however, there seems little doubt that the measurement of plasma LDH-5 is a useful blood-borne indicator of stress.

These investigations have produced convincing evidence for the skeletal muscle origin for the increase in plasma LDH and hence, also in LDH-5. The increase in circulating LDH-5 is consistent with it having come from LDH containing between 50% and 68% of that isozyme (see Fig. 49.2, Table 49.3), precisely the amount found in skeletal muscle. Unlike in pigs (AR Jones, unpublished data) the distribution of the isozymes in deer liver is different, and there is no likelihood of the increases in LDH shown in these experiments having a hepatic origin.

Measurements of corticosteroids, especially cortisol, have been frequently used as a means of attempting to measure an animal's response to stressors or disturbance (Fulkerson and Jamieson 1982; Hennesy et al. 1982; Kenny and Tarrant 1987). We have measured plasma cortisol in stressed fallow deer and found no correlation between it and plasma LDH-5. We are convinced as to the value of LDH-5 determinations to assess stress even though it is not correlated with plasma cortisol. The release of LDH-5 from muscles is likely to have a very different time course to the release of cortisol. It is noticeable (Fig. 49.3) that, although all the cortisol levels are high (in excess of 15 ng/ml), there is no relation between cortisol and duration of stress. However, in view of the probably rapid release of corticosteroids, detectable after 5 min in rats (Sakellaris and Vernikos-Danellis 1975) and sheep (Fell and Shutt 1989), these results are not surprising. We would suggest that cortisol levels in plasma may be quantitatively related to the duration of stress for a short time, perhaps 30 min; thereafter, it is likely that elevated levels will only give qualitative indications of stress. LDH-5 probably does not appear in the plasma until stress has exceeded 60 to 90 min and may plateau after 200 min (Fig. 49.1a). However, it should be emphasized that cortisol levels vary in individual animals on a seasonal and diurnal basis and can change in relation to nonstressful events such as feeding (Delahunt and Mellsop 1987; Follenius and Brandenburger 1980).

The use of percentage LDH-5 in plasma as a stress measure rather than absolute values, although easier and cheaper, carried with it the inevitable consequence that, especially at high levels, changes in percentage will progressively under-represent the LDH-5 increase. This problem can be overcome in part by transforming the percentage data, or completely by taking the extra step of determining activities. Of more concern are the possible releases of LDH-5 from muscles due to exercise and bruising. The same reservations hold for the use of CPK (Rose 1986). It is worth noting that the extent of exercise and bruising in the various groups represented in Fig. 49.1 was very unequal, and yet the slopes of percentage LDH-5 increase are very similar. We also have evidence that mice stressed by being placed in solitary confinement show raised plasma LDH-5 but a significantly lower level of activity than the controls housed in groups of four (AR Jones, unpublished data). In this case it is also unlikely that bruising played any part in elevating LDH-5. Nonetheless, the interactions of stress, bruising, and exercise on circulating enzymes of muscles origin need further investigation.

Acknowledgements

The Mammal Conservation Trust kindly permitted us to sample at their catch-ups in which we were guided and assisted by J. Pemberton. R. Putman and J. Langbein supplied us with data on individual deer for which we are most grateful. Financial support for some parts of the work came from the RSPCA and the Animal Health Trust and from support to SEP through a University of Reading Studentship.

References

Baldock NM, Sibly RM, Penning PD (1988) Behaviour and seasonal variation in heart rate in domestic sheep, *Ovis aries*. Anim Behav 36:35–43

Baldock NM, Penning PD, Sibly RM (1987) A system for recording sheep ECG in the field using a miniature 24-hour tape recorder. Comput Electronics Agric 2:57–66

Bergmeyer HU (1974) Methods of Enzymic Analysis (2nd English ed), vol 2. Verlag-Chemie, Academic Press, New York

Cannon WB (1929) Bodily Changes in Pain, Hunger, Fear and Rage: An Account of the Function of Emotional Excitement. Appleton, New York

Cunnah D, Jassop DS, Besser GM, Rees LH (1987) Measurement of circulating corticotrophin-releasing factor in man. J Endocrinol 113:123–131

Delahunt JW, Mellsop G (1987) Hormone changes in stress. Stress Med 3:123–134

Fell LR, Shut DA (1989) Behavioural and hormonal responses to acute surgical stress in sheep. Appl Anim Behav 22:283–294

Follenius M, Brandenburger G (1980) Evidence of a delayed feedback effect on the midday plasma cortisol peak in man. Horm Metabol Res 12:638–639

Fulkerson WJ, Jamieson PA (1982) Pattern of cortisol release in sheep following administration of synthetic ACTH or imposition of various stressor agents. Aust J Biol Sci 35:215–222

Griffin JFT, Bisset LR, Fisher MW (1988) Influence of management stress on immunity in farmed deer. Proc Deer Seminar for Vets 5:145–163

Hennessy DP, Coghlan JP, Hardy KJ, Wintour EM (1982) Development of the pituitary-adrenal axis in chronically cannulated ovine fetuses. J Dev Physiol 4:329–352

Humphries RE, Smith RH, Sibly RM (1989) Effects of human disturbance on the welfare of park fallow deer. Deer 7:458–463

Jones AR, Price SE (1990) Can stress in deer be measured? Deer 8:25–27

Kallweit E, Unshelm J, Wemken K, Smidt D (1981) Effects of transportation and preslaughter holding on some blood parameters and on meat quality in fattening bulls. In Hood DE, Tarrent PV (eds) Problem of Dark-Cutting in Beef. Commission of the European Communities, Brussels, pp 404–409

Kenny FJ, Tarrant PV (1987) The reaction of young bulls to short-haul road transport. Appl Anim Behav Sci 17:209–227

Latner AL, Skillen AW (1968) Isozymes in Biology and Medicine. Academic Press, New York

McAllum HJF (1985) Stress and postcapture myopathy in red deer. R Soc N Z Wellington Bull 22:65–72

Reddy MVV, Kastenschmidt LL, Cassens RG, Briskey EJ (1971) Studies on stress-susceptibility: the relationship between serum enzyme changes and the degree of stress-susceptibility. Life Sci 10: 1381–1391

Rose RJ (1986) Endurance exercise in the horse—a review. Part II. Br Vet J 142:542–552

Sakellaris PC, Vernikos-Danellis J (1975) Increased rate of response of the pituitary adrenal system in rats adapted to chronic stress. Endocrinology 97:597–602

Sibly RM, Humphries RE (1989) Effects of stress and disturbance in deer. In Brown MB, Goldspink CR (eds) Management, Conservation and Interpretation of Park Deer. British Deer Society, Manchester, pp 27–38

Smith RF, Dobson H (1990) Effect of preslaughter experience on behaviour, plasma cortisol and muscle pH in farmed red deer. Vet Rec 126:155–158

Sybesma W, Eikelenboom G (1978) Methods of predicting pale, soft, exudative pork and their application in breeding programmes—a review. Meat Sci 2:79–90

Sybesma W, Hessel-deHeer JCM (1967) LDH_5 and meat quality in pigs. Proc 13th Europ Meeting Meet Res Workers, Rotterdam, Section C10:1–9

Warris PD, Kestin SC, Brown SN, Wilkins LJ (1984) The time required for recovery from mixed stress in young bulls and the prevention of dark cutting beef. Meat Sci 10:53–68

50

Seasonal Pattern of Births, Female Reproductive Success and Causes of Neonate Mortality of Farmed Red Deer and Fallow Deer in Northern New Zealand

G.W. Asher and M. Langridge*

Breeding female populations on seven red deer (*Cervus elaphus*) and five fallow deer (*Dama dama*) farms in northern New Zealand (latitude 37°–38° S) were monitored daily during the calving/fawning seasons (November–January) of 1980/81–1983/84. All neonates were tagged and weighed within 24 h of birth, and most were identified to their dams. Causes of neonate mortality were assessed by post-mortem examination of all retrieved specimens. Females of breeding age (> 15 months) were weighed immediately prior to the rut preceding each birth season, and linear regression analyses were performed to determine the influences of dam age and prerut (nonpregnant) live weight on neonate birth date and birth weight. Birth records were obtained for 932 red calves born to 1005 hinds (91.8% calving rate) and 853 fallows fawns born to 960 does (88.9% fawning rate); 115 red calves (12.5% of calves born) and 161 fallow fawns (18.9%) died before weaning in March (at 2 to 4 months of age). The resultant weaning rates (offspring weaned/females present at parturition) were 80.4% and 71.2%, respectively. The distribution of birth dates were skewed for both species, with the right-hand tail representing late conceptions. Between-farm and year corrections on birth data (i.e., alignment to median birth date) indicate that fallow deer had a more synchronous birth pattern, with 83.7% of fawns born over a 21-day period (i.e., 10 days either side of median), compared with 67.6% for red deer. However, fallow deer births generally started 2 to 3 weeks later than red deer. Mortality diagnoses of 102 red calves (88.7% of known deaths) and 161 fallow fawns (100%) indicate that dystocia, starvation, and misadventure accounted for 40.2%, 34.3%, and 13.7% respectively of red calf deaths; whereas nonviability (failure to walk and suckle), infectious agents, starvation, dystocia, and misadventure accounted for 24.9%, 22.3%, 19.3%, 14.3%, and 11.2% respectively of fallow fawn deaths. Neonate mortality increased with decreasing birth weight. Sex-corrected birth weights of neonates were positively related to birth date and dam prerut live weight for both species. Dam age (primiparous vs multiparous) had a significant effect on birth date for red deer and birth weight for fallow deer.

*Ruakura Agricultural Centre, MAFTech North, Private Bag, Hamilton, New Zealand

51

Matching of Pasture Growth Rate, Animal Requirements, and Animal Performance to Establish Stocking Policies for Farming Red Deer Under New Zealand Grassland Farming Conditions

Peter G. Floyd*

This paper illustrates a range of pasture growth rates from different locations in New Zealand. The animal feed requirements for breeding hinds, venison, and velvet stags, and the growth rates of the same are discussed. The matching of these basic resources to achieve a profitable farming venture under various stocking policies and farm management practices and conditions are shown. The results illustrate the significance of making the best business decisions based on the biology of the deer and the natural growing features of pasture. The best results are achieved on properties where the pasture supply is available during the peak demand periods such as during lactation. Where pasture supply is inadequate, supplementary feeding or changing the balance of numbers of hinds and stags can significantly benefit the overall results.

*1 B Brocas Ave., Hamilton, New Zealand

52
Profitability of Farming Red Deer in New Zealand

Peter G. Floyd*

This study is based on experiences over the past 10 years, analyzing the performance of over 300 deer farms, under a wide range of climatic and physical conditions. The financial returns or profitability of farming deer in New Zealand vary widely depending on a large number of factors. This paper outlines these factors and summarizes the financial returns from a range of farming practices and policies and stocking rates. A study of the comparative returns per acre and a comparison with other forms of livestock farming at similar stocking rates is summarized. Returns on capital invested in land, buildings, and livestock, as well as an analysis of the capital and land area required to return an annual net income of $NZ 50 000, from various farming systems are presented. The results show that the farming of red deer (*Cervus elaphus*) to be the most profitable form of livestock farming under a range of farming policies and environments. It is shown that the most consistent financial returns have been achieved from a policy of farming stags for venison and velvet with a slaughter date of 21 to 25 months. These returns have always been superior under high stocking rates and when the breeding hing prices have been below $NZ1 500 each. This policy is best combined with the farming of a breeding herd; however, not all situations and farming conditions are suitable.

*1 B Brocas Ave., Hamilton, New Zealand

53
Responses of Farmed Red Deer to Being Handled

Sarah E. Price and Alick R. Jones*

In the time that deer farming has been carried out in Britain, there has been limited selective breeding for temperament on farms with established herds. On many farms the deer have been brought straight from the hill in Scotland or from deer parks. It is apparent that despite a lack of prior experience they are adaptive animals and can be handled with ease using the correct facilities and expertise. The behavioral and physiologically responses were measured in a group of red deer (*Cervus elaphus*) during handling on a lowland arable farm. These animals had prior experience of handling and were used to the proximity of humans. Behavioral observation of the group in field was done for 2 days previous to and 2 days following handling. There were no significant changes, but there was a trend for some measures of restlessness, i.e., increased locomotion including fence-walking on the first day following handling but not the second day. Single blood samples were taken from the deer in the same sequence in which they were handled. These were assayed for cortisol, lactic acid, total LDH (lactate dehydrogenase), and LDH-5. Cortisol levels were significantly raised from control levels, but no difference was seen between those at the beginning, middle, or end of the handling session. LDH-5 was similar to resting levels in the beginning and middle batch but raised in the final batch.

*Department of Pure and Applied Zoology, Reading University, Whiteknights, P.O. Box 228, Reading, RG6 2AJ, United Kindom

54
A Study on the Biological Characteristics of Sika Deer

Rongqian Qin and Shangren Li*

A series of studies on the biological characteristics of penned sika deer (*Cervus nippon*) were conducted to provide ways to raise productive efficiency and select improved varieties. The results are as follows: The average weight of the adult male was 122.56 kg and that of the female was 71.19 kg. The average height of the male was 106.7 cm and that of the female was 90.9 cm. The average weight of the three-branched antler was 2.59 kg. The total length was 45.04 cm. The period of antler loss is 4 to 6 years. The average dressing percentage of adult males was 54.32%. The difference of the weight of the testis between the periods of intensive sexuality and antler growth was significant. The duration of pregnancy was 237.17 days. The weight of the newborn deer was 6.13 kg. The day of the estrous cycle was 12.51 and the duration of the estrus was 12–36 h. The growth rates of body weight were 104.97% (first month), 70.76% (second month), and 45.83% (third month), respectively. The VFAs of the rumen contents of the adult females were C_2, 64.56% ± 3.85%; C_3, 18.05% ± 2.01%; C_4, 13.4% ± 2.51%; iso-C_5, 1.39% ± 0.49%; and C_5, 2.46% ± 0.26%.

*Department of Animal Science, Jilin Agricultural University, Changchun, Jilin, People's Republic of China

55 Hungarian Red Deer and Its Potential Performance as a Farm Deer

Peter Horn and László Sugár*

The body size of red deer (*Cervus elaphus*) decreased over centuries in continental Europe. However, one of the largest deer in body as well as antler size and weight survived in southwestern Hungary. The average carcass weight of mature red deer stags is now 200 to 220 kg, and that of hinds is 100 to 120 kg. The average antler weight of mature stags is about 8 to 9 kg. Most of the world record deer trophies were harvested here. The Hungarian red deer was chosen by New Zealand specialists to improve venison and velvet production, and about 300 Hungarian red deer are now involved in the genetic improvement of New Zealand deer. Farming red deer is a new branch of animal husbandary in Hungary. The present wild population of about 90 000–100 000 must be reduced by 30% to 50% in a few years because of damage to plant agriculture and forests. The experience gained so far regarding growth rates and reproduction with farmed red deer utilizing rolling country grassland areas in Hungary is encouraging. Deer farming seems to be a competitive alternative to sheep and cattle farming in certain regions of Hungary.

*Pannonian Agricultural University, Faculty of Animal Science, H-7401 Kaposvar, P.O. Box 16, Hungry

Part 4
Deer Products

56
Venison and Other Deer Products

Ken R. Drew

Abstract

Game production is becoming an important industry in several countries. The main products from deer are meat, velvet antler, and leather. Venison from farmed deer has been shown to differ very little from wild-shot venison when the carcasses are of equal weight. The meat from young deer is extremely lean, having a fat content in the range of 5% to 10% of carcass weight and a very high lean/fat ratio. Electrical stimulation of carcasses will reduce the pH to 6.0 within 1 to 2 h compared with 6 h in unstimulated carcasses. Chilled venison, when vacuum packaged and held at −1° C, had a shelf life in excess of 12 weeks. Controlled atmosphere packaging did not further improve shelf life or product quality. Velvet antler is a valuable deer product used in oriental medicine for treating a wide range of human conditions. The product is often given to children as a protection against problems such as influena. Deer skin leather is a particularly durable product which is now being used in a wide range of fashion garments.

Key words: Carcass, deer, deer skins, fat, meat quality, meat packaging, velvet antler, venison

Introduction

The role of animal products in the human diet has been the subject of a voluminous literature in recent years. The National Research Council (NRC 1988) stated that "Animal products contribute more than one-third of the calories and between a third and all of the other major nutrient in the food supply." Red meat comprises almost half of the energy and protein contributed by animal products.

World trade in game meat products is now greater than 30 000 tons per year (Luxmoore 1989), yet little of that will be from farmed or closely managed animals. New Zealand is one of the few countries to have recently established the full domestication of deer for farming, and the production of ranched venison is expected to increase from about 4000 tons in 1989 to 20 000 tons by the mid-1990s (Fennessy and Taylor 1989). Commercially, the deer is a multipurpose animal providing venison to an international market, velvet antler for the oriental medicine trade and other less important products such as hides, pizzles, ligaments, and teeth.

The farming of deer in low input systems for sustainable agricultural production should have a favorable future in a society which is becoming increasingly sensitive to environmental pollution, animal manipulation, and feed additives.

Venison

Venison or deer meat can be derived from feral or farm-raised deer, and from several different species or strains of deer. The quality of the product may be greatly influenced by commercial treatment such as pre- and postslaughter handling, packaging, storage, transport, and cooking preparation. If science had been commissioned to produce a "designer" red meat that had all the best attributes of our traditional farm animals and none of the perceived bad features, then the successful result would have been something remarkably like venison.

Carcass Characteristics of Feral and Farmed Deer

Differences in carcass composition between some feral ungulates and the same species in domestication were considered by Crawford (1968), who concluded that the marked differences in fatty acid composition were dietary in origin. Carcasses from feral, grass-fed, and feedlot-raised red deer (*Cervus elaphus*) in New Zealand were analyzed for differences in composition. Figure 56.1 indicates that variation in carcass fatness was mainly a function of carcass weight and not environment. Because there were no significant environmental differences the following common regression equation was calculated (71 animals ranging in weight from 23.7 kg to 143.6 kg and from 1 to 9 years of age):

$Y = 0.191X - 2.49$ where $Y = \%$ carcass fat and $X =$ carcass weight (kg), the SE of the slope and intercept were 0.008 and 0.58 respectively, and $r^2 = 0.89$.

The feral deer certainly have less fat than farmed or feedlot-raised deer because the feral deer were much smaller at equal age. In relation to carcass weight both feral and farmed deer in their first winter (6 month of age) are fatter than expected. This may have been due to lack of rumen development and reliance on milk as a

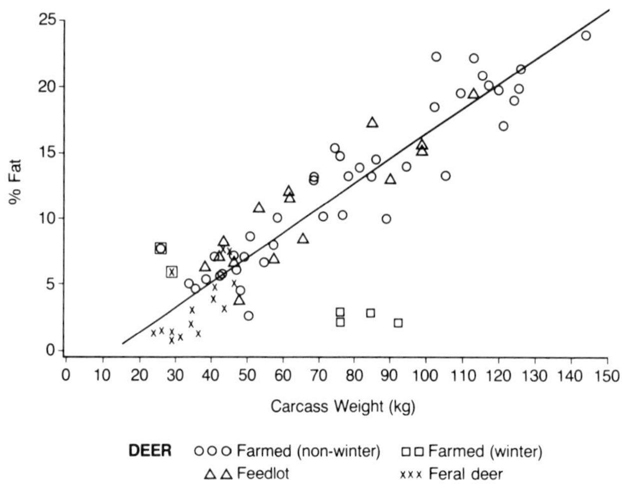

FIGURE 56.1. Carcass weight and fatness in red deer stags.

TABLE 56.1. Composition of leg meat from feral and farmed red deer ($n = 6$ per group)

	Feral	Farmed	
Age (months)	27	12	27
Carcass weight (kg)	43.1	40.8	75.7
Lean meat (g/100 g)	95.6	95.6	88.0
Fat (g/100 g)	3.3	3.3	10.9
Minerals (g/100 g)	1.1	1.1	1.1
Saturated and mono-unsaturated fatty acid (S) (g/100 g)	2.8	3.0	10.6
Polyunsaturated fatty acids (P) (g/100 g)	0.5	0.3	0.3
Tryglycerids/phospholipids	3.4	5.6	17.8
P/S ratio	0.178	0.100	0.028

major dietary component. Deer are very seasonal animals in their growth patterns and show marked mobilization of fat reserves during the breeding season (Drew 1985). The very low winter carcass fat content (Fig. 56.1) is typical of the very limited adipose tissue stored in deer during the winter season. The slope of the regression line indicates that 1 kg of carcass gain comprises 0.19 kg fat, a figure that is a little lower than that published by Drew (1985). Rate of fat deposition in deer is less than half of that shown in sheep by Fennessy and Greer (1982).

There are substantial differences between the meat composition of 27-month-old feral and farmed animals, but this is a function of differences in carcass weight (Table 56.1). Twelve-month-old farmed animals are very similar in weight to 27-month-old feral deer, and their carcass compositions are almost identical. There are some differences between the groups of deer in polyunsaturated (P) fat levels and these are more fully documented by Manley and Forss (1979). Feral animals tend to be higher in P fatty acids than farmed (Table 56.1). Most of the P is in the phospholipid fraction, and although environmental effects were present, they were not quantitatively important. Because of the substantial adipose tissue depot in the 27-month-old farmed deer, their P:S ratio is about one-sixth of that in feral deer.

Possible differnces in the taste of venison derived from feral, farmed, and feedlot deer were researched by Forss et al. (1979). These researchers found that differences between carcasses from the same feed environmental were substantial, and generally concluded that, "it is unlikely that meat from deer not older than 27 months and on different diets would be distinguished when eaten." Brittin et al. (1981) found that beef was juicier, much higher in fat and of similar tenderness, when compared with meat from feral male deer.

Farmed Deer Carcass Composition

Carcass characteristics of different speies and strains of farmed, entire, male deer are compared with Aberdeen Angus cattle in Table 56.2. All deer less than 26 months of age have a separable lean content which ranges from 72.7% to 76% of the carcass weight. A comparable figure for cattle is 62%, although varition will be wide between breeds and sexes. Seman and McKenzie-Parnell (1989) quoted figures of separable lean (% hot carcass weight) that ranged from 48.0% in pork to 59% for beef, veal, and chicken. Mutton and lamb were intermediate. Mature deer get very fat at the end of summer, and the lean content is reduced to 60%–70%. In contrast, the deer fat levels are low in animals less than 2 years of age.

TABLE 56.2. Farmed deer carcass composition

Species	Age	Weight (kg)	Carcass weight: Lean (%)	Carcass weight: Fat (%)	Carcass weight: Bone (%)	Lean/fat	Lean/bone
Fallow[a]	13–25 months	24–40	73.9	9.1	13.6	8.1	5.4
Red	26 months	62.6	72.7	7.0	20.3	10.4	3.6
	9 and 10 years	129.5	70.9	14.2	14.9	5.0	4.8
NZW[b]	26 months	83.0	72.7	4.2	23.5	17.3	3.1
	4 and 5 years	115.3	66.8	14.7	18.5	4.5	3.6
Hybrid (NZW/red)	26 months	78.0	73.8	5.4	22.0	13.7	3.3
	4 and 5 years	115.5	60.9	19.7	19.4	3.1	3.1
Hybrid (elk/red)	11 months	67.6	76.0	4.7	19.3	16.2	3.9
Angus cattle[c] (bull)	2 years	250.0	62.0	21.5	14.0	2.9	4.4

[a]Gregson and Purchas (1985)
[b]From an NZ population of mixed elk/red breeding
[c]Maiga (1974)
NZW, New Zealand wapiti

Carcass bone content in cattle is generally lower than in deer with the exception of fallow deer. Bone content in carcasses of traditionally farmed species can range from 17% in lamb to 28% in veal with beef, pork, and chicken in between (Seman and McKenzie-Parnell 1989). The lean/fat ratio, as a measure of edible soft tissue, shows that young deer are very high when compared with beef. Calculations from the data of Seman and McKenzie-Parnell (1989) yield lean/fat ratios of 1.6, 1.9, 2.6, 3.2, 3.9, and 5.1 for mutton, pork, beef, lamb, chicken, and veal, respectively. All these figures are much lower than those for young deer which range from 8.1 to 16.2. Two types of young deer stand out in Table 56.1 as extremely good for carcass production. The first is fallow deer (*Dama dama*) with a low bone content and very high lean/bone ratio. The other is the elk (*Cervus elaphus canadensis*)/red hybrid, which at only 11 months of age can produce a 68 kg carcass with 76% lean, 4.7% fat, and a bone content of 19%.

Factors Affecting Venison Quality

The two most important features of meat are its tenderness when it is eaten and its color when it is purchased. Venison quality has been the subject of recent significant work.

Electrical Stimulation and Meat Quality

Postslaughter electrical stimulation of carcasses is now widely used to prevent toughness in meat that is caused by cold or thaw shortening. Experiments from Chrystall and Devine (1983) and Drew et al. (1988) have shown that if non-stimulated deer carcasses are subjected to 0° to 2° C temperatures within 2 h of slaughter then tough meat will be produced. The critical pH in lamb muscle required to avoid cold shortening is about 6.0 (Chrystall et al. 1984). In venison experiments it took at least 6 h at 10° C to achieve this pH in nonstimulated carcasses compared with 1 to 2 h when stimulation was used. When carcasses were held at 10° C for 2 h after slaughter then 0° C for a further 22 h, electrically stimulated carcasses were 50% more tender than non-stimulated ones. It is possible to substitute carcass storage at an elevated temperature (e.g., 10° C) for electrical stimulation in order to achieve tender venison, but this delays carcass cutting where regulations require a deep bone temperature of not higher than 7° C before cutting is permitted.

Skinning and Evisceration

Carcass contamination during processing will shorten the shelf life of the meat, and deer hair is frequently found on skinned carcasses. The inverted deer dressing system developed by A.W.A. Milmech in New Zealand (Drew 1989), involves holding the carcass on a rail by the front legs and mechanically pulling the skin from the head and over the back legs. The procedure prevents hand or knife contact with the high-value saddle/hind leg part of the carcass. There is considerable varition in the microbiological loading on venison carcass at different plants (Table 56.3). The Invermay plant uses inverted dressing while the other two use traditional procedures. Although all average counts are low and all plants have worked efficiently, the range in microbial counts is much greater from the commercial plants than from Invermay. Operator skill has a major influence in the quality from traditional plants but less so when inverted dressing is used. The bacteriological loading on the farmed deer carcasses compared very favorably with beef where bacteriological counts of 500 to 1000/cm^2 are common (Nottingham and Wyborn 1975) and feral venison can have more than 10 000/cm^2 (Sumner et al. 1977).

Packaging

Venison from New Zealand is mainly vacuum packed when transported as a chilled product.

TABLE 56.3. Microbial counts (expressed as colony forming units/cm^2 with range in parentheses) from farmed deer carcasses processed at three New Zealand plants

Slaughter plant	Carcass site: Shoulder	Middloin	Leg
Invermay	1 (0–11)	1 (0–24)	1 (0–16)
Plant B	33 (0–53 000)	30 (0–2000)	ND
Plant C	10 (0–10 000)	21 (0–158 000)	7 (0–2000)

ND, not determined

Modern technologies such as controlled atmosphere packaging have been researched using venison (Seman et al. 1988, 1989). Table 56.4 shows the effect of the packaging system on venison quality when the product was stored for up to 18 weeks at 0° C. There was a trend towards increasing tenderness with length of storage in all treatment, but this was quantitatively small and all samples were quite tender. Color stability of displayed meat is an important matter, and vacuum-packed venison was found to be slightly better than other storage systems, although all packaging systems produced meat with a very short display life after 18 weeks of storage. Venison color has been shown by Stevenson et al (1989) to be very highly correlated with acceptability. Since venison is rather dark in color due to the high iron content, some consumers will see it as unattractive. Color deterioration with increasing storage time will exaggerate that perception. Relatively long periods of storage with venison do not seem to have much effect on odor (Table 56.4) and this may be due to the very low fat levels and, therefore, a reduced tendency towards rancidity.

TABLE 56.4. Packaging system and storage time as they influence venison quality (data from Semen et al. 1988)

	Storage time at 0° C (weeks)			
	1	6	12	18
Venison tenderness (kPa/cm)				
Vacuum	27.8	26.1	25.0	25.9
CO_2-UHB[a]	32.1	25.3	26.9	25.7
CO_2-foil[b]	24.4	26.2	26.4	25.9
Color stability[c]				
Vacuum	5	4	3	2
CO_2-UHB[a]	5	3	2	1
CO_2-foil[b]	3	3	2	2
Odor[d]				
Vacuum	3	3	2	2.5
CO_2-UHB[a]	3	3	2.9	2.2
CO_2-foil[b]	3	3	3	2

[a] CO_2 flush ultra-high-barrier film
[b] CO_2 flush polyester aluminium foil laminate
[c] Days of display to reach color acceptability score of 2 (purchase with reservations)
[d] Scored on 3-point scale: 3, no off odors; 2, off odor—not objectionable; 1, off odor—objectionable

Shelf Life of Chilled and Frozen Venison

Drew and Fennessy (1986) found that there was a considerable decrease in toughness of vacuum-packed venison leg and loin cuts during the first 6 weeks of storage at 1° C but no change thereafter (Fig. 56.2). Changes in tenderness during the first 6 weeks of storage do differ from what is presented in Table 56.4 with the latter showing only small changes from weeks 1 to 6 in tenderness. This difference is likely to be due to differences in storage temperature and the time from packaging to first sampling. In practical terms, well-prepared and packaged chilled venison does not deteriorate noticeably over 12 weeks of storage, but undesirable changes begin to occur after that time. The shelf life of frozen venison is long if it is well-prepared and packaged. Storage at −12° C and −18° C of whole muscle cuts and "retail-ready" portions of venison for up to 28 months has few measurable effects on product quality, except that there was an indication of noticeable rancidity in "retail-ready" portions packed in permeable film after 8 months of storage (J.M. Stevenson et al., unpublished data).

Nutritive Value of Venison

Nutrition-related health problems arise in affluent societies from over consumption of fat, saturated fatty acids, and cholesterol (NRC 1988). Red meat is a major source of these items,

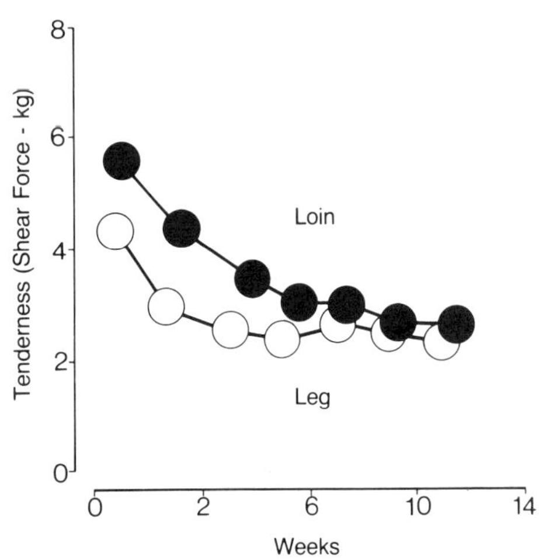

FIGURE 56.2. Changes in venison tenderness with holding time at 1° C.

TABLE 56.5. Nutrient composition per 100 g of untrimmed venison meat from red deer (data from Drew and Semen 1987)

	Loin	Leg
Protein (g)	24.7	23.8
Fat (g)	3.3	3.0
Water (g)	70.8	71.2
Minerals (g)	1.4	1.9
Energy (kJ)	545	519
Cholesterol (mg)	66	74
Calcium (mg)	5	3
Magnesium (mg)	25	29
Sodium (mg)	51	47
Potassium (mg)	352	367
Copper (μg)	190	216
Iron (μg)	3820	3900
Zinc (μg)	2820	2510
Selenium (μg)	2.2	2.2

Mean carcass weight, 69.4 kg; $n = 10$

and consumers are seeking information on nutrient contents of many meats. Drew and Seman (1987) have studied venison nutrient content and compared it with meat from traditional livestock. Venison loin and leg are extremely high in protein and iron while being very low in fat, energy, and cholesterol (Table 56.5). It clearly ranks as a preferred health food for the 1990s.

Breeds of Farmed Deer

Fallow, red, wapiti, hybrid, and to some extent axis deer are all farmed for venison. Fallow deer mature rapidly and need to be slaughtered as yearlings if overfatness is to be avoided. Overfatness can also be a problem in red deer if the animals are slaughtered in summer at more than 2 years of age. Wapiti and red/wapiti hybrids mature more slowly and do not usually show much summer fat deposition until 3 years of age and older (Drew and Hogg 1990). A wide variety of cervids will hybridize and produce fertile offspring (Dratch and Fennessy 1985). The use of a wapiti bull mated to a red deer hind will produce a large, fast-growing calf with excellent carcass characteristics at an early age (Pearse 1988). The hybrid bulls are now in great demand in New Zealand as sires over red hinds to produce one-quarter wapiti progeny. These animals will become important venison producers in future years. The half-bred wapiti/red animal at an early age (11 months) has a bigger and better carcass than a 2-year-old red deer (Table 56.6). Lean content of 76% is extremely high, fat is very low, and the hybrid does not appear to have a large bone content.

Velvet Antler

The partly grown and immature antlers from several species of deer are a very valuable product in traditional oriental medicine. Kong and But (1985) have comprehensively reviewed the products of deer for medicinal purposes. New Zealand developed a sophisticated velvet antler industry because of its good internal communication, a strong industry structure, and a business community prepared to invest in the cooking/drying technology required to produce top-quality velvet antler. Prices to farmers have fluctuated considerably since 1980, but the return has never been less than about $60US/kg for the best grade. Since red deer stags will produce 2.5 kg/year, this represents a good return to the farmer.

The use of velvet antler in a medical practice has been described by Yoon (1989) who says that he gives about 70% of his velvet antler to children in helping to prevent such problems as influenza.

TABLE 56.6 Tissue carcass composition of 26-month-old red stags and 11-month-old elk/red hybrid stags expressed as percentage of carcass weight

	Lean (%)	Fat (%)	Bone (%)
Red deer (mean carcass weight 63 kg; $n = 53$)	72.7	7.0	20
Elk/red hybrid (mean carcass weight 68 kg; $n = 8$)	76.0	4.7	19

The product is used in small doses (e.g., 4 g) and usually with herbal medicines. Extracts of velvet antler are also used as acupuncture injections having special effects on problems such as sciatic neuritis, impotence, shoulder pain, neck stiffness, and the after-effects of paralysis. Extracts of velvet antler in the form of pantocrin and rantocrin have been made in the USSR for a very long time and are used in human medicine.

The New Zealand Game Industry Board has established a good working relationship with the Korean Pharmaceutical Traders Association and now has a comprehensive velvet antler grading system in place to allow buyers to establish their preferences. Strict procedures are in place to see that analgesic drugs are used when velvet is cut from stags so that they feel no pain. The procedure must be done under the supervision of a veterinarian. After removal, the velvet is cooled, wrapped and frozen, pending sale or drying for further processing. Because of the high price currently being paid for velvet antler, many New Zealand deer farmers are holding stags back from slaughter in order to take a velvet antler cut; therefore, export volumes are going up rapidly.

Deer Skins

Deer leather from well-prepared skin produces a particularly high quality leather suitable for fashion garments. The strength of the leather (60–90 N/mm) is much greater than that of sheep leather (25–35 N/mm). Therefore, deer skin can be processed to a very thin leather (e.g., 0.4 mm) which is excellent for high fashion garments (Clark and Webster 1985). Skin can be damaged during the yarding and holding of animals preslaughter. Flailing feet and hard antlers are commonly seen in the deer yard and these do physical damage to skins which precludes their use for top-quality leather garments. As deer farming grows around the world, the quantity and quality of deer skins should become increasingly important in the making of leather goods.

References

Brittin HC, Armes CL, Ramsey CB, Simpson CD (1981) Palatability of meat from mule deer. J Food Sci 46:1805–1808

Chrystall BB, Devine CE (1983) Electrical stimulation of deer carcasses. N Z J Agric Res 26:89–92

Chrystall BB, Devine CE, Ellery S, Wade L (1984) Low voltage electrical stimulation of lamb; its effect on muscle pH and tenderness. N Z J Agric Res 27:513–523

Clark JE, Webster RM (1985) Leather production from New Zealand deer skins. In Fennessy PF, Drew KR (eds) Biology of Deer Production. R Soc N Z Wellington Bull 22:303–306

Crawford MA (1968) Fatty-acid ratios in free-living and domestic animals. Lancet, June 22:1329–1333

Dratch P, Fennessy PF (1985) Directions in deer breeding. 2. Efficiency. Deer Farmer 23:31–33

Drew KR (1989) Technologies to improve the shelf life and presentation of fresh venison. Proc Ruakura Deer Industry Conf, MAF, Hamilton, New Zealand, pp 27–30

Drew KR (1985) Meat production from farmed deer. In Fennessy PF, Drew KR (eds) Biology of Deer Production. R Soc N Z Wellington Bull 22:285–290

Drew KR, Fennessy PF (1986) Venison research—carcass features, processing and packaging. In Wilson PR (ed) Proc Deer Course for Vets, vol 3. Palmerston North, New Zealand, pp 17–34

Drew KR, Hogg BW (1990) Comparative carcass production from red, wapiti and fallow deer. Proc 8th Conf of the Australian Assoc Animal Breeding Genetics, Simon Printing Coy, Palmerston North, pp 491–496

Drew KR, Seman DL (1987) The nutrient content of venison. Proc Nutr Soc N Z 12:49–55

Drew KR, Crosbie SF, Forss DA, Manley TR, Pearse AJ (1988) Electrical stimulation and aging of carcasses from red, fallow and New Zealand Wapiti-type male deer. J Sci Food Agric 43:245–259

Fennessy PF, Greer GJ (1982) Progeny testing of selected lean and fat rams. Proc N Z Soc Anim Prod 42:137–140

Fennessy PF, Taylor PG (1989) Deer farming in Oceania. In Hudson RJ, Drew KR, Baskin LM (eds) Wildlife Production Systems. Cambridge University Press, Cambridge, pp 309–322

Forss DA, Manley TR, Platt MP, Moore VJ (1979) Palatability of venison from farmed and feral red deer. J Sci Food Agric 30:932–935

Gregson JE, Purchas RW (1985) The carcass composition of male fallow deer. In Fennessy PF, Drew KR (eds) Biology of Deer Production. R Soc N Z Wellington Bull 22:295–298

Kong YC, But PPH (1985) Deer—the ultimate medicinal animal (antler and deer parts in medicine). In Fennessy PF, Drew KR (eds) Biology of Deer Production. R Soc N Z Wellington Bull 22:311–326

Luxmoore RA (1989) International trade. In Hudson RJ, Drew KR, Baskin LM (eds) Wildlife Production Systems. Cambridge University Press, Cambridge, pp 28–49

Maiga AM (1974) Physical and chemical composition of the carcass of the domestic bovine as influenced by breed, sex, level of feed intake and stage of growth. PhD Thesis, Cornell University, Ithaca, New York

Manley TR, Forss DA (1979) Fatty acids of meat lipids from young red deer (*Cervus elaphus*). J Sci Food Agric 30:927–931

Nottingham PM, Wyborn R (1975) Microbiology of beef processing. II. Chilling and aging. N Z J Agric Sci 18:23–27

National Research Council (NRC) (1988) Designing foods. National Academy Press, Washington, D.C.

Pearse AJ (1988) Wapiti and hybrids—special management needs. In Wilson PR (ed) Proc Deer Course for Vets, vol 5. Palmerston North, New Zealand, pp 164–177

Seman DL, McKenzie-Parnell JM (1989) The nutritive value of meat as a food. In Purchas RW, Butler-Hogg BW, Davies AS (eds) Meat Production and Processing. N Z Soc Anim Prod, Occasional Publication 11:13–28

Seman DL, Drew KR, Littlejohn RP (1989) Packaging venison for extended chilled storage: comparison of vacuum and modified atmosphere packaging containing 100% carbon dioxide. J Food Protection 52:886–893

Seman DL, Drew KR, Clarken PA, Littlejohn RP (1988) Influence of packaging, method and length of chilled storage on microflora, tenderness and colour stability of venison loins. Meat Sci 22:267–282

Stevenson JM, Seman DL, Wetherall IL, Littlejohn RP (1989) Evaluation of venison colour by an objective method using CIELAB values. J Food Sci 54:1661–1662

Sumner JL, Perry IR, Reay CA (1977) Microbiology of New Zealand feral venison. J Sci Food Agric 28:829–832

Yoon P (1989) Some mysteries of the East revealed. Deer Farmer 58:17–23

57

Body Composition of Female Svalbard Reindeer: The Effects of Age and Reproductive Status

Nicholas J.C. Tyler

Abstract

The influence of age and reproductive status on body composition was studied in 38 female Svalbard reindeer, *Rangifer tarandus platyrhynchus*, shot in late autumn and early winter as part of an investigation of the causes of periodic low fecundity in this subspecies. Nonlactating females were on average heavier and fatter than lactating females. Unusual for this genus, total dissectible fat (TDF, kg) was inversely related to age in both nonlactating ($P < 0.05$) and lactating females ($P < 0.002$). Moreover, the effect of age on fat reserves tended to be greater in lactating compared with nonlactating females ($0.05 < P < 0.10$). Svalbard reindeer are evidently less able to compensate for the burden of lactation than other subspecies of *Rangifer* despite excellent forage quality in summer. A pronounced decline in TDF in the oldest lactating females may reflect increased investment in their last offspring.

Key words: Adipose tissue, fat reserves, fertility, lactation, reindeer, reproduction

Introduction

Female wild reindeer and caribou, *Rangifer tarandus* spp., are typically highly fertile. Mean rates of fertility or fecundity in mainland (subarctic) populations exceed 76% and show relatively little variation between years (Table 57.1). By contrast, in high arctic populations, rates of fertility or fecundity are low (mean <48%) and annually highly variable (Table 57.1). The demographic consequences of this variation can be considerable; in Svalbard periodic low fecundity is the principal factor limiting the annual rate of increase of reindeer (Tyler 1987a); however, the cause for this is not known.

Fertility in female mountain reindeer and barren-ground caribou, like many other cervids, is related to body weight and fat reserves (Mitchell and Brown 1974; Dauphiné 1976; Hamilton and Blaxter 1980; Reimers 1983; Sæther and Haagenrud 1983; Leader-Williams 1988). Albon et al. (1986) showed that differences in fertility between age classes in red deer hinds, *Cervus elaphus*, were due to independent effects of age-specific differences in body weight, body (jaw) size, fat reserves, and reproductive history. This chapter describes the influence of age and reproductive history on body weight and fat reserves in female Svalbard reindeer shot early in winter during, or shortly after, the rut. The data form part of a larger study which investigates the physiological mechanism and potential adaptive significance of natural variation in fecundity in this unusual subspecies.

Materials and Methods

Size and Body Composition

Body size, body weight, and fat content were measured in 38 female Svalbard reindeer (19

TABLE 57.1. Rates of fertility or fecundity (calves or fetuses per 100 adult females) in reindeer/caribou from subarctic and high-arctic populations

Locality	Latitude	Pregnancy or birth rate (%) Range	Mean	CV	Years	Source
Subarctic						
Norway, Hardangervidda	60° N	57–87[a]	76.3	10.7	11	T. Skogland, personal communication
Snøhetta	62° N	60–86[a]		12.0	9	T. Skogland, personal communication
Knutshø	62° N	90–100[a]		11.5	6	T. Skogland, personal communication
Canada, Newfoundland	48° N	72–91[b]	84.4	7.9	10	Bergerud (1971)
Northwest Territories	60° N	73–100[c]	88.3	15.7	3	Dauphiné (1976)
Beverly	64° N	79–100[c]			8	Thomas and Barry (1990)
High arctic						
Svalbard reindeer	78° N	9–73[b]	47.4	44.7	11	Tyler (1987a)
Peary caribou	75° N	16–92[c]	31.5	64.0	4	Thomas (1982)

[a]Pregnancy rate from counting antlers (see Skogland 1985)
[b]Birth rate
[c]Pregnancy rate from shot sample
CV, coefficient of variation
Years, number of years in which data were collected

lactating and 19 nonlactating) shot in Nordenskiøld Land, Svalbard (78° N, 16° E), in September ($n = 5$), October ($n = 10$), and November ($n = 26$) 1980, 1981, and 1988, respectively.

All specimens were weighed whole to 0.1 kg (less blood lost from the wound). The rumen was emptied and its contents weighed by difference; ingesta-free body weight (IFBW) was calculated to 0.1 kg for each specimen by subtraction. The length of one metatarsus was measured according to Dauphiné (1976) to 1 mm.

The total mass of dissectible fat (TDF, kg) was measured according to Tyler (1987b). Age of specimens was determined by counting annulations in the cementum of first incisor teeth (Reimers and Nordby 1968; Leader-Williams 1979).

Analysis

The overall median collection date was 21 November; measurements of weight (IFBW, TDF) were corrected, using linear regression models, as if all the animals had been shot on this date (Albon et al. 1986), and only corrected values were used in the analysis.

The body size of reindeer from different collections was compared using Kruskal-Wallis analysis of variance (Siegel 1956). Two sample comparisons were made using Mann-Whitney *U* tests (Siegel 1956). Linear regression models were calculated using the least square method and compared using the *t* statistic (Kleinbaum and Kupper 1978). The null hypothesis was rejected at *P* (two-tailed) < 0.05 in all tests.

Results

There were no significant differences in body sizes of reindeer between collections as measured length of metatarsus ($P > 0.2$); therefore, the data were combined for all years. The median ages of specimens in the combined sample were: lactating females, 7 years (range, 3–12) and nonlactating females, 7 years (range, 2–11).

Nonlactating females were significantly heavier and had significantly larger fat reserves than lactating females (Table 57.2). IFBW tended to increase with age in nonlactating females and to decrease age in lactating females, but neither relationship was significant.

By contrast, the total weight of dissectible fat in

TABLE 57.2. Body composition of female Svalbard reindeer (age, 2 to 12 years) in November

	Ingesta-free body weight (kg)			Total dissectible fat (kg)			Fat content (% live weight)		
	NL		L	NL		L	NL		L
Median	62.8	***	53.5	14.0	***	11.3	20.0	**	17.2
Range	57.6–68.5		44.0–61.4	11.3–17.0		6.8–15.1	15.9–28.5		10.4–21.9
n	18		18	19		19	16		18

NL, nonlactating; L, lactating; ***$P < 0.001$; **$P < 0.01$

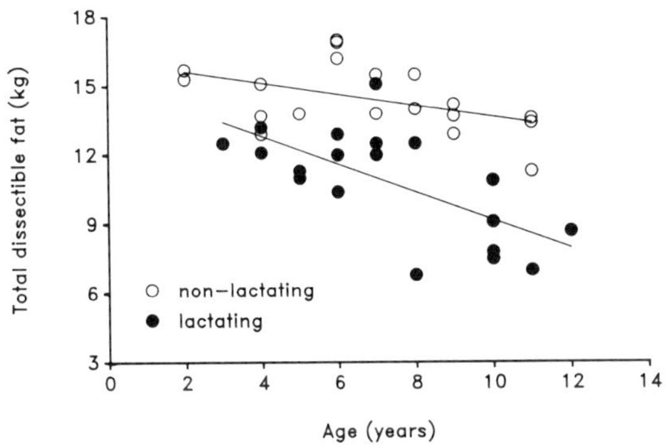

FIGURE 57.1. Influence of age on fat reserves in female Svalbard reindeer (age, 2 to 12 years) in November.

the carcasses was inversely related to age in both nonlactating females ($m = -0.285$, $r^2 = 23.8$, $P = 0.047$) and lactating females ($m = -0.606$, $r^2 = 44.8$, $P < 0.002$). Furthermore, the effect tended to be greater in lactating compared with nonlactating females ($T = 1.622$, $df = 34$, $0.05 < P < 0.1$) (Fig. 57.1).

Discussion

These results demonstrate that in Svalbard reindeer, like several other species of cervids, females which have nursed calves throughout the summer are likely to weigh less and have smaller reserves of fat early in winter, after the rut, than females with no calves. They also reveal several important differences between Svalbard reindeer and other subspecies of reindeer/caribou.

All the animals collected in this study, whether lactating or not, had large subcutaneous deposits of fat. Accumulation of exceptionally large reserves of fat in summer and autumn is an important characteristic of Svalbard reindeer (Reimers et al. 1982). This fat can provide a significant supplement to the animals' meager winter diet (Mathiesen et al. 1984) and may also be important in both sexes for enhancing reproductive success (Tyler 1987b; Tyler and Blix 1990). In contrast to this, females in other subspecies of *Rangifer* are typically lean, and a substantial proportion (25% to 90%) carry no rump fat early in winter after the rut (Dauphiné 1976; Leader-Williams and Ricketts 1982). However, although the Svalbard reindeer had much more fat than other *Rangifer*, and might therefore be said to have been in excellent physiological condition, the size of their reserves was individually highly variable (Table 57.2).

Differences in body weight and fat reserves between lactating and nonlactating females are much larger in Svalbard reindeer compared with other subspecies of *Rangifer*. Nonlactating barren-ground caribou cows (*R.t. groenlandicus*) tend to be heavier and fatter in November than lactating females, but no statistically significant differences have been found (Dauphiné 1976; Adamczewski et al. 1987). On South Georgia, nonlactating female reindeer (*R.t. tarandus*) were significantly heavier and fatter at the rut than lactating females but the difference, nevertheless, was small; few females (< 10%) had subcutaneous reserves of fat and the peak weight of kidney fat, their principal reserve, was less than

100 g (Leader-Williams and Ricketts 1982). In Svalbard reindeer, by contrast, the *difference* in median TDF between the two groups was 2.7 kg (Table 57.2). Large differences in body weight and fat reserves between lactating and nonlactating females indicate that Svalbard reindeer are less able to compensate the burden of lactation than other female *Rangifer*. Rates of production of ruminal volatile fatty acids, which reflect the quality of food eaten, are much higher in Svalbard reindeer (549 mM/kg$^{0.75}$·day) compared with reindeer in north Norway (178 mM/kg$^{0.75}$·day) or to caribou in Alaska (185 mM/kg$^{0.75}$·day (White and Staaland 1983). Hence, in Svalbard, fattening is probably limited by the brevity of summer rather than the quality of summer forage.

Strong inverse relationships between age and fat reserves (Fig. 57.1) have not previously been reported in reindeer/caribou. Leader-Williams and Ricketts (1982) found no differences in fat reserves across age classes in female Norwegian reindeer on South Georgia. Dauphiné (1976) found a small decline in fat reserves with age in barren-ground caribou, but only in animals aged $\geqq$ 10 years. The progressive decline in fat reserves with increasing age in nonlactating female Svalbard reindeer reflects the cummulative effect of past lactations together with debilitating effects of age, such as increased tooth wear (Tyler 1986).

It is not clear, however, why the effect of age on fat reserves was more pronounced in lactating females. Scottish red deer hinds show the opposite trend. In these the effect of age on autumn fat reserves is most pronounced in nonlactating females (Mitchell et al. 1977), presumably reflecting the fact that all lactating hinds aged $\geqq$ 3 years are lean in autumn. Nevertheless, there is indirect evidence that the proportion of resources which red deer hinds invest in their calves during lactation increases with the age of the dams (Clutton-Brock 1984), and it would be useful to test this in Svalbard reindeer by examining the relationships between fat reserves in calves and their mothers' age and condition.

Acknowledgements

This study was supported, in part, by the Norwegian MAB Program, the Norwegian Research Council for Science and the Humanities, and a Special Research Grant from the ECO-Sciences division of NATO.

References

Adamczewski JZ, Gates CC, Hudson RJ, Price MA (1987) Seasonal changes in body composition of mature female caribou and calves (*Rangifer tarandus groenlandicus*) on an arctic island with limited winter resources. Can J Zool 65:1149–1157

Albon SD, Mitchell B, Huby BJ, Brown D (1986) Fertility in female red deer (*Cervus elaphus*): the effects of body composition, age and reproductive status. J Zool, London (A) 209:447–460

Bergerud AT (1971) The Population Dynamics of Newfoundland Caribou. Wildl Monog No 25

Clutton-Brock TH (1984) Reproductive effort and terminal investment in iteroparous animals. Am Natural 123:212–229

Dauphiné TC Jr (1976) Biology of the Kaminuriak population of barren-ground caribou. Part 4. Growth, reproduction and energy reserves. Can Wildl Service Rep Ser No 38

Hamilton WJ, Blaxter KL (1980) Reproduction in farmed red deer. 1. Hind and stag fertility. J Agr Sci Cambridge 95:261–273

Kleinbaum DG, Kupper LL (1978) Applied Regression Analysis and Other Multivariate Methods. Duxbury Press, Massachusetts

Leader-Williams N (1988) Reindeer on South Georgia: the Ecology of an Introduced Population. Cambridge University Press, Cambridge

Leader-Williams N (1979) Age determination of reindeer introduced into South Georgia. J Zool, London 188:501–515

Leader-Williams N, Ricketts C (1982) Seasonal and sexual patterns of growth and condition of reindeer introduced into South Georgia. Oikos 38:27–39

Mathiesen SD, Orpin CG, Blix AS (1984) Rumen microbial adaptation to fibre digestion in Svalbard reindeer. Can J Anim Sci (Supp) 64:261–262

Mitchell B, Brown D (1974) The effects of age and body size on fertility in female red deer (*Cervus elaphus* L.). Proc Int Union Game Biol (13E) 11:89–98

Mitchell B, Staines BW, Welch DW (1977) Ecology of red deer: a Research Review Relevant to Their Management in Scotland. Institute of Terrestrial Ecology, Cambridge

Reimers E (1983) Reproduction in wild reindeer in Norway. Can J Zool 61:211–217

Reimers E, Nordby Ø (1968) Relationship between age and tooth cementum layers in Norwegian reindeer. J Wildl Manage 32:957–961

Reimers E, Ringberg T, Sørumgård R (1982) Body composition of Svalbard reindeer. Can J Zool 60:1812–1821

Skogland T (1985) The effects of density-dependent resource limitations on the demography of wild reindeer. J Anim Ecol 54:359–374

Siegal S (1956) Nonparametric statistics for the behavioral sciences. McGraw-Hill Kogakshua, Tokyo

Sæther BE, Haagenrud H (1983) Life history of the moose (*Alces alces*): fecundity rates in relation to age and carcass weight. J Mammal 64:226–232

Thomas DC (1982) The relationship between fertility and fat reserves of Peary caribou. Can J Zool 60:597–602

Thomas DC, Barry SJ (1990) Age-specific fecundity of the Beverly herd of barren-ground caribou. Rangifer (Special Issue) 3:257–263

Tyler NJC (1987a) Natural limitation of the abundance of the high arctic Svalbard reindeer. PhD thesis, Cambridge University, Cambridge

Tyler NJC (1987b) Body composition and energy balance of pregnant and non-pregnant Svalbard reindeer during winter. Symp Zool Soc London 57:203–229

Tyler NJC (1986) The relationship between the fat content of Svalbard reindeer in autumn and their death from starvation in winter. Rangifer (Special Issue) 1:311–314

Tyler NJC, Blix AS (1990) Survival strategies in arctic ungulates. Rangifer (Special Issue) 3:211–230

White RG, Staaland H (1983) Ruminal volatile fatty acid production as an indicator of forage quality in Svalbard reindeer. Acta Zoologica Fennica 175: 61–63

58
Evaluation of Carcass Composition Changes to Fallow Bucks Castrated Prepuberally

Robert C. Mulley and Anthony W. English

Abstract

A trial was conducted to compare, by dissection of individual muscles, bones and fat, carcass changes to intact fallow bucks (*Dama dama*) and bucks castrated prepuberally and killed at 17 months of age. Intact bucks grew faster than castrated bucks, but much of the weight difference resulted from the increased weight of the head and bones of intact deer (25%). Most of the muscles of castrates (77/94) were lighter than the same muscles in their intact counterparts, but the major differences occurred in the commercially less useful forequarter muscles. The total weight and proportion to carcass weight of the hindquarter were not changed significantly by castration. The muscle bone ratio was not significantly different between intacts and castrates, but the muscle fat ratio was significantly reduced by castration. The total weight of muscle on intact and castrated bucks of the same body weight was not significantly different, but there was a greater proportion of muscle on the hindquarter of the castrate. The results of this work demonstrate that the castrate offers a viable alternative to intact bucks being killed for venison during the fallow deer breeding season when carcass bruising due to fighting is difficult to avoid. Prepuberal castration was shown to eliminate antler spikes and aggressive behavior, and the small carcass changes associated with castration were easily compensated by increased ease of management.

Key words: Castration, carcass composition, fallow deer, growth rate

Introduction

Body form and composition are modified by castration, resulting in carcass characteristics that differ from the intact animal (Butterfield 1988). The use of castration as a husbandry procedure with farmed deer has been tested on red deer (Blaxter et al. 1974; Drew et al. 1978; Kay et al. 1981; Tan and Fennessy 1981; Drew 1985) and fallow deer (Mulley and English 1985; Asher 1988; Adam 1988; Mulley 1989). These studies have shown that castration increased the ease with which animals could be handled and managed, but reduced growth rates and increased carcass fatness compared with intact animals (Table 58.1). Therefore, the need to castrate surplus bucks destined for slaughter has been balanced between ease of animal management and production loss. Red deer stags appear to be less aggressive towards one another when confined in yards, compared with fallow bucks during the breeding season. Carcass bruising and, occasionally, death of animals have been problems associated with the yarding and slaughter of fallow bucks during the breeding season, and castration of animals to be killed at this time of year may afford some advantages to deer farmers (Mulley and English 1985; Fennessy and Drew 1987).

The decision to castrate or not should be made after evaluation of growth and carcass composition differences between intact and castrated animals. Conclusions from previous studies have been based on growth rate changes and

TABLE 58.1. A comparison of carcass characteristics for red and fallow deer from published literature

	Age (months)	Live weight (kg)	Fasted live weight (kg)	Dressed carcass weight (kg)	Dressing proportion	Muscle fat (%)	Authors
Red deer							
Stag	15	67.2	57.3	36.3	0.63	1.2	Kay et al. (1981)
	15	83.8	—	43.8	0.52[a]	—	Drew et al. (1978)
	27	118.1	—	67.6	0.57[a]	—	Drew et al. (1978)
	27	—	—	—	—	3.0	Tan and Fennessy (1981)
Castrate	15	58.4	50.6	30.4	0.60	1.0	Kay et al. (1981)
	16	76.1	—	40.6	0.53[a]	—	Drew et al. (1978)
	27	98.4	—	55.8	0.57[a]	—	Drew et al. (1978)
	27	—	—	—	—	3.9	Tan and Fennessy (1981)
Hind	17	43.1	37.2	22.0	0.59	1.8	Kay et al. (1981)
Fallow deer							
Buck	17	—	43.8	26.8	0.61	2.1	Mulley (1989)
	20	—	46.8	30.4	0.65	—	Mulley and English (1985)
Castrate	17	—	40.8	24.7	0.61	10.0	Mulley (1989)
	20	—	41.7	24.2	0.58	—	Mulley and English (1985)
Doe	Adult	—	41.0	24.0	0.59	12.6	Mulley (1989)

[a]Dressing proportion of unfasted live weight
Dressing percentage = dressing proportion × 100

gross carcass differences, without distinguishing where those changes occurred. The present trial was conducted to compare, by dissection and weighing of individual muscles, bones and fat, carcass differences between intact fallow bucks and fallow bucks castrated prepuberally and killed at 17 months of age during the fallow deer rut.

Materials and Methods

Twenty 6-month-old fallow bucks were stratified on weight and allocated to one of two treatment groups. Bucks in one group were left intact, while bucks in the remaining group were castrated by applying elastic rubber ligatures (Elastrator, Blenheim, New Zealand). The two groups of ten bucks were run as one mob and given unlimited access to improved pasture and water for the duration of the trial. All animals were weighed once monthly from commencement of the trial until slaughter at 17 months of age. A multivariate repeated measures analysis (Gilmour 1988) was used to examine the effect of castration on growth rate (g/day) over the 11 monthly growth periods, simultaneously.

In May 1987, when the animals were 17 months old, five intact animals were matched with five castrated animals of equal weight according to their weight at the beginning of the trial. These animals were fasted for 16 h before slaughter and transported to an abattoir in single animal crates, in which they were killed. For this group of ten animals, two animals per day were slaughtered to allow examination and dissection of each animal for total body and carcass composition. A recovery sheet of all body components such as major organs, blood, hide, head, and gut fill was compiled at slaughter, and then each carcass was chilled for 24 h before dissection of the left side of each carcass into component muscle, bone, and fat (Butterfield and May 1966). The right side of each carcass was divided into primal cuts (leg, shoulder, saddle, neck and

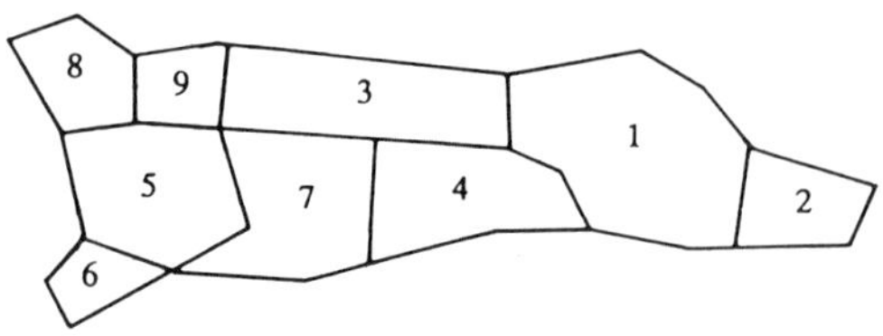

FIGURE 58.1. Diagrammatic representation of the standard muscle groups as described by Butterfield and Berg (1966).

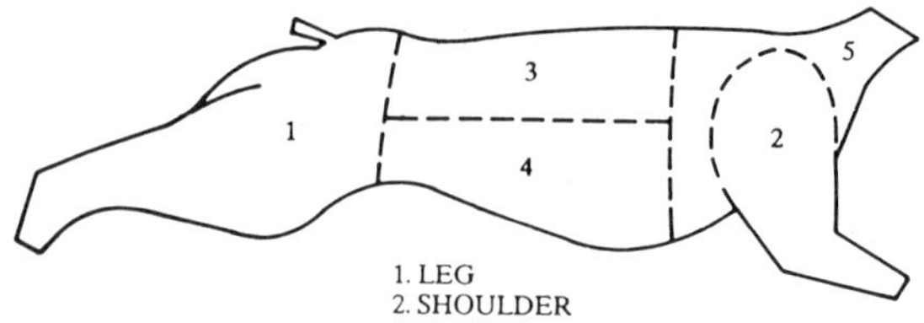

FIGURE 58.2. Diagrammatic representation of the five primal carcass cuts. Right-hand side of each carcass was sectioned into these cuts for weighing.

chuck, flap) with each cut being weighed separately.

Dissected muscles were trimmed from tendons and aponeuroses at the level of the last vestige of muscle tissue so that there may have been some tendenous tissue left on the muscle, but no muscle tissue was removed with the tendon. The surfaces of the muscles were cleaned of all fat and connective tissues, weighed and grouped into one of nine standard muscle groups for statistical analysis (Fig. 58.1). Selected muscles from both the forequarter and hindquarter of each carcass were tested for tenderness using a Warner-Bratzler shear (Ramsbottom and Strandine 1948). A portion of the M. glutaeus medius (one of the largest hindquarter muscles) from each animal was minced and stored frozen at -20° C for subsequent chemical fat analysis using a Foss-Let density apparatus (Usher et al. 1973).

The remaining ten animals (five from each group) were transported to an abattoir in one group and slaughtered on the same day following a 16 h overnight fast. The carcasses were chilled for 72 h before again being broken into primal cuts for weighing (Fig. 58.2).

Results

From the beginning of the trial when they were six months old (May 1986) until slaughter at 17 months of age (May 1987), intact bucks grew faster than castrated bucks and they were 3.0 kg heavier at slaughter ($P < 0.05$) (Table 58.2). Intact bucks grew antler pedicles 8 to 10 cm long and antler spikes 6 to 15 cm long compared with zero pedicle development in castrated bucks.

As a result of castration, the weight and proportion to live weight of the thymus and rumenoreticulum contents were significantly increased and the weights of the head (-0.52 kg), liver, kidney, large intestine, seminal vesicles, and distal forelimbs were significantly reduced. Relative to live weight, castration significantly reduced the proportion of the head, seminal vesicles, and blood weights.

The left side of each cold carcass was sectioned

TABLE 58.2. Effect of castration on growth compared with intact fallow bucks

Treatment group	n	Birth weight December, 1985 (kg)		Live weight May, 1986 (kg)		Fasted 16 h live weight May, 1987 (kg)		Cold carcass weight (kg)		Dressing proportion	
		$\bar{x}$	SE	$\bar{x}$	SE	$\bar{x}$	SE	$\bar{x}$	SE	$\bar{x}$	SE
Intact	10	4.2	0.13	26.7	0.9	43.8	0.6	26.8	0.5	0.61	0.01
Castrate	10	4.4	0.24	26.5	0.9	40.8	0.6	24.7	0.3	0.61	0.01

TABLE 58.3. Mean side weight, hindquarter weight, and forequarter weight for 17-month-old intact and castrated fallow bucks

	Intact		Castrate	
	$\bar{x}$	SE	$\bar{x}$	SE
Three-rib hindquarter (kg)	7.12	0.15	6.96	0.08
Ten-rib fourquarter (kg)	6.12	0.23	5.44*	0.10
Total left side (kg)	13.24	0.29	12.40	0.16

*Castration effect, $P < 0.01$

into a three-rib hindquarter and a ten-rib forequarter (Table 58.3). Castration significantly reduced the weight and proportion of the forequarter ($P < 0.01$) but did not significantly change the weight or proportion of the hindquarter. Castration also decreased the proportion of muscle and increased the proportion of fat relative to total body weight, but had no effect on bone proportion. The net effect of proportionately decreased muscle and increased fat in castrated animals was that there was no significant difference between the two groups of animals when comparing half-carcass weights, nor for the proportion of the half-carcass to live weight. As a consequence the muscle bone ratio was not significantly different between groups but the muscle fat ratio was significantly reduced by castration ($P < 0.001$).

The muscle weight difference between intact and castrated bucks was due to a reduction of both hindquarter and forequarter total muscle weights in the castrates, with hindquarter muscle weights being significantly different ($P < 0.01$). However, the proportion of hindquarter muscle to total muscle in castrates compared with intact bucks was significantly greater ($P < 0.001$). Castration also reduced the total weight of muscles in each of the nine standard muscle groups (Fig. 58.1) but changed the proportion of muscle in only four of the groups. Total muscle weight for groups 1 and 2 muscles was proportionately larger in castrates than in intacts, while total muscle weight in groups 8 and 9 muscles was proportionately smaller in castrates than in intact bucks. Castration also produced a weight reduction in 77 of the 94 individual muscles dissected from each side, and this effect was distributed over the whole carcass (Table 58.4). Proportional increases in muscle weight relative to total muscle weight occurred in 21 muscles represented in muscle groups 1 to 6 in castrated animals, while proportional decreases in muscle weight occurred in 20 muscles represented in muscle groups 3, 5, 6, 8, and 9. There were no

TABLE 58.4. Carcass distribution of weight and proportion changes to individual muscles resulting from castration of 17-month-old fallow bucks

		Castration			
		Weight		Proportion	
Muscle group	Number of muscles	Increase	Decrease	Increase	Decrease
1	20	—	12	11	—
2	10	—	10	1	—
3	9	—	8	3	3
4	7	—	3	3	—
5	13	—	12	2	3
6	11	—	9	1	1
7	5	—	5	—	—
8	4	—	4	—	3
9	15	—	14	—	10
Total	94	0	77	21	20

TABLE 58.5. Primal cuts as mean weight and proportion of total half carcass weight for 17-month-old intact and castrated fallow bucks; all weights expressed as "bone-in" weights

	Weight (kg)				Proportion			
	Intact		Castrate		Intact		Castrate	
	$\bar{x}$	SE	$\bar{x}$	SE	$\bar{x}$	SE	$\bar{x}$	SE
Leg	5.2190	0.0886	5.0350	0.1097	0.4120	0.0018	0.4300	0.0000
Shoulder	2.3310	0.3930	2.0920*	0.4510	0.1840	0.0040	0.1800	0.0076
Saddle	1.9240	0.0627	1.7900	0.0440	0.1500	0.0045	0.1520	0.0058
Neck and chuck	1.9910	0.0408	1.5420*	0.0971	0.1580	0.0376	0.1320	0.0058
Flap	1.1241	0.0792	1.2710	0.0904	0.0960	0.0040	0.1080*	0.0058
Total half carcass	12.7060	0.1631	11.7300	0.2364				

*Castration effect, $P < 0.05$

differences in muscle tenderness between muscles from bucks in either of the treatment groups.

Weights of fat dissected from subcutaneous, intermuscular, kidney, channel, omental, and mesenteric fat partition were collected separately. There were significant ($P < 0.001$) increases in the weight of fat deposited in all fat depots in castrated animals and also significant increases in the chemical fat content of muscles and carcass fat depots. Although small weight changes occurred in four of the five primal cuts as a result of castration, only the weight of the flap relative to the total half carcass weight was proportionately changed ($P < 0.05$) (Table 58.5).

Discussion

The results of this trial define the body and carcass composition of intact fallow bucks at about the age and weight that they are usually slaughtered for venison in Australia and New Zealand. The results demonstrate that the carcasses of castrates are a suitable alternative to those from intact bucks slaughtered during the breeding season. Although castrates grew more slowly and had lighter carcasses than intact bucks, much of the difference was accounted for by the significantly heavier head and bones of intacts (25% of weight difference). Most of the muscles of castrates were lighter than the same muscles in their intact counterparts (77/94), but the differences were small and of the little commercial significance. The major differences occurred in the commercially less useful forequarter muscles. When the selvage from the muscles of the leg were removed to produce what is commercially known as "Denver leg steak," there were no differences in total leg muscle weight between the two treatment groups. While the producer may lose up to 2 kg per carcass by castrating, the actual "bone-in" weight of the commercially important primal cuts (saddle and leg, Table 58.5) is not reduced. Kay et al. (1981) also came to this conclusion for 15- and 27-month-old red deer stags.

An important management consideration is that castrates do not have to be yarded for removal of antler spikes if castration is carried out before pedicle development, and the behavioral changes that result from castration equal anything that can be produced chemically and are permanent. Future research to establish the per hectare cost of producing venison from intact and castrated bucks will also help to resolve the question of whether castration of deer is cost effective or not.

Increased carcass fatness in castrated deer has been reported in red deer (Blaxter et al. 1974; Drew et al. 1978; Tan and Fennessy 1981; Fennessy and Drew 1987) though not in young animals (Kay et al. 1981). In a previous study of 20-month-old fallow bucks and castrates (Mulley and English 1985), there were no apparent gross carcass differences, but chemical analysis of muscle and carcass dissections were not performed. The results of the present study indicate that castration has increased carcass fatness when carcasses from 17-month-old intact and castrate animals are compared. However, many intact fallow bucks are slaughtered for the venison market in summer condition and are known to be fatter (McCall 1985) than castrates killed from late fall through winter. Over-fatness can be a problem in older castrates and intact red deer stags (Drew et al. 1978) and fallow bucks (Asher 1985), but while fallow bucks are being slaughtered at a young age, over-fatness is unlikely to be a problem.

Fallow bucks have again been shown to produce a carcass with a high muscle bone ratio, a high dressing out proportion, and low levels of fat compared to other domestic species. Castration was shown to eliminate aggressive behavior, and the small carcass changes associated with castration are easily compensated for by increased ease of management. Fresh fallow venison year-round can be achieved by careful management that includes castration of a proportion of surplus bucks to be killed during the breeding season.

Acknowledgements

This work was supported by the Australian Special Rural Research Council (ASRRC) and several state deer farming associations. Grateful thanks are due to R.M. Butterfield and P. Martin from the University of Sydney, Department of Veterinary Anatomy, for performing carcass dissections.

References

Adam JL (1988) Growth. In Allen PL, Asher GW (eds) Progressive Fallow Farming. New Zealand MAF, Ruakura Agricultural Centre, Hamilton, pp 47–55

Asher GW (1988) Antlers. In Allen PL, Asher GW (eds) Progressive Fallow Farming. New Zealand MAF, Ruakura Agricultural Centre, Hamilton, pp 40–46

Asher GW (1985) Meat production from fallow deer. In Fennessy PF, Drew KR (eds) Biology of Deer Production. R Soc N Z Wellington Bull 22:299–301

Blaxter KL, Kay RNB, Sharman GAM, Cunningham JMM, Hamilton WJ (1974) Farming the Red Deer. Department of Agriculture and Fisheries for Scotland, Her Majesty's Stationery Office, Edinburgh

Butterfield RM (1988) New Concepts of Sheep Growth. Department of Veterinary Anatomy, University of Sydney, Australia

Butterfield RM, Berg RT (1966) A classification of bovine muscles based on their relative growth patterns. Res Vet Sci 7:326–332

Butterfield RM, May NDS (1966) Muscles of the Ox. University of Queensland Press, St. Lucia, Australia

Drew KR (1985) Meat production from farmed deer. In Fennessy PF, Drew KR (eds) Biology of Deer Production. R Soc N Z Wellington Bull 22:285–290

Drew KR, Fennessy PF, Greer GJ (1978) The growth and carcass characteristics of entire and castrate red stags. Proc N Z Soc Anim Prod 38:142–144

Fennessy PF, Drew KR (1987) Meat production from farmed deer. The Federal Deerbreeder. Aust Deer Farm Fed, 6:16–21

Gilmour AR (1988) REG—A generalised linear models programme, 3rd ed. Misc Bull 1, Div Agr Serv, Department of Agriculture, NSW, Australia

Kay RNB, Sharman GAM, Hamilton WJ, Goodall ED, Pennie K, Coutts AGP (1981) Carcass characteristics of young red deer farmed on hill pasture. J Agric Sci Cambridge 96:79–87

McCall JE (1985) The carcass composition and meat quality of male fallow deer. MScAg thesis, Massey University, Palmerston North, New Zealand

Mulley RC (1989) Reproduction and performance of farmed fallow deer (*Dama dama*). PhD thesis, University of Sydney, Australia

Mulley RC, English AW (1985) The effects of castration of fallow deer (*Dama dama*) on body growth and venison production. Anim Prod 41:359–361

Ramsbottom JM, Strandine EJ (1948) Comparative tenderness and identification of muscles in wholesale beef cuts. Food Res 13:315–330

Tan GY, Fennessy PF (1981) The effect of castration on some muscles of red deer (*Cervus elaphus* L.). N Z J Agr Res 24:1–3

Usher CD, Green CJ, Smith CA (1973) The rapid estimation of fat in various foods using the FOSSLET density apparatus. J Food Tech 8:429–437

59

Legal Status of Exotic Deer Throughout the United States

R. Terry Ervin, Stephen Demarais, and David A. Osborn

Abstract

The legal status of exotic deer throughout the United States was determined from a questionnaire mailed to state wildlife, meat inspection, and agricultural marketing agencies during 1989. The response rate varied among state agencies ($P < 0.01$), with the largest number of questionnaires returned by wildlife agencies and the smallest number of questionnaires returned by agricultural marketing agencies. Knowledge of the legal status of exotic deer production was compared among state agencies. In most instances, wildlife agencies responded with "don't know" less often than agricultural marketing and meat inspection agencies, indicating that wildlife agencies may be the best source of information concerning exotic deer production. Most wildlife agencies (64%) perceive exotic deer as potential competitors with native wildlife (i.e., pests). An important conclusion that can be drawn from this questionnaire concerns the general lack of knowledge among state agencies which either are or could be directly involved in regulating the exotic deer industry.

Key words: Exotic deer, game ranching, deer farming, legislation

Introduction

Commercial deer production is an established industry in New Zealand and the United Kingdom, but remains a relatively unexplored industry in the United States, despite the substantial number of ranches stocking exotic deer. Production and marketing potential for the exotic deer industry in the United States could be limited by regulations imposed at the state level. The perception of exotic deer as an ecologically viable product could vary among states and among agencies within a state due to the controversial nature of this resource (Demarais et al. 1990) and thus impact the future regulatory climate. Our objectives were to determine the legal status, agency perception, and availability of production-related programs related to exotic deer throughout the United States.

Methods

A questionnaire requesting information on the legal status of exotic deer was mailed to three agencies or departments in all 50 states during 1989: wildlife, meat inspection and agricultural marketing. Within some states, the meat inspection and agricultural marketing sources of information were departments or branches within the same agency. An initial mailing of 150 surveys was followed 4 weeks later with a second mailing to nonrespondents. The results reported in this paper represent only those responses received from state agencies as specified in the survey. Results were summarized by state agency/ department.

The questionnaire referred to "exotic deer" as nonnative deer and antelope, such as blackbuck antelope (*Antilope cervicapra*), and axis (*Axis*

axis), fallow (*Cervus dama*), red (*Cervus elaphus*) and sika deer (*Cervus nippon*). No attempt was made to collect data on any other exotic animals.

Many responses were grouped according to whether an agency knew or did not know the answer to a question. The chi-square test for independence was used to determine whether departments differed in their knowledge with respect to the questions asked in the survey. In cases where only one agency within a state answered a question it was assumed that this agency's response was correct. When responses differed between agencies within a state the disagreement was noted. These results are listed according to state and response in Table 59.1.

Results

Survey Response Rate

Surveys were returned from at least one of the targeted agencies for all states. The response rate differed among state agencies ($P < 0.01$), with the largest number of questionnaires returned by wildlife agencies and the smallest number of questionnaires returned by agricultural marketing agencies. The following summarizes the response rate: 90% of wildlife agencies returned their surveys, representing 41% of all surveys returned; 74% of meat inspection agencies returned their surveys, representing 33% of all surveys returned; and 59% of marketing agencies returned their surveys, representing 26% of all surveys returned.

Legal Status of Exotic Deer

When asked the legal status of exotic deer in their states, the responses were as follows: (1) game, 6 states; (2) domestic, 17 states; (3) unclassified, 4 states; (4) both game and domestic, 2 states; (5) don't know, 1 state; and (6) conflicting answers from within states, 20 states. Some state agencies suggested that "nongame" and "exotic" categories be added to the survey.

Regulations Governing the Shipment of Exotic Deer Within and into States

When asked if there were state regulations governing shipment of exotic deer within their states, the answers were as follows: (1) no, 10 states; (2) yes, 20 states; and (3) don't know, 1 state. The remaining 19 states reported disagreement among agencies. State agencies differed ($P < 0.05$) in their knowledge of state regulations dealing with the transportation of exotic deer within a state. More wildlife agencies (95.6%) knew about regulations than meat inspection (78.4%) and marketing (72.4%) agencies.

When asked if there were regulations governing shipment of exotic deer into their states, the answers were as follows: (1) no, 2 states; (2) yes, 32 states; (3) don't know, 1 state; and (4) disagreement among agencies within a state, 15 states. State agencies differed ($P < 0.05$) in their knowledge with respect to state regulations dealing with the importation of exotic deer into a state. More wildlife agencies (93.3%) knew about regulations than meat inspection (82.9%) and marketing (67.9%) agencies.

Agencies Having Jurisdiction over Exotic Deer Within States

Wildlife agencies had jurisdiction over exotic deer in 27 states. Meat inspection and marketing agencies were combined to represent an agricultural agency, representing 9 states. Some states separated jurisdictional control of exotic deer between wildlife and agriculture agencies, representing 10 states. Disagreement occurred within 4 states in which agency X claimed agency Z had jurisdiction, while agency Z claimed agency X had jurisdiction.

State-Sponsored Programs

When asked whether exotic deer production qualified for a state-sponsored tax credit program, responses indicated that 38 states do not, 3 states do, 8 states had agencies which "don't know," and 1 state had disagreement among agencies. States with tax credit possibilities were Minnesota, Texas, and Vermont. Respondents in Minnesota indicated that a tax credit exists if deer ranching qualifies as an agricultural enterprise. Respondents in Vermont indicated that credits are available for normal farm operations. Although tax credits for agricultural production encompassing deer

TABLE 59.1. A survey of the legal status of exotic deer in the United States

	AK	AL	AR	AZ	CA	CO	CT	DE	FL	GA	HI	IA	ID	IL	IN	KS	KY	LA	MA	MD	ME	MI	MN	MO	MS	MT	NC	ND	NE	NH
Legal status																														
Game		X									X	X					X													
Domestic				X	X		X				X							X	X				X	X				X		
Unclassified			X					X					X																	
Game and domestic																				X										
Don't know															X															
Disagreement	X					X			X					X			X				X	X			X	X	X		X	X
Regulated shipments within state																														
No			X			X				X								X							X	X		X	X	
Yes	X	X		X	X				X		X	X				X	X		X	X										X
Don't know															X															
Disagreement							X	X					X	X							X	X	X	X			X			
Regulated shipments into state																														
No																														
Yes	X	X	X	X	X						X	X	X			X	X	X	X	X	X			X	X	X		X		X
Don't know															X															
Disagreement						X	X	X	X	X				X								X	X				X		X	
Agency with jurisdiction																														
Wildlife	X	X	X	X	X	X	X		X			X			X			X	X	X				X	X	X	X			X
Agricultural																					X	X						X	X	
Wildlife/agri-cultural										X	X		X			X	X													
Disagreement								X						X											X					
Tax credit program																														
No	X	X		X	X			X	X	X	X	X	X	X		X	X	X		X	X	X		X	X	X	X	X	X	X
Yes																							X							
Don't know			X			X	X								X				X											
Disagreement																														
Tax exemption program																														
No	X	X		X	X			X	X	X	X	X	X	X		X	X	X		X	X	X		X	X	X	X	X	X	X
Don't know			X			X	X								X				X											
Disagreement																							X							

TABLE 59.1. (*Continued*)

	NJ	NM	NV	NY	OH	OK	OR	PA	RI	SC	SD	TN	TX	UT	VA	VT	WA	WI	WV	WY
Legal status																				
Game					X													X		
Domestic				X		X		X			X	X	X			X	X			
Unclassified																				X
Game and domestic		X																		
Disagreement	X		X				X		X	X				X	X				X	
Regulated shipments within state																				
No								X											X	
Yes		X	X		X				X		X	X				X				X
Don't know																				
Disagreement	X			X		X	X			X			X	X	X		X	X		
Regulated shipments into state																				
No													X						X	
Yes	X	X	X	X	X		X		X	X	X	X		X		X				X
Don't know																				
Disagreement						X		X							X		X	X		
Agency with jurisdiction																				
Wildlife	X	X			X	X		X		X		X			X					X
Agricultural				X					X		X					X	X			
Wildlife/agri-cultural			X				X						X	X				X		
Disagreement																			X	
Tax credit program																				
No	X	X	X			X	X	X	X	X	X	X		X			X	X	X	
Yes												X				X				
Don't know				X	X															X
Disagreement															X					
Tax exemption program																				
No	X	X	X			X	X	X	X	X	X	X		X	X	X		X	X	
Don't know				X	X								X				X			X
Disagreement																				

TABLE 59.1. (*Continued*)

	AK	AL	AR	AZ	CA	CO	CT	DE	FL	GA	HI	IA	ID	IL	IN	KS	KY	LA	MA	MD	ME	MI	MN	MO	MS	MT	NC	ND	NE	NH
Cost share program																														
No	X	X		X	X	X		X	X	X	X	X	X	X		X	X	X		X	X	X	X	X	X	X	X	X	X	X
Don't know			X				X								X				X											
Sold live in state																														
No		X						X												X										
Yes				X	X	X	X		X		X	X		X		X		X	X		X	X	X	X	X	X	X		X	X
Don't know			X										X		X															
Disagreement	X									X							X											X		
Opportunities on public lands																														
No		X	X	X	X	X	X	X	X	X		X	X			X		X	X	X	X	X	X	X	X	X	X	X	X	X
Yes											X						X													
Don't known															X															
Disagreement	X													X																
Processed meat sold in state																														
No		X																		X					X		X			
Yes	X			X	X	X	X	X	X	X	X	X		X		X	X	X				X	X	X		X			X	X
No response			X										X		X				X		X							X		
Meat function required																														
No		X				X	X	X	X		X		X		X			X							X		X	X		X
Yes			X							X		X				X	X				X	X		X		X				
Don't know				X															X	X										
Disagreement	X				X									X									X						X	
Facilities regulated																														
No		X				X				X		X			X					X		X			X		X	X		
Yes				X					X		X		X				X	X	X		X			X						
Don't know			X													X														
Disagreement	X				X		X	X						X									X			X			X	X
States which have recent or pending state legislation	X			X	X	X				X	X			X							X								X	

TABLE 59.1. (*Continued*)

	NJ	NM	NV	NY	OH	OK	OR	PA	RI	SC	SD	TN	TX	UT	VA	VT	WA	WI	WV	WY
Cost share program																				
No	X	X	X			X	X	X		X	X	X		X	X	X		X	X	
Don't know				X	X				X				X				X			X
Sold live in state																				
No									X	X									X	X
Yes		X	X	X	X	X	X	X			X	X	X	X	X	X	X			
Don't know																				
Disagreement	X																	X		
Opportunities on public lands																				
No	X		X	X	X	X	X	X			X	X		X		X	X	X	X	X
Yes													X		X					
Don't know																				
Disagreement		X							X	X										
Processed meat sold in state																				
No										X										
Yes		X	X	X	X	X	X	X	X		X	X	X		X	X	X	X	X	X
No response	X													X						
Meat inspection required																				
No	X		X			X		X		X	X			X				X	X	
Yes		X			X							X	X			X				
Don't know									X											
Disagreement				X			X								X		X			X
Facilities regulated																				
No				X	X			X			X		X				X		X	
Yes		X							X					X		X				
Don't know																				
Disagreement	X		X			X	X			X		X			X			X		X
States which have recent or pending state legislation				X	X		X		X		X		X							

ranching where allowed, none of the respondents from these 3 states clarified whether exotic deer production was considered a "normal" agricultural or farm enterprise. State agencies did not differ ($P > 0.05$) in their knowledge concerning the availability of a tax credit program. Wildlife (66.7%), meat inspection (58.3%), and marketing (51.7%) agencies were equally aware of this topic.

When asked whether exotic deer production qualifies for a state-sponsored tax exemption program, responses indicated that 39 states do not, 10 states had agencies which "don't know," and 1 state had disagreement among agencies. Agencies did not differ ($P > 0.05$) in the knowledge concerning the availability of a tax exemption program. Wildlife (66.7%), meat inspection (52.8%), and marketing (58.6%) agencies were equally aware of this topic.

When asked whether exotic deer production qualifies for a state-sponsored cost-sharing program responses indicated that 40 states do not, and 10 states had agencies which "don't know." State agencies did not differ ($P > 0.05$) in their knowledge concerning the availability of a state-sponsored cost-sharing program. Wildlife (66.7%), meat inspection (52.8%), and marketing (58.6%) agencies were equally aware of this topic.

Exotic Deer Sold Live

When asked whether exotic deer may be sold live within their states, the responses indicated that 34 states allowed exotic deer to be sold live, 7 states do not, 3 states had agencies which "don't know," and 6 states had disagreement among agencies. State agencies did not differ ($P > 0.05$) in their knowledge concerning the sale of exotic deer or their products. Wildlife (86.7%), meat inspection (72.2%), and marketing (75.9%) agencies were equally aware of this topic.

Exotic Deer Hunting Opportunities on Public Lands

When asked if exotic deer hunting opportunities are available on public lands in their states, the responses indicated that 40 states do not have exotic deer hunting opportunities on public lands, 4 states do have opportunities, 1 state had agencies which "don't know," and 5 states had disagreement among agencies. State agencies differed ($P < 0.005$) in their knowledge concerning the availability of exotic deer hunting on public lands. Wildlife agencies (100%) knew most often whether this opportunity existed, compared with meat inspection (69.4%) and marketing (75.9%) agencies.

Processed Exotic Deer Meat Sold in State

When asked if processed exotic deer meat may be sold in their states, the responses indicated that 5 states (Alabama, North Carolina, Maryland, Mississippi, and South Carolina) did not allow the sale of processed exotic deer meat from supermarkets, restaurants, farmer's markets, telephone/mail-order, and/or meat markets, while the remaining 37 states responding to this question did allow the sale of processed exotic deer meat in at least one of the markets as long as inspection standards are met. Oregon and Virginia indicated that although meat can be sold in these states, this meat must be imported and not slaughtered in the state.

State or Federal Inspection of Exotic Deer Meat in the State

When asked if there is a requirement for state or federal inspection of exotic deer meat in their states, the responses indicated that 14 states required either a state or federal inspection of the meat; 22 states did not require inspections; 4 states had agencies which "don't know," and 10 states had disagreement among agencies. Results indicated that state agencies differed ($P < 0.005$) in their knowledge concerning the inspection requirements for exotic meat. Meat inspection agencies (97.3%) knew more often whether this requirement existed compared with wildlife (59.1%) and marketing (69.0%) agencies.

State Regulations Governing Exotic Deer Production Facilities

When asked if there are special state regulations governing exotic deer production facilities, the responses are as follows: 13 states had regulations which in some way govern production facilities; 17 states did not have regulations; 2 states had agencies which "don't know"; and 18 states had disagreements among agencies. Elaboration of responses from several states reveals the following: (1) facilities must allow reasonable freedom of movement (California); (2) exotics can be processed in state-licensed slaughter facilities but they must be kept separate by space and time, and must be labeled "not for sale" (Illinois); (3) specifications are held on fence height, shelters, and food and water systems which prevent exotics from being in contact with other wildlife (Missouri); and (4) facilities must be escape-proof and of adequate size to allow freedom of movement (New Jersey and Wyoming). State agencies did not differ ($P > 0.05$) in their knowledge concerning production facilities. Meat inspection (81.1%), wildlife (86.4%), and marketing (75.9%) agencies were equally aware of this topic.

Agency Perception of Exotic Deer

State agencies differed ($P < 0.001$) in their perceptions of exotic deer. Of the 42 wildlife agencies responding to the survey, 32 (76%) considered them to be potential competitors with native wildlife (i.e., pests). No wildlife agencies thought of them as having attractive qualities. Meat inspection (18%) and marketing (31%) agencies also recognized the potential of exotic deer as competitors with native wildlife, however, some meat inspection (33%) and marketing (31%) agencies indicated positive perceptions of exotics.

Recent or Pending State Legislation

Fifteen states had either recent or pending state legislation which affected the status of exotics in the state. Elaboration of responses from several states revealed the following: (1) regulations are being modified in Nebraska whereby only mule (*Odocoileus hemionus*) and white-tailed (*Odocoileus virginianus*) deer will be required to be covered by a permit (exotic deer will be exempt); (2) fallow deer (*Cervus dama*) have recently been approved for commercial farming in Maryland (4/6/88), and the department of agriculture is finalizing rules for licensing and regulating fallow deer farming and processing; (3) California is currently revising its regulations regarding exotic animals and deer may no longer be kept as pets, while breeding and exhibitors must specify and document their experience breeding or raising deer; and (4) permits are required for elk (*Cervus elaphus*) farming in Alaska.

Interest in Receiving Survey Results

State agencies had equally high ($P > 0.05$) interest levels for receiving the results of the survey. Of the agencies responding, 84% (36 states) of the wildlife agencies, 79% (22 states) of the marketing agencies, and 86% (31 states) of the meat inspection agencies wanted to receive the survey results.

Discussion

Long-term security for investment in the exotic deer industry could be related to the regulatory "climate" of a particular state. One method of gauging this "climate" could be to examine the perception of exotic deer by the agency with regulatory authority. The agency which most often had jurisdiction over exotic deer also most often had a negative perception of this resource. Potential investors should consider this aspect on a state-by-state basis.

The survey results identified several topics which could be addressed by private and/or public sources with an interest in the exotic deer industry. Those sources interested in promoting expansion of the industry should note the low number of states offering state-sponsored incentive programs. Only 4 states offered one or more of the tax credit, tax exemption, or cost-sharing programs listed in the survey. The lack of specific legal classification for exotic deer in 11 states would be of interest both to sources promoting the industry and to those sources interested in retarding industry expansion.

Perhaps the single most important fact developed from this questionnaire concerns the general lack of knowledge among state agencies which either are or could be directly involved in regulating the exotic deer industry. This lack of knowledge was most evident among the meat inspection and agricultural marketing agencies, although lack of knowledge appeared to be related to the type of question asked. The wildlife agency was most often the most knowledgeable, particularly concerning questions related to regulations. The high level of interest in receiving the results of this survey could be attributed to a general need for information by the respective agencies.

The exotic deer industry has expanded within the United States and this will probably continue as long as the market for domestic venison remains high. The political "climate" will be a factor in the extent of expansion. Fifteen states had recent or pending legislation which would affect the legal states of exotic deer. The results of these and future political activities will determine the future status of exotic deer production as an alternative agricultural enterprise within the United States.

Acknowledgements

This is Texas Tech University College of Agricultural Sciences publication no. T-1-319.

Reference

Demarais S, Osborn DA, Jackley JJ (1990) Exotic big game: a controversial resource. Rangelands 12: 121–125

60

Harvest Considerations for Free-Ranging Venison Production in Texas

David A. Osborn, Stephen Demarais, and David M. Hughes, Jr

Abstract

Axis deer (*Axis axis*) ($n = 108$), fallow deer (*Dama dama*) ($n = 92$), and sika deer (*Cervus nippon*) ($n = 93$) were harvested from common range in the Edwards Plateau region of Texas during seasonal (summer and winter) sampling periods from December 1987 through August 1989. All deer were female and estimated to be 1.5, 2.5, or $\geqq 3.5$ years of age. Free-ranging female fallow and sika deer retained past 1.5 years of age for venison production utilized resources with little increase in value. Free-ranging female axis deer gained carcass weight with additional age; however, this weight gain may not have justified the continued investment of forage resources. Higher mean carcass weights of female sika deer during winter compared with summer may have reflected changes in body fat reserves and should be considered when scheduling venison harvests.

Key words: Axis deer, dressing percentage, fallow deer, Edwards Plateau, kidney fat index, sika deer, Texas

Introduction

Commercial harvest of free ranging exotic deer in Texas provides supplemental income to ranchers and is an effective means of population control. Ranchers can maximize net returns from exotics sold for meat by harvesting individuals at an optimal age. Harvest should take place when production costs (e.g., additional investment of forage resources) and other market factors indicate that producing heavier carcass weights (the weight for which the rancher is paid), higher dressing percentages (the proportion of the animals' live weight which is carcass), or improved venison quality will not be cost-effective. The objectives of this research were: (1) to assess the effects of age and season on selected body weight measurements and kidney fat deposits of axis, fallow, and sika deer in the Edwards Plateau region of Texas, and (2) to discuss age- and season-related harvest considerations relative to venison quality and potential economic return from free-ranging production.

Methods

Data were collected from up to 108 axis deer, 92 fallow deer, and 93 sika deer. Deer were harvested by the Texas Wild Game Cooperative, All That's Deer, Inc., and Texas Tech University from four privately owned ranches located in Kerr and Real Counties, Texas, during seasonal—summer, 15 July to 15 August; winter, 15 December to 15 January—sampling periods from December 1987 through August 1989. These ranches were predominantly rangeland used for the production of domestic livestock and wildlife and were considered representative of the Edwards Plateau region (Landers 1987). Deer were shot in the head with a large caliber rifle and transported to a portable slaughter facility for processing. All deer were female and classified as either 1.5, 2.5, or ≧ 3.5 years of age based on tooth eruption and wear criteria (Graf and Nichols 1966; Chaplin and White 1969; Duff 1969). The mean and standard deviation of animals' ages in the ≧ 3.5 years of age group were: axis deer, 5.4 and 1.5; sika deer, 5.1 and 1.5; and fallow deer, 5.5 and 1.6.

Live weight, defined as the weight of a freshly harvested animal minus blood loss from the wound, was recorded using a spring scale. An electronic scale was used to determine carcass weight prior to chilling, defined as the weight of an eviscerated animal minus the head, hide, and feet. All weights were recorded to the nearest 0.45 kg. Dressing percentage (dressing %) was calculated by dividing the carcass weight by live weight for each animal and multiplying by 100. Riney's (1955) kidney fat index (KFI) was used as an index to body fat (Finger et al. 1981).

Analysis of variance in a completely randomized design with a factorial arrangement of age and season was used to determine the effects of age and season on live weight, carcass weight, dressing %, and KFI of axis, fallow, and sika deer. Data were pooled across ranches and years. Analyses of KFI for each deer species and dressing percentage for fallow deer were conducted on rank-transformed data to remedy violations of normality and/or homogeneity of variances among the residuals for the raw data, as indicated by the Kolmogorov-Smirnov test and Bartletts' test. When the F-test on main effects was significant ($P < 0.05$), differences among means were identified using Fisher's least significant difference test.

Results and Discussion

Age Class Effects

Age class affected ($P < 0.05$) live weights of axis deer, fallow deer, and sika deer (Table 60.1). Live weights of axis and fallow deer increased ($P < 0.05$) through the ≧ 3.5 year age class, whereas sika deer did not increase ($P > 0.05$) beyond the 2.5 year age class. Carcass weight changed ($P < 0.05$) with age class, but only in axis deer (Table 60.2), with the highest value at the ≧ 3.5 age class. Carcass weights of fallow deer and sika deer did not increase ($P > 0.05$) after 1.5 years of age (Table 60.2). The percentage of dressing changed ($P < 0.05$) with age class, but only in sika deer (Table 60.3). The percentage of dressing in sika deer did not increase beyond the 2.5 year age class. Age class affected ($P < 0.05$) KFI, but only in fallow deer, with a decline in KFI beyond the 2.5 year age class (Table 60.4).

The increased live weights of free-ranging female axis and fallow deer through the ≧ 3.5

TABLE 60.1. Effects of age class and season on live weight of female axis, fallow, and sika deer from the Edwards Plateau region of Texas, December 1987 to August 1989

		Age class											
		1.5			2.5			≧ 3.5			Combined[a]		
Species	Season	n	$\bar{x}$	SE	n	$\bar{x}$	SE	n	$\bar{x}$	SE	n	$\bar{x}$	SE
Axis deer	Summer	10	37.2	1.1	16	41.5	0.7	28	47.9	0.8	54	44.0	0.8
	Winter	12	40.2	1.2	9	45.1	1.4	33	47.7	0.6	54	45.6	0.7
	Combined[b]	22	38.8A	0.9	25	42.8B	0.7	61	47.8C	0.5	108	44.8	0.5
Fallow deer	Summer	16	30.1	0.9	6	35.2	1.4	27	37.6	0.9	49	34.9	0.8
	Winter	4	33.9	1.2	11	36.2	0.8	28	38.9	0.8	43	37.7	0.6
	Combined[b]	20	30.9A	0.8	17	35.9B	0.7	55	38.3C	0.6	92	36.2	0.5
Sika deer	Summer	8	31.9	1.4	8	37.0	1.9	32	39.1	0.8	48	37.5	0.8
	Winter	6	34.0	1.0	11	40.3	1.6	28	39.0	0.8	45	38.7	0.7
	Combined[b]	14	32.8A	0.9	19	38.9B	1.3	60	39.0B	0.6	93	38.1	0.5

[a]Season effect on live weight of axis deer ($P < 0.05$)

[b]Age effect on live weight ($P < 0.0001$); means within the "combined" row for each variable with the same letter are not different ($P > 0.05$)

TABLE 60.2. Effects of age class and season on carcass weight of female axis, fallow, and sika deer from the Edwards Plateau region of Texas, December 1987 to August 1989

		Age class											
		1.5			2.5			≧3.5			Combined[a]		
Species	Season	n	$\bar{x}$	SE	n	$\bar{x}$	SE	n	$\bar{x}$	SE	n	$\bar{x}$	SE
Axis deer	Summer	10	19.9	0.9	16	22.4	0.5	28	25.0	0.5	54	23.3	0.4
	Winter	12	20.0	0.9	9	22.9	0.8	32	23.8	0.4	53	22.8	0.4
	Combined[b]	22	20.0A	0.6	25	22.6B	0.5	60	24.3C	0.3	107	23.0	0.3
Fallow deer	Summer	17	14.9	0.6	6	15.8	0.9	27	16.5	0.5	50	15.8	0.3
	Winter	4	15.3	1.0	10	16.0	0.8	28	17.1	0.5	42	16.7	0.4
	Combined[b]	21	15.0	0.5	16	15.9	0.6	55	16.8	0.4	92	16.2	0.3
Sika deer	Summer	8	16.0	1.2	8	17.8	1.4	30	18.1	0.5	46	17.7	0.5
	Winter	6	18.3	0.8	11	20.5	1.3	28	18.4	0.5	45	18.9	0.5
	Combined	14	17.0	0.8	19	19.4	1.0	58	18.2	0.4	91	18.3	0.3

[a]Season effect on carcass weight of sika deer ($P < 0.05$)
[b]Age effect on carcass weight ($P < 0.0001$); means within the "combined" row for each variable with the same letter are not different ($P > 0.05$)

TABLE 60.3. Effects of age class and season on dressing percentage of female axis, fallow, and sika deer from the Edwards Plateau region of Texas, December 1987 to August 1989

		Age class											
		1.5			2.5			≧3.5			Combined[a]		
Species	Season	n	x	SE	n	$\bar{x}$	SE	n	$\bar{x}$	SE	n	$\bar{x}$	SE
Axis deer	Summer	10	53.2	1.1	16	54.0	0.6	28	52.3	0.6	54	52.9	0.4
	Winter	12	49.7	1.1	9	50.9	1.1	32	49.8	0.6	53	50.0	0.5
	Combined	22	51.3	0.9	25	52.9	0.6	60	50.9	0.4	107	51.5	0.5
Fallow deer	Summer	16	49.8	1.3	6	44.6	1.4	27	43.9	0.9	49	45.9	0.8
	Winter	4	45.3	2.5	10	44.4	1.7	28	44.0	0.9	42	44.2	0.7
	Combined	20	48.9	1.2	16	44.5	1.1	55	43.9	0.6	91	45.0	1.0
Sika deer	Summer	8	49.8	2.4	8	47.9	1.7	30	46.5	0.8	46	47.3	0.7
	Winter	6	54.1	3.0	11	50.5	1.7	28	47.0	0.8	45	48.8	0.9
	Combined[b]	14	51.7A	1.9	19	49.4A	1.3	58	46.7B	0.6	91	48.0	1.0

[a]Season effect on dressing percentage of axis deer ($P < 0.0001$)
[b]Age effect on dressing percentage ($P < 0.01$); means within the "combined" row for each variable with the same letter are not different ($P > 0.05$)

year age class are consistent with reports for female white-tailed deer (*Odocoileus virginianus*) in Texas (Knowlton et al. 1979), mule deer (*O. hemionus*) in Colorado (Anderson et al. 1974), and red deer (*Cervus elaphus*) in New Zealand (Caughly 1971). Growth rate in deer (Wood et al. 1962; Kay 1985; Adam 1989) and herbivores in general (Bogart and Taylor 1983) declines following puberty. After maturation, changes in live weight may reflect fat deposits, water stores, rumen fill, and reproductive status (Bogart and Taylor 1983; Kelly et al. 1987). The lack of change in sika deer beyond the 2.5 year age class could have been due to either earlier maturation or a confounding effect of fat deposition, water stores, or rumen fill.

Carcass weight and percentage of dressing data indicate that free- ranging female axis, fallow, and sika deer do not produce significant additional muscle mass beyond the 1.5 year age class. Comparisons with similar data are limited due to the preponderance of males used in other studies.

TABLE 60.4. Effects of age class and season on kidney fat index (KFI) of female axis, fallow, and sika deer from the Edwards Plateau region of Texas, December 1987 to August 1989

		Age class											
		1.5			2.5			≧3.5			Combined[a]		
Species	Season	n	$\bar{x}$	SE	n	$\bar{x}$	SE	n	$\bar{x}$	SE	n	$\bar{x}$	SE
Axis deer	Summer[b]	10	26.6	4.9	16	41.3	9.9	28	50.3	5.3	54	43.2	4.3
	Winter[b]	12	64.1	14.2	9	69.1	21.2	34	44.9	5.4	55	53.1	5.7
	Combined	22	47.0	8.9	25	51.3	10.0	62	47.3	3.8	109	48.2	3.6
Fallow deer	Summer	17	71.6	7.1	6	35.7	9.0	27	34.5	3.9	50	47.2	4.1
	Winter	4	107.3	7.5	11	82.1	24.0	26	56.9	6.8	41	68.6	8.0
	Combined[b]	21	78.4A	6.6	17	65.7B	16.5	53	45.5B	4.2	91	56.9	4.4
Sika deer	Summer	8	75.1	7.3	8	66.3	22.7	32	66.4	10.8	48	67.8	8.1
	Winter	7	108.6	25.7	12	124.9	25.3	25	102.5	8.8	44	109.6	9.3
	Combined	15	90.7	12.9	20	101.4	18.5	57	82.2	7.5	92	87.8	6.5

[a]Season effect on KFI of fallow deer ($P < 0.001$) and sika deer ($P < 0.01$)
[b]Season and age-season interaction effects on KFI of axis deer ($P < 0.05$); season effect on KFI for 1.5 year ($P < 0.05$) and ≧ 3.5 year ($P < 0.01$) age classes
[c]Age effect on KFI ($P < 0.05$); means within the "combined" row for each variable with the same letter are not different ($P > 0.05$)

Significant differences in body size and growth patterns existed between the sexes in free-ranging mule deer (Anderson et al. 1974) and farmed red deer (Kelly et al. 1987). Heavier carcass weights in axis deer and greater percentage of dressing in sika deer beyond the 1.5 year age class could have been due to fat deposition, although this was not indicated by the KFI data.

Only fallow deer exhibited an age-related change in fat deposition, as indexed by KFI. The lower KFI in the ≧ 3.5 year age class could have been associated with the demands of lactation. The energy required for the lactation period of red deer hinds is 43% of their annual energy requirement (Kelly et al. 1987).

Age-related variations in body characteristics of the free-ranging deer sampled in this study could be different from those present under intensively farmed conditions, due to differences in forage quantity and/or quality. These differences could have been further impacted by our sampling of animals during the two most stressful periods of the year in Texas.

Effect of the Season of Harvest

Season of harvest affected ($P < 0.05$) live weight, but only in axis deer (Table 60.1), with live weights being generally greater during winter. Carcass weight was greater ($P < 0.05$) for winter-harvested sika deer and did not differ ($P > 0.05$) between seasons for axis deer or fallow deer (Table 60.2). The percentage of dressing was greater ($P < 0.05$) for summer harvested axis deer (Table 60.3). Winter KFI values were greater ($P < 0.05$) than summer KFI values for all three species (Table 60.4).

Seasonal increases and decreases in the body weight of deer are primarily associated with reproductive activity but also may represent voluntary shifts in feeding activity in response to metabolic changes and/or forage availability and quality during critical stress periods (Fowler et al. 1967; Short 1975; Mautz 1978; King and Murphy 1985; Schwartz et al. 1987). Weight changes associated with season differ between sexes in Cervidae (Kelly et al. 1987; Schwartz et al. 1987), and these changes are generally more abrupt for males than females (Wood et al. 1962; Adam 1989).

Our data indicated free-ranging female fallow and sika deer had higher KFI values and potentially more body fat during winter than during summer. The lack of a consistent seasonal pattern in axis deer could have been due to their year-round reproductive cycle (Howery et al.

1989) and/or their tendency for limited fat deposition (Graf and Nichols 1966).

Harvest Recommendations

Optimum harvest time depends on a variety of criteria, including market demand, producer supplies, and carcass characteristics. Specific considerations relative to the carcass include seasonal- and age- related variations in quality (i.e., fat content) and the costs associated with animal maintenance.

The increased carcass fat content associated with maturity in red deer stags (Kelly et al. 1987) would not be a harvest consideration for free-ranging fallow and sika does in the Edwards Plateau region during summer and winter. Seasonal changes in body fat content could be a consideration when scheduling harvest for axis, fallow, and sika, with winter levels being higher than summer. However, data are not currently available which would allow clear incorporation of our seasonal fat deposition results into the harvest decision-making process. Additionally, the importance of our reported age-and season-related fat content variations would have to be evaluated relative to other factors, such as market demand and producer supplies.

The cost of maintaining a free-ranging animal on rangelands is an important economic component in the scheduling of harvest. Assuming a grazing fee of $5.00 per year and a stocking rate of 1 animal unit per 8.1 ha, which are fairly generalized values for the Edwards Plateau region (R.Q. Landers, personal communication), an animal unit month of grazing costs of $8.30 US. Assuming that an "average" exotic deer is equivalent to 1/6 animal unit, it would cost $16.60 just for the forage costs to maintain an animal for an additional year. The costs would be even greater if discounting procedures (i.e., net present value) and lost opportunity costs were considered.

Producers in the Edwards Plateau region are paid by carcass weight at the time of harvest. Assuming an average price to producers of $4.40 per kg, the average increased value for each additional year of growth was calculated. The carcass value of axis deer would increase by $11.44 from the 1.5 to the 2.5 year age class and by $7.48 from the 2.5 to the ≧ 3.5 year age class. Because there were no statistically significant increases in fallow and sika deer carcass weights, no increased value to these species could be assigned for an additional year of growth.

The increased axis deer carcass value associated with an additional year of growth does not even cover the forage costs for that animal. Taking into consideration discounting procedures and lost opportunity costs would further imbalance this relationship for axis deer.

We conclude that for venison production there is no economic justification for maintaining free-ranging axis, fallow, and sika deer beyond the 1.5 year age class. It must be understood that our conclusions apply only to free-ranging female axis, fallow, and sika deer harvested during winter and/or summer in the Edwards Plateau region of Texas. Conclusions may differ for free-ranging females during other seasons, and for males or farm-raised deer.

Acknowledgements

We thank the owners and managers of the Y.O., Bowman, Johnson, and Two-Dot ranches for their hospitality and assistance. We also thank the Texas Wild Game Cooperative, All That's Deer, Inc., J.J. Jackley, M. Bailey, and many students from Texas Tech University for assistance in data collection. We thank G.W. Davis, J.J. Jackley, and R.T. Ervin for manuscript review. Funding for this research was provided by the Exotic Wildlife Association, various private landowners, Texas Tech University, and the Houston Livestock Show and Rodeo Association. This is manuscript T-9-593 of the College of Agricultural Sciences, Texas Tech University.

References

Adam JL (1989) Growth. In Allen PL, Asher GW (eds) Progressive Fallow Farming: Proceedings of a Course on Fallow Deer Farming. Ruakura Agricultural Centre, Hamilton, New Zealand, pp 47–55

Anderson AE, Medin DE, Bowden DC (1974) Growth and morphometry of carcass, selected bones, organs, and glands of mule deer. Wildl Monogr 39, Wildlife Society, Washington, D.C.

Bogart R, Taylor RE (1983) Scientific Farm Animal Production. Burgess Publishing, Minneapolis, Minnesota

Caughley G (1971) Demography, fat reserves and body size of a population of red deer (*Cervus elaphus*) in New Zealand. Mammalia 35:369–383

Chaplin RE, White RWG (1969) The use of tooth eruption and wear, body weight and antler characteristics in the age estimation of male wild and park fallow deer (*Dama dama*). J Zool (London) 157:125–132

Duff KR (1969) Tooth eruption as a guide to ageing Japanese sika deer (*Cervus nippon*) in Dorset. Deer 2:566–567

Finger SE, Brisbin Jr IL, Smith MH (1981) Kidney fat as a predictor of body condition in white-tailed deer. J Wildl Manage 45:964–968

Fowler JF, Newsom JD, Short HL (1967) Seasonal variation in food composition and weight in male and female white-tailed deer. Proc Annu Conf SE Assoc Game Fish Comm 21:24–31

Graf W, Nichols Jr L, (1966) The axis deer in Hawaii. Bombay Nat Hist Soc 63:630–734

Howery LD, Pfister JA, Demarais S (1989) Seasonal reproductive activity of 4 exotic ungulates in Texas. J Wildl Manage 53:613–617

Kay RNB (1985) Body size, patterns of growth, and efficiency of production in red deer. In Fennessey PF, Drew KR (eds) Biology of Deer Production. R Soc N Z Wellington Bull 22:411–421

Kelly RW, Fennessy PF, Moore GH, Drew KR, Bray AR (1987) Management, nutrition, and reproductive performance of farmed deer in New Zealand. In Wemmer CM (ed) Biology and Management of the Cervidae. Smithsonian Institution Press, Washington, D.C. pp 450–460

King JR, Murphy ME (1985) Periods of nutritional stress in the annual cycles of endotherms: fact or fiction. Am Zool 25:955–964

Knowlton FF, White M, Kie JG (1979) Weight patterns of wild white-tailed deer in southern Texas. In Drawe LD (ed) Proc First Welder Wildl Found Symp. Welder Wildlife Foundation, Contribution B-7, Corpus Christi, Texas, pp 55–64

Landers RQ (1987) Native vegetation of Texas. Rangelands 9:203–207

Mautz WW (1978) Sledding on a bushy hillside: the fat cycle in deer. Wildl Soc Bull 6:88–90

Riney T (1955) Evaluating condition of free-ranging red deer, with special reference to New Zealand. N Z J Sci Tech 36:428–463

Schwartz CC, Regelin WW, Franzmann AW (1987) Seasonal weight dynamics of moose. Swedish Wildl Res (suppl) 1:301–310

Short HL (1975) Nutrition of southern deer in different seasons. J Wildl Manage 39:321–329

Wood AJ, Cowan IMcT, Nordan HC (1962) Periodicity of growth in ungulates as shown by deer of the genus *Odocoileus*. Can J Zool 40:593–603

Part 5
Reproduction

61
Reproduction in Deer: Adaptations for Life in Seasonal Environments

Andrew S.I. Loudon and Barry R. Brinklow

Abstract

Deer became a distinct family in the early Miocene era, evolving in high latitudes at a time of global cooling. Many species that now live in temperate latitudes have well-developed physiological mechanisms to suit them to a seasonally changing environment. In addition, some species that live at low tropical latitudes with a less marked seasonal environment also show a degree of inherent rhythmicity. Deer living in temperate latitudes show overt seasonal fluctuations in reproduction, metabolism, appetite, growth, and pelage growth and moult. These seasonal cycles are closely linked temporally and are difficult to separate experimentally. The predictable changes in the length of the photoperiod have been shown to be the major environmental factor controlling seasonal cycles via the secretion of the hormone melatonin during the night. Treatment with melatonin induces phase shifts in all of these rhythms. Recent studies have shown, however, that the seasonal red deer (*Cervus elaphus*) exhibits an overt rhythm in its metabolism, coat growth, and associated endocrine parameters when kept on constant midsummer day lengths from birth. This rhythm is similar to that of animals on natural day lengths and indicates that the rhythm is endogenous. In some tropical species such as the Axis deer (*Axis axis*), stags exhibit a circannual cycle of antler growth, testis growth and regression, and body weight change. These cycles are not synchronized to the environment, even when animals are kept in temperate latitudes or treated with melatonin. Thus, tropical species may retain many of the endogenous rhythmicity evolved by their temperate ancestors.

Key words: Appetite, circannual rhythms, gonadotrophins, growth, puberty, reproduction, seasonality

Introduction

Distribution and Timing of Breeding Seasons

Extant species of deer currently account for 40 out of the 176 species of the Ruminantia. The Cervidae became a distinct family within the Ruminantia some 40 million years ago, appearing in the early Miocene Eurasian fauna. The family evolved in the northern hemisphere at a time of general global cooling and increased seasonality of climate and food supply. Their current distribution extends from the cold polar regions (up to latitude 80° north in the case of the reindeer) to tropical regions. In general, there is a close correlation between latitude and the timing of the breeding season, and species that occur between 20° north and 20° south experience little seasonality in their natural habitats and remain aseasonal breeders even after transfer to temperate environments (Whitehead 1972 and Zuckerman 1953, reviewed by Lincoln 1985).

A number of species exhibit distinct geographical variations in the pattern and timing of births. Thus, the southern Indian races of the Axis deer exhibit little or no synchrony of breeding or of the antler cycle (Phillips 1928; Morris 1935) while northern Indian and Nepalese races exhibit a seasonal birth peak

(Schaller 1967; Mishra and Wemmer 1987). Similarly, the most widely distributed species, the American white-tailed deer (*Odocoileus virginianus*), which has a distribution extending from 55° north to 18° south is highly seasonal in the temperate zone and in the tropical zone breeds throughout the year (Blouch 1987). Clearly, latitudinal differences in the timing of the breeding season are a reflection of selection processes which have ensured that periods of high energy requirements, such as late pregnancy and early lactation, coincide with periods of maximal food supply.

Evidence from studies of wild red deer living in impoversihed habitats in northern Britain indicate that small natural differences in the timing of birth are associated with significant differences in survival (Clutton-Brock et al. 1982). Late-born claves have a lower probability of survival through the winter as they have low body weights due to shortened lactation (Clutton-Brock et al. 1982). Similarly, early-born calves are significantly lighter and have a lower probability of surviving over the summer (Blaxter and Hamilton 1980; Clutton-Brock et al. 1982). Thus, birth seasons in seasonal habitats are compressed by processes that select against a genetic predisposition to produce young outside the optimal period.

Growth and survival of deer born in seasonal habitats depends directly upon maternal milk yield (Loudon and Kay 1984, Robbins et al. 1987). In female deer (as in other ruminants), the energy requirements of lactation exceed those of pregnancy (Arman et al. 1974; Loudon and Kay 1984; Robbins et al. 1987), and it is at this stage of the reproductive cycle that the food requirements of the mother are normally maximal. In seasonal species, such as the red deer, maternal food intake at peak lactation may exceed that of nonlactating animals by a factor of more than two (Arman et al. 1974). Most species of deer have low fat reserves relative to domesticated ruminants. Consequently, they are not able to utilize substantial body reserves to support lactation and must rely on current food intake for the production of milk. In free-ranging red deer, the type and amount of vegetation available to the mother has a profound effect on her grazing behavior and milk production and the growth of her calf (Loudon et al. 1984). An exception appears to be the Svalbard reindeer (*Rangifer tarandus*), on the Island of Spitzerbergen, which occupies the most extreme seasonal environment of any deer and accumulates substantial fat reserves (up to 29% of body weight; Tyler 1987) which may be mobilized during lactation. Thus, the timing of the breeding season is ultimately constrained by energetic considerations, in particular the temporal pattern of food availability and its affects on maternal food intake, milk yield, and the consequent growth of the calf.

In the larger deer species of approximately 100 kg or more, peak lactation and maternal food intake do not occur until at least 40 days after birth (Loudon et al. 1983; Robbins et al. 1987). As a consequence, animals such as red deer with a calving date in early June (northern hemisphere) normally exhibit peak energy requirements in late June/early July. At this time of year, food intake and daily grazing times of lactating hinds are maximal (Loudon et al. 1984; Heydon et al. this volume). However, in agricultural systems, maximal grass production rates occur several weeks before the peak in food requirement of the lactating hind. In contrast, in their natural habitat of forests and forest-edge the phenology and development of wild grasses occur later in the year and at a more appropriate time for peak food intake. The mismatch of plant production and animal food requirements has profound effects on the economics of farmed red deer. For this reason, considerable research effort has been devoted to the study of the physiological mechanisms underlying the timing of the breeding season in domesticated deer.

Adaptations for Life in Seasonal Environments

Reproduction

Role of Gonadotrophins

In animals with relatively long gestation lengths such as deer, the timing of mating must be synchronized to occur at a time of year different from that best suited to the production of young. Proximal environment cues are used by the animals to time the endocrine changes necessary for mating and conception. In domestic ruminants, seasonal changes in the secretion of the gonadotrophins luteinizing hormone (LH) and follicle stimulating hormone (FSH) control the timing of the onset and termination of the breeding season. Both hormones are known to be regulated by the hypothalamic decapeptide, gonadotrophin releasing hormone (GnRH). Luteinizing hormone is secreted in discrete pulses (Katangole et al. 1971), and each episode is a direct consequence of an episodic release of GnRH (Clarke and Cummins 1982). In the red deer stag, seasonal changes in the pulsatile secretion of LH and the release of FSH are associated with a seasonal cycle of development and regression of the testis (Lincoln 1971) and testosterone secretion (Suttie et al. 1984). Lincoln (1985) summarized the role of LH and GnRH in the control of the seasonal sexual cycle in male deer. For this reason, the following discussion refers predominantly to female deer.

LH Secretion During the Breeding Season

Experiments on sheep have demonstrated that the feedback of ovarian steroids controls the pulsatile secretion of GnRH from the hypothalamus. In sheep during the breeding season, the sustained rise in estrogen secretion from developing follicles in the ovary causes an increase in LH pulse frequency which ultimately results in a surge of LH from the pituitary and ovulation (Baird 1978; Karsch et al. 1979). Following ovulation, the GnRH pulse generator is "clamped" by exposure to ovarian steroids, principally progesterone, and episodic secretion of LH is reduced (Goodman and Karsch, 1980). There is, however, a dramatic increase in the amplitude of LH pulses at this time (Goodman and Karsch 1980, 1981). Following luteolysis and the decline in progesterone, the frequency of LH pulses increases, causing follicular growth and the next ovulation. Thus, during the estrous cycle, ovarian steroids feed back on the hypothalamus to modulate both the frequency and amplitude of LH secretion.

Changes in LH concentrations during the estrous cycle have been published for several deer species, including roe deer (*Capreolus capreolus*; Schams et al. 1980, white-tailed deer (Plotka et al. 1980), and fallow deer (*Dama dama*; Asher et al. 1986). In none of these studies were blood samples collected with sufficient frequency to monitor changes in episodic LH secretion. Recently, Loudon et al. (1990) described the pattern of LH secretion over the estrous cycle in the Père David's deer hind (*Elaphurus davidianus*) with 15-min samplings of catheterized animals during both the follicular and luteal phases of estrous cycles synchronized by the removal of progesterone implants. The frequency of LH pulses during the follicular phase was 0.6/h, declining to 0.24/h during the luteal phase at a time of elevated progesterone concentrations (see Fig. 61.1). Mean LH concentrations declined shortly before the onset of the LH surge which may have been due to the negative feedback effects of increased plasma estrogen on the LH pulse generator (Kesner et al. 1981). The LH surge had a duration of approximately 15 h which is similar to that reported for fallow deer. These data broadly accord with data from studies of domesticated sheep (McLeod et al. 1982, 1983), but the LH surge has a duration of approximately double that of cattle (Rahe et al. 1980, Walters and Schallenberger 1984) and goats (Bon Durant et al. 1981).

Asher et al. (1986) described the events around estrus and the LH surge in fallow deer. Estrus occurred 48 to 56 h following the removal of an intravaginal progesterone releasing device and within 4 h of the LH surge and ovulation. Female fallow deer ceased sexual behavior within 1 h of mating, although unmated females were receptive to males for up to 8 h (Asher et al. 1986). This is in contrast to red deer in which estrus can extend for to up to 24 h (Guiness et al. 1971), and hinds frequently mate more than once with the same or different males. In fallow deer, plasma estradiol-17β does not rise significantly at the time of the LH surge. Also in fallow, white-tailed (Plotka et al. 1980) and red deer (Kelly et al. 1985), there is little correlation between the timing of estrus and plasma estrogen concentrations. This contrasts with domestic sheep, where there is a sustained rise in plasma estrogen levels at the time of the LH surge and estrus (Karsch et al. 1979). The mechanism(s) whereby estrus ceases abruptly in fallow deer is unknown

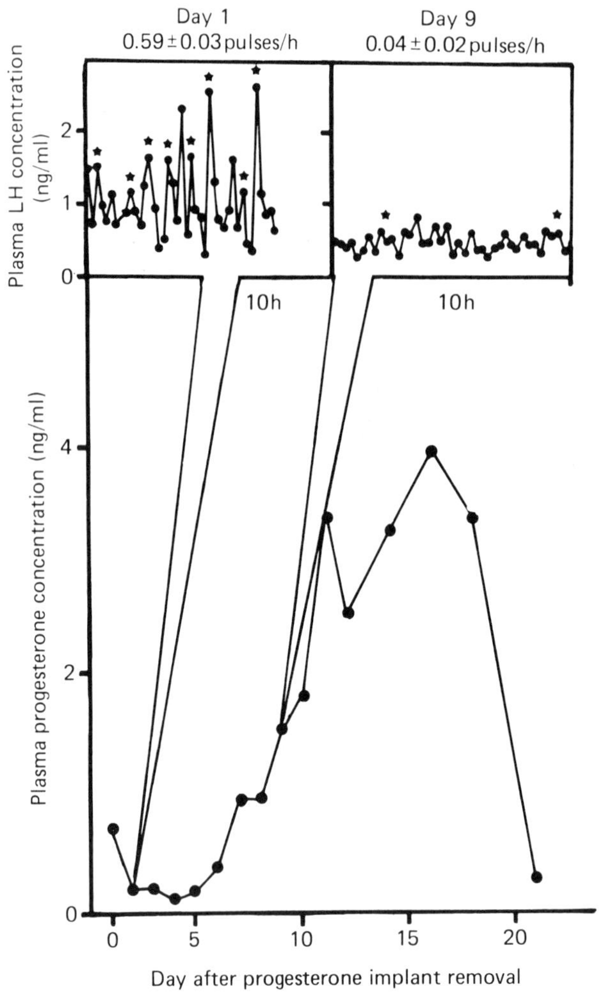

FIGURE 61.1. Diagram shows mean plasma progesterone concentrations over an estrous cycle for a Père David's deer hind after synchronization with progesterone implants (withdrawn on day 0). The panels above show pulsatile LH profiles over 10 h for the same animal during the follicular phase (28 h after implant removal) and luteal phase (9 days after implant removal) of the cycle. Stars indicate pulses as defined in Loudon et al. (1990). Summary statistics for follicular and luteal phases for $n = 5$ animals are also shown; data from Curlewis et al. (1988a, b) and Loudon et al. (1990).

and further comparative data on the hormonal events around estrus in deer are needed.

The Transition to and from Seasonal Anestrus

Very few data are available concerning the mechanisms regulating the transition to and from the breeding season in female deer. In seasonally polyestrus deer unmated females are normally only able to undergo repeated estrus cycles for less than 6 months of the year—red or Père David's deer, approximately 160 days (Loudon et al. 1989), fallow deer, approximately 120 days (Asher 1985).

In sheep, LH pulse frequency is greatly reduced during seasonal anestrus (Scaramuzzi and Baird 1977; Yuthasastrakosol et al. 1975). In the absence of adequate gonadotrophin secretion, ovarian follicles do not mature (McNatty et al. 1981). During anestrus, exogenous GnRH or LH treatment results in follicle maturation and ovulation (McNatty et al. 1981; McLeod et al. 1982). This indicates that the failure to ovulate during seasonal anestrus is primarily due to a deficiency of gonadotrophin support. The transition to seasonal anestrus in sheep is believed to be caused by an abrupt change in the steroid feedback effects of ovarian estrogen on the hypothalamic GnRH pulse generator and hence LH secretion (Legan et al. 1977). During the breeding season, estrogen stimulates GnRH pulsatile activity. However, at the onset of seasonal anestrus, estrogen begins to suppress the GnRH pulse generator. These transitions are mediated by seasonal changes in day length via the secretion of melatonin (Bittman and Karsch 1984; see below). During seasonal anestrus, LH pulse frequency in intact animals is very low and does not change until approximately 1 week before the onset of the breeding season (I'Anson and Legan 1988). Thus, in sheep, seasonal anestrus apparently is relatively uniform, and sensitivity to estradiol feedback undergoes abrupt changes, dramatically increasing and decreasing at the onset and termination of the breeding season.

The episodic secretion of LH during anestrus in the Père David's deer does not conform to the pattern described in sheep (Fig. 61.2; Curlewis et al. 1991; McLeod et al., Brinklow et al. this volume). LH pulse frequency was low in early anestrus (February, 3.3 pulses/18 h), similar to the frequency observed in anestrous domestic sheep. However, in mid-anestrus (April), LH pulse frequency rose to 8.4 pulses/18 h, declining slightly by late anestrus (June) to 6.3 pulses/18-h period (see Fig. 61.2). Indeed, in late anestrus, LH pulse frequency approached that observed during the follicular phase of the breeding season (Loudon et al. 1990). These seasonal changes in LH pulse frequency were associated with significant changes in the response to a fixed dose of GnRH with a smaller response in early than mid- and late anestrus (Curlewis et al. 1991). In addition, the induction of ovulation by exogenous GnRH administration after progestegen priming was significantly more effective in late than early anestrus (Brinklow et al. this volume). Such variations during anestrus have not been reported for other species, and the results suggest that in this species animals enter a state of "deep" anestrus immediately after the end of the breeding season. The midpoint of the anestrous period is associated with a substantial increase in LH pulse frequency. The reduction in episodic secretion of LH immediately after transition to anestrus may be caused by an abrupt change in steroid feedback, as observed in sheep. The mechanism(s) whereby episodic secretion of LH is gradually restored during anestrus is unknown.

Although ovulation can be induced during anestrus by exogenous GnRH, animals return to anestrus following the withdrawal of GnRH treatment and the demise of the induced corpus luteum (McLeod et al., 1991). This even occurs when animals are induced to ovulate close to the onset of the breeding season (see Brinklow et al this volume). The timing of the onset of the breeding season in GnRH-treated and control animals was not significantly different, and there was evidence in both groups of a transitory rise in plasma progesterone concentrations before

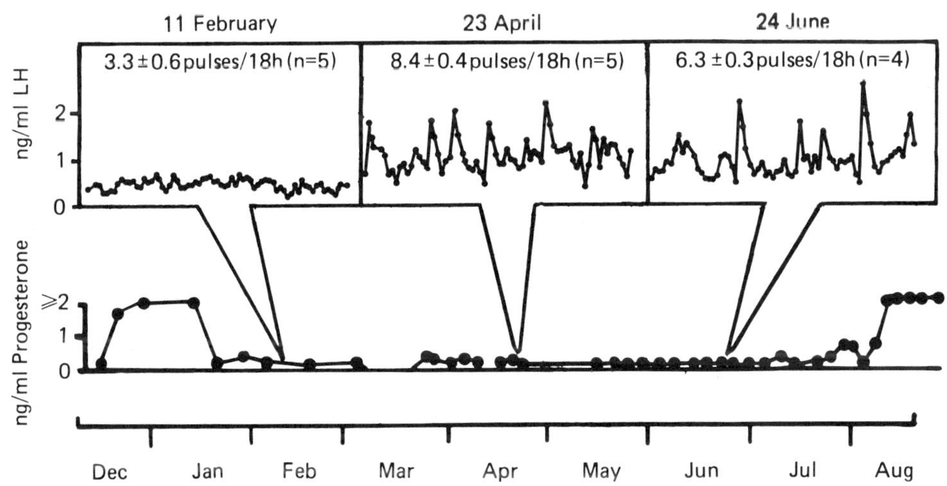

FIGURE 61.2. Diagram shows a plasma progesterone profile over the period of seasonal anestrus for a single Père David's deer hind with the panels above showing 8-h LH profiles in early (11 Feb), mid-(24 April), and late (23 June) anestrus; data from Curlewis et al. (1988a, b) and McLeod et al., 1991.

the first estrous cycle of the breeding season (McLeod et al. 1990, Curlewis et al. 1988a). Thus, while transitions to and from the breeding season may be primarily a consequence of seasonal changes in the secretion of gonadotrophins, manipulations of these hormones by treatment with exogenous GnRH have no effect on the central seasonal time-keeping mechanism that is responsible for switching on the GnRH pulse generator at the beginning of the breeding season.

Metabolism and Growth

Studies of enhoused animals have demonstrated substantial seasonal changes in metabolic rate in seasonally breeding deer and domesticated sheep (Silver et al. 1969; Simpson et al. 1978; Blaxter and Boyne 1982; Renecker and Hudson 1986). In the primitive Soay sheep there is a sinusoidal seasonal change of approximately 26% in estimated metabolizable energy requirements for maintenance (Argo and Smith 1983). These changes in metabolic rate are associated with a marked seasonal cycle of voluntary food intake (VFI) and body fat reserves and, in the young animal, growth rate (Pollock 1975; Milne et al. 1978; Kay 1979; Brown et al. 1978; Forbes et al. 1979a, b; Milne et al. 1987; Loudon et al. 1989). Changes in metabolic rate precede those of appetite, suggesting a causal relationship. The seasonal deposition of fat in the autumn and early winter in wild deer is a reflection of endogenous growth processes that occur even when animals are maintained under conditions of ad libitum food intake. Seasonal changes in VFI and body weight are probably a direct consequence of an underlying seasonal rhythm in metabolic rate and energy requirements.

Seasonal appetite rhythms in deer persist independently of pregnancy and lactation or competition between males during the rut. There are, however, sex differences in the amplitude and timing of the annual VFI cycle. In red deer stags, the cycle has an almost three fold amplitude (Kay 1979, Suttie and Kay 1985) in contrast to one and one half to two-fold observed in females (Curlewis et al. 1988b). Volunary food intake in enhoused stags declines abruptly at the onset of the normal rutting season (Kay 1979); in females, VFI declines more slowly in the autumn (Curlewis et al. 1988b). Similar sex differences have been reported for the black-tailed deer (*Odocoileus hemionus*) by Bandy et al. (1970). Female Père David's deer (*Elaphurus davidianus*) exhibit a VFI cycle with a more than three fold amplitude (Loudon et al. 1989), although no comparable data exist for males.

Castration of red deer stags greatly diminishes the amplitude of the seasonal VFI cycle (Kay 1979), with appetite cycles in castrates resembling those of female deer. However, where VFI cycles of intact and castrated red deer stags are corrected for metabolic body weight (body weight to the log power 0.75; Kleiber 1961), the differences in amplitude of the VFI cycle disappear (see Fig. 61.3). Thus, it appears that differences in the amplitude of the appetite cycle may to a large extent be due to differences in body size. That the primary mechanisms regulating seasonal changes in food intake are independent of gonadal steroids has also been demonstrated by workers studying seasonal metabolic cycles in other mammals such as Syrian hamsters (Bartness and Wade 1984; Wade and Bartness 1984).

Linkage of Seasonal Rhythms of Reproduction and Growth

In deer, seasonal physiological rhythms are clearly synchronized to seasonal environmental changes. In autumn rutting male deer, such as the red and fallow deer and the North American Odocoilids, the seasonal peak in food intake and body condition usually occurs just prior to the onset of sexual and aggressive behavior in the late summer and early autumn (Wood et al. 1962, Suttie and Kay 1985, Asher et al. 1987). In male red deer during the rut, feeding time is greatly reduced, declining in some individuals to zero for short periods (Clutton-Brock et al. 1982), and adult stags lose most of the fat reserves accumulated over the summer (Mitchell et al. 1976). Similarly, in females, the endogenous seasonal cycle of food intake persists, even in nonbreeding animals, such as Odocoileus (Bandy et al. 1970) and red deer (Curlewis et al. 1988b) and is usually coincident with the normal timing of peak lactation in mid-summer; for Odocoileus see Sadleir (1980) for red deer Loudon et al. (1983). During

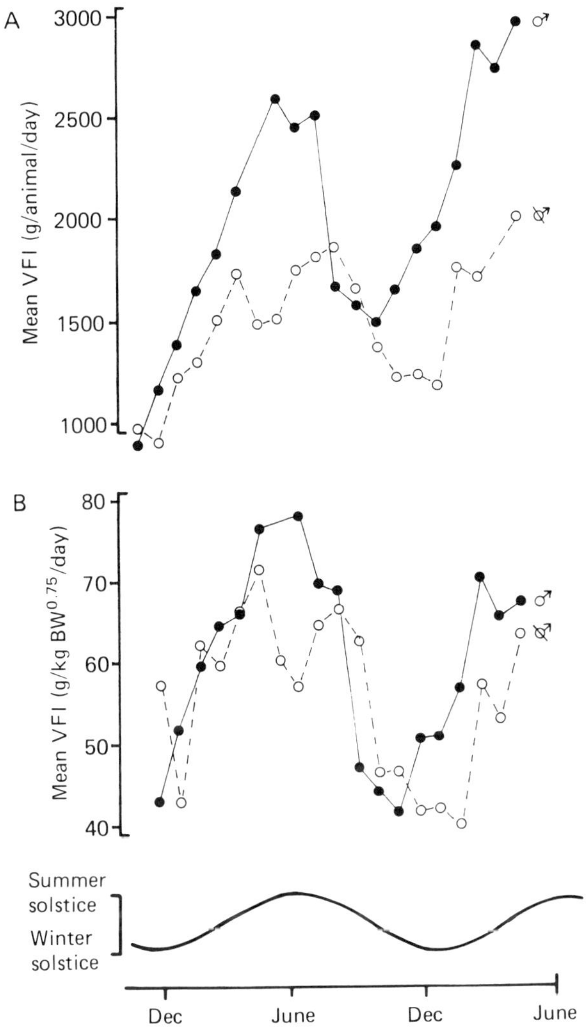

FIGURE 61.3. Mean voluntary food intake of intact (●, $n = 2$) or castrate (○, $n = 2$) red deer stags fed ad libitum pelleted concentrate diet on natural photo-periods. (A) Expressed as food intake per animal per day and (B) intake corrected for metabolic weight; after Kay (1979).

lactation, female deer frequently lose large amounts of body condition; for red deer, Mitchell et al. (1976); reindeer, Dauphine (1976); roe deer, Ellenberg (1978). In wild red deer on impoverished habitats in northern Britain, the metabolic strain of lactation can delay or even block the onset of the normal breeding season (Mitchell and Lincoln 1973; Loudon et al. 1983, Heydon et al. this volume). Thus, in both sexes, the endogenous seasonal growth and appetite cycles are associated with the annual cycle of energy expenditure resulting from reproduction.

Further evidence indicating a link between growth and reproductive cycles comes from studies of female Père David's deer. The species originates from China and is related to the red deer; recently they have been hybridized by artificial insemination (Asher et al. 1988a). The onset of the breeding season and the timing of the rut in the Père David's deer occurs in midsummer, some 3 to 4 months before red deer (Schaller and Hamner 1978; Curlewis et al. 1988a). Gestation length is approximately 282 days (Wemmer 1983), 1.5 months longer than in the red deer. Therefore, the peak in calving usually precedes that of red deer and many other north temperate zone cervids by at least 2 months (Bedford 1951, Schaller and Hamner 1978). In a comparative study of enhoused female Père David's and red deer, the seasonal cycles of food intake, the timing of the onset and termination of estrous cycles, growth and moult of the coat, and seasonal changes in plasma prolactin and triiodothyronine concentrations of female Père David's deer are phase advanced by approximately 2 months in comparison with the red deer (Loudon et al. 1989). Thus a 2-month earlier calving season is reflected in a similar advance in the seasonal and growth processes. Data summarized from this work are shown in Fig. 61.4. The timing of the seasonal prolactin cycle in the Père David's deer is unusual as this hormone normally exhibits a peak at mid-summer and a nadir at mid-winter (Lincoln 1985; Loudon et al. 1989). Although the breeding season was advanced by 3 months in the Père David's deer the total duration of the season (i.e., period over which animals exhibited estrous cycles) was virtually identical in the 2 species (160 vs 161 days for the red and Père David's deer respectively).

In red and Père David's deer calves weaned at 3 months of age there were no apparent species differences in the seasonal pattern of prolactin secretion, VFI, and growth over the first year of life. However, at 11 months of age there was a significantly earlier peak in prolactin secretion in

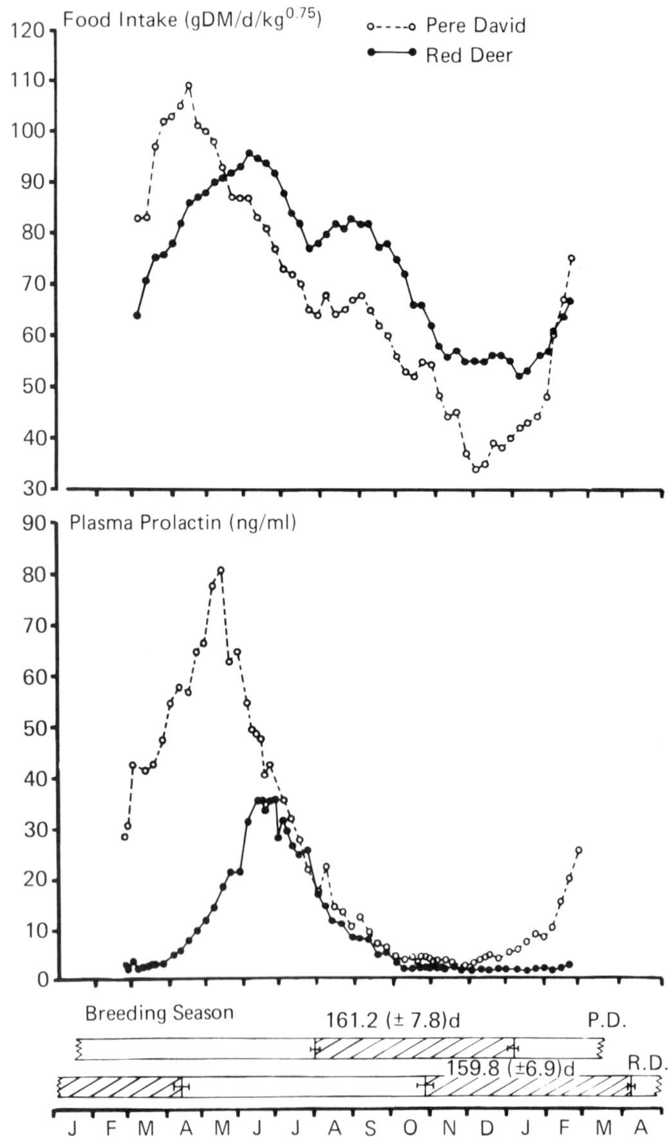

FIGURE 61.4. Seasonal changes in mean weekly food intake and plasma prolactin concentrations for six adult female red deer (●) and six adult female Père David's deer (○) over a period of 15 months. The data are presented after quadratic smoothing. The timing of the onset and duration of the breeding season are indicated by the hatched bars; upper bar Père David's deer and lower bar red deer; from Loudon et al. (1989).

the Père David's deer. The seasonal VFI cycle was also advanced by this stage and an early decline relative to the red deer was also associated with a 3-month advance in the timing of puberty. A consequence of the similarity of the food intake and growth cycles of the two species in the first year is that Père David's deer, which are normally conceived 2 months before red deer, are exposed to a longer "growing season" before seasonal inappetance is established in the first winter after birth. From this, it is clear that the phase differences in the timing of seasonal cycles of adult deer develop with age (Loudon et al. 1989).

Role of Prolactin in Seasonal Physiological Changes

Plasma concentrations of prolactin are highly seasonal in many mammals. In general, concentrations are maximal during the long photoperiods of midsummer and minimal during midwinter. There is clear experimental evidence from a number of different species that seasonal changes in prolactin are directly involved in the control of the coat growth and moult cycle, e.g., in the mink (Martinet et al. 1984) and hamster (Duncan and Goldman 1984). In red deer, manipulations of prolactin are associated with changes in the growth and moult of the coat (Curlewis et al. 1988b; Milne et al. 1990).

In white-tailed deer (Mirachi et al. 1978), red deer (Scuttie et al. 1984), and roe deer (Schams and Barth 1982) concentrations of prolactin are maximal in midsummer. The early peak in the Père David's deer appears to be a notable exception, perhaps associated with unusually early calving (Loudon et al. 1989; see above). There is no obvious relationship between the seasonal prolactin profiles and the timing of the onset of the breeding season, which differs significantly between different species of deer (Lincoln 1985). In eutherian mammals, the secretion of pituitary prolactin is thought to be primarily under the inhibitory control of hypothalamic dopamine secretion. There is also evidence from sheep that dopamine plays a role in the suppression of LH secretion during seasonal anestrus (Meyer and Goodman 1986), and it might therefore be the case that hypothalamic dopamine plays a dual role in regulating both prolactin and LH secretion.

Two recent studies provide evidence of a possible relationship between prolactin secretion, the timing of the breeding season and also cycles of food intake and pelage growth in deer. Red deer hinds treated in mid-summer during anestrus with the dopamine agonist bromocroptine exhibited a depression in prolactin and VFI and disruption of the normal pattern of winter coat growth (Milne et al. 1990). However, the onset of the breeding season in the treated group was not significantly affected. In a further experiment (Curlewis et al. 1988b) red deer hinds were treated during the breeding season in late winter for an 8.5-month period with a slow release preparation of bromocriptine. Treatment greatly delayed the seasonal rise in prolactin secretion and reduced the amplitude of the peak. It also had profound effects, including the suppression of food intake, delay in the moult of the winter coat, growth of the summer coat, and even the growth of the subsequent winter coat, nearly a year later. In addition, the timing of the onset of seasonal anestrus was delayed by 55 days. Thus, although it is unlikely that dopamine or prolactin play a major role in the suppression of pulsatile secretin of LH in deer, the data do indicate that a large number of seasonally changing parameters were simultaneously affected by administration of bromocriptine.

Environmental Control Mechanisms

Changes in the duration of daylength provide the most predictable environmental signal for cueing seasonal physiological processes. In sika deer stags, Goss (1969a, b; Goss and Rosen 1973) demonstrated that artificial manipulations of photoperiod can alter the timing of the antler cycle. Red deer also respond to changing photoperiod, both in changes in reproductive activity and pattern of food intake and growth (Suttie and Kay 1985). Where red deer stags have been kept on an artificial 6-month photoperiods in which the annual changes in day length have been compressed into two cycles per year, males generate two sets of antlers per year, undergo two periods of rutting and sexual activity, and exhibit two distinct appetite and body weight cycles

(Kay 1979; Simpson et al. 1978, 1983). In red deer females, manipulations of photoperiod can also cause an early onset to the breeding season as well as advancing seasonal pelage changes (Webster and Barrell 1985).

Seasonal changes in reproduction in mammals are regulated by changes in the duration of secretion of melatonin from the pineal gland during the hours of darkness (see Reiter 1980). In one of the earliest experiments using melatonin in deer, Bubenik (1983) treated intact adult white-tailed bucks orally each day from midsummer through to late autumn. The timing of the daily dosing was designed to coincide with a period approximately 2 h before the onset of darkness and the subsequent natural rise in plasma melatonin levels. Oral dosing thereby extended the daily function of elevated melatonin concentrations in the blood since the rumen provides a slow-release mechanism. This treatment induced an early onset of seasonal aggressive and sexual activity and also caused premature cleaning and hardening of the antlers, an androgen dependent process. Oral dosing (Adam et al. 1989, 1986; Asher et al. 1987; Milne et al. 1990; Heydon et al. this volume) and continuous slow-release implants of melatonin have also been employed (Lincoln et al. 1984; Asher et al. 1988b; Heydon et al. this volume) have been shown to be effective in manipulating seasonality, but the mechanism may be different (see below).

Treatment with melatonin during the summer has been shown to advance the onset of the breeding season in male red deer (Lincoln et al. 1984), white-tailed deer (Bubenik 1983), and fallow deer (Asher et al. 1987). Studies demonstrating advances in the timing of the onset of estrous cyclicity of female red and fallow deer are summarized in Table 61.1. The interval from the timing of the onset of treatment to the first estrous cycle ranges from 71 to 96 days, a shorter interval in experiments in which treatments commenced later in the summer. This gives a slope of the response to melatonin of less than unity (Fig. 61.5) with the regression line for red deer intercepting with the data for fallow deer. A group of lactating animals treated with melatonin implants and deliberately maintained on a low plane of nutrition have been excluded from this analysis, as undernutrition influences the date of first estrus (see Heydon et al., this volume). The above data indicate that the melatonin signal acts to cue an underlying seasonal rhythm to environmental photoperiod, rather

TABLE 61.1. Summary of published data for the advancement of the onset of the breeding season by melatonin treatment in 6 studies with red deer and one with fallow deer.

Physiological status	Onset of breeding season Control	Treated	Days advance	Interval (days)[a]	Treatment type	Reference
Prepubertal	24 Apr	29 Mar	26	81	Daily injection	Webster and Barrell 1985[b,e]
	24 Apr	23 Mar	32	75	Short days	
Lactating	17 Oct	15 Sept	32	88	Daily oral	Adam et al. 1986[c]
Nonlactating	21 Oct	16 Sept	35	89	Daily oral	Adam et al. 1986[c]
Nonlactating	8 Oct	19 Sept	19	78	Daily oral	Milne et al. 1990[c]
	5 Oct	13 Sept	22	76	Daily oral	Milne et al. 1990[c]
Nonlactating	16 Oct	2 Oct	14	71	Daily oral	Heydon (unpublished data)[c]
Lactating[d]	16 Oct	28 Sept	18	96	Implants	
Pregnant nonpregnant	23 Apr	28 Feb	54	110	Implants	Asher et al. 1988[b,c]

[a] Interval from onset of treatment to onset of breeding
[b] Southern hemisphere dates
[c] Breeding season as determined from progesterone profiles
[d] Animals on a low plane of nutrition
[e] Breeding based on mating dates

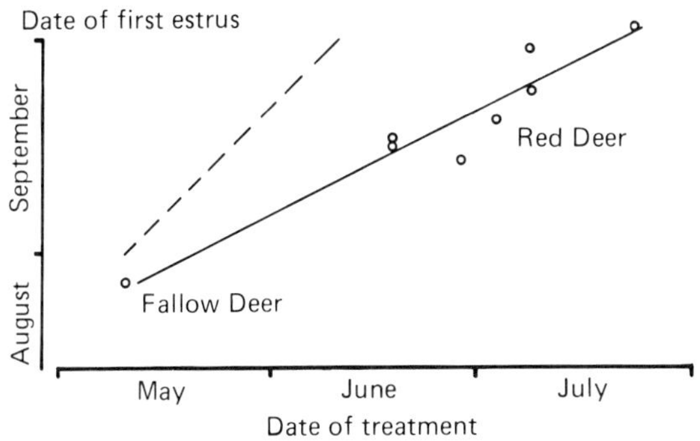

FIGURE 61.5. Relationship between date of onset of the breeding season to time of treatment with melatonin for red deer and a single study of the fallow deer; the studies are described in Table 61.1. Solid line represents the regression analysis of the first seven studies in Table 61.2, excluding the data for the fallow deer. Slope 0.50, $r\,0.88$, $P < 0.01$. The slope of 1 is indicated with the dotted line. For comparative purposes, data from New Zealand are presented with respect to northern hemisphere dates.

than serving simply to "throw" a photoperiodic switch that results in the restoration of gonadotrophin secretion after a fixed period of time.

A range of other physiological parameters is also affected by melatonin treatment. Treatment in summer results in a rapid decline in prolactin secretion (red deer: Webster and Barrell 1985, Milne et al. 1990), has significant effects on the seasonal body weight cycle (fallow: Asher et al. 1988b; red deer: Adam et al. 1986, Milne et al. 1990) and on the growth and moult of the coat (red deer: Webster and Barrell 1985, Milne et al. 1990). Milne et al. (1990) orally dosed red deer hinds with melatonin from July for a 120-day period and reported that melatonin treatment caused a decline in food intake approximately 4 to 6 weeks in advance of control animals. Treatment with melatonin shifts seasonal cycles in pelage growth and body condition and has potentially serious consequences for the animal. In the experiment of Milne et al. (1990) above, following melatonin treatment over the summer, deer were maintained out of doors on pastures in northeast Scotland over the winter. A number of the treated animals died, possibly due to poor coat development and low body condition. Further, Asher et al. (1988b) reported that fallow does treated with melatonin during late pregnancy have a high incidence of lactational failure and, as a result, calves suffered high perinatal mortality. These data also suggest that the initiation of lactation (perhaps involving prolactin) in seasonal mammals such as fallow deer may be under photoperiodic control. The above data indicate that great care must be taken before the widespread commercial use of melatonin is established for application in farmed deer.

Endogenous Rhythms

Female sheep do not need to experience a decline in photoperiod after the summer solstice to commence breeding at the normal time in the autumn. In one experiment (Robinson et al. 1985) adult ewes were maintained on artificial fixed summer solstitial photoperiods from midsummer and not permitted to experience a decline in day length. These ewes commenced cycling at the same time as controls exhibiting no change in the duration of the daily melatonin signal (Robinson et al. 1985). Further evidence for a change in reproductive response irrespective of the prevailing melatonin signal comes from studies of pinealectomized ewes that were infused with an artificial melatonin profile designed to mimic long photoperiods (Wayne et al. 1988). In these experiments, the "long-day" profile initially caused a suppression of LH typical of anestrus. Subsequently, however, LH levels rose again despite a constant "long-day" melatonin signal.

In our recent unpublished experiments, we have demonstrated that pre-pubertal hinds transferred to summer solstice photoperiods at their first winter solstice showed a rapid and early rise in plasma prolactin concentrations, an early moult to summer coat, and a marked advance in the seasonal appetite rhythm (see Fig. 61.6). Subsequently, appetite and prolactin declined, and the animals grew winter coats significantly in advance of controls on natural photoperiods despite the fact that these animals were kept throughout on long photoperiods. In addition, the timing of puberty in this group was advanced by 61 days compared with controls. A similar phase advance in the breeding season of adult sheep exposed to long photoperiods has been demonstrated by Malpaux et al. (1989). Thus, the use of the term "short-day" breeder with respect

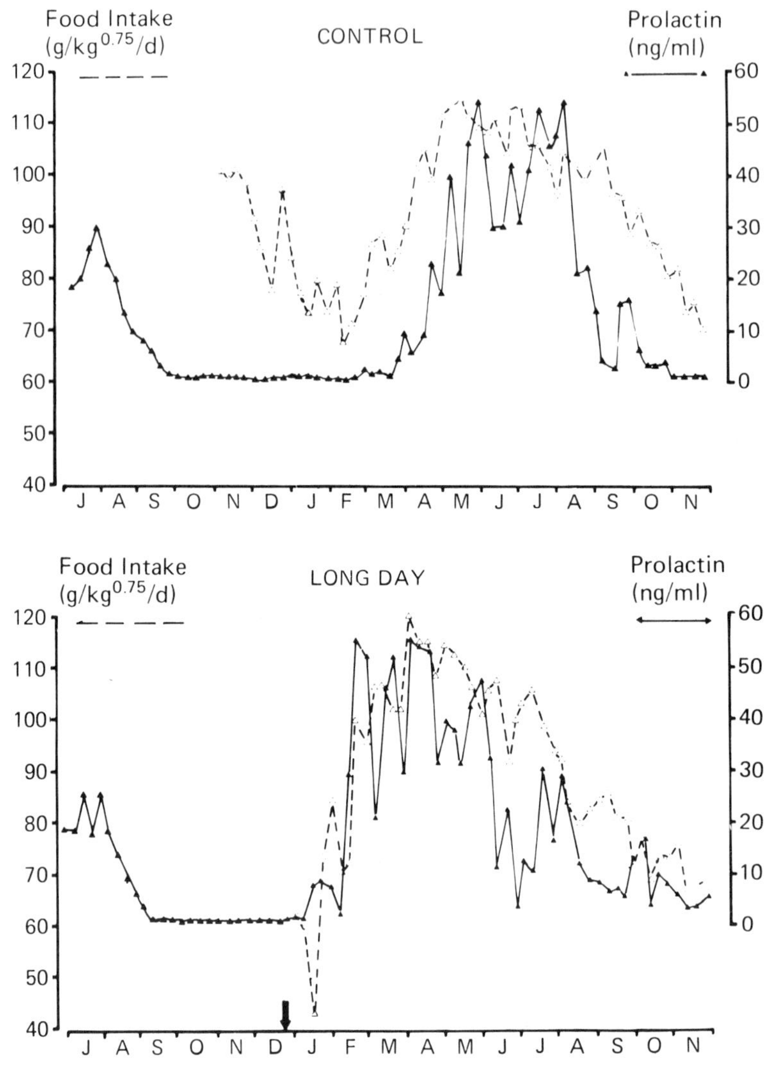

FIGURE 61.6. Seasonal cycle of prolactin (▲) and voluntary food intake (△) as X24 kg dry matter per day/kg$^{0.75}$ live weight) from birth to puberty in a group of hind calves ($n = 7$) maintained under natural photoperiodic conditions (top panel) and a group ($n = 6$) exposed to midsummer photoperiods for the subsequent 10-month period (18 h light; 6 h dark) from the first winter solstice of life as indicated by the arrow.

to many seasonal deer species is no more than a descriptive term and provides no insight into the underlying physiological mechanisms determining the timing of breeding seasons.

The environmental factors responsible for the termination of the breeding season in sheep have been investigated by Robinson and Karsch (1984). These investigators demonstrated that ewes maintained on winter solstice photoperiods from midwinter did not affect the timing of the onset of seasonal anestrus, demonstrating that an increase in photoperiod was not necessary to end the breeding season. In deer, however, Adam et al. (1989) orally dosed red deer hinds during the breeding season (February) with melatonin in late afternoon and maintained treatment for 15 months. This regimen prevented the establishment of anestrus, and plasma prolactin levels remained low and within the typical midwinter range throughout the period of treatment. Thus, it seems that in red deer hind (but not domesticated sheep), increasing photoperiods after the winter solstice serves two roles. First, the increase in day length effectively terminates the breeding season (i.e., estrus cycles) and, second, acts as a timing signal which initiates processes that result in the onset of the next breeding season and other associated changes several months later (i.e., the "long-day" refractory process; see Nicholls et al. 1988). The key photoperiodic events in the timing of the endogenous annual cycle in red deer may occur during the spring and early summer and not during the period of shortening day lengths in the autumn. This phenomenon is revealed in the experiment in which hinds were "locked" onto midsummer day lengths from mid-December (see Fig. 61.6). Such a treatment simultaneously advanced the phase of a number of different seasonal rhythms, including food intake, prolactin level, coat growth, and the timing of puberty. In red deer, Adam et al. (1989) have shown that melatonin treatment started in midwinter may trap animals at a fixed stage of their annual cycle and prevent long-day stimulated anestrus and the subsequent onset of long-day photorefractoriness from occurring.

Studies of long-lived seasonal mammals have demonstrated that physiological rhythms can persist under constant photoperiodic conditions. Sika stags maintained for a very long time on fixed photoperiods, and therefore presumably experiencing an unvarying daily melatonin signal, underwent cyclical growth and shedding of antlers (Goss and Rosen 1973). Endogenous rhythms clearly persist in the absence of any photoperiodic entrainment since pinealectomized or ganglionectomized male deer continue to exhibit long-term changes in food intake and reproduction (Brown et al. 1978; Plotka et al. 1979, 1981; Lincoln 1985). Brinklow and Loudon (1990) have recently demonstrated that endogenous rhythms of food intake, body weight, plasma prolactin level, and coat growth are present in animals kept on long days from birth. Similarly, in sheep Howles et al. (1982) demonstrated a free-running rhythm of for testis size with a period of approximately 55 weeks in rams held on constant long, short, or equitorial photoperiods. Studies of adult ewes maintained on constant photoperiods have also demonstrated endogenous long-term rhythms of LH and prolactin (Karsch et al. 1989), although a recent report by Jackson et al. (1990) has shown that some animals maintained on equatorial photoperiods tended to breed continuously.

Tropical Deer

There is very little published information concerning the reproductive biology of tropical and subtropical deer. Nineteen of the 40 deer species are found between latitudes 20° north and 20° south (Whitehead 1972). At these latitudes, deer may cue the timing of births to coincide with local periods of maximum food abundance, although species living in the same geographical area (such as the hog, sambar, and barking deer of northern and central India) exhibit significant differences in the timing of breeding (Mishra and Wemmer 1987). It is not known how the timing of breeding is cued in such species.

A herd of Axis deer from southern India kept in Britain at 52° north (Loudon and Curlewis 1988) exhibited no obvious calving season. Individual stags grew and shed their antlers approximately once every 13 months, and these antler cycles were associated with cycles in testis size, the secretion of testosterone, neck girth, and body weight (Loudon and Curlewis 1988). However, within the axis deer herd studied, the individual antler cycles were not synchronized to one another, and each animal appeared to exhibit its own free-running rhythm. Thus, in this aseasonal species there is clear evidence of cyclical changes in testicular function and an associated antler cycle, as in north temperate species of seasonally breeding deer (Lincoln 1985). Melatonin implanted in the summer into stags whose antlers were growing had no effect on the rate of growth of the antler, dimensions of the neck or size of the testis (Loudon and Curlewis 1988). This is in marked contrast to the effect of similar implants administered to red deer stags at the same time of year and stage of the antler cycle (Lincoln et al. 1984). Thus, a tropical species such as the axis deer may undergo persistent life-long circannual physiological rhythms that are not entrained by photoperiod. Such animals provide further evidence that seasonality in temperate deer is an endogenous process in which changes in day length and melatonin entrain annual physiological cycles.

In the course of their evolution, deer have developed a suite of physiological adaptations to cope with life in temperate environments with relatively predictable annual variations in food supply. Tropical and subtropical representatives of this family currently present one of the great challenges facing the biologist with an interest in deer. In an earlier review, Lincoln (1985) drew attention to the paucity of knowledge of these animals. It is clear from a few studies of captive populations that tropical species may well have retained many of the adaptations of their northern temperate relatives. Many of the tropical species live in habitats that are severely threatened and their survival requires vigorous action by conservationists. Our knowledge of deer over the past 10 years has increased dramatically as a result of the large body of research carried out on a few species. This has been spurred on in particular by the rapidly developing deer-farming industry. It is to be hoped that in the next 10 years, some of this knowledge could be applied to the problems of captive breeding of endangered representatives of this widespread and highly adaptable ungulate family.

Acknowledgements

We thank C. Argo and R. Kay for permission to quote from their unpublished analysis of voluntary food intake of red deer stags, the staff of the Whipsnade research facility for their assistance with the work on deer quoted within this review and B. McLeod for comments on an earlier draft. We thank the Agriculture and Food Research Council for financial support.

References

Adam CL, Moir CE, Shiach P (1989) Melatonin can induce year-round ovarian cyclicity in red deer (*Cervus elaphus*). J Reprod Fert 87:401–408

Adam CL, Moir CE, Atkinson T (1986) Induction of early breeding in red deer (*Cervus elaphus*) by melatonin. J Reprod Fert 76:569–573

Argo CM, Smith JS (1983) Relationships of energy requirements ans seasonal cycles of food intake in Soay rams. J Physiol 343:22–24P

Arman P, Kay RNB, Goodall ED, Sharman GAM (1974) The composition and yield of milk from captive red deer. J Reprod Fert 37:67–84

Asher GW (1985) Oestrous cycle and breeding season of farmed fallow deer (*Dama dama*). J Reprod Fert 75:521–529

Asher GW, Adam JL, Otway W, Bowmar P, van Reenan G, Mackintosh CG, Dratch P (1988a) Hybridization of Père David's deer (*Elaphurus davidianus*) and red deer (*Cervus elaphus*) by artificial insemination. J Zool London 215:197–203

Asher GW, Barrell GK, Adam JL, Staples LD (1988b) Effects of subcutaneous melatonin implants on reproductive seasonality of farmed fallow deer (*Dama dama*). J Reprod Fert 84:679–691

Asher GW, Barrell GK, Peterson AJ (1986) Hormonal changes around oestrus of farmed fallow deer (*Dama dama*). J Reprod Fert 78:487–496

Asher GW, Day AM, Barrell GK (1987) Annual liveweight and reproductive changes of male fallow deer (*Dama dama*) and the effect of daily oral administration of melatonin in summer on the attainment of seasonal fertility. J Reprod Fert 79:353–362

Baird DT (1978) Pulsatile secretion of LH and ovarian estradiol during the follicular phase of the sheep estrous cycle. Biol Reprod 18:359–364

Bandy PJ, McCowan I, Wood AK (1970) Comparative growth in four races of black-tailed deer (*Odocoileus hemionus*) Part 1. Growth in body weight. Can J Zool 48:1404–1410

Barthess TJ, Wade GN (1984) Photoperiodic control of body weight and energy metabolism in Syrian hamsters (*Mesocricetus auratus*): role of pineal gland, melatonin gonads and diet. Endocrinology 114:492–498

Bedford, 12th Duke of (1951–2) Pere David's deer. The history of the Woburn herd. Proc Zool Soc 121:327–333

Bittman EL, Karsch FJ (1984) Nightly duration of pineal melatonin secretion determines reproductive response to inhibitory day length in the ewe. Biol Reprod 30:585–593

Blaxter KL, Boyne AW (1982) Fasting and maintenance metabolism of sheep. J Agric Sci Cambridge 99:611–620

Blaxter KL, Hamilton WJ (1980) Reproduction in farmed red deer 2. Calf growth and mortality. J Agric Sci Cambridge 95:275–284

Blouch RA (1987) Reproductive seasonality of the white-tailed deer on the Columbian Llanos. In: Wemmer C (ed) Biology and Management of

Cervidae. Smithsonian Institution Press, Washington D.C., pp 339–343
Bon Durant RH, Darien BJ, Munro CJ, Stabenfeldt GH, Wang P (1981) Photoperiod induction of fertile oestrus and changes in LH and progesterone concentrations in yearling dairy goats (*Capra hircus*). J Reprod Fert 63:1–9
Brinklow BR, Loudon ASI (1990) Development of seasonal rhythms in a long-lived ungulate: The red deer (*Cervus elaphus*). J Interdisc Cycle Res 21:173–175
Brown RD, Cowan RL, Kavanaugh EJF (1978) Effect of pinealectomy on seasonal androgen titres, antler growth and feed intake in white-tailed deer. J Anim Sci 47:435–440
Bubenik G (1983) Shift of seasonal cycle in white-tailed deer by oral administration of melatonin. J Exp Zool 225:155–156
Clarke IJ, Cummins JT (1982) The temporal relationship between gonadotrophin-releasing hormone (GnRH) and luteinizing hormone (LH) secretion in ovariectomized ewes. Endocrinology 111:1737–1739
Clutton-Brock TH, Guiness FE, Albon SD (1982) Red Deer: Behaviour and Ecology of Two Sexes. Edinburgh University Press, Edinburgh
Curlewis JD, McLeod BJ, Loudon ASI (1991) LH secretion and response to GnRH during seasonal anoestrus of the Père David's deer hind. J Reprod Fert 91:131–138
Curlewis JD, Loudon ASI, Coleman AMP (1988a) Oestrous cycles and the breeding season of the Père David's deer hind (*Elaphurus davidianus*) J Reprod Fert 82:119–126
Curlewis JD, Loudon ASI, Milne JD, McNeilly AS (1988b) Effects of chronic long-acting bromocriptine treatment on liveweight, voluntary food intake, coat growth and breeding season in non-pregnant red deer hinds. J Endocrinol 119:413–420
Dauphine TC (1976) Biology of the Kaminuriak population of barren-ground caribon. Part 4. Growth, reproduction and energy reserves. Can J Wildl Serv Rep Ser 38:1–71
Duncan MJ, Goldman BD (1984) Hormonal regulation of the annual pelage colour cycle in the Djungarian hamster (*Phodopus sungorus*) 11: role of prolactin. J Exp Zool 230:97–103
Ellenberg H (1978) The population ecology of roe deer in Central Europe. Spixiana (Suppl 2)
Forbes JM, El Shahat AA, Jones R, Duncan JGS, Boaz TG (1979a) The effect of day length on the growth of lambs. 1. Comparisons of sex, level of feeding, shearing and breed of sire. Anim Prod 29: 33–42
Forbes JM, Driver PM, Brown WB, Scanes CG (1979b) The effect of day length on the growth of lambs. 2. Blood concentrations of growth hormone, prolactin, insulin, thyroxine and the effect of feeding. Anim Prod 29:43–51
Goodman RL, Karsch FJ (1981) The hypothalamic pulse generator: a key determinant of reproductive cycles in sheep. In: Follett BK, Follett DE (eds) Biological Clocks in Seasonal Reproductive Cycles. J Wright, Bristol, UK
Goodman RL, Karsch FJ (1980) Pulsatile secretion of luteinizing hormone: differential suppression of ovarian steroids. Endocrinology 107:1286–1290
Goss RJ (1969a) Photoperiodic control of antler cycles in deer. I Phase shift and frequency changes. J Exp Zool 170:311–324
Goss RJ (1969b) Photoperiodic control of antler cycles in deer. II Alterations in amplitude. J Exp Zool 171:223–234
Goss RJ, Rosen JK (1973) The effect of latitude and photoperiod on the growth of antlers. J Reprod Fert Suppl 19:111–118
Guiness FE, Lincoln GA, Short RV (1971) The reproductive cycle of the female red deer, *Cervus elaphus* L. J Reprod Fert 27:427–438
Howles CM, Craighton J, Haynes NB (1982) Long-term testicular volume and plasma prolactin in rams reared for three years in constant photoperiod. J Reprod Fert 65:439–446
I'Anson H, Legan SJ (1988) Does the first LH surge of the breeding season initiate the first full-length cycle in the ewe? J Reprod Fert 82:761–767
Jackson GL, Hansen H, Kao C (1990) Continuous exposure of Suffolk ewes to an equitorial photoperiod disrupts expression of the annual breeding season. Biol Reprod 42:63–73
Karsch FJ, Robinson JE, Woodfill CJ, Brown MB (1989) Circannual cycles of luteinizing hormone and prolactin secretion in ewes during prolonged exposure to a fixed photoperiod: evidence for an endogenous rhythm. Biol Reprod 41:1034–1046
Karsch FJ, Foster DL, Legan SJ, Ryan KD, Peter GK (1979) Control of the preovulatory endocrine events in the ewe: Interrelationships of estradiol, progesterone and luteinizing hormone. Endocrinology 105:421–425
Katangole CB, Naftolin F, Short RV (1971) Relationship between blood levels of luteinizing hormone and testosterone in bulls, and the effects of sexual stimulation. J Endocrinol 50:457–466
Kay RNB (1979) Seasonal changes of appetite in deer and sheep. ARC Res Rev 5:13–15
Kelly RW, McNatty KP, Moore GH (1985) Hormonal changes about oestrus in female red deer. In: Fennessy P, Drew K (eds) The Biology of Deer Production. R Soc N Z Wellington Bull 22:181–184

Kesner JS, Convey EM, Anderson CR (1981) Evidence that estradiol induces the preovulatory LH surge in cattle by increasing pituitary sensitivity to LHRH and then increasing LHRH release. Endocrnology 108:1386–1391

Kleiber M (1961) The Fire of Life: an Introduction to Animal Energetics. Wiley, New York p 454

Legan SJ, Karsch FJ, Foster DL (1977) The endocrine control of seasonal reproductive function in the ewe: a marked change in response to the negative feedback action of estradiol on luteinizing hormone secretion. Endocrinology 101:818–824

Levine JE, Pau K-Y P, Ramirez VD, Jackson GL (1982) Simultaneous measurement of luteinizing hormone-releasing hormone and luteinizing hormone release in unanaesthetised ovariectomised sheep. Endocrinology 111:1449–1455

Lincoln GA (1985) Seasonal breeding in deer. In: Fennessey P, Drew K (eds) The Biology of Deer Production. R Soc N Z Wellington Bull 22:

Lincoln GA (1971) The seasonal reproductive changes in the red deer stag (*Cervus elaphus*). J Zool 163:105–123

Lincoln GA, Kay RNB (1979) Effect of season on the secretion of LH and testosterone in intact and castrated red deer stags (*Cervus elaphus*). J Reprod Fert 55:75–80

Lincoln GA, Fraser HM, Fletcher TJ (1984) Induction of early rutting in male red deer (*Cervus elaphus*) by melatonin and its dependence on LHRH. J Reprod Fert 72:339–343

Lincoln GA, Fraser HM, Fletcher TJ (1981) Antler growth in male red deer (*Cervus elaphus*) after active immunization against LHRH. J Reprod Fert 66:703–708

Loudon ASI, Curlewis JD (1988) Cycles of antler and testicular growth in an aseasonal tropical deer (*Axis axis*). J Reprod Fert 83:729–738

Loudon ASI, Kay RNB (1984) Lactational constraints on a seasonally breeding mammal: the red deer. Symp Zool Soc London 51:233–252

Loudon ASI, McLeod BJ, Curlewis JD (1990) Pulsatile secretion of LH during the periovulatory and luteal phases of the oestrous cycle in the Père David's deer hind (*Elaphurus davidianus*). J Reprod Fert 89:663–670

Loudon ASI, Milne JA, Curlewis JD, McNeilly AS (1989) A comparison of the seasonal hormone changes and patterns of growth, voluntary food intake and reproduction in juvenile and adult red deer (*Cervus elaphus*) and Père David's deer (*Elaphurus davidianus*) hinds. J Endocr 122:733–745

Loudon ASI, Darroch AD, Milne JA (1984) The lactation performance of red deer on hill and improved species pastures. J Agric Sci Cambridge 102:149–158

Loudon ASI, McNeilly AS, Milne JA (1983) Nutrition and lactational control of fertility. Nature 302:145–147

Malpaux B, Robinson JC, Wayne NL, Karsch FJ (1989) Regulation of the onset of the breeding season of the ewe: Importance of long days and of an endogenous reproductive rhythm. J Endocr 122: 269–278

Martinet L, Allain D, Weiner C (1984) Role of prolactin in the photoperiodic control of moulting in the milk (*Mustela vison*). J Endocrinol 103:9–15

McLeod BJ, Haresign W, Lamming GE (1983) Induction of ovulation in seasonally anoestrous ewes by continuous infusion of low doses of GnRH. J Reprod Fert 68:489–495

McLeod BJ, Haresign W, Lamming GE (1982) Response of seasonally anoestrus ewes to small dose multiple injections of GnRH with and without progesterone pre-treatments. J Reprod Fert 65:223–230

McLeod BJ, Brinklow BR, Curlewis JD, Loudon ASI (1991) Efficacy of intermittent or continuous administration of GnRH in inducing ovulation in early and late seasonal anaestrus in the Père David's deer hind. J Reprod Fert 91:229–238

McNatty KP, Gibb M, Dobson C, Thurley DC (1981) Evidence that changes in luteinizing hormone secretion regulate the growth of the preovulatory follicle in the ewe. J Endocrinol 90:375–389

Meyer SL, Goodman RL (1986) Separate neural systems mediate the steroid- dependent and steroid-independent suppression of tonic luteinizing hormone secretion in the anestrus ewe. Biol Reprod 35:562–571

Milne JA, Loudon ASI, Sibbald AM, Curlewis JD, McNeilly AS (1990) Effects of melatonin and a dopamine agonist and antagonist on seasonal changes in voluntary food intake, reproductive activity and plasma concentrations of prolactin and tri-iodothyronine in red deer hinds. J Endocrinol 125:241–249

Milne JA, Sibbald AM, McCormack HA, Loudon ASI (1987) The influence of nutrition and management of the growth of red deer calves from weaning to 16 months of age. Anim Prod 45:511–522

Milne JA, Macrae JC, Spence AM, Wilson S (1978) A comparison of the voluntary of the voluntary intake and digestion of a range of forages at different times of the year by sheep and red deer. Br J Nutr 40: 347–357

Mirachi RE, Howland BE, Scanlon PF, Kirkpatrick RL, Sanford LM (1978) Seasonal variation in plasma LH, FSH, prolactin and testosterone concentrations in adult male white-tailed deer. Can J Zool 56:121–127

Mishra HJ, Wemmer CM (1987) The comparative breeding ecology of four cervids in Royal Chitwan National Park. In: Wemmer CM (ed) Biology and Management of the Cervidae. Smithsonian Institution Press, Washington DC., pp

Mitchell B, Lincoln GA (1973) Conception dates in relation to age and condition in two populations of red deer in Scotland. J Zool London 171:141–152

Mitchell B, McCowan D, Nicholson IA (1976) Annual cycles of body weight and condition in Scottish red deer. J Zool London 180:107–127

Morris RC (1935) Growth and shedding of antlers in sambar (*Cervus unicolor*) and cheetal (*Axis axis*) in south India. J Bombay Nat Hist Soc 37:485

Nicholls TJ, Goldsmith AR, Dawson A (1988) Photorefractoriness in birds and comparison with mammals. Physiol Rev 68:133–176

Phillips WWA (1928) A guide to the mammals of Ceylon. Spolia Zeylancia 14:1–50

Plotka ED, Seal US, Letellier MA, Verme LJ, Ozoga JJ (1981) Effects of pinealectomy on seasonal phenotypic changes in white-tailed deer (*Odocoileus virinialis borealis*), In: Matthews CD, Seamark RF (eds) Pineal Function. Elsevier/North-Holland, Amsterdam, pp 45–56

Plotka ED, Seal US, Verme LJ, Ozoga JJ (1980) Reproductive steroids in deer III, luteinizing hormone, estradiol and progesterone around estrus. Biol Reprod 22:576–581

Plotka ED, Seal US, Letellier MA, Verme LJ, Ozoga JJ (1979) Endocrine and morphologic effects on pinealectomy in white-tailed deer. In: Alexander J (ed) Animal Models for Research on Contraception and Fertility Research. Harper and Row, Hagerstown, pp 452–466

Pollock AM (1975) Seasonal changes in appetite and sexual condition in red stags maintained on six-month photoperiod. J Physiol 244:95–96

Rahe CH, Owens RE, Fleeger JL, Newton HJ, Harms PG (1980) Pattern of plasma luteinizing hormone in the cyclic cow: dependence upon the period of the cycle. Endocrinology 107:498–503

Reiter RJ (1980) The pineal and its hormones in the control of reproduction in mammals. Endocrine Rev 1:109–131

Renecker LA, Hudson RJ (1986) Seasonal energy expenditure and thermoregulatory responses of moose (*Alces alces*). Can J Zool 64:322–327

Robbins CT, Oftedal OT, O'Rourke KI (1987) Lactation, early nutrition and hand-rearing of wild ungulates with special reference to deer. In: Wemmer C (ed) Biology and Management of the Cervidae. Smithsonian Institution Press, Washington, D.C., pp

Robinson JE, Karsch FJ (1984) Refractoriness to inductive day lengths terminates the breeding season of Suffolk-cross ewes. Biol Reprod 31:656–663

Robinson JE, Wayne NL, Karsch FJ (1985) Refractoriness to inhibitory daylengths initiates the breeding season of the Suffolk ewe. Biol Reprod 32:1024–30

Sadleir RMFS (1980) Milk yield of the black-tailed deer. J Wildlife Management 44:472–478

Scaramuzzi RJ, Baird DT (1977) Pulsatile release of luteinizing hormone and the secretion of ovarian steroids in sheep during anestrus. Endocrinology 101:1801–1806

Schaller GB (1967) The Deer and the Tiger: a Study of Wildlife in India. University of Chicago Press, Chicago

Schaller GG, Hamner A (1978) Rutting behavior of the Pere David's deer (*Elaphurus davidianus*). Zoologische Garten 48:1–15

Schams D, Barth D (1982) Annual profiles of reproductive hormones in peripheral plasma of the male roe deer (*Capreolus capreolus*). J Reprod Fert 66:463–468

Schams D, Barth D, Karg H (1980) LH, FSH and progesterone concentration in peripheral plasma of the female roe deer, (*Capreolus capreolus*) during the rutting season. J Reprod Fert 60:109–114

Silver H, Colovos NF, Holter JB, Haynes HH (1969) Fasting metabolism of white-tailed deer. J Wildlife management 33:490–498

Simpson AM, Suttie JM, Kay RNB (1983) The influence of artificial photoperiod on the growth, appetite and reproductive status of male red deer and sheep. Anim Reprod Sci 6:291–299

Simpson AN, Webster AJF, Smith JF, Simpson CA (1978) The efficiency of utilization of dietary energy for sheep (*Ovis ovis*) and red deer (*Cervus elaphus*). Comp Biochem Physiol 59A:95–99

Suttie JM, Kay RNB (1985) Influence of plane of winter nutrition on plasma concentrations of prolactin and testosterone and their association with voluntary food intake in red deer stags (*Cervus elaphus*). Anim Prod Sci 8:247–258

Suttie JM, Lincoln GA, Kay RNB (1984) The endocrine control of antler growth in red deer stags. J Reprod Fert 71:7–15

Tyler NJC (1987) Body composition and energy balance of pregnant and non-pregnant Svalbard reindeer during winter. Symp Zool Soc Lond 57:203–226

Wade GN, Bartness TJ (1984) Seasonal obesity in Syrian hamsters: effects of age, diet, photoperiod and melatonin. Am J Physiol 247:R328–R334

Walters DL, Schallenberger E (1984) Pulsatile secretion of gonadotrophins, ovarian steroids and ovarian oxytocin during the periovulatory phase of the oestrous cycle in the cow. J Reprod Fert 71:503–512

Wayne NL, Malpaux B, Karsch FJ (1988) How does melatonin code for day length in the ewe: duration of nocturnal melatonin release for coincidence of melatonin with a light sensitive period? J Biol Reprod 39:66–75

Webster JR, Barrell GK (1985) Advancement of reproductive activity, seasonal reduction in prolactin secretion and seasonal pelage changes in pubertal red deer hinds (*Cervus elaphus*) subjected to artificially shortened daily photoperiod or daily melatonin treatment. J Reprod Fert 73:223–260

Wemmer C (1983) Systematic position and anatomical traits. In: Beck B, Wemmer C (eds) The Biology and management of an Extinct Species—Père David's Deer. Noyes Publications, N.J., pp 15–25

Whitehead GK (1972) Deer of the World. Constable, London

Wood AJ, Cowan I McT, Nordan HC (1962) Periodicity of growth in ungulates as shown by deer of the genus Odocoileus. Can J Zool 40:593–603

Yuthasastrakosol P, Palmer WM, Howland BE (1975) Luteinising hormone, oestrogen and progesterone levels in peripheral serum of anoestrus and cyclic ewes as determined by radioimmunoassay. J Reprod Fert 43:57–65

Zuckerman S (1953) The breeding season of mammals in captivity. Proc Zool Soc Lond 122:827–950

62

Influence of Plane of Nutrition on Seasonal Changes in Food Intake and Reproduction in Grazing Red Deer

Matthew J. Heydon, Andrew S.I. Loudon, John A. Milne, and Barry R. Brinklow

Abstract

The influence of herbage availability and endogenous appetite changes on food intake and the timing of breeding were investigated in lactating, nonlactating and melatonin-treated red deer hinds (*Cervus elaphus*) on long and short sward pastures. On the high-sward pasture, intake of melatonin-treated hinds was significantly lower than controls. Lactation and level of food intake did not affect the time of onset of estrous cycles, but on a restricting low-sward pasture fewer lactating animals cycled. Differences in appetite were expressed through time spent grazing and diet selection.

Key words: Alkane, food intake, lactation, melatonin, red deer, reproduction

Introduction

Enhoused red deer exhibit pronounced seasonal rhythms in metabolism and reproduction under the control of photoperiod (Kay 1979; Suttie and Simpson 1985). These changes correspond to seasonal changes in herbage availability, suggesting such cycles are an adaptation to predictable changes in resources (Kay and Staines 1981). Few studies have investigated the importance of these adaptations in free-ranging deer, particularly the direct influence of herbage availability on metabolism and reproduction. This interaction is particularly relevant to the Scottish red deer, as many populations occupy impoverished open moorland habitats (Kay and Staines 1981). In both wild (Mitchell et al. 1976; Albon et al. 1986) and farmed (Hamilton and Blaxter 1980) populations, fertility and onset of breeding have been correlated with body weight, suggesting herbage availability exerts a major influence on reproduction. Studies of domestic ruminants have demonstrated correlations between sward characteristics and diet selection and herbage intake (Allden and Whittaker 1970; Arnold 1975), suggesting the influence of appetite may be subordinate to herbage availability.

We conducted two experiments to investigate the interaction between appetite and food availability. The first considered the effect of herbage availability on forage intake. The second investigated the effect of sward height on food intake, behavior, and the onset of the breeding season in lactating and nonlactating hinds. In both experiments, exogenous melatonin was administered to some hinds to advance the breeding season and autumn appetite.

Materials and Methods

Both experiments were performed at the Macaulay Land Use Research Institute's Glensaugh Research Station (Laurencekirk, Kincardineshire, U.K.) between July 1987 and October 1988. The origin of the red deer herd is described by Blaxter et al. (1974). The hinds used in experiments were accustomed to intensive handling.

In Experiment 1, 16 nonlactating hinds were treated with 10 mg/day melatonin (Sigma Chemical, Poole, Dorset, U.K.) orally at 1600 h from 24 July to 8 October. Sixteen untreated, nonlactating hinds were controls. Between 8 and 27 October, eight hinds from each group were allocated to either a low (LP) or a high (HP) herbage availability pasture. Sward heights and herbage mass were 3.4 cm and 894 kg DM/ha for LP, and 5.5 cm and 1573 kg DM/ha for HP. Pastures were a perennial rye grass/white clover mixture with the same management history. Sward heights were managed by manipulating stocking density.

In Experiment 2, three groups of hinds were kept on a pasture maintained at 4 to 5 cm (LP) from July to October: a eight lactating hinds (group LL); seven lactating hinds treated with two 0.5 g subcutaneous melatonin implants (Sigma Chemical) on 23 June (group LLM), and eight nonlactating hinds (group LNL). In addition, eight lactating hinds were kept on a pasture maintained at 8 to 10 cm (group HL).

Onset of the breeding season was determined from plasma progesterone levels in samples collected by jugular venipuncture at twice weekly intervals. The onset was defined as the first value < 1.6 nmol/l followed by two or more samples with concentrations $\geqq 1.6$ nmol/l. Anestrous progesterone concentrations were near or below the detection limit of the assay (0.3 nmol/l).

Herbage intake and digestibility were measured using the n-alkane technique (Mayes et al. 1986). Each measurement consisted of 11 days on which hinds were dosed with pellets containing 170 mg dotriacontane (C32). From day 7 to day 12 fecal samples were collected. These were bulked for each hind to give a mean estimate of intake over the measurement period. Plant material representative of that eaten by the hinds was collected for analysis of the diet alkane content. In Experiment 2 a single measurement of calf intake was made during September using a 100 mg C32 pellet.

Observations of the time spent grazing, ruminating, or in other activities were made following each intake determination. In Experiment 1, activity was noted every 10 min during daylight hours, and in Experiment 2 every 5 min between approximately 0300 and 2300 h. Bite rate was estimated from repeated determinations of the time taken for 20 bites. Bite weight was derived from the daily herbage intake divided by predicted total number of bites per day.

Hind live weight was recorded weekly. Calf body weight was measured twice weekly during Experiment 2 as an index of lactational performance. Herbage mass was estimated by clipping herbage to ground level using electric clippers, and sward height was measured with a sward stick.

Analysis of varience was applied to the results, and differences between means were examined using Student's *t*-test.

Results

Experiment 1

Treatment with melatonin advanced the date of first estrus by 14.7 days (controls, 16 October ± 2.1 days; melatonin-treated, 1 October ± 1.2 days; $P < 0.001$). Herbage intakes (g DM per day/kg$^{0.75}$) of melatonin-treated hinds on the HP (70.9 ± 2.7) were lower ($P < 0.05$) than those of the control group (90.0 ± 6.0), but higher ($P < 0.0001$) than the intakes of either group on LP (control 51.2 ± 3.4; melatonin-treated 51.3 ± 2.8). There were no differences between the groups on LP. Groups on the LP spent longer times grazing and being active ($P < 0.05$) and less time ruminating ($P < 0.001$). Bite rates did not differ despite intake and sward height differences ($P < 0.05$).

Experiment 2

The mean onset of estrous cyclicity for hinds treated with melatonin (28 September ± 4.0 days) was earlier ($P < 0.05$) than untreated lactating hinds on the LP (16 October ± 5.2 days). Only 5 out of 8 LL hinds exhibited estrous cycles compared with all animals in the other groups ($P < 0.01$). The mean dates of onset in groups HL and LNL were 13 October ± 1.6 days and 19 October ± 2.7 days respectively.

There were consistent differences in the sward

height (LP 4.4 cm ± 0.2; HP 10.1 ± 0.5) and herbage mass (LP 466 kg DM/ha ± 95; HP 1659 ± 189) throughout the experiment. In vivo digestibilities were different between pastures (HP = 0.77 and LP = 0.71; $P < 0.01$) and between lactating and non-lactating hinds on LP (LL, LLM = 0.70, LNL = 0.73; $P < 0.05$) but did not change with time. Sward height and digestibility were positively correlated ($n = 0.80$, $P < 0.001$).

Herbage intake, corrected for differences in metabolic body weight, were higher ($P < 0.001$) for the hinds garzing the HP than the LP on all but the first comparison (Fig. 62.1). Melatonin treatment did not affect food intake in lactating hinds on the LP, and intakes declined as the summer progressed in both groups. In contrast, intakes of nonlactating hinds on LP were lower ($P < 0.01$) than those of lactating hinds. Herbage

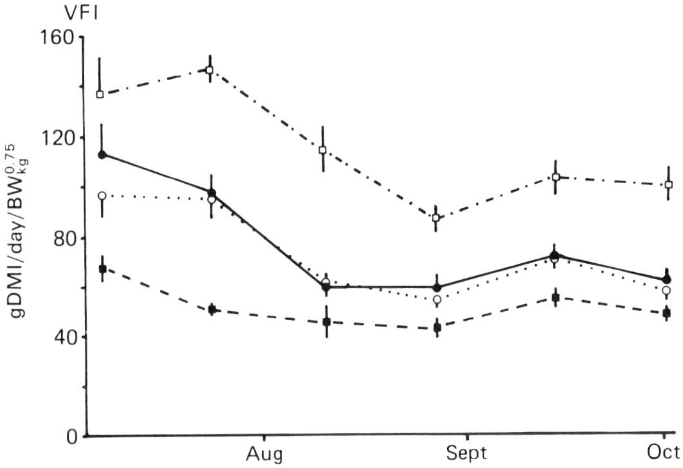

FIGURE 62.1. Mean (⊥ SEM) herbage intakes of red deer hinds in Experiment 2 between July and October for groups lactating (LL; ●); lactating with melatonin implants (LLM; ○) and nonlactating (LNL; ■) on a low sward pasture, and lactating on a high sward pasture (HL; □).

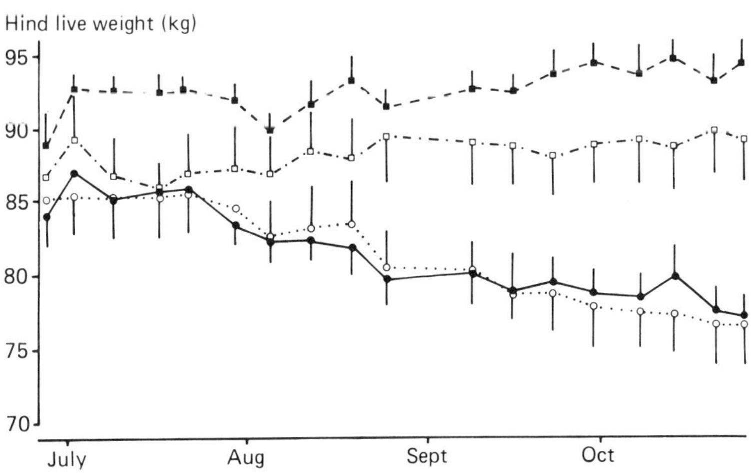

FIGURE 62.2. Changes in live weight ($\bar{x} \pm$ SEM) of red deer hinds in Experiment 2 betwen July and October for groups lactating (LL; ●); lactating with melatonin implants (LLM; ○) and nonlactating (LNL; ■) on a low sward pasture, and lactating on a high sward pasture (HL; □).

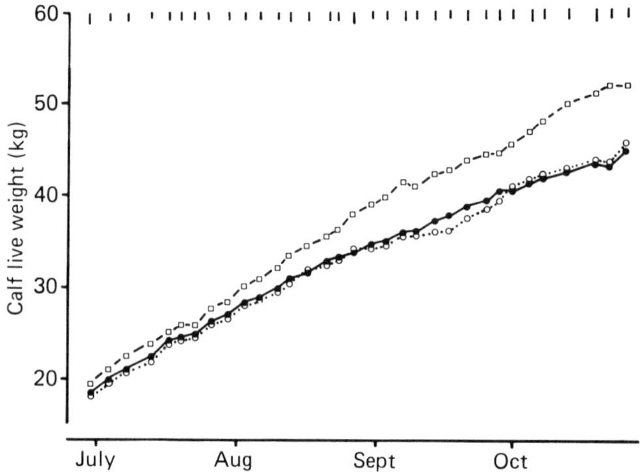

FIGURE 62.3. Changes in mean live weight ($\bar{x} \pm$ SEM) of red deer calves in Experiment 2 between July and October of hinds lactating (LL; ●) and lactating with melatonin implants (LLM; ○) on a low sward pasture, and lactating on a high sward pasture (HL; □). Vertical bars represent one pooled SEM; $n = 23$.

intakes by calves measured during September were higher ($P < 0.05$) for animals on the LP treatment (35.4 ± 1.72 g DM/day/kg$^{0.75}$) than on the HP (30.0 ± 1.06).

Lactating hinds on LP grazed for a greater proportion of the observed period ($P < 0.001$) than hinds on HP (51% and 39%). Bite rate was higher ($P < 0.001$) on the LP (82.1 and 61.6 bites/min, for LL and HL), but estimated bite weight was lower (36.3 and 85.8 mg DM/bite for LL and HL, respectively; $P < 0.001$). Proportion of time spent grazing (45%) and estimated bite weight (31.0 mg DM/bite) of group LNL were lower than of group LL ($P < 0.01$ and $P < 0.05$, respective). Bite rate and weight were significantly correlated to both sward height and mass ($P < 0.01$ and $P < 0.001$, respectively).

Hind live weight remained stable throughout the study for groups HL and LNL. In contrast, it showed a progressive declined for the lactating groups on LP (Fig. 62.2). There were no differences in calf live weights of either group on the LP ($P > 0.05$). Calves were heavier ($P < 0.05$) on HP from week 8 through to the end of the study (Fig. 62.3).

Discussion

The lower intakes of melatonin-treated hinds on HP in Experiment 1 are consistent with the effects observed in enhoused animals (Milne et al. 1990), suggesting that the autumnal appetite decline (Loudon et al. 1989) is exerting an influence on the intake of grazing hinds. The absence of any differences in intake on the shorter sward indicates that expression of an endogenous appetite cycle can be suppressed by limiting herbage availability. The longer grazing time of LP groups reveals that animals were unsuccessful in compensating for limited food availability by extending the grazing period. This is consistent with observations of sheep-grazing behavior (Allden and Whittaker 1970). Failure to observe differences in the feeding behavior of HP groups despite a significant difference in intake may be due to insufficient length of observation, particularly at night.

In Experiment 2, hinds lactating on LP were unable to meet their nutrient requirements, as reflected by weight losses and reduced calf growth. Poorer milk production of hinds grazing impoverished pastures (Loudon et al. 1984) may explain the higher herbage intakes of calves on LP. Despite low herbage mass, lactating hinds maintained higher food intakes than nonlactating hinds, although these intakes were lower than the HL hinds. The ability of nonlactating hinds to maintain high body weights on very short grass pasture may explain why similar animals on Rhum spend a greater proportion of their time utilizing heather moorland avoiding unnecessary competition with lactating hinds, who concentrate their efforts on the better quality pastures of the glen bottoms (Clutton-Brock et al. 1982).

Differences in the grazing time, bite weight, and bite rate of lactating hinds on the two pastures reflect previously observed effects of sward height and mass on sheep grazing behavior (Allden and Whittaker 1970). The larger bite weight and lower diet digestibility of lactating than nonlactating hinds on LP suggests the former were less selective in diet choice. This is supported by a positive correlation between sward height and digestibility.

The failure of melatonin implants to induce differences in nutritional parameters on a restricting pasture is consistent with the results of Experiment 1. In contrast, the lower herbage intakes and poorer body condition compared with the HL group and the suckling stimulus of the calf compared with the LNL group, did not prevent an advance in the onset of the breeding season in the melatonin-treated group. In addition, there were no differences in the onset of estrous cyclicity between nontreated groups irrespective of live weight, intake or lactational status. This contradicts the evidence of some previous studies which suggested nutrition (Loudon et al. 1983) and lactation (Adam et al. 1985) have a significant effect on the time of onset of the breeding season.

The failure of three out of eight hinds in group LL to cycle may reflect the low body weight of these hinds. This is supported by studies of both wild (Mitchell et al. 1976; Albon et al. 1986) and farmed (Hamilton and Blaxter 1980) red deer in which a strong correlation has been demonstrated between fertility and body weight/condition. The failure to cycle when in poor condition has adaptive significance for wild deer in marginal habitats such as Scotland by avoiding pregnancy when in poor condition, thus increasing the probability of surviving the subsequent winter.

Acknowledgements

This study was supported by a studentship from the Natural Environment Research Council (U.K.). Special thanks to S. Busby, M.M. Gallagher, and A. Sneddon for assistance with field work and P.M. Colgrove, C.S. Lamb, R.W. Mayes, and A.M. Sibbald for assistance with analysis.

References

Adam CL, Moir CE, Atkinson T (1985) Plasma concentrations of progesterone in female red deer during the breeding season, pregnancy and anoestrus. J Reprod Fert 74:631–636

Albon SD, Mitchell B, Huby BJ, Brown D (1986) Fertility in female red deer: the effects of body composition, age and reproductive status. J Zool London (A) 209:447–460

Allden WG, Whittaker IAMcD (1970) The determinants of herbage intake by grazing sheep: the interrelationships of factors influencing herbage intake and availability. Aust J Agric Res 21:755–766

Arnold GW (1975) Herbage intake and grazing behaviour in ewes of four breeds at different physiological states. Aust J Agric Res 26:1017–1024

Blaxter KL, Kay RNB, Sharman GAM, Cunningham JMM, Hamilton WJ (1974) Farming the Red Deer, London, Her Majesty's Stationery Office

Clutton-Brock TH, Iason GR, Albon SD, Guiness FE (1982) Effects of lactation on feeding behaviour and habitat use in wild red deer hinds. J Zool London 198:227–236

Hamilton WJ, Blaxter KL (1980) Reproduction in farmed red deer 1. Hind and stag fertility. J Agric Sci Cambridge 95:261–273

Kay RNB (1979) Seasonal changes of appetite in deer and sheep. ARC Research Rev 5:13–15

Kay RNB, Staines BW (1981) The nutrition of red deer. Nutr Abst Rev B 51:601–622

Loudon ASI, McNeilly AS and Milne JA (1983) Nutrition and lactation control of fertility in red deer. Nature 302:145–147

Loudon ASI, Darroch AD, Milne JA (1984) The lactation performance of red deer on hill and improved species pastures. J Agric Sci Cambridge 102:149–158

Loudon ASI, Milne JA, Curlewis JD, McNeilly AS

(1989) A comparison of the seasonal hormone changes and patterns of growth, voluntary food intake and reproduction in juvenile and adult red deer and Père David's deer hinds. J Endocrinol 122:733–745

Mayes RW, Lamb CS, Colgrove PM (1986) The use of dosed and herbage n-alkanes as markers for the determination of herbage intake. J Agric Sci Cambridge 107:161–170

Milne JA, Loudon ASI, Sibbald AM, Curlewis JD, McNeilly AS (1990) The effects of melatonin and a dopamine agonist and antagonist on seasonal changes in voluntary intake, reproductive activity and plasma concentrations of prolactin and tri-iodothyronine in red deer hinds. J Endocrinol 125:241–249

Mitchell B, McCowan D, Nicholson IA (1976) Annual cycles of body weight and condition in Scottish red deer. J Zool London 180:107–127

Suttie JM, Simpson AM (1985) Photoperiodic control of appetite, growth, antlers and endocrine status of red deer. In Fennessy PF, Drew KR (eds) Biology of Deer Production. R Soc N Z Wellington Bull 22:429–432

63
Patterns of LH Secretion in Cycling and Seasonally Anestrous Père David's Deer Hinds

Bernard J. McLeod, Jonathon D. Curlewis, and Andrew S.I. Loudon

Abstract

Changes in patterns of episodic luteinizing hormone (LH) secretion were monitored during the follicular and luteal phases of the estrous cycle, and during seasonal anestrus on five occasions. The LH response to a dose of gonadotrophin-releasing hormone (GnRH) capable of evoking release of an LH episode was also determined in the luteal phase of the cycle, and at the same five intervals during seasonal anestrus. In the breeding season, mean LH concentrations did not differ significantly between the follicular and the luteal phase of the estrous cycle, but the mean frequency of LH episodes was higher ($P < 0.001$) during the follicular phase. During seasonal anestrus, mean LH concentrations were significantly higher ($P < 0.05$) in midanestrus than in early or late anestrus. In addition, mean LH episode frequencies were higher in both mid- (April, May, $P < 0.01$) and late (June, $P < 0.05$) anestrus, than in early anestrus (February, March). The mean amplitude of endogenous LH episodes did not differ significantly between stage of the estrous cycle or over the anestrus period, but the mean LH response to exogenous GnRH was significantly lower ($P < 0.05$) in early anestrus than in midanestrus or late anestrus, at which times it was comparable to that recorded in the luteal phase. The patterns of LH monitored in Père David's (*Elaphurus davidianus*) deer hind during the estrous cycle were similar to those occurring in domestic ruminants, except that the magnitude of LH episodes was not increased in the luteal phase. The differences in endogenous and GnRH-induced LH release recorded over the anestrus period, which have not previously been observed in a seasonal breeding species, indicate an early period of deep anestrus.

Key words: Anestrus, estrous cycle, GnRH, LH, Père David's deer, pulsatile secretion

Introduction

The gonadotrophic hormone luteinizing hormone (LH) is secreted in discrete pulses and is released as a direct consequence of pulsatile release of gonadotrophin-releasing hormone (GnRH) from the hypothalamus (Clarke and Cummins 1982). In domestic ruminants, different reproductive states are associated with different patterns of LH secretion, particularly of episodic LH release. For example, the frequency with which LH episodes occur in the ewe is much higher during the breeding season than during seasonal anestrus (Baird and Scaramuzzi 1976; Yuthasatrakosol et al. 1977), and this is reflected in differences in mean plasma LH concentrations (Legan and Karsch 1979; Webster and Haresign 1983). Similarly, periods of prepubertal (Foster et al. 1975; Schams et al. 1981), postpartum (Lamming et al. 1981; Wright et al. 1983), and lactational (Peters et al. 1981, Wright et al. 1983) acyclicity in cows and sheep, are associated with a low frequency of LH episodes. There is an increase in LH episode frequency following regression of the corpus luteum in the estrous cycle in the cow (Walters and Schallenberger 1984), the ewe (Baird 1978), and the goat (Mori and Kano 1984) that is associated with the final stages of preovulatory follicle development prior to ovulation. In addition, episode frequency is greater during the follicular phase than during the luteal phase of the estrous cycle (Goodman and Karsch 1980; Rähe et al. 1980).

There is very little information available on patterns of LH secretion in deer, although changes in mean LH concentrations during the estrous cycle have been described for roe deer (*Capreolus capreolus*; Schams et al. 1980), white-tailed deer (*Odocoileus virginianus*; Plotka et al. 1980) and fallow deer (*Dama dama*; Asher et al. 1986). However, the frequency of blood sampling employed in these studies did not allow LH episodes to be defined. The objective of the present study was to monitor changes in mean plasma concentrations of LH over the estrous cycle and during seasonal anestrus, and to relate these changes to patterns of episodic LH secretion. In addition, LH release in response to a dose of GnRH capable of evoking an LH episode was compared at different stages of anestrus and in the luteal phase of the estrous cycle.

Methods

Animals and Management

Tame, adult, postpubertal Père David's deer hinds, in which the time of onset of the breeding season had been determined from profiles of plasma progesterone concentrations, were maintained on pasture at Whipsnade Park, Bedfordshire, U.K. (latitude 50° 30′N). During periods of intensive blood sampling, the animals were housed indoors under conditions of natural photoperiod and temperature and were fed hay and pelleted lucerne.

Experiment 1—Estrous Cycle

In midbreeding season (September), five hinds (mean live weight 162 ± 6.3 kg) were treated with progesterone implants to synchronize estrus. Five implants (Sil-Estrus; Ceva, Sydney, Australia), each containing 375 mg progesterone in a silicone elastomer matrix, were inserted subcutaneously (under local anesthesia) in the axilla region and left in situ for 11 days. At the time of implant removal, the hinds were injected intramuscularly with prostaglandin (1 ml, Estrumate; Coopers Animal Health, Crewe, Cheshire, U.K.) and were fitted with an indwelling jugular vein catheter. Starting 16 h after implant removal, blood samples for LH determination were taken at 15-min intervals for 36 h, and then at 3-h intervals for a further 48 h. A second period (10 h) of blood-sampling at 15-min intervals was carried out 9 days after implant removal to assess patterns of LH secretion during the luteal phase. On the following day the animals were subjected to a GnRH challenge. Four injections of 2.0 μg GnRH (Lutal; Fabwerke Hoechst, Frankfurt, West Germany) in 2.0 ml of sterile saline, were given at 2-h intervals. Blood samples for LH determination were taken at 15-min intervals, from immediately before the first, until 2 h after the last GnRH injection. Blood samples for progesterone determination were taken three times weekly while the progesterone implants were in situ, and daily for 10 days after implant withdrawal. The hinds were checked twice daily (0800 h and 1700 h) for signs of behavioral estrus, using the procedure described by Loudon et al. (1990).

Experiment 2—Seasonal Anestrus

Five hinds (mean live weight 173 ± 5.2 kg) were subjected to five periods of intensive blood sampling (samples collected at 15-min intervals for 18 h) undertaken at 4 to 5 week intervals (11 February, 17 March, 23 April, 27 May, and 24 June) throughout the period of seasonal anestrus. Indwelling jugular vein catheters were inserted the day before each blood-sampling period. The day following each of these sampling periods, the hinds were subjected to a GnRH challenge identical to that described in Experiment 1. Blood samples were taken at 15-min intervals, from immediately before the first, until 2 h after the last injection. From early December until mid-August, blood samples for progesterone analysis were taken twice weekly by jugular venepuncture. The time of cessation and onset of the breeding season, was determined from these progesterone profiles.

Hormone Analyses

Plasma LH concentrations were measured using a heterologous ovine double-antibody radioimmunoassay (McLeod et al. 1982), which has been validated for deer LH (Loudon et al. 1990). Within this study, the limit of sensitivity of the assay was 0.2 ng NIH-LH-S24 equivs/ml plasma, and the intra- and interassay coefficients of variation were 8.6% and 9.8% respectively. Plasma progesterone concentrations were determined using the radioimmunoassay method of Curlewis et al. (1988), previously validated for this species. The intra- and interassay coefficients of variation were 3.9% and 10.4%, respectively, and the limit of sensitivity of the assay was 0.1 ng/ml plasma.

Analysis of Data

An LH episode was defined as (1) an increase in plasma LH concentrations of at least four times the coefficient of variation of duplicate pairs, in which (2) there was not more than one sample point between the peak and the preceding trough, and (3) there were no less than two sample points between the peak and the succeeding trough. Assessment of episodic patterns of secretion during the follicular phase of the estrous cycle was restricted to four animals, due to failure of the indwelling catheter in the remaining hind. This animal was blood-sampled at 3-h intervals by jugular venepuncture to monitor the preovulatory surge release of LH. A preovulatory LH surge was defined as a sustained (> 5.0 h) elevation in LH concentrations above 4 ng/ml, with a peak concentration during this time that exceeded 10 ng/ml (McLeod et al. 1985). The onset of the peak was defined as the time when LH concentrations first exceeded 4 ng/ml.

The response to GnRH was calculated as the difference in plasma LH concentrations between the sample taken immediately before injection of GnRH and the highest concentration recorded within 30 min of injection. The onset of seasonal anestrus was defined as the date when plasma progesterone concentrations fell below 0.3 ng/ml following the last luteal phase of the breeding season, and the onset of the breeding season was defined as the date of the first increase in progesterone concentrations above 1 ng/ml for more than two consecutive samples.

Differences in mean LH concentrations and in mean amplitude of LH episodes were compared by ANOVA, and differences in LH episode frequency were assessed by chi-squared analysis. All values are presented as mean ± SEM.

Results

Experiment 1—Estrous Cycle

Plasma Progesterone Concentrations and Estrus Behavior

Over the period that the progesterone implants were in situ, mean plasma progesterone concentrations were 5.6 ± 0.5 ng/ml. By 16 h after implant removal, mean concentrations had fallen to 0.7 ± 0.1 ng/ml, and 24 h later they were below 0.2 ng/ml. Four of the five hinds were in full standing estrus when checked at 0800 h on the third day after implant removal, and the remaining animal was in estrus at 0800 h the following day. Plasma progesterone concentrations remained low (< 0.2 ng/ml) until the fourth day after estrus, when they increased significantly ($P < 0.05$) in all five hinds. The mean maximum progesterone concentrations recorded in the subsequent luteal phase were 3.6 ± 0.1 ng/ml.

Plasma LH Concentrations

Mean LH concentrations during the follicular phase (0.57 ± 0.09 ng/m) did not differ significantly from those in the luteal phase (0.74 ± 0.13 ng/ml; Fig. 63.1). Although LH episodes were recorded in all hinds at both stages of the estrous cycle, these occurred at a higher ($P < 0.001$) frequency during the follicular phase (7.08 ± 1.08 and 2.88 ± 0.23 episodes/12 h for the follicular and luteal phases, respectively). There were no significant differences in the mean amplitude of LH episodes recorded in the follicular phase (0.99 ± 0.14 ng/ml) with those monitored in the luteal (1.05 ± 0.21 ng/ml) phase of the estrous cycle.

The preovulatory LH surge ($n = 5$) had a mean time of onset of 59.4 ± 3.3 h after removal of the progesterone implants, a mean duration of 15.3 ± 1.2 h, and reached a mean maximum concentration of 17.1 ± 2.85 ng/ml. When the follicular phase LH profiles were standardized

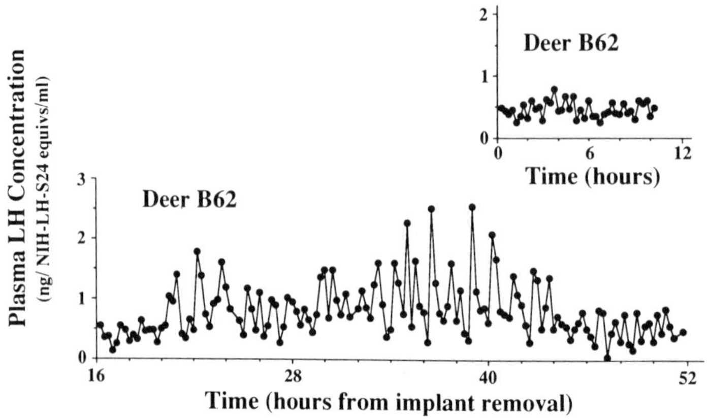

FIGURE 63.1. Plasma LH concentrations in an individual Père David's deer hind (Deer B62) during the follicular phase of a progesterone-synchronized estrous cycle and (inset) during the subsequent luteal phase. Blood samples were taken at 15-min intervals.

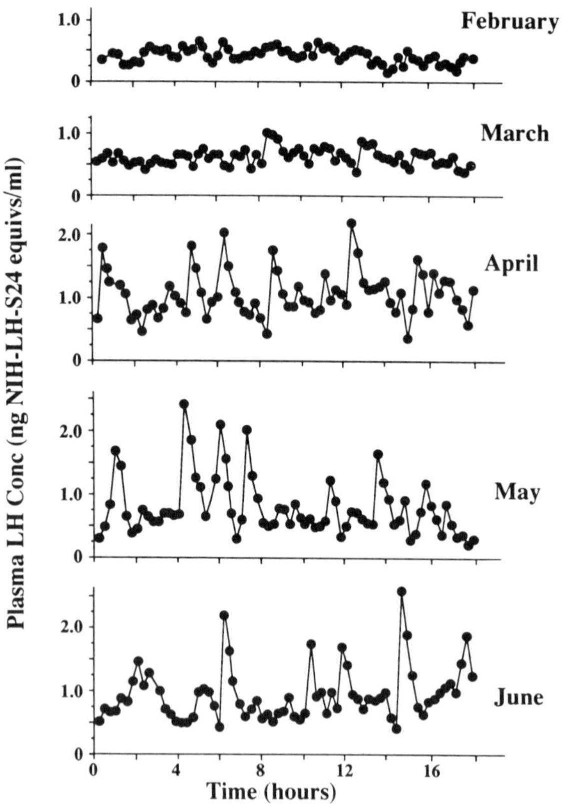

FIGURE 63.2. Plasma LH concentrations in individual Père David's deer hinds over the period of seasonal anestrus. Blood samples were taken at 15-min intervals.

about the time of onset of the preovulatory surge, there was a significant effect of time on mean LH concentrations over the period leading up to the surge. Mean LH concentrations were 0.42 ± 0.04, 0.53 ± 0.04, 0.88 ± 0.09, and 0.80 ± 0.05 ng/ml over the periods 0 to -6 h, -6 to -12 h, -12 to -18 h, and -18 to -24 h before the onset of the LH surge.

Experiment 2—Seasonal anestrus

Plasma Progesterone Concentrations and Seasonal Breeding Activity

The mean date for the start of the anestrous period (the end of the last luteal phase of the breeding season) was January 2 (± 4.7 days). Plasma progesterone concentrations then remained low (< 0.2 ng/ml) until mid- to late July when a short luteal phase (3 to 8 days) was recorded in all hinds. The start of the first normal-length luteal phase of the breeding season occurred on 4 August (± 5.2 days). Therefore, the first LH profile (11 February) had been carried out at a mean time of 40 ± 4.7 days after the end of the breeding season, and the last profile (24 June) had been at a mean time of 41 ± 5.2 days before the start of the next breeding season.

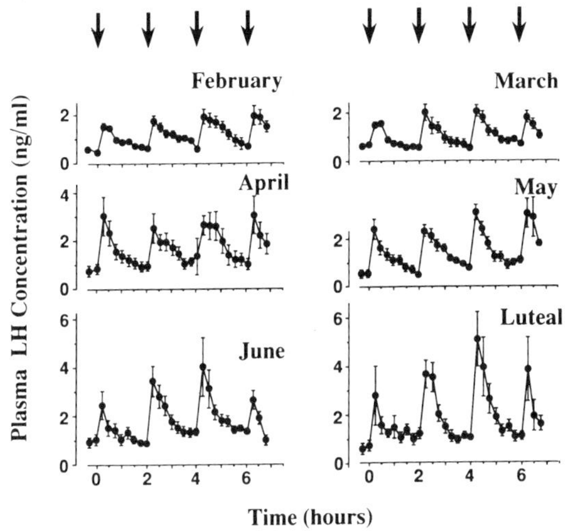

FIGURE 63.3. Mean LH concentrations in response to four i.v. injections of 2.0 μg GnRH given at 2-h intervals on five occasions throughout the period of seasonal anestrus, and in the luteal phase of the estrous cycle. Blood samples were taken at 15-min intervals. The times of GnRH injections are shown by the arrows.

Plasma LH Concentrations

There were significant changes in mean LH concentrations over the period of seasonal anestrus, with plasma concentrations being higher ($P < 0.05$) in midanestrus (April, 0.85 ± 0.12 ng/ml) than in early anestrus (February, 0.53 ± 0.05 ng/ml; Fig. 63.2). In addition, LH episode frequency was higher in midanestrus (April, 5.6 ± 0.85 episodes/18 h, $P < 0.001$) and in late anestrus (4.6 ± 0.83 and 4.2 ± 0.24 episodes/12 h in May and June, $P < 0.05$), than that in early anestrus (2.2 ± 0.39 and 2.4 ± 0.41 episodes/12 h in February and March). The mean amplitude of LH episodes recorded at each stage of anestrus did not change significantly. The mean amplitude of the LH response to GnRH was 1.21 ± 0.12, 1.24 ± 0.16, 1.77 ± 0.21, 1.91 ± 0.10, and 1.99 ± 0.31 ng/ml, in February, March, April, May, and June, respectively (Fig. 63.3). The magnitude of the response was significantly higher ($P < 0.05$) in mid- (April, May) and late (June) anestrus, than in early (February, March) anestrus.

Discussion

The mean frequency with which LH episodes occurred during the follicular phase of the estrous cycle is similar to that reported for sheep and cattle (Baird 1978; Yuthasastrakosol et al. 1977; Rahe et al. 1980). This is not surprising as there is convincing evidence that it is this high frequency of LH episodes that initiates pre-ovulatory follicle development. For example, the administration of repeated injections of low doses of GnRH to artificially impose a high frequency of LH episodes will induce ovulation in several species including deer (Fisher et al. 1989; McLeod et al. 1991). The decline in mean LH concentrations immediately prior to the onset of the LH surge, also noted previously in cattle and sheep (Walters and Schallenberger 1984; McLeod et al. 1988), is probably attributable to the effects of increased estrogen secretion at this time (Baird 1978). In contrast to the ewe in which there is a dramatic increase in LH episode amplitude during the luteal phase compared with that in the follicular phase (Baird and Scarramuzzi 1976; Goodman and Karsch 1980), there was no significant difference in the mean amplitude of LH episodes with stage of estrous cycle in these hinds. It is possible that high amplitude LH episodes do occur in deer, but at such a low frequency that none occurred during the 10-h sampling period undertaken in the present study. It is of interest to note that the mean amplitude of LH episodes recorded in mid- and late anestrus was significantly higher ($P < 0.01$) that that in the luteal phase of the cycle.

The magnitude of the LH surge was similar to that recorded in fallow deer (Asher et al. 1986) and in cattle (Rahe et al. 1980; Walters and Schallenberger 1984), but three to six times lower than occurs in sheep and goats (Baird and Scarramuzzi 1976; Bon Durant et al. 1981).

In the ewe, mean LH concentrations are higher during the breeding season than during anestrus, and LH levels fall abruptly at the onset of seasonal anestrus (Legan and Karsch 1979; Webster and Haresign 1983). However, there is little information on changes in the pattern of episodic secretion over the period of anestrus. Studies based on ovariectomized, estrogen-implanted (Walton et al. 1980) and in ovary-intact (I'Anson and Legan 1988) ewes suggest that there is an abrupt increase in LH episode frequency immediately before the first ovulation of the breeding season. In contrast, we found in the hind that there was an increase in LH episode frequency in midanestrus and that episode frequency at this time approached that recorded during the follicular phase of the estrous cycle (0.47 ± 0.05 and 0.59 ± 0.03 episodes/h for midanestrus (April) and the follicular phase, respectively). This does not preclude the occurrence of a further, abrupt increase in episode frequency prior to the first ovulation of the breeding season, particularly as episode frequency had decreased again by late anestrus in these deer. Nevertheless, the pattern of LH secretion recorded in this study is at variance with the mechanism postulated to control seasonality in the ewe, namely, that there is a photoperiod-induced change in sensitivity to estradiol negative-feedback on LH secretion at the onset and cessation of the breeding season (Legan et al. 1977; Karsch et al. 1985). The pattern of LH secretion recorded in the present study could be explained by a change (in midanestrus) in the mechanism imposing this increased sensitivity to estradiol. For example, the onset of anestrus may be due to refractoriness to the otherwise stimulatory effects of a short-day photoperiod, and the maintenance of acyclicity later in the anestrous period to the suppressive effect of a long-day photoperiod. The changes in LH secretion monitored between early and midanestrus would then reflect the transition between these two mechanisms. Evidence in support of a period of increased pituitary activity during the anestrous period is given by reports of a single ovulation occurring in some breeds of sheep (Thimonier and Mauleon 1969).

To date, there have been few studies of seasonal changes in the LH response to low doses of exogenous GnRH that include comparisons between different stages of anestrus. It has been reported that the LH response to GnRH administration in the ovariectomized ewe is lower during seasonal anestrus than during the breeding season, but also that a physiological dose of GnRH can evoke an LH response in anestrous ewes (Bittman et al. 1985). In the current study, there were significant differences in the LH response to GnRH administered in early anestrus with that to the same dose given in mid- or late anestrus. Curlewis et al. (1991) suggested that this, along with the changes in endogenous LH release, indicated that seasonal anestrus in the Père David's deer hind is not a uniform state, but is characterized by an early period of deep anestrus. In support of this suggestion, we have demonstrated that more hinds will ovulate in response to a prolonged period of GnRH administration in late anestrus than to the same treatment given in early anestrus (McLeod et al. 1991).

In summary, the patterns of LH secretion recorded in the Père David's deer hind during the estrous cycle are comparable to those reported for domesticated ruminants. The single exception is that the amplitude of LH episodes was not increased during the luteal phase compared with the follicular phase. The changes in LH secretion that occurred over the period of seasonal anestrus have not previously been recorded in a seasonal breeding species, and they suggest that seasonal anestrus is characterized by an early period of deep anestrus.

Acknowledgements

We thank A. Beasey and B. Eves for care and husbandry of the animals, NIADDK and National Hormone and Pituitary Program for the ovine pituitary hormones, Hoechst Pharmaceuticals for the GnRH preparations, and the AFRC for financial support.

References

Asher GW, Barrell GK, Peterson AJ (1986) Hormonal changes around oestrus in farmed fallow deer, *Dama dama*. J Reprod Fert 78:487–496

Baird DT (1978) Pulsatile secretion of LH and ovarian oestradiol during the follicular phase of the sheep estrous cycle. Biol Reprod 18:359–364

Baird DT, Scarramuzzi RJ (1976) Changes in the secretion of ovarian steroids and pituitary luteinizing hormone in the preovulatory period in the ewe; the effect of progesterone. J Reprod Fert 70:237–245

Bittman EL, Kaynard AH, Olster DH, Robinson JE, Yellon SM, Karsch FJ (1985) Pineal melatonin mediates photoperiodic control of pulsatile luteinizing hormone secretion in the ewe. Neuroendocrinology 40:409–418

Bon Durant RH, Darien BJ, Munro CJ, Stabenfeldt

GH, Wang P (1981) Photoperiod induction of fertile oestrus and changes in LH and progesterone in yearling dairy goats. J Reprod Fert 63:1–9
Clarke IA, Cummins JT (1982) The temporal relationship between gonadotrophin releasing hormone (GnRH) and luteinizing hormone (LH) secretion in ovariectomized ewes. Endocrinology 111:1737–1739
Curlewis JD, McLeod BJ, Loudon ASI (1991) LH Secretion and response to GnRH during seasonal anoestrus of the Père David's deer hind. J Reprod Fert 91:131–138
Curlewis JD, Loudon ASI, Coleman APM (1988) Oestrous cycles and the breeding season of the Père David's deer hind (*Elaphurus davidianus*). J Reprod Fert 82:119–126
Fisher MW, Fennessey PE, Davis GH (1989) A note on the induction of ovulation in lactating red deer hinds prior to the breeding season. Anim Prod 49:134–138
Foster DL, Lemons JA, Jaffe RB, Niswender GD (1975) Sequential patterns of circulating luteinizing hormone and follicle-stimulating hormone in the female sheep from early post-natal life through the first estrous cycles. Endocrinology 97:985–994
Goodman RL, Karsch FJ (1980) Pulsatile secretion of luteinizing hormone; differential suppression by ovarian steroids. Endocrinology 107:1286–1290
I'Anson H, Legan SJ (1988) Changes in LH pulse frequency and serum progesterone concentrations during the transition to breeding season in the ewe. J Reprod Fert 82:341–345
Karsch FJ, Bittman EL, Foster DL, Goodman RL, Legan SJ, Robinson JE (1985) Neuroendocrine basis of seasonal reproduction. Rec Prog Hormone Res 40:185–225
Lamming GE, Wathes DC, Peters AR (1981) Endocrine patterns of the post partum cow. J Reprod Fert (Suppl) 30:155–170
Legan SJ, Karsch FJ (1979) Neuroendocrine regulation of the estrous cycle and seasonal breeding in the ewe. Biol Reprod 20:74–85
Legan SJ, Karsch FJ, Foster DL (1977) The endocrine control of seasonal reproductive function in the ewe; a marked change in response to the negative feedback action of estradiol on luteinizing hormone secretion. Endocrinology 101:818–824
Loudon ASI, McLeod BJ, Curlewis JD (1990) Pulsatile secretion of LH during the periovulatory and luteal phases of the oestrous cycle in the Père David's deer hind (*Elaphurus davidianus*). J Reprod Fert 89:663–670
McLeod BJ, Brinklow BR, Loudon ASI, Curlewis JD (1991) Efficacy of intermittent or continuous administration of GnRH in inducing ovulation early and late in the period of seasonal anoestrus in Père David's deer hind. J Reprod Fert 91:229–238
McLeod BJ, Haresign W, Peters AR, Humke R, Lamming GE (1988) The development of subcutaneous-delivery preparations of GnRH for the induction of ovulation in acyclic sheep and cattle. Anim Reprod Sci 17:33–50
McLeod BJ, Peters AR, Haresign W, Lamming GE (1985) Plasma LH and FSH responses and ovarian activity in prepubertal heifers treated with repeated low doses of GnRH for 72 h. J Reprod Fert 74:589–596
McLeod BJ, Haresign W, Lamming GE (1982) The induction of ovulation and luteal function in seasonally anoestrous ewes treated with small-dose multiple injections of GnRH. J Reprod Fert 65:215–221
Mori Y, Kano Y (1984) Changes in plasma concentrations of LH, progesterone and oestradiol in relation to the occurrence of luteolysis, oestrus and time of ovulation in the Shiba goat. J Reprod Fert 72:223–230
Peters AR, Lamming GE, Fisher MW (1981) A comparison of plasma LH concentration in milked and suckled post partum cows. J Reprod Fert 62:567–573
Plotka ED, Seal US, Verme LJ, Ozoga JJ (1980) Reproductive steroids in deer. III Luteinizing hormone, estradiol and progesterone around estrus. Biol Reprod 22:575–581
Rahe CH, Owens RE, Fleeger JL, Newton HJ, Harms PGH (1980) Patterns of plasma luteinizing hormone in the cyclic cow; dependence upon the period of the cycle. Endocrinology 107:489–503
Schams D, Schallenberger E, Gombe S, Karg H (1981) Endocrine patterns associated with puberty in male and female cattle. J Reprod Fert (Suppl) 30:103–110
Schams D, Barth F, Karg H (1980) LH, FSH and progesterone concentrations in peripheral plasma of the female roe deer (*Capreolus capreolus*) J Reprod Fert 60:109–114
Thimonier J, Mauleon P (1969) Variations saisonnieres du comportement d'oestrus et des activities ovarienne et hypophysaire chez les ovins. Annales de Biologie Animale et Biochemie Biophysiologie 9:233–250
Walters DL, Schallenberger E (1984) Pulsatile secretion of gonadotrophins, ovarian steroids and ovarian oxytocin during the periovulatory phase of the oestrous cycle in the cow. J Reprod Fert 71:503–512
Walton JS, Evins JD, Fitzgerald BP, Cunningham FJ (1980) Abrupt decrease in daylength and short-term changes in the plasma concentrations of FSH, LH

and prolactin in anoestrous ewes. J Reprod Fert 59:163–171
Webster GM, Haresign W (1983) Seasonal changes in LH and prolactin concentration in ewes of two breeds J Reprod Fert 67:465–471
Wright PJ, Clarke IJ, Findlay JK (1983) The induction of fertile oestrus in seasonally anoestrous ewes using a continuous low dose administration of gonadotrophin releasing hormone. Aust Vet J 60:254–255
Yuthasastrakosol P, Palmer WM, Howland BE (1977) Release of LH in anoestrous and cyclic ewes. J Reprod Fert 50:319–321

64

Reproductive Success of Female Fallow Deer in Relation to Age and Condition

Jochen Langbein and Rory Putman

Abstract

While the effects of age and condition on female reproductive success have been widely reported in a number of deer species, they have received little study in fallow deer (*Dama dama*). Beginning in 1986, we established in many British deer parks marked fallow populations with a range of densities and resource availabilities. Annual live capture of the deer provided physical data in relation to age and sex, and subsequent observations were made of marked animals to assess individual fawning and rearing success, as well as mating and parturition dates.

Despite wide variation in mean live body weights between parks, reproductive rates of adult does were very high at all sites; however, fecundity of yearling does was consistently lower than that of adults, with differences greatest at parks offering restricted resources. The winter body-weight threshold, at which 50% of yearling produced fawns, was determined as 32 kg. Yearlings which did reproduce were shown to produce lighter offspring than adults, and to fawn an average of 11 days later. Near maximum reproduction was attained in park populations kept at 6 deer/ha or less, as long as at high densities animals were offered adequate supplementary feeding and access to shelter during winter.

Key words: Body weight, birth weight, fallow deer, parturition date, reproductive success

Introduction

The effects of age and condition on female reproductive success have been investigated in a number of deer species including among others red deer (*Cervus elaphus*), reindeer (*Rangifer tarandus*), and white-tailed deer (*Odocoileus virginianus*). Marked differences in reproductive rates were observed for animals from different populations at different levels of condition (Clutton-Brock et al. 1983, Mitchell and McCowan 1986, Ratcliffe 1987; Reimers 1983, Skogland 1985; Teer 1984), between those which produced young the previous year and others which had not, and also between yearling and adult females in populations with restricted resources. Previous studies of reproductive success in fallow deer have been restricted to general descriptions of reproductive performance based on data derived from culled animals (Sterba and Klusak 1984; Chapman and Chapman 1975), or have dealt predominantly with the physiology of puberty and estrus cycles (Fisher 1982; Sterba and Klusak 1984; Asher 1986).

The life-time reproductive success of deer (number, quality, and sex of surviving progeny) is influenced by a wide array of factors. In this chapter we will concentrate on aspects of fallow deer reproduction which have not been well documented to date (fawning rate, parturition date, and off-spring birth weight and growth), and evaluate each with respect to the age, body weight, and reproductive history of the mothers.

Methods

Over 5000 fallow deer were caught between 1986 and 1990 in a range of British deer parks

(Langbein 1990). Captures took place annually between November and March; animals caught were weighed, aged, and tagged, and the females' reproductive status checked.

Reproductive success of known-aged animals was monitored for marked individuals in eight main study parks located throughout England and Wales. Herd sizes ranged from 30 to 1200 deer, kept at over-winter stocking densities from 2.2 to 6.0 per hectare. Visits were made at least every 5 to 10 days to each site from the end of May to August every year. On each visit does were assessed visually for progression of pregnancy, using a telescope to read ear-tags, and for those pregnant the date of parturition or onset of lactation (udders increase conspicuously in size after parturition) was estimated. Neonatal loss of fawns could be assessed from condition of udders, which regressed quickly in dams whose fawns died during the summer.

Searches were made during June and July to tag and sex a sample of newly born fawns, and to obtain "birth weights" of any judged to be less than 3 days of age. These searches provided information on birth dates and neonatal losses, as well as allowing matching of fawns with their dams of known age through later observation of suckling bouts. Reproductive condition of individual females during the summer could frequently be related to observations made of the same animal during studies of the previous autumn's rut (Langbein and Thirgood 1989).

In this presentation does are described as yearlings if they were less than 2 years old at the previous rut (though in fact 2 years old by the end of gestation), while those referred to as adult were already over 2 years old at the rut. Milk does are those that had reared fawns in the previous summer; yeld does are those that had not produced a fawn the previous year or had lost it soon after parturition.

Results

Fawning Rate

Fawning rates of adult females varied little between populations and exceeded 82% at all parks but one. Yearlings' rates showed greater variation, ranging from 12.5% to 94.3%, and with only one exception were lower than those of adult does in the same park (Table 64.1; Wilcoxon's matched pairs test $P < 0.05$).

Body weights of female fallow also varied widely between the study sites, and the extent to which differences in fawning rates relate to differences in the mother's weight the previous winter was explored. Data for all sites were combined, and animals were then grouped into weight classes such that each contained 5% of the

TABLE 64.1. Fallow deer fawning success and summer offspring survival (1987–89) in eight British deer parks. (Parks are arranged in order of increasing average adult female live weight)

Park code	Mean winter live weight		Yearlings			Adult does			Percentage of does rearing fawns in 2 consecutive years	
	Yearlings (kg)	Adult does (kg)	*n*	% fawned	% with live fawns in August	*n*	% fawned	% with live fawns in August	*n*	%
A	32.8	37.9	8	12.5	12.5	18	94.4	94.4	14	71.4
B	34.7	38.8	44	86.4	86.4	56	82.1	78.6	13	69.4
E	35.0	40.8	6	66.7	66.7	31	77.4	71.0	id	–
H	37.9	41.4	6	66.7	50.0	24	91.7	70.8	11	63.6
G	35.6	42.2	106	88.6	83.0	108	88.8	86.1	66	90.9
I	36.8	44.2	35	94.3	88.6	75	96.0	96.0	68	92.6
J	40.5	44.3	38	78.9	76.3	80	85.0	82.5	37	94.6
K	40.2	46.4	15	86.7	80.0	35	94.3	91.4	14	92.8

id: insufficient data

entire sample. The proportion of animals in each weight class that produced live fawns was calculated and plotted against weight, to explore whether or not there was evidence for a threshold weight for successful reproduction; analyses were carried out separately for yearlings and adults. Fawning success of adult does exceeded 75% in each 5-percentile of the available sample (total $n = 240$), and no direct link of fawning rate and body weight was apparent. (From Table 64.1 it is clear that even Park A, with the lowest average body weights, showed adult fawning rates $> 90\%$).

Fawning rates of yearlings in contrast did show a relationship to body weight (Fig. 64.1) Among the lightest group of animals (29 to 31.5 kg) none produced a fawn, but fawning rate rose quickly to 100% for animals of 33 kg and above. A threshold mean winter live weight at which 50% of yearlings are likely to produce a fawn, could be calculated (after Hamilton and Blaxter 1980; Albon et al. 1986) as 32 kg. The fawning rate remained at 100% for all 70 yearlings weighing between 33 and 42 kg but in fact fell slightly for the few heavier animals.

Studies of ungulate reproduction have often assessed the reproductive success of milk and yeld females separately, since the costs of raising a calf one year may reduce fecundity at the following rut. Table 64.1 shows the percentage of fallow does known to have raised fawns in two consecutive summers at each park, the remainder having either failed to fawn or lost their young early in at least 1 year. The number of individually recognizable does that reared fawns to weaning in 2 consecutive years only fell below 90% in those four parks which also exhibited lighter than average body weights. Park I provided us with the most individuals monitored for over 2 consecutive years, with 96% ($n = 24$) observed to have produced in 3 consecutive years, and 83% ($n = 6$) for 4 years. Milk does thus frequently attained very high conception rates, which in most parks exceeded those recorded for adult yeld does.

As tagging began less than 5 years ago few known aged animals are yet over 6 years old. However, among tagged animals whose tooth eruption and wear pattern was examined after death, no does with I1-incisors worn down to 5 mm or less were found to be under 6 years old. Therefore, selecting from our (live) adult does all those with incisors less or equal to 5 mm, and separately those with I1s over 7 mm distinguishes among our adult sample a group of "old," as well as a relatively "young" does. Ninety-five percent of the "old" does ($n = 41$), and 90% of the "young" adult does ($n = 179$) produced fawns. However, the "old" group suffered a 15% summer fawn-mortality compared with only 1.2% recorded

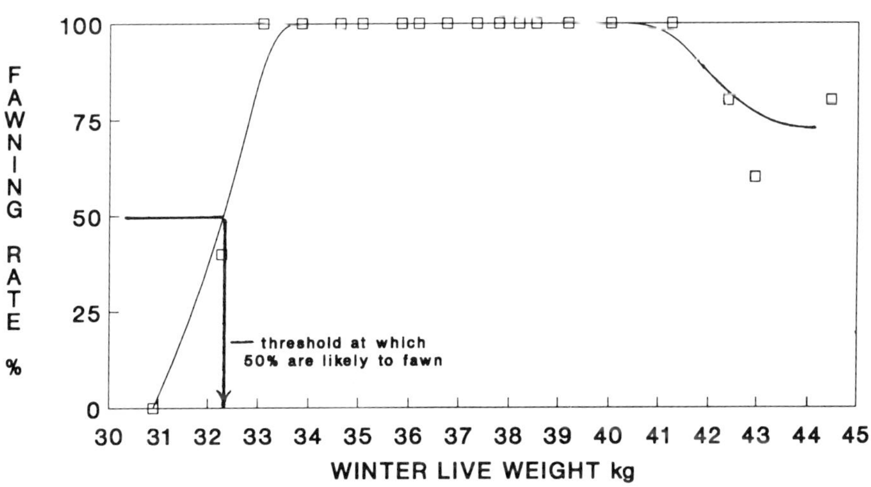

FIGURE 64.1. Fawning success of yearling fallow does, in relation to body weight. Combined data from eight British parks; each data point shown represents the percentage which fawned among five does in a given weight class.

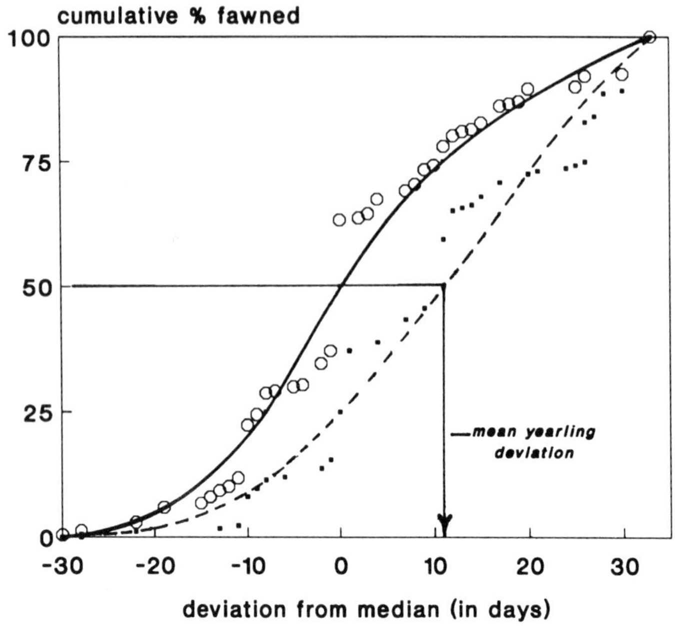

FIGURE 64.2. Temporal pattern of parturition dates of adult and yearling does during June and July, based on data from seven British parks. Individual parturition dates were all calculated as the number of days deviation from the median parturition date in that year for adults at their park of origin. Curves are fitted through consecutive quartiles; ■, yearlings ($n = 175$); ○, adult does ($n = 247$).

losses in the case of the younger group, giving respective weaning rates of 80% and 89%. On average less than 5% of yearling and adult mothers lost their fawns within 1 to 2 months of birth. Losses of fawns during the winter among the eight populations included in this presentation were also negligible.

Parturition Date

The dates individual does gave birth were assessed in seven of the study parks in each of 2 years. The date by which 50% of the does were known to have produced fawns varied between parks from 11 June to 6 July for adults and 16 June and 17 July for yearlings. A difference in median parturition dates for yearling and adult does was shown in 11 of the 14 park-years, with yearlings having the later median in all but one case. Yearlings fawned an average of 11 days later than adults (Fig. 64.2). That yearlings in fact tend to be mated later than adults was confirmed by observations made of tagged does during the rut; for example, during 1988 in park G, only 1 (5%) of 20 matings seen by 16 October involved a yearling doe, but this rose to 38% of 29 matings observed after 17 October (see Langbein 1990).

Birth Weights

Based on a total of 84 fawns weighed at less than 3 days old in all parks combined, females were lighter (mean 4.3 kg) than males (mean 4.7 kg, $t = 2.25$, $P < 0.03$). These results support the larger body measurements of males recorded for wild fallow fawns by Braza et al. (1988) in Donana, Spain. When using only birth weights of those fawns for which we were also able to determine the age of the mother, our sample sizes were reduced considerably; nevertheless, significant weight differences could be demonstrated between fawns born to yearlings and those born to adult dams—(female fawns: yearling dams, mean = 3.6 kg, $n = 6$, adult dams, mean = 4.3 kg, $n = 11$, $t = -3.37$, $P < 0.005$; male fawns: yearling dams, mean = 3.73 kg, $n = 7$, adult dams, mean 4.8 kg $n = 13$, $t = -3.98$, $P < 0.001$). No significant differences in offspring birth weights could be shown between parks, despite the highly significant variation in body weights of does.

Growth

How important differences in birth dates and birth weights are to the lifetime reproductive success of females will depend on whether these differences have long-lasting effects on the survival and reproductive chances of their offspring. Effects of birth date were examined by comparing the body weights reached by fawns at about 5 months (obtained during live-capture operations in November/December). Male fawns born between 1 and 7 June (week 1) were significantly heavier by the beginning of winter than those born 8 to 14 June (week 2; $t = 2.3$, $P < 0.05$), as well as being heavier than those born 15 to 21 June (week 3; $t = 3.8$, $P < 0.01$). Differences between fawns born in weeks 2 and 3 were not significant for either sex ($P > 0.1$).

Regressions of live weight in November/December (at 5 to 6 months) on birth weight were not significant within any single site, although sample sizes were small. When combining data for all sites, positive significant relationships did emerge—(males: $R^2 = 31\%$, $n = 24$, $P < 0.01$; females: $R^2 = 14\%$, $n = 21$, $P = 0.1$). The considerable residual variation may reflect effects introduced by combining data for several sites, in terms of differing environmental conditions (and consequently also the maternal investment) experienced after birth.

Discussion

The reproductive rate of female red deer (Hamilton and Blaxter 1980; Clutton-Brock et al. 1983; Albon et al. 1986), white-tailed deer (Teer 1984), and reindeer (Reimers 1983) has been demonstrated to be closely linked to condition, particularly as indicated by body weight. In our studies the mean body weight of fallow does was found to vary widely among study sites within years, reflecting differences in the climate (particularly length of growing season and rainfall), food supplementation, and stocking densities (Langbein 1990). Despite this, little variation in adult does' parturition rates was evident. Indeed the high fecundities and rearing success of yearlings as well as adult "milk" and "yeld" does indicated that some of our populations were reproducing at close to their full potential, despite being kept at relatively high densities (e.g., Parks G, I, J, and K: range 2.7 to 6.0 fallow/ha). Although conception rates of wild and penned fallow in other studies are also generally high, (>85% Chapman and Chapman 1975; Asher 1986), it is notable that weaning rates on farms often fall below 70% (Asher 1986; Vigh-Larsen 1988).

Most females in our populations well exceeded body-weight thresholds required for ovulation and were able to reach these again while weaning fawns during the next rut. However, in resource-restricted populations yearling body weights were significantly lower, as was fecundity of yearlings and does with fawns at heel. Similar results suggesting that the first responses to restriction in resources are seen in reduction of yearling pregnancy and pregnancy among currently lactating hinds, have been shown by Ratcliffe (1987) in red deer and Reimers (1983) in reindeer (see review by Hanks 1981). We estimated threshold winter live weight at which 50% of yearling does are likely to produce fawns at 32 kg. This is in agreement with English (1984) and Asher (1986) who suggest that estrus activity in yearling does on fallow farms in Australia and New Zealand commences at 28 to 30 kg, while Vigh-Larsen (1988) recorded a 50% yearling fawning rate on a Danish farm where yearlings averaged 32.6 kg live weight at the end of winter. These thresholds are surprisingly low, as we already recorded equivalent weights for many female fawns at 6 months of age (in parks J, I, and K; Langbein 1990).

Yearlings even in our "heaviest" herds produced fawns significantly later than older does. In contrast Asher (1986) only found such differences in some New Zealand farms, which raised yearlings and older does separately, and attributed them to that variation in management. Our results with yearling and adult does in mixed herds suggest that differences are more fundamental and pubertal does are less responsive to the environment stimuli for the induction of estrus than are adult animals. For red deer on Rhum it has been demonstrated that a 2-week delay in birth may have a significant detrimental effect on survival of calves the following winter (Clutton-Brock et al. 1987). The later birth dates of fawns born to yearling mothers demonstrated here (of on average 11 days) may not decrease their chances of survival significantly, given the more temperate conditions of England, but combined with their lighter birth weights do put them at a clear disadvantage when compared with fawns born to adult does.

In parks (where fawns are not usually separated from the remainder of the herd for separate rearing) fawns that enter winter relatively light are unable to catch up this deficit in relation to heavier cohort members during the following year (Langbein 1990), and their reproductive chances as yearlings are thus also likely to be impaired. In addition to any effects of late parturition on offspring, Clutton-Brock et al. (1983) suggest that for red deer reproductive performance of the mother is related to the previous year's conception date, with a 1% decline in fecundity for every day's delay in the conception of the previous calf. Fecundity of 2-year-old fallow does in our studies was generally as high as that of older does, despite their later parturition dates. Also an advancement in the mean parturition date in successive years could be demonstrated for some cohorts, for example, for the 1985 cohort in Park G median parturition dates were 23 June in 1987 and 20 June in 1988 $n = 48$).

A further difference in the success of fawns may relate to maternal behavior and lactation rate. Our previous work has shown that 2-year-old fallow does have reached body weights very close to those of fully grown animals (Langbein 1990), and thus it seems likely that lactation rates of does producing their first fawn at that time fall little short of those of older does. Suckling bouts by fawns with "yearling" dams were on average shorter than those with adults (for fawns < 4 weeks old, as well as those > 4 weeks old), but differences were not statistically significant (see Langbein and Putman 1989). However, it was noted that in the parks showing low mean female body weights (A, B) few suckling bouts occurred after September, while in parks with higher weights (G, J) suckling was observed frequently during the rut in October/November.

Reproductive rates of fallow recorded by us in British parks are generally higher, and the summer fawn losses lower, than those reported on many deer farms (English 1984; Asher and Adam 1985; Vigh-Larsen 1988). Vigh-Larsen emphasized that the unsatisfactory reproductive success of fallow recorded on Danish farms seemed to be largely due to inadequate supplementary feeding of deer kept at very high densities, as well as the insufficient shelter available on farms in general. In conclusion, studies in parks clearly demonstrate that fallow deer can achieve their full reproductive potential even at high stocking densities, as long as they receive appropriate levels of supplementary feed and are provided with shelter during the months when natural forage availability and quality are limited.

Acknowledgements

We would like to thank the owners and staff of all our study parks for their assistance throughout this work. The research was funded by the Science and Engineering Research Council and the British Department of the Environment.

References

Albon SD, Mitchell B, Huby BJ, Brown D (1986) Fertility in female red deer: the effects of body composition, age and reproductive status. J Zool London 209:447–460

Asher GW (1986) Studies on the reproduction of farmed fallow deer. Ph.D. thesis, University of Canterbury, New Zealand

Asher GW, Adam JL (1985) Reproduction of farmed red and fallow deer in northern New Zealand. In Fennesy PF, Drew KR (eds) Biology of Deer Production. R Soc N Z Wellington Bull 22:217–224

Braza F, San Jose C, Blom A (1988) Birth measurements, parturition dates, and progeny sex ratio of *Dama dama* in Donana, Spain. J Mammal 69:607–610

Chapman D, Chapman NG (1975) Fallow Deer. Terence Dalton, Lavenham, Suffolk, U.K.

Clutton-Brock TH, Major M, Albon SD, Guinness FE (1987) Early development and population dynamics in red deer. I. Density dependent effects on juvenile survival. J Anim Ecol 56:53–67

Clutton-Brock TH, Guinness FE, Albon SD (1983) The cost of reproduction to red deer hinds. J Anim Ecol 52:367–383

English AW (1984) Fallow deer—their biology, behavior, and management in Australia. Deer Refresher Course for Vets Proc No 72. University of Sydney, Australia, pp 285–304

Fisher K (1982) Rut, breeding season, conception, birth, and plasma testosterone levels in young fallow deer. Acta Endo 99 (Suppl. 246): 14

Hanks J (1981) Characterization of population condition. In Fowler CW, Smith TD (eds) Dynamics of Large Mammal Populations. Wiley, New York, pp 47–73

Hamilton WJ, Blaxter KL (1980) Reproduction in farmed deer: I. Hind and stag fertility. J Agrc Sci, Cambridge 95:261–273

Langbein J (1990) Effects of density and age on body condition, reproductive performance, behavior, and survival of fallow deer. Ph.D. thesis, University of Southampton, Great Britain

Langbein J, Putman RJ (1989) Effects of density and age on the performance of fallow deer. In Baxter-Brown M, Goldspink CR (eds) Management, Conservation and Interpretation of Park Deer. British Deer Society, Reading, pp 5–21

Langbein J, Thirgood SJ (1989) Variation in mating systems of fallow deer in relation to ecology. Ethology 83:195–214

Mitchell B, McCowan D (1986) Performance and population dynamics in relation to management of red deer *Cervus elaphus* at Glenfeshie, Inverness-shire, Scotland. Biological Conservation 37:237–267

Ratcliffe PR (1987) The management of red deer in the commercial forests of Scotland related to population dynamics and habitat changes. Ph.D. thesis, University of London, England

Reimers E (1983) Reproduction in wild reindeer in Norway. Can J Zool 61:211–217

Skogland T (1985) The effects of density-dependent resource limitations on the demography of wild reindeer. J Anim Ecol 54:359–374

Sterba O, Klusak K (1984) Reproductive Biology of Fallow deer, *Dama dama*. Acta Scientarium Naturalium, Brno, Czeckoslowakia

Teer JG (1984) White-tailed deer population management: lessons from the Llano Basin, Texas. In Halls LK (ed) White-Tailed Deer: Ecology and Management. Stackpole Books, Harrisburg, PA, pp 261–292

Vigh-Larsen F (1988) Deer farming in Denmark, with special emphasis on the management and handling of fallow deer. In Reid HW (ed) The Management and Health of Farmed Deer. Kluwer Academic Publishers, Dordrecht, pp 61–70

65
Impact of Melatonin on Time of Breeding in Farmed Red Deer

Clare L. Adam

Abstract

Seasonal breeding and late puberty, under the influence of pineal melatonin secretion, limit the productivity of farmed red deer (*Cervus elaphus*). These studies recorded responses to exogenous melatonin given daily in an afternoon feed. Given for 3 months starting midsummer, melatonin advanced by 5 weeks breeding and conception dates in adult and pubertal hinds. Given for several months starting late winter to cyclic hinds, melatonin induced continued estrous cyclicity, and to anestrous hinds, which had previously been exposed for 1 month to continuous light or for 6 weeks to artificial long days, melatonin advanced by 4 months the onset of the next breeding season. Conceptions were thereby achieved in midsummer. Given for 12 months starting late winter to 8-month-old hinds, melatonin delayed puberty, but if such hinds had previously been exposed to artificially long days for 5 months or to continuous light for 1 month, puberty was advanced by 6 or 2 months, respectively. Calves resulting from out-of-season conceptions and born in winter or spring achieved live-weight targets early in the production year. Strategic melatonin administration, therefore, offers the potential to control puberty, breeding, and calving dates to advantage.

Key words: Breeding season, calving season, melatonin, puberty, red deer

Introduction

The characteristic features of seasonal breeding and relatively late attainment of puberty can impose limitations on the productivity of red deer farmed for meat production. Seasonal estrous cycles commence in October (northern hemisphere) for mature and pubertal 16-month-old hinds, and can continue until March. Nevertheless most conceptions normally occur in October, resulting in a June (midsummer) calving period, which maximizes neonatal survival in the wild. This pattern does not necessarily match the pattern of feed availability in the farm situation, and, furthermore, calves born in June neither reach puberty nor attain an economic slaughter weight until their second year. There would be obvious production advantages in being able to manipulate calving dates to suit prevailing management conditions, to advance puberty, to reduce the birth-to-slaughter interval, and to provide a less seasonal supply to the venison market.

It is generally accepted that the decreasing photoperiod of the autumn, mediated by the diurnal secretory pattern of the pineal hormone melatonin, is the primary cue for timing the breeding season in red deer, as demonstrated for sheep (Bittman et al. 1983), but it is not clear to what extent prior exposure to long day length may be a necessary prerequisite for a reproductive response to short day length (Yellon and Foster 1985). This paper presents the results of a series of research studies conducted over the past 6 years which (1) evaluated ovarian responses of adult and pubertal hinds to exogenous melatonin administered at different times of the year, with or without prior exposure to long days, (2) investigated the feasibility of out-of-season conception and calving, and (3) assessed the performance of calves born out-of-season.

Methods

Experiments were conducted with red deer, average adult live weight 94 kg (hinds) and 150 kg (stags), from the Rowett herd in Aberdeen, Scotland (57° N). Deer either grazed ryegrass/clover pastures or received a diet of hay and barley indoors. All calves were suckled naturally by their dams.

Melatonin was administered to deer in natural photoperiod (indoors or out) by inclusion in a supplementary afternoon feed (1500–1600 h) at an individual dose rate of 5 mg (hinds), 10 mg (stags), or 2.5 mg (prepubertal hinds). This protocol was known to induce a short-day diurnal melatonin profile in the blood (C.L. Adam, unpublished data). To assess ovarian activity, blood samples were taken two or three times weekly for plasma progesterone determination by radioimmunoassay (Adam et al. 1985). Calves were weighed at birth and at fortnightly intervals thereafter.

Results

Melatonin Administration in Summer

Given for 3 months starting in June (mid-anestrus), melatonin advanced the onset of seasonal estrous cyclicity in lactating and non-lactating hinds (Adam et al. 1986) and in yearling pubertal hinds (Adam 1985) by 5 weeks ($P < 0.001$) (Table 65.1). Stags given melatonin over the same period showed a 5-week advance in onset of rutting behavior and were allowed to mate with these groups of hinds. Conception rates at first estrus were similarly high between treated and untreated animals (89% and 93%, respectively, for mature hinds; 100% for all pubertal hinds); thus, the subsequent calving period was effectively advanced for the former by 5 weeks to April/May. Similar results have now been repeated consistently in commercial UK farm conditions (C.L. Adam, unpublished data).

Melatonin Administration from Winter Onwards: Mature Hinds

Given for 16 months starting in February (late breeding season) to hinds still showing estrous cyclicity, melatonin prevented the normal transition to anestrus (100%) and induced continued cyclicity for periods of 1 year or more (Adam et al. 1989a). In the same trial, given over the same period to hinds, which had previously been induced into premature anestrus by exposure for 1 month to continuous artificial light, melatonin advanced by 132 days (4 months) the onset of the next breeding season (80%) and again induced continued estrous cyclicity for periods up to 1 year or more (Fig. 65.1).

The above experimental treatments were repeated in 1989 with the aim of assessing the fertility of the out-of-season estrous cycles. Ten mature hinds and one stag were exposed to a 6-week pretreatment period of artificial long days (16 h light: 8 h dark), which induced

TABLE 65.1. Date of onset of breeding in hinds fed melatonin in the summer ($P < 0.001$ for all three groups)

Physiological status	Date of onset of breeding			
	Melatonin-fed $\bar{x} \pm$ SEM	*n*	Controls $\bar{x} \pm$ SEM	*n*
Lactating[a]	15 September ± 1.6 days	12	17 October ± 2.2 days	9
Nonlactating[a]	16 September ± 2.4 days	6	21 October ± 1.5 days	5
Pubertal[b]	26 September ± 3.8 days	9	31 October ± 1.8 days	12

[a]From Adam et al. (1986)
[b]From Adam (1985) and unpublished data

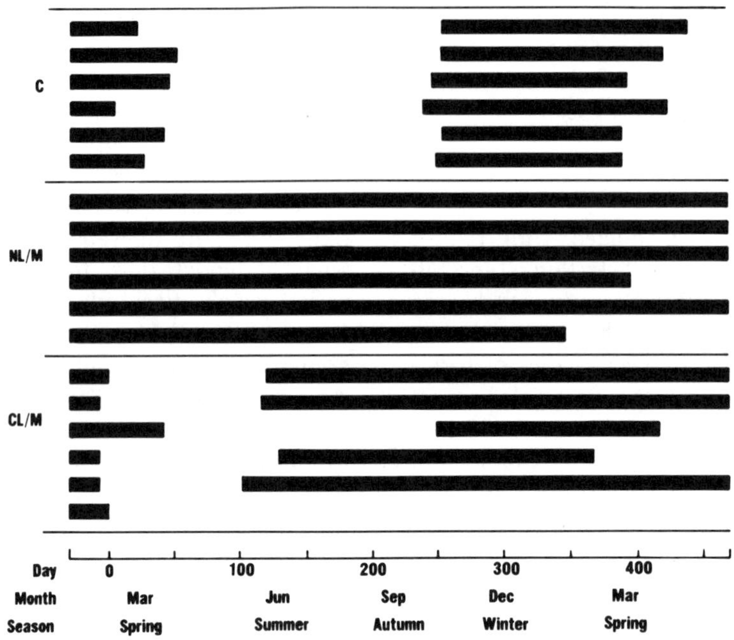

FIGURE 65.1. Incidence of ovarian cyclicity (solid bars) in individual control hinds (C: $n = 6$) and in hinds fed melatonin from February onwards (days 1–475) having been pretreated in natural daylength (NL/M: $n = 6$) or for 1 month in continuous light (CL/M: $n = 6$); adapted from Adam et al. (1989a).

premature anestrus and premature antler growth, respectively, followed by daily melatonin administration from February onwards. Onset of breeding activity was advanced by 4 months in both sexes to May/June, and 40% of the hinds conceived at this time. An additional ten normally cycling hinds given melatonin over the same period showed continued cyclicity (90%) and 60% conceived when allowed to mate in June. Melatonin administration continued during pregnancy in the summer and autumn and was discontinued at the winter solstice in December. These ten hinds subsequently calved in January/February 1990.

Melatonin Administration from Winter Onwards: Pubertal Hinds

Melatonin given for 12 months starting in February to 8-month-old prepubertal hinds (Adam et al. 1989b) delayed puberty, preventing ovarian cyclicity commencing at the normal age of 16 months in the autumn (100%). However, given from February to hinds that had been in artificially long days (16 h light: 8 h dark) for the previous 5 months, and to hinds that had been in continuous artificial light for the previous 1 month, melatonin advanced puberty to 10 months (50%) and 14 months (25%) of age, respectively (Fig. 65.2).

Growth of Calves Born Out-of-Season

Calves born at pasture in April/May (spring) to hinds that had conceived on average 5 weeks early, following melatonin administration in the previous summer, had a similar mean birth weight and grew at a similar mean rate as calves born in June (summer) (Adam and Moir 1987) at, respectively, 9.0 kg (SEM 0.10) and 407 g/day (SEM 9.8) to 10 weeks of age ($n = 66$). Although growth rates subsequently decreased earlier in the calender year for early born calves, they maintained a significant live-weight advantage

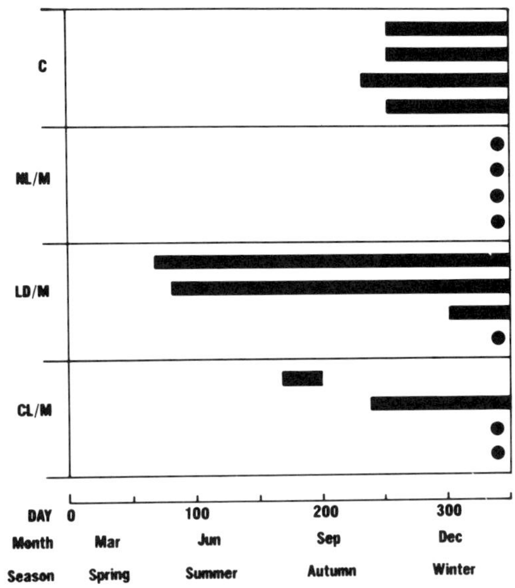

FIGURE 65.2. Incidence of pubertal ovarian cycles (solid bars) in individual control hinds (C: $n = 4$) and in hinds fed melatonin from February onwards (days 1–340) at 8 months of age having been pretreated in natural daylength (NL/M: $n = 4$), for 5 months in artificially long days (16 h light, 8 h dark) (LD/M: $n = 4$), or for 1 month in continuous light (CL/M: $n = 4$). ●, no ovarian cycles detected by day 340; data from Adam et al. (1989b).

throughout, remaining on average about 7 kg heavier than summer-born calves at the onset of winter in November.

Nine calves born indoors in January/February (late winter) 1990 had normal mean birth weight for the herd, 9.2 kg (SEM 0.21), but below average live-weight gains to 10 weeks of age, 303 g/day (SEM 15.3). Nonetheless, they averaged 32 kg (SEM 1.8) live weight in April when put out to pasture with their dams, in contrast with June-born calves, which did not achieve this weight until August (4 months later) or indeed April/May-born calves which achieved this weight in June/July (2 to 3 months later).

Discussion

Exogenous melatonin clearly has profound effects on breeding activity in red deer. The timing of the ovarian response to stimulation by melatonin in red deer evidently depends on initial reproductive status and recent photoperiodic history, and any refractoriness to prolonged melatonin administration is slow to develop. Continued year-round cyclicity was induced in hinds showing estrous cycles at the start of treatment, but there was a characteristic delay (3 to 4 months) in the response time shown by hinds which were initially anestrus and/or had been in long day length (whether natural in summer, or artificial) or continuous light. This delay may reflect either the time taken to develop sensitivity to the melatonin (short-day) signal following exposure to long days, or the duration of time required to reset the reproductive axis from the anestrous state, or indeed both. Interestingly, whether hinds had been previously exposed to continuous light, which may effectively have removed all photoperiodic input as in pinealectomy, or to long day photoperiodic stimulation, they showed a similar response to subsequent melatonin administration. The duration of the response time to melatonin shown by anestrous hinds apparently increases with earlier treatment initiation dates, i.e., both earlier in the anestrous period and earlier before the summer solstice (Adam et al. 1989c; Asher 1990) thus implicating the possible importance of varying "depth" of anestrus over the spring and summer months and of a requirement for prior exposure to long (or increasing) day length before a response to short days (melatonin) may be elicited. An increased understanding is required of the underlying mechanisms involved in decoding the melatonin signal, and the recent studies of Morgan et al. (1989), which have identified and characterized high-affinity melatonin receptors in the pars tuberalis of the sheep pituitary (another short-day breeder), represent a considerable advance.

The results for the prepubertal hinds demonstrate a requirement for prior experience of long day length before ovarian activity can be stimulated by melatonin administration. Above a threshold live weight at 16 months of age (Hamilton and Blaxter 1980), puberty is normally stimulated by the short days of autumn; however, the preceding long days of summer may be equally critical for initial entrainment or priming the immature animal to develop sensitivity to short day length, as in sheep (Yellon and Foster 1985). The pubertal live weight may, of course, be achieved earlier than 16 months, and the above results show that hinds are indeed capable of responding positively at a younger age to the melatonin (short-day) stimulus following long-day exposure.

Various studies have confirmed the advancement by up to 2 months of seasonal breeding in red deer hinds given melatonin during the summer months (Nowak et al. 1985; Webster and Barrell 1985; Fisher et al. 1988; Asher 1990). Conception dates and hence parturition dates are effectively advanced by up to 2 months in the late spring (Adam et al. 1986; Asher 1990). However, the induction of breeding activity beyond this at other times of the year, or indeed the abolition of reproductive seasonality by alternative strategic administration of melatonin as described, offers even greater potential for out-of-season breeding.

The effective reversal of mating and calving periods, with hinds giving birth in the late winter months (January–February), has not previously been reported. This dissociation of the annual reproductive cycle from the seasonal cycle of photoperiod opens intriguing new areas for investigation. It is well established, for example, that voluntary food intake is normally depressed by the short days of winter (Kay 1979). Others (Curlewis et al. 1988; Loudon et al. 1989) have been unable to separate seasonal cycles of appetite and reproduction in red deer, and our winter-calving hinds indeed demonstrated a voluntary food intake more appropriate to the demands of lactation than to the prevailing photoperiod. They consumed about 3 kg dry matter (DM) daily, equivalent to 31 MJ metabolizable energy (ME), the same as hinds lactating indoors in summer and considerably more than the 1.6 to 1.8 kg DM (16 to 18 MJ ME) normally consumed daily by hinds in winter (Adam 1986). Lactogenesis was evidently not inhibited by the short day length, but lactation (yield and composition) may have been affected since the winter-born calves apparently grew indoors some 25% slower to 10 weeks of age compared with summer-born calves outdoors. However, the nutritional status of the dam may account for some of this discrepancy as summer-born calves sucking hinds indoors also grow at a slower rate (18%) than summer-born calves at pasture (C.L. Adam, unpublished data). Of course, photoperiod may have a more direct effect on the calves themselves; calves born in the long days of summer experience decreasing photoperiod through to the winter (with live-weight gain declining), whereas calves born in the short days of winter first experience increasing day length followed by prolonged long day exposure.

The judicious administration of melatonin to red deer offers a practical tool to overcome reproductive seasonality and manipulate breeding and calving dates to advantage. There is also the potential to advance puberty by up to 6 months, thereby advancing recruitment of hinds to the productive herd. Calves born in late winter or spring achieve live-weight targets earlier in the production year than do summer-born calves (e.g., 32 kg live weight in April or June compared with the normal August weight) and offfer the potential to produce finished carcasses for venison production with a shorter birth-to-slaughter interval and at different times of the calendar year. Although it is technically feasible to manipulate breeding dates to calve red deer at a time of year suitable to management conditions, or to divide the herd in order to spread calving and finishing dates more evenly over the year, critical economic evaluations would be required before the true commercial value of these findings may be assesed.

In conclusion, quite apart from its potential commercial impact as a manipulative tool, melatonin provides a powerful means of exploring experimentally a whole range of phenomena such as precocious puberty, advance and reversal of breeding and calving seasons, and the interactions between season (photoperiod), appetite, and physiological state, all of which have a profound influence on the biological productivity of red deer.

Acknowledgements

I thank C. Kyle, P. Young, T. Atkinson, and the staff of the Duthie Farm for their invaluable assistance in carrying out these trials and J. Robinson for advice on preparing the manuscript.

References

Adam CL (1986) Feeding. In Alexander T (ed) Management and Diseases of Deer. Vet Deer Society, London pp 25–37

Adam CL (1985) Recent developments for lowground red deer farming. Annu Rep Rowett Res Inst 40:34–50

Adam CL, Moir CE (1987) A note on the effect of birth date on the performance of suckled red deer calves and their dams on low-ground pasture. Anim Prod 44:330–332

Adam CL, Moir CE, Shiach P (1989a) Melatonin can induce year-round ovarian cyclicity in red deer (*Cervus elaphus*). J Reprod Fert 87:401–408

Adam CL, Moir CE, Shiach P (1989b) Effect of melatonin on puberty in female red deer. J Reprod Fert Abstract Series 3, Abstract 110

Adam CL, Moir CE, Shiach P (1989c) Plasma prolactin concentrations in barren, pregnant and lactating red deer (*Cervus elaphus*) given melatonin to advance the next breeding season. Anim Reprod Sci 18:77–86

Adam CL, Moir CE, Atkinson T (1986) Induction of early breeding in red deer (Cervus elaphus) by melatonin. J Reprod Fert 76:569–573

Adam CL, Moir CE, Atkinson T (1985) Plasma concentrations of progesterone in female red deer (*Cervus elaphus*) during the breeding season, pregnancy and anestrus. J Reprod Fert 74:631–636

Asher GW (1990) Effect of subcutaneous melatonin implants on the seasonal attainment of puberty in female red deer (*Cervus elaphus*). Anim Reprod Sci 22:145–159

Bittman EL, Dempsey RJ, Karsch FJ (1983) Pineal melatonin secretion drives the reproductive response to daylength in the ewe. Endocrinology 113:2276–2283

Curlewis JD, Loudon ASI, Milne JA, McNeilly AS (1988) Effects of chronic long-acting bromocriptine treatment on body weight, voluntary food intake, coat growth and breeding season in non-pregnant red deer hinds. J Endocrinol 119:413–420

Fisher MW, Fennessy PF, Milne JD (1988) Effects of melatonin on seasonal physiology of red deer. Proc N Z Soc Anim Prod 48:113–116

Hamilton WJ, Blaxter KL (1980) Reproduction in farmed red deer. 1. Hind and stag fertility. J Agric Sci Cambridge 95:261–273

Kay RNB (1979) Seasonal changes of appetite in deer and sheep. Agric Res Council Rev 5:13–15

Loudon ASI, Milne JA, Curlewis JD, McNeilly AS (1989) A comparison of the seasonal hormone changes and patterns of growth, voluntary food intake and reproduction in juvenile and adult red deer (*Cervus elaphus*) and Pére David's deer (*Elaphurus davidianus*) hinds. J Endocrinol 122:733–745

Morgan PJ, Williams LM, Davidson G, Lawson W, Howell HE (1989) Melatonin receptors on ovine pars tuberalis: characterisation and autoradiographical localization. J Neuroendocrinol 1:1–4

Nowak R, Elmhirst RN, Rodway RG (1985) A note on the effect of melatonin feeding on the initiation of ovarian activity and on plasma prolactin levels in lactating and non-lactating red deer hinds. Anim Prod 40:515–518

Webster JR, Barrell GK (1985) Advancement of reproductive activity, seasonal reduction in prolactin secretion and seasonal pelage changes in pubertal red deer hinds (*Cervus elaphus*) subjected to artificially shortened daily photoperiod or daily melatonin treatments. J Reprod Fert 73:255–260

Yellon SM, Foster DL (1985) Alternate photoperiods time puberty in the female lamb. Endocrinology 116:2096–2097

66

Duration of Melatonin Treatment Affects the Timing of Puberty in Young Red Deer Hinds

Mark W. Fisher, Lynne M. Meikle, and Peter F. Fennessy

Abstract

The timing of puberty in farmed red deer (*Cervus elaphus*) hinds can be altered by the strategic administration of melatonin, and the present study was undertaken to investigate the importance of the duration of treatment. Six groups, each of five hinds aged about 12 months were used. The control group was untreated throughout while the remaining five groups received approximately 30, 60, or 90 days of melatonin from subcutaneous implants. Treatment began in three groups in spring (on 30 November for 30, 60, or 90 days), and in two groups in summer (on 30 December for 60 days and on 30 January for 30 days) so that each duration of treatment either started (30 November) or finished (approximately 1 March) at the same time. Treatment begun in spring lasting 60 or 90 days resulted in all hinds displaying ovarian activity about 20 March, representing about a 4-week advancement in the onset of puberty. Melatonin administered for 30 days in spring and 30 or 60 days in summer resulted in only a proportion (two or three of five) of hinds responding with early ovarian activity. In addition, melatonin treatment significantly advanced the seasonal patterns in live-weight gain and coat growth in some of the earlier treated groups, and reduced prolactin secretion in all treated groups. These results indicate that approximately 60 days of exogenous melatonin treatment is required to induce an alteration in the timing of puberty in red deer hinds, providing that it is given at an appropriate time of the year.

Key words: Hinds, live weight, melatonin, pelage, progesterone, prolactin, puberty, red deer.

Introduction

The onset of puberty in farmed red deer (*Cervus elaphus*) hinds typically occurs at about 16 to 17 months of age in the autumn. For red deer at Invermay, the threshold weight for attainment of puberty ascertained from calving data is 65 to 75 kg (Kelly et al. 1982). The timing of puberty can be altered by the strategic administration of exogenous melatonin, the pineal hormone which entrains circannual physiological rhythms with external environmental rhythms (Webster and Barrell 1985; Asher 1990; Fisher et al. 1990). Characteristically, a latent period of at least 70 days occurs between the initial exposure to melatonin and the onset of ovarian activity (Fisher et al. 1990); however, the duration of treatment required is unknown.

We have previously demonstrated that the most appropriate time for the initiation of melatonin treatment in order to advance the onset of puberty is late spring in November (Fisher et al. 1990). The present study was designed to extend these observations to determine the duration of treatment required to influence puberty. In addition, plasma prolactin concentrations were measured to provide an endocrine indicator of the efficacy of melatonin treatment (Webster and Barrell 1985; Adam et al. 1987, 1989).

Materials and Methods

Animals and Treatment

Thirty pasture-farmed yearling red deer hinds aged about 12 months at the beginning of the experiment (30 November 1988) were randomly allocated to six groups of five hinds. One group was left untreated (controls) while the remaining five groups were treated with two subcutaneous melatonin (Regulin) implants (Regulin, Melbourne, Australia) each containing 18 mg melatonin. These implants maintain plasma melatonin concentrations at or above normal night-time values for at least 30 days (Asher 1990; Webster et al. 1991), and so they were administered on one, two, or three occasions at 30-day intervals, to give a total of approximately 30, 60, or 90 days of treatment, respectively. Treatment began on 30 November (group designated Nov 30, Nov 60, or Nov 90), on 30 December (group Dec 60), or 30 January (group Jan 30) so that each duration of treatment either started (30 November) at the same time or was designed to run out at about the same time, i.e., early March assuming 30-day melatonin elevations (Asher 1990; Webster et al. 1991). Implants were placed subcutaneously at the base of the ear. The hinds were maintained as one group throughout the experiment until 13 April, which is approximately the onset of the normal breeding season in yearling hinds at Invermay (Fisher et al. 1986), when they were randomly allocated to one of two single-sire mating groups, each with an untreated adult stag. The two stags were interchanged 21 days later (4 May) so all hinds had access to both stags which were removed after another 20 days (24 May).

Measurements

Hind live weights were recorded at 2- to 3-week intervals throughout the study. Blood samples were collected via jugular venepuncture, twice weekly from 23 January to 13 April 1989, and plasma progesterone concentrations were determined to monitor ovarian activity. Progesterone concentrations were determined as described by Jopson et al. (1990), with the onset of ovarian activity taken as the first, short ovarian cycle (Fisher et al. 1990). Plasma prolactin concentrations were determined on every second sample using an enzyme-linked immunosorbent assay. This assay used rabbit antiserum to ovine prolactin (courtesy D.F.M. Van de Wiele, Research Institute for Animal Husbandry, Schoonoord, The Netherlands) and ovine prolactin (NIADDK-o-prl-16) for conjugation to bovine thyroglobulin prior to absorption onto microtiter plates and as assay standards. Dilutions of hind plasma and purified cervine prolactin were parallel to the ovine prolactin standards. Assay sensitivity was less than 2.5 ng/ml and inter- and intraassay coefficients of variation 12.4% and 7.3%, respectivley, for a pool containing 40.7 ng/ml.

Whenever the hinds were weighed or blood-sampled, coat condition was assessed as either summer (reddish-brown), moulting, or winter (grayish-brown) as reported by Ryder (1977). A month after the stags were removed (23 June), pregnancy was diagnosed by real-time ultrasound scan (Aloka, Japan) using a rectal probe while the animals were mechanically restrained. Individual calving dates were recorded by observing the hinds daily.

Statistical Analysis

The data are given as the mean ± SEM and were examined by analysis of variance and Student's t-test. Live weight on 9 November 1989 was confounded by pregnancy status; therefore, it was adjusted by using a covariate for the nonpregnant hinds. The relationship between live weight and the onset of ovarian activity was analyzed using linear regression analysis.

Results

Ovarian Activity

None of the control hinds displayed any ovarian activity by 13 April, in contrast to some hinds (two to five of five) in the treated groups (Fig. 66.1). In two groups, Nov 90 and Nov 60, all animals displayed ovarian activity. The timing of the onset of ovarian activity (mean 20 March, range 27 February to 4 April) in those animals responding was similar in all treatment groups (Table 66.1).

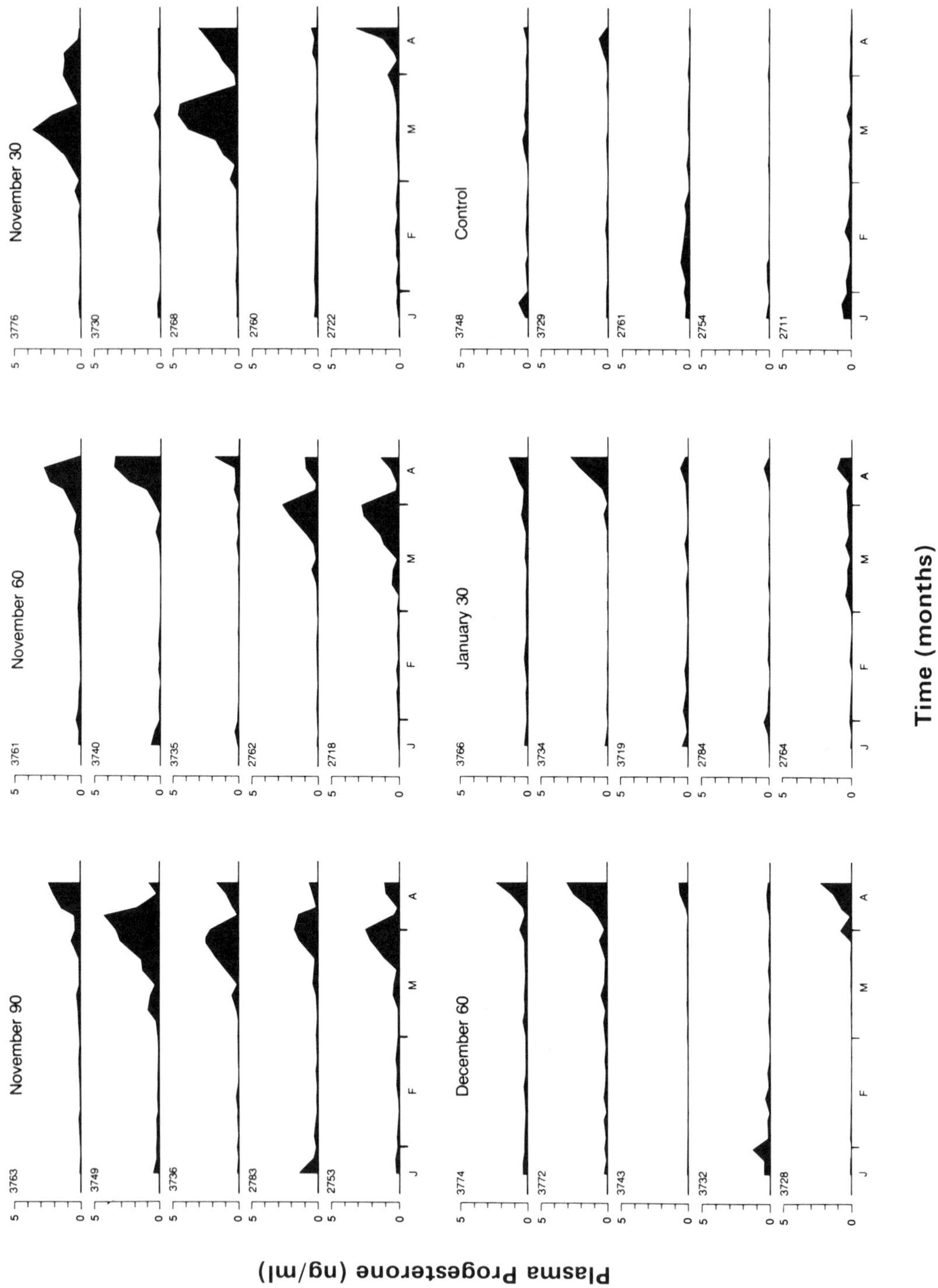

FIGURE 66.1. Ovarian activity as indicated by plasma progesterone concentrations in hinds untreated or treated with melatonin.

TABLE 66.1. The effect of duration of melatonin treatment on ovarian activity prior to the onset of the normal breeding season

Group (month treatment began and duration in days)	Hinds with ovarian activity ($n = 5$ per group)	Onset of ovarian activity of those responding		Mean interval from start of treatment to onset of ovarian activity of those responding (days)
		Mean	Range	
November 90	5	16 March	9 March–28 March	106
November 60	5	21 March	9 March–4 April	111
November 30	3	10 March	27 February–28 March	100
December 60	3	30 March	28 March–31 March	90
January 30	2	28 March	—	57
Control	0	—	—	—

Fertility and Calving

Of the 30 hinds, 27 were diagnosed pregnant at 30 to 60 days of gestation. Of the three nonpregnant animals, one each from Nov 60, Dec 60, and control groups, only the control hind failed to display any ovarian activity prior to the introduction of stags to the hinds. All hinds diagnosed as pregnant subsequently calved, with 17/27 (63%) corresponding to conception within the first 21 days of stag introduction. All calved between 23 November and 31 December and the mean date of calving (14 December) was similar for all treatment groups.

Live Weight

The onset of ovarian activity was not significantly related to live weight at any time. Melatonin treatments begun in November (90, 60, and 30 days) advanced the seasonal pattern of live-weight change. There were lower rates of gain during late summer/early autumn but higher rates during late autumn/winter, so that ultimately live weight was unaffected (Table 66.2).

Coat Growth

Melatonin treatment tended to advance the timing of the summer coat moult (Fig. 66.2), but

TABLE 66.2. Mean live-weight gains and live weight in hinds untreated or treated with melatonin

Group	Live weight of hinds on 30 November 1988 (kg)	Liveweight Gain (kg)		Live weight of hinds[a] on 9 November 1989 (kg)
		Late summer/ early autumn (30 January–6 April)	Late autumn/ winter (6 April–27 July)	
November 90	75.0	2.2*	5.8**	99.8
November 60	74.2	−1.2***	4.4*	97.1
November 30	75.6	1.2**	2.8	98.2
December 60	74.4	4.2	0.8	100.1
January 30	77.0	3.8	1.2	100.6
Control	75.2	5.6	−0.4	98.4
SED	2.99	1.35	1.87	4.17

[a]Live weights adjusted for presence of nonpregnant hinds
$*P < 0.05$; $**P < 0.01$; $***P < 0.001$ for comparisons with control hinds

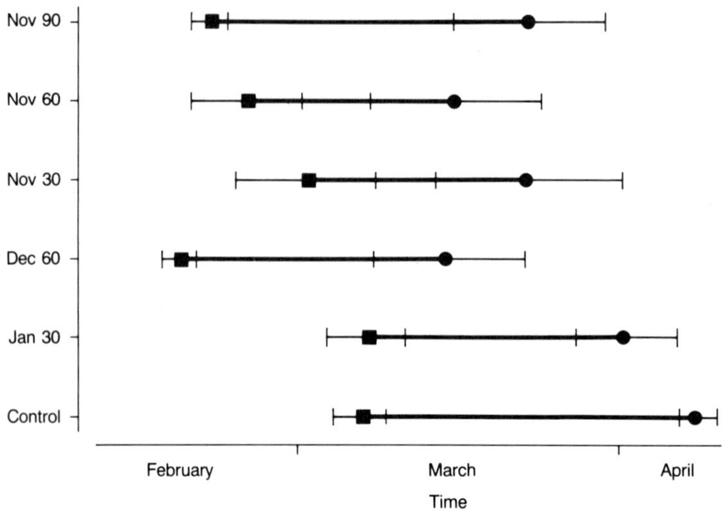

FIGURE 66.2. Mean ± SEM (light lines plus outer vertical lines) time of the start of the summer coat moult (■) and the appearance of a complete winter coat (●) separated by the period of moulting (heavy line).

only the Nov 90 and Dec 60 groups moulted significantly earlier ($P < 0.05$ and $P < 0.01$, respectively) than the controls. The timing of the complete appearance of the winter coat and the period between the start and the completion of the moult (mean 26.2 days) were similar in all groups.

Plasma Prolactin

Mean plasma prolactin concentrations (Fig. 66.3) in untreated control hinds showed a decline through summer and autumn (116 ng/ml on 23 January to 4.7 ng/ml on 10 April). Melatonin treatment significantly reduced

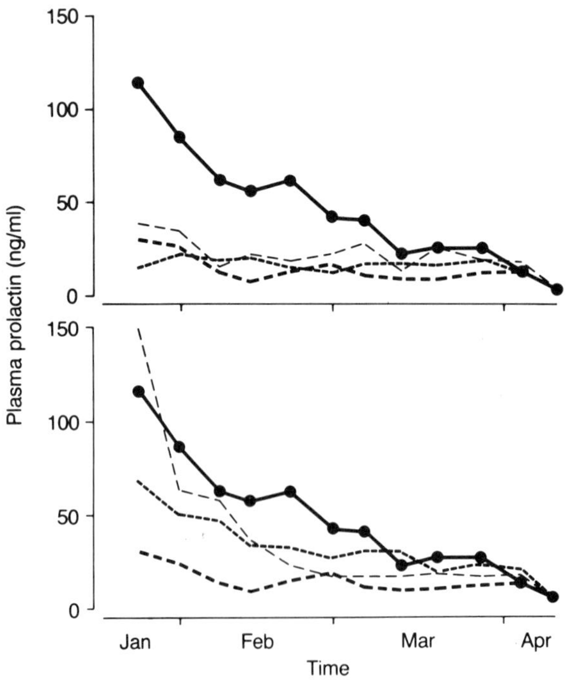

FIGURE 66.3. Mean plasma prolactin concentrations in untreated (heavy line) and melatonin-treated (broken lines) hinds. The upper panel represents treatments of 90 (———), 60 (--), and 30 (——) days duration that began at the same time (30 November) while the lower panel represents 90 (———), 60 (--), and 30 (——) days of treatments that began in November, December, and January respectively, so that they finished at the same time (approximately 1 March). Note that the untreated and 90-day treatment groups are common to both graphs.

plasma prolactin concentrations in all groups until 20 March when concentrations in control hinds had fallen to 27.6 ng/ml and concentrations in the treated hinds averaged 18.2 ng/ml. November 90-, 60-, and 30-day treatments were similar, averaging 0 to 36 ng/ml throughout sampling and were significantly lower than the control animals from 23 January until 20 February. The reduction in prolactin concentrations was less evident in the December 60 and January 30 groups, being significantly lower than the control group on fewer occasions.

Discussion

In the ewe, the species of domestic ruminant most extensively studied, there is a lag phase of approximately 55 to 65 days between photoperiod or melatonin treatments and a neuroendocrine response (Bittman et al. 1983). It is difficult to extrapolate from studies investigating the physiological control of seasonality to the present study utilizing a pharmacological treatment; however, it appears that melatonin treatment must also be administered for approximately 60 days. Although plasma melatonin concentrations were not measured in the present experiment, Asher (1990) reported that these implants administered to yearling hinds raised plasma concentrations in excess of normal night-time levels for at least 30 days. All hinds in each group in the present study required the 60- or 90-day treatments to respond with early ovarian activity. This is in accordance with work in ewes where longer (36 to 40 days or more) treatment periods have usually been associated with a more consistent ovarian response. Likewise, shorter periods of melatonin treatment have resulted in transitory ovarian activity, short luteal phases, and inadequate progesterone secretion (Nowak and Rodway 1987; Wigzell et al. 1988; Stellflug et al. 1988). In accordance with our earlier work (Fisher et al. 1990) it appears that the time melatonin is administered is important in determining the timing of puberty, and from the limited data in this study, it would appear that the time of treatment initiation was more important than the time of treatment completion. Thus, melatonin must be administered at an appropriate time of year (late spring) and for an appropriate duration (at least 60 days) in order to influence the timing of puberty in all treated hinds, assuming that hinds have reached an appropriate threshold or critical live weight necessary for the attainment of puberty (Hamilton and Blaxter 1980; Kelly et al. 1982). All hinds in the present experiment were well grown (a mean of 75.2 kg at the start), and it is likely that photoperiod and the pineal gland are the primary determinants of the timing of puberty, at least in well-grown animals.

The stag can influence the onset of the breeding season (Moore and Cowie 1986; McComb 1987; Fisher and Fennessy 1990); therefore, to avoid any uncertainty in the interpretation of the results, stags were precluded from the present experiment until the normal breeding season. Similarly, social cues can influence the onset of the breeding season in ewes (Wayne et al. 1989; Sunderland et al. 1990), thus sexual activity in some hinds may have influenced such activity in others since all were maintained as one group until stag introduction. Thus, it is possible that the results noted in the present experiment may have been more apparent if untreated and treated hinds were kept separately.

The small, but significant, changes in the timing of body and coat growth recorded in the present experiment are similar to those recorded earlier (Fisher et al. 1990) and reinforce the reproductive results obtained. Thus, these aspects of seasonal physiology in the young deer are influenced most by longer periods of exposure to melatonin given from late spring. Similarly, prolactin secretion was affected by treatment, as in other studies (Webster and Barrell 1985; Adam et al. 1987, 1989; Milne et al. 1990). The spring treatments were most effective in reducing prolactin concentrations regardless of treatment duration, whereas the reduction brought about by the summer treatments was probably masked by the normal seasonal decline seen in control animals.

Acknowledgements

The Invermay staff who have assisted with data collection, analysis, and presentation, especially R.J. Young, A.J. Whaanga, R.E. Labes, T.R.

Manley, P.D. Johnstone, and B.A. Shackell are very gratefully acknowledged. L.D. Staples and Regulin Ltd. (now Schering Agrochemicals Ltd.), Melbourne, Australia, are thanked for the generous supply of melatonin implants. L.K. Lewis, Lincoln University, Canterbury, New Zealand, is gratefully thanked for the prolactin determinations.

References

Adam CL, Moir CE, Shiach P (1989) Plasma prolactin concentrations in barren, pregnant and lactating red deer (*Cervus elaphus*) given melatonin to advance the next breeding season. Anim Reprod Sci 18:77–86

Adam CL, Atkinson T, Moir CE (1987) Melatonin lowers plasma prolactin levels in female red deer (*Cervus elaphus*). J Pineal Res 4:13–20

Asher GW (1990) Effect of subcutaneous melatonin implants on the seasonal attainment of puberty in female red deer (*Cervus elaphus*). Anim Reprod Sci 22:145–159

Bittman EL, Dempsey RJ, Karsch FJ (1983) Pineal melatonin secretion drives the reproductive response to daylength in the ewe. Endocrinology 113:2276–2283

Fisher MW, Fennessy PF (1990) A note on melatonin-treated red deer stags advancing the onset of the calving season in hinds. Anim Prod 51:213–216

Fisher MW, Fennessy PF, Johnstone PD (1990) The timing of melatonin treatment affects the seasonal onset of ovarian activity, coat growth and live weight in young red deer hinds. Anim Reprod Sci 23:49–59

Fisher MW, Fennessy PF, Suttie JM, Corson ID, Pearse AJT, Davis GH, Johnstone PD (1986) Early induction of ovulation in yearling red deer hinds. Proc N Z Soc Anim Prod 46:171–173

Hamilton WJ, Blaxter KL (1980) Reproduction in farmed red deer. 1. Hind and stag fertility. J Agric Sci Cambridge 95:261–273

Jopson NB, Fisher MW, Suttie JM (1990) Plasma progesterone concentrations in cycling and in ovariectomised red deer hinds: the effect of progesterone supplementation and adrenal stimulation. Anim Reprod Sci 23:61–73

Kelly RW, Fennessy PF, Moore GH, Drew KR, Bray AR (1982) Management, nutrition, and reproductive performance of farmed deer in New Zealand. In Wemmer CM (ed) Biology and Management of the Cervidae. Smithsonian Institution Press, Washington, D.C., pp 450–460

McComb K (1987) Roaring by red deer stags advances the date of oestrus in hinds. Nature 330:648–649

Milne JA, Loudon ASI, Sibbald AM, Curlewis JD, McNeilly AS (1990) Effects of melatonin and a dopamine agonist and antagonist on seasonal changes in voluntary intake, reproductive activity and plasma concentrations of prolactin and tri-iodothyronine in red deer hinds. J Endocrinol 125:241–249

Moore GH, Cowie GM (1986) Advancement of breeding in non-lactating adult red deer hinds. Proc N Z Soc Anim Prod 46:175–178

Nowak R, Rodway RG (1987) Length of melatonin exposure and onset of ovarian activity in anoestrous ewes. J Reprod Fert 80:343–347

Ryder ML (1977) Seasonal coat changes in grazing red deer (*Cervus elaphus*). J Zool London 181:137–143

Stellflug JN, Fitzgerald JA, Parker CF, Bolt D (1988) Influence of concentration, duration and route of administration of melatonin on reproductive performance of spring-mated Polypay and Polypay-cross ewes. J Anim Sci 66:1855–1863

Sunderland SJ, O'Callaghan D, Boland MP, Roche JF (1990) Social cues can alter the timing of reproductive transitions in ewes. J Reprod Fert Abstr Ser 5:28 (Abstract)

Wayne NL, Malpaux B, Karsch FJ (1989) Social cues can play a role in timing onset of the breeding season of the ewe. J Reprod Fert 87:707–713

Webster JR, Barrell GK (1985) Advancement of reproductive activity, seasonal reduction in prolactin secretion and seasonal pelage changes in pubertal red deer hinds (*Cervus elaphus*) subjected to artificially shortened daily photoperiod or daily melatonin treatments. J Reprod Fert 73:255–260

Webster JR, Suttie JM, Corson ID (1991) Effects of melatonin implants on reproductive seasonality of male red deer (*Cervus elaphus*). J Reprod Fert 92:1–11

Wigzell S, Robinson JJ, Wallace JM, Aitken RP (1988) Duration of melatonin treatment and ovarian activity in ewes. Anim Prod 46:510 (Abstract)

67

Field Studies of Advancing the Breeding Season in Farmed Red Deer Using Melatonin Implants: Two Dosage Regimes and the Influence of the Stag

Peter R. Wilson

Abstract

Two experiments were conducted on red deer (*Cervus elaphus*) farms to further investigate melatonin implants for advancing the onset of the breeding season. A three-treatment regime of 2 × 18 mg melatonin implants (Regulin) at a 45-day interval was compared with a three-treatment regime using 2 × 18 mg melatonin implants at 30-day intervals commencing 8 December. Hinds were joined on 1 to 6 March with melatonin-treated stags. Ultrasound fetal age predictions were undertaken on a sample of each group on each property 25 May to 8 June. The advancement of median-predicted calving dates averaged 12 days for the three-treatment hinds and 15 days for the hinds receiving two treatments when compared with untreated control hinds.

A second experiment was conducted to determine whether advancement of estrus in mixed-age hinds could be achieved by exposure to stags with advanced rutting behavior following melatonin treatment. Stags were treated with melatonin commencing 8 December and joined with hinds 1 to 6 March. The median predicted calving dates of hinds exposed to treated stags were 8 to 10 days earlier than those mated to untreated control stags. Prior exposure to vasectomized treated stags resulted in no further advancement of calving. These studies showed that a two-treatment melatonin implant regime resulted in the same advancement of calving as the previously reported three-treatment regime and that stags induced to rut early resulted in advancement of estrus in hinds.

Key words: Advanced breeding, melatonin, red deer, "stag-effect"

Introduction

Previous studies have shown that melatonin administration to red deer hinds by the appropriate regime either orally (Adam and Atkinson 1984; Nowak et al. 1985), by injection (Webster and Barrell 1985), or subcutaneous implants (G.K. Barrell and L.D. Staples, personal communication; Fisher et al. 1988; Fisher and Fennessy 1991; Fisher et al. 1990; Wilson et al. 1991), resulted in an advancement of the onset of the breeding season and mean or median calving date. Currently, the most practical method for administration of melatonin on commercial deer farms is by subcutaneous implants (Wilson 1989).

Several variations of treatment commencement dates, dose rates, and treatment frequency have been employed in red deer. The greatest advance in median or mean calving date in red deer occurred in response to a 30-day interval, three-treatment regime using 2 × 18 mg melatonin implants per treatment commencing between 27 November and 16 December (Wilson et al. 1991). Other workers observed up to 16 days advance in mean calving date in red deer (G.K. Barrell and L.D. Staples, personal communication). In fallow deer, Asher et al. (1988) reported a 52-day advance in mean fawning date after a 30-day interval, three- treatment regime using 1 × 18 mg melatonin implant per treatment commencing 10 November. For commercial and practical reasons a regime involving fewer treatments while achieving the maximum advance would be desirable.

In earlier reports of advanced breeding, most authors noted that untreated control hinds calved earlier than expected (G.K. Barrell and L.D. Staples, personal communication; Fennessy and Fisher 1988; Wilson et al. 1991). In most of those trials, untreated control hinds were joined with stags given a melatonin treatment regime to advance the onset of the rut. That observation suggested that exposure to an early-rutting stag prior to the breeding season induced estrus and ovulation in hinds or that estrous hinds induced early estrus in herd mates. McComb (1987) reported an estrus synchronization and estrus advancement effect in red deer hinds exposed to artificially transmitted stag-roaring sounds prior to introduction of a stag at the commencement of the breeding season on 1 April. Data of Fisher and Fennessy (1991) indicate the mean calving date of adult red deer hinds exposed prior to the normal mating season to a melatonin-treated stag was 12 days earlier than in hinds mated with untreated stags. The commercial application and management implications of this finding are of considerable interest to the New Zealand deer farmer.

The present studies were designed to investigate the efficacy of a two-treatment compared with a three-treatment melatonin implant regime in 1-year-old red deer hinds and to evaluate the influence of early rutting stags on the onset of estrus in adult red deer hinds on large commercial deer farms.

Materials and Methods

Experiment 1—Two- and Three-Treatment Regimes in 1-Year-Old Hinds

One-year-old hinds from three commercial red deer farms were given 2 × 18 mg melatonin implants ("Regulin," Schering Agrochemicals) by subcutaneous implantation on either each of three occasions at 30-day intervals (three-treatment group) or on two occasions at a 45-day interval (two-treatment group; Table 67.1). Treatment commenced 8 December. Untreated hinds acted as controls. On each farm the necessary number of stags was treated with

TABLE 67.1. Summary of melatonin treatment regimes and mating programs for 1-year-old red deer hinds

Farm no.		1			2			3	
Locality		Hawkes Bay			Gisborne			Taumaranui	
Group	T2	T3	C	T2	T3	C	T2	T3	C
Number of hinds	18	19	18	50	118	50	54	197	65
Treatment dates[a]	8 Dec	8 Dec		8 Dec	8 Dec		8 Dec	8 Dec	
(Hinds and stags)	20 Jan	7 Jan		20 Jan	7 Jan		20 Jan	7 Jan	
		9 Feb			7 Feb			7 Feb	
Weight at commencement (kg)	NRC	NR	NR	67.7	67.4	67.9	NR	NR	NR
Weight at joining (kg)	NR	NR	NR	NR	NR	NR	Average 72.8		
Date stags joined	6 Mar			1 Mar			1 Mar		
No of stags	1 to combined group			1	3	1	2	6	1
Stag treated (T) or untreated (C)		T		T	T	C	T	T	C
Date stags replaced with untreated stag	25 Apr			12 Apr			12 Apr		
Date stag withdrawn	20 May			12 May			12 May		
Date ultrasound examination	8 Jun			25 May			25 May		

T2, two melatonin treatments; T3, three melatonin treatments; C, Untreated control; NR, not recorded
[a]Two × 18 mg melatonin implants per treatment

implants in a regime identical to that for the three-treatment hind group. On farm 2 (Table 67.1) control, two- and three- treatment groups were mated together, whereas on farms 2 and 3 untreated control hinds were isolated from treated hinds and mated with untreated stags. Hinds from farm 2 were transported on 22 February to farm 3 and were mated on that farm. For clarity of presentation, that group retains its identity with the farm of origin (i.e., group 2). All deer were managed on pasture throughout the trial.

On either 25 May or 8 June a sample of hinds (see Table 67.3) from each group were examined rectally by ultrasound to record pregnancy and fetal and uterine dimensions for fetal age estimation according to the method of Bingham et al. (1990) which was validated by Wilson and Bingham (1990). Median-predicted calving dates were calculated for each group.

Experiment 2—Influence of Melatonin-Treated Stags

Hinds and stags more than 2 years of age from two properties were used for this study (Table 67.2). On farm 1, three stags were vasectomized, and given a three-treatment regime of 2 × 18 mg Regulin implants. They were joined with 150 randomly selected mixed-age hinds on 16 February (VTS group). These stags were replaced 6 March by intact melatonin- treated stags and those in turn were replaced by untreated stags 25 April. A second group of 150 hinds were joined with stags treated as for experiment 1 above (TS group), which were replaced 25 April by untreated stags. A third group of 300 mixed-age hinds were joined 6 March with untreated stags which were also replaced 25 April. All hinds were mated in single sire groups; mating was completed 20 May by withdrawal of stags.

On farm 2, 947 mixed-age hinds were joined in multiple sire groups on 1 March with stags treated with melatonin as above. One hundred and thirty-seven mixed-age hinds were joined with untreated stags to act as controls. All stags were replaced by untreated stags 12 April. On the dates indicated in Table 67.2, 20 to 27 hinds from each group were examined for pregnancy and fetal aging by ultrasound, and median-predicted calving dates were calculated.

TABLE 67.2. Summary of melatonin treatment regimes and mating programs for stags mated to untreated adult hinds

Farm no.		1			2	
Location		Hawkes Bay			Taumaranui	
Stag status	Treated Vasectomized	Treated Intact	Untreated Control		Treated Intact	Untreated Control
No. of stags	3	7	—		42	—
Date treated[a]	8 Dec	8 Dec	—		8 Dec	—
	7 Jan	7 Jan	—		7 Jan	—
	9 Feb	9 Feb			7 Feb	—
Date stags joined	16 Feb	6 Mar	25 Apr	Hind groups TS (n = 947)	1 Mar	12 Apr
Hind group VTS[b] (n = 150)		6 Mar	25 Apr	Hind groups C (n = 137)		1 Mar
Hind group TS(n = 150)			6 Mar			12 Apr
Hind group C(n = 300)			25 Apr			
Date stags withdrawn	20 May all groups				12 May all groups	
Mating management	S	S	S		M	M
Hind: stag ratio	50:1	50:1	50:1		40:1	50:1
Date ultrasound examination		8 Jun			25 May	

[a]Two × 18 mg melatonin implants
[b]Stags replaced those joined earlier
VTS, treated vasectomized followed by treated entire stags; TS, treated entire stags; C, untreated control stags; S, single sire groups, M, multisire group

Results

Experiment 1

The overall mean pregnancy rates of untreated and two-treatment one-year-old hinds were 83.5% whereas the mean pregnancy rate in three-treatment hinds was 89.3% (Table 67.3). In treated groups, first calves were predicted to be born 5 to 37 days earlier than those of untreated control groups. The two-treatment regime resulted in an advance of predicted median calving date compared with controls of 12 to 20 days (average 15 days) while three treatments resulted in an advance of predicted median calving date of 8 to 18 days (average 12 days).

Experiment 2

Pregnancy rates for hinds exposed to melatonin-treated stags without prior exposure to vasectomized treated stags were 100% and 96%, whereas for hinds mated to control stags, pregnancy rate was 87% and 100% for farms 1 and 2, respectively (Table 67.4). Hinds given exposure to vasectomized-treated stags prior to an entire treated stag had a pregnancy rate of 100%. The predicted median calving dates of hinds mated with treated stags were 10 and 8 days earlier than those mated to control stags on farms 1 and 2, respectively. Predicted median calving date of hinds exposed to vasectomized melatonin-treated stags for 21 days prior to introduction of

TABLE 67.3. Pregnancy rates and predicated calving dates for melatonin treated and control 1-year-old hinds

Farm no.		1			2			3	
Group	T2	T3	C	T2	T3	C	T2	T3	C
No. ultrasound scanned	15	13	12	20	24	23	20	27	25
% Pregnant	93	92	100	65	83	70	95	82	88
Predicted date first calf	4 Nov	10 Nov	19 Nov	24 Oct	8 Nov	30 Nov	12 Nov	1 Nov	17 Nov
Predicted date last calf	21 Dec	23 Dec	27 Dec.	After[a] 11 Dec	After[a] 10 Dec	After[a] 13 Dec	After[a] 11 Dec	After[a] 13 Dec	After[a] 14 Dec
Predicted median calving date	18 Nov	21 Nov	1 Dec	18 Nov	30 Nov	8 Dec	22 Nov	16 Nov	4 Dec
Days advanced median calving	13[b]	10[b]		20[c]	8[c]		12[c]	18[c]	

[a]At date of scanning, pregnancy of late calvers was not sufficiently advanced to allow age prediction equations to be used
[b]Mated in contact with control hinds
[c]Mated apart from control hinds
T2, two melatonin treatments 45 days apart; T3, three melatonin treatments 30 days apart; C, untreated control

TABLE 67.4. Predicted pregnancy rates and calving dates for mixed-age hinds exposed to vasectomized and/or intact treated stags, or untreated control stags

Farm no.		1		2	
Hind group	VTS	TS	C	TS	C
No. scanned	20	24	23	24	23
% Pregnant	100	100	87	96	100
Predicted date first calf	6 Nov	29 Oct	16 Nov	1 Nov	14 Nov
Predicted date last calf	11 Dec	19 Dec	22 Dec	After 16 Dec[a]	After 13 Dec[a]
Predicted median calving date	16 Nov	15 Nov	25 Nov	22 Nov	30 Nov
Days advanced median calving	9	10	—	8	—

[a]At date of scanning, pregnancy of late calvers was not sufficiently advanced for ultrasound age prediction equations to be used
VTS, hinds exposed to vasectomized treated stags followed by entire treated stags; TS, hinds exposed to entire treated stags; C, hinds exposed to untreated stags

melatonin-treated intact stags resulted in a predicted 9-day advance in median calving date compared with controls. The predicted onset of calving of hinds mated with treated stags was 18 and 13 days earlier than those mated with untreated stags on farms 1 and 2, respectively.

Discussion

Advancement of calving in red deer has been shown to be possible by artificial shortening of daily photoperiod (Webster and Barrell 1985) and has been observed in lactating red deer hinds after daily oral or injectable melatonin treatment (Adam and Atkinson 1984; Webster and Barrell 1985; Adam et al. 1986). However, for practical purposes long-acting melatonin implants, Regulin, (Staples 1989) have many advantages over daily manipulations. Regulin implants, normally given by a three-treatment regime, advance the breeding season in deer by up to 52 days (Asher et al. 1988; Fisher et al. 1988; Fisher and Fennessey 1991; Fisher et al. 1990; Wilson et al. 1991). To achieve acceptance by the farmer, regimes to advance calving must be economical and practical.

The present data based on fetal age estimation utilize ultrasound scanning and measurements of fetal and uterine dimensions. Bingham et al. (1990) and Wilson and Bingham (1990) confirmed that the mean error of fetal age estimation was 0.97 days, and the predicted median calving dates of six monitor groups involving 132 deer varied 0, 0, 0, 0, 1, and 4 days from actual median calving dates. Therefore, this technique provides a reliable estimate of median calving date, and the conclusions made in the present study can be accepted with confidence.

In previous trials using 1-year-old hinds, untreated control groups were mated in-contact with melatonin-treated stags. Fisher and Fennessy (1991) suggest that the advancement of calving in melatonin treated hinds may be a combination of the effect of melatonin on the hind together with an influence of the early rutting stag. In Experiment 1, control hinds on farm 1 were mated with treated stags, but on farms 2 and 3 control hinds were mated with untreated stags. There was a difference in median calving date advancement between the three farms, but median calving date of controls on farms 2 and 3 were 3 to 7 days after that on farm 1, again suggesting a stag effect.

In all 1-year-old groups on farm 1, pregnancy percentage, determined by ultrasound examination, was satisfactory. On farm 2, pregnancy rates were lower than normal, probably because body weights were below target, but possibly because the mating period was shorter. Hinds from farm 3 were transported to farm 2 immediately prior to the introduction of stags, and this may have resulted in a lower conception rate, and, therefore, depression of the overall effectiveness of the melatonin treatment in advancing the median calving date. The overall 12 to 15 day average advancements of median calving dates were lower than the 22-day average advancement in the median calving date reported elsewhere by Wilson et al. (1990). This may be attributable to the more restricted mating period in the present study, causing a more condensed calving pattern, and an earlier median calving in control deer. However, this study has confirmed that melatonin administration to hinds by a 45-day interval, two-treatment regime using Regulin is as effective in advancing the median calving date as a 30-day interval, three-treatment regime reported elsewhere by Wilson et al. (1991).

The predicted commencement of calving of yearling hinds was as early as 24 October. This indicated that some treated yearling hinds were in estrus as early as 3 March (farm 2). In other groups, matings were as early as 10 March (farm 3) or 13 March (farm 1). Untreated control hinds commenced mating from 29 March (farm 1) to 11 April (farm 2) which is considered normal for this age group of farmed red deer (P.R. Wilson, unpublished data).

The important implication of Experiment 1 is that the two-treatment regime requires less handling of hinds and a lower labor requirement, and fewer implants, resulting in a lower cost; however, the calving span of melatonin-treated groups was longer than for control groups, particularly on farm 2. This can be influenced by the farmer by earlier withdrawal of the stag at the end of mating, but that may result in a slightly lower pregnancy rate. An alternative is to use ultrasound pregnancy-testing techniques early in pregnancy as described by Revol and Wilson

(1989) and Bingham et al. (1990) to distinguish early- and late-calving hinds to be managed separately.

Confirmation of the effect of treated stags shown by experiment 2 explains the observations of calving earlier than expected in untreated control deer mated in contact with melatonin-treated hinds in earlier trials (Wilson et al. 1991; G.K. Barrell and L.D. Staples personal communication). Earlier work by Moore and Cowie (1986) indicated a slight advancement of mean calving date of hinds exposed to vasectomized stags prior to the rut. The stags used in that study were not treated with melatonin to advance the rut and may not have been rutting as vigorously as treated stags; however, that study gave an indication that the stag may indeed have an influence on the date of onset of estrus.

The 8- to 10-day advance in median calving date in adult hinds mated with melatonin-treated stags shown by Experiment 2 is similar to a 12-day mean calving date advance shown by Fisher and Fennessy (1991). They reported a mean calving date of 21 November in hinds mated with a treated stag compared with 4 December for those mated with an untreated stag. In the present study, median calving dates for groups mated with treated stags were 15 to 22 November and for control hinds, 25–30 November. The striking feature of the data of Fisher and Fennessey (1991) was the very condensed calving pattern of hinds mated with a treated stag (19 days). This pattern was not observed in the present study where the shortest calving span was 35 days and the longest approximately 60 days.

McComb (1987) proposed that the mechanism for advancement of estrus in hinds was the roaring of stags which evoked an early onset of estrus in hinds rather than pheromonal or other behavioral influences such as chasing or herding. However, in that study entire stags were joined 1 April, which is considered to be the normal time for onset of estrus in farmed red deer hinds, thus precluding determination of possible influences prior to the normal breeding season. In the present study and in that of Wilson et al. (1991), it was noted by farmers that roaring by treated stags commenced considerably earlier than normal, and the author noted that aggressive behavior of treated stags was often considerable at the time of administration of the third melatonin treatment (9 February). The "ram effect" observed with ewes, resulting in the induction of ovulation after exposure to a ram following a period of isolation is attributable to a pheromonal response, but behavioral factors, including vocalization, exert some influence (Martin et al. 1986). Further investigation is needed to determine whether the influence of the stag is due to vocalization or physical behavior in deer.

The failure of prior exposure of hinds to vasectomized melatonin-treated stags to advance the onset of the breeding season further than that caused by intact treated stags alone is not surprising in view of the fact that in sheep a period of isolation from the male is necessary for the "ram effect" to occur (Martin et al. 1986). However, Moore and Cowie (1986) reported an advancement of the breeding season after exposure of hinds to a vasectomized (untreated) stag prior to the normal rutting season. The calving pattern in response to treated intact stags suggests that the influence of the stag may be very rapid in some circumstances. On farm 1, first matings occurred as early as 9 March, only 3 days after introduction of treated stags. On farm 2, matings first commenced 11 March, 10 days after the stags were introduced. Data of Wilson et al. (1991) also suggest a considerable variation between stags and their ability to advance estrus in hinds. However, the ultrasound-scanned hinds were randomly selected from all mating groups so stag variation was included in the estimation of median calving date in the present study.

In this study and that of Fisher and Fennessy (1991) the influence of the stag has been shown on mixed-aged hinds. It is not known whether the stag influences the onset of estrus in 1-year-old hinds, but the data of Wilson et al. (1991) strongly suggest a similar stag influence on that age group.

There are practical advantages in using the influence of the stag to advance the onset of the breeding season in mixed-aged hinds. For commercial herds it is not practical to muster and yard mixed-aged hinds from mid-November until late January because of the risk of mismothering newborn calves and the risk of injuries to small deer. Therefore, melatonin implantation regimes in adult hinds similar to those used in yearling hinds are not advisable. There remains some concern about possible influences of melatonin on the onset of lactation in late pregnant deer. Asher et al. (1988) reported failure to rear fawns by fallow does (*Dama dama*) treated with melatonin prior to lactation, suggesting that melatonin directly or indirectly inhibited lactogenesis. Nowak et al. (1985) observed a reduction of blood prolactin concentration in melatonin-treated hinds. However, Adam et al. (1989) reported no negative effect on subsequent lactation of pregnant red deer hinds that were treated with melatonin despite a reduction in prolactin concentrations. The ability to advance the onset of the breeding season simply by melatonin treatment of stags precludes these potential problems. However, the farmer must accept that the advancement of estrus resulting from the "stag effect" is only approximately one-half of that achievable with melatonin treatment of hinds in combination with mating with treated stags. Early weaning and optimum nutrition may be necessary to achieve the maximum response to either technique (Wilson 1989).

Acknowledgements

The willing help of staff of Ben Nevis, Opou and Otuiti Stations is gratefully acknowledged. Linton Staples is thanked for his advice and B. Revol and K. Killorn are thanked for assistance with ultrasound pregnancy diagnoses. The study was funded by Regulin Ltd., Melbourne, and Agland Holdings Ltd., Waipukurau. Melatonin implants were supplied by Regulin Ltd., now Schering Agrochemicals Ltd.

References

Adam CL, Atkinson T (1984) Effect of feeding melatonin to red deer (*Cervus elaphus*) on the onset of the breeding season. J Reprod Fert 72:463–466

Adam CL, Moir CE, Shiach P (1989) Plasma prolactin concentrations in barren, pregnant and lactating red deer (*Cervus elaphus*) given melatonin to advance the breeding season. Anim Reprod Sci 18:77–86

Adam CL, Moir CE, Atkinson T (1986) Induction of early breeding in red deer (*Cervus elaphus*) by melatonin. J Reprod Fert 76:569–573

Asher GW, Barrell GK, Adam JL, Staples LD (1988) Effects of subcutaneous melatonin implants on reproductive seasonality of farmed fallow deer (*Dama dama*). J Reprod Fert 84:679–691

Bingham CM, Wilson PR, Davies AS (1990) Real time ultrasonography for pregnancy diagnosis and estimation of fetal age in farmed red deer. Vet Rec 126:102–106

Fennessy PF, Fisher MW (1988) Advancing the breeding season in red deer. Proc Deer Course for Vets 5. Deer Branch, N Z Veterinary Assoc, pp 17–27

Fisher MW, Fennessy PF (1991) A note on melatonin treatment of red deer stags to advance the onset of the calving season in hinds. Anim Prod 51 (in press)

Fisher MW, Fennessy PF, Milne JD (1988) Effects of melatonin on seasonal physiology of red deer. Proc N Z Soc Anim Prod 48:113–116

Fisher MW, Fennessy PF, Johnston PD (1990) The timing of melatonin treatment affects the seasonal onset of ovarian activity, coat growth and liveweight in young red deer hinds. Anim Reprod Sci 23:47–59

McComb K (1987) Roaring by red deer stags advances the date of estrus in hinds. Nature 330:648–649

Martin GB, Oldha CM, Cognie Y, Pearce DI (1986) The physiological responses of anovulatory ewes to the introduction of rams—a review. Livestock Prod Sci 15:219–248

Moore GW, Cowie GM (1986) Advancement of breeding in non-lactating adult red deer hinds. Proc N Z Soc Anim Prod 46:175–178

Nowak R, Elmhirst RN, Rodway RG (1985) A note on the effect of melatonin feeding on the initiation of ovarian activity and on plasma prolactin levels in lactating and non-lactating red deer hinds. Anim Prod 40:515–581

Revol B, Wilson PR (1989) Ultrasonographic pregnancy diagnosis in red deer. Proc Deer Course for Vets 6. Deer Branch, N Z Veterinary Assoc, pp 36–53

Staples LD (1989) The development of "Regulin" from concept to international marketplace. Aust J Biotech 3:164–175

Webster JR, Barrell GK (1985) Advancement of reproductive activity, seasonal reduction in prolactin secretion, and seasonal pelage changes in pubertal red deer hinds (*Cervus elaphus*) subjected to shortened daily photoperiod or daily melatonin treatments. J Reprod Fert 73:255–260

Wilson PR (1989) Advanced calving in deer: practical aspects. Proc Deer Course for Vets 6. Deer Branch, N Z Veterinary Assoc, pp 54–68

Wilson PR, Bingham CM (1990) Pregnancy diagnosis and prediction of calving dates in red deer from 44 to 110 days gestation using real-time ultrasound scanning. Vet Rec 126:133–135

Wilson PR, Walker IH, Bond DB, Middleberg A, Staples LD (1991) Field evaluation of melatonin implants to advance the breeding season in one-year-old red deer hinds. N Z Vet J 39:23–28

68

Induction of Ovulation in Père David's Deer Hinds at Two Stages of Seasonal Anestrus

Barry R. Brinklow, Bernard J. McLeod, Andrew S.I. Loudon, and Jonathon D. Curlewis

Abstracts

Père David's deer hinds were treated with gonadotrophin-releasing hormone (GnRH) administered either intermittently by IV injections (2 μg GnRH at 2-h intervals) for 4 days or as a continuous subcutaneous infusion (1 μg/h) for 14 days after progestagen pretreatment early (February to March) or late (May to June) in the period of seasonal anestrus. Both treatments at both stages of anestrus induced ovulation indicated by an increase in progesterone concentration exceeding 0.5 ng/ml for at least 7 days accompanied by behavioral estrus and an luteinizing hormone (LH) surge. Treatment was significantly more effective in inducing ovulation in late (9/12 animals) than early (3/13 animals) anestrus ($P < 0.01$), and the onset of estrus behavior and LH surge were significantly earlier in late than early anestrus ($P < 0.05$). These results indicate that there is a period of deeper anestrus following the end of the breeding season.

Key words: Anestrus, GnRH, ovulation, Père David's deer

Introduction

The Père David's deer (Mi Lu; *Elaphurus davidianus*) is a seasonally breeding deer originally from East Asia, where they are said to have become extinct early in the twentieth century, being maintained as a captive population in Europe. The species breeds seasonally, and females in the U.K. will, if unmated, undergo repeated estrous cycles from July to December (Curlewis et al. 1988). This breeding season is similar to that of domestic sheep in which the control of estrous cycling in the female is via the feedback effects of estradiol on the secretion of gonadotrophin-releasing hormone (GnRH; Legan and Karsch 1980), giving rise to differences in LH pulsatile secretion between the breeding season and seasonal anestrus (Scarramuzzi and Baird 1977; McLeod et al. 1982a, b, I'Anson and Legan 1988). Curlewis et al. (1991) recently reported that in the Père David's deer, in contrast to the sheep, there is a change in pattern of luteinizing hormone (LH) secretion over seasonal anestrus, indicating a period of deep anestrus with low LH pulse frequency and a lower LH response to administered GnRH in early anestrus. LH pulse frequency and response to GnRH subsequently increase in mid- and late seasonal anestrus.

In seasonally anestrus sheep and goats ovulation can be induced by repeated injections of GnRH to artificially impose a high frequency of LH episodes (McLeod et al. 1982a, b; Knight et al. 1988). Though increased LH pulsatility is thought to be the major factor inducing the later stages of preovulatory follicle development

(Rahé et al. 1980; Baird 1978), continuous infusions of GnRH resulting in an elevation of basal LH concentrations have also been shown to induce ovulation in anestrus sheep (McLeod et al. 1983, 1988), though this treatment is not effective in cattle (Jagger et al. 1987).

The present study was designed to assess the pituitary and ovarian responses to GnRH administered either intermittently or continuously in early and late anestrus at times when the endogenous LH pulse frequently was low or high.

Materials and Methods

Adult hinds were maintained on pasture at Whipsnade Wild Animal Park, Bedfordshire, U.K. (latitude 51° 30′ N). During periods of intensive blood sampling they were enhoused under conditions of natural photoperiod and fed pelleted lucerne and hay ad libitum with continuous access to fresh water. All hinds were bled twice weekly throughout the studies in order to determine the onset and cessation of seasonal anestrus and the incidence of ovulation.

Experiment 1

In late February 1987, six hinds were initially treated with progesterone for 14 days via 5 SC implants each containing 275 mg in the flank area (Sil Estrus, Ceva, Sydney, Australia). From 2 to 4 h after implant removal the animals were treated with 2.0 μg GnRH (Lutal, Fabwerke Hoechst, Frankfurt, Germany) by IV injection at 2-h intervals for 4 days. Blood samples for plasma LH determination were withdrawn before each injection and at 15-min intervals 0–8, 16–24, 40–48, and 64–72 h after the first GnRH injection. The foregoing regime was repeated in early June with 12 days of progesterone pretreatment. Implant removal and replacement were carried out under xylazine sedation (20 mg, Bayer Pharmaceuticals, Leverkusen, Germany), and jugular cannulae for sampling and injection were inserted at the time of implant removal. The effect of progesterone treatment alone was investigated in two further 12-day periods in early April and early May. In addition to the above, five hinds were monitored for plasma progesterone to establish the timing of normal seasonal anestrus.

Experiment 2

In this experiment the effect of continuously administered GnRH was investigated in early and late anestrus. In late February and again in late April (1988) 13 adult hinds were pre-treated with progestagen intravaginal sponges (two per animal Chronogest [fluogesterone acetate] Intervet, Cambridge, U.K.) for 12 days. At the time of sponge removal seven animals were sedated (20 mg xylazine) and osmotic minipumps (Alzet 2ML2, Alza Corp Palo Alto, Calif., USA) were inserted SC in the flank and left for 14 days. The minipumps were preincubated in sterile saline at 37° C for 16 h before insertion and delivered GnRH at 1.0 μg/h in 3.8 μl saline. Blood samples for plasma LH determination were collected from all 13 hinds by jugular venepuncture at 2-h intervals 16–22, 46–52, and 76–82 h after minipump insertion. The animals were randomly reallocated to GnRH-treated and control groups between the early and late anestrus periods.

Hormone Assays

Plasma LH concentrations were measured using the ovine double antibody radioimmunoassay of McLeod et al. (1982a) and validated for deer by Loudon et al. (1991). In Experiment 1, plasma progesterone concentrations were determined on solvent-extracted samples according to the method described by Curlewis et al. (1988). In Experiment 2, plasma progesterone was determined by the direct radioimmunoassay of McNeilly and Fraser (1987) validated for deer by McLeod et al. (1991).

Data Analysis

A preovulatory LH surge was defined as an increase in plasma concentration above 4 ng/ml for greater than 5 h with a peak value exceeding 10 ng/ml. The onset of the surge was defined as the time when the plasma LH concentration first exceeded 4.0 ng/ml (McLeod et al. 1985). Plasma progesterone concentrations above 0.5 ng/ml were considered evidence of an active corpus luteum. Differences in mean LH concentrations were assessed by analysis of variance and in the incidence of estrus, preovulatory LH surge, and ovulation using chi-squared analysis.

Results

Experiment 1

There was a remarkable degree of synchrony between the animals' plasma progesterone in both the time of termination of the breeding season (6 January ± 3.2 days; $n = 11$) and the subsequent time of onset of the next breeding season (20 July ± 2.2 days, $n = 6$; Fig. 68.1).

In the group of treated animals (below) the progesterone implants elevated plasma concentrations to a mean of 2.55 ± 0.15 ng/ml, comparable with normal luteal phase values. The profiles show that elevated plasma progesterone concentrations indicative of ovulation followed by formation of a corpus luteum occurred in only one of six animals in early anestrus (B54) and five of six in late anestrus. Administration followed by withdrawal of progesterone implants in April and May did not result in any elevation of plasma progesterone above 0.5 ng/ml for at least 12 days after implant removal.

In the five untreated hinds, the onset of the breeding season (defined by the first increase in plasma progesterone concentration above 0.5 ng/ml for 7 days or longer) was preceded by a transient increase in progesterone concentration. Similar increases were recorded in three of six treated animals of which two had undergone a GnRH-stimulated estrous cycle.

An example of the profile of plasma LH concentrations in one GnRH-treated animal (B54 in late anestrus) showing the response to the GnRH injections over the periods of intensive (15-min) blood sampling and the LH surge is shown in Fig. 68.2. Two of the six animals exhibited behavioral estrus and an LH surge in early and five of six in late anestrus. Of these, one animal (B51) in early anestrus did not have an extended elevation of plasma progesterone

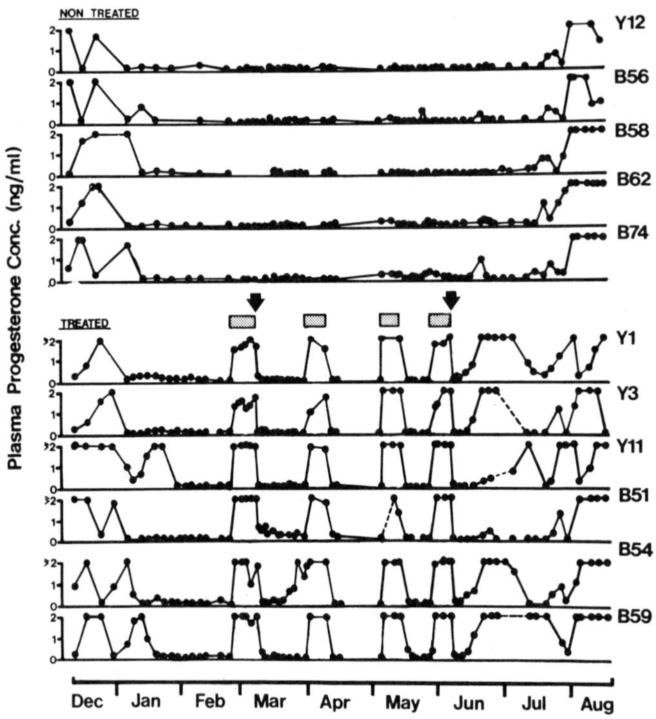

FIGURE 68.1. Plasma progesterone concentrations in Père David's deer hinds (Experiment 1) that were untreated or were subjected to periods of progesterone treatment with and without subsequent treatment with repeated injections of GnRH (2.0 μg/injection) at 2-h intervals for 96 h. Periods of progesterone treatment shown by solid bars, times of GnRH treatment by arrows.

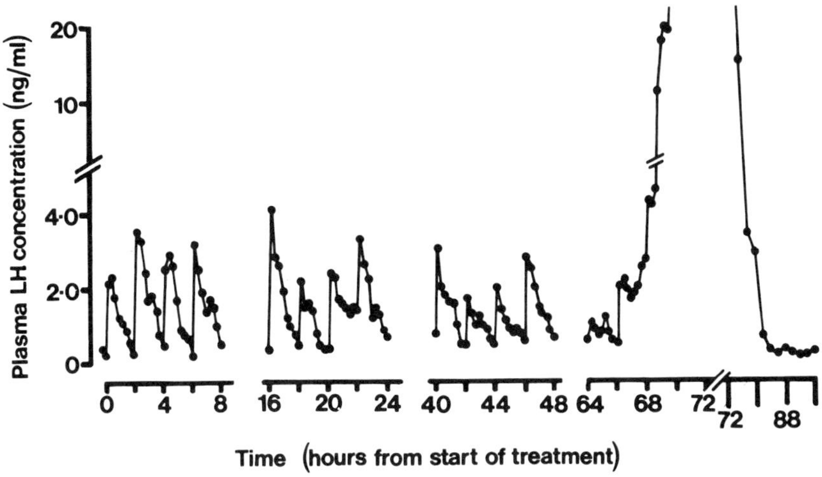

FIGURE 68.2. Plasma LH concentrations over four periods of frequent blood sampling (samples taken at 15-min intervals) in a Père David's deer hind treated with repeated injections of GnRH (2.0 μg GnRH at 2-h intervals) beginning at time 0 for 5 days.

indicative of a competent corpus luteum. There was no significant difference in the mean peak LH response to GnRH injections between the four intensive sampling periods in either early or late anestrus; however, the mean overall response was greater in late (4.03 ± 0.28 ng/ml) than early (3.12 ± 0.26 ng/ml) anestrus ($P < 0.05$). In addition the time of onset of the LH surge from the beginning of GnRH treatment was earlier in June than in March (59.4 ± 0.25 vs 85.8 ± 3.76 h), as was the time of observation of behavioral estrus (62.4 ± 0.98 vs 91.0 ± 1.0 h).

Experiment 2

Blood sampling began after the onset of seasonal anestrus in all animals and no spontaneous elevations of progesterone occurred indicating the onset of the subsequent breeding season until after mid-July (Fig. 68.3). The profiles show ovulation indicated by elevated progesterone concentrations occurred in two of seven animals in early anestrus, both of which remained elevated for more than 2 months indicating the presence of persistent corpora lutea. These two animals were removed from the second part of the experiment as their plasma progesterone levels were still elevated at the time of the second treatment. Two further animals had transient elevations above 0.5 ng/ml (Bk2 and Y3). In late anestrus, four of six GnRH treated animals had indications of ovulation. There were no elevations in plasma progesterone in any of the animals treated with progestagen alone.

In this experiment, with 2-h sampling for 9-h periods, there was no significant difference in plasma LH concentrations between early and late anestrus in either the GnRH-treated or the nontreated animals, though the GnRH-treated animals had significantly higher levels than those untreated (early 2.57 ± 0.73 vs 0.48 ± 0.09; late 2.18 ± 0.49 vs 0.69 ± 0.15; $P < 0.001$). Behavioral estrus was detected in three of seven GnRH-infused animals in early and three of six in late anestrus, being significantly earlier in late anestrus (46.7 ± 0.67 vs 66.2 ± 0.32 h from pump insertion; $P < 0.05$). No animals treated with progestagen alone showed evidence of behavioral estrus.

Discussion

This study clearly demonstrates that both intermittent and continuous GnRH treatments are capable of inducing ovulation accompanied by estrus behavior, a preovulatory LH surge, and a competent progesterone-producing corpus luteum. Pooling the results from experiments 1 and 2, significantly more hinds ovulated in

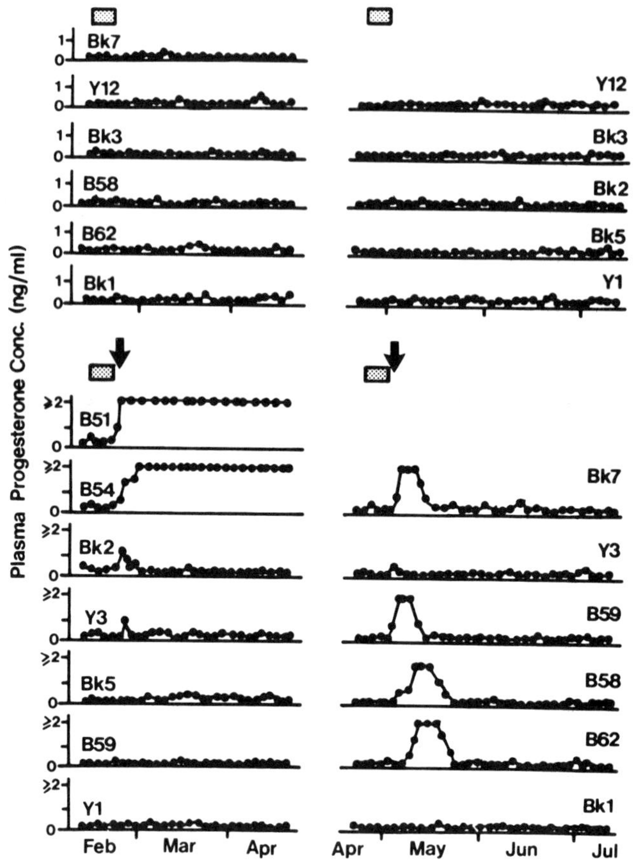

FIGURE 68.3. Plasma progesterone concentrations in Père David's deer hinds (Experiment 2) treated with a continuous infusion of GnRH (1.0 μg/h) or untreated following a period of progestagen pretreatment. Periods of progestagen treatment are shown by hatched bars, times of GnRH treatment by solid arrows.

response to GnRH treatment in late than in early seasonal anestrus ($P < 0.01$). Furthermore, the response in LH surge and estrus behavior was 20 to 30 h earlier in late than early anestrus. In Experiment 1, the LH response to GnRH injections was significantly greater in late than in early anestrus, comparable to the results of Curlewis et al. (1991).

The data presented here and that of Curlewis et al. (1991), which show endogenous LH pulse frequency to be significantly lower in early than mid- and late anestrus, indicate that in the Père David's deer there is a period of deep anestrus at the end of the breeding season which changes as anestrus progresses. The reduced pituitary responsiveness to GnRH in early anestrus resulting in reduced LH secretion would give rise to reduced follicular development resulting in smaller follicles. This may explain the increased time to induced ovulation in early anestrus and the smaller proportion of animals responding.

The timing of the LH surge relative to the expression of estrus behavior was comparable to that of Père David's hinds in the breeding season (Loudon et al. 1991). The practical significance of differences in the timing of the responses to GnRH treatment administered either by injection or continuous infusion is difficult to interpret as the two studies were carried out in different years. Differences of this order have been described in sheep of the same breed treated with GnRH in different years (Hunter et al. 1986; McLeod et al. 1988).

Two hinds in Experiment 2 exhibited persistent corpora lutea after GnRH treatment. Similar periods of consistently elevated plasma progesterone have been previously reported in this species at the onset of the breeding season by Curlewis et al. (1988).

The high degree of synchrony in the termination and onset of the breeding season seen in Experiment 1 may indicate that social factors are involved in the timing of the breeding season in these animals kept in the absence of males. In addition we invariably observe an earlier onset of breeding season (as judged by mating behavior and calving dates) in the main herd of Père David's deer at Whipsnade with stags present, than in hinds kept in the absence of males.

At the beginning of the breeding season in the untreated animals in Experiment 1, all five showed a transient elevation of progesterone. It is interesting to note that two of six treated animals which ovulated in response to GnRH in late anestrus also showed a similar transient elevation of progesterone. This may indicate that the induction treatment, even quite late in seasonal anestrus, did not affect the normal ovarian mechanisms associated with onset of the breeding season.

In conclusion, the data presented support the existence of a period of deep anestrus at the end of the breeding season in Père David's deer. Induction of ovulation late in anestrus probably does not advance the onset of the normal breeding season which may occur at the normal time. Induction of ovulation in Père David's deer is effective using both intermittent and continuous administration of GnRH after progestagen priming.

Acknowledgements

We thank A. Beasey and B. Eves for care of the animals, A. Coleman and J. Boyle for assistance with hormone assays, Hoechst, Germany, for the GnRH preparation, NIADDK National Hormone and Pituitary Program for the LH standard, and A. McNeilly of the MRC Unit of Reproductive Biology in Edinburgh for provision of progesterone antisera and labelled tracer.

References

Baird DT (1978) Evidence in vivo for the two-cell hypothesis of oestrogen synthesis by the sheep Graafian follicle. J Reprod Fert 50:183–185

Curlewis JD, McLeod BJ, Loudon ASI (1991) LH secretion and response to GnRH during seasonal anoestrus of the Père David's deer hind. J Reprod Fert 91:131–138

Curlewis JD, Loudon ASI, Coleman APM (1988) Oestrous cycles and the breeding of the Père David's deer hind (*Elaphrus davidianus*). J Reprod Fert 82:119–126

Hunter MG, Southee JA, McLeod BJ, Haresign W (1986) Progesterone pre-treatment has a direct effect on GnRH induced preovulatory follicles to determine their ability to develop into normal corpora lutea in anoestrus ewes. J Reprod Fert 76:349–363

I'Anson H, Legan SJ (1988) Changes in LH pulse frequency and serum progesterone concentrations during the transition to breeding season in the ewe. J Reprod Fert 82:341–345

Jagger J, Peters AR, Lamming GE (1989) Hormone responses to low-dose GnRH treatment in postpartum beef cows. J Reprod Fert 80:263–269

Jagger J, Peters AR, McLeod BJ, Lamming GE (1987) Pituitary and ovarian responses to prolonged administration of low doses of GnRH in prepubertal heifers. Anim Reprod Sci 18:111–124

Knight CH, McLeod BJ, Haresign W, Wilde CT (1988) Exogenous GnRH induces ovulation in seasonally anoestrus lactating goats. J Reprod Fert 83:679–686

Legan SJ, Karsch FJ (1980) Photoperiodic control of seasonal breeding in ewes: modulation of the negative feedback action of estradiol. Biol Reprod 23:1061–1068

Loudon ASI, McLeod BJ, Curlewis JD (1990) Pulsatile secretion of LH during the periovulatory and luteal phases of the oestrous cycle in Père David's deer hind (*Elaphurus davidianus*). J Reprod Fert 89: 663–670

McLeod BJ, Brinklow BR, Curlewis JD, Loudon ASI (1991) Efficacy of intermittent or continuous administration of GnRH in inducing ovulation in early and late seasonal anoestrus in the Père David's deer hind. J Reprod Fert 91:229–238

McLeod BJ, Haresign W, Peters AR, Humke R, Lamming GE (1988) The development of subcutaneous-delivery preparations of GnRH for the induction of ovulation in acyclic sheep and cattle. Anim Reprod Sci 17:33–50

McLeod BJ, Peters AR, Haresign W, Lamming GE (1985) Plasma LH and FSH responses and ovarian activity in prepubertal beef heifers treated with repeated injections of low doses of GnRH for 72 h. J Reprod Fert 74:589–596

McLeod BJ, Haresign W, Lamming GE (1983) Induction of ovulation by continuous infusion of low doses of GnRH. J Reprod Fert 68:489–495

McLeod BJ, Haresign W, Lamming GE (1982a) The induction of ovulation and luteal function in seasonally anoestrous ewes treated with small-dose multiple injectins of GnRH. J Reprod Fert 65:215–221

McLeod BJ, Haresign W, Lamming GE (1982b) Response of seasonally anoestrous ewes to small-dose multiple injections of GnRH with and without progesterone pre-treatment. J Reprod Fert 65:223–230

McNeilly AS, Fraser HM (1987) Effect of gonadotrophin-releasing hormone agonist-induced suppression of LH and FSH on follicle growth and corpus luteum function in the ewe. J Endocrinol 115:273–282

Rahe CH, Owens RE, Fleeger JL, Newton HJ, Harms PGH (1980) Patterns of plasma luteinizing hormone in the cyclic cow; dependence upon the period of the cycle. Endocrinology 107:489–503

Scarramuzzi RJ, Baird DT (1977) Pulsatile release of luteinizing hormone and the secretion of ovarian steroids in sheep during anestrus. Endocrinology 101:1801–1806

69
Sexual Attraction in White-Tailed Deer

Mark D. Whitney, Daniel L. Forster, Karl V. Miller, and R. Larry Marchinton

Abstract

We conducted two experiments to determine the source of sexually attractive odors in white-tailed deer (*Odocoileus virginianus*). In Experiment 1, estrous, midcycle, and anestrous doe urines, buck urine, a commercial deer lure, and sterile saline were tested to determine their attractiveness to free-ranging deer on study areas in Georgia and Florida. All urine samples were obtained directly from the bladder. A randomized incomplete block design with paired 1 m^2 tracking pits (10-m apart) was used. Deer visitation rates to test areas were highest for buck urine but were not statistically different for the compounds tested ($P < 0.05$). In Experiment 2, bladder urine or vaginal secretions from estrous and midcycle females and a control were applied to ovariectomized females. Attractiveness to an epididymectomized, teaser male was determined by recording the number of approaches and time associated with each doe. With both measures, attractiveness of estrous vaginal secretions was significantly greater than the control ($P < 0.03$). Neither midcycle vaginal secretions, midcycle urine, nor estrous urine elicited significantly more responses than the control. Our data indicate that sexually attractive odors in female white-tailed deer originate in the reproductive tract. The possibility that these odors may become associated with voided urine during urination must be explored.

Key words: Attractant, estrus, pheromone, reproduction, urine, white-tailed deer

Introduction

In the United States, numerous urine-based scent products are marketed for attracting white-tailed deer (*Odocoileus virginianus*). However, there has been little research published that provides a scientific basis for this. Urine is an accurate indicator of physiological status because of its function in removing metabolic waste (Albone 1984). White-tailed deer behavior suggests that urine could be a source of pheromones signalling estrus (Moore and Marchinton 1974).

A second possible source of a signal indicating reproductive status is mucus or other secretions from the female reproductive tract. Paleologou (1977) concluded that a sex pheromone emanated from the cervicovaginal mucus of estrous cows (*Bos taurus*). McDonald (1980) noted hypertrophy of the vaginal epithelium, and Schilling and Zust (1968) observed increased viscosity of vaginal mucus in estrous cattle. Similar occurrences in white-tailed deer (K.V. Miller, unpublished data) make vaginal secretions a likely source of a sex pheromone; however, the roles of urine and vaginal secretions in deer reproductive behavior have not been established. The objectives of our two experiments were to determine: (1) whether deer urine acts as an attractant to bucks, and (2) if estrous urine or estrous vaginal secretions contain a pheromone that elicits sexual behavior in males.

Methods

Experiment 1

Experiment 1 was conducted on the Bishop F. Grant Forest (area I) in Putnam County, Georgia, and Tall Timbers Research Station (area II) in Leon County, Florida. Deer densities were estimated at 12 to 14 animals/km^2 and 10 to 12 animals/km^2 on the two study areas, respectively. White-tailed deer were the only ungulates inhabiting the study areas. Limited sport hunting was allowed on area I but prohibited on area II.

From October 1986 through February 1987, attractiveness of anestrous doe urine, estrous doe urine, midcycle doe urine, buck urine, a commercial buck lure, and sterile saline (0.25 N) was studied. Doe urine was collected via catheterization from captive females whose estrous cycles were monitored daily with epididymectomized bucks (Knox 1987). Urine was collected during estrus as well as between estrus periods (but not within 2 days of standing estrus). Anestrous doe urine also was collected during August and September before the does' first estrous cycle of the season. Buck urine was taken from the bladders of antlered bucks harvested during the rut. Samples were combined within classes and stored at $-80°$ C.

Pressurized containers were charged with 180 ml of solution containing 1 part urine and 3 parts sterile saline (0.25 N); commercial deer lure was transferred undiluted. Automatic dispensers were used to release 1.2 cc ($\bar{x} \pm SD = 1.21 \pm 0.07$ cc; $n = 10$) of test mixture into the center of a 1-m^2 tracking medium every 6 min ($\bar{x} \pm SD = 16.06 \pm 0.37$; $n = 10$) for 48 h.

A balanced, incomplete block statistical design was used. Treatments were assigned to 100-m radius plots in pairs and placed 10 m apart (Forster 1988). Each treatment was paired with all others for six 3-day periods during fall and winter 1986. Pits were observed at 24, 48, and 72 h and number of tracks/pit recorded. Pits were smoothed after each observation. Visitations were expressed as a percentage of deer-visits/pit per period and rates transformed (arcsine) before statistical analyses were conducted (Forster 1988). Visitations were analyzed by period on both sites because of significant treatment × period interactions.

An index of track size was calculated by multiplying the mean length and width of a sample of five tracks. Differences in track size/treatment per period were tested using the largest tracks recorded for each visitation. Treatment observations then were pooled across periods and mean sizes of all track sets were used for comparisons.

We assumed that the number of tracks/visit was related to time spent in the tracking pit. The mean number of tracks/visit was tested among treatments as an index of "interest" in the treatment. Multiple visitations were excluded, and observations were pooled by treatment and period to test for differences in track numbers and size.

Experiment 2

All research for Experiment 2 took place at the University of Georgia, School of Forest Resources' Whitehall Deer Research Facility in Athens, Georgia. Urine was collected by the same methods used in Experiment 1, except that midcycle doe urine was not collected within 6 days of standing estrus. Vaginal secretions were collected by inserting a Tampax tampon (Tambrands, Lake Success, N.Y.) into the vagina subsequent to urine collection. The tampon remained in the animal for 15 min and then was removed. One-half of each tampon was removed for use in another study. Tampons noticeably contaminated with urine were discarded. Tampons were wrapped in aluminium foil and stored at $-40°$ C until ready for use. Prior to the trials, samples were thawed to ambient temperature.

Four ovariectomized does and one epididymectomized male were used as test animals. The ovariectomies were performed to eliminate behavioral and/or physiological cues of estrus.

Two does were used in each test: one of two does received either estrous or midcycle urine or vaginal mucus as a treatment. The second doe was used as a control. Treated and control tampons were attached with adhesive tape to the does' rumps above the base of the tail. Urine treatments were prepared by soaking one-half of a tampon in 10 ml of urine. Vaginal secretion treatments were applied by attaching the tampon used in collecting the sample. Control animals had a "blank" tampon attached to their rumps. Animals receiving one of the four treatments were not used again for 48 h. The does were selected randomly from the eligible animals.

Twenty trials were run using each treatment except midcycle vaginal secretions (17 trials). During each trial, the two does and the buck were moved to an observation pen. Trials were 30 min in duration, and activities of all three deer were recorded. The number of courtship approaches (Brown and Hirth 1978) the buck made to each doe was counted, and the time he associated with each female was recorded in seconds.

Approach data were analyzed using the analysis of variance randomized complete block design. A square root transformation was used to normalize the approach data. Time data were analyzed using a paired *t*-test. Statistical analyses for Experiments 1 and 2 were conducted using Statistical Analytical Systems (SAS Institute, 1985).

Results

Experiment 1

Visitation rates did not differ ($P > 0.05$) among any of the treatments or the control on either site (Tables 69.1 and 69.2). Deer tracks occurred in 21.1% and 26.3% of the observations on study areas I and II, respectively. A total of 245 deer visits was recorded with multiple visits occurring a minimum of 62 times (25.3%).

Number of tracks/visit was highly variable, ranging from 1 to 43 ($\bar{x} = 8.4$; SD = 6.7), on area I and 1 to 41 ($\bar{x} = 9.1$; SD = 7.0), on area II. Differences in mean number of tracks/visit were not detected ($P > 0.10$) among treatments on either area; however, deer lure received fewest tracks/visit on both areas, and buck urine the overall highest. Buck urine ranked 2 and 1 on areas I and II, respectively. This suggests that at least some deer visiting the pits showed more interest in and spent more time "assessing" the information contained in buck urine than the other treatments.

Although the sex of whitetails visiting tracking pits could not be determined with certainty, tracks of adult males have been shown to be larger than other sex and age classes (P.K. Swiderek, unpublished data). Track size indices ranged from 14.4 to 42.7 ($n = 929$) on area I and 11.1 to 46.2 ($n = 1222$) on area II. Track sizes did not differ significantly ($P \geqq 0.10$) among treatments and periods, but mean track sizes/visit were larger in tracking pits treated with buck urine on both study areas. In addition, of the largest 25% of tracks observed, the highest percentage was in pits treated with buck urine, suggesting that buck urine may have elicited a greater response from adult males.

A high proportion of total visits occurred during early sampling periods (Fig. 69.1). The tendency for a reduction in deer visitations may have been due to habituation. Precipitation was not correlated significantly with visitation rates on either area ($P > 0.10$, Spearman's correlation coefficient). Low number of visitations on area I during periods 4 and 5 may have been due to behavioral changes associated with two either-sex hunts (25–29 November and 3 December). Decreased activity following the rut also may explain this reduction in visits. Most rutting activity occurred during the middle of November on area I, which corresponded closely with peak visitations. The second rise in visitations in late December and January may have been due to a second peak in breeding activity. Visitations rates also were greatest on area II just prior to peak rut and declined gradually throughout the breeding season. A younger age structure on area I may explain why fewer visitations were recorded than on area II even though deer densities were similar on both areas.

Experiment 2

In trials comparing estrous urine, midcycle urine, or midcycle vaginal secretions with the control, there were no significant differences ($P > 0.05$) detected in either mean number of approaches to the doe, or in the time he associated with her (Table 69.3). However, in trials comparing estrous vaginal secretions and a control, mean number of approaches were 3.55 and 2.45 for the vaginal secretions and the control, respectively. Mean times associated with each doe were 1025.9 s and 774.1 s for the vaginal secretions

TABLE 69.1. Tracking pit visitation by white-tailed deer on area I (Bishop F. Grant Memorial Forest), showing number of observations (O) and number (V) and percentage (%) of visits

	Treatment																	
	Anestrous urine			Estrous urine			Midcycle urine			Buck urine			Deer lure			control		
Sampling date	O	V	%	O	V	%	O	V	%	O	V	%	O	V	%	O	V	%
16–18 October	15	6	40.0	15	6	40.0	15	4	26.7	15	6	40.0	15	6	40.0	9	5	55.6
29–31 October	15	2	13.3	15	5	33.3	15	2	13.3	15	6	40.0	15	3	20.0	15	2	13.3
21–23 November	15	3	20.0	15	2	13.3	15	3	20.0	12	1	8.3	15	3	20.0	15	4	26.7
9–11 December	15	1	6.7	15	0	0.0	15	1	6.7	15	0	0.0	15	3	20.0	15	1	6.7
21–23 December	15	3	20.0	15	2	13.3	15	0	0.0	12	3	25.0	12	4	33.3	15	1	6.7
12–14 January	15	6	40.0	15	4	26.7	12	3	25.0	15	4	26.7	15	3	20.0	15	2	13.3
Total	90	21	23.3	90	19	21.1	87	13	14.9	84	20	23.8	87	22	25.3	84	15	17.9

TABLE 69.2. Tracking pit visitation by white-tailed deer on area II (Tall Timbers Research Station), showing number of observations (O) and number (V) and percentage (%) of visits

	Treatment																	
	Anestrous urine			Estrous urine			Midcycle urine			Buck urine			Deer lure			control		
Sampling date	O	V	%	O	V	%	O	V	%	O	V	%	O	V	%	O	V	%
4–6 November	15	6	40.0	15	7	46.7	15	4	26.7	15	5	33.3	15	2	33.3	15	5	13.3
3–5 December	15	10	66.7	9	8	88.9	12	6	50.0	15	9	60.0	15	3	20.0	15	4	26.7
16–18 December	15	8	53.3	12	6	50.0	9	3	33.3	15	6	40.0	15	2	16.7	15	2	13.3
2–4 January	15	5	33.3	15	3	20.0	15	2	13.3	15	7	46.7	15	0	0.0	15	0	0.0
25–27 January	12	3	25.0	15	4	26.7	15	2	13.3	15	2	13.3	12	3	25.0	15	2	13.3
15–17 February	15	2	13.3	15	1	6.7	15	2	13.3	15	1	6.7	15	0	0.0	15	0	0.0
Total	87	34	39.1	81	29	35.8	81	19	23.5	90	30	33.3	87	10	11.5	90	13	14.4

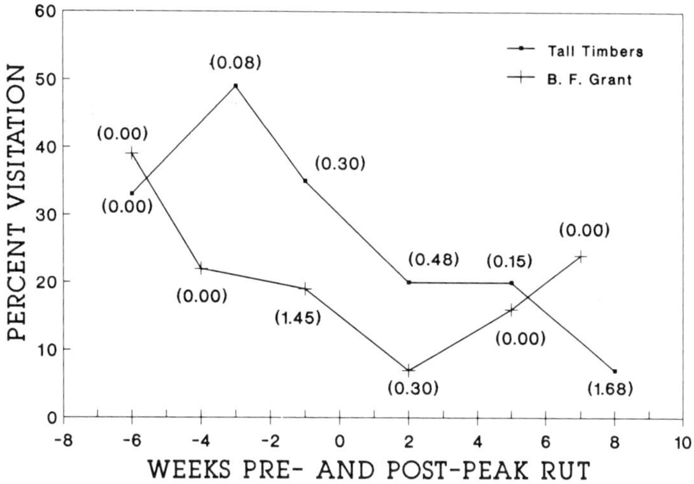

FIGURE 69.1. White-tailed deer visitations to tracking pits on area I (B.F. Grant) and area II (Tall Timbers). Sampling periods are presented by weeks before and after peak rutting activity on each area (i.e., B.F. Grant, 23–29 November, and Tall Timbers, 21–27 December). Mean daily precipitation (cm) is indicated for each sampling period.

and the control, respectively. Approach means were significantly different ($F = 7.25$, $P < 0.01$), as were mean times associated with each doe ($T = 2.30$, $P < 0.03$). For each response variable, the male indicated a preference for the estrous vaginal secretions (Table 69.3).

TABLE 69.3. Mean number of courtship approaches and mean time associated with treated female white-tailed deer versus controls

Treatment	Mean number of approaches	Mean time in seconds
Estrous urine	2.35	940.1
Control	1.95	859.9
Midcycle urine	1.95	933.1
Control	1.85	866.9
Estrous vaginal secretions	3.55	1,025.9
Control	2.45*	774.1**
Midcycle vaginal secretions	2.70	893.8
Control	1.53	906.2

$^*P \leq 0.01$; $^{**}P \leq 0.03$; otherwise, no significant difference

Discussion

The results of Experiment 1 indicate that deer were attracted to treatment sites as evidenced by what were judged to be high visitation rates. High visitation rates may have been due to factors other than the treatments presented (e.g., freshly exposed soil). Nevertheless, visitation rates among treatments and control were not statistically different. Mean number of tracks/visit or track sizes did not differ significantly among treatments, but on both areas highest values were associated with buck urine, suggesting more interest and response by adult bucks.

Results of Experiment 2 support those of Experiment 1 as the male showed no statistical preference for estrous or midcycle urine over the control. Likewise, no preference was displayed for midcycle vaginal secretions. However, a significant difference was evident in both response variables when estrous vaginal secretions were compared with the control. The combined results of these experiments indicate that the female reproductive tract is a source of a pheromone that elicits male courtship behavior and also could act as a sexual attractant.

Pheromones signaling estrus also have been demonstrated in the reproductive tract of other mammals including the rhesus monkey (*Macaca mulatta*) (Michael and Keverne 1970; Michael et al. 1971), the golden hamster (*Mesocricetus auratus*) (Johnston 1974, 1975; Singer et al. 1976), and the domestic cow (*Bos taurus*) (Klemm et al. 1987; Paleologou 1977). Possibly, urine could act as a medium of transfer that exposes these pheromones to the environment. All urine samples used in our experiments were taken directly from the bladder. Voided urine can be distinctly different from bladder urine (Albone 1984). During passage through the urogenital tract, the urine could acquire mucosal constituents that contain a pheromone indicating reproductive status. However, it is our contention that the urine itself is not the source of a sexually attractive pheromone.

Urine traditionally has been implicated to be the source of reproductive information in white-tails because of the buck's response known as flehmen. Flehmen introduces urine to the vomeronasal organ (Estes 1972). However, the vomeronasal organ accesses a part of the brain that regulates the reproductive physiology of an animal (Winans and Scalia 1970; Scalia and Winans 1976), and would not elicit the immediate response needed to ensure successful breeding. In contrast, the main olfactory system accesses parts of the brain associated with immediate behavioral responses (Scalia and Winans 1976). Therefore, we suggest that stimulation of the main olfactory epithelium, rather than the accessory olfactory epithelium of the vomeronasal organ, may be used in detecting a pheromone that elicits approach and copulation.

Apparently, urine does not have to be present for a buck to detect an estrous doe. Estrous odors, arising from the reproductive tract and detected through olfaction, are enough to elicit courtship behavior in the buck. However, additional research, including tests using voided urine, are necessary to determine if these odors are present in urine after passage through the urogenital tract.

Olfaction has been shown to be important in detection of estrus in other mammalian species. Male golden hamsters with their olfactory bulbs removed lost their ability to detect estrous females (Murphy and Schneider 1970). In domestic sheep (*Ovis aries*), olfactory ablations affected the detection of estrus in ewes, by rams (Lindsay 1965; Fletcher and Lindsay 1968), whereas blocking access to the vomeronasal organ does not impair estrus detection in cows (Izard 1983). Urinalysis through vomeronasal stimulation during flehmen likely serves to stimulate the reproductive physiology of the male to ensure that both sexes are in peak breeding condition at the appropriate time.

The role(s) of pheromones in the communication of reproductive information in white-tailed deer is not completely understood. Although our studies indicate that the constituents of bladder urine are not used to communicate sexual receptivity, behavioral patterns of white-tailed deer clearly indicate that voided urine is an important communicative mechanism in this species. Additional bioassays and chemical analyses are warranted to understand the types of information communicated.

Acknowledgements

We thank the numerous students of the School of Forest Resources, University of Georgia, who assisted in sample collection. We acknowledge support from Wellington Outdoors, Madison, Ga., McIntire-Stennis Project GEO-0030-MS-H, and Tall Timbers Research Station.

References

Albone ES (1984) Mammalian Semiochemistry. Wiley, New York

Brown BA, Hirth DH (1978) Breeding behavior in white-tailed deer. Proc Welder Wildl Found Symp 1:83–95

Estes RD (1972) The role of the vomeronasal organ in mammalian reproduction. Mammalia 36:314–341

Fletcher RD, Lindsay DR (1968) Sensory involvement in the mating behavior of domestic sheep. Anim Behav 16:410–414

Forster DL (1988) Attractiveness of conspecific bladder urine in white-tailed deer. MSc thesis, University of Georgia, Athens

Izard MK (1983) Pheromones and reproduction in domestic animals. In Vandenbergh JG (ed) Pheromones and Reproduction in Mammals. Academic Press, New York, pp 253–285

Johnston RE (1975) Sexual excitation function of hamster vaginal secretion. Anim Learning Behav 3:161–166
Johnston RE (1974) Sexual attraction function of golden hamster vaginal secretion. Behav Biol 12:111–117
Klemm WR, Hawkins GN, De Los Santos E (1987) Identification of compounds in bovine cervico-vaginal mucus extracts that evoke male sexual behavior. Chem Senses 12:77–87
Knox WM (1987) Recurrent estrous cycles and reproductive steroids in white-tailed deer. MSc thesis, University of Georgia, Athens
Lindsay DR (1965) The importance of olfactory stimuli in the mating behavior of the ram. Anim Behav 13:75–78
McDonald LE (1980) Veterinary Endocrinology and Reproduction. Lea and Febiger, Philadelphia
Michael RP, Keverne EB (1970) Primate sex pheromones of vaginal origin. Nature 225:84–85
Michael RP, Keverne EB, Bonsall RW (1971) Pheromones: isolation of male sex attractants from a female primate. Science 172:964–966
Moore WG, Marchinton RL (1974) Marking behavior and its social function in white-tailed deer. In Geist V, Walther F (eds) The Behaviour of Ungulates and Its Relation to Management. IUCN Publ. 24. Morges, Switzerland, pp 447–456
Murphy Mr, Schneider GE (1970) Olfactory bulb removal eliminates mating behavior in the male golden hamster. Science 167:302–304
Paleologou AM (1977) Detecting oestrus in cows by a method based on bovine sex pheromones. Vet Rec 100:319–320
SAS Institute (1985) SAS/STAT™ guide for personal computers, 6th ed. Cary, N.C.
Scalia F, Winans SS (1976) New perspectives on the morphology of the olfactory system: olfactory and vomeronasal pathways in mammals. In Doty RL (ed) Mammalian Olfaction, Reproductive Processes, and Behavior. Academic Press, New York, pp 7–28
Schilling E, Zust J (1968) Diagnosis of oestrus and ovulation in cows by pH- measurements intra vaginam and by apparent viscosity of vaginal mucus. J Reprod Fert 15:307–311
Singer AG, Agosta WC, O'Connell RJ, Pfaffman C, Bowen DV, Field FH (1976) Dimethyl disulfide: an attractant pheromone in hamster vaginal secretion. Science 191:948–950
Winans SS, Scalia F (1970) Amygdaloid nucleus: new afferent input from the vomeronasal organ. Science 170:330–332

70
Artificial Insemination of Farmed Chital Deer

Gillian E. Mylrea, Anthony W. English, Robert C. Mulley, and Gareth Evans

Abstract

Estrus and ovulation were synchronized in 20 parous chital hinds (*Axis axis*) by inserting an intravaginal silicone elastomer, progesterone-releasing device (CIDR G) for 12 days. All hinds received 300IU PMSG (Folligon) at the time of CIDR removal. Nineteen hinds retained their CIDRs for the 12-day period and were allocated to one of four treatment groups. Hinds were inseminated intracervically at either 48 or 58 h after CIDR removal with either fresh or frozen-thawed semen. Each inseminate contained 135×10^6 motile spermatozoa. The overall pregnancy rate to artificial insemination, assessed by ultrasonography at day 53, was 63%. Fresh semen inseminations resulted in more calvings than frozen-thawed semen (44.4% vs 10%; $P < 0.10$), with an overall calving rate of 26%.

Key words: Axis deer, chital deer, estrus synchronization, intracervical artificial insemination

Introduction

Chital deer are one of four deer species farmed in Australia for venison production. Chital deer are a tropical species originating from India and Sri Lanka. They do not have the well-defined breeding season exhibited by the temperate species, with females able to conceive at any time of the year.

Australia's chital deer population is derived from four deer introduced into Queensland in 1863 (Bentley 1978). Several thousand chital deer are currently estimated to exist in this wild population and are the only source of chital deer for farms in Australia. Hence, the introduction of new genetic material into Australia would increase the gene pool and reduce the possibility of inbreeding. Importation of semen for use in artificial insemination (AI) programs would enable the introduction of genetic material without the high costs and risks associated with live animal transport and reduce the risks of introducing exotic diseases or parasites. This chapter describes the first attempt at artificial insemination in chital deer.

Methods

Semen Collection and Processing

Semen was collected from stags by electroejaculation using the Colorado electroejaculator (Australian Livestock Genetics, Australia) with a ram probe. All stags were sedated with 3 mg/kg xylazine hydrochloride (Rompun, Bayer, Australia) and 3 mg/kg ketamine hydrochloride (Ketamine, Parnell Laboratories, Australia) injected IM. The pattern of stimulation was at intervals of 3 s "on" and 3 s "off" for up to 120 s with voltage gradually increased over time. Samples were usually obtained after approximately 60 to 90 s of stimulation.

The ejaculate processed for freezing was collected during February 1989. At the time of collection, the sample was evaluated for volume, motility, live-to-dead ratio, and concentration (Evans and Maxwell 1987). The semen sample was then diluted 1:2 (semen:diluent) in a 3% egg yolk-Tris-citrate diluent (Evans and Maxwell 1987) to a final concentration of 135×10^7 live spermatozoa/ml and then cooled to 5° C over 1.5 h. The cooled semen was then frozen in pellets (0.1 ml/pellet) in cavities in blocks of dry-ice. The frozen pellets were then transferred into liquid nitrogen for storage.

The fresh semen ejaculate was collected 30 min prior to artificial insemination. The semen was evaluated as for the frozen semen and diluted 1:2 in a 2.5% egg yolk-citrate diluent (Evans and Maxwell 1987) to a final concentration of 135×10^7 live spermatozoa/ml and maintained at 30° C in a water bath until required for insemination.

The pellets of frozen semen were thawed rapidly in a dry test tube held in a water bath at 37° C and immediately inseminated.

Estrus Synchronization

All hinds in the trial ($n = 20$) were parous (3 to 8 years old) and had been weaned 4 weeks prior to estrus synchronization. Estrus was synchronized in the hinds following the insertion of an intravaginal silicone elastomer, progesterone-releasing device (CIDR G, Riverina Artificial Breeders, Australia) for 12 days. At the time of CIDR withdrawal, all hinds received 300 IU PMSG (Folligon, Intervet Australia) as an IM injection. A vasectomized teaser stag was run with the hinds from the time of CIDR removal until the time of insemination.

Insemination

One of the hinds failed to retain her CIDR and was withdrawn from the trial. The remaining hinds were randomly allocated to one of four treatment groups and were inseminated intracervically at either 48 or 58 h after CIDR removal with either fresh or frozen-thawed semen. Hinds were sedated with 3 mg/kg xylazine hydrochloride and 3 mg/kg of ketamine hydrochloride IM and placed in a cradle in sternal recumbency so that the hind limbs were elevated.

A duck-billed speculum was inserted into the vagina and a light source directed through the open speculum into the anterior vagina to visualize the cervix. Mucus present in the vagina was aspirated so that the cervix was clearly visible. A plastic inseminating pipette with a tapered tip, bent at an angle of approximately 30° and connected to a 1.0 ml syringe, was used for the inseminations. The pipette delivered only a single dose and was reloaded for each insemination. An attempt was made to introduce the pipette as deeply as possible into the cervix using gentle pressure and manipulation. The depth of penetration was measured, and then the semen was deposited into the cervix by depressing the plunger on the syringe. Once inseminated, the hind was placed in sternal recumbency in a small, dark pen and subcutaneously given 2 ml yohimbine hydrochloride (Reverzine, Parnell Laboratories, Australia).

Each inseminate, whether fresh or frozen contained 135×10^6 motile spermatozoa. The post-thaw motility of the frozen semen was 50%. Hence, to ensure that the same number of spermatozoa were inseminated in all females, the volume of frozen semen inseminated (0.2 ml) was twice the volume of fresh semen inseminated (0.1 ml).

A crayon-harnessed stag was run with the hinds from 18 to 25 days after AI. Observations were made twice daily to detect crayon mating marks (i.e., return to service). On day 53 after AI all hinds were pregnancy tested by the use of ultrasonography. Hinds were restrained in a drop-floor crush, and the ultrasound probe was placed on the ventral abdomen just anterior to the mammary gland.

Results

Five of the 19 hinds (26%) artificially inseminated were recorded as having crayon-mating marks between 19 and 24 days after insemination (mean 22 ± 2.1 days; Table 70.1). Ultrasound pregnancy diagnosis on day 53 indicated that 12 hinds were pregnant with the remaining 7 not detectably pregnant. Only 5 hinds calved to the artificial insemination (26%) with a mean gestation length of 240 ± 1.7 days.

Overall, fresh semen inseminations resulted in more calvings than frozen-thawed semen insemi-

TABLE 70.1. Pregnancy diagnosis and calving results of cervical artificial insemination in chital hinds

Time of insemination post-CIDR removal (h)	48	48	58	58
Semen	Fresh	Frozen	Fresh	Frozen
Number of hinds	5	5	4	5
Nonreturn to service	5	3	4	2
Positive ultrasound, day 53	4	1	4	3
Calving to AI	3	0	1	1

nations (44.4% vs 10%; $P < 0.10$), but there were no differences between the time of insemination nor was there an interaction between the type of semen or the time of insemination (Table 70.2).

Discussion

This study has demonstrated that cervical artificial insemination of farmed chital deer using either fresh or frozen-thawed semen can be successful, although there is considerable room for improvement. AI has been studied in red deer (*Cervus elaphus*) (Krzywinski and Jaczewiski 1978), wapiti (*C. e. canadensis*) (Haigh et al. 1984), white-tailed deer (*Odocoileus virginianus*) (Jacobson et al. 1989), reindeer (*Rangifer tarandus*) (Dott and Utsi 1973) and fallow (*Dama dama*) (Asher et al. 1988) with highly variable results. The methods of estrus synchronization/detection and insemination in these studies varied widely.

The discrepancy between the number diagnosed to be pregnant on ultrasonography and the number actually calving may have resulted from embryonic mortality after day 53 of gestation. Chital deer normally give birth to a single calf. Ovulation rates in naturally cycling and estrus-synchronized (using CIDR Gs only) hinds are rarely multiple (G.E. Mylrea, unpublished data). The use of PMSG in this study may have increased the ovulation rate, resulting in multiple conceptions. Chital hinds may be unable to carry multiple fetuses to term with fetal losses resulting. Asher (1987) reported a fallow doe (a monovular species) which had been treated with PMSG aborting twin fetuses during the midtrimester of pregnancy. A study in fallow deer reported the possibility of 15% embryonic mortality 35 days after cervical AI in fallow deer (Asher et al. 1988). However, the accuracy of the methods used to diagnose pregnancy in that study were questioned, raising the possibility of false positive diagnoses.

The ultrasound technique used in this study to diagnose early pregnancy was based on the technique used successfully in fallow deer (Mulley et al. 1987). This technique has not been validated in chital deer. Hence, the difference in conception rates based on ultrasound versus calving rates may have resulted from false positives on ultrasound. Therefore, ultrasonography as a method of early pregnancy diagnosis in chital deer needs to be validated in chital deer to ensure that it is an accurate method of pregnancy diagnosis.

It was only possible to pass the pipette into the uterus in 26% (5/19) of hinds in this study. Only in one of the 10 hinds (10%) inseminated with

TABLE 70.2. Effect of depth of insemination on pregnancy and calving rates in chital hinds

Depth of insemination (cm)	Fresh diluted semen			Frozen-thawed semen		
	Hinds inseminated	Hinds positive ultrasound, day 53	Hinds calving	Hinds inseminated	Hinds positive ultrasound, day 53	Hinds calving
Up to 1.5 into cervix	1	1	1	1	0	0
1.5–3.0 into cervix	3	3	1	7	2	0
3.0–4.0 into cervix	1	0	0	1	1	0
Into uterus	4	4	2	1	1	1

frozen-thawed semen could the pipette be passed directly into the uterus, and this hind was the only one in the group to calve. In the group inseminated with fresh semen it was possible to deposit the semen directly into the uterus in 4 of the 9 hinds (21%). All 4 hinds were detected pregnant on ultrasound but only 2 calved. Two other hinds in this group calved to inseminations that were deposited only 1 and 3 cm respectively into the cervix. Although numbers were too small to enable statistical analysis of these data, it would appear that the depth of insemination of fresh semen did not have a major influence on pregnancy rate or calving rate, as calves resulted from both shallow and deep cervical inseminations. However, the depth of deposition of frozen-thawed semen into the genital tract did appear to be an important factor influencing fertility. Studies in sheep and goats have indicated that fresh semen was of higher fertility than frozen-thawed semen and that fertility of frozen-thawed semen increases with increasing depth of deposition, with deposition directly into the uterus giving the best results (Lightfoot and Salamon 1970; Ritar and Salamon 1983). Thus, it would seem that to achieve acceptable conception rates in chital deer, particularly with frozen-thawed semen, one might require intrauterine insemination via laparoscopy.

There are many factors that may affect the success of an AI program. These include the method of estrous synchronization, semen quality, timing of insemination, method of insemination, and animal handling management. Information on the role of each of these factors in chital deer is largely unknown because of the small amount of research done on this species. Clearly, further studies are required to identify which components of the AI program are affecting its success and to investigate ways of improving such areas.

References

Asher GW (1987) Induction of oestrus and ovulation in farmed fallow deer (*Dama dama*) by using progesterone and PMSG treatment. J Reprod Fert 81:113–118

Asher GW, Adam JL, James RW, Barnes D (1988) Artificial insemination of farmed fallow deer (*Dama dama*): fixed time insemination at a synchronized oestrous. Anim Prod 47:487–492

Bentley A (1978) An Introduction to the Deer of Australia. The Roetong Trust Service Fund, Forests Commission, Victoria

Dott HM, Utsi MNP (1973) Artificial insemination of reindeer (*Rangifer tarandus*). J Zool London 170: 505–508

Evans G, Maxwell WMC (1987) Salamon's Artificial Insemination of Sheep and Goats. Butterworths, Sydney

Haigh JC (1984) Artificial insemination of two white-tailed deer. JAVMA 185:1446–1447

Haigh JC, Shadbolt MP, Glover GJ (1984) Artificial insemination of wapiti (*Cervus elaphus*). Proc Am Assoc Zoo Vets, Louisville, Kentucky, p 173

Jacobson HA, Bearden HJ, Whitehouse DB (1989) Artificial insemination trials with white-tailed deer. J Wildlife Management 53:224–227

Krzywinski A, Jaczewski Z (1978) Observations on the artificial breeding of red deer. Symp Zool Soc London 43:271–287

Lightfoot RJ, Salamon S (1970) Fertility of ram spermatozoa frozen by the pellet method. J Reprod Fert 22:385 398

Mulley RC, English AW, Rawlinson RJ, Chapple RS (1987) Pregnancy diagnosis of fallow deer by ultrasonography. Aust Vet J 64:8 257–258

Ritar AJ, Salamon S (1983) Fertility of fresh and frozen-thawed semen of the angora goat. Aust J Biol Sci 36:49–59

71
Changes in Live Weight and the Reproductive Tract of Farmed Red Deer Stags from 6 to 27 Months of Age

James R. Webster, James M. Suttie, Barry A. Veenvliet, Ian D. Corson, and Robert E. Labes

Abstract

Groups of farmed red deer stags ($n = 49$) were weighed and slaughtered, and the reproductive tracts dissected, at 2- or 3-month intervals from age 6 to 27 months. The penis, testes, epididymides, seminal vesicles, ampullae, and prostate gland were weighed separately, and testes sections were examined using light microscopy. At 6 months of age reproductive tracts were immature. The seminiferous tubules were mostly solid and contained spermatogonia only. At 9 months the testes had doubled in size, and spermatocytes were seen indicating that spermatogenesis had begun; at 12 months, elongated spermatids were present. A dramatic increase in the weight of the reproductive tract occurred between 12 and 15 months of age at which time the seminiferous tubules were mature and fully active. This was coincident with the timing of the seasonal rut in older stags. Over winter, from 15 to 21 months of age, weight of the reproductive tracts decreased, and live weights fell slightly. Live weights began to increase again at 24 months of age, and this was associated with the resumption of reproductive tract development. At the time of the second rut in March (27 months of age), the testes and accessory glands attained their greatest weights recorded. Live weights at slaughter increased during spring and summer and fell in autumn. We conclude that reproductive development of farmed red deer stags from 6 to 27 months of age consists of two phases of growth which occur during spring and early summer each year. The first phase includes the onset of spermatogenesis indicating that puberty occurs between 9 and 15 months of age. After puberty is attained, the stags exhibit a decline in reproductive state over winter followed by recrudescence in spring, similar to the reproductive cycle of adult stags.

Key words: Growth, male, puberty, red deer, reproduction, season

Introduction

The physical development of the male reproductive tract during puberty has been studied in wild Scottish red deer (*Cervus elaphus*). Lincoln (1971a) found that spermatogenesis began at about 10 and was complete at 15 months of age. Testicular and accessory gland (seminal vesicles, ampullary gland, and prostate) weight increased slowly from when the testes descended into the scrotum at around 3 months until about 14 months of age, reaching a peak during the breeding season (rut) at 15 months. After the rut the testes and accessory glands decreased in weight, and thereafter an annual rhythm of seasonal development and regression ensued, such as was apparent in adult stags (Lincoln 1971b).

The stags studied by Lincoln (1971a, b) were very small and at 15 months were 55 to 60 kg or only about half the weight of typical well-grown farmed red deer stags of similar age and genotype in New Zealand (Suttie 1981). As nutrition and growth are known to play a role in pubertal development (Foster et al. 1988) we have studied the temporal reproductive development in well-grown New Zealand red deer stags from 6 to 27 months of age.

Materials and Methods

Farmed red deer stag calves ($n = 49$) born in December 1984 (southern hemisphere) were randomly allocated to one of nine groups ($n = 5$ or 6 per group) which were slaughtered at 2- to 3-month intervals from 6 to 27 months of age. The animals were held indoors under natural day length and fed a pelleted barley/soya meal based concentrate ration ad libitum throughout the study.

After a 24-h fast the animals were weighed and slaughtered in a research abattoir. The reproductive tracts were recovered then they were dissected, and the testes, epididymides, seminal vesicles, ampullary glands, prostate body, and the bulbourethral glands were weighed. Dissection followed Lincoln (1971b) and Wallace and Birtles (1985). Sections of testis from each animal were fixed in Bouin's solution and stained with hematoxylin and eosin. They were examined at a magnification of X450. Antler status was recorded for each animal.

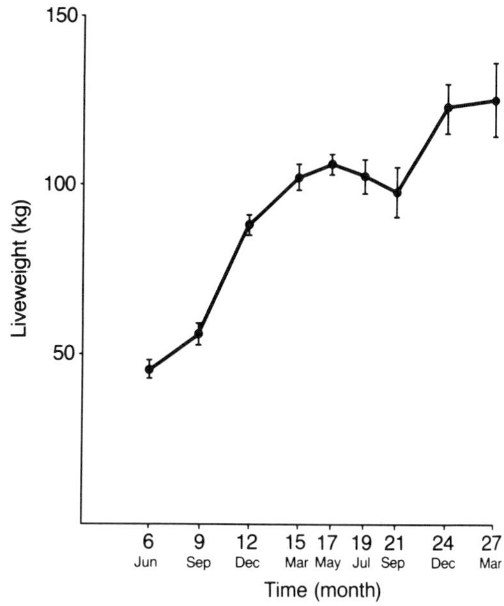

FIGURE 71.1. Live weight of slaughtered red deer stags ($\bar{x} \pm$ SEM).

Results

The mean weight of each group increased to 17 months of age then fell. The last group to be slaughtered (at 27 months of age) was the heaviest (Fig. 71.1). Testis weight increased from June to March with a marked increase from December to March (Fig. 71.2). The testes weight of stags from groups slaughtered from March to September then progressively decreased in size. The epididymides and seminal vesicle weights followed a similar pattern to that of the testes (Fig. 71.3). Peak weights of both were found from the stags killed at 15 and 27 months.

In contrast, the weight pattern of the ampullae and prostate differed (Fig. 71.4) both from each other and from the pattern of the testes, epididymides, and seminal vesicles. The ampullae increased in weight rapidly in samples from June to March (6 to 15 months) and again from September to peak in March (21 to 27 months). The weight of the prostate also increased until March (15 months) and then showed a more

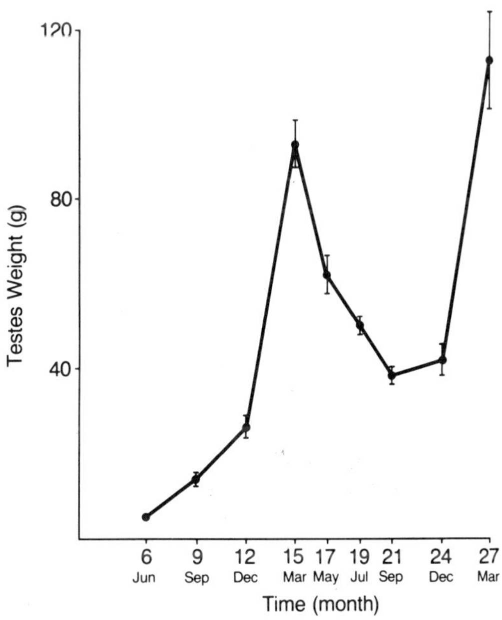

FIGURE 71.2. Mean ($\pm$ SEM) testes weight of slaughtered stags.

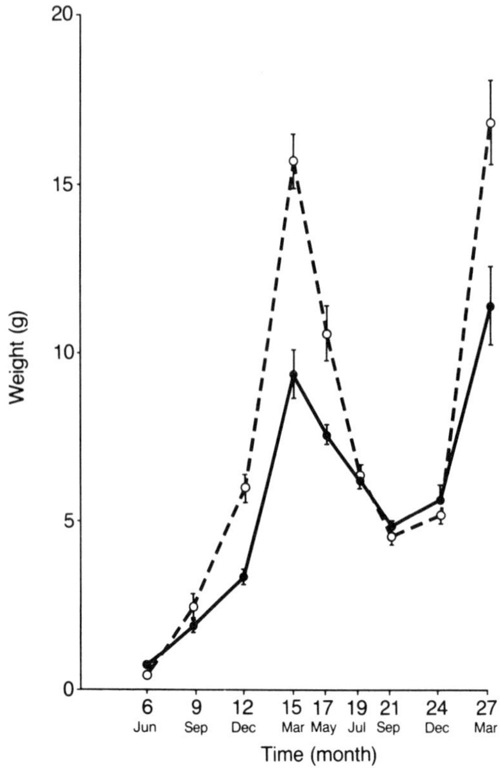

FIGURE 71.3. Mean (±SEM) epididymides (•) and seminal vesicle (○) weight of stags.

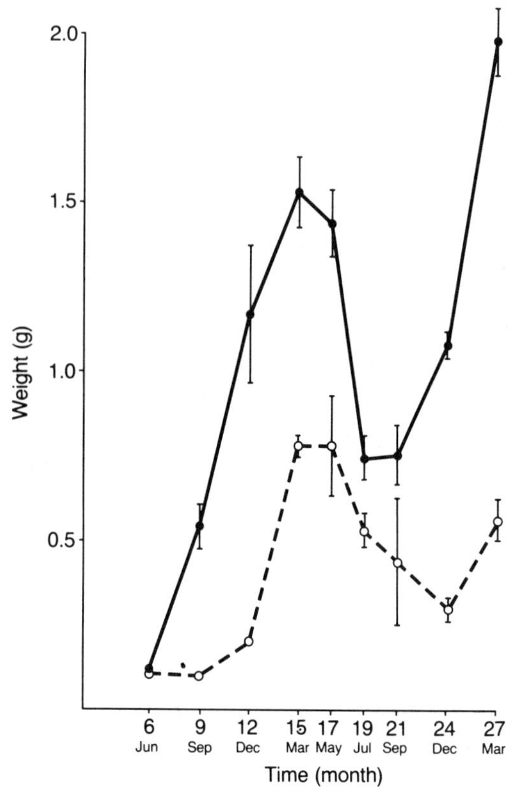

FIGURE 71.4. Mean (±SEM) ampullary gland (•) and prostate weight (○) of stags.

prolonged decrease until December. Mean weight at 27 months was less than at 15 to 17 months. Bulbourethral glands weighed between 0.15 and 0.40 g and did not vary seasonally or during development.

The testes of 6-month-old stags were characterized by small closed seminiferous tubules with few spermatogonia (Fig. 71.5). In the 9-month-old stags the number of spermatogonia had increased, although the tubules were still closed. There were maturing spermatids in the testes of the 12-month-old stags. The testes of 15-month-old stags showed all stages of spermatogenesis with maturing spermatids in the tubules which now had a well-pronounced lumen.

The stags had pedicles only in September at 9 months of age and were in velvet antler in December at 12 months of age. They were in hard antler from March to September (15 to 21 months), in velvet in December (24 months), and in hard antler in March (27 months).

Discussion

The weight gain pattern of the stags was normal for red deer farmed in New Zealand (Fennessy et al. 1981). The group slaughtered during March (15 months of age) achieved a maximum live weight of 107 ± 2.0 kg which is the normal live weight for yearling stags (Fennessy et al. 1981). The stags slaughtered at 27 months weighed 124 ± 11 kg which is a little smaller than the expected live weight of 130 kg at that age. The stags began pedicle development within the weight range expected for farmed New Zealand red deer of 50 to 60 kg (Suttie and Kay 1983).

In general, the pattern of development and regression of the reproductive organs followed closely that published by Lincoln (1971a). In the present study, the testes were heavier at all stages compared with those of wild Scottish red deer (Lincoln 1971a), however, the seasonal decrease in weight from 15 months to 24 months of age

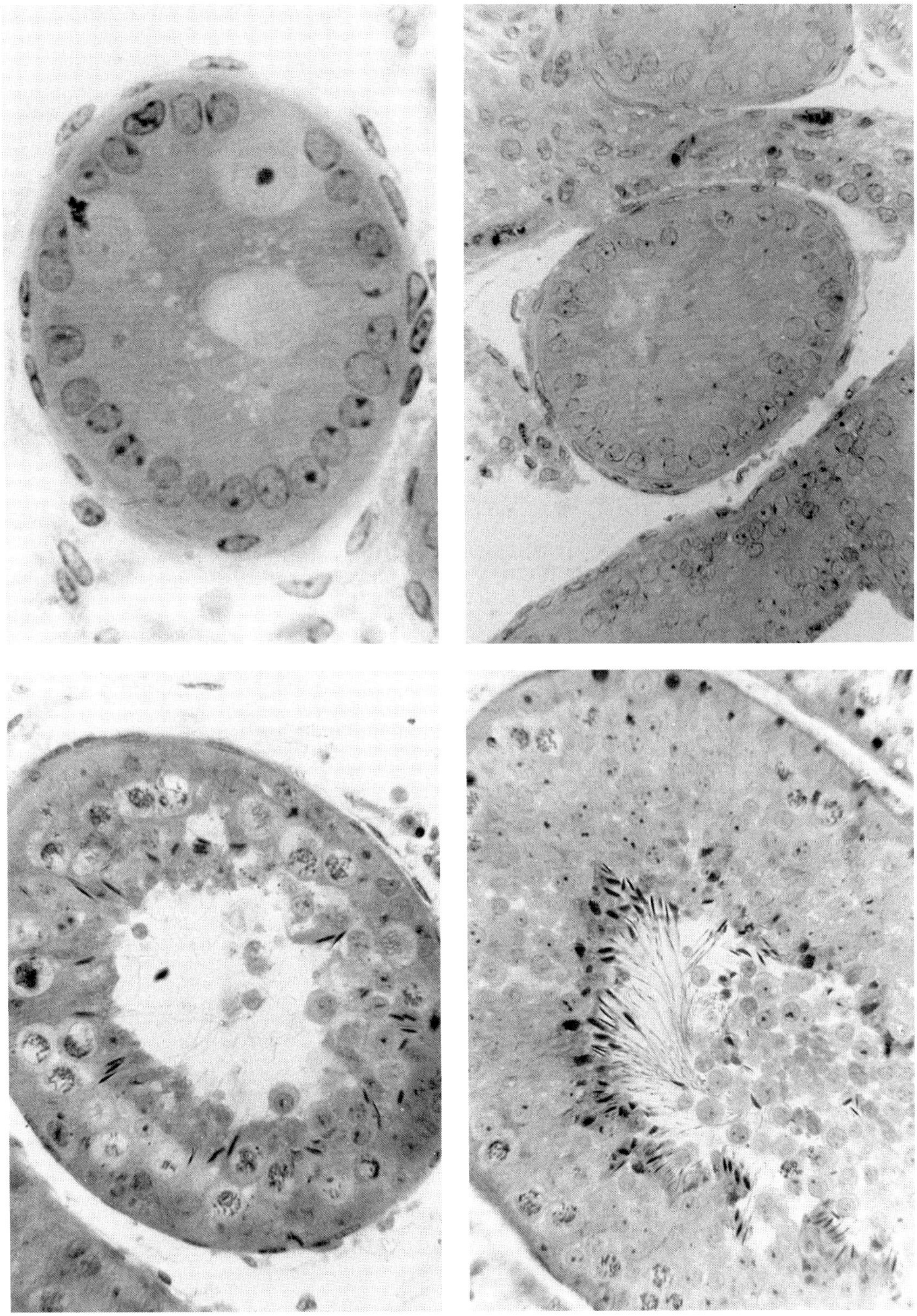

FIGURE 71.5. Histological sections of testes from typical stags at 6 (top left), 9 (top right), 12 (lower left), and 15 (lower left) months of age.

was about 50% in both studies. This indicates that the New Zealand deer proportionately were following the same seasonal pattern. The presence of sperm in the testes of the 12-month old stags in the present study was earlier than reported by Lincoln (1971a). This probably indicates that the larger New Zealand deer were advanced in reproductive development compared with their wild Scottish counterparts.

The epididymides, prostate, and ampullary glands were of similar size and had a similar pattern of development and regression compared with Lincoln (1971a). In contrast the seminal vesicles of the N Z stags were larger and showed a greater regression in size after the 15-month rut than reported by Lincoln (1971a). Lincoln (1971a) did not report bulbourethral glands in the stags from his study. The present study confirms the finding of Wallace and Birtles (1985) that they exist in red deer, but indicates that they are small and do not show a clear seasonal or developmental rhythm.

Antler cycles were in synchrony with the rhythm of testicular development and regression such that velvet antler grew when the testes were small or developing. Antlers were hard when testes were large or regressing, and antler casting occurred when the testes were completely regressed (Lincoln 1971b).

The present study confirms that farmed red deer in New Zealand have a well-defined developmental pattern of reproductive organ growth. Full physical reproductive maturity is reached at 15 months of age, although some sperm may be present as early as 12 months of age. It is unlikely that the stags are fertile at 12 months due to the lack of development of the epididymides and accessory glands indicating successful ejaculation would be unlikely.

References

Fennessy PF, Moore GH, Corson ID (1981) Energy requirements of red deer. Proc N Z Soc Anim Prod 41:167–143

Foster DL, Ebling FJP, Vannerson LA, Bucholtz DC, Wood RI, Micka AF, Suttie JM, Veenvliet BA (1988) Modulation of gonadotropin secretion during development by nutrition and growth. 11th Int Congress on Anim Reprod and Artificial Insemination, Dublin, Ireland, p 101–108

Lincoln GA (1971a) Puberty in a seasonally breeding mammal, the red deer stag (*Cervus elaphus* L). J Reprod Fert 25:41–54

Lincoln GA (1971b) The seasonal reproductive changes in the adult red deer stag (*Cervus elaphus* L). J Zool London 163:105–123

Suttie JM (1981) The influence of nutrition and photoperiod on the growth and development and endocrine status of captive red deer and Soay rams. PhD thesis, University of Aberdeen, Scotland

Suttie JM, Kay RNB (1983) The influence of nutrition and photoperiod on the growth of antlers of young red deer. In Brown RD (ed) Antler Development in Cervidae. Caeser Kleberg Wildl Res Inst, Kingsville, Texas, pp 61–71

Wallace V, Birtles MJ (1985) The accessory sex glands of the red deer stag. In Fennessy PF, Drew KR (eds) The Biology of Deer Production. R Soc N Z Wellington Bull 22:193–196

72
Influence of Sex, Habitat, and Genotype on the Growth Patterns of White-Tailed Deer

Paul Leberg, Michael H. Smith, and I. Lehr Brisbin, Jr.

Abstract

Growth patterns were examined using age-weight data from 21 264 white-tailed deer (*Odocoileus virginianus*) collected over a 22-year period on the Savannah River Site, located near Aiken, South Carolina, USA. The Richards growth equation was used to evaluate the effects of sex, habitat, and genetic variation on asymptotic weight and the length of the growth period. Male deer grew for a longer period of time and obtained greater adult weights than did females. This variation in growth patterns is probably due to differences in life history strategies of the two sexes. Contrary to previous investigations, no associations were detected between genotypic variation at nine electrophoretic loci, or overall heterozygosity, and growth patterns of deer. Growth patterns of deer from upland habitat were different than those of deer from an adjacent swamp. Both sexes tended to grow faster and attain greater body weights in the uplands than in the swamp; however, asymptotic weight of males was the only growth parameter that differed significantly between the habitat types. Differences in growth functions among habitats may be related to differences in density, habitat quality, or the genetic composition of the populations. These results suggest that patterns of growth, which can affect levels of productivity of cervid populations, can differ significantly over relatively small geographic areas.

Key words: Body size, growth, heterozygosity, Richards growth curve, sexual dimorphism, white-tailed deer

Introduction

The growth pattern of an organism can affect many aspects of its ecology. Growth rate and adult size are often highly correlated with other life history traits such as fecundity and mortality rates (Clutton-Brock et al. 1982; Stearns and Koella 1986). Furthermore, growth patterns can affect the outcome of management practices designed to optimize the biomass or trophy quality of individuals (Leberg et al. 1991). Because variation in growth patterns can affect the ecology of organisms, it is important to determine what factors affect individual growth in natural populations.

A wide variety of factors are thought to affect the growth and size of cervids, including habitat and individual levels of genetic variation (Brisbin and Lenarz 1984; Chesser and Smith 1987); however, a comprehensive study of the effects of these factors on cervid growth in a natural population is lacking. Our objective is to evaluate the effects of sex, habitat, and genetic variation on the growth of white-tailed deer.

Study Area and Methods

The Savannah River Site (SRS) is an 80 972 ha government reserve located near Aiken, South Carolina, USA. The site can be divided into two major habitats, swamp and upland (Urbston 1967). The swamp is characterized by bottomland hardwoods and islands of pine, while the upland is composed of pine plantations with bottomland hardwoods in stream floodplains.

The SRS deer herd originated from remnant populations located in the swamp. With the protection afforded by the creation of the SRS in the early 1950s, the herd expanded rapidly throughout the site. An annual controlled harvest, usually held during October to December, was initiated in 1965. Starting in 1965, age and weight data have been collected from all harvested deer. Special collections and accidental deaths of deer have been used to obtain data at other times of the year. All animals were weighed to the nearest 0.45 kg and aged to year class based on tooth eruption and wear (Severinghaus 1949). Age in months was determined using the mean parturation date for the SRS deer herd of 1 June (Rhodes et al. 1986).

Weight and age data were fit using the generalized Richards growth equation (Richards 1959). The Richards equation, as reparameterized by Brisbin et al. (1986), is

$$W_t = \{W_\infty^{1-m} - (W_\infty^{1-m} - W_0^{1-m})* \exp^{-2t/T*(1+m)}\}^{1/(1-m)}$$

where W_t is body mass (kg) at age t (mo), W_∞ is asymptotic body weight, W_0 is the mass at birth (kg), T is the growth period (mo), and m is a shape parameter (Richards 1959; Brisbin et al. 1986). Changes in the numerical value of the shape parameter, m, change the position of the inflection point of the curve. Higher values of m indicate that the point of inflection of the growth curve occurs later in the growth period (Richards 1959). The growth period (T) represents the period of time required to complete the "majority" of individual growth (Brisbin et al. 1986). The proportion of asymptotic mass attained by age T is partially dependent on m. Based on the estimates of m from this study, deer attain approximately 90% to 97% of asymptotic mass by age T.

The flexibly shaped Richards function accounts for more variation in age-mass relationships and produces less biased estimates of growth parameters than sigmoid models with fixed inflection points such as the von Bertalanffy, Gompertz, and logistic curves (Leberg et al. 1989). The Weibull function, another flexible curve that has been used to model deer growth (Fuller et al. 1989), fits our data as well as did the Richards equation. However, we use the Richards function in this analysis of growth because a past simulation study demonstrated the robustness of the various parameters of the Richards curve to the vagaries of check-station data (Leberg et al. 1989). Such information is not available for the Weibull function. Furthermore, the Richards curve has been reparameterized so that all of its parameters have meaningful biological interpretations (Brisbin et al. 1986); this is not the case with the parameters of the Weibull function.

The Richards equation was fit to age-weight data using the NLIN subroutine of SAS (1985). The secant method (DUD) of iteration was used to generate estimates for all four of the parameters of the Richards equation. For deer collected after 1974, the effect of multiple locus heterozygosity on growth was determined by comparing the growth curves of individuals that were heterozygous for 0, 1, 2, or 3 or more electrophoretic loci. These estimates of heterozygosity were based on genotypes of the same nine polymorphic loci examined by Scribner et al. (1989). These loci were resolved using standard electrophoretic techniques (Manlove et al. 1975). The individual effects of each of these loci on growth were also assessed by comparing growth curves of individuals that were either homozygous or heterozygous at each locus. Differences in the effects of habitat on growth were assessed by comparing growth parameters estimated for deer collected from the swamp and upland habitat.

Because there is strong sexual dimorphism in the growth of deer (Chesser and Smith 1987; Fuller et al. 1989) separate curves were fit to data from males and females to evaluate the effects of genotype and habitat. Due to intensive harvest efforts, few SRS deer reached the older age classes. Less than 10% of the sample was over 2.5 years of age (Leberg et al. 1989). To increase sample sizes of these older age classes, samples were combined across years. There was some annual variation in growth patterns (P.L. Leberg, unpublished data), but because the size of these effects was small and did not affect the outcome of the analyses reported here, samples have been combined across years in all analyses.

An F-test (White and Brisbin 1980) was used to determine if growth curves differed between sexes, habitats, multiple-locus heterozygosity classes, and single-locus heterozygosity. If differences were detected between curves, a second F-test was used to evaluate the significance of differences between individual parameter estimates (White and Brisbin 1980). Because the amount of growth completed by age T is partially dependent on the value of m, the significance of estimates of T was evaluated in two ways. In the first case, estimates of T were compared between curves for which m was estimated separately. In the second case, estimates of T were compared between curves in which m was fixed at the value estimated from a Richards equation fit to the pooled data. Both comparisons yielded similar results; therefore, only results of the former comparison will be discussed.

Simulations suggest that estimates of m and W_0 from data collected at hunter check-stations may be imprecise and biased (Leberg et al. 1989; P.L. Leberg, unpublished data). Generally, with this type of data, W_0 and m are usually overestimated; however, by allowing these parameters to vary, the estimates of W_∞ and T are improved (Leberg et al. 1989; P.L. Leberg unpublished data). Although all four parameters were estimated from the data, only estimates of W_∞ and T will be discussed. These two parameters are generally insensitive to the problems of nonuniform sampling and aging errors associated with check-station data collection (Leberg et al. 1989).

Results

Analyses were based on weights of 10 143 male and 11 121 female deer collected from 1965 to 1987. Mean fall weights of white-tailed deer increase rapidly over the first several years of life and then approach an asymptote (Table 72.1). Some differences were observed in the growth of the two sexes. Within each age class, males were larger than females. Males also continued to increase in weight after females had ceased

TABLE 72.1. Sample sizes (n), mean body weights ($\bar{x}$), and their standard errors (SE) for male and female deer from upland and swamp habitats harvested between October and December on the Savannah River Site between 1965 and 1987 (approximately 92% of the total sample). Only weights of individuals collected during the fall are presented to make comparisons with other studies straightforward

Age (years)		Males Upland	Males Swamp	Males Total	Females Upland	Females Swamp	Females Total
0.5[a,b]	n	3130	517	3647	2868	510	3378
	$\bar{x}$	29.2	28.3	29.1	25.7	25.2	25.7
	SE	0.1	0.2	0.1	0.1	0.2	0.1
1.5[a]	n	2295	432	2727	2271	407	2678
	$\bar{x}$	51.9	49.7	51.6	41.6	41.2	41.6
	SE	0.1	0.3	0.1	0.1	0.2	0.1
2.5[a,b]	n	1833	320	2153	2473	478	2951
	$\bar{x}$	66.8	63.9	66.4	46.5	45.5	46.3
	SE	0.2	0.7	0.2	0.1	0.2	0.1
3.5[a,b]	n	812	137	949	922	212	1134
	$\bar{x}$	77.1	73.9	76.6	48.9	47.3	48.6
	SE	0.4	0.7	0.3	0.2	0.4	0.2
4.5[a]	n	200	59	259	265	97	362
	$\bar{x}$	80.9	75.2	79.6	50.1	49.5	49.9
	SE	0.7	1.3	0.6	0.4	0.5	0.3
≥ 5.5	n	63	27	90	75	54	129
	$\bar{x}$	80.2	78.1	79.6	49.6	49.8	49.7
	SE	1.5	2.2	1.3	0.7	0.8	0.5

[a]Weights of males were significantly different between habitats based on a t-test ($P < 0.05$)
[b]Weights of females were significantly different between habitats based on a t-test ($P < 0.05$)

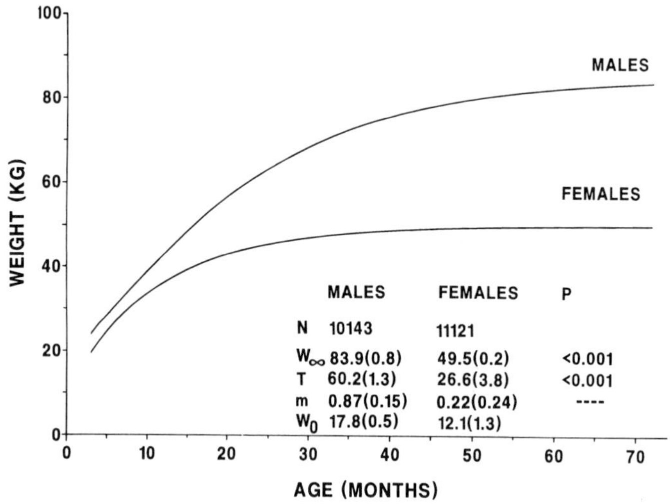

FIGURE 72.1. Growth functions, sample sizes (N) and estimates of asymptotic weight (W_∞), growth period (T), growth curve shape (m), and birth weight (W_0) for male and female white-tailed deer from the Savannah River Site, located near Aiken, South Carolina. Estimates of asymptotic standard errors are in parentheses. Probability (P) that W_∞ and T are the same for both curves was determined using an F-test. Differences between m and W_0 were not evaluated because estimates of these parameters were known to be biased and imprecise.

growing. Male and female regression equations were different ($f_{4,21256} = 7222.6$, $P < 0.001$). Estimates of both asymptotic weight and growth period were significantly higher for male than female deer (Fig. 72.1).

There were no differences in growth curves among deer of either sex with different numbers of heterozygous loci ($P > 0.10$). Furthermore, there was no evidence that being heterozygous or homozygous at any of the nine loci examined had any effect on growth patterns of either sex ($P > 0.10$). These relationships were not affected by the habitat from which the deer were collected.

Deer from the two habitats had different growth equations ($f_{4,10135} = 40.1$, $P < 0.001$ for males; $f_{4,11113} = 12.3$, $P < 0.001$ for females). Males and females from the swamp tended to have lower asymptotic weights and longer growth periods than those in the uplands (Fig. 72.2). However, asymptotic weight of males was the only growth parameter that differed between habitats (Fig. 72.2). These same trends are seen in mean weights of deer harvested in fall hunts (Table 72.1). In each age class, but 5.5 years, males in the upland were significantly larger than those in the swamp. While females also tended to weigh more in the upland than in the swamp, these differences in weight were smaller than those in males and not as consistent across age classes (Table 72.1).

Discussion

Weights of deer on the SRS fall well within the ranges previously reported (Sauer 1984). Although much of this data was based on field-dressed weights, a conversion factor (Sauer 1984) allowed comparison of weights of SRS deer to those from other studies. Deer in this study were considerably smaller than deer from Minnesota and from a high quality range in New York. South Carolina deer were generally larger than those from Texas and Florida. Although this general north-south gradient in deer weights has been previously noted (Sauer 1984; Fuller et al. 1989), the mechanism responsible for this cline is still unresolved (Fuller et al. 1989).

Male and female deer on the SRS exhibited marked sexual dimorphism in growth patterns. Males obtained twice the adult mass of females

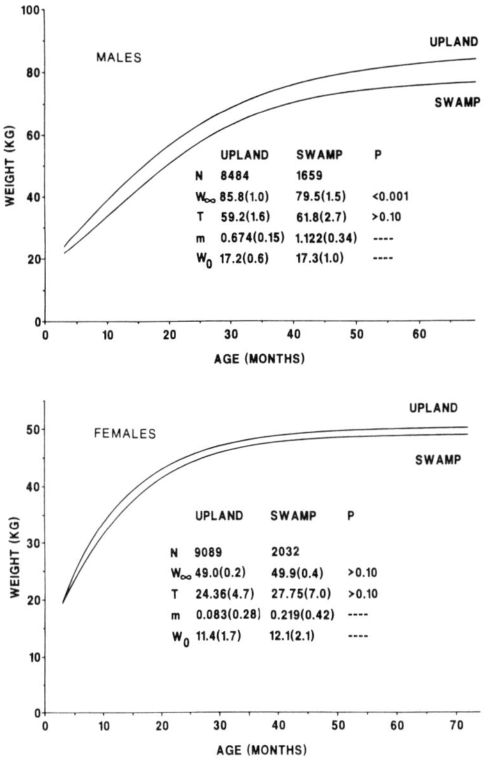

FIGURE 72.2. Growth functions, sample sizes (N) and estimates of asymptotic weight (W_∞), growth period (T), growth curve shape (m) and birth weight (W_0) for male and female white-tailed deer from upland and swamp habitats on the Savannah River Site, located near Aiken, South Carolina. Estimates of asymptotic standard errors are in parentheses. Probability (P) that W_∞ and T are the same for both curves was determined using an F-test. Differences between m and W_0 were not evaluated because estimates of these parameters were known to be biased and imprecise.

and grew for a considerably longer period of time. Sexual dimorphism in adult size and length of growth period have been noted previously in this species (Roseberry and Klimstra 1975; Chesser and Smith 1987; Fuller et al. 1989) and in many other cervids (Clutton-Brock et al. 1982). Although the causes of sexual dimorphism in mammals are incompletely understood, they are probably the result of different selection pressures on the sexes (Ralls 1977; Clutton-Brock et al. 1982). Growth patterns exhibited by white-tailed deer fit the predictions of a proposed model of cervid growth (Clutton-Brock et al. 1982). Because males contribute little to offspring care or production, they can devote more energy to biomass accumulation. Larger body size is further favored by sexual selection because greater size may influence the ability of males to compete for mates (Hirth 1977; Ozoga 1972). Females reach adult weight early in life, allowing the allocation of more energy to reproduction and parental care later in life.

There were no detectable differences in the growth patterns of deer with different numbers of heterozygous loci or with different genotypes at individual loci. Positive associations between heterozygosity and growth have been reported in a large number of organisms (Mitton and Grant 1984). In the white-tailed deer on the SRS, positive associations between heterozygosity and traits related to individual growth have been documented several times. For instance, fetuses with higher heterozygosity have higher growth rates than fetuses with reduced heterozygosity (Cothran et al. 1983; Leberg et al. 1990). Adult body weight and antler size also have been observed to be positively associated with individual heterozygosity (Cothran et al. 1983; Scribner et al. 1989).

More closely related to this study, Chesser and Smith (1987) reported that more heterozygous deer grew more slowly but attained greater adult mass than less heterozygous individuals. The major difference between the current study and that of Chesser and Smith is that they used a growth equation with fixed shape. Subsequent simulations have suggested that the equation they used produces biased and imprecise estimates of growth parameters (Leberg et al. 1990). Furthermore, Brisbin et al. (1986) have shown that conclusions based on statistical comparisons of fixed-shaped sigmoid equations can be altered when a flexible-shaped growth curve is used.

Differences in statistical approaches between the two studies do not explain why we found no relationship between heterozygosity and growth patterns, when such relationships exist between genotype and other traits related to growth. If heterozygosity did influence postnatal growth in deer, its effects may not have been detected because they are generally small and may be temporally variable. Effects of heterozygosity on both body weight and antler size appear to vary from year to year (Scribner et al. 1989). This interaction between heterozygosity and some temporally varying factor on growth would go undetected in our analysis, because the paucity of older-aged individuals necessitated pooling data among cohorts. Other sources of variation, such as measurement error, would also make it difficult to detect a small effect of heterozygosity on deer growth.

Regardless of the explanation for our failure to observe a relationship between postnatal growth and heterozygosity, it appears that individual heterozygosity does not account for much of the variation in postnatal growth of SRS deer. This should not be construed as evidence that genetic variation is not important in growth, because there is ample evidence to the contrary. There is considerable additive genetic variation in the growth patterns of many organisms (Falconer 1960); however, these additive effects are not detected with protein electrophoresis.

Previous research has reported morphological differences between deer from the swamp and uplands of the SRS (Brisbin and Lenarz 1984). The present study indicated that there was substantial variation between the growth patterns of deer from these same habitats. However, asymptotic mass of males was the only individual growth parameter to differ between habitats. The reason for these differences between habitats that are immediately adjacent to one another is unclear. Greater weights in the upland habitat as compared with the swamp may be due to differences in forage quality. These growth differences may also be a function of herd density. Dapson et al. (1979) thought that deer in the swamp were much closer to carrying capacity than in the upland areas because they exhibited many reproductive characteristics that were indicative of overpopulation. Additionally, the population in the swamp is hunted less extensively than in the uplands. There is also evidence to suggest that deer from the two habitats may belong to different genetic units (Ramsey et al. 1979; Smith et al. 1991). The differences in growth between the upland and the swamp may be due to genetic differences in the populations. If there is a genetic basis for these differences in growth, it may be due to random genetic drift or habitat-specific selection pressures (Brisbin and Lenarz 1984). None of these explanations of habitat-specific growth patterns is easily eliminated with the present data.

It is also not clear why male growth patterns show more habitat-specific variation than do those of females. If there are large differences in density between the habitats, the reduced size of males in the swamp may be due to increased competition between males for mates. In cervids, as population size increases, so does intrasexual competition, resulting in males devoting more energy to combat and display and less energy to growth (Clutton-Brock et al. 1982). Alternatively, the differences in the response of male and female growth functions to habitat effects may also be explained by increased competition for forage. If quality forage is rarer in the swamp than in the upland, deer in the swamp would be expected to have a harder time obtaining energy for growth and reproduction than deer in the upland. Does mature more rapidly and at smaller sizes than males, thus allowing them to devote more energy to offspring production. As a result, resource competition may affect the ability of females to produce offspring, rather than inhibiting growth as appears to be the case with males.

Factors affecting growth patterns may also be directly or indirectly affecting other components of population productivity such as fecundity and/or mortality in the two habitats (Johns et al. 1977; Dapson et al. 1979). Adult reproduction, for example, is highly correlated with body weight (Nixon 1971; Clutton-Brock et al. 1982), so higher reproductive rates would be expected in deer from the upland sites. Dapson et al. (1979) and Rhodes et al. (1986) reported higher pregnancy rates and numbers of fetuses per doe in the upland habitat. Higher maternal and neonate weights appear to lead to decreased fawn mortality (Cook et al. 1971; Clutton-Brock et al. 1982). Because growth rates and age-specific weights are higher in the uplands, fawn mortality rates in the upland populations should be lower than those in the swamp. In fact, based on life table analyses, fawns in the uplands have higher survival rates than those in the swamp (Dapson et al. 1979).

The influence of habitat on the growth patterns of deer in the SRS herd appears to be greater than the influence of genetic variation. Furthermore, habitat-specific effects on growth appear to be greater in males than females. These results suggest that changes in habitat can affect growth patterns over a small geographic area. Because of the close association between growth and other life history features, managers should recognize that even small differences in local habitat can have large effects on the productivity of deer herds.

Acknowledgements

We thank P.E. Johns and employees of the Savannah River Site and the U.S. Forest Service who helped with data collection. Special thanks are extended to C. McCreedy and N. Matthews for critically reading earlier drafts of this manuscript. This research was conducted in conjunction with programs of the Institute of Ecology and was supported by contract DE-AC09-76SR00-819 between the University of Georgia and the U.S. Department of Energy.

References

Brisbin IL Jr, Lenarz MS (1984) Morphological comparisons of insular and mainland populations of southeastern white-tailed deer. J Mammal 65:44–50
Brisbin IL Jr, White GC, Bush PB (1986) PCB intake and the growth of waterfowl: multivariate analyses based on a reparameterized Richard's sigmoid model. Growth 50:1–11
Chesser RH, Smith MH (1987) Relationship of genetic variation to growth and reproduction in white-tailed deer. In Wemmer CM (ed) Biology and Management of the Cervidae. Smithsonian Institution Press, Washington, D.C., pp 168–177
Clutton-Brock TH, Guiness FE, Albon SA (1982) Red Deer. Behavior and Ecology of Two Sexes. University of Chicago Press, Chicago
Cook RS, White M, Trainer DO, Glazener WC (1971) Mortality of young white-tailed deer fawns in south Texas. J Wildlife Management 35:47–56
Cothran EG, Chesser RK, Smith MH, Johns PE (1983) Influences of genetic variability and maternal factors on fetal growth in white-tailed deer. Evolution 37:282–291
Dapson RW, Ramsey PR, Smith MH, Urbston DF (1979) Demographic differences in contiguous populations of white-tailed deer. J Wildlife Management 43:889–898
Falconer DS (1960) Introduction to quantitative genetics. Ronald Press, New York
Fuller TK, Pace RM III, Markl JA, Coy PL (1989) Morphometrics of white-tailed deer in north-central Minnesota. J Mammal 70:184–188
Hirth DH (1977) Social Behavior of White-Tailed Deer in Relation to Habitat. Wildl Monogr No 41. Wildlife Society, Washington D.C.
Johns PE, Baccus R, Manlove MN, Pinder JE III, Smith MH (1977) Reproductive patterns, productivity, and genetic variability in adjacent white-tailed deer populations. Proc Southeastern Assoc Fish Wildl Agencies 31:167–172
Leberg PL, Smith MH, Brisbin IL Jr, Scribner KT (1991) Optimization strategies in a deer harvest program. 18th Proc Congr Int Union of Game Biol (in press)
Leberg PL, Smith MH, Rhodes OE Jr (1990) The association between heterozygosity and the growth of deer fetuses is not explained by effects of the loci examined. Evolution 44:454–459
Leberg PL, Brisbin IL Jr, Smith MH, White GC (1989) Factors affecting the analysis of growth patterns of large mammals. J Mammal 70:275–283
Manlove MN, Avise JC, Hillestad HO, Ramsey PN, Smith MH, Straney DO (1975) Starch gel electrophoresis for study of population genetics in white-tailed deer. Proc Southeastern Assoc Fish Wildl Agencies 30:487–492
Mitton JB, Grant MC (1984) Associations among protein heterozygosity, growth rate, and developmental homeostasis. Annu Rev Ecol Syst 15:479–499
Nixon CM (1971) Productivity of white-tailed deer in Ohio. Ohio J Sci 71:218–229
Ozoga JJ (1972) Aggressive behavior of white-tailed deer at winter cuttings. J Wildlife Management 36:861–868
Ralls K (1977) Sexual dimorphism in mammals: avian models and unanswered questions. Am Nat 111: 917–938
Ramsey PR, Avise JC, Smith MH, Urbston DF (1979) Biochemical variation and genetic heterogeneity in South Carolina deer populations. J Wildlife Management 43:136–142
Rhodes OE Jr, Novak JM, Smith MH, Johns PE (1987) Assessment of the degree of fawn breeding in a South Carolina deer herd. Proc Southeastern Assoc Fish Wildl Agencies 40:430–437
Rhodes OE Jr, Scribner KT, Smith MH, Johns PE (1986) Factors affecting number of fetuses in a white-tailed deer herd. Proc Southeastern Assoc Fish Wildl Agencies 39:380–385
Richards FJ (1959) A flexible growth function for empirical use. J Exp Bot 10:290–300
Roseberry JL, Klimstra WD (1975) Some morphological characteristics of the Crab Orchard deer herd. J Wildlife Management 39:48–58

SAS (1985) SAS users guide: statistics. SAS Institute, Cary, North Carolina
Sauer PR (1984) Physical characteristics. In Halls LK (ed) White-Tailed Deer. Ecology and Management. Stackpole Books, Harrisburg, Pa., pp 73–90
Scribner KT, Smith MH, Johns PE (1989) Environmental and genetic components of antler growth in white-tailed deer. J Mammal 70:284–291
Severinghaus CW (1949) Tooth development and wear as criteria of age in white-tailed deer. J Wildlife Management 13:195–216
Smith MH, Willis KB, Johns PE (1991) Spatial-genetic variability in a white-tailed deer herd. 19th Proc Congr Int Union Game Biol (in press)
Stearns SC, Koella JC (1986) The evolution of phenotypic plasticity in life-history traits: predictions of reaction norms for age and size at maturity. Evolution 40:893–914
Urbston DF (1967) Herd dynamics of a pioneer-like deer population. Proc Southeastern Assoc Fish Wildl Agencies 21:42–57
White GC, Brisbin IL Jr (1980) Estimation and comparison of parameters in stochastic growth models for barn owls. Growth 44:97–111

73

Temporal Relationship Between Estrus, the Preovulatory LH Surge, and Ovulation in Fallow Deer

G.W. Asher[1], M.W. Fisher[2], J.F. Smith[1], H.N. Jabbour[1] and C.J. Morrow[1]

To determine the precise timing of ovulation relative to the onset of estrus and the preovulatory LH surge in fallow deer (*Dama dama*), 20 (> 5 years old) does were allocated to two treatments designed to synchronize estrus on or about 17 May. Treatment 1 does (prostaglandin-induced estrus) received intravaginal progesterone-releasing devices for 13 days followed by an injection of 1500 mg coprostenol on day 12 (15 May) of the subsequent estrous cycle. Treatment 2 does (progesterone-induced estrus) received CIDRs for a 13-day period, with withdrawal on 15 May. All does were run with crayon-harnessed bucks (10:1 ratio) from the start of synchronization treatment (1800 h 15 May). Five does per treatment were blood sampled via in-dwelling jugular cannulae every 2 h for 70 h from coprostenal administration/CIDR withdrawal, and the plasma was analyzed for LH. Does within each treatment were randomly allocated to one of four ovarian examination times (12, 16, 20, or 24 h after onset of estrus). Laparoscopy was repeated at 12-h intervals until ovulation (hemorrhagic rupture point) was recorded. Does failing to exhibit estrus were examined 72 h after synchronization treatment. Seventeen does exhibited estrus at a mean (± SEM) interval from synchronization of 44.6 ± 3.6 h for Treatment 1 ($n = 9$) and 34.1 ± 2.5 h for Treatment 2 ($n = 8$). The incidence of ovulation was 0% at 12, 16, and 20 h ($n = 4$, 4 and 4 observations), 50% at 24 h ($n = 10$), and 100% at 28, 32, and 36 h ($n = 4$, 5, and 3). The onset of the preovulatory LH surge ($n = 8$) occurred at the onset of estrus, with maximum LH concentrations (30 ng/ml) occurring 4 to 8 h later. Of three does not exhibiting estrus, two (Treatment 2) had active corpora lutea (indicating ovulation during CIDR insertion) and one (Treatment 1) had a large unruptured follicle 72 h after prostaglandin treatment. The data indicate that ovulation in fallow deer occurs ~ 24 h after the onset of estrus and 16 to 20 h after the peak of the preovulatory LH surge.

[1]Ruakura Agricultural Centre, MAFTech North, Private Bag, Hamilton, New Zealand and [2]Invermay Agricultural Centre, MAFTech South, Private Bag, Mosgiel, New Zealand

74

Reproductive Behavior of Captive Chital Deer in Australia: Seasonal Changes and a Comparison of Activity During Restricted and Unrestricted Mating Periods

Rosalie S. Chapple*

The behavior of captive chital (*Axis axis*) stags and hinds in Australia was observed. Activity profiles were recorded for 6 to 12 h every week over 1 year. The oldest and dominant stag spent 10.5% of his time throughout the year engaged in rutting activities, which was significantly more than the younger stags, who spent 2.6% to 5.3% of their time engaged in such activities. Hinds in estrus were mostly (70%) attended by the dominant stag. Rutting was significantly less intense than that of seasonal breeders and was dispersed throughout the year, with two major peaks of activity—one in August/September coincident with a birth peak and prior to antler casting, and the second coinciding with antler hardening early in the year. In a separate study of the same deer, a dominant stag (6 years old) and a subordinate stag (2 years old) were joined with 35 hinds for 6 weeks, after 5 months of separation. The mating activity of this herd was compared with that during unrestricted mating. During unrestricted mating, when there was rarely more than one hind in estrus at a time, there was no evidence of harem formation or of any other mating strategy. However, signs of harem formation were evident during the 6-week restricted mating period. Adaptation is therefore apparent in the mating strategy of the chital stag, and can be made use of to increase mating efficiency so that many hinds may be mated within several days. The time spent courting a hind was reduced from at least 1 day during unrestricted mating, to less than 10 h during restricted mating. During the short mating period, aggression and homosexual activities were frequent between the hinds, but were rare during unrestricted mating.

*Department of Animal Health, University of Sydney, Camden, New South Wales 2570, Australia

75
Mating Tactics in the Male Roe Deer

Olof Liberg, Annkristin Axen, Agneta Johansson, and Kjell Wahlstrom*

Male roe deer (*Capreolus capreolus*) defend large resource-holding territories for part of the year. The function of these territories has been unclear, as has the question of whether roe deer are mono- or polygamous. On the basis of data from extensive telemetry and visual observations during three mating seasons in a continuing population study of roe deer in south-central Sweden, we show that roe males are both highly polygamous and that territory defense is the dominant mating tactic. Also, two subordinate tactics were identified, territory satellites and peripheral scroungers. No males younger than 3 years old held territories, but the subordinate tactics were performed by both young (1 and 2 years old) males and by older former territory holders. Territorial males got access to and courted intensively 2 to 5 females per season. All of the few matings observed were performed by territorial males. They spent 50% to 68% of their time on sexual behavior during the rut, while corresponding figures for nonterritorial males were 0% to 35%. Territory turnover was fast. All recorded holders in 1987 were replaced by 1989. Some females made short visits to distant territory males during their estrus. Both courtship and matings were observed during these visits.

*Zoology Department, University of Stockholm, 10691 Stockholm, Sweden

76

Effects of Strategic Short-term and Long-term Melatonin Treatment on Estrous Cyclicity in Fallow Deer

C.J. Morrow, G.W. Asher, and H.N. Jabbour*

Preliminary data are presented for an ongoing trial designed to determine the effects of strategic treatment with SC melatonin implants (Regulin; Regulin Australia, Melbourne, Australia) on seasonal estrous cyclicity and live-weight changes of fallow does (*Dama dama*). Sixteen adult (> 3 years old) does were allocated to one of four treatment groups, balanced for live weight, in October 1988. All does were in their third trimester of pregnancy. Group 1 does (summer implantation; 120-day treatment period) each received double melatonin implants on four occasions at 28-day intervals from 2 November (i.e., 30 to 40 days prior to parturition); group 2 does (continuous implantation; 18-month treatment period) received implants at 28-day intervals from 2 November, with the implantation schedule continuing until May 1990; group 3 does (winter implantation; 120-day treatment period) will receive implants on four occasions at 28-day intervals from 31 July; and group 4 does (controls) have received no implants. All does are being run continuously with single vasectomized bucks fitted with ram mating harnesses to mark estrous does. Harness crayons are replaced twice weekly and observations to record mating marks are conducted daily. All does are being blood sampled by jugular venepuncture twice weekly from 2 November 1988 to 30 May 1990, and plasma will be analyzed for concentrations of progesterone. Live weights are recorded at 2-week intervals. All 16 does fawned in December 1988. However, six of eight does treated during late pregnancy (groups 1 and 2) failed to initiate lactation and subsequently lost their fawns. The remaining eight does all reared their fawns successfully. Does in groups 1 and 2 exhibited first estrus of the 1989 breeding season 6 to 8 weeks earlier than groups 3 and 4. The mean ($\pm$ SEM) dates of first estrus were 1 March $\pm$ 7.2 days, 22 February $\pm$ 7.7 days, 22 April $\pm$ 1.3 days, and 20 April $\pm$ 1.9 days for groups 1, 2, 3, and 4 respectively. The data presently (July 1989) indicate that three does in group 1 are still exhibiting 19- to 25-day estrous cycles, all does in group 2 have become anestrous after exhibiting only four estrous cycles, and all does in groups 3 and 4 are still cyclic half-way through their normal breeding season. Melatonin treatment has resulted in marked live-weight loss in group 1 and 2 does relative to groups 3 and 4. Continuous melatonin implantation (group 2) has been associated with the most severe reductions in live weight, and this may explain the rapid reversion to an anestrous state.

---

*Ruakura Agricultural Centre, MAFTech North, Private Bag, Hamilton, New Zealand

77
Effects of Melatonin Implants in White-Tailed Deer

Harry A. Jacobson and Stephen J. Waldhalm*

Melatonin implants were given to seven adult female white-tailed deer (*Odocoileus virginianus*) 15–17 September 1987. Progesterone pessaries administered 6–20 October 1987 and prostaglandin F_2 alpha administered at pessary removal failed to induce preseason estrus. First estrus was observed about 2 weeks earlier than expected for one of the seven but was delayed 1 month or later than expected for the remaining does. Three of the seven does failed to have normal vernal molts in either 1988 or 1989 and molted from winter coat to winter coat in September 1988. Two of the three delayed vernal molts in late June and early August 1989. A fourth doe maintained her same winter coat from the time of implant through early August of 1989 when she underwent a delayed vernal molt. The remaining three does underwent normal molts in 1988. Six of the seven does gave birth in 1988, and no impediment of lactation was noted. Fawns nursed by one doe, like their dam, had a delayed (August) vernal molt in 1989. All other fawns had normal vernal molts in May. A melatonin implant was given an adult buck on 18 May 1989. His winter molt was completed by 4 July; velvet rub-out was completed on 10 July, and sperm presence was confirmed when electroejaculated on 24 July. All three events occurred at least 2 months earlier than normal.

*Department of Wildlife and Fisheries and College of Veterinary Medicine, Mississippi State University, Mississippi State, MS 39762 USA

78

Synchronization and Superovulation in Captive White-Tailed Deer

S.J. Magyar, C.M. Hodges, and D.C. Kraemer*

The object of this study was to evaluate two methods of synchronization and two methods of synchronization-superovulation in captive white-tailed deer (*Odocoileus virginianus*). Thirteen hand-raised does and one hand-raised vasectomized male white-tailed deer were studied during three breeding seasons (1986–87, 1987–88, 1988–89). The does were synchronized using one or the other of two treatment methods. One treatment method used a single IM injection consisting of 25 mg prostaglandin F_2. The other method of synchronization used 60 mg medroxyprogesterone acetate (MAP) impregnated in an intravaginal sponge. The does on synchronization-superovulation treatments received either 28 mg FHS-P in a decreasing dose regimen plus a single IM injection of 25 mg prostaglandin F_2, or MAP-impregnated intravaginal sponges plus 1000 IU PMSG. Synchronization of estrous in captive white-tailed deer is possible using either prostaglandin F_2 or a progesterone-impregnated intravaginal sponge. Does receiving prostaglandin F_2 cycled 60% of the time, while the does with MAP intravaginal sponges cycled 75% of the time. Superovulation of captive white-tailed deer using FSH-P and PMSG in our does has proven to be unsatisfactory, with only one doe having a response.

*Department of Veterinary Physiology and Pharmacology, Texas A&M University, College Station, TX 77843, USA

79

Studies on Superovulation and Embryo Recovery in Farmed Red and Fallow Deer

H.N. Jabbour, G.W. Asher, J.G.E. Thompson, H.R. Tervit, and C.J. Morrow*

The effects of various gonadotrophin regimens on ovarian ovulatory responses, endocrine changes and recovery/fertilization rates of ova in red (*Cervus elaphus*) and fallow deer (*Dama dama*) were examined in 1989. Fifty mature females of each species were treated during their breeding (CIDR) season with intravaginal progesterone-releasing devices for 14 days (insertion day = day 0) and one of five doses of ovine FSH (0, 0.25, 0.5, 0.75, 1.0 units). All animals received a dose of 200 IU PMSG on day 11 and eight doses of FSH administered at 12-h intervals starting at PMSG administration. After CIDR removal, all hinds/does were run with fertile males (10:1 ratio). They also received intravaginal inseminations (20 to 30 × 10^6 motile thawed spermatozoa/insemination) on four occasions at 12-h starting 24 h after CIDR removal. Four animals in each treatment were blood sampled every 2 days from CIDR insertion until day 24 and every 2 h for 70 h after CIDR removal. Plasma samples were analyzed for progesterone, estradiol-17B, androstenedione, and LH. Ova were recovered by flushing the uterus during midventral laparotomy under general anesthesia 7 days after CIDR removal. The numbers of corpora lutea (CL) and unruptured follicles (> 5 mm) were recorded. Both species exhibited a curvilinear pattern of ovarian response to increasing doses of FSH. The highest numbers of CL were observed following treatment with 0.5 units of FSH. Ova recovery rates were low for both breeds (red, 32.7% ± 5.1%; fallow, 30.6% ± 5.1%) with no differences between treatment groups. The mean fertilization rate (± SEM) of ova recovered from red hinds was 50.2% ± 8.2%, with the majority of the embryos at the morula stage. None of the ova recovered from fallow deer had cleaved. Present indications are that sperm transport, and hence fertilization, may be adversely affected by high follicular secretion of estradiol. It is apparent that studies are needed to define the optimal site and time of semen deposition within the reproductive tract to produce satisfactory fertilization rates.

*Ruakura Agricultural Centre, Private Bag, Hamilton, New Zealand

80

Real-time Ultrasonography for Pregnancy Diagnosis and Fetal Age Estimation in Red Deer from Conception to 40 Days' Gestation

Brigitte Revol and Peter R. Wilson*

Ultrasonography has proven to be a simple, accurate, and cost-effective method of pregnancy diagnosis in farmed red deer (*Cervus elaphus*) in New Zealand. Previous studies have shown the accuracy of pregnancy detection from 30 days of gestation and that fetal age estimation equations could be used to predict calving from 42 to 110 days of gestation. However, no data were available on the signs of pregnancy and accuracy of pregnancy determination from conception to 30 days of gestation. This report presents data on ultrasonography of early pregnancy in red deer. Mating dates were recorded for 35 mixed-age farmed red deer hinds. Rectal ultrasound examinations using a 5-mHz rectally placed ultrasound transducer were undertaken at intervals from 0 to 42 days of gestation. Images were recorded on video tape and signs of early pregnancy and fetal development were recorded. By 7 days of gestation, a 5-mm vesicle could be observed and by day 14, edema of uterine horns was apparent in some cases. A comma-shaped fetus 6 mm long, fetal membranes, and placentomes could be observed at day 24. The heart beat could be observed on day 28 when the fetus was 10 mm long. Limb buds were observed on day 31 and by day 37 the head with nose and eyes was clearly distinguishable. Fetal movements were observed on day 42. While the accuracy of pregnancy detection from day 30 was acceptable (97%), the accuracy between 20 and 30 days was 80% and that for 7 to 20 days was 60%.

*Veterinary Clinical Sciences Department, Massey University, Palmerston North, New Zealand

81

Breeding Performance of Female Chinese Muntjac Deer in England

Norma G. Chapman*

Chinese muntjac (*Muntiacus reevesi*) are aseasonal breeders with a postpartum estrus. Long-term observations on a total of 13 captive females kept in paddocks with sexually mature males showed that most females conceive before the age of 10 months and thereafter have the potential to produce a single fawn at intervals of about 7 months. The minimum interbirth period was 211 days: for 53% the mean was 216 days. This productivity can be maintained over many years (e.g., 22 fawns by the age of 14 years). Births occurred in all months of the year, and the sex ratio was close to parity, situations which also apply to the feral populations in England as evidenced by investigations on hundreds of carcasses.

*Larkmead, Barton Mills, Suffolk IP28 6AA, U.K.

82
Variation in Mating Systems of Fallow Deer in Relation to Ecology

Jochen Langbein and Simon J. Thirgood*

Previous explanations of variation in ungulate mating systems have concentrated on differences between species in relation to their ecology. Comparisons within species, however, whether between populations living in different environments, or within populations in different circumstances, avoid the problems of major differences between species masking subtle responses to environmental variation. The aims of this study were: (1) to define more clearly the different types of mating systems exhibited by fallow deer (*Dama dama*) and (2) to examine which factors of the population structure and vegetational environment are most important in influencing the mating system adopted by the population. In addition, we briefly examined individual reproductive success of males adopting the different mating strategies both within and between populations. Variation in mating systems of fallow deer was studied in wild and park populations and could be broadly distinguished into seven categories: following, harems, dominance groups, stands, temporary stands, multiple stands, and leks. The ecological factors which exerted the greatest influence on mating systems were buck density, doe density, and habitat structure and tree cover, resulting in pronounced changes in the degree and type of territoriality observed. Individual male reproductive success was highly skewed in all systems. A greater range of concurrent mating strategies was observed in wild populations than in parks.

*Department of Biology, University of Southampton, Southampton, SO9 5NH, U.K.

83
Variation in Conception Dates in Black-Tailed Deer in Western Washington

Kenneth J. Raedeke and Cynthia A. Kunz*

The season of birth in deer in temperate climates is generally considered to coincide with the most favorable period of the year for survival of the fawns. The dates in conception in black-tailed deer (*Odocoileus hemionus columbianus*), calculated from embryos collected during the hunting season were studied in two populations from 1980 to 1988. Substantial variations between years were noted in one of the populations. Four potential factors have been postulated as possible causes of this annual variation: (1) disturbance due to hunting during the breeding season; (2) changes in female age structure; (3) availability of sexually mature males for breeding; and (4) variations in physiological condition of does. The first three factors have been rejected as possible causes of the dramatic shifts noted. The fourth factor, influenced by annual changes in climatic conditions and hence forage productivity, is the subject for continued research.

*College of Forest Resources AR-10, University of Washington, Seattle, WA 98195, USA

84
Aspects of Placentation in White-Tailed Deer

Patrick F. Scanlon*

The placenta of ruminants consists of placentomas of which the caruncles comprise the maternal component. Among ruminants the Cervidae species generally seem to have a relatively small number of placentomas. White-tailed deer (*Odocoileus virginianus*) taken by hunters were examined in early pregnancy following implantation. Numbers of placentomas ranged from 4 to 14; the modal value was 6. All caruncles tended to be attached to the fetal membranes regardless of the number of fetuses present (1 to 3) and also tended to be evenly distributed between both uterine horns.

*Department of Fisheries and Wildlife Sciences, Virginia Polytechnic Institute and State University, Blacksburg, VA 24061, USA

85
Establishing Early Pregnancy by the Evaluation of the Serum Progesterone Level

Zoltan Zomborszky[1], László Sugár[1], and Tibor Feher[2]

The levels of serum progesterone were studied in red deer (*Cervus elaphus*) hinds by radioimmunoassay (RIA) in order to establish early pregnancy or infertility. Twelve yearlings, nine 2-year-olds, and 13 older hinds took part in this experiment. Six of the seven infertile hinds had 0.3 nmol/l progesterone level; however, in one the value was 1.9 nmol/l. At the same time (8/11/88), progesterone values were measured in pregnant hinds: (1) One of the seven yearlings had 1.6 nmol/l on the estimated 21st day of pregnancy; values of the other six ranged from 2.6 to 9.6 nmol/l on days 23 to 43 of pregnancy. (2) One of the nine 2-year-old hinds had 2.2 nmol/l on day 13 of pregnancy; the others were sampled between day 15 and 48 of pregnancy and had levels of 5.1 to 11.2 nmol/l. (3) The progesterone values of the older pregnant hinds were between 2.6 and 6.7 nmol/l measured on the estimated days 16 to 52 of pregnancy.

[1]Pannonian Agricultural University, Faculty of Animal Science, H-7401 Kaposvar, P.O. Box 16, and [2]Endocrinological Laboratory, Clinic of Internal Medicine, Semmelweis University of Medicine, H-1083, Budapest, Koranyi S. 2/a., Hungary

86
Circadian and Circannual Melatonin Secretion in Finnish Semidomesticated Reindeer

Eija Eloranta[1], Jouni Timisjarvi[1], Mauri Nieminen[2], Juhani Leppaluoto[1], and Olli Vakkuri[1]

Mammalian species from temperate and subarctic areas display distinct annual fertility cycles and their reproductive physiology responds to cues from the environment. The daily photoperiod is the major environmental factor repeated with great regularity from year to year. There is compelling evidence to suggest that in small laboratory animals the pineal gland transfers the photoperiodic information entering the eyes to the hypothalamo-hypophyseal-gonadal axis via melatonin. The pineal gland is required for photoperiodic regulation of seasonal breeding. The reindeer is a short-day breeder and the association of melatonin with its reproductive cycle was investigated in this study. Ten female reindeer (*Rangifer tarandus tarandus*) and five calves were studied. The endogenous circadian and circannual rhythms of melatonin were investigated using 4-h samplings for 24 h every 3 months in a natural photoperiod. Serum melatonin levels exhibited a pronounced seasonal rhythm ($F = 5.88$, $P < 0.001$). In autumn (12L:12D), in winter (0L:24D), and in spring (12L:12D) the circadian rhythm in melatonin secretion was seen in hinds with acrophase at night ($P < 0.001$, compared with day). However, in summer (24L:0D) no circadian rhythm occurred. The amplitude in melatonin secretion was at its maximum in autumn. In prepubertal calves, the circadian melatonin rhythm was recorded in autumn and spring, but in autumn the peak levels were lower than those of adults ($P < 0.05$). In winter and in summer the melatonin levels were equal, and no rhythm was recorded in calves.

[1]Department of Physiology, University of Oulu, SF-90220 Oulu and [2]Finnish Game and Fisheries Research Institute, Reindeer Research Unit, SF-96100 Rovaniemi, Finland

Part 6
Genetics

87
Application of Biochemical Genetics to Deer Management: What the Gels Tell

Peter A. Dratch and Josephine M. Pemberton

Abstract

Since the first electrophoretic study of deer was reported in *Nature* in 1961, there have been over 100 papers published on the subject. Most of the early studies fall into two categories: (1) those surveying a few individuals from several species, generally utilizing blood samples from zoo animals, and (2) those examining many individuals from one or a few species, often using tissue samples obtained from carcasses during hunting seasons. The interspecific studies have identified several species-specific markers in proteins from serum and red blood cells. The cumulative results of the major intraspecific studies have revealed great differences in the amount of variation between species. White-tailed deer (*Odocoileus virginianus*) and reindeer (*Rangifer tarandus*) are the most polymorphic species, while European fallow deer (*Dama dama*) and Père David's deer (*Elaphurus davidianus*) show the least variation. Where polymorphism has been found, usually it has indicated significant spatial and temporal differentiation; the gene frequency data verify that deer species are seldom panmictic. Using the standard measures of electrophoretic variation (average heterozygosity, percentage of polymorphic loci, and average number of alleles per locus), significant differences between populations within a species have also been widely reported. Most electrophoretic studies of deer have assumed that polymorphic loci are neutral in relation to natural selection; however, when there has been an attempt to test for selection, several parameters of fitness have been associated with variation at electrophoretic loci in white-tailed deer, red deer (*Cervus elaphus*), and reindeer. The intensification of deer management in the past decade has brought new genetic problems as well as the application of advanced biomedical techniques in addressing them. Plasma proteins, isozymes, and mitochondrial DNA have provided markers that document hybridization both in captive breeding programs and deer in the wild. Laboratories now conduct routine deer blood-typing. As deer genomes diverge, and some species receive increased protection while others are bred for commercial purposes, biochemical and molecular techniques are bound to have a growing role in deer management.

Key words: Biochemical genetics, DNA, electrophoresis, isozymes, polymorphism, proteins

Introduction

During the past three decades, well over 100 papers have been published on biochemical genetics of deer. The literature on this subject has never been reviewed nor has a comprehensive bibliography been published. The need for such a review is now pressing for a number of reasons. The deer farming industry continues to grow worldwide, and there are now laboratories conducting commercial deer blood-typing to confirm the genetic background and parentage of particular animals. In addition, there has been an increasing attempt on the part of wildlife managers to incorporate genetic information in plans to conserve particular species or populations.

The current interest in deer genetics is reflected in the concern of the Deer Specialist Group of the International Union of the Conservation of Nature, a new standing committee on deer within the International Society for Animal Genetics as well as a session devoted to genetics at the Second International Symposium on Biology of Deer. The number of scientific articles on the biochemical genetics of deer has more than doubled in the past decade. This review has attempted to include all of those papers, either published or in press at the time of the conference. There probably have been a few oversights, and the authors would appreciate knowing of any omitted references.

The review is divided into two parts, each directed primarily at a different audience. The first part discusses some of the generalizations that emerge from this deer genetics literature that are pertinent to deer management, both of captive and wild populations. The second part is organized as species accounts of the electrophoretic research done to date, so that scientists working in this expanding field can readily retrieve papers from disparate sources.

Amount of Genetic Variation in Deer

Electrophoretic studies comparing deer species have identified abundant protein and DNA differences that can be utilized for forensic identification (Dilworth and McKenzie 1970; Hoyem and Thorson 1970; Bunch et al. 1979a, b; Hindmarsh and Schuster 1979; Lawton and Sutton 1982; McClymont et al. 1982; Mardini 1984; Pex and Wolfe 1985; McCormick et al. 1988; Murphy et al. 1988) as well as to refine deer phylogenies (Baccus et al. 1983; Miyamoto et al. 1990; Cronin 1991b).

In contrast, surveys of biochemical variation within populations of moose (*Alces alces*), elk, Pére David's deer, and fallow deer found low polymorphism in each species (Ryman et al. 1977; Cameron and Vyse 1978; Ryder et al. 1981; Pemberton and Smith 1985). These results prompted speculation that large mobile mammals such as deer might be intrinsically low in variation, conforming to the environmental grain hypothesis (Selander and Kaufman 1973). Additional research has found more polymorphic loci in moose (Ryman et al. 1980), elk (Baccus et al. 1983; Dratch and Gyllensten 1985), and fallow deer (Hartl et al. 1986; Randi and Apollonio 1988; Herzog 1988). In other species where little or no variation has been found in particular populations such as Norway red deer, Iceland reindeer, and Maryland sika deer, founder effects and human-controlled breeding can often explain loss of genetic variation (Ryman et al. 1980; Røed et al. 1985; Feldhamer et al. 1982).

Intraspecific biochemical variation is quantified in three ways: heterozygosity (the likelihood that any individual animal will have different alleles from each parent at a particular locus), percentage of polymorphic loci, and the average number of alleles per locus. Using all three of these indices, Smith et al. (1991b) evaluated ten deer species and found that the amount of variation does not differ significantly from that of mammals in general (Nevo et al. 1984).

While body size generally is not a good predictor of genetic variability (Wooten and Smith 1985), other possibilities have just begun to be examined. Hartl and Reimoser (1988) tested the prediction made by Harrington (1985) at the first Deer Biology Conference that the r-selected deer species—those with high reproductive rates which live in subclimax habitats—would be more polymorphic than K-selected species. The high heterozygosity they found in roe deer fits Harrington's prediction, as do published data on polymorphism from several other deer species. This idea, along with the hypothesis that habitat generalists will be more polymorphic than specialists, warrants examination in future studies.

There are two practical conclusions that can be drawn from the research on the extent of biochemical variation in deer. All species are invariant at some loci, and these are the systems which should be used as species specific markers in forensic work. Secondly, when an adequate number of loci are surveyed, deer have levels of electrophoretically detectable variation typical of mammals (Ryman et al. 1980; Smith et al. 1991b). Low electrophoretic variation in a deer population often does not characterize that species generally; it is more likely attributable to the breeding history of that particular population.

Population Subdivision

If there is a single consistent generalization from the electrophoretic studies of wild deer populations, it is that these species are genetically subdivided. When several populations have been studied and biochemical polymorphism has been identified, significant gene frequency differences between populations have also been found. Population differentiation has been documented in red deer (McDougall and Lowe 1968; Gyllensten et al. 1983), white-tailed deer (Manlove et al. 1976; Kennedy et al. 1987; Karlin et al. 1989), moose (Ryman et al. 1980; Chesser et al. 1982b), elk (Dratch and Gyllensten 1985), reindeer (Røed 1986), caribou (Røed and Whitten 1986), roe deer (Hartl and Reimoser 1988), and mule deer (Smith et al. 1991b). Further research has shown that gene frequencies change over time (Smith et al. 1984). Despite these fluctuations, in some cases such as red deer and elk, gene frequency differences reflect the subspecies designations initially made on the basis of morphology (Gyllensten et al. 1983; Dratch and Gyllensten 1985).

What is the biochemical evidence that a location may harbor more than one population? The first suggestion of subdivision is that the genotype frequencies from the sampling area do not meet Hardy-Weinberg expectations due to a scarcity of heterozygotes. The reason for this underpins any electrophoretic analysis of natural populations. After a population subdivides, the gene frequencies in the subsequent populations can become progressively different over time, either due to genetic drift or natural selection. Because the populations are no longer a single breeding unit, the likelihood that a male and female with different genotypes will meet and breed decreases. The degree of population differentiation can be quantified and compared with other populations and species using the *F*-statistics developed by Wright (1978). Genetic analysis has shown that populations can be differentiated within a few miles, particularly where there are different habitats (Smith et al. 1984).

For the manager of wild populations, an electrophoretic study from tissues of harvested animals can provide a cost-effective method of determining breeding units (Gyllensten et al. 1982; Scribner et al. 1985). For the manager of captive deer, the same methods can ascertain whether a group of animals is inbred (Pemberton 1988). The important conclusion from these studies of gene frequency is that populations are the reservoirs of genetic variation. By using biochemical and molecular methods, that variation can be monitored, and management decisions can be taken so it is maintained.

Hybridization

One area where biochemical and molecular methods have shown obvious utility in recent years is in the detection of hybridization in deer. Natural matings between mule deer and white-tailed deer have been documented in a zone of introgression from Alberta to Texas (McClymont et al. 1982; Carr et al. 1986; Cronin et al. 1988). Gavin and May (1988) have found hybridization between black-tailed deer and the rare Columbian white-tailed deer. Hybrids between red deer and introduced sika in Ireland (Harrington 1979) and red deer and elk (Dratch 1987a) have been confirmed with electrophoretic tests on controlled crosses. Putative red/rusa (*Cervus timorensis*) and red/sambar (*Cervus unicolor*) hybrids remain to be definitively tested. Hybridization of red deer and Père David's deer by artificial insemination was confirmed electrophoretically (Asher et al. 1988); the two have previously been classified as different genera. A fixed isozyme difference was used to confirm hybridization between the European fallow and the rare Mesopotamian fallow deer on New Zealand farms (Dratch 1987b).

Results from study of the New Zealand wapiti population provide one example of how electrophoresis can be used to trace the extent of hybridization. Seventeen captive elk from the United States were released in remote Fiordland, New Zealand, in 1905. Within 25 years, hybridization with the encroaching red deer population was suspected, particularly because of the strange combination of roaring and bugling heard during the rutting season. A joint expedition led by Olaus Murie in 1949 confirmed limited hybridization on the basis of morphology (Murie 1951). When blood samples from 200 animals captured in Fiordland since 1975 were analyzed for red deer protein markers, 95% were found to be hybrids (Dratch 1987c).

Biochemical markers suggest that one common definition of a species—animals that are reproductively isolated under natural conditions—requires a proviso when applied to the Cervidae. If humans have been responsible for removing geographic or ecological barriers, deer can be wild yet not under natural conditions. However, electrophoresis has a more important role to play than simply identifying hybrids. Studies focusing on disparate populations and analyzing them for a large number of loci can identify genetic discontinuities as well as particular localities low in variation. In this way biochemical genetics can contribute important data in addressing taxonomic and conservation questions.

In the case of hybridization mentioned, electrophoresis verified the cause of atypical morphological features recognized by other biologists. Laboratories in New Zealand and Canada have analyzed thousands of deer for evidence of hybridization, and in Canada the primary purpose is to control hybridization and protect wild populations. While we now have the means to detect hybridization after the fact, the ability to control it may prove a decisive factor in the success of deer farming.

Natural Selection

Perhaps the most contentious issue surrounding biochemical genetics is whether natural selection is acting on electrophoretic loci in wild populations. The prevailing view has been that these loci are selectively neutral most of the time, and that electrophoretic variation is maintained by population dynamics and genetic drift (King and Jukes 1969; Kimura 1983). It is widely acknowledged, however, that natural selection is very difficult to prove, particularly when relying on culled animals for samples. Moreover, surveys which depend on electrophoresis to provide diagnostic markers for particular populations often have an unstated interest in those marker loci not being subject to selective pressures.

Electrophoretic studies addressing natural selection have either examined the effects of individual loci or associations between multilocus heterozygosity and some characteristics associated with fitness. In studies of the Savannah River Plant white-tailed deer population, multilocus heterozygosity has been associated with testes size in fawns (Urbston 1976), number of fetuses (Johns et al. 1977; Rhodes et al., Chapter 89, this volume), rate of fetal development (Cothran et al. 1983; Maffei and Woolf 1987; Leberg and Smith 1990) antler growth (Smith et al. 1983, 1991a; Scribner et al. 1984; Scribner and Smith 1990), adult female body size and fat levels (Cothran et al. 1983), and with conception time and body size (Chesser and Smith 1987). Some of these associations with heterozygosity may not persist with time as Leberg et al. (Chapter 72, this volume) have shown for postnatal growth in Savannah River deer.

Single locus effects have been found for β-hemoglobin and sorbitol dehydrogenase in white-tailed deer (Baccus et al. 1977; Chesser et al. 1982a; Smith et al. 1983), transferrin in reindeer (Røed 1987), and transferrin, mannose phosphate dehydrogenase, and isocitrate dehydrogenase in red deer (Pemberton et al. 1988, 1991). The latter research represents a new departure for two reasons. First, live deer were bled and their fate followed; second, a mechanism is suggested whereby electrophoretic variation is maintained within populations rather than across them. Utilizing the long-term study of deer on part of the Island of Rhum where the population has quadrupled and food is limiting, Pemberton et al. (1988) have shown that particular alleles are strongly associated with calf survival. Their follow-up paper which focuses on adult females shows selective effects at that same loci, but for different genotypes. For example, those females carrying a particular mannose phosphate isomerase allele have the poorest chance of survival, but if they do survive the juvenile period, they have a high lifetime reproductive output. Their calves, which carry the same mannose phosphate isomerase allele, often die young.

This evidence of countervailing selection at the biochemical level provides a mechanism for the high levels of polymorphism found in natural populations. To test further for the effects of selection, electrophoretic and radiotelemetry studies could be combined to examine deer populations subjected to natural predation.

New Techniques, New Target Species

The trend in electrophoretic studies of deer has been the elucidation of previously cryptic variation as a result of greater resolution available with improved methods. The use of polyacrylamide as a gel medium was crucial in revealing many alleles for transferrin and other plasma proteins. Isoelectric focusing showed additional hemoglobin variation that is still not fully understood. The first DNA study of deer was published only 7 years ago (Lima-de-Faria et al. 1984, 1986; Scherton et al. 1987); and the first to look at a wild population at the molecular level are even more recent (Carr et al. 1986; Cronin et al. 1988). Analysis of mitochondrial and minisatellite DNA, as well as DNA amplification by the polymerase chain reaction (PCR) offer powerful forensic tools in identifying species from small or old samples (Thommasen et al. 1989; Cronin et al. 1991a; Waye and Haigh, Chapter 88, this volume).

There is little doubt that new methods will be applied to deer, especially as these animals become more valuable either as livestock or as rare trophies. Before we set our sights on mapping the genome of red deer or mule deer, it should be noted that wild populations of half of the world's 40 deer species have not been studied at all for biochemical variation. Moreover, it is many of these deer species that are the most threatened. Smith et al. (1986) have pointed the way in this regard, applying their knowledge of the North American white-tails in a study of the South American populations and the brockets, *Mazama americana* and *M. gouazoubira*. The results suggest not only that the taxonomy might need revision, but that there is variation in these deer that can be used as a genetic baseline if their numbers or habitat diminish.

Decades of research have produced a sophisticated set of methods for studying genetic variation in deer at the biochemical and molecular level. We now have a better idea of what protein and isozyme systems are likely to be polymorphic (Table 87.1), as well as how to develop sampling strategies for unstudied species. Generalizations about any species are tenuous without looking at many loci in several populations. As deer become more valuable there will likely be increased incentive to look for genetic markers, not only for hybridization but for growth and reproduction. Hopefully, those conducting the next decades of research will continue to devote some of their energy to the rare and wild deer of the world.

Electrophoretic Variation in Deer Species

This record of the cervid family is divided according to subfamily. It uses the nomenclature of Allendorf and Utter (1979), numbering loci beginning with the most cathodal. The naming and numbering of comparable loci were altered to fit this convention. The full names of all abbreviated proteins and isozymes are found in Table 87.1.

Subfamily Cervinae

Red Deer (*Cervus elaphus*)

Red deer are native to Europe and are still common there. Several subspecies, some with ranges as far east as the Himalayas, are now rare (Whitehead 1972). European red deer have been introduced to many other parts of the world and are the most commonly farmed deer species in Britain and New Zealand. Extensive electrophoretic studies of European red deer have revealed abundant polymorphism.

Among plasma proteins, transferrin is commonly polymorphic. Most populations have two to four transferrin alleles, except in Norway where no variation was found (Lowe and McDougall 1961; McDougall and Lowe 1968; Kravchenko and Kravchenko 1971; Bergmann 1976; Wegge 1978; Gyllensten et al. 1980, 1983; Herzog 1986; Tate and Dratch 1988; Stratil et al. 1990). Kravchenko and Kravchenko (1971) also reported two other polymorphic serum proteins, Pa and Sd-2, although they were not fully described. Variation has also been reported at EST-1 (Pemberton et al. 1988), α-2-globulin (Tate and Dratch 1988), and vitamin D binding protein GC (Van de Weghe et al. 1982). Most recently, Stratil et al. (1990) have confirmed variation at

TABLE 87.1. Proteins and isozymes studied in deer. N_L denotes the number of loci studied and P_{SP} the number of species in which polymorphism has been reported

Abbreviation	N_L	P_{SP}	Name	Enzyme Commission Number
AAT	2	3	Aspartate aminotransferase	2.6.1.1
ACP	2	4	Acid phosphatase	3.1.3.2
AK	2	2	Adenylate kinase (myokinase)	2.7.4.3
ALB	1	1	Albumin	
ADA	2	1	Adenosine deaminase	3.5.4.4
ADH	1	1	Alcohol dehydrogenase	1.1.1.1
AMY	1	0	Amylase	3.2.1.1
BG	1	0	Alpha-2-B glycoprotein	
CK	4	0	Creatine kinase	2.7.3.2
CAT	1	1	Catalase	1.11.1.6
DIA	1	1	Diaphorase	1.6.2.2
EST	2	1	Esterase	3.1.1.1
FH	1	0	Fumerate hydratase	4.2.1.2
GAPDH	1	0	Glyceraldehyde-3-phosphate dehydrogenase	1.2.1.12
GC	1	1	Vitamin D binding protein	
GD	1	1	Glucose-6-phosphate dehydrogenase	1.1.1.49
GDH	1	3	Glucose dehydrogenase	1.1.1.47
GLO	1	1	Glyoxalase	4.4.1.5
GPI	2	4	Glucose-6-phosphate isomerase	5.3.1.9
GOT (=AAT)	2	3	Glutamate oxaloacetate transaminase	
GPD	2	3	Alpha-glycerophosphate dehydrogenase	1.1.1.8
α-HB	1	1	Alpha-hemoglobin	
β-HB	1	1	Beta-hemoglobin	
IDH (=ICD)	2	4	Isocitrate dehydrogenase	1.1.1.42
LDH	2	6	Lactate dehydrogenase	1.1.1.27
MDH	2	3	Malate dehydrogenase	1.1.1.37
ME (=MOD)	2	4	Malic enzyme	1.1.1.40
MNR	1	1	Menadione reductase	
MPI	2	7	Mannose-6-phosphate isomerase	5.3.1.8
NP	1	1	Nucleoside phosphorylase	2.4.2.1
PA	1	1	Postalbumin	
PEP	3	3	Peptidase	3.4.11
PGM	3	6	Phosphoglucomutase	2.7.5.1
PER	1	1	Peroxidase	1.11.1.7
PGD	1	2	Phosphogluconate dehydrogenase	1.1.1.44
PI	8	1	Protease inhibitors PI1-PI8	
PPB	1	0	Plasma protein-B	
SDH (=IDDH)	1	1	Sorbitol dehydrogenase	1.1.1.14
SOD (=TO)	3	5	Superoxide dismutase	1.15.1.1
TRF	1	5	Transferrin	

most of these loci and have found additional polymorphism at several protease inhibitors (PI-2, PI-3, PI-4, PI-5, and PI-8).

Initially, no hemoglobin variation was detected by electrophoresis (Maughan and Williams 1967) or isoelectric focusing (Lawton and Sutton 1981, 1982), but two more recent studies report extensive and complex hemoglobin variation (Herzog 1986; Pemberton et al. 1988).

In enzyme surveys of many different European populations, polymorphism has been found at ACP-1, ACP-2, one esterase locus, GPI-1, IDH-2, LDH-2, MDH-2, ME-1, MPI, PGM-1, PGM-2, and SOD-1 (Baccus et al. 1983; Gyllensten et al. 1983; Dratch and Gyllensten 1985; Hartl et al. 1986, 1990; Van Wieren 1989). A whey protein in red deer milk is also polymorphic (McDougall and Stewart 1976). Previously reported polymorphism at IDH-2, MPI, and transferrin have been used by Pemberton et al. (1988, 1991) to test whether particular alleles confer selective advantage.

Electrophoretic data have been used to compare red deer with other artiodactyls (Lowe and McDougall 1961; Maughan and Williams 1967; McDougall and Lowe 1968; Lawton and Sutton 1982; Baccus et al. 1983), especially the closely related wapiti (Dratch and Gyllensten 1985; Dratch 1987a). Polymorphism within red deer has also been used in parentage testing of farmed red deer (Tate and Dratch 1988).

Wapiti (*Cervus elaphus*)

The natural range of wapiti is from North America, where they are commonly called elk, to eastern Asia. Present populations are patchily distributed, and many in the United States are the result of introductions from Yellowstone National Park (Bryant and Maser 1982). North American wapiti have also been introduced to other parts of the world including Europe (Lever 1985). Wapiti hybridize with European red deer and hybrids are fertile. Some biologists still contend that elk should be considered a separate species, *C. canadensis*. On New Zealand deer farm hybridization is being used to increase the size of the farmed red deer and optimize carcass weight in relation to breeding female size. Asian wapiti are common on Chinese deer farms. All the studies reported relate to North American wapiti.

No polymorphism has been found in either serum proteins (Johnson 1968; Cameron and Vyse 1978, 1979; McClymont et al. 1982; Baccus et al. 1983) or hemoglobin (Butcher and Hawkey 1977; Cameron and Vyse 1978; Bunch et al. 1976). An enzyme survey of Yellowstone Park wapiti revealed only one polymorphic locus, IDH-2 (Cameron and Vyse 1978, 1979), but subsequent surveys including other populations have found additional variation at MPI-1, PEP-B, PGM-1, and SOD-1 (Baccus et al. 1983; Dratch and Gyllensten 1985). Electrophoretic data have been used to compare wapiti with other large grazing animals (Baccus et al. 1983), especially the closely related red deer, where electrophoresis has been employed to identify hybrids (Dratch and Gyllensten 1985; Dratch 1987a). Cronin (1991b) has compared the mtDNA of elk with that of other cervids.

Sika Deer (*Cervus nippon*)

Many subspecies of sika deer, each native to a particular region or island in Asia have been given subspecies names; some are now very rare (Whitehead 1972). Sika are common on Chinese deer farms and are now being farmed in several other countries as well. Sika have been introduced to many other parts of the world (Whitehead 1972). They hybridize with the much larger red deer and introgression of red deer stocks is a problem (Lowe and Gardiner 1974). All reported electrophoretic studies have examined introduced populations and the results probably do not characterize all variation in the species.

Kravchenko et al. (1971) report transferrin polymorphism in Russian sika deer. In 15 British sika, transferrin and other serum proteins were monomorphic (McDougall and Lowe 1968). No hemoglobin polymorphism was detected electrophoretically (Lowe and McDougall 1961; Maughan and Williams 1967; Taylor and Easley 1977) but sequencing studies indicated considerable alpha chain heterogeneity (Taylor and Easley 1977). Subsequent isoelectric focusing studies have found hemoglobin variation (Butcher and Hawkey 1977; Lawton and Sutton 1982), but it has not been studied in detail. The only survey of enzyme loci in sika found no polymorphism among 19 loci screened in an introduced population in the northeastern United States (Feldhamer et al. 1982).

Père David's Deer (*Elaphurus davidianus*)

The world population of Père David's deer is descended from a small number of animals assembled at Woburn Park, England, at the beginning of this century. These deer were themselves the remnants of a population held for at least 2000 years in a hunting park in Peking (Whitehead 1972).

No polymorphism has been detected in the

plasma proteins (Ryder et al. 1981; Woodruff and Ryder 1986) or the hemoglobin (Butcher and Hawkey 1977; Lawton and Sutton 1981, 1982) of this species. In enzyme surveys, only one locus, a plasma esterase, is reported to be polymorphic (Lawton and Sutton 1982; Woodruff and Ryder 1986). Fixed protein differences have been used to confirm hybridization of Père David's and red deer by artificial insemination (Asher et al. 1988).

European Fallow Deer (*Dama dama*)

European fallow deer are patchily distributed throughout Europe and have been introduced to many other parts of the world (Whitehead 1972). Held in many ornamental deer parks in Europe, fallow are now the most common farmed deer species in Australia and Europe excluding Britain, where as in New Zealand they rank second to red deer. A number of studies have failed to find any polymorphism in transferrin and other serum proteins (Lowe and McDougall 1961; McDougall and Lowe 1968; Gunther and Scheil 1982; Pemberton and Smith 1985; Hartl et al. 1986), though recently transferrin variation has been reported in German fallow (Herzog 1989). Both electrophoresis and isoelectric focusing detected no hemoglobin variation (Maughan and Williams 1967; Lawton and Sutton 1982; Pemberton and Smith 1985; Hartl et al. 1986). In enzyme surveys, no polymorphism was found by Munday et al. (1974), Lawton and Sutton (1982), Gunther and Scheil (1982), or Pemberton and Smith (1985), but Hartl et al. (1986) found one polymorphic system, catalase. Randi and Appollonio (1988) screened 51 loci and found catalase and peroxidase variation, although the genotype frequencies for peroxidase alleles were far from Hardy-Weinberg predictions.

Mesopotamian Fallow Deer (*Dama mesopotamica*)

Native to Iran, Iraq, Israel, Jordan, and Syria (Whitehead 1972), Mesopotamian fallow are now extremely rare. A number are in captivity in Europe, Israel, and New Zealand; the remainder are in reserves in Iran. Farming interest centers on using Mesopotamian bucks to introduce variability into European fallow, with which this species easily hybridizes.

Gunther and Scheil (1982) screened ten serum and red cell loci in four Mesopotamian deer along with European fallow deer. Fixed differences from European fallow were indicated at AMY and GPI, and GLO was polymorphic. Electrophoresis has been used to confirm hybridizations between European and Mesopotamian fallow (Dratch 1987b) as well as to distinguish fallow deer venison from that of other species (Murphy et al. 1988).

Other Cervinae Species

There have been no electrophoretic studies of natural populations of the other Cervinae species. These deer are mostly native to Asia (Whitehead 1972). There is some farming of the Indian axis or chital deer (*Axis axis*) and the Indonesian rusa deer. In comparative studies, hemoglobin has been screened in small samples of zoo animals of Indian barasingha (*Cervus duvauceli*) and Indian hog deer (*Axis porcinus*), as well as chital and rusa, with variation reported for chital (Naik et al. 1964) and barasingha (Lawton and Sutton 1981). Naik et al. (1964) also tested GD in axis and found that the enzyme was NAD dependent in this species. Lawton and Sutton (1982) found no intraspecific polymorphism at the SOD-1 locus in barasingha, rusa, or axis deer. No electrophoretic data have been reported for tufted deer (*Elaphodus cephalophus*), Duhl's or Bawean deer (*Axis kuhlii*), Calamian deer (*Axis calamianensis*), sambar (*Cervus unicolor*), Eld's deer (*Cervus eldi*), or Thorold's deer (*Cervus albirostris*). These are species of India, Southern Asia, China, and Indonesia.

Subfamily Odocoileinae

White-Tailed Deer (*Odocoileus virginianus*)

White-tailed deer range from Canada to northern South America and are the most common deer species in the United States (Whitehead 1972). Many different subspecies have been named. Extensive electrophoretic studies have been carried out on this species and, on a multilocus basis, it is the most polymorphic deer species so far studied. The earliest work on serum protein variation was by Cowan and Johnston (1962) and Van Tets and Cowan (1966), but since protein bands were not identified, comparison with later work is not possible. In subsequent studies, transferrin has usually proven polymorphic; seven alleles were detected by Quinteros and Miller (1969), three in studies from South Carolina (Manlove et al. 1975; Ramsey et al. 1979; Breshears et al. 1988) and two by Miller et al. (1965) and LeResche et al. (1974). Though finding no variation in transferrin, Seal and Erickson (1969) reported polymorphism in an unnamed fetal serum protein. Unlike transferrin, albumin variants are rare or absent in white-tailed deer (Seal and Erickson 1969; Manlove et al. 1976; McClymont et al. 1982; Baccus et al. 1983; Smith et al. 1984, 1986; Sheffield et al. 1985). Rare, naturally occurring hybrids between white-tailed deer and mule deer can be detected by their heterozygous albumin pattern (McClymont et al. 1982; Carr et al. 1986). The albumin difference between white-tailed deer and mule deer can be electrophoretically detected in fecal material (Scribner and Warren 1984).

White-tailed deer hemoglobin is highly polymorphic and has been the subject of intensive research, initially because of the observation that red blood cells in this species assume sickle-cell shapes (Weisberger 1964). Several studies have aimed to relate sickling to hemoglobin type. Kitchen et al. (1964, 1966, 1967) found as many as six adult hemoglobin types and two fetal hemoglobins in Florida white-tails. Huisman et al. (1968) studying penned deer, described 17 phenotypes, the result of seven β-hemoglobin alleles; Huisman later described the structural basis of alpha chain hemoglobins in white-tails (1974). Other studies have found hemoglobin polymorphism in white-tail populations from Iowa (Miller et al. 1965), South Carolina (Manlove et al. 1975; Ramsey et al. 1979; Chesser et al. 1982a; Breshears et al. 1988), and Minnesota (Seal and Erickson 1969). In allozyme surveys, populations from several areas in the species' range have been surveyed and polymorphism has been detected at 71% of loci. The following number of alleles have been detected for the protein loci listed: four alleles, EST-2, PGM-2, SDH; three alleles, AAT-1, GPD-2, LDH-1, MDH-1, MPI, PEP-1, PGM-1; two alleles AAT-2, α-HB, ACP, AK-2, ADH, EST-4, GD, GDH, LDH-2, ME-1, ME-2, PEP-2, PEP-3, PGD, and SOD (Manlove et al. 1975; Johns et al. 1977; Price et al. 1979; Ramsey et al. 1979; Baccus et al. 1983; Smith et al. 1984, 1986; Sheffield et al. 1985; Kennedy et al. 1987; Maffei and Woolf 1987; Breshears et al. 1988). Mitochondrial DNA variation has been found in white-tailed deer from Montana (Cronin et al. 1988) and Minnesota (Cronin et al. 1991b), and used in combination with isozymes to examine spatial heterogeneity. A number of studies of white-tailed deer have tested whether selection is acting on electrophoretic loci (Baccus et al. 1977; Chesser et al. 1982a), as well as whether multilocus heterozygosity is correlated with fitness-related characteristics (Johns et al. 1977; Smith et al. 1983; Cothran et al. 1983; Chesser and Smith 1987)

Mule Deer/Black-Tailed Deer (*Odocoileus hemionus*)

Mule deer are native to the western United States, except along the coast from California to Alaska, where they are replaced by conspecific black-tailed deer (Wallmo 1981). Early studies by Cowan and Johnston (1962) and Van Tets and Cowan (1966) suggested serum protein polymorphism but did not identify the proteins involved. More recently, McClymont et al. (1982) found no albumin variation in 100 mule deer screened. Differences in albumin phenotypes have been used to differentiate mule deer from white-tailed deer as well as from naturally occurring hybrids (McClymont et al. 1982; Scribner and Warren 1984; Stubblefield et al. 1986; Cronin et al. 1988). Bunch et al. (1976) found evidence of two hemoglobin alleles in 36 individuals. Isozyme surveys have screened 34 loci and reported polymorphism at ADA, ME-2, MNR, NP, PEP-2, and PGD (Baccus et al. 1983; Smith et al. 1990).

Cameron and Vyse (1979) compared mule deer with elk and white-tails, and Baccus et al. (1983) included mule deer in their interspecific comparison of large grazing mammals. Studies of mule deer mitochondrial DNA (Carr et al. 1986; Cronin et al. 1988) have been used to confirm hybridization with white-tails in Texas and Montana. Cronin (1991a) has found a discordance in the mitochondrial and nuclear genetic relationships in black-tailed deer and mule deer, with the latter having mtDNA haplotypes more similar to those in white-tailed deer.

Reindeer/Caribou (*Rangifer tarandus*)

This species has a circumpolar distribution, and many subspecies have been described. Subspecies native to Alaska and Canada are known as caribou, and those occurring elsewhere are called reindeer. Reindeer have been introduced to many arctic and antarctic regions from which they were absent (Whitehead 1972). Besides large wild populations, reindeer are herded by arctic tribes and are the most numerous domesticated deer species.

Reindeer transferrin is the most polymorphic deer locus that has been described. More recent studies have undoubtedly used gels of superior resolving power (e.g., polyacrylamide) than earlier ones. In a major series of papers, covering reindeer in Scandinavia, Svalbard (Spitzbergen), Alaska, and Canada, Røed and coworkers have identified 25 different transferrin alleles (Røed 1985a, 1985b, 1985c; Røed et al. 1985, 1986; Røed and Whitten 1986; Røed et al. 1987). No other polymorphic serum proteins were found by Gahne and Rendel (1961), Braend (1964a), or Storset et al. (1978), but Nadler et al. (1967) noticed variation in an unnamed serum protein.

Polymorphism has been described in haptoglobin and a second general protein (Røed 1985a, 1986), although none of the studies that have analyzed albumin have found polymorphism (Gahne and Rendel 1961; Nadler et al. 1967; Storset et al. 1978; Røed 1985a, 1985c, 1986). Baccus et al. (1983) found a fixed mobility difference between reindeer and caribou albumin. No hemoglobin polymorphism has been found, either using electrophoresis (Gahne and Rendel 1961; Storset et al. 1978; Røed 1985a, 1985c, 1986) or isoelectric focusing (Butcher and Hawkey 1977; Lawton and Sutton 1981, 1982). In enzyme surveys, polymorphism has been found at AAT-2, ACP-2, EST, GDH, GPI-2, LDH-2, MPI-1, MPI-2, SOD-2, and SOD-4 (Storset et al. 1978; Baccus et al. 1983; Røed 1985a, 1985c, 1986). Baccus et al. (1983) also note fixed differences between reindeer and caribou at GPI-1 and IDH-2.

In some studies, reindeer have been screened for comparison with other artiodactyls (Butcher and Hawkey 1977; Lawton and Sutton 1981; Baccus et al. 1983; Cronin 1991a), but in the majority of cases intraspecific variation, especially in transferrin, has been used to reveal population differentiation (Braend 1964b; Zhurkevich and Fomicheva 1976; Storset et al. 1978; Soldal and Staaland 1980; Røed 1983, 1985a, 1985b, 1985c, 1986; Røed et al. 1985, 1986; Røed and Whitten 1986). Many interesting points have emerged from these studies. A number of studies have reported that Svalbard reindeer have just two transferrin alleles, which are different from all alleles present in Scandinavia, the nearest mainland (Storset et al. 1978; Soldal and Staaland 1980; Røed 1985c).

Røed et al. (1986) and Røed and Whitten (1986) have shown that one of the Svalbard alleles exists in Alaskan and Canadian native carbou, implying that the Svalbard reindeer originated in the Canadian arctic rather than from Europe. Transferrin variation in the introduced reindeer in Iceland provides a good example of founder effect when compared with variation in the Norwegian parent herds (Røed et al. 1985). There are consistent genetic differences between wild and domestic reindeer herds in Scandinavia (Røed 1983, 1985b, 1986). Røed (1987) has also reported an association between transferrin genotype and body size in domesticated reindeer.

Moose (*Alces alces*)

Moose range through Canada, Alaska, Russia, and Northern Europe, where they are sometimes known as European elk. Most studies of serum proteins, in both European and North American moose, have failed to find any polymorphism (Braend 1962; Nadler et al. 1967; Shubin 1969; Seal and Karns as reported in LeResche et al. 1974; Wilhelmson et al. 1978; McClymont et al. 1982). However, Gyllensten et al. (1980) found a variant transferrin allele segregating at low frequency in a large sample of Scandinavian moose. No polymorphism has been found in hemoglobin, either by electrophoresis (Wilhelmson et al. 1978) or by isoelectric focusing (Butcher and Hawkey 1977; Lawton and Sutton 1982). In enzyme surveys, polymorphisms have been reported at GPI-1, MDH-2, MPI-2, PEP-B, and PGM-1 (Ryman et al. 1977, 1980; Wilhelmson et al. 1978; Chesser et al. 1982a, b; Baccus et al. 1983). All of these polymorphisms were found in European moose; other areas of the species' range are much less well studied to date. Electrophoretic and molecular data on moose have been used in interspecific comparisons (Butcher and Hawkey 1977; Lawton and Sutton 1981; Baccus et al. 1983; Cronin 1991a) and to demonstrate population differentiation (Ryman et al. 1977, 1980; Chesser et al. 1982b).

Roe Deer (*Capreolus capreolus*)

This species is distributed throughout Europe and in Asia from the Soviet Union to China (Whitehead 1972). Despite its vast distribution, there are relatively few electrophoretic studies of roe deer. No transferrin or albumin variation has been detected in roe deer (Gyllensten et al. 1983; Baccus et al. 1983). Hemoglobin appeared monomorphic on both electrophoresis (Maughan and Williams 1967) and isoelectric focusing (Lawton and Sutton 1981, 1982). In the early enzyme surveys, roe deer were compared with other artiodactyls. Polymorphism is reported at MPI-2 and PEP-2 (Lawton and Sutton 1981; Baccus et al. 1983). Hartl and Reimoser (1988) studied five Austrian roe populations and at MPI and PEP 2 found seven additional polymorphic loci: ACP-2, AK-2, DIA-2, LDH-2, ME- 2, PGM-1, and PGM-2.

Pampas Deer (*Ozotoceros bezarticus*)

Pampas deer live in the plains regions of Southern America. In a sample of 36 individuals, Quinteros et al. (1971) found 17 different transferrin phenotypes, and although the number of alleles involved was not indicated, this locus is evidently highly polymorphic.

Red Brocket (*Mazama americana*)

In the only electrophoretic study of this South American species, Smith et al. (1986) screened 52 individuals across 19 enzyme loci and found that AAT-1, AAT-2, GDH, GPD-1, β-HB, IDH-2, LDH-1, LDH-2, MPI, PEP-B, and PGM-1 were polymorphic. Cronin (1991a) also compared the mitochondrial DNA of red brocket with other Cervidae.

Other Odocoilinae Species

Smith et al. (1986) compared a single specimen of the South American brown brocket deer (*Mazama gouazoubira*) with other deer species at 19 loci. It appeared to be heterozygous at LDH-2, MPI, and PGM-1. There have been no reports of electrophoretic studies of the seven remaining Odocoileinae species, the little red brocket (*Mazama chunyi*), huemals (*Hippocamelus bisulcus* and *H. antisensis*), marsh deer (*Blastocerus dichotomus*), and pudu (*Pudu pudu* and *P. amphistophiles*), all of which are native to South America.

Subfamily Hydropotinae

Chinese Water Deer (*Hydropotes inermis*)

Chinese water deer are widespread in China and Korea, but both of the published genetic studies examine the introduced British population of this species (Whitehead 1972). Three of 21 deer screened by electrophoresis had a variant hemoglobin type (Maughan and Williams 1967; Maughan 1969). A single specimen was compared with other artiodactyls for isoelectric point differences in hemoglobin and SOD-1 (Lawton and Sutton 1981, 1982).

Reeves's Muntjac (*Muntiacus reevesii*)

Although again a species native to China, both studies of Reeves muntjac relate to the population introduced to Britain. In a study of 40 feral animals, Maughan and Williams (1967) detected two hemoglobin alleles and also noted that the hemoglobin was sickled (Maughan 1969). Hemoglobin and superoxide dismutase from two Reeves's muntjac were compared with other artiodactyls by isoelectric focusing (Lawton and Sutton 1982).

Other Muntiacinae Species

All muntjac species are native to various parts of India, China, and Southeast Asia (Whitehead 1972). Hemoglobin and SOD-1 from one Indian muntjac (*Muntiacus muntjak*) were included in the Lawton and Sutton (1982) isoelectric focusing survey. Hemoglobin from a muntjac of unidentified species was also screened by Naik et al. (1964). There are no studies relating to Roosevelt's muntjac (*M. rooseveltorum*), the black muntjac (*M. crinifrons*), or Fea's muntjac (*M. feae*).

Acknowledgements

We would like to acknowledge the encouragement of Bill Mutch, Robert Smith, and Michael Tate in early stages of this review and thank Stephen Busack, John Linnell, and Michael Smith for suggestions which improved the final manuscript.

References

Allendorf FW, Utter FM (1979) Population genetics. In Hoar WS, Randall DJ, Brett JR (eds) Fish Physiology, 8. Academic Press, New York, pp 407–454

Asher GW, Adam JL, Otway W, Bowmar P, van Reenan G, Mackintosh CG, Dratch P (1988) Hybridization of Pere David's deer (*Elaphurus davidianus*) and red deer (*Cervus elaphus*) by artificial insemination. J Zool London 215:197–203

Baccus R, Ryman N, Smith MH, Reuterwall C, Cameron D (1983) Genetic variability and differentiation of large grazing mammals. J Mammal 64:109–20

Baccus R, Hillstead H, Johns P, Manlove M, Marchinton RL, Smith MH (1977) Prenatal selection in white-tail deer. Proc Annu Conf SE Assoc Fish, Wildl Agen 31:173–179

Bergmann F Von (1976) Contributions to the study of infrastructures of red deer. Part II First investigations of the elucidation of genetic structure of red deer populations employing protein Z. Judgwiss 22:28–35 (in German)

Braend M (1964a) Polymorphism in the serum proteins of reindeer. Nature 203:764

Braend M (1964b) Genetic studies on serum transferrins in reindeer. Hereditas 52:181–188

Braend M (1962) Studies on blood and serum groups in the elk (*Alces alces*). Ann N Y Acad Sci 97: 296–305

Breshears DD, Smith MH, Cothran EG, Johns PE (1988) Genetic variability in white-tailed deer. Heredity 60:139–146

Bryant LD, Maser C (1982) Classification and distribution. In Thomas JW, Toweill DE (eds) The Elk of North America: Ecology and Management. Stackpole Books, Harrisburg, P., pp 1–60

Bunch TD, Meadows RW, Foote WC, Egbert LN, Spillett JJ (1976) Identification of ungulate hemoglobins for law enforcement. J Wildlife Management 40:517–22

Butcher PD, Hawkey CM (1977) A comparative study of haemoglobins from the Artiodactyla by isoelectric focusing. Comp Bioch Physiol 56B:335–339

Cameron DG, Vyse ER (1979) Comparative genetics of elk and other cervidae of the Yellowstone Park region. In Boyce MS, Hayden-Wing LD (eds) North American Elk: Ecology, Behavior and Management. University of Wyoming Press, Laramie, pp 34–37

Cameron DG, Vyse ER (1978) Heterozygosity in Yellowstone Park elk, *Cervus canadensis*. Biochem Gen 16:651–657

Carr SM, Ballinger SW, Derr JN, Blankenship LH, Bickham JW (1986) Mitochondrial DNA analysis of hybridization between sympatric white-tailed deer and mule deer in West Texas. Proc Natl Acad Sci, USA 83:9576–9580

Chesser RK, Smith MH (1987) Relationship of genetic variation to growth and reproduction in white-tailed deer. In Wemmer CM (ed) Biology and Management of the Cervidae. Smithsonian Institution Press, Washington, D.C., pp 168–177

Chesser RK, Smith MH, Johns PE, Manlove, MN, Straney DO, Baccus, R (1982a) Spatial, temporal and age-dependent heterozygosity of β-hemoglobin in white-tailed deer. J Wildlife Management 46:983–490

Chesser RK, Reuterwall C, Ryman N (1982b) Genetic differentiation of Scandinavian moose *Alces alces* populations over short geographical distances. Oikos 39:125–130

Cothran EG, Chesser RK, Smith MH, Johns PE (1983) Influences of genetic variability and maternal factors on fetal growth in white-tailed deer. Evolution 37:282–291

Cowan IMcT, Johnston PA (1962) Blood serum protein variations at the species and subspecies level in deer of the genus Odocoileus. Systematic Zool 11:131–138

Cronin MA (1991a) Mitochondrial and nuclear relationships of white-tailed deer, mule deer and black-tailed deer (*Odocoileus*). Can J Zool (in press)

Cronin MA (1991b) Mitochondrial DNA phylogeny of deer (Cervidae). J Mammal (in press)

Cronin MA, Palmisciano DA, Vyse ER, Cameron DG (1991a) Mitochondrial DNA in wildlife forensic science: species identification of tissues. Wildl Soc Bull (in press)

Cronin MA, Nelson ME, Pac DF (1991b) Spatial

heterogeneity of mitochondrial DNA and allozymes among populations of white-tailed and mule deer. J Hered (in press)

Cronin MA, Vyse ER, Cameron DG (1988) Genetic relationships between mule deer and white-tailed deer in Montana. J Wildlife Management 52:320–328

Dilworth TG, McKenzie JA (1970) Attempts to identify meat from game animals by starch gel electrophoresis. J Wildlife Management 34:917–921

Dratch PA (1987a) A marker for red deer-wapiti hybrids. Proc N Z Soc Anim Prod 46:179–182

Dratch PA (1987b) Deer blood-typing: applications and developments. Proc of a Deer Course of Veterinarians 4:28–32

Dratch PA (1987c) The Fiordland wapiti: their history in their genes. Deer Farmer 40:35–37

Dratch PA, Gyllensten U (1985) Genetic differentiation of red deer and North American elk (wapiti). In Fennessy PF, Drew KR (eds) The Biology of Deer Production. R Soc N Z Wellington Bull 22:37–40

Feldhamer GA, Morgan RP, McKeown PE, Chapman JA (1982) Lack of polymorphism in liver and muscle enzymes from sika deer (*Cervus nippon*). J Mammal 63:512–514

Gahne B, Rendel J (1961) Blood and serum groups in reindeer compared with those in cattle. Nature 192:529–530

Gavin TA, May B (1988) Taxonomic status and genetic purity of Columbian white-tailed deer. J Wildlife Management 52:1–10

Gunther A, Scheil H-G (1982) Serologische Untersuchungen beim Europäischen und beim Mesopotamishen Damhirsch. 56 Hauptversammlung der Deutschen Gesellschaft fur Säugetierkunde, Salzburg

Gyllensten U, Ryman N, Reuterwall C, Dratch P (1983) Genetic differentiation in four European subspecies of red deer (*Cervus elaphus* L.). Heredity 51:561–580

Gyllensten U, Ryman N, Reuterwall C (1982) Allozyme differentiation within and between subspecies of red deer and the management of red deer genetic resources in Sweden. Naturvardsverkert Rapport snv pm 1525 (in Swedish)

Gyllensten U, Reuterwall C, Ryman N, Stahl G (1980) Geographical variation of transferrin allele frequencies in three deer species from Scandinavia Hereditas 92:237–241

Harrington R (1979) The hybridization of red deer (*Cervus elaphus* L.) and Japanese sika deer (*C. nippon nippon* Temminck, 1838). Proc 14th Intl Congress Game Biologists, Irish Wildl Pubs, pp 559–571

Harrington R (1985) Evolution and distribution of the Cervidae. In Fennessy PF, Drew KR (eds) The Biology of Deer Production. R Soc N Z Wellington Bull 22:3–11

Harris MJ, Huisman THJ, Hayes FA (1973) Geographic distribution of hemoglobin variants in the white-tailed deer. J Mammal 54:270–274

Hartl GB (1986) Genetische Variabilität beim Rotwild-Answirkungen anthropogener Einflüsse auf den Genpool von Wildtierpopulation. CIC-Rotwildtagung, Graz, pp 423–431

Hartl GB, Reimoser F (1988) Biochemical variation in rue deer (*Capreolus capreolus*): are r-strategists among deer genetically less variable than K-strategists? Heredity 60:221–227

Hartl GB, Willing R, Lang G, Klein F, Koller J (1990) Genetic variability and differentiation in red deer (*Cervus elaphus* L) of central Europe. Genet Sel Evol 22:289–306

Hartl GB, Schleger A, Slowak M (1986) Genetic variability in fallow deer. *Dama dama* L. Anim Gen 17:335–341

Herzog S (1989) Genetic polymorphism of transferrin in fallow deer, *Cervus dama* L. Anim Genetics 20:421–426

Herzog S (1988) Polymorphism and genetic control of erythrocyte 6-phosphogluconate dehydrogenase in the genus. Anim Genetics 19:291–294

Herzog S (1986) Biochemisch-genetische Unterschungen and Rotwild (*Cervus elaphus* L.) CIC-Rotwildtagung, Graz pp 432–433

Hindmarsh M, Schuster NM (1979) Forensic techniques for the identification of venison and other meats to establish the species of origin Department of Agriculture, Victoria. Australia Agric Note Series 15, Symp on Deer Farming in Victoria, pp 259–263

Hoyem T, Thorson B (1970) Myoglobin electrophoretic patterns in identification of meat from different animal species. J Agric Food Chem 18:737–739

Huisman THJ (1974) Structural aspects of fetal and adult hemoglobins from nonanemic ruminants. Ann N Y Acad Sci 241:392–410

Huisman THJ, Dozy AM, Blunt MH (1968) The hemoglobin heterogeneity of the Virginia white-tailed deer: a possible genetic explanation. Arch Biochem Biophys 127:711–717

Johns PE, Baccus R, Manlove M, Pinder J, Smith MH (1977) Reproductive patterns, productivity and genetic variability in adjacent white-tailed deer populations. Proc Annu Conf SE Assa Fish, Wildl Agen 31:167–172

Johnson ML (1968) Application of blood protein electrophoretic studies to problems of mammalian taxonomy: Syst Zool 17:23–30

Karlin AA, Heidt GA, Sugg DW (1989) Genetic

variation and heterozygosity in white-tailed deer in Arkansas. Am Mid Nat 121:273–284

Kennedy PK, Kennedy ML, Beck ML (1987) Genetic variability in white-tailed deer (*Odocoileus virginianus*) and its relationship to environmental parameters and herd origin. Genetica 74:189–201

King JL, Jukes TH (1969) Non-Darwinian evolution. Science 164:788–798

Kimura M (1983) The Neutral Theory of Molecular Evolution. Cambridge University Press, Cambridge, England

Kitchen H, Putnam FW, Taylow WJ (1967) Hemoglobin polymorphism in white-tailed deer: subunit basis. Blood 29:867–877

Kitchen H, Putnam FW, Taylor WJ (1966) The structural basis for the polymorphic hemoglobins of white-tailed deer associated with sickled and nonsickled erythrocytes. Int Symp Comp Hemoglobin Structure, Thessaloniki, 11–13 April 1986, pp 73–82

Kitchen H, Putnam FW, Taylor WJ (1964) Hemoglobin polymorphism: its relation to sickling of erythrocytes in white-tailed deer. Science 144:1237–1239

Kravchenko DN, Kravchenko RS (1971) Polymorphous systems of blood serum proteins in *Cervus elaphus* L. Cytology and Genetics 4:311–315 (in Russian)

Kravchenko RS, Galkin VS, Kravchenko DN (1971) Polymorphic proteins in the blood plasma of spotted deer. Collected Scientific Work of the Laboratory of Deer Keeping, 3rd ed. Gorno-Altaisk, p 131

Lawton ME, Sutton JG (1982) Species identification of deer blood by isoelectric focusing. J Forensic Sci 22:361–366

Lawton ME, Sutton JG (1981) Species identification of deer blood by isoelectric focusing. Deer 5:285–288

Leberg PL, Smith MH (1990) The association between heterozygosity and growth of deer fetuses is not explained by effects of the loci examined. Evolution 44:454–458

LeResche RE, Seal US, Karns PD, Franzmann AW (1974) A review of blood chemistry of moose and other Cervidae with emphasis on nutritional assessment. Naturaliste Canadien 101:263–290

Lever C (1985) Naturalized Mammals of the World. Longman, New York

Lima-de-Faria A, Arnason U, Widegren B, Isaksson M, Essen-Moller J, Jaworska H (1986) DNA cloning and hybridization in deer species supporting the chromosome field theory. Biosystems 19:185–212

Lima-de-Faria A, Arnason U, Widegren B, Essen-Moller J, Isaksson M, Olsson E, Jaworska O (1984) Conservation of repetitive DNA sequences in deer species studied by Southern blot transfer. J Molec Evol 20:17–24

Lowe VPW, Gardiner AS (1974) A re-examination of the subspecies of red deer (*Cervus elaphus*) with particular regard to stocks in Britain. J Zool London 174:185–201

Lowe VPW, McDougall EI (1961) Serum B-globulin types in red deer and other species and their stability in the presence of bacteria. Nature 192:983–984

Maffei MD, Woolf A (1987) The influence of genic heterozygosity on the potential productivity of white-tailed deer. Trans Ill Acad Sci 80:309–316

Manlove MN, Smith MH, Hillestad HO, Fuller SE, Johns PE, Straney DO (1976) Genetic subdivision in a herd of white-tailed deer as demonstrated by spatial shifts in gene frequencies. Proc 30th Annu Conf SE Game Fish Comm 30:487–492

Manlove MN, Avise JC, Hillestad HO, Ramsey PR, Smith MH, Straney DO (1975) Starch gel electrophoresis for the study of population genetics in white-tailed deer. Proc 29th Annu Conf SE Game Fish Comm 29:392–403

Mardini A (1984) Species identification of tissues of selected mammals by agarose gel electrophoresis. Wildl Soc Bull 12:149–251

Maughan E (1969) Further studies on the haemoglobin types in deer. Deer 1:314

Maughan E, Williams JRB (1967) Haemoglobin types in deer. Nature 215:404–450

McClymont RA, Fenton M, Thompson JR (1982) Identification of cervid tissues and hybridization by serum albumin. J Wildlife Management 46:540–544

McCormick RJ, Moore TD, Field RA, Hepworth WG (1988) Identification of deer meat in cooked products. Wildl Soc Bull 16:433–436

McDougall EI, Lowe VPW (1968) Transferrin polymorphism and serum proteins of some British deer. J Zool London 155:131–140

McDougall EI, Stewart JC (1976) The whey proteins of the milk red deer (*Cervus elaphus* L.) A homologue of bovine B-lactoglobulin. Biochemical J 153:647–55

Miller W, Haugen AO, Roslien DJ (1965) Natural variation in the blood proteins of white-tailed deer. J Wildlife Management 29:717–723

Miyamoto MM, Kraus F, Ryder OA (1990) Phylogeny and evolution of antlered deer determined from mitochondrial sequences. Proc Natl Acad Sci USA 87:6127–6131

Munday BL, Corbould A, Goodsall G (1974) Identification of muscle tissue from mammals. J Wildlife Management 38:884–886

Murie OJ (1951) Hybridization. In Poole AL (ed) Preliminary Reports of The New Zealand-American Fiordland Expedition. N Z Dept Sci Ind Res Bull 103:26

Murphy BR, Smith LM, Lee DJ, Terre DR (1988) Forensic discrimination of exotic and native ungulates. Wildl Soc Bull 16:24–27

Nadler CF, Hughes GE, Harris KE, Nadler NW (1967) Electrophoresis of the serum protein of transferrins of *Alces alces* (elk), *Rangifer tarandus* (reindeer), and *Ovis dalli* (Dall sheep) from North America. Comp Biochem Physiol 23:149–157

Naik SN, Bhatia HM, Baxi AJ, Naik PV (1964) Haematological study of Indian spotted deer (Axis deer). J Expt Zool 155:231–236

Nevo E, Beiles A, Ben-Schlomo R (1984) The evolutionary significance of genetic diversity: ecological, demographic and life history correlates, In Mann GS (ed) Evolutionary Dynamics of Genetic Diversity. Springer-Verlag, Berlin, pp 13–21

Oates DW, Dent NL, Pearson K (1979a) Differentiation of white-tailed and muie deer tissue by isoelectric focusing. Wildl Soc Bull 7:113–116

Oates DW, Pearson K, Dent N (1979b) Differentiation of white-tailed and mule deer blood and tissue by isoelectric-focusing. Nebraska Game and Parks Commission, Lincoln, Nebraska

Pemberton JM (1988) Genetics of deer. Publ Vet Deer Assoc 3:3–20

Pemberton JM (1983) Blood grouping of deer and its potential in deer management. Some Research on Deer and Its Potential in Deer Management, Symp Br Deer Soc and the Nature Conservancy Council, London, pp 31–34

Pemberton JM, Smith RH (1985) Lack of biochemical polymorphism in British fallow deer. Heredity 55:199–207

Pemberton JM, Albon SD, Guinness FE, Clutton-Brock TH (1991) Countervailing selection in different fitness components in female red deer. Evolution 45:93–103

Pemberton JM, Albon SD, Guinness FE, Clutton-Brock TH, Berry RJ (1988) Genetic variation and juvenile survival in red deer. Evolution 42:921–934

Pex JO, Wolfe JR (1985) Phenotyping phosphoglucose isomerase in the West coast cervids for species identification and individualization. J Forensic Sci 30:114–118

Price PK, Cartwright M, Rogers MJ (1979) Genic variation in white-tailed deer from Arkansas. Proc Arkansas Acad Sci 33:64–66

Quinteros IR, Miller WJ (1969) New transferrin phenotypes of white-tailed deer. Analecta Veterinaria 1:93–98 (in Spanish)

Quinteros IR, Muller AO, Miller WJ, Bischoff JR (1971) Transferrin phenotypes of Argentine deer. (*Ozotoceros bezoarticus* Celer) Analecta Veterinaria 3:107–114 (in Spanish)

Ramsey PR, Avise JC, Smith MH, Urbston DF (1979) Biochemical variationand genetic heterogeneity in South Carolina deer populations. J Wildlife Management 43:136–142

Randi F, Appollonio M (1988) Low biochemical variability in European fallow deer (*Dama dama* L.): natural bottlenecks and the effects of domestication. Heredity 61:405–410

Røed KH (1987) Transferrin variation and body size in reindeer. *Rangifer tarandus* L. Hereditas 106:67–71

Røed KH (1986) Genetic variability in Norwegian wild reindeer. (*Rangifer tarandus* L.) Hereditas 104:293–298

Røed KH (1985a) Genetic variability in Norwegian semi-domestic reindeer. (*Rangifer tarandus* L.) Hereditas 102:177–184

Røed KH (1985b) Genetic differences at the transferrin locus in Norwegian semi-domestic and wild reindeer. (*Rangifer tarandus* L.) Hereditas 102:199–206

Røed KH (1985c) Comparison of the genetic variation in Svalbard and Norwegian reindeer. Can J Zool 63:2038–2042

Røed KH (1983) Enzyme polymorphism in one wild and two semi-domestic reindeer (*Rangifer tarandus* L.) herds of southern Norway. Acta Zoologica Fennica 175:81–83

Røed KH, Whitten KR (1986) Transferrin variation and evolution of Alaskan reindeer and caribou. *Rangifer tarandus* L. Rangiger (Special Issue 1):247–251

Røed KH, Mossing T, Nieminen M, Rydberg A (1987) Transferrin variation and genetic structure of reindeer populations in Scandinavia. Rangiger 7:12–21

Røed KH, Staaland H, Broughton E, Thomas DC (1986) Transferrin variation in caribou (*Rangifer tarandus* L.) on the Canadian Arctic islands. Can J Zool 64:94–98

Røed KH, Soldal AV, Thórisson S (1985) Transferrin variability and founder effect in Iceland reindeer. *Rangifer tarandus* L. Hereditas 103:161–164

Ryder OA, Brisban PC, Bowling AT, Wedemeyer EA (1981) Monitoring genetic variation in endangered species. In Scudder GGE, Reveal JL (eds) Evolution Today. Proc 2nd Intl Congr of Systematics and Evol Biol, Carnegie-Mellon Univ, Pittsburgh, pp 417–424

Ryman N, Baccus R, Reuterwall C, Smith MH (1981) Effective population size and loss of genetic variability in moose and white-tailed deer under different hunting regimes. Oikos 36:257–266

Ryman N, Reuterwall C, Nygren K, Nygren T (1980) Genetic variation and differentiation in Scandinavian moose (*Alces alces*): Are large mammals monomorphic? Evolution 34:1037–1049

Ryman N, Beckman G, Brun-Petersen G, Reuterwall C (1977) Variability of red cell enzymes and genetic implications of management policies in Scandinavian moose. (*Alces alces*). Hereditas 85:157–162

Scherton H, Arnason U, Lima-de-Faria A (1987) The chromosome field theory tested in muntjac species by cloning and hybridization. Hereditas 107:175–184

Scribner KT, Smith MH (1991) Genetic variability and antler development. In Bubenik Ga, Bubenik AB (eds) Horns, Pronghorns and Antlers. Evolution, Morphology and Social Significance of Cranial Appendages in Ruminants. Springer-Verlag, NY, pp 460–473

Scribner KT, Warren RJ (1984) Electrophoretic identification of white-tailed and mule deer feces: a preliminary assessment. J Wildlife Management 48:656–865

Scribner KT, Smith MH, Johns PE (1989) Environmental and genetic components of growth in white-tailed deer. J Mammal 70:284–291

Scribner KT, Wooten MC, Smith MH, Johns PE (1985) Demographic and genetic characteristics of white-tailed deer populations subjected to different harvest methods. In Beasom SL, Robertson SF (eds) Proc Symp on Game Harvest Management. Caesar Kleberg, Wildlife Research Institute, Kingsville, Texas, pp 197–212

Scribner KT, Smith MH, Johns PE (1984) Age, condition and genetic effects of the incidence of spiked bucks. Proc Annu Conf SE Game Fish Commun 38:23–32

Seal US, Erickson AW (1969) Hematology, blood chemistry and protein polymorphisms in the white-tailed deer (*Odocoileus virginianus*). Comp Biochem Physiol 30:695–713

Selander RK, Kaufman DW (1973) Genic variability and strategies of adaptation in animals. Proc Natl Acad Sci USA 70:1875–1877

Sheffield SR, Morgan RP, Feldhamer GA, Harman DH (1985) Genetic variation in white-tailed deer (*Odocoileus virginianus*) populations in Western Maryland. J Mammal 66:243–255

Shubin PN (1969) Genetics of transferrin of reindeer and European elk. Genetika 5:37 (in Russian) Translations in Soviet Genetics 5:26–29

Shubin PN, Turubanov MN (1976) Transferrin polymorphism in northern deer. Doklady Akademii Nauk SSSR 193:697–699 (in Russian) Translations in Soviet Genetics 12:448–450

Smith MH, Willis KB, Johns PE (1991b) Spatial-genetic variation in a white-tailed deer herd. Proc 19th Cong Int Union Game Biol, Trondheim, Norway (in press)

Smith MH, Scribner KT, Johns PE, Rhodes OE (1991a) Genetics and antler development. Proc 18th Cong Int Union Game Biol, Krakow, Poland (in press)

Smith MH, Scribner KT, Carpenter LH, Garrott RA (1990) Genetic characteristics of Colorado mule deer (*Odocoileus hemionus*) and comparisons with other cervids. Southwest Natural 35:1–8

Smith MH, Branan WV, Marchinton RL, Johns PE, Wooten MC (1986) Genetic and morphologic comparisons of red brocket, brown brocket and white-tailed deer. J Mammal 67:103–111

Smith MH, Baccus R, Hillestad HO, Manlove MN (1984) Population genetics. In Halls LK (ed) White-Tailed Deer: Ecology and Management. Stackpole Books, Harrisburg, PA, pp 119–128

Smith MH, Chesser RK, Cothran EG, Johns PE (1983) Genetic variability and antler growth in a natural population of white-tailed deer. In Brown RD (ed) Antler Development in Cervidae. Caesar Kleberg Wildlife Research Institute, Kingsville, Texas, pp 365–387

Smith MH, Hillestad HO, Manlove MN, Marchinton RL (1976) Use of population genetics data for the management of fish and wildlife populations. In Sabol K (ed) Transactions 41st N Am Wildl Nat Res Conf, Washington D.C., pp 119–123

Soldal AV, Staaland H (1980) Genetic variation in Norwegian reindeer. In Reimers E, Gaare E, Skjenneberg S (eds) Proc 2nd Int Reindeer/Caribou Symp, Røros, Norway, 1979, Direktoratet for vilt og ferskvannfisk, Trondheim, pp 396–402

Storset A, Olaisen B, Wika M, Argmov BJ (1978) Genetic markers in the Spitzbergen reindeer. Hereditas 88:113–115

Stratil A, Glasnák V, Bobák P, Čížová D, Gábrišová E, Kaláb P (1990) Variation of some serum proteins of red deer, *Cervus elaphus* L. Anim Genetics 21:285–293

Stubblefield SS, Warren RJ, Murphy BR (1986) Hybridization of free-ranging white-tailed and mule deer in Texas. J Wildlife Management 50:688–690

Tate ML, Dratch PA (1988) Inherited protein variation and parentage testing in farmed red deer. Proc N Z J Anim Prod 48:181–185

Taylor WJ, Easley CW (1977) Multiple hemoglobin a-chains in the sika deer (*Cervus nippon*). Biochem Biophys Acta 492:126–135

Thommasen HV, Thomson MJ, Shutler GG, Kirby LT (1989) Development of DNA fingerprints in wildlife forensic science. Wildl Soc Bull 17:321–326

Turubanov MN, Shubin PN (1971) New alleles of the transferrin locus in reindeer. Genetika 7:171–173 (in Russian) Translations in Soviet Genetics 7:265–266

Van de Weghe A, Van Zeveren A, Bouquet Y (1982) The vitamin D binding protein in domestic animals. Comp Biochem Physiol 73:977–982

Van Tets P, Cowan IMcT (1966) Some sources of variation in the blood sera of deer (Odocoileus) as revealed by starch gel electrophoresis. Can J Zool 44:631–647

Van Wieren (1989) Genetische variatie in het Veluwse edelhert *Cervus elaphus*. Lutra 32:193–200

Wallmo OC (1981) Mule deer and black-tailed deer distribution and habitats. In Wallmo OC (ed) Mule Deer and Black-Tailed Deer of North America. Stackpole Books, Harrisburg, P.a., pp 1–25

Wegge P (1978) Blood serum electrophoresis of insular and mainland red deer in Norway. Second Int Theriological Congr, Brno, abstract

Weisberger AS (1964) The sickling phenomenon and heterogeneity of deer hemoglobin. Proc Soc Expt Biol Med 117:276–280

Whitehead GK (1972) Deer of the World. Constable, London

Wilhelmson M, Juneja RK, Bengtsson S (1978) Lack of polymorphism in certain blood proteins of European and Canadian moose (*Alces alces*). Naturaliste Canadien 105:445–449

Woodruff DS, Ryder OA (1986) Genetic characterization and conservation of endangered species: Arabian oryx and Pere David's Deer. Isozyme Bull 19:33

Wooten MC, Smith MH (1985) Large mammals are genetically less variable? Evolution 39:210–212

Wright S (1978) Variability with and among natural populations. Evolution and the Genetics of Populations, vol 4. University of Chicago Press, Chicago

Zhurkevich NM, Fomicheva II (1976) Genetic polymorphisms of transferrings of blood serum in reindeer (*Rangifer tarandus* L.) indigenous to north eastern Siberia. Genetika 12:56–65

88
DNA Probes for Identification of Cervidae

John S. Waye and Jerry C. Haigh

Abstract

DNA fingerprinting provides the potential for genetic individualization within a given species. Current DNA fingerprinting methods utilize cross-reacting human probes to detect highly polymorphic sequences found in animal species. However, since the DNA fingerprints are not species-specific, additional tests are necessary to determine the species of origin. Species identification can be accomplished using conventional protein makers, although the usefulness of such markers may be limited by the quantity or quality of the specimen. An alternative approach is the use of species-specific DNA makers. Highly repetitive satellite DNAs are characterized by high levels of intraspecies homogeneity and interspecies heterogeneity, ideal features for differentiating among animal species. We report the isolation and characterization of satellite DNA probes suitable for sensitive and specific identification of cervid species.

Key words: Cervidae, DNA probes, satellite DNA, species identification

Introduction

The use of human DNA probes to generate individual-specific "DNA fingerprints" for animal species represents an important application of DNA technology (e.g., Jeffreys and Morton 1987; Georges et al. 1988; Thomassen et al. 1989). However, there are limits to the types of information that can be obtained by DNA fingerprinting. In particular, DNA fingerprints do not provide any information as to the species of origin. In fact, the DNA fingerprints of human genomic DNA appear very similar to those of animal and plant species. As such, there may be a need for two types of markers, one to identify the species and another to discriminate among the members of the species.

There are several features of satellite DNA that can be exploited for the development of sensitive and specific tests of species identity. Satellite DNAs are ubiquitous components of eukaryotic genomes and may be used to identify virtually any species. In addition, satellite DNAs are highly repetitive and can be detected with a high degree of sensitivity. Lastly, satellite DNAs are subject to rapid genetic turnover in a concerted fashion (Dover 1982), such that there is an accumulation of species-specific sequence variation and a general tendency towards intraspecies homogeneity and interspecies heterogeneity.

Satellite DNA sequences have been studied in a wide range of animal and plant species, including several species of Cervidae (Lima-de-Faria et al. 1984; Bogenberger et al. 1987). These studies have indicated that satellite DNA sequences are "conserved" among different cervid species, yet exhibit subtle interspecies differences in restriction endonuclease (RE) cleavage periodicity, and nucleotide sequence homology.

In this report, we describe a general strategy for the isolation of satellite DNA probes from virtually any species, requiring no prior information regarding the species or the satellite DNA family. Using an approach similar to that of Lima-de-Faria et al. (1984), we isolated a series of satellite DNA probes suitable for the identification of cervid species. Beginning with white-tailed deer (*Odocoileus virginianus*), a satellite DNA probe was identified and cloned based solely on its RE cleavage periodicity. Subsequently, the cloned white-tailed deer probe was used to identify related satellite DNAs of several different cervid species.

Methods

Genomic DNA Samples

Whole blood samples were collected from 65 animals, and each sample was split. The animals belonged to the following species: white-tailed deer (*Odocoileus virginianus*), mule deer (*Odocoileus hemionus*), fallow deer (*Dama dama*), moose (*Alces alces americanus*), wapiti (*Cervus elaphus canadensis*), Japanese sika deer (*Cervus nippon*), and plains bison (*Bison bison bison*). Genomic DNA was isolated from 0.7 ml volumes of blood using standard methods (Maniatis et al. 1982).

Isolation of Satellite DNA Clones from White-Tailed Deer

Genomic DNA was isolated from white-tailed deer, digested with a number of different REs, and analyzed by agarose gel electrophoresis. The RE *Rsa*I produced intense satellite bands of ~250 and ~500 basepairs (bp). The ~250 bp fragment was excised from a preparative 1.5% low melting point agarose gel, ligated to *Hinc*I-digested plasmid pUC19, and introduced into *Escherichia coli* JM101. Recombinant clones were screened by colony hybridization (Maniatis et al. 1982) using the uncloned "satellite band" (labeled with ^{32}P-dCTP) as a hybridization probe. One of the positively hybridizing clones, designated pWTD3, contained a single copy of the white-tailed deer ~250 bp *Rsa*I satellite DNA sequence.

DNA Hybridizations

DNA samples were digested with REs according to the manufacturers' specifications, subjected to electrophoresis through 1.2% agarose slab gels, and transferred to nylon membranes (Hybond-N, Amersham) using the method of Southern (1975). Membranes were prehybridized for 1 to 2 h at 65° C in a solution of 6 × SSC, 5 × Denhardt's solution, 0.5% SDS and 100 μg/ml denatured salmon sperm DNA (Maniatis et al. 1982), and hybridized for at least 2 h at 65° C in the same solution containing 10% dextran sulfate and 10^6 cpm/ml ^{32}P-labeled probe. Following hybridization, the membranes were washed at normal stringency (65° C, 0.1 × SSC/0.1% SDS) or high stringency (75° C, 0.1 × SSC/0.1% SDS). Autoradiography was conducted for 2 to 6 h at −70° C using XAR5 x-ray film (Kodak) and Lightning Plus intensifying screens (Dupont).

Results and Discussion

Satellite DNA Repeats

In general, satellite DNAs are characterized by high levels of intraspecies homogeneity and interspecies heterogeneity, hallmark features of what has been described as horizontal or concerted evolution. It is thought that the individual satellite DNA repeat units of different species accumulate mutations independently and that these mutations are somehow spread throughout the many repeat units of the tandem arrays and fixed in the population (Dover 1982). As such, any mutations that accumulate after the separation of two species can become identifying features of the satellite DNA and the individual species.

There are several molecular features that can be used to distinguish one satellite DNA from another, and hence one species from another. Satellite DNAs are most easily distinguished based on the size and sequence composition of the tandemly repeated units, features that can be indirectly evaluated using REs. For example, the satellite DNA of species no. 1 may consist of a tandemly repeated 1000 bp repeat unit that happens to contain a single recognition site for the RE *Xba*I. The same satellite DNA of a closely related species, species no. 2, may consist of a 1000 bp repeat unit that contains two *Xba*I sites, located 500 bp apart. Upon digestion with *Xba*I, the satellite DNA of species no. 1 will appear as a highly repeated restriction fragment of 1000 bp, whereas that of species no. 2 will appear as a repeated 500 bp fragment. Therefore, based solely on their RE cleavage periodicities, these satellite DNAs can be distinguished from each other.

Identification and Cloning of Satellite DNA from White-Tailed Deer

To identify RE cleavage periodicities of white-tailed deer satellite DNA, genomic DNA was digested with a number of different REs (13 in total) and analyzed using agarose gel electrophoresis. Following ethidium bromide staining, the genomic DNA fragments were visualized by UV fluorescence (data not shown). Many of the RE digestions appeared as featureless smears of unresolved fragment lengths. However, there were several REs that yielded discrete "satellite bands" superimposed upon the unresolved smears. In particular, the RE *Rsa*I yielded two intense satellite bands of ~250 bp and ~500 bp. The more intense ~250 bp satellite band was isolated from a preparative agarose gel, labeled with ^{32}P-dCTP, and used as a Southern hybridization probe against white-tailed deer genomic DNA. The results of the analysis indicate that the uncloned ~250 bp *Rsa*I satellite band is very homogeneous (Fig. 88.1). This is demonstrated by its ability to detect the corresponding ~250 bp *Rsa*I fragment of genomic DNA with a high degree of specificity and sensitivity. Moreover, indistinguishable results were obtained when the cloned ~250 bp *Rsa*I probe (pWTD3) was subsequently used to probe the same membrane (data not shown).

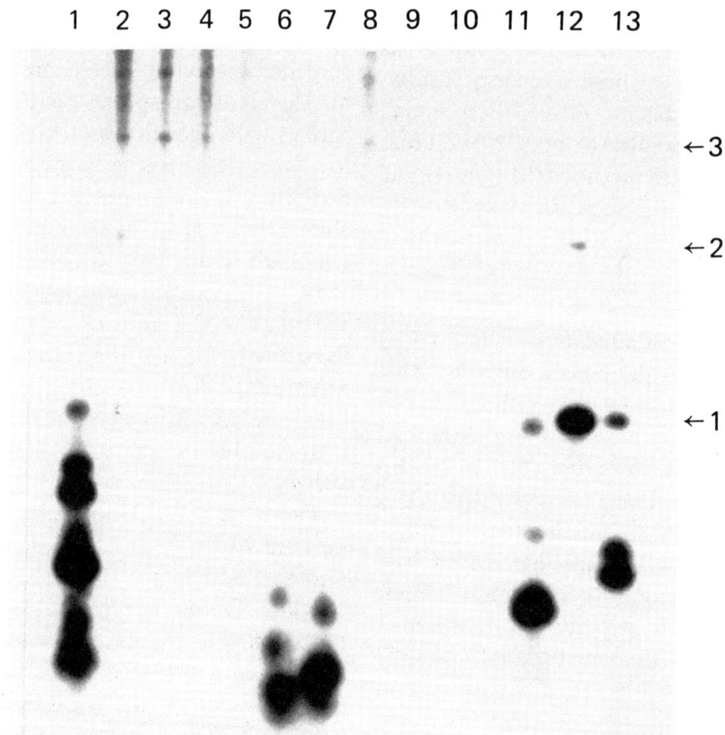

FIGURE 88.1. Restriction cleavage periodicities of white-tailed deer satellite DNA detected using the uncloned ~250 bp white-tailed deer probe and normal wash stringency. Lane 1, *Alu*I; 2, *Hind*III; 3, *Bam*HI; 4, *Dra*I; 5, *Dde*I; 6, *Hae*III; 7, *Hinf*I; 8, *Eco*RI; 9, *Bgl*II; 10, *Pst*I; 11, *Rsa*I; 12, *Xba*I; 13, *Sst*I. Size markers are indicated in multiples of the 1000 bp repeat unit.

The RE cleavage periodicities detected by Southern hybridization with the ~250 bp *Rsa*I probe provide insight into the overall organization of this white-tailed deer satellite DNA. At varying intensities, several of the REs revealed a ladder of hybridization fragments that increase at ~1000 bp increments (Fig. 88.1). This is most obvious for *Xba*I, having a predominant hybridization fragment of ~1000 bp and less intense hybridization fragments of ~2000 bp and ~3000 bp. Overall, these results indicate that this particular satellite DNA is probably organized as tandem ~1000 bp repeat units that contain single *Xba*I sites in the majority of repeat units. The patterns detected with other REs indicate that there are multiple cleavage sites per repeat unit for *Alu*I, *Hae*III, *Hinf*I, *Rsa*I and *Sst*I. Conversely, cleavage sites for *Hind*III, *Bam*HI, *Dra*I, *Dde*I, *Eco*RI, *Bgl*II, and *Pst*I occur infrequently in the repeat units, such that the tandem arrays are left intact by RE digestion.

The restriction cleavage periodicities identified for this white-tailed deer satellite DNA appear very similar to those detected in white-tailed deer using a cloned Indian muntjac (*Muntiacus muntjac vaginalis*) satellite DNA probe (Bogenberger et al. 1987). In that particular study, the Indian muntjac probe cross-hybridized with a wide range of cervids, each of which could be distinguished based on RE cleavage periodicity. Therefore, it would seem probable that the white-tailed deer satellite DNA described in this report belongs to the same satellite DNA family as that previously characterized for the Indian muntjac (Bogenberger et al. 1987).

Analysis of Cervid and Bovid Species with pWTD3

To evaluate the possible utility of pWTD3 as a species-specific marker, a blind study was conducted using 130 blood samples from 65 animals (six cervids, one bovid). DNA samples were digested with *Sst*I and separated using agarose gel electrophoresis. The visual patterns of satellite bands in the ethidium-bromide stained samples were sufficient to divide the samples into five distinct groups (data not shown). Upon subsequent hybridization with pWTD3, the visible satellite bands were detected as hybridization fragments. The five distinct hybridization patterns correspond to the following: (1) white-tailed deer and mule deer, (2) wapiti and Japanese sika deer, (3) fallow deer, (4) moose, and (5) plains bison. Analysis with several other REs failed to differentiate white-tailed deer from mule deer, or wapiti from Japanese sika deer.

Figure 88.2 shows the five Southern hybridization patterns detected using pWTD3. Of the six REs evaluated in this study, *Hae*III is the most discriminating since it cleaves the satellite repeat units of each group and generates readily distinguishable hybridization patterns. The hybridization results demonstrate the utility of these sequences for evaluating evolutionary relationships between species. Based on the results of the blind study, one would correctly predict that white-tailed deer and mule deer are very closely related, as are wapiti and Japanese sika deer. It is also notable that the plains bison DNA sample (the only bovid species included in the study) hybridized very weakly to the pWTD3 probe. This reflects the greater evolutionary distance separating plains bison and white-tailed deer relative to the distances separating white-tailed deer from the other cervids.

Strategies for Species Identification

Perhaps the simplest approach for species identification is to isolate genomic DNA and visually assess the pattern of satellite DNA fragments generated by RE digestion. However, this approach can only be used if there is sufficient DNA for visual examination and the samples are not of mixed origin. In addition, it is possible that the satellite DNAs of completely unrelated species may fortuitously exhibit a similar or identical RE cleavage periodicity. As such, this method should be considered a presumptive test for species identity.

A more definitive test of species identity involves the use of satellite DNAs as qualitative species-specific hybridization probes. For such tests, the probe should hybridize predominantly, if not exclusively to genomic DNA of the corresponding species. To evaluate the hybridization specificity of pWTD3 for white-tailed deer (and mule deer), genomic DNA samples of cervid and bovid species were manually spotted (dot

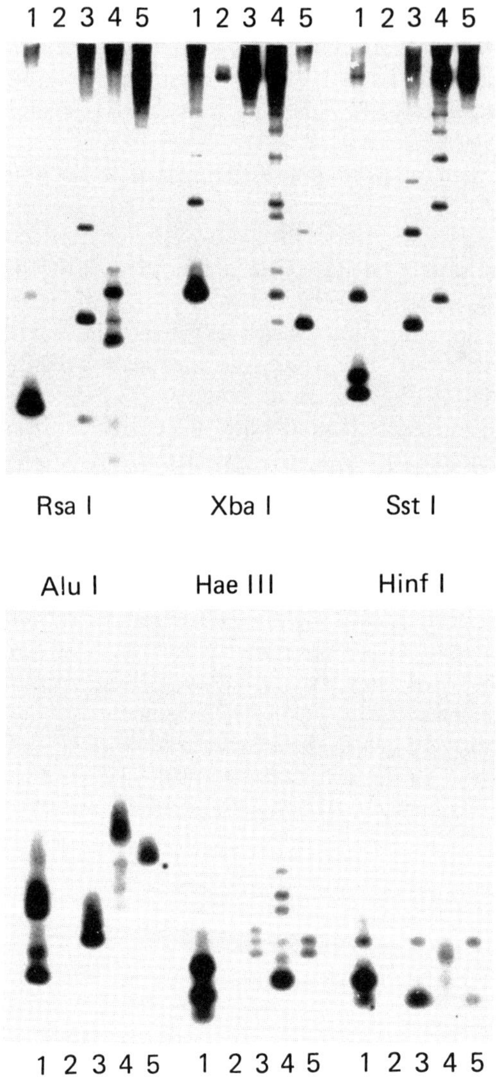

FIGURE 88.2. Restriction cleavage periodicities of cervid and bovid satellite DNAs detected using the pWTD3 probe under normal wash stringency. Lane 1, white-tailed deer; 2, plains bison; 3, wapiti; 4, moose; 5, fallow deer.

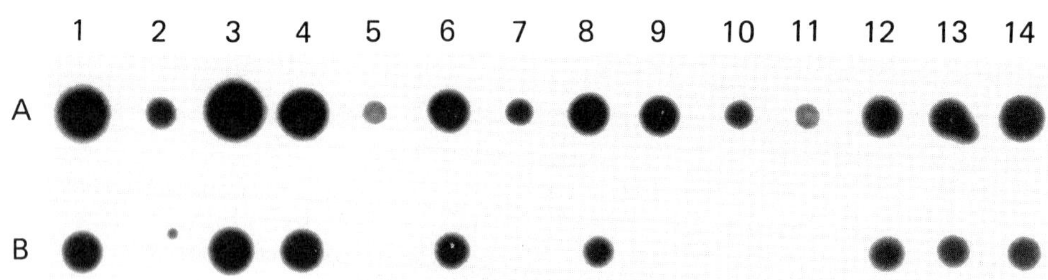

FIGURE 88.3 Cervid and bovid genomic DNAs analyzed by dot blot hybridization with pWTD3. Panels A and B were washed at normal and high wash stringencies, respectively. White-tailed deer, numbers 1, 4, 6, 8, 13, 14; mule deer, 3, 12; Wapiti, 2, 5, 9; fallow deer, 7, 10; plains bison, 11. The small "offline" hybridization signal of sample 2 (panel B) is a hybridization artifact, not a bone fide positive signal.

blotted) onto a nylon membrane and hybridized to pWTD3 probes. The results indicate that pWTD3 is capable of cross-hybridizing with heterologous cervid and bovid species under conditions of normal wash stringency, but is largely specific for white-tailed deer and mule deer under higher stringency wash conditions (Fig. 88.3). Accordingly, it may be possible to use pWTD3 for qualitative testing procedures.

Satellite DNA probes can also be used to identify species based on their RE cleavage periodicities as detected by Southern hybridization (Fig. 88.2). Although this method is more time-consuming than the above strategies, it provides a more definitive indication of species identity. Moreover, this method can be used to identify many different species using a single hybridization probe (e.g., identification of moose or wapiti with the white-tailed deer probe).

Lastly, it may be possible to devise strategies for the amplification of species-specific satellite DNAs using the polymerase chain reaction (PCR) (Saiki et al. 1988). This approach has been used for rapid amplification of satellite DNA that is specific for the human Y chromosome (Witt and Erickson 1989). Using PCR-based strategies, it may be possible to identify individual animal species from samples that contain as few as single cells.

Acknowledgements

The authors thank the Marimba Farm and Forestry Farm Zoo (Saskatoon, Sask.) for providing the samples used in the blind study. DNA probes are freely available upon request to J.S.W.

References

Bogenberger JM, Neitel H, Fittler F (1987) A highly repetitive component common to all cervidae: its organization and chromosomal distribution during evolution. Chromosoma 95:154–161

Dover G (1982) Molecular drive: a cohesive mode of species evolution. Nature 299:111–116

Georges M, Lequarre A-S, Castelli M, Hanset R, Vassart G (1988) DNA fingerprinting in domestic animals using four different minisatellite probes. Cytogenet Cell Genet 47:127–131

Jeffreys AJ, Morton DB (1987) DNA fingerprints of dogs and cats. Animal Genet 18:1–15

Lima-de-Faria A, Arnason U, Widegren B, Essen-Moller J, Isaksson M, Olsson E, Jaworska H (1984) Conservation of repetitive DNA sequences in deer species studied by Southern blot transfer. J Molec Evol 20:17–24

Maniatis T, Fritsch EF, Sambrook J (1982) Molecular cloning: a laboratory manual. Cold Spring Harbor Laboratory, Cold Spring Harbor, New York

Saiki RK, Gelfand DH, Stoffel S, Scharf SJ, Higuchi R, Horn KB, Mullis KB, Erlich HA (1988) Primer-directed enzymatic amplification of DNA with a thermostable DNA polymerase. Science 239:487–491

Southern EM (1975) Detection of specific sequences among DNA fragments separated by gel electrophoresis. J Molec Biol 98:503–517

Thomassen HV, Thomson MJ, Shutler GG, Kirby LT (1989) Development of DNA fingerprints for use in wildlife forensic science. Wildl Soc Bull 17:321–326

Witt M, Erickson RP (1989) A rapid method for detection of Y-chromosomal DNA from dried blood specimens by the polymerase chain reaction. Hum Genet 82:271–274

89
Prenatal Reproductive Losses in White-Tailed Deer

Olin Eugene Rhodes, Jr., Michael H. Smith, and Ronald K. Chesser

Abstract

Estimates of the percentages of ovulated eggs that do not become visible embryos were made using data from white-tailed deer (*Odocoileus virginianus*) on the Savannah River Site (SRS; 11%) and in deer from herds across North America (13%). Does on the SRS lose an average of 16.4 offspring per 100 females of all age classes due to loss of ovulated eggs. Genetic heterozygosity, at seven polymorphic loci, and the kidney fat index were not related to egg loss of does on the SRS. Consistency in the relationship between the numbers of fetuses and numbers of eggs ovulated in deer herds across North America suggests that egg loss may not be directly related to environmental conditions for white-tailed deer.

Key words: Ovary, corpora lutea, reproduction, white-tailed deer

Introduction

The productivity of a deer herd partially depends upon the number of ovulated eggs and the percentage of eggs that never complete development. The latter reduces the number of fawns, but is not normally considered as important in modifying the productivity of a deer herd. The number of fetuses in white-tailed deer varies according to the level of genetic heterozygosity of the mother (Chesser and Smith 1987). These differences in fetal number could be due to the number of eggs ovulated or the proportion of ovulated eggs that are fertilized and develop into fetuses. An examination of the frequency distributions of different-aged females that ovulate different numbers of eggs which develop into viable fetuses could provide an understanding of environmental and genetic limitations on reproduction in this southeastern herd, on the Savannah River Site (SRS) near Aiken, South Carolina.

Our objectives were to (1) describe prenatal loss of eggs, embryos, or fetuses of white-tailed deer on the SRS and estimate the magnitude of such losses for white-tailed deer across North America; (2) estimate age-specific losses in productivity of SRS white-tailed deer resulting from prenatal mortality, and (3) test the hypotheses that fetal number or prenatal egg loss is significantly related to multilocus genetic heterozygosity of female white-tailed deer on the SRS.

Materials and Methods

Female white-tailed deer were collected during fall hunts in 1983 ($n = 759$; 1 September to 31 December) and 1984 ($n = 540$; 1 October to 31 December) as well as in winter-spring collections during 1985 ($n = 48$; 1 January to 1 June). These data result from the breeding periods of 1983 and 1984 and were pooled unless otherwise specified. Females were aged using tooth eruption and wear (Servinghaus 1949), and both ovaries were collected and stored in 10% buffered formalin (Wobeser et al. 1980). The uterus of each female was inspected for signs of swelling (indicating pregnancy), and visible fetuses were collected. One incidence of fetal reabsorption was observed. Fetuses were taken from 89 (1983), 137 (1984), and 38 (1985) females. In 1984 and 1985 kidneys and their peripheral fat were collected from pregnant females, and kidney fat index (KFI) values were calculated (Finger et al. 1981). Samples of liver, muscle, and blood tissues were taken from each doe for subsequent electrophoretic analyses.

Ovaries were inspected for corpora lutea using macroscopic ovarian analysis (Trauger and Haugen 1965). Corpora lutea were classified as either corpora lutea of esterous (CLE) or pregnancy (CLP) based on size without a priori information regarding the female's reproductive status (pregnant or nonpregnant). Corpora lutea were classified as CLPs in 100% of the pregnant females, and only data from pregnant females were analyzed.

Pregnant females were surveyed for genetic variation at seven polymorphic loci (Manlove et al. 1976). Females were recorded as either homozygous or heterozygous at a locus on the basis of the banding patterns of their gene products. Individual deer were categorized into multilocus heterozygosity classes, ranging from 0, 1, 2, 3, and $\geqq 4$, based on the total number of their heterozygous loci.

A survey of the literature was conducted to find information that included both mean numbers of CLPs and fetuses per female for North American white-tailed deer herds. Data from 23 herds were available for analysis. The mean number of CLPs per female was regressed on mean number of fetuses per female for these herds using least squares regression. The theoretical expectation for the relationship between numbers of CLPs and numbers of fetuses is a line with an intercept of 0 and slope of 1, assuming that no prenatal loss or identical twins occur. Negative deviations from this expectation (slope < 1) can be used to estimate the average prenatal loss of eggs, embryos, or fetuses across herds. Our estimate of prenatal loss is probably a slight underestimate as there are no estimates of the rate of occurrence of identical twins in the selected herds. Identical twinning would increase the slope of the relationship.

Fisher's exact test was used to test for differences in the distributions of females with 1 or $\geqq 2$ fetuses or CLPs among age classes of females that lost eggs, did not lose eggs, and all females combined. To test for differences in the distributions of females that lost or did not lose eggs among age classes ($\geqq 1.5$) *G*-tests of independence were used; these statistics were also used to determine if females that lost or did not lose eggs and females that ovulated one or two or more eggs were distributed independently with regard to their multilocus heterozygosities. Kidney fat index values were log transformed, and an analysis of variance (ANOVA) was calculated to test for differences in mean KFIs between age classes of females ($\geqq 1.5$ years) that lost or did not lose eggs. Embryonic loss was calculated for each age class by dividing total number of fetuses by total number of CLPs and subtracting this result from 1. Univariate statistics, regression analyses, *G*-tests of independence, Fisher's exact tests, and ANOVA were performed using SAS (SAS Institute 1985). Hierarchical log-linear models were calculated using BMDP (Dixon and Brown 1983). Statistical significance was set a priori at $P \leqq 0.05$.

Results

The majority of singleton offspring born to yearling (1.5 years) and adult ($\geqq 2.5$ years) female white-tailed deer on the SRS was not due to the ovulation of a single egg, but because of the failure of one or more ovulated eggs to become an embryo (Fig. 89.1). There was no detectable loss of ovulated eggs in the fawn (0.5 years) females (Fig. 89.1). Most fawn females which breed ovulate only one egg, and females that lost their only offspring were not detectable in our sample.

Females 1.5 years old or older ovulate an average of 1.98 ova per female and a much greater proportion of these females ovulate two eggs rather than one or three (Fig. 89.1). Fourteen of the sampled females ovulated three eggs, and of those only seven had three fetuses (fawns, 0 of 0; 1.5 years, 0 of 2; 2.5 years, 2 of 4; and $\geqq 3.5$ years, 5 of 8). Percentages of eggs lost by

FIGURE 89.1. Age-specific percentages of pregnant white-tailed does that had one, two, or three corpora lutea of pregnancy (CLP). Age-specific percentages of females with a single fetus are also provided.

TABLE 89.1. Age-specific reproductive success of pregnant white-tailed does on the Savannah River Site (1983–84)

	Age class (years)				
	0.5	1.5	2.5	≧ 3.5	Sum
Total fetuses	16	117	194	137	464
Total CLPs[a]	16	138	212	154	520
Embryonic loss (%)[b]	0	15	8	11	11[e]
Percent population female ($n = 1326$)	29.4	25.1	27.9	17.6	100
Productivity per 100 females[c]	12.7	41.9	51.6	32.6	138.8
Adjusted productivity per 100 females[d]	12.7	49.5	56.4	36.6	155.2

[a]Corpora lutea produced pregnancies
[b]Percentage of eggs that did not become embryos
[c]Productivity per 100 females based on an average of 40% female fawn breeding (Rhodes et al. 1991)
[d]Productivity adjusted for embryonic losses calculated as percentage of female population times mean fetal number times 100 for age classes > 0.5 years assuming 100% breeding for all adult does
[e]Average embryonic loss calculated across all age classes > 0.5 years

females that ovulated two or three times were 10.3% and 21.4%, respectively. Proportions of females with one vs two fetuses or CLPs were not significantly different among age classes (≧ 1.5 years) of females that lost eggs, did not lose eggs, and all females combined.

Numbers of fetuses and CLPs detected in females within each age class are provided in Table 89.1. Distributions of females that lost or did not lose eggs were not significantly different among age classes (≧ 1.5 years). The percentage of the female population represented by females within each age class was estimated from data that included both pregnant and nonpregnant females collected in 1983 and 1984–85 combined (Table 89.1).

Productivity per 100 females was estimated based on the age-specific contributions of offspring from females of all age classes. The proportion of females represented by each age class multiplied by the mean fetal number for each age class, by the proportion of females that bred in

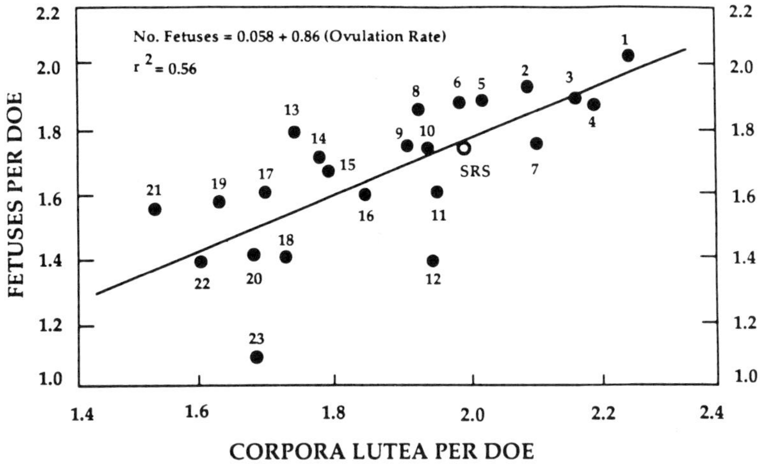

FIGURE 89.2. Linear regression ($P < 0.001$) of the mean number of corpora lutea of pregnancy per doe and the mean number of fetuses per doe for data from 23 North American deer herds including deer from the Savannah River Site (SRS). SRS data were not used in the regression analysis. Standard errors of the intercept and slope were 0.32 and 0.17, respectively. Numbers associated with the data points correspond with the following sources: (1) Haugen 1975; (2) Harder and Peterle 1974, Bell and Peterle 1975; (3) Ransom 1967; (4) Roseberry and Klimstra 1970; (5) Shaw and McLaughlin 1951; (6) Ransom 1966; (7) Nixon 1965, 1971; (8) Ransom 1967; (9) Wilson and Sealander 1971; (10) Anonymous 1951, Laramie and Silver 1968, White and Sealander 1971; (11) Kie and White 1985; (12) Servinghaus and Cheatum 1956; (13) Banasiak 1961; (14) Hilyer 1982; (15) Wilson and Sealander 1971; (16) Barron and Harwell 1973; (17) Wilson and Sealander 1971; (18) Mansell 1974; (19) White 1973; (20) Nobel 1974; (21) Teer et al. 1965; (22) Woolf and Harder 1979; (23) Osborne 1976.

each age class, and by 100 estimated the age-specific contributions of fetuses per 100 females (Table 89.1). The mean number of CLPs per female within each age class was substituted for mean fetal number per female in the previous relationship to estimate adjusted productivity per 100 females. Adjusted productivity is an estimate of the potential productivity of 100 females of all age classes if no egg loss had occurred. An average of 16.4 eggs per 100 females of all age classes did not become visible embryos in this herd.

The regression of mean number of CLPs per doe on mean number of fetuses per doe was significant ($P \leq 0.01$), and our estimate of average egg loss in North American white-tailed deer herds was 13.8% (Fig. 89.2). Estimates of fetal numbers per doe ranged from 1.11 (Osborne 1976) to 2.07 (Haugen 1975) and estimates of CLPs per doe ranged from 1.53 (Teer et al. 1965) to 2.27 (Haugen 1975).

Pregnant females were variable for an average of 2.2 loci of the seven examined. A hierarchical log-linear model was used to determine if fetal number was independent of multilocus heterozygosity among three sampling peiods—1974–76 (Chesser and Smith 1987), 1983 and 1984–85 (Table 89.2). The three-way interaction between sampling period, number of fetuses, and multilocus heterozygosity was not significant. The two-way test of independence of fetal number and multilocus heterozygosity (pooling across sampling periods) was almost significant ($P = 0.06$). The relationship between multilocus heterozygosity and fetal number was significant using data from 1974–76 ($P = 0.01$; Chesser and Smith 1987), but not using data from 1983 ($P = 0.76$), 1984 ($P = 0.46$), or 1983 and 1984

TABLE 89.2. Number and percentage (in parentheses) of female white-tailed deer in each multilocus heterozygosity class (0, 1, 2, 3, ≧ 4) are presented for does with one and two or more fetuses within each of three sampling periods (years) and for all sampling periods combined

Years	Fetuses (*n*)	Multilocus heterozygosity class 0	1	2	3	≧ 4
1974–76	1	8	16	10	13	12
		(14)	(27)	(17)	(22)	(20)
	≧ 2	6	27	55	35	20
		(4)	(19)	(38)	(25)	(14)
1983	1	4	6	6	7	2
		(16)	(24)	(24)	(28)	(8)
	≧ 2	6	15	25	22	7
		(8)	(20)	(33)	(29)	(10)
1984	1	2	11	11	13	9
		(4)	(24)	(24)	(28)	(20)
	≧ 2	4	38	41	33	13
		(3)	(29)	(32)	(26)	(10)
Combined	1	14	33	27	33	23
		(11)	(25)	(21)	(25)	(18)
	≧ 2	16	80	121	90	40
		(5)	(23)	(35)	(26)	(11)

combined ($P = 0.42$). Females that lost or did not lose eggs and females that ovulated one or two or more eggs were distributed independently with regard to their multilocus heterozygosities (data from 1983 and 1984 combined). Mean KFI values were not significantly different between age classes of females (≧ 1.5 years) that lost and did not lose eggs.

Discussion

White-tailed deer on the SRS lose an average of 11% of eggs ovulated by females of all age classes. This is probably an underestimate due to our inability to detect the loss of eggs in females that ovulate one egg. A large proportion of the pregnant fawns ovulate only one egg, and undetected egg loss in this age class alone could be substantial; an average of 40% of the fawn females breed annually on the SRS (Rhodes et al. 1991). Egg losses in fawns have been reported for several herds (21%, Roseberry and Klimstra 1970; 13%, Nixon 1971; 9%, Haugen 1975).

Although the numbers of females in the SRS deer herd that did or did not lose eggs are distributed independently of age class (1.5, 2.5, or ≧ 3.5 years), yearling females have the greatest egg losses at 15% (Table 89.1). Scanlon (1991) reported that yearling females that ovulated two eggs lost higher percentages of eggs than did adult females. However, other studies have reported lower egg loss rates in yearling females than in adult females (Roseberry and Klimstra 1970; Nixon 1971). Barron and Harwell (1973) found no significant differences in loss rates among age classes of female white-tailed deer, but significant yearly variation in adult loss rates does occur.

Fawn females ovulate fewer eggs and have fewer fetuses than do other females. There are no significant differences in numbers of eggs ovulated or numbers of fetuses per female among age classes ≧ 1.5 years for females that lose eggs, do not lose eggs, and all females combined. Females that lose eggs average 1.09 fetuses and 2.13 CLPs per doe, and females that do not lose eggs average 1.93 fetuses and 1.93 CLPs per female. Most females that have a single fetus do so due to the loss of an egg(s) and not because of the ovulation of a single egg. Females ≧ 1.5 years have two fetuses more often than one, and very few of these females have three offspring. Egg losses in females that ovulated three eggs are high (21.4%). Does on the SRS generally produce two offspring, and attempts to produce more than two often fail, whereas single offspring are generally the result of a failure of eggs to become fertilized and/or develop.

The productivity per 100 females of all age classes observed during the 1983 and 1984 breeding seasons is slightly higher at 138.8 fetuses than the average over years (1965–84) on the SRS at 126 fetuses (Rhodes et al. 1991). Our estimate of potential losses in productivity per 100 females, due to egg loss, is 16.4 offspring, which is conservative given our inability to detect loss in does that ovulate a single egg. Fluctuations in age-specific egg losses over a range of different age structures could produce different overall losses in productivity than those estimated for our herd in which age-specific differences were not significant. There are a number of white-tailed deer herds in North America with documented age-specific differences in numbers of CLPs per pregnant female (Roseberry and Klimstra

1970). Even if age-specific egg loss occurs in other herds, it cannot account for much of the variation in fetal number (Fig. 89.2).

Fetal number was not independent of multilocus heterozygosity in female white-tailed deer on the SRS during the mid 1970s (Chesser and Smith 1987). Females with fewer heterozygous loci had one rather than two or more fetuses more often than did females with intermediate numbers of heterozygous loci. Although our data (1983 and 1984) show the same trend for the relationship between fetal number and multilocus heterozygosity as did the data of Chesser and Smith (1987; Table 89.2), the differences are not quite significant ($P = 0.06$), and this relationship is not significant using data for either 1983 or 1984 independently or combined.

Numbers of eggs ovulated (one or two or more) and numbers of females that lost or did not lose eggs are also not significantly related to multilocus heterozygosity. Loss of eggs leading to the production of one rather than two or more fetuses in less heterozygous females is a potential mechanism through which the result of Chesser and Smith (1987) could be obtained. However, the relationship between multilocus heterozygosity and fetal numbers is not temporally consistent, and there are no data on ovulation rates from 1974–76 with which to test whether egg loss is related to multilocus heterozygosity. Although there is no evidence that egg loss is related to multilocus heterozygosity in the SRS deer herd during the breeding seasons of 1983 and 1984, there are data that support a relationship between ovulation rate and multilocus heterozygosity in white-tailed deer from Illinois (Maffei and Woolf 1987). They report that the occurrence of corpora albicantia is positively correlated with multilocus heterozygosity ($P = 0.037$).

Fertilization frequencies (fetuses per female/eggs ovulated per female) of less than one occur often in mammalian species (Asdell 1946). Physiological mechanisms, such as failure of fertilized eggs to implant or to survive more than several days after implantation, and blockage of the ovaduct have been suggested to explain egg loss (Hesselton and Jackson 1974). Increased rates of egg loss in deer from poor vs high quality range (Cheatum and Morton 1946) or "flushing" of ova in response to sudden increases in habitat quality (Ransom 1967) have also been proposed. In addition to these causative mechanisms, there is undoubtedly some error in the correct specification of corpora lutea of pregnancy vs those of a previous heat.

The results of the regression analysis of mean CLPs per doe on mean fetuses per doe (Fig. 89.2) do not support the hypothesis that egg losses in white-tailed deer are directly related to environmental quality. The coefficient of determination ($r^2 = 0.56$) for the regression is quite high considering that the effects of a wide range of environmental and physical conditions, productivity rates, and techniques of ovarian analysis are represented in these data. The proportion of eggs lost by does reproducing with high and low ovulation rates and under a wide range of environmental and physical conditions is relatively constant. The loss rate for the SRS herd is very close to the predicted value calculated for all other herds. The relationship across herds would not be expected if environmental factors were directly related to egg loss or if females reproducing at varying levels of productivity lost different proportions of eggs.

The proportions of eggs lost by breeding females could be the result of physiological or genetic factors that operate at a relatively constant level in many different environments. Whereas data for the SRS deer support this general hypothesis, genetic heterozygosity and KFI fail to explain losses in fertility for individual deer. However, other physiological and genetic factors need to be considered in future studies. Considering the consistency of the relationship between number of fetuses and CLPs, the simplest explanation involves the probability of bringing sperm and egg together, and this hypothesis will need to be rejected before more complicated explanations are invoked.

Acknowledgements

This research was supported by contract DEAC09-76SR00819 between the University of Georgia's Institute of Ecology and the Department of Energy. Other groups cooperated in conducting this research including the employees of EI du Pont de Nemours Company, Clemson University, and the South Carolina Wildlife and Marine Resources Department who granted permission to conduct the research. Special thanks are extended to P.E. Johns and our colleagues at the Savannah River Ecology Laboratory who contributed their labor and ideas toward the completion of this work.

References

Anonymous (1951) Summary of Deer Research. New Hampshire Fish and Game Department

Asdell SA (1946) Patterns of Mammalian Reproduction. Comstock Publishing, New York

Banasiak CF (1961) Deer in Maine. Maine Game Div Bull 6

Barron JC, Harwell WF (1973) Fertilization rates of south Texas deer. J Wildl Manage 37:179–182

Bell RL, Peterle TJ (1975) Hormone implants control reproduction in white-tailed deer. Wildl Soc Bull 3:152–156

Cheatum EL, Morton GH (1946) Breeding season of white-tailed deer in New York. J Wildl Manage 10:249–263

Chesser RK, Smith MH (1987) Relationship of genetic variation to growth and reproduction in the white-tailed deer. In Wemmer CM (ed) Biology and Management of the Cervidae. Smithsonian Institution Press, Washington, D.C., PP 168–177

Dixon WJ, Brown MB (1983) Biomedical Computer Programs. University of California Press, Berkeley

Finger SE, Brisbin IL, Jr., Smith MH, Urbston DF (1981) Kidney fat as a predictor of body condition in white-tailed deer. J Wildl Manage 45:964–968

Harder JD, Peterle TJ (1974) Effect of diethylstillbestrol on reproductive performance of white-tailed deer. J Wildl Manage 38:183–196

Haugen AO (1975) Reproductive performance of white-tailed deer in Iowa. J Mammal 56:151–159

Hesselton WT, Jackson LW (1974) Reproductive rates of white-tailed deer in New York State. N Y Fish Game J 21:135–152

Hilyer BM (1982) Productivity in the Fort Rucker, Alabama white- tailed deer herd. MSc thesis, Auburn University, Auburn, AL

Kie JG, White M (1985) Population dynamics of white-tailed (*Odocoileus virginianus*) on the Welder Wildlife Refuge. Southwestern Naturalist 30:105–118.

Laramie HA, Silver H (1968) The White-Tailed Deer in New Hampshire. New Hampshire Fish Game Dep Survey Report 10

Maffei MD, and Woolf A (1987) The influence of genic heterozygosity on the potential productivity of white-tailed deer. Trans Ill Acad Sci 80:309–316

Manlove NM, Smith MH, Hillestad HO, Fuller SE, Johns PE, Straney DO (1976) Genetic subdivision in a herd of white-tailed deer as demonstrated by spatial shifts in gene frequencies. Proc Annu Conf SE Game Fish Wildl Agen 30:487–492

Mansell WD (1974) Productivity of white-tailed deer on the Bruce Peninsula, Ontario. J Wildl Manage 38:808–814

Nixon CM (1971) Productivity of white-tailed deer in Ohio. Ohio J Sci 71:217–225

Nixon CM (1965) White-tailed deer growth and productivity in eastern Ohio. Game Res Ohio 3:123–136

Nobel RE (1974) Reproductive Characteristics of the Mississippi White-Tailed Deer. Mississippi Game Fish Commission, Jackson, MS

Osborne JS (1976) Population dynamics of the Blackbeard Island white tailed deer. MSc thesis, University of Georgia, Athens

Ransom AB (1967) Reproductive biology of white-tailed deer in Manitoba. J Wildl Manage 31:114–123

Ransom AB (1966) Breeding seasons of white-tailed deer in Manitoba. Can J Zool 44:59–62

Rhodes OE Jr, Smith MH, Scribner KT, Johns PE (1991) Factors affecting productivity of a white-tailed deer herd. Proc 18th Int Union Game Biologists, Krakow, Poland (in press)

Roseberry JL, Klimstra WD (1970) Productivity of white-tailed deer on Crab Orchard National Wildlife Refuge. J Wildl Manage 34:23–28

SAS Institute (1985) SAS Users Guide Statistics, 5 ed. SAS Institute Inc, North Carolina

Scanlon PF (1991) Patterns of embryonic loss in white-tailed deer. Proc 18th Int Union Game Biologists, Krakow, Poland (in press)

Servinghaus CW (1949) Tooth development and wear as a criteria of age in white-tailed deer. J Wildl Manage 13:195–216

Servinghaus CW, Cheatum EL (1956) Life and times of the white-tailed deer. In Taylor WP (ed) The Deer of North America. Stackpole Publishing, Harrisburg, Pa., Wildlife Management Institute, Washington, D.C., pp 57–186

Shaw SP, McLaughlin CL (1951) The management of white-tailed deer in Massachusetts. Mass Div Fish Game Res Bull 13

Teer JG, Thomas JW, Walter EA (1965) Ecology and management of white-tailed deer in Llano Basin of Texas. Wildl Monogr 15, The Wildlife Society, Washington, D.C.

Trauger DL, Haugen AO (1965) Corpora lutea variations of white-tailed deer. J Wildl Manage 29:487–492

White M (1973) The white-tailed deer of the Arkansas National Wildlife Refuge. Texas J Sci 24:451–489

Wilson SN, Sealander JA (1971) Some characteristics of white-tailed deer reproduction in Arkansas. Proc 25th Annu Conf SE Assoc Game Fish Commis 25:53–65

Wobeser GA, Spraker TR, Harms VL (1980) Collection and field preservation of biological materials. In Schemnitz SD (ed) Wildlife Techniques Manual. Wildlife Society, Washington, D.C., pp 537–552

Woolf A, Harder JD (1979) Population dynamics of a captive white-tailed deer herd with emphasis on reproduction and mortality. Wild Monogr 67, The Wildlife Society, Washington, D.C.

90
Genetic Variability and Secondary Productivity

Michael H. Smith and Ronald K. Chesser*

Differences in genetic variability or heterozygosity, as determined by electrophoresis, have been particularly well studied within and among cervid species. For example, heterozygosity has been estimated for over 15 000 white-tailed deer (*Odocoileus virginianus*) in South Carolina. Higher levels of genetic variability appear to be positively related to measures of secondary productivity, including growth rates, body size, fat storage, and reproductive effort. All of these relationships have been documented for white-tailed deer in South Carolina, but some of them have also been observed for this species from other localities for across-species comparisons. These relationships imply that individuals with higher estimated heterozygosities have more energy available for productivity than genetically less variable individuals, and they may (1) obtain more and better food, (2) better assimilate the food that is eaten, and/or (3) have a more efficient maintenance metabolism. There is support for hypotheses (1) and (3) but not (2) from research conducted on noncervids. Because characteristics such as number of offspring, body size, and antler size are all important to productivity and the population biology of cervids, it is important to determine whether heterozygosity is just correlated with them or causes their variation. Heterozygosity is an important characteristic of cervids that should be of interest to managers and population biologists. Maintenance of genetic variability should be a goal of wildlife managers as well as conservation biologists, because of its potential impact on productivity and population dynamics.

*Savannah River Ecology Laboratory, Drawer E, Aiken, South Carolina 29802, USA

91

Effect of Social Structure on Genetic Structure of White-Tailed Deer in the Central Adirondack Mountains

Nancy E. Mathews and William F. Porter*

Behavioral and genetic characteristics of social groups of white-tailed deer (*Odocoileus virginianus*) were studied on the Huntington Wildlife Forest, Newcomb, New York. Previous studies have focused on either social behavior or genetic characteristics and have hypothesized that where a social structure is developed, there should also be genetic subdivisions. To date, few empirical data exist to support this hypothesis. This study focuses on two hypotheses: (1) social groups of white-tailed deer represent genetically distinct subunits within the population, and (2) adjacent social groups on summer range and groups that use the same winter range are more genetically similar. Eight social groups are defined based on movement patterns for 45 radio-collared does. Genetic analyses, based on starch-gel electrophoresis, indicate that social groups are genetically distinct subunits, and clusters of social groups using the same winter range may represent an interbreeding unit. We found no association between genetic distance and geographic distance on summer ranges. However, we found a significant association between genetic distance and winter yard membership. Because social groups have a high degree of overlap on summer range, we suspect that individual males breed females in several groups, homogenizing genetic structure. This homogenization would account for the lack of genetic and geographic association between otherwise distinct groups of females on summer range.

*State University of New York, College of Environmental Science and Forestry, Syracuse, New York 13210; Texas Cooperative Fish and Wildlife Research Unit, Texas Tech University, Lubbock, Texas 79409 USA

92
Establishment of Mesopotamia Fallow Deer in New Zealand and Their Hybridization with European Fallow Deer

Whitley Otway*

Mesopotamia fallow deer (*Dama dama*) are an endangered species with some 200 in the world, mainly in Iran. As their breeding and habitat is under threat, it was decided to establish a purebred herd in New Zealand. In November 1985, the first Mesopotamia buck and doe were imported from Opel Zoo, Frankfurt, Germany to Charnley Park, New Zealand. The doe had conceived in quarantine in England and gave birth to a female fawn at Charnley Park in May 1986. The buck adapted rapidly to the change in hemispheres and naturally mated the pure doe and 23 European fallow does that year. Hybridization with European fallow deer had been known to occur, so backcrossing the pure Mesopotamia buck to F1 hybrids, and then further backcrossing to the three-quarter bred, etc., was decided on to produce "pure" Mesopotamia females. Two further pure Mesopotamia bucks have been obtained, but no more females are available. The hybrid males (F1) are fertile and have produced offspring from European fallow does that are significantly heavier than pure European fallow fawns. This is of advantage in venison production. In 1988, an unsuccessful attempt was made to superovulate the pure-bred Mesopotamia doe and mate her to the pure-bred buck.

*Charnley Park, Kaukapakapa, New Zealand

93
Hybridization of Père David's and Red Deer

Whitley Otway*

Hybrids of Père David's (*Elapharus davidianus*) and red deer (*Cervus elaphus*) have occurred very rarely by natural mating at Woburn Abbey, England. Such hybrids would be ideal for deer farming on account of their size, antler quality, and potential for earlier calving. In November 1983, the first Père David's deer to be imported into New Zealand arrived at Charnley Park. The two stags and four hinds acclimatized rapidly to the change in hemispheres and unsuccessful attempts were made in 1984 and 1985 to produce hybrids by natural mating of the Père David stags with New Zealand red hinds. In April 1986, semen was collected from the Père David's stag by electroejaculation and inseminated into 20 "prepared" red hinds. The semen was placed in utero following laparotomy. Two hybrid fawns resulted, and one was successfully reared. This F1 hybrid has antlers which are intermediate between the two species. This F1 Père David's/red hybrid was mated in March 1988, at age 15 months with 20 pure red hinds, resulting in 14 fawns. Morphological characteristics of these fawns from birth to 15 months of age show considerable variation. The study is continuing to investigate growth rate, calving dates of progeny, and antler growth and quality. In conclusion, we found that the F1 Père David/red hybrid deer is fertile. The Père David's/red hybrid and the F1/red deer hybrids are suitable for venison production. Antler velvet production and quality, together with calving dates of the F1/red deer will be reported as the study proceeds.

*Charnley Park, Kaukapakapa, New Zealand

94
Genetic Variation, Parentage Testing, and Hybrid Identification in New Zealand Red Deer

Mike Tate*

A total of 450 red deer (*Cervus elaphus*) from four New Zealand feral herds were tested for genetic variation at ten blood protein loci. The herds were chosen to be representative of the range of European stock acclimatized to New Zealand. Animals captured from these feral herds have provided the initial stock for many New Zealand deer farms. The genetic variation found in this study suggests a parentage test based on eight of the loci would detect about 80% of any mismatches in deer farm pedigree records ($P_E = 0.8$). The red deer in the study were compared electrophoretically at the ten loci with 20 wapiti (*Cervus elaphus manitobensis*), 20 sika deer (*Cervus nippon*), and 20 Père David deer (*Elaphus davidianus*). The wapiti showed alleles that did not occur in the New Zealand red deer sample at four loci (in two cases the differences were fixed), the sika at three loci (one fixed difference) and the Père David deer showed fixed differences to red deer at five loci. In all the species other loci showed significant allele frequency differences to red deer. Identifying hybrids with these species has become necessary to maintain pure herds of farmed red deer in New Zealand. The presence of alleles that do not occur in New Zealand red deer provides the ability to screen biochemically for hybridization.

*Invermay Agricultural Centre, Private Bag, Mosgiel, New Zealand

95
Importance of Habitat Considerations in Interpreting Electrophoretic Data for White-Tailed Deer Populations

Robert E. Vanderhoof and Raymond K. Anderson*

Liver samples were collected from hunter-harvested white-tailed deer (*Odocoileus virginianus*) on Chambers Island ($n = 53$) and mainland Door County, Wisconsin ($n = 240$) during 1981–83. Five loci were chosen for horizontal starch-gel electrophoretic analysis on the basis of their reported polymorphism in white-tailed deer: sorbitol dehydrogenase (SDH), esterase (ES-2), phosphoglucomutase (PGM-2), malate dehydrogenase (MDH-1), and lactate dehydrogenase (LDH-2). Genotypic frequencies differed significantly ($P < 0.05$) between island and mainland herds at three of five loci studied. Significant deviations from Hardy-Weinberg equilibrium were observed for one of five loci from island deer and four of five loci for mainland deer. The proportion of heterozygotes was significantly higher for island deer at three of five loci. The proportion of loci for which an individual was heterozygous was also significantly higher for the island deer; *F*-statistics revealed significant genetic subdivision between island and mainland deer herds. Recruitment rate, body weight, antler points, main beam length, and main beam diameters were all significantly smaller for deer on Chambers Island than for deer on the mainland.

*Department of Wildlife and Fisheries, Mississippi State University, P.O. Drawer LW, MS State, MS 39762 and College of Natural Resources, University of Wisconsin-Stevens Point, Stevens Point, Wisconsin 54481, USA

Part 7
Nutrition and Physiology

96
Nutritional in Relation to Season, Lactation, and Growth of North Temperate Deer

Robert G. White

Abstract

Timing of reproduction in north temperate cervids has evolved in response to factors selecting for optimum birth weight, newborn survival, and growth to adult body size. Although melatonin controls proximal timing of reproduction, nutrition and range quality have a modulating effect. How plane of nutrition, maternal body condition, lactation, and offspring behavior interact to determine ovulation rate and conception success is at the center of understanding density-dependent effects in wild populations. Summer nutrition is involved; however, carry-over effects from winter and stochastic losses of newborn to predators interact in the field. Birth weight is highly correlated with maternal body weight, and in caribou, maternal protein reserves. Young animals strive to achieve body weight set-points, but the actual trajectory and ability to carry out compensatory growth may depend on lactation and grazing. Lactation initiates rate of growth to sexual maturity. Factors limiting milk synthesis rate involve both prepartum body condition and food availability postpartum. Growth rate of nursing-ruminating offspring (i.e., > 8 weeks of age) is highly linearly related to milk intake. The results strongly support the hypothesis that rumen digestion meets maintenance requirements, while milk nutrients escaping modification by rumen fermentation are preferentially used for growth.

Key words: Caribou, deer, grazing, growth, lactation, range, reindeer, seasonality

Seasonality and the Timing of Birth in Relation to Weather and Nutrition

In north temperate cervidae, timing of birth is highly constrained to a brief period in association with snow-melt and/or just prior to the emergence of new plant growth; however, genetic and ecological processes that cause this rather simple biological event are not fully understood. Proximal timing of reproduction by melatonin is now well recognized, but year-to-year and site-to-site variations in snow-melt and date of plant "green-up" are highly variable and are associated with the date of parturition. Long-term trends of snowmelt patterns have been shown (Vibe 1967). In a North American example, Barter Island, in northeastern Alaska, a 3-week trend to earlier melt-off since 1950 has been noted (Foster 1989; Eastland and White 1991). This area is of considerable interest to biologists as it is adjacent to the calving grounds of caribou (*Rangifer tarandus granti*) of the Porcupine herd (187 000 in 1989; S.G. Fancy unpublished data). In the past decade peak calving has occurred 4 days earlier (S.G. Fancy, personal communication), which is presently 3 to 8 June (Fancy et al. 1989). Whether this change has occurred in response to snowmelt is not known; however, the earlier calving presumably reflects genetic plasticity in the timing of birth for caribou. Whether this also reflects variation in gestation length is conjectural (Skogland 1984) and not well supported. Furthermore, Eastland et al. (1989) reported that patterns of snowmelt affected selection of calving sites by caribou which may be related to nutrition and predator avoidance.

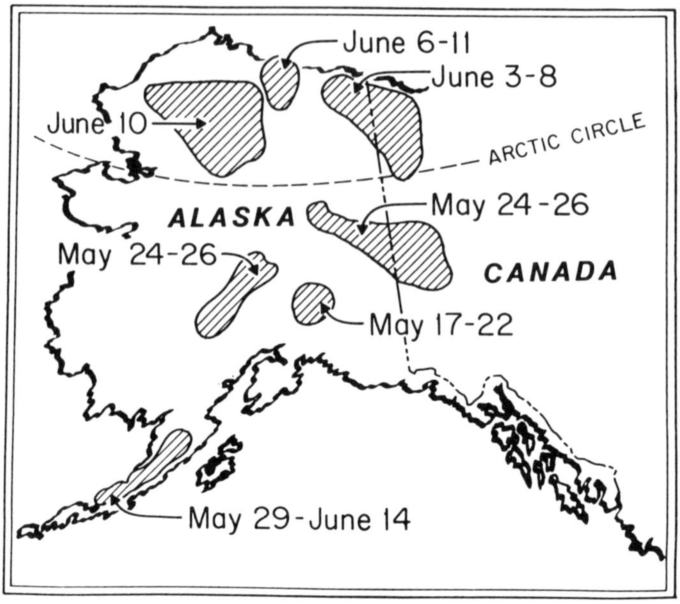

FIGURE 96.1. Example of variation in between-population peak calving dates of Alaskan caribou (Eastland and White 1991). Estimates were made by employees of Alaska Department of Fish and Game and the U.S. Fish and Wildlife Service.

Peak calving dates also correlate with density-dependent effects due to nutrition (Espmark 1980; Skogland 1984, 1985; Clutton-Brock et al. 1987). Among caribou herds in Alaska peak calving varies by more than 4 weeks (10 May to 10 June; Fig. 96.1). These results could suggest calving is latitudinally based; however, it is more likely to be in relation to seasonal phenology; the most southern population calves as late as the most northern. Latitudinal (Bischoff 1957; Ransom 1966; Robinette et al. 1977) and altitudinal (Weber 1966; Robinette et al. 1977; Bowyer 1991) trends in birth date and synchrony have been reported for cervids as in other ruminants (Rutberg 1987; Bunnell 1982). In a northern example, Skogland (1989) has shown that peak calving in Norwegian reindeer (*R.t. tarandus*) is highly correlated with timing of new plant growth, and the implication is that females time birth to maximize parental investment through lactation. The differences in calving date can be attributed to genetic origin of the populations as well as to a nutritional effect due to plant phenology (Skogland 1989). Variation in date of peak calving also has implications for animal performance (Clutton-Brock and Albon 1989). Reimers (1983a) demonstrated that female reindeer calving early (1 May) are 30 kg heavier than those calving late (29 May), the response over these populations was linear (Fig. 96.2), and populations varied widely in their ecology and population density. These data are from the same reindeer populations reported by Skogland (1983, 1984, 1989), who showed that fetal weight increased as a power function of maternal weight in late gestation (Fig. 96.2). Similar results occur in caribou, but Allaye-Chan and White (1991) showed that fetal weight at term is correlated with maternal body protein and not with body fat. In red deer (*Cervus elaphus*) birth weight is affected by an interaction between population density and overwinter calf survival and spring weather (Clutton-Brock et al. 1987). Birth weight may vary among years, probably in response to yearly ecological effects (Albon et al. 1983; Albon, this volume) and is correlated with many future life-history traits (Clutton-Brock and Albon 1989). Reproduction rate in *Odocoileus* is also

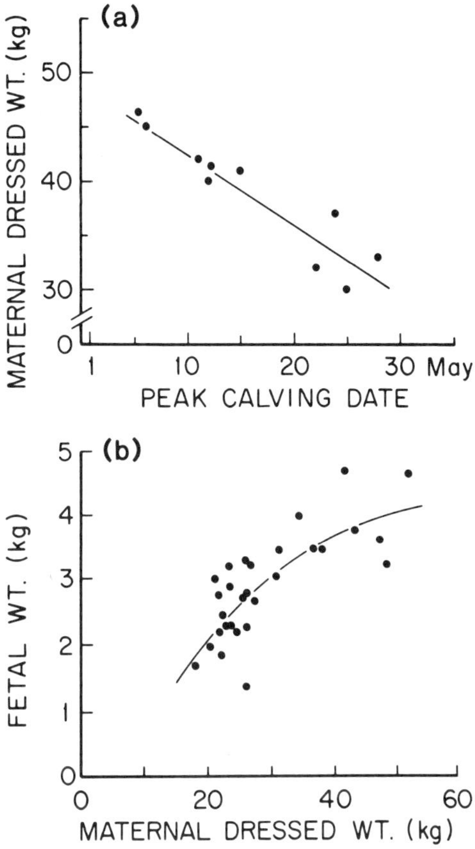

FIGURE 96.2. Decrease in maternal weight of pregnant Norwegian reindeer in relation to peak calving date (a). Data are from discrete reindeer herds (Reimers 1983a) which also show a significant effect ($r^2 = 0.63$, $P < 0.001$) of maternal body size on fetal weight (b) as reported by Skogland (1984) for three reindeer herds.

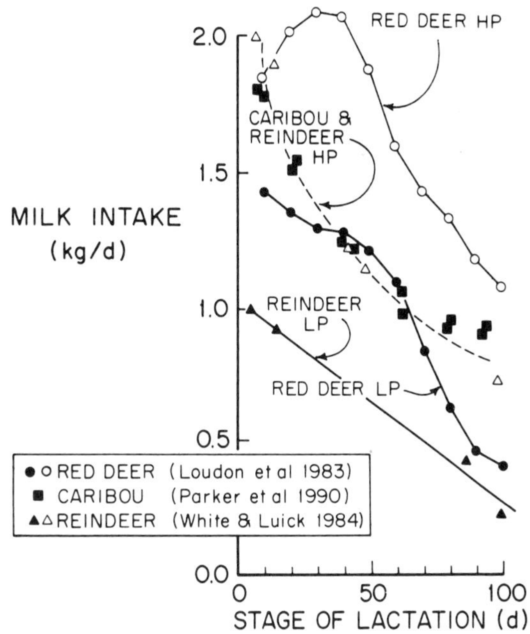

FIGURE 96.3. Effect of plane of nutrition on the lactation curve for red deer, reindeer, and caribou. Milk production (kg/day per animal) in red deer was estimated by calf weighing, whereas that for reindeer and caribou was determined using the double-isotope technique (Holleman et al. 1975, 1988).

related to weather conditions (Picton 1984); however, a clear relationship to nutrition was not made.

Lactation

Peak production of milk and the shape of the lactation curve are both affected by nutrition in cervids. Therefore, timing of calving also may have important implications for milk production (White and Luick 1984) as well as birth weight. If plane of nutrition during late pregnancy is low, and maternal weight and fat reserves lessened, total milk production is reduced, and its peak occurs within the first week of lactation (Rognmo et al. 1983). If maternal body condition and plane of nutrition are high, then in some deer species total milk production is higher (Loudon and Kay 1984; Carl and Robbins 1988, but see also Parker et al. 1990), and Loudon and Kay (1984) showed that the peak is 3 weeks postcalving (Fig. 96.3). Data shown in Fig. 96.3 are taken from different studies that were not matched for absolute plane of nutrition. Data on peak production of milk for reindeer and caribou fed high-plane rations (White and Luick 1984; Parker et al. 1990) suggest that natural selection has not resulted in major differences in the shape of their lactation curves even though caribou and reindeer vary in life-history traits. In particular, reindeer mature earlier than caribou, and the peak calving date in reindeer precedes caribou by 3 to 4 weeks at the same latitude (e.g., 28 April for semi-domestic reindeer and 25 May for caribou of the Delta herd, at the Large Animal Research Station, Fairbanks, Alaska).

TABLE 96.1 Comparison of early winter (November) mean (± SE) body weight (kg) of caribou (originating from the Delta herd in interior Alaska) and reindeer calves maternally raised under the same environmental conditions at the Large Animal Research Station, University of Alaska Fairbanks

Sex	Caribou			Reindeer			P
	n	$\bar{x}$	SE	n	$\bar{x}$	SE	
Female	3	69	1.9	5	49	2.5	< 0.001
Male	2	74	3.0	2	55	5.0	0.10–0.05

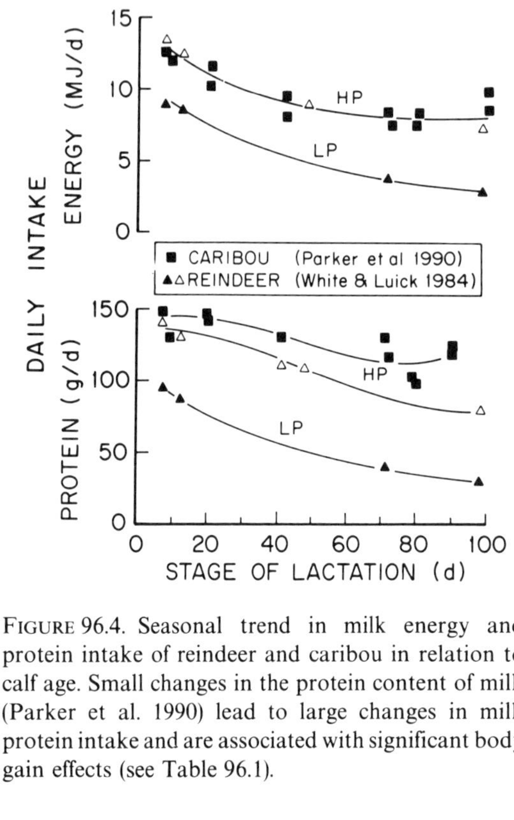

FIGURE 96.4. Seasonal trend in milk energy and protein intake of reindeer and caribou in relation to calf age. Small changes in the protein content of milk (Parker et al. 1990) lead to large changes in milk protein intake and are associated with significant body gain effects (see Table 96.1).

Because lactation curves of reindeer and caribou are similar, the possibility that milk quality, and particularly protein content, may have major implications for calf growth deserves further attention. Milk dry matter, protein and energy increase inversely with milk production in some cervids (Loudon and Kay 1984; White and Luick 1984). Under the same comparative high plane of nutrition, caribou calves achieve higher body weights by winter than do reindeer (Table 96.1). Although the volume of milk production is similar (Fig. 96.3), milk protein concentration in caribou is higher than in reindeer. Higher protein intake by caribou calves (Fig. 96.4), therefore, may explain their greater weaning weights (Table 96.1). No comparative foraging studies have been made, so that differences in foraging rate cannot be discounted. Future research on lactation should focus on the relative roles of maternal genetic/hormonal control versus that due to the mammary gland's own autocrine control system (sensu Wilde and Peaker 1990).

Forage Protein to Energy Ratio and Partitioning of Absorbed Nutrients

Seasonal changes in the ratio of forage protein energy have implications for growth and changes in body condition of deer on a proximal (White et al. 1987) and evolutionary basis (Geist 1971, 1986). We question "to what extent does the dietary energy/protein ratio effect utilization of body reserves for lactation in caribou and reindeer?" These studies suggest that metabolizable energy intake (MEI) tends to limit milk production in caribou and reindeer for most of lactation when dietary crude protein is at least 10% (Allaye-Chan and White 1991). Our interpretation is that maternal fat reserves and MEI interact to maximize milk production where protein content of forage and projected protein intake of individuals is extremely high (White et al. 1987). Matching milk synthesis to energy supply presumably occurs through hormonal control over the partitioning of nutrients between the mammary gland and other maternal tissues. In sheep with experimentally induced high plasma concentrations of growth hormone (GH) accompanied by low insulin (I), a low I:GH ratio is correlated with high rates of lactogenesis (Hart et al. 1978). This association holds when lactogenesis is increased by cold stress associated with winter shearing (Symonds et al. 1990). The hypothesis is that low insulin concentration minimizes glucose uptake by peripheral tissues, sparing it for uptake by the mammary gland (Lomax et al. 1979). High GH levels may increase lipolysis of adipose tissue during lactation (Johnson and Hart 1986), presumably to rapidly supply precursors of milk fat that would spare and be more efficiently used than ruminal acetate for the synthesis of milk lipids. The role of metabolic intermediates, particularly propionic and acetic acids, in controlling metabolic partitioning could act indirectly through affects on insulin secretion, but a direct role by mass action of acetate on mammary lipogenesis versus use of propionate on maternal tissue cannot be dismissed. Variation in metabolic partitioning due to endocrines (I:GH ratio) occurs during lactation and fetal development without accompanying changes in plasma concentrations of glucose and nonesterified fatty acids (Symonds et al. 1988a, 1988b, 1990). Alternatively, control by mass action would be accompanied by changes in plasma concentrations of these intermediates.

Forage Quality and Quantity

For wild deer, the question as to whether food quality or availability drives the early lactation curve is speculative. Behavioral studies indicate that most deer undergo a brief period of perinatal hypophagia followed by a maximization of time spent foraging; increasing daylight hours at northern latitudes may facilitate this effect. These data may be interpreted as showing that deer are maximizing intake; however, the extent to which they meet requirements varies between years and between habitats. The hypothesis that deer track closely plant phenology (Klein 1970) explains their ability to obtain high quality food; however, if metabolizable energy is limiting, then forage of higher biomass also must be ingested (White and Trudell 1980). Under all but the most extreme circumstances, quality of food is high in spring and early summer. Between-year differences, driven by variation in abiotic conditions such as snowmelt, temperature, and moisture, control both initiation of plant growth and plant productivity so that plant biomass per se may limit eating rate, particularly in deer populations that can be limited in feeding time by insect harassment (Fancy 1986). Albon (Chapter 2 this volume) emphasized that between-year abiotic effects have extreme implications for red deer survival and growth. This has direct implications for many cervids as low forage biomass may limit eating rate, as reported for reindeer and elk (*Cervus canadensis*) (Trudell and White 1981; Wickstrom et al. 1984; Spalinger et al. 1986). Nonetheless, not all deer species eat in relation to biomass, as exemplified by studies on mule deer (*O. hemionus*) and black-tailed deer (*O.h. columbianus*) (Wickstrom et al. 1984; Spalinger et al. 1988).

Studies on the effect of restricted grazing and competition for food resources on food intake in cervids are few. Some of our early studies in northern Alaska (White and Trudell 1980) suggest that under restricted grazing (i.e., lower forage biomass), food intake is maintained by a combination of diet shifting and consumption of dead plant material of low nutritional value. This analysis (Fig. 96.5) shows that a continuous 3-day grazing event in a limited area reduces plant biomass and results in a decline in daily dry matter intake from 99 to 83 $g/kg^{0.75}$. A significant replacement of willows (*Salix* sp.) by lichens plus dead graminoid leaves prevents an extreme reduction in food intake, but also results in an increase in rumen retention time. This increased retention time increases diet digestibility, which results in no change in digestible dry matter intake. Thus, MEI is maintained; however, intake of crude protein would be reduced by 50% (Fig. 96.5). Clearly, this outcome could have major implications for production as at this time of the year caribou are depositing protein or are lactating. Mobilization of body protein may not be possible for synthesis of milk protein (Reid et al. 1966; Coppock et al. 1968; Kaufmann 1979; White et al. 1989), and a decline in milk production would be predicted.

Reproduction and Maternal Body Condition

Demands of lactation in concert with variation in forage quality and availability affect maternal body condition in late summer, which may be important to both timing of rut and reproductive success via ovulation, conception, and fetal retention and development. For species such as caribou and reindeer, subject to extreme insect harassment, then, the energetic costs of fighting insects (Russell 1976; Reimers 1980) and time taken from foraging (Fancy 1986) combine to

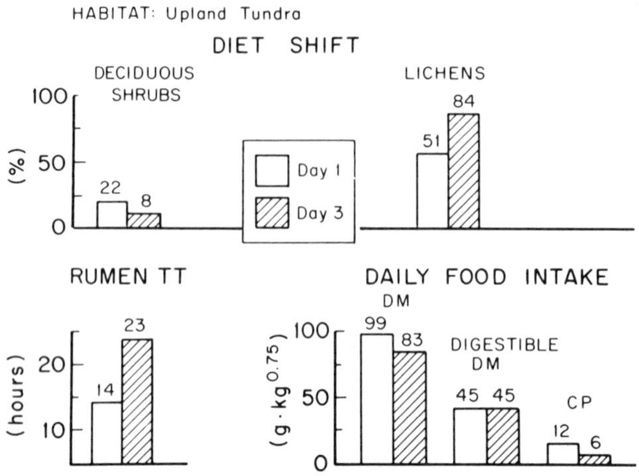

FIGURE 96.5. Projected changes in diet, rumen turnover time (TT), daily dry matter (DM) intake, digestible dry matter intake and crude protein (CP) intake in reindeer tethered on upland tundra for a continuous 3 days of grazing. Diet was assessed by visual observation, rumen TT using ^{51}CrEDTA and digestibility from dietary components suspended in the rumen in sacco for the retention time (i.e., 14 or 23 h). Data are reanalyzed from White and Trudell (1980) and Moore (1982).

exert a powerful control over body condition of females in the fall. Theoretical and empirical studies show that conception success is related to maternal weight for deer species in environments without predators (Hamilton and Blaxter 1980; Reimers 1983a; Albon et al. 1983, 1986; Clutton-Brock and Albon 1989); however, evidence for heavily predated species such as caribou is equivocal (Davis et al. 1991) but gaining support (Cameron et al. 1991). High postnatal calf losses due to predation can result in an increase in body condition of nonlactating females, which in turn can result in a higher conception success (i.e., it raises herd fecundity and may possibly maintain early breeding with consequent early calving). A clear effect of density-dependent conception success in relation to body weight has been shown for red deer in which the body weight required for median fertility increases linearly with density (Clutton-Brock and Albon 1989). Whether this is a social factor or is a reflection of the lack of "flushing," i.e., a rising plane of nutrition which induces ovulation (Lindsay 1976), is not known. In caribou, conception success appears more closely associated with maternal fat reserves than body weight (Thomas 1982; Allaye-Chan and White 1991). Early calving is usually associated with low animal density, high birth weight, good calf survival, and therefore a high plane of nutrition. Nevertheless, variation in calving dates in deer species suggests advantages exist to later calving, or at least late calving may not be detrimental to short-term fitness. Although not reported for deer, behavioral studies in muskoxen (*Ovibos moschatus*) and Dall's sheep (*Ovis dalli*) noted that late-born individuals spend more time suckling than early-born individuals (Jingfors 1984; J.L. Rachlow and R.T. Bowyer, unpublished data). These studies suggest to me that offspring born into more advanced vegetation biomass can benefit by compensatory growth from an increased maternal plane of nutrition, perhaps accompanied by a more efficient maternal nutrient partitioning to lactation with or without an increased expenditure of maternal reserves.

The hypothesis that plane of nutrition in summer, as dictated by habitat condition, can affect both maternal body condition and reproductive success in reindeer has been stressed by Reimers (1983b). In contrast, Skogland (1984) using data from the same reindeer population argues that winter range ecology, through carry-over effects, has the most important nutritional control over fecundity as there were no measurable differences in summer range (Skogland 1983). To some extent these counterviews can be explained by the compensatory growth phenomenon (Reimers 1983b; Renecker and Samuel 1991). The degree to which compensatory growth is possible depends on summer nutrition which must be excellent to elicit a full response. In the Norwegian studies, summer range differences may be large yet not necessarily statistically different. I have shown that small differences in range biomass and quality can have major effects on individual productivity through multiplier effects (White 1983).

Compensatory Growth and Body Weight Set-Points

In recent years a reemphasis has been placed on the ability of young animals to grow to meet body mass and body dimensional set-points (Renecker and Samuel 1991) and to undergo compensatory growth. Provided plane of nutrition is high, young animals undergo compensatory growth in summers following low growth periods when the set-point was not achieved (Fig 96.6). If summer nutrition combines with factors that limit daily food intake, then compensation cannot be made and carry-over effects from the previous winter would be expressed. The extent that compensatory growth can occur in the nursing calf that is ruminating deserves further study. Milk supply controls growth rates of deer calves (Loudon and Kay 1984; White and Luick 1984) in the preruminant stage of development. Loudon and Kay (1984) speculate on the importance of milk in maintaining an optimal ratio of amino-nitrogen

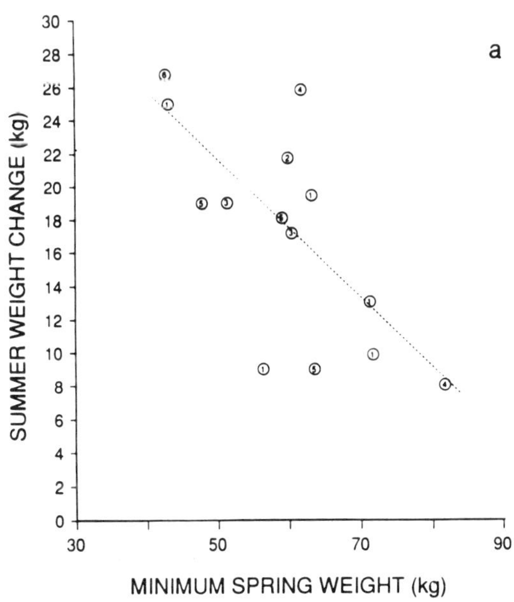

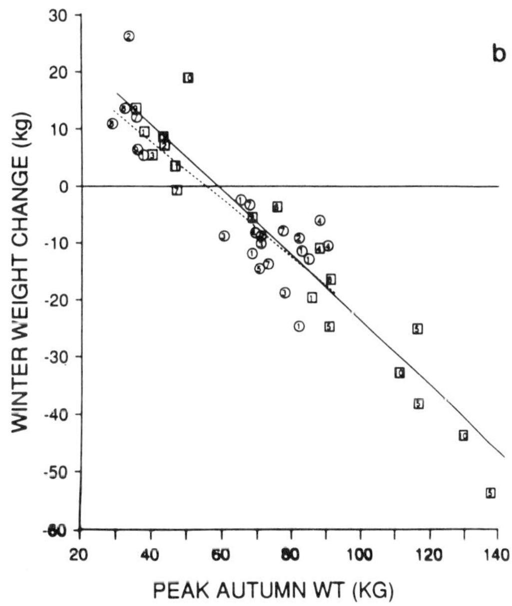

FIGURE 96.6. Weight change of hand-reared, female mule deer (animals were 5 to 72 months of age) during spring and summer in relation to minimum spring weight (Renecker and Samuel 1991). Equation is as follows: $Y = -0.403X + 41.35$; $SE_b = 0.122$, $r^2 = 0.47$, $P < 0.01$ (a). Winter weight change of hand-reared mule deer (animals were 5 to 72 months of age) fed ad libitum a pelleted alfalfa-concentrate ration (b). Equations are as follows: for females, $Y = -0.515X + 28.53$, $SE_b = 0.067$, $r^2 = 0.74$, $P < 0.01$; for males $Y = -0.594X + 34.51$, $SE_b = 0.041$, $r^2 = 0.93$, $P < 0.01$. Observations with the same number represent an individual animal in different years. Data are reproduced courtesy of the *Canadian Journal of Zoology*.

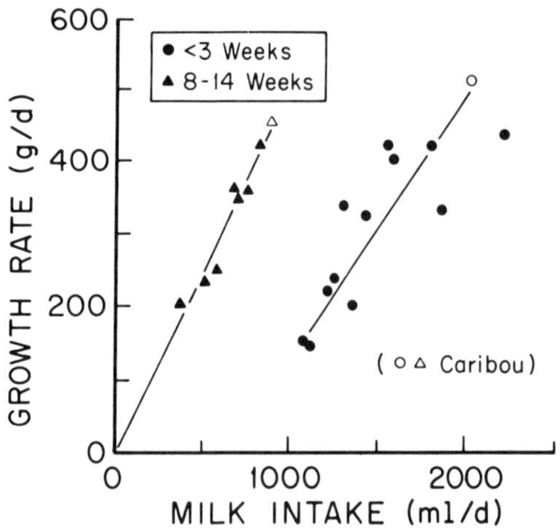

FIGURE 96.7. Relations between calf growth (g/day) and milk intake (ml/day) of reindeer (●, ▲) and caribou (0, △). Data are from White and Luick (1984) and Parker et al. (1990). The left shift in the relationship with age may reflect preferential use of milk nutrients for growth once rumen function and forage nutrients can support maintenance requirements.

metabolizable energy for maintenance and growth. It is surprising that relations between growth and milk intake of the preruminant are so variable, e.g., those reported for reindeer (Fig. 96.7) and red deer (Loudon and Kay 1984). If milk production were to assist in growth compensation in calves, then the relation between calf growth and milk intake in the ruminant (late summer) stage of development should be demonstratable. Relations between growth rate and milk intake in reindeer (Fig. 96.7) show that the function changes with age, probably as forage intake and rumen metabolic by-products add to the nutrient intake of calves and alter the amino-nitrogen metabolizable energy ratio (sensu Loudon and Kay 1984). After 8 weeks the regression line passes through the origin (Fig. 96.7), and the fit is very good. The hypothesis advanced is that nutrient supply from milk is partitioned differentially to that of ruminal origin. In the preruminant calf, milk nutrients are used for both maintenance and growth; however, as milk production declines and the rumen develops and makes a significant contribution to nutrient supply, I postulate that milk nutrients are used at maximum efficiency for growth while rumen metabolites are used to meet maintenance requirements. Hence, in temperate systems the level of milk intake in late lactation may be critical to maintaining growth rates of calves as food quality declines. This phenomenon could promote compensatory growth, and also maintain growth into winter. On extreme low planes of nutrition, typical where range-carrying capacity is exceeded, females could prolong lactation and undergo a breeding pause. Maternal investment theory (Trivers and Willard 1973) predicts that maximum benefit would occur if the supported offspring is a male. Prolonged lactation would maximize first-year growth and thereby project male offspring to a growth curve reaching higher mature adult body weight than males of the same cohort weaned at the usual time (i.e., immediately prerut).

Future Studies

In adapting to a highly seasonal environment, deer species are necessarily constrained in their timing of reproduction and growth phases and thus cannot change when encountering more nonseasonal and agricultural systems that may typify farmed deer production in New Zealand, Australia, and China and which is developing in North America. In this chapter I have emphasized the way lactation might respond to ecological changes and how nutrient-partitioning of the adult female might be affected. A different "strategy" in nutrient partitioning may occur in young and adult males in which natural selection has resulted in high growth efficiency. Fennessey et al. (1991) hypothesize that females have a more fat-biased metabolism than males. Developing management systems to overcome genetic constraints presents a challenge to agricultural biologists and current research emphasis is on understanding mechanisms underlying these life-history traits. Such research undoubtedly will lead to management breakthroughs. For the wildlife biologist the same research effort will improve our understanding of animal ecology and evolution. Current research on seasonal rhythms will be beneficial to both researchers. The more northern cervids, which are perhaps most highly constrained in their physiological responses, have been studied in depth. A comparative study with white-tailed (*Odocoileus virginianus*) and mule deer, which show local adaptations to temperate through tropical systems, will give even further insight into which selective forces have been most influential in the origin of the various physiological mechanisms that underlie life-history traits.

Acknowledgements

I would like to thank R.T. Bowyer and D.R. Klein for their review of this manuscript and A.C. Allaye-Chan and W.A. Eastland for inclusion of unpublished data. L.A. Renecker provided input on recent research on compensatory growth. Funding for my studies was made possible through grants from the Canadian Wildlife Service, the U.S. Fish and Wildlife Service (RWO no. 30), and a grant to the Large Animal Research Station by the Institute of Arctic Biology. I would like to thank the Mississippi State University for a travel grant to present this paper.

References

Albon SD, Mitchell B, Huby BJ, Brown D (1986) Fertility in female red deer (*Cervus elaphus*): the effects of body composition, age and reproductive status. J Zool London A 209:447–460

Albon SD, Mitchell B, Staines BW (1983) Fertility and body weight in female red deer: a density-dependent relationship. J Anim Ecol 52:969–980

Allaye-Chan AC, White RG (1991) Seasonal and between year variation in body condition in caribou of the porcupine herd. In Butler C, Mahoney S (eds) Proc 4th North Am Caribou Workshop. Newfoundland and Labrador Wildlife Division, Department of Environment and Lands, St. John's Nfld., pp 103–108

Bischoff AI (1957) The breeding season of some California deer herds. Calif Fish Game 43:91–96

Bowyer RT (1991) Timing of parturition and lactation in southern mule deer. J Mammal (in press)

Bunnell FL (1982) The lambing period of mountain sheep: synthesis, hypotheses, and tests. Can J Zool 60:1–14

Cameron RD, Smith WT, Fancy SG (1991) Comparative body weights of pregnant/lactating and nonpregnant female caribou. In Butler C, Mahoney S (eds) Proc 4th North Am Caribou Workshop. Newfoundland and Labrador Wildl Div, Department of Environment and Lands, St. John's, Nfld., pp 109–114

Carl GR, Robbins CT (1988) The energetic cost of predator avoidance in neonate ungulates: hiding versus following. Can J Zool 66:239–246

Clutton-Brock TH, Albon SD (1989) Red Deer in the Highlands. BSP Professional Books, Oxford

Clutton-Brock TH, Guinness FE, Albon SD (1987) Interactions between population density and maternal characteristics affecting fecundity and juvenile survival in red deer. J Anim Ecol 56:857–871

Coppock CE, Tyrrell HF, Merrill WG, Reid JT (1968) Protein reserves in cows. Proc Cornell Nutr Conf Feed Manuf, pp 86–94

Davis JL, Adams LG, Valkenburg P, Reed DJ (1991) Relationships between body weight, early puberty, and reproductive histories in central Alaskan caribou. In Butler C, Mahoney S (eds) Proc 4th North Am Caribou Workshop. Newfoundland and Labrador Wildlife Division, Department of Environment and Lands, St. Johns, pp 115–142

Eastland WA, White RG (1991) Potential effects of global warming on calving caribou. In Weller G (ed) Proc Int Conf on the Role of Polar Regions in Global Change (in press)

Eastland WG, Bowyer RT, Fancy SG (1989) Effects of snow cover on selection of calving sites by caribou. J Mammal 70:824–828

Espmark Y (1980) Effects of maternal pre-partum undernutrition on early mother-calf relationships in reindeer. In Reimers E, Gaare E, Skjenneberg S (eds) Proc 2nd Int Reindeer/Caribou Symp, Norway 1979. Direktoratet for vilt og ferskvanns-fisk, Trondheim, pp 485–496

Fancy SG (1986) Daily energy budgets of caribou: a simulation approach. PhD thesis, University of Alaska, Fairbanks

Fancy SG, Pank LF, Whitten KR, Regelin WL (1989) Seasonal movements of caribou in arctic Alaska as determined by satellite. Can J Zool 67:644–650

Fennessy PE, Thompson JM, Suttie JM (1991) Season and growth strategy in red deer: evolutionary implications and nutritional management. In Renecker LA (ed) Proc 2nd Int Wild Ranching Symp (in press)

Foster JL (1989) The significance of the date of snow disappearance on the arctic tundra as a possible indicator of climate change. Arct Alp Res 21:66–70

Geist V (1986) The paradox of the great Irish stags. Nat. History 3:54–64

Geist V (1971) Mountain Sheep—A Study in Behavior and Evolution. University of Chicago Press, Chicago

Hamilton WJ, Blaxter KL (1980) Reproduction in farmed red deer. 1. Hind and stag fertility. J Agric Sci 95:261–273

Hart IC, Chadwick PME, Simmonds AD (1978) The effect of intravenous bovine growth hormone on human pancreatic growth hormone-releasing factor on milk production and plasma hormones and metabolites in sheep. J Endocrinol 105:189–196

Holleman DF, White RG, Lambert PJ (1988) Analytical procedures for estimating milk intake and yield in steady-state and nonsteady-state systems. J Dairy Sci 71:1189–1197

Holleman DF, White RG, Luick LR (1975) New isotope methods for estimating milk intake and yield. J Dairy Sci 58:1814–1821

Jingfors K (1984) Observations of cow-calf behavior in free-ranging muskoxen. Biol Pap Univ Alaska Spec Rep 4:105–109

Johnson ID, Hart IC (1986) Manipulation of milk yield with growth hormone. In Haresign W, Cole DJA (eds) Recent Advances in Animal Nutrition. Butterworths, London, pp 105–123

Kaufmann W (1979) Protein utilization. In Broster WH, Swan H (eds) Feeding Strategy for the High Yielding Dairy Cow. Granada Publishing, London, pp 90–113

Klein DR (1970) Tundra ranges north of the boreal forest. J Range Manage 23:8–14

Lindsay DR (1976) The usefulness to the animal producer of research findings in nutrition on reproduction. Proc Aust Soc Anim Prod 11:217–224

Lomax MA, Baird GD, Mallinson CB, Symonds HW (1979) Differences between lactating and non-lactating dairy cows in concentration and secretion rate of insulin. Biochem J 180:281–289

Loudon ASI, Kay RNB (1984) Lactational constraints on a seasonally breeding mammal: the red deer. In Peaker M, Vernon RG, Knight CH (eds) Physiological Strategies in Lactation. Symp Zool Soc London 51:233:252

Moore JT (1982) Factors affecting grazing patterns of reindeer on summer tundra ranges. MSc thesis, University of Alaska, Fairbanks 144 pp

Parker KL, White RG, Gillingham MP, Holleman DF (1990) Comparison of energy metabolism in relation to daily activity and milk consumption by caribou and muskox neonates. Can J Zool 68:106–114

Picton HD (1984) Climate and the prediction of reproduction of three ungulate species. J Appl Ecol 21:869–879

Ransom AB (1966) Breeding season of white-tailed deer in Manitoba. Can J Zool 44:59–62

Reid JT, Moe PW, Tyrell HF (1966) Energy and protein requirement of milk production. J Dairy Sci 49:215–223

Reimers E (1983a) Reproduction in wild reindeer in Norway. Can J Zool 61:211–217

Reimers E (1983b) Growth rate and body size differences in Rangifer, a study of causes and effects. Rangifer 3:3–15

Reimers E (1980) Activity patterns: the major determinant for growth and fattening in Rangifer. In Reimers E, Gaare E, Skjenneberg S (eds) Proc 2nd Int Reindeer/Caribou Symp, Norway 1979. Direc-toratet for vilt og ferskvannstisk, Trondheim, pp 466–474

Renecker LA, Samuel WM (1991) Growth and seasonal weight changes as they relate to spring and autumn set points in mule deer. Can J Zool 69:744–747

Robinette WL, Hancock NV, Jones DA (1977) The Oak Creek Mule Deer Herd. Utah State Div Wildl Res Publ 77–15

Rognmo A, Markussen KA, Jacobsen E, Blix AS (1983) Effects of improved nutrition in pregnant reindeer on milk quality, calf birth weight, growth, and mortality. Rangifer 3:10–18

Russell DE (1976) Computer simulation of *Rangifer* energetics. M.F. Thesis. University of British Columbia, Victoria

Rutbeg AT (1987) Adaptive hypotheses of birth synchrony in ruminants: an interspecific test. Am Naturalist 130:692–710

Skogland T (1989) Comparative social organization of wild reindeer in relation to food, mates and predator avoidance. Advances in Ethology 29. Paul Parey, Scientific Publishers, Berlin.

Skogland T (1985) The effects of density-dependent resource limitation on the demography of wild reindeer. J Anim Ecol 54:359–374

Skogland T (1984) The effects of food and maternal conditions on fetal growth and size in wild reindeer. Rangifer 4:39–46

Skogland T (1983) The effects of density-dependent resource limitation on size in wild reindeer. Oecologia 60:156–168

Spalinger DE, Hanley TA, Robbins CT (1988) Analysis of the functional response in foraging in the Sitka black-taild deer. Ecology 69:1166–1175

Spalinger DE, Robbins CT, Hanley TA (1986) The assessment of handling time in ruminants: the effect of plant chemical and physical structure on the breakdown of plant particles in the rumen of mule deer and elk. Can J Zool 64:312–321

Symonds ME, Bryant MJ, Lomax MA (1990) Metabolic adaptation during lactation in winter-shorn sheep. J Agric Sci Cambridge 114:201–205

Symonds ME, Bryant MJ, Lomax MA (1988a)

Metabolic adaptation during pregnancy in winter-shorn sheep. J Agric Sci Cambridge 11:137–145
Symonds ME, Bryant MJ, Shepherd DAL, Lomax MA (1988b) Glucose metabolism in shorn and unshorn pregnant sheep. B J Nutrition 60:249–263
Thomas DC (1982) The relationship between fertility and fat reserves in Peary caribou. Can J Zool 60:597–602
Trivers RL, Willard DE (1973) Natural selection of parental ability to vary the sex ratio. Science 179: 90–92
Trudell J, White RG (1981) The effect of forage structure and availability on food intake, biting rate, bite size and daily eating time of reindeer. J Appl Ecol 18:63–81
Vibe C (1967) Arctic animals in relation to climatic fluctuations. Meddelr Gronland 170:1–277, 153–162, and 181–192
Weber AJ (1966) Regional differences in fawning times of North Carolina deer. J Wildl Manage 30:843–845
White RG (1983) Foraging patterns and their multiplier effects on productivity of northern ungulates. Oikos 40:377–384
White RG, Luick JR (1984) Plasticity and constraints in the lactational strategy of reindeer and caribou. In Peaker M, Vernon RG, Knight CH (eds) Physiological Strategies in Lactation. Symp Zool Soc London 51. Academic Press, London, pp 215–232
White RG, Trudell J (1980) Vegetation use and forage consumption by reindeer and caribou in Arctic tundra. Arctic Alpine Res 12:511–529
White RG, Holleman DF, Tiplady BA (1989) Seasonal body weight, body condition, and lactational trends in muskoxen. Can J Zool 67:1125–1133
White RG, Holleman DF, Hubbert ME, Staaland H (1987) Herbivores in cold climates. In Hacker JB, Tennouth JH (eds) The Nutrition of Herbivores. Academic Press, Sydney, pp 465–486
Wickstrom ML, Robbins CT, Hanley TA, Spalinger DE, Parish SM (1984) Food intake and foraging energetics of elk and mule deer. J Wildl Manage 48:1285–1301
Wilde CJ, Peaker M (1990) Autocaine control in milk secretion. J Agric Sci Cambridge 114:235–238

97

Reevaluation of the Basal Metabolic Cycle in White-Tailed Deer

Peter J. Pekins, William W. Mautz, and John J. Kanter

Abstract

We investigated the seasonal cycle of basal metabolism in white-tailed deer (*Odocoileus virginianus*) by measuring basal metabolic rate (BMR) of four nonpregnant and four pregnant does from winter to summer. Metabolism was measured monthly with indirect respiration calorimetry from January–July/August, 1989. Mean monthly BMR of nonpregnant does ranged from 85.5 to 89.0 kcal/kg body weight$^{0.75}$/day. The mean BMR of nonpregnant does was 87.1 kcal. Individual, nonpregnant BMR, averaged across all months, varied from 80.7 to 93.5 kcal with minimal monthly or seasonal change. Metabolism of pregnant does increased curvilinearly from January to parturition. Postpartum BMR averaged 86.4 kcal, similar to nonpregnant does. Our data are similar to prior winter measurements of BMR, but are substantially lower than previously reported summer BMRs. Our data do not support the proposed existence of an inherent seasonal cycle in basal metabolism in white-tailed deer. Experimental temperatures and animal activity within metabolism chambers may have affected previous summer measurements.

Key words: BMR, deer, energy, gestation, metabolism, season, white-tailed deer

Introduction

Overwinter survival of white-tailed deer in northern regions of its range in dependent on a combination of physiological and behavioral adaptations. Behavioral strategies may include occupation of mature softwood stands (deer yards) (Gill 1957), selection of favorable microhabitats (Moen 1976), and decreased activity (Gill 1957). Physiological adaptations include a highly insulative pelage, deposition and subsequent catabolism of significant fat reserves (Mautz 1978), voluntary decrease in food consumption (Long et al. 1965), and reduction in basal metabolic rate (BMR) (Silver et al. 1969). A reduced BMR during winter, versus summer, theoretically lowers energy requirements during the resource-limited winter.

Seasonal metabolic cycles have been reported for moose (*Alces alces*) (Regelin et al. 1985; Renecker and Hudson 1986), elk (*Cervus canadensis*) (Pauls et al. 1981), and roe deer (*Capreolus capreolus*) (Weiner 1977), although metabolic depression from summer to winter is extremely variable (10% to 50%). In white-tailed deer, the seasonal metabolic cycle (> 50% amplitude) is based primarily upon a limited number of laboratory measurements and experimental deer (Silver et al. 1969, 1971; Thompson et al. 1973; Holter et al. 1975). However, Kanter (1989) recently found no difference in winter and summer BMR of four yearling does. Importantly, Kanter (1989) only used data from bedded animals at thermoneutral temperatures. Previous studies in metabolic chambers did not account for posture or activity and may have exposed deer to temperatures below the thermoneutral zone during summer, both of which elevate metabolic above BMR.

Discrepancies in results and methods of previous studies warranted further investigation of the metabolic cycle in deer under laboratory conditions indicative as possible of basal metabolism. The objective of this study was to further clarify whether a basal metabolic cycle exists in white-tailed deer by measuring basal metabolism of the same deer from winter to summer.

Methods

Eight does ($\geqq 1.5$ years old in January) were used to investigate the presence of a seasonal cycle in basal metabolism. Deer were maintained on ad libitum pelleted grain (Thompson 1972) and housed in outdoor enclosures exposing them to daily and seasonal weahter. Four deer were not bred the previous fall to compare metabolism of pregnant and nonpregnant animals. Animals were measured monthly from January–July/August, 1989, except pregnant deer during June.

Metabolism was measured with indirect calorimetry by measuring oxygen consumption in an open-flow system. Deer were confined in a wire-mesh pen ($0.8 \times 1.5 \times 1.2$ m) within a temperature-controlled chamber at thermoneutral temperatures determined by Kanter (1989). Data were collected from 2100 to 0700 h after a 12 h acclimation period in the chamber, preceded by a 48 h fast outdoors to ensure a postabsorptive state (Silver et al. 1969); water was provided ad libitum. Oxygen consumption and percent of time deer were standing was averaged every 20 min by a microcomputer that monitored electronic sensors and continuously recorded chamber and line temperature, barometric pressure, flow rate, oxygen concentration, water vapor pressure, and deer posture (bedded or standing). Analysis included only intervals with 100% bedded time, immediately preceded by two continuous intervals of 100% bedded time (i.e., last 20 min of a continuous 60-min bedded period). Previous trials indicated that this lag time was necessary to allow gas concentrations to stabilize after animal activity. Oxygen consumption was converted to kcal/kg $BW^{0.75}$/day assuming 4.7 kcal/LO_2. Analysis of variance ($P < 0.05$) was used to test for differences in metabolism between groups and months.

Results

All deer lost mass ($\bar{x} = 5.5$ kg) from January–March, maintained mass (± 1.5 kg) during April and May, and gained mass during summer. Two pregnant does lost 8 and 10 kg each from January–March. Pregnant does produced two singletons and two sets of twins. One set was born the night after a May trial, and presumably died of exposure within 24 h. It is unknown if lack of maternal care (no milk was in the stomachs), or premature birth, possibly related to the 72 h fast and chamber trial, contributed to the fawn

TABLE 97.1. Average monthly basal metabolism (kcal/kg $BW^{0.75}$/day) of four pregnant and four nonpregnant adult white-tailed deer from winter to summer in New Hampshire, 1989

Chamber temp (°C)	Nonpregnant Date	Mass (kg) $\bar{x}$	Metabolism $\bar{x}$	SE[a]	Pregnant Date $\bar{x}$	Mass (kg) SE	Metabolism $\bar{x}$	SE	% Gestational Period[b] $\bar{x}$	SE
2.0	16–19 Jan	56.5	87.3	5.4	20–26 Jan	69.0	76.0	2.0	35	4
2.0	15–20 Feb	53.5	85.5	4.5	22–28 Feb	63.5	79.7	4.5	48	2
7.5	22–25 Mar	52.5	86.3	4.1	27–30 Mar	63.0	91.0	4.0	70	3
11.0	20–25 Apr	53.0	89.0	2.1	26–29 Apr	64.0	107.0	6.0	85	3
15.0	24–31 May	53.0	86.5	2.6	15–18 May	65.0	117.5	7.8	94	3
21.0	19–21 Jun	54.0	86.5	2.5		—	—	—	—	—
25.0	1–4 Aug	56.0	88.7	3.7	24–27 Jul	67.0	86.4	4.1	0	0

[a] SE calculated from individual means
[b] Gestational period assumed as 200 days

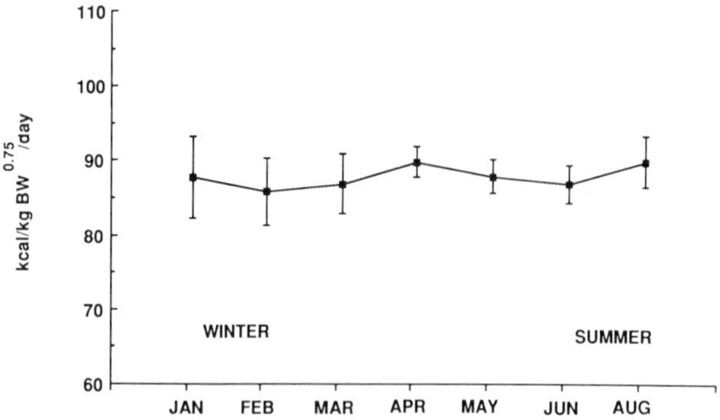

FIGURE 97.1. Basal metabolism of four nonpregnant white-tailed deer from winter to summer in New Hampshire, 1989.

mortality. Data analysis included at least five 20-min intervals per deer each month.

Individual BMR of nonpregnant deer, averaged across all months, ranged from 80.7 (SE = 4.0) to 93.5 (SE = 2.4) kcal/kg $BW^{0.75}$/day; no monthly or seasonal pattern was evident in highest or lowest BMR. Mean monthly BMRs of nonpregnant does were not different and ranged from 85.5 (SE = 4.5) to 89.0 (SE = 2.1) kcal/kg $BW^{0.75}$/day (Table 97.1, Fig. 97.1). The overall mean BMR of nonpregnant deer was 87.1 (SE = 2.6) kcal/kg $BW^{0.75}$/day.

Mean pregnant metabolism was not different ($P < 0.05$) from nonpregnant metabolism during January–March. Metabolism of pregnant deer increased curvilinealy from January to parturition with major gestational costs occurring after March (Fig. 97.2). Energetic cost of pregnancy at 94% of the gestational period was about 1.5 times that in January. Mean postpartum BMR (July) was similar to mean nonpregnant BMR (Table 97.1), and was not different ($P < 0.05$) from pregnant metabolism in January. Individually, postpartum BMR of two deer was similar to their metabolism in January, whereas two deer exhibited 10% and 27% higher summer metabolism. The latter animal exhibited similar metabolism during February and July.

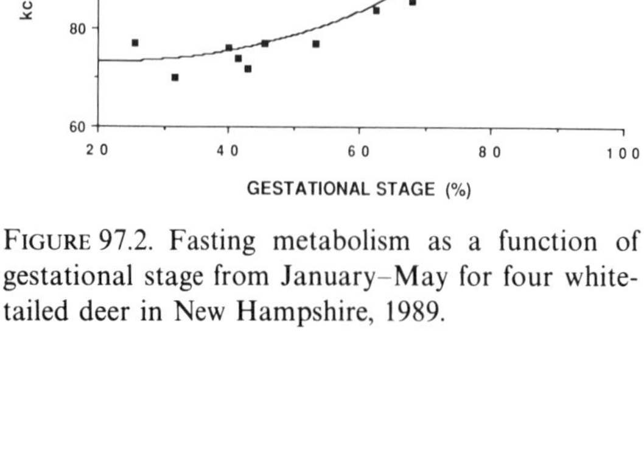

FIGURE 97.2. Fasting metabolism as a function of gestational stage from January–May for four white-tailed deer in New Hampshire, 1989.

Discussion

Results from nonpregnant does indicated no presence of a seasonal basal metabolic cycle. Similarly, pregnant does revealed no differences between January and July metabolism, although one deer exhibited a substantially higher metabolism during July. However, her metabolism in February and July were similar, as were those of the other three deer. Although the metabolism of pregnant and nonpregnant does during January were not significantly different ($P < 0.05$), metabolism of pregnant deer was about 14% lower. This may relate to the pregnant deer's larger size (Table 97.1), presumably attributed to higher fat content in the body, which theoretically influences (decreases) feed intake and fasting metabolism (Arnold 1985; Hudson and Christopherson 1985). It should be recognized that all deer in this study were on the same nutritional plane, and differences could be random. However, the potential effect of relative fat deposition in relation to voluntary reduction in feed intake and seasonal basal metabolism is unknown, but seems pertinent to understanding and predicting energy requirements of free-ranging deer.

Metabolic costs in winter were similar to those of past studies (Silver et al. 1969, 1971; Thompson et al. 1973; Kanter 1989), whereas our summer metabolic values are substantially lower (about 60%) than others, except for Kanter (1989). Arguably, elevated summer values in earlier studies reflect unmonitored activity and possibly thermal stress within the metabolic chamber. Minimal estimates of the cost of standing over lying in deer is 20% (Weiner 1977), however, Mattfeld (1974) and Mautz and Fair (1980) found increases nearer 60%. Mautz et al. (1985) found a mean 88% increase for deer in the same chamber used in this study, presumably because nonbedded activity includes movement beyond standing. Kanter (1989) estimated the lower critical temperature of deer in summer pelage as 20° to 25° C, meaning that previous studies, where experimental summer temperatures equaled 16° to 21° C, may have induced thermal stress and subsequent rise in heat production. Kanter (1989) reported a 40% rise in metabolism above BMR at 16° C during summer. Thus, significant activity and/or thermal stress could explain the disparity in reported summer metabolisms.

Although our data point to a lack of any inherent basal metabolic cycle or rhythm, it is important to recognize the similarity in winter metabolism reported in all studies. Hence, bioenergetic models and activity budgets built upon the basal energy requirements of deer during winter are little affected by our results. It is understandable, though ironic, that the resource limitations of winter likely led to the interpretation of a depressed winter metabolism. As Renecker and Hudson (1986) interpreted, a more accurate description would be an elevated summer metabolism in relation to the predicted interspecies BMR (70 kcal/kg $BW^{0.75}$ per day; Kleiber 1975). Obviously, models addressing seasons other than winter may need to be adjusted depending upon assumptions of basal metabolism.

Winter survival in an energy-demanding, resource-limited environment that imposes a negative energy balance on deer cannot be explained, in part, by a genetically inherited metabolic adaptation. Therefore, the relative importance of other adaptations, both physiological and behavioral, are magnified in the context of winter bioenergetics of deer. Although quantitative assessments of winter energy expenditure by deer may require no change, the relative importance of other physiological adaptations are different and require further investigation. In particular, the potential influence of body composition on feed intake, activity, and fasting metabolism during winter.

Acknowledgements

This study was funded through the McIntire-Stennis research program and is Scientific Contribution No. 1684 from the New Hampshire Agricultural Experiment Station. We thank numerous graduate and undergraduate students who assisted with husbandry and handling of the deer, and J.A. Litvaitis and D.L. Verbyla for constructive reviews of this paper.

References

Arnold GW (1985) Regulation of forage intake. In Hudson RJ, White RG (ed) Bioenergetics of Wild Herbivores. CRC Press, Boca Raton, Fla., pp 81–101

Gill JD (1957) Review of deer yard management 1956. Maine Department of Inland Fish and Game Bulletin 5

Holter JB, Urban WE Jr, Hayes HH, Silver H, Skutt HR (1975) Ambient temperature effect on physiological traits of white-tailed deer. Can J Zool 53:679–685

Hudson RJ, Christopherson RJ (1985) Maintenance metabolism. In Hudson RJ, White RG (eds) Bioenergetics of Wild Herbivores. CRC Press, Boca Raton, Fla, pp 121–142

Kanter JJ (1989) The influence of temperature and season on metabolic rate of white-tailed deer. MSc thesis, University of New Hampshire, Durham

Kleiber M (1975) The Fire of Life. Robert E. Greiger, Huntington, N.Y.

Long TA, Cowan RL, Strawn GD, Wetzel RS, Miller RC (1965) Seasonal fluctuations in feed consumption of the white-tailed deer. Pa State Univ, College of Agric, Prog Rep 262

Mattfeld GF (1974) The energetics of winter foraging by white-tailed deer—a perspective on winter concentration. PhD thesis, State University of New York, Syracuse

Mautz WW (1978) Sledding down a bushy hillside: the fat cycle in deer. Wildl Soc Bull 6:88–90

Mautz WW, Fair J (1980) Energy expenditures and heart rate for activities of white-tailed deer. J Wildl Manage 44:333–342

Mautz WW, Pekins PJ, Warren JA (1985) Cold temperature effects on metabolic rate of white-tailed, mule, and black-tailed deer in winter coat. In Fennessy PF, Orew KR (eds) The Biology of Deer. R Soc N Z Bull 22:453–457

Moen AN (1976) Energy conservation by white-tailed deer in the winter. Ecology 57:192–198

Pauls RW, Hudson RJ, Sylvin S (1981) Energy expenditure of free-ranging wapiti. Alberta Wildl Prod Res Bull 3:87–91

Renecker LA, Hudson RJ (1986) Seasonal energy expenditure and thermoregulatory responses of moose. Can J Zool 64:323–327

Regelin WL, Schwartz CC, Franzman AW (1985) Seasonal energy metabolism of adult moose. J Wildl Manage 49:388–393

Silver H, Holter JB, Colovos NF, Hayes HH (1971) Effects of falling temperatures on heat production in fasting white-tailed deer. J Wildl Manage 35: 37–46

Silver H, Colovos NF, Holter JB, Hayes HH (1969) Fasting metabolism of white-tailed deer. J Wildl Manage 33:490–438

Thompson CB (1972) Nutrition of white-tailed deer fawns. MSc thesis, University of New Hampshire, Durham

Thompson CB, Holter JB, Hayes HH, Silver H, Urban WE Jr (1973) Nutrition of white-tailed deer. I. Energy requirements of fawns. J Wildl Manage 37:301–311

Weiner J (1977) Energy metabolism of the roe deer. Acta Theriol 22:3–24

98
Nutritional Strategies and Winter Survival of European Roe Deer in Norway

Øystein Holand and Hans Staaland

Abstract

The European roe deer (*Capreolus capreolus*) are the smallest free-ranging cervid in the northern temperate region. Recently, they have expanded their distribution north of the Arctic circle. Due to their low body mass (25 to 35 kg) they depend, presumably, on high quality forage to cover their energy demands. However, in Norway available winter staple is mostly low-quality forage from dwarf shrubs like blueberry (*Vaccinum myrtillus*), and when snow cover exceeds about 0.5 m the animals are forced to feed on twigs and buds from deciduous and evergreen trees. Apparently the roe deer have developed morphological, physiological, and behavioral traits that enable them to survive winter on such forage. Among these winter adaptations are a 40% increase in alimentary dry matter content, reduced food intake, and increased total mean retention time of gastrointestinal fill. Based on in vitro studies with rumen inoculum from roe deer culled at different seasons, there is apparently little cellulolytic activity during all seasons. Amount of fat deposits, the midsummer rut, and delayed implantations are additional strategies for winter survival.

Key words: Nutritional strategies, roe deer, winter survival

Introduction

Since the turn of the century, European roe deer have expanded their range from Sweden through the southeastern corner of Norway, and it is now a numerous game species in most areas in the southern and middle part of Norway. At the same time red deer (*Cervus elaphus*) have expanded from the west coast through southern Norway, and moose (*Alces alces*) have increased from a small base to a number sustaining a yearly harvest of about 25 000 animals. The reasons for these expansions are obscure. The larger species, by virtue of their body size, can be presumed to be preadapted to a strict seasonal environment, whereas the smaller roe deer should be constrained by its small body to survive on high-quality winter forage. This is in accordance with the hypothesis of nutritional constraints arising from body size (Demment and Van Soest 1985). Nevertheless the roe deer, except for large die-offs in years with adverse winter climatic conditions, has expanded and now exists in large populations. In this chapter we review the evidence that this is possible due to physiological and reproductive traits enabling these animals to survive winters on low-quality forage.

Body Composition

The field-shot animals have a well-defined annual cycle in whole body fat concentration, with no intersexual asynchrony (Fig. 98.1). Maximum fat reserves in roe deer are relatively small compared with most northern cervids

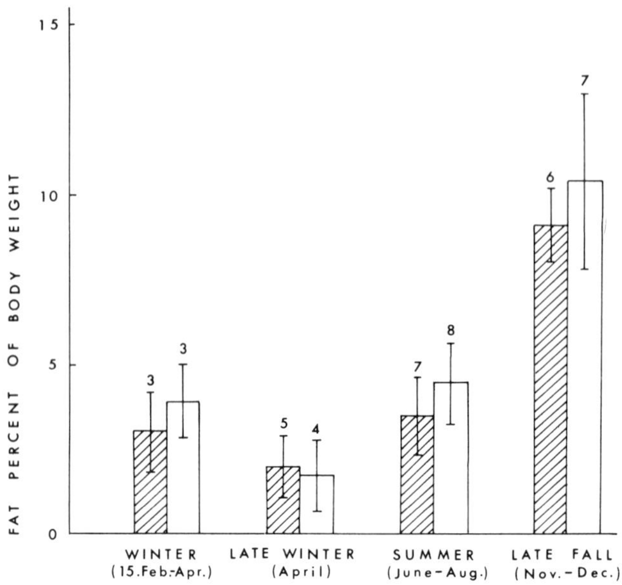

FIGURE 98.1. Seasonal variation in fat concentration (±SD, number of animals) in Eureopean roe deer ▨ females; □, males. (Data from Holand 1991a).

(Robbins et al. 1974; Reimers et al. 1982; Torbit et al. 1985), however, all other northern cervids deplete their energy reserves in fall due to the demanding rutting season. In particular males enter the winter in poor physical condition. On the contrary, the European roe deer have their rutting period in July and early August and have ample time to recover physically before the onset of winter.

The roe deer probably originated in the Middle East with a relative mild winter climate (von Raesfeld 1978). The early rut, coupled with delayed implantation found in European roe deer, has not evolved as adaptations to severe winter conditions. However, we believe that early rut is beneficial to winter survival in Fennoscandia. The body reserves of the roe deer would at best cover the energy demands for about 3 weeks (Holand 1990). The principal source of energy in winter is undoubtedly the food supply.

Digestion, Turnover, and Capacity

Winter rumen contents of European roe deer showed a cell wall content of about 50% (Ø. Holand, unpubl. data). This is in the same range as observed in mule deer (*Odocoileus hemionus*) (Hobbs et al. 1983). Although analyses of chemical compounds in the rumen are subject to error when attempts are made to relate them to contents of forages (Hungate 1966), the high cell wall content indicates a winter diet quality way below what is encountered on summer range. Therefore, we expect a seasonal change in the nutritional strategy of roe deer.

In vitro studies indicated no rumen microbial adaptation to a high fiber winter diet. The IVDMD of typical winter forage plants were 30% to 40% irrespective of season and type of inocula (Table 98.1). Typical summer forage plants were digested significantly better (+3.5 to +6.3 percentage units) with inocula from summer-killed roe deer compared with inocula from animals sampled in winter. This is in contrast to the work of Dissen and Hartfiel (1985), who suggested relative high digestability of plants rich in cellulose. Hofmann (1982), however, points out that the cellulolytic breakdown in European roe deer is hardly detectable. Hofmann (1985) suggests that extragastric microbial digestion plays an important role in the winter period, however, a longer total mean retention time in winter could lead to better utilization of this diet than that expressed from in vitro results. This was confirmed by

TABLE 98.1. Mean in vitro dry matter digestibility of forages using inocula of European roe deer from summer and winter. (Data from Holand 1991b)

Plant species	Plant part	Month	Winter (Feb) ($n = 5$)		Summer (June & Aug) ($n = 6$)	
			$\bar{X}$	SD	$\bar{X}$	SD
Potential summer forages						
Oxalis acetosella	Leaves	June	60.6	1.9	66.6	1.2**
Filipendula ulmaria	Leaves	June	56.4	0.8	62.7	1.1**
Chamaenrion augustif ifolium	Leaves	June	56.4	0.9	60.8	0.8**
Sorbus aucuparia	Leaves	June	54.7	1.9	58.2	1.3**
Salix caprea	Leaves	June	52.0	1.2	56.5	1.0**
Rubus idaeus	Leaves	June	52.3	1.6	56.3	0.8**
Potential winter forages						
Vaccinium myrtillus	Green shoots	Feb	36.0	0.7	35.2	0.7 NS
Pinus sylvestris	Needles	Feb	34.5	1.2	33.9	1.3 NS
Salix caprea	Twigs	Feb	33.7	1.7	31.9	2.0 NS
Sorbus aucuparia	Twigs	Feb	33.4	0.5	32.4	2.5 NS
Picea abies	Needles	Feb	32.7	0.9	31.6	1.4 NS
Betula pubescens	Twigs	Feb	28.6	0.6	27.6	1.1 NS

NS, no seasonal significant difference ($P > 0.05$); **, seasonal significant difference ($P < 0.01$)

the calculation of the fractional digestibility based on nonabsorbed, particulate-phase marker (cerium-141 chloride) using a single-dose injection (Holleman and White 1989) which yielded a value of 55% for blueberry stems (Table 98.2). Forage with a digestibility of 64% is required to meet the energetic maintenance cost in roe deer (Weiner 1977). In vivo studies performed by Drozdz (1979) indicate a digestibility coefficient of natural winter forages (twigs of trees and shrubs) slightly above 40%. He suggested that easily digestible food (i.e., herb layer plant, grass, or leaves) must be an important part of the winter diet.

Field-shot roe deer in the southeastern part of Norway had a 40% higher total alimentary

TABLE 98.2. Food intake and digesta passage in roe deer given blueberry stems ad libitum under winter conditions ($n = 6$)

	$\bar{x}$	SD
Body weight (kg)	24.8	3.4
Food intake (g DM/kg$^{0.75}$ per day)	36.9	2.8
Total mean retention time (h)	31.1	2.5
Transit time (h)	9.9	2.0
Rumen turnover time (h)	15.5	1.4
Alimentary fill (g DM/kg)	15.6	3.9
DM digestability (%)	54.2	2.9

tract fill (on a dry matter basis) in winter compared with summer (Table 98.3). This was mainly due to increased rumen fill. The omasum and the large intestine (including cecum) were also significantly larger in winter compared with summer. This supports the hypothesis that roe deer increase rumen fill in winter as a compensation for a low-quality diet, indicating that they tend towards a mixed feeding strategy in winter (Kay et al. 1980). Gasaway and Coady (1974) also reported an increase of the rumenoreticulum content in Alsaskan moose (*Alces alces*) under winter conditions in the same magnitude as found in roe deer. These observations contrast with those for other northern ruminants, where a relative lower rumen fill was found in winter (White et al. 1987). At the same time, a decrease in dry matter intake coupled with increased total mean retention time, using single-dose injection (Holleman and White 1989), was found in winter for our animals fed a concentrate diet (Table 98.4). When fed a staple winter food (i.e., blueberry stems), the intake dropped drastically and TMRT increased (Table 98.2). This contrasts with the hypothesis that concentrate selectors should excrete lignified and poorly digested forages rapidly, thereby reducing the limitation on voluntary intake

TABLE 98.3. Alimentary fill (g) in European roe deer collected in summer and winter. (Data from Holand 1991c)

	Winter (15 Dec–Apr) ($n = 27$)		Summer (June–Nov) ($n = 36$)		
	$\bar{x}$	SD	$\bar{x}$	SD	
Body weight (kg)	24.9	2.7	26.1	3.4	NS
Rumenoreticulum					
Wet weight	1972	502	1498	413	**
Dry weight	370	109	243	89	**
Omasum					
Wet weight	44	22	26	17	**
Dry weight	10	5	5	4	**
Abomasum					
Wet weight	76	40	98	52	NS
Dry weight	13	7	14	7	NS
Small intestine					
Wet weight	324	112	371	124	NS
Dry weight	45	15	51	17	NS
Cecum					
Wet weight	178	44	140	43	**
Dry weight	29	8	22	8	**
Large intestine					
Wet weight	382	123	287	97	**
Dry weight	83	26	58	24	**
Total					
Wet weight	2977	513	2421	591	**
Dry weight	553	109	393	117	**

NS, no seasonal significant difference ($P > 0.05$); **, seasonal significant difference ($P < 0.01$)

TABLE 98.4. Seasonal change in food intake and digesta passage in roe deer given a concentrate diet ad libitum ($n = 6$)

	Summer		Winter		
	$\bar{x}$	SD	$\bar{x}$	SD	
Body weight (kg)	24.8	3.4	24.3	1.8	NS
Food intake (g DM/kg$^{0.75}$ per day)	57.6	2.4	47.9	7.2	**
Total mean retention time (h)	19.6	2.0	24.2	1.3	**
Transit time (h)	6.8	1.6	7.6	0.8	NS
Rumen turnover time (h)	8.7	1.9	13.1	2.1	**
Alimentary fill (g DM/kg)	11.9	5.3	12.3	3.9	NS
DM digestibility (%)	86.7	3.6	87.1	2.7	NS

NS, no seasonal significant difference ($P > 0.05$); **, seasonal significant difference ($P < 0.01$)

imposed by gut capacity (Hobbs et al. 1983; Renecker and Hudson 1990). Perzanowski (1978) found a negative relationship between food intake and fiber content in the diet, indicating bulk limitation by the indigestible fraction, which again limits dry matter intake.

Discussion

The European roe deer is the smallest free-ranging wild cervid in Fennoscandia. Bucks reach a mean body weight of about 30 kg, and females weight 2 to 3 kg less in late fall (Ø. Holand unpublished data). Calves go into their first winter with a body weight of about 20 kg. All other cervids enter their first winter with body weights exceeding adult roe deer bucks, e.g., reindeer (*Rangifer tarandus*) weigh 40 to 50 kg (Reimers et al. 1983), red deer 50 kg (R. Langvatn, personal communication), and moose 150 kg or more (Sæther 1985). Compared with northern cervids, the roe deer calves are relatively close to adult weight at the onset of winter (Ø. Holand, unpublished data). Attaining a large body size is of vital importance in order to survive the first winter. Roe deer in Fennoscandia have a significantly higher body weight than roe deer in continental Europe (Stubbe and Passarge 1979). This is in accordance with the hypothesis of nutritional constraints arising from body size (Demment and Van Soest 1985).

The roe deer is small, it has an unfavorable scaling relative to common snow depths, and its feet are poorly adapted to movement in deep snow; this may reduce the area available for foraging (Cederlund 1981). Thus, snow depths of more than 0.5 m prevent the roe deer from reaching the best quality forage, such as blueberry stems (Cederlund et al. 1980). Throughout winter the roe deer is frequently forced to feed on low-quality forage of twigs and branches from deciduous and evergreen trees. This diet resembles that of moose (Cederlund et al. 1980), which are at least five times larger than roe deer. Although the roe deer shows a more selective feeding behavior than moose, this would not fully compensate for the difference in capacity of handling a high fiber diet which is normally closely related to body size (Hanley 1982). The classification of roe deer as a concentrate selector (Hofmann 1985) does not take into account the influence of a highly repeatable seasonal environment as found in large parts of Fennoscandia. Hofmann's classification scheme is only applicable to northern cervids, if limited to the annual period of plant growth as pointed out by Klein (1985).

Adverse snow conditions in combination with low ambient temperature coincide with large numbers of die-offs of roe deer in Fennoscandia (Hagen 1958; Borg 1970; Hjeljord 1980). Snow depth will, to a large degree, depend on canopy closure and the texture of the landscape. During severe snow conditions roe deer can often be found in bluffs where small snow-free areas can be found. Because of their small size, the roe deer's absolute energy demands are low, and they can subsist in these small favorable microhabitats for relatively long periods. It is well known that in snowy winters roe deer congregate and frequently move to agricultural land to feed on crops. Supplementary feeding has also increased over the past decades. The roe deer is also attracted to newly logged clear-cuts where food (e.g., aboreal lichens) is easily accessible. Human modification of the landscape is an important aspect of winter roe deer survival and the increased population dispersal obseved during the past few decades in Norway.

Conclusion

The European roe deer in most parts of Fennoscandia adapt to a changing habitat and shift from a concentrate-selector strategy in summer, i.e., a low gut fill, a high intake of easily digestible forage, and a short retention time, to a bulk-limiting situation in winter, i.e., an increased gut fill, a relative low intake of a diet rich in fiber, and a long retention time. Our findings suggest, in agreement with Baker and Hobbs (1987) and Renecker and Hudson (1990), a highly dynamic relationship between intake, fill, and turnover influenced by season and diet. The winter nutritional strategy observed is a small ruminants' response to a winter diet rich in fiber. In addition, we hypothesize that a selective feeding behavior, a winter depletion of body fat reserves, a midsummer rut, and an opportunistic utilization of human-made landscape are the keys to winter survival of European roe deer in most parts Fennoscandia.

References

Baker DL, Hobbs NT (1987) Strategies of digestion: digestive efficiency and retention time of forage diets in montane ungulates. Can J Zool 65:1978–1984

Borg K (1970) On mortality and reproduction of roe deer during the periode 1948–1969. Swed Wildl Res 7:121–149

Cederlund G (1981) Daily and seasonal activity pattern of roe deer in a boreal habitat. Swed Wildl Res 12:3–41

Cederlund G, Ljungquist H, Markgren G, Staelfelt F (1980) Seasonal differences between moose and roe deer at Grimsø in central Sweden; results of rumen content analyses. Swed Wildl Res 11:169–247

Demment MW, Van Soest PJ (1985) A nutritional explanation for body-size patterns of ruminants and nonruminant herbivores. Am Nat 125:641–672

Dissen J, Hartfiel W (1985) Beobachtungen zum Äsungverhalten sowie Untersuchungen zum Nährstoffverdaulichkeit von Rehwild. Z Jagdwiss 31:83–91

Drozdz A (1979) Seasonal intake and digestibility of natural foods by roe-deer. Acta Theriol 24:137–170

Gasaway WC, Coady JW (1974) Review of energy requirements and rumen fermentation in moose and other ruminants. Nat Can 101:227–262

Hagen Y (1958) Vinterernæringen hos rådyr og elg. Jeger og Fisker 10:453–464

Hanley TA (1982) The nutritional basis for food selection by ungulates. J Range Manage 35:146–151

Hjeljord O (1980) Viltbiologi. Landbruksforlaget, Oslo

Hobbs NT, Baker DL, Gill BR (1983) Comparative nutritional ecology of montane ungulates during winter. J Wildl Manage 47:1–17

Hofmann RR (1985) Digestive physiology of the deer—their morphophysiological specialisation and adaptation. In Fennessy PF, Drew KR (eds) Biology of Deer Production. R Soc N Z Wellington Bull 22: 393–407

Hofmann RR (1982) Vergleich der einzelnen Wildwiederkauer. Deutsche Jagd-Zeitung 2:4–5

Holand Ø (1990) Body reserves and winter survival in European roe deer. 19th Int Congr Game Biol, Trondheim, pp 187–191

Holand Ø (1991a) Seasonal variation in body composition of European roe deer (submitted to Can J Zool).

Holand Ø (1991b) In vitro dry matter disappearance using roe deer inocula from winter and summer (submitted to J Range Manage)

Holand Ø (1991c) Winter digestive strategy of a concentrate selector in Norway: the European roe deer (in prep)

Holleman DF, White RG (1989) Determination of digesta volume and passage rate from non-absorbed particulate phase markers using single dosing method. Can J Zool 67:488–494

Hungate RE (1966) The Rumen and Its Microbes. Academic Press, New York

Kay RNB, Engelhardt WV, White RG (1980) The digestive physiology of wild ruminants. In Rucklebush Y, Thivend P (eds) Digestive Physiology and Metabolism in Ruminants. AVI, Westport, pp 743–761

Klein DR (1985) Population ecology: the interaction between deer and their food supply. In Fennessy PF, Drew KR (eds) Biology of Deer Production. R Soc N Z Wellington Bull 22:13–22

Perzanowski K (1978) The effect of winter food composition on roe-deer energy budget. Acta Theriol 23:451–457

Reimers E, Klein D, Sørumgård R (1983) Calving timme, growth rate and body size of Norwegian reindeer on different range. Artic Alpine Res 15:107–118

Reimers E, Ringberg T, Sørumgård R (1982) Body composition of Svalbard reindeer. Can J Zool 60:1812–1821

Renecker LA, Hudson RJ (1990) Digestive kinetics of moose (*Alces alces*), wapiti (*Cervus elaphus*) and cattle. Anim Prod 50:51–61

Robbins CT, Moen AN, Reid JT (1974) Body composition of white-tailed deer. J Anim Sci 38:871–876

Stubbe C, Passarge H (1979) Das Rehwild. Verlag Neumann–Neudamm, Berlin

Sæther BE (1985) Annual variation in carcass weight of Norwegian moose in relation to climate along a latitudinal gradient. J Wildl Manage 49:977–983

Tilley JMA, Terry RA (1963) A two-stage technique for the invitro digestion of forage crops. J Br Grl Soc 18: 104–111

Torbit SC, Carpenter LH, Alldredge AW Swift DM (1985) Mule deer body composition—a comparison of methods. J Wildl Manage 49:86–91

Von Raesfeld F (1978) Das Rehwild Naturgeschichte, Hege und Jagd. Verlag Paul Parey, Hamburg and Berlin

Weiner J (1977) Energy metabolism of the roe-deer. Acta Theriol 22:3–24

White RG, Holleman DF, Hubbert ME, Staaland H (1987) Herbivores in cold climates. In Hacker JB, Ternout JH (eds) The nutrition of Herbivores. Academic Press, Sydney, Australia, pp 465–486

99
A Pattern of Plasma Potassium Concentrations in Adult Red Deer Stags

Neville B. Jopson and Peter F. Fennessy

Abstract

This study investigated the relationship between blood electrolytes and the annual reproductive cycle in red deer stags. Mature farmed red deer stags ($n = 36$) were slaughtered around the rut (February to May), at antler casting (September), full velvet antler (January), and the following rut (April), and blood samples and compositional data were collected. Blood samples were collected immediately after death. Blood was also collected monthly from a group of eight stags from June to April. Plasma and red blood cell sodium and potassium concentrations were measured.

Analyses of blood revealed a pattern in plasma potassium in the slaughtered stags that was in phase with the annual reproductive cycle. Concentrations rose above the normal reference range (about 3.5 to 5.4 mmol/l) to reach a peak of 10.6 ± 0.69 (mean $\pm$ SD) in mid-March and returned to normal as antler casting approached in late August. There was a second peak in the live stags (10.6 ± 1.31 mmol/l) in mid-July which coincided with a late winter resurgence in rutting activity. However, an elevated plasma potassium concentration was not recorded over the second rut in either slaughtered or live animals. The data indicate that stags entered the first rut in poor condition with little fat reserves because of a spring drought, whereas stags in the second year were in much better condition. We suggest the elevated concentrations of plasma potassium may be an indication of muscle catabolism in the rut.

Key words: Hyperkalemia, potassium, red deer, seasonal changes

Introduction

Homeostatic control of the blood electrolytes, especially potassium (K^+) and sodium (Na^+), is regarded as a feature of many mammalian systems (Hays and Swenson 1970). Cell volume, pressure, and membrane potential depend on electrolyte concentrations remaining at set levels (Ganong 1973). Muscle (especially cardiac muscle) and nervous tissue function rely on the maintenance of high and low concentrations of sodium and potassium, respectively, in the extracellular space. The reverse situation is true of the intracellular space. Close correlations exist between exercise-induced extracellular increases in K^+ in the dog, increase in muscle blood flow, increase in oxygen consumption, and increase in muscle performance (Hirche et al. 1980). Larger increases in extracellular K^+ result in disorders in muscle and nervous tissues and unless reversed will ultimately result in death (Henry et al. 1974).

Reports on normal blood concentrations of K^+ and Na^+ in cervids show some variation between and within species (Chapman 1977; McAllum 1981). In general, most cervids, including the red deer, have normal plasma concentrations of about 5 mmol/l K^+ and 140 mmol/l Na^+ (Wilson and Pauli 1983); however, there are noteworthy reports of hyperkalemia in red deer.

McAllum (1985) discussed postcapture myopathy in wild red deer, a stress-related condition which results in the death of a number of deer within a few days of capture. Many hematological and biochemical parameters were elevated in myopathic deer, including plasma K^+ which reached an average of 19.1 mmol/l, 0.5 h postcapture (four times the normal level). Nonmyopathic deer (those studied which survived the first few days of captivity) were also hyperkalemic at 0.5 h postcapture with an average plasma K^+ concentration of 9.6 mmol/l. In an earlier report McAllum (1981) recorded a mean plasma K^+ concentration of 9.8 mmol/l for a group of 18-month-old red deer stags but did not comment on the elevated K^+ concentration.

This study describes dramatic changes in blood electrolytes and investigates their possible relationship with changes in body composition of red stags through the annual cycle.

Materials and Methods

Fifty-six 4- and 5-year-old red stags located at the Invermay Agricultural Centre in the South Island of New Zealand were used. Twenty-five were treated with 36 mg melatonin (Regulin, Regulin, Victoria, Australia) on 19 December 1988 and again on 20 January 1989. Stags (total, $n = 36$) were slaughtered around the rut (February, March, April and May 1989), at antler casting (September 1989), full velvet antler (January 1990), and the following rut (April 1990) (February and March 1989, $n = 8$, all other slaughters $n = 4$). Blood samples were collected within 2 min of death. From June 1989, blood samples were collected monthly from a group ($n = 8$) of live stags (i.e., the repeat-sampled group).

The stags were weighed monthly except during the period of October to December. Carcasses and visceral organs from the slaughtered stags were dissected and weighed as part of a whole carcass analysis study. Fat was trimmed from the omentum, kidneys, and pelvic cavity, and pooled to produce the omental, kidney, knob, and channel fat (OKKC fat).

The concentration of K^+ and Na^+ in the plasma and red blood cells (calculated from the totals for whole blood and plasma and using the measured packed cell volume) were determined using a flame photometer (Radiometer, FLM3 flame photometer, Copenhagen, Denmark). Data were analyzed using an analysis of variance and Student's *t*-test as was appropriate.

Results

No changes in plasma potassium, live weight, hot carcass weight, or internal fat could be attributed to melatonin treatment; therefore, the results for the two groups have been combined.

Plasma K^+

The normal reference range for plasma K^+ concentrations for red deer is between 3.5 and 5.4 mmol/l (MAF Quality Management 1990). Plasma collected from slaughtered animals in the period from February to May had K^+ concentrations well above the normal reference values reaching a peak in March (Fig. 99.1) of 10.6 ± 0.69 mmol/l (mean $\pm$ SD). Thereafter, plasma K^+ concentration declined by about 50% over the final three slaughters (4.6 ± 1.18, 6.0 ± 0.45, and 5.3 ± 0.67 for September, January and April, respectively). Monthly sampling began in June on the live animals. K^+ concentrations peaked in mid-July at 10.6 ± 1.31 mmol/l. This was at a time that coincided with a small resurgence in rutting behavior. However, the plasma K^+ concentration had dropped below 5 mmol/l when sampled just over 1 week later; thereafter, values remained low.

Such an elevation of plasma potassium was not recorded in the period around the rut in 1990 in either slaughtered or repeat-sampled live red stags. However, plasma samples collected at slaughter, a mixture of arterial and venous blood, usually had similar or higher K^+ concentrations than venous blood collected the previous day from live animals (0.0, 1.5, and 0.3 mmol/l difference for September, January and April slaughters, respectively).

Plasma Na^+

Plasma Na^+ did not vary greatly throughout the experiment, with an average concentration of 144 ± 0.4 (mean $\pm$ SEM). Any variation in

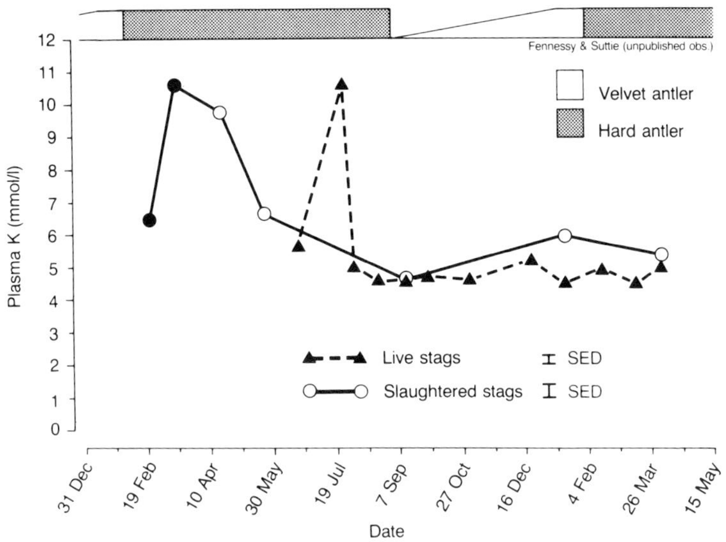

FIGURE 99.1. Potassium concentration in plasma from red stags in relation to the annual antler cycle (solid symbols $n = 8$; open symbols $n = 4$).

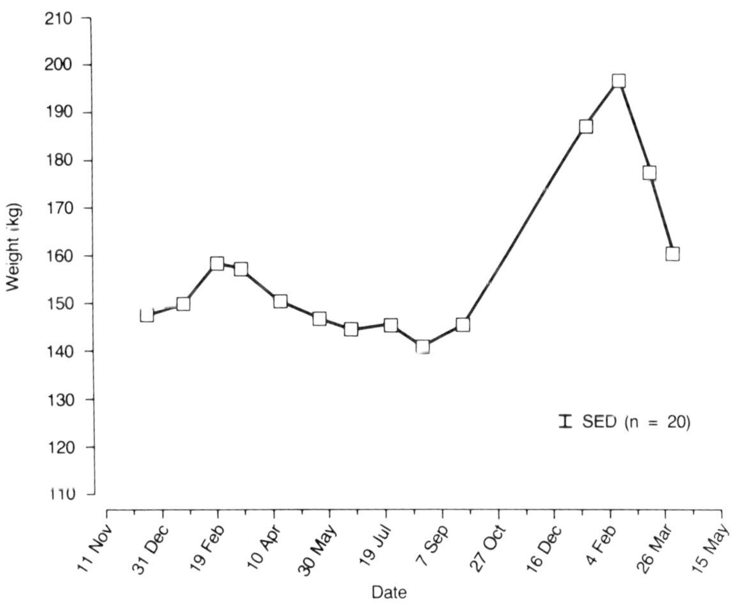

FIGURE 99.2. Pattern of live-weight change in a group of red deer stags.

plasma Na^+ was not correlated with trends in plasma K^+.

Live Weight

At the beginning of the study (December 1988), the stags weighed 148 ± 14.6 kg (mean ± SD) and grew to reach a peak in February of 158 ± 11.4 kg (Fig. 99.2). From this peak the stags gradually lost weight reaching a nadir of 141 ± 9.2 kg in August 1989. Live weight was gained rapidly in spring, reaching a peak of 196 ± 16.0 kg in February 1990. Weight was then lost rapidly at an average rate of 0.77 kg/day over

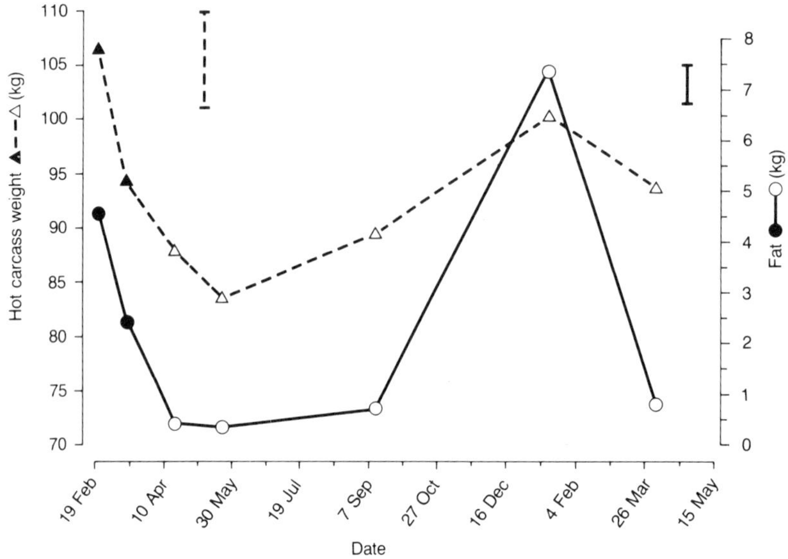

FIGURE 99.3. Seasonal patterns in hot carcass weight and internal fat depots in red deer stags (fat depots represent the combined weight of the omentum and kidney, knob and channel fat; error bars indicate SED; solid symbols $n = 8$; open symbols $n = 4$).

6 weeks (this compares with 0.14 kg/day for the same period the previous year). At the end of the study in April 1990, the stags weighed 160 ± 10.3 kg. For the seven stags for which full data are available, the live-weight change from the winter trough to prerut peak was 33 ± 9.5 kg and 62 ± 8.9 kg for years 1 (1988/89) and 2 (1989/90), respectively. The slaughtered stags had similar live weights, at each time of slaughter, to that of the group of stags presented in Fig. 99.2.

Carcass Weight and Fat Depots

Hot carcass weight fell rapidly from 106 ± 10.8 kg in February to a nadir of 84 ± 13.0 kg in May 1989 (Fig. 99.3). Carcass weight rose from the postrut trough slowly at first, then more rapidly to reach a peak of 100 ± 10.0 kg in January 1990. Loss of carcass weight was slower over the second rut, falling to 94 ± 4.9 kg in April 1990, an average loss of 0.08 kg/day (compared with 0.25/day for the previous year).

The OKKC fat followed a similar pattern to that of hot carcass weight, falling rapidly at first, then slowly from a February peak of 4.6 ± 1.79 kg to a May trough of 0.4 ± 0.27 kg (Fig. 99.3). There was virtually no change in fat over winter with a difference of only 0.3 kg of OKKC fat between the May and September groups. OKKC fat was deposited rapidly over spring reaching a peak of 7.4 ± 1.27 kg in January, approximately 60% more than the peak of the previous year. OKKC fat loss was dramatic over the second rut falling on average 6.6 kg to reach 0.8 ± 0.40 kg in April.

Discussion

The principle danger in hyperkalemia in several mammalian species is cardiac arrest. Electrocardiographic abnormalities may be observed when the serum potassium reaches about 7 mmol/l and are usually severe when the serum potassium is greater than 8 mmol/l. Death usually occurs when the serum potassium reachs about 7 mmol/l (Tasker 1980). According to our data and that of McAllum (1985), red deer stags are able to tolerate elevations in plasma K^+ which would be potentially lethal in humans, horses, cattle, and dogs (Tasker 1980). Initially the pattern of plasma K^+ concentrations appeared to correlate well with that of neck girth, a secondary sexual characteristic. However, a second peak in plasma K^+ was recorded in July in the eight animals sampled from June 1989 through April 1990, a peak which coincided with a late winter behavioral resurgence in rutting behavior. There was also no difference between the control and melatonin-treated stags; if the changes in plasma K^+ were related to the sexual cycle, then such a change could reasonably have been expected considering the influence of melatonin on seasonality in red stags (Fisher et al. 1988). Furthermore, no elevation was recorded in plasma K^+ in the year 2, either in blood collected from slaughtered or live animals.

The greatest difference between these two successive seasons was in the level of nutrition of the stags. The spring and summer of 1988/89 was one of the driest on record at Invermay. Only half the normal rainfall fell in September to November 1988, and air temperatures were on average 2° C higher than a 35-year rolling average for Invermay. The stags went into the rut in poor condition, approximately 20 kg below their expected live weight in a more normal season. Fat reserves were also lower than would normally be expected for that time of year; this is apparent from comparisons of the stags in the present work with those of Drew (1985). In the latter study, stags had 7.2 kg OKKC fat compared with only 4.6 kg in the prerut slaughter of 1989. Consequently, muscle mass must have been catabolized to provide the energy needed to meet the stags' energy requirements through the rut. This rapid loss of protein is indicated by the greater rate of fall in hot carcass weight in year 1 compared with year 2 which approached significance ($P < 0.1$). The 21% drop in hot carcass weight in year 1 is also greater than the 14% to 17% fall in carcass weight reported for wild red deer stags by Mitchell et al. (1976). The indicated loss of protein began before fat stores were exhausted, as would be expected if deer were to follow the three-phase fasting model presented for rats, penguins, and geese (Cherel et al. 1988). The stags may have shifted from phase 2 (a period of economy with most of the energy requirement being derived from lipid while energy derived from proteins remains at low and steady values) to phase 3 (protein utilization and increase in the rate of body mass loss). The shift to phase 3 is not the single consequence of lipid exhaustion as there are normally significant fat stores still available (Cherel et al. 1988). Throughout the fast, stags mainly catabolize fat to meet energy requirements (Drew 1985). *Ad libitum* food intake over the rut is very low (Fennessy et al. 1991), meaning that once fat reserves reach a low level, the stag has no option but to catabolize protein, rapidly increase food intake, or die.

Two possible mechanisms for release of intracellular K^+ can be excluded with some confidence. Prolonged exercise or stress has resulted in profound increases in extracellular electrolytes in a number of species including humans (Rose et al. 1970), horses (Keenan 1979; Rose et al. 1980), zebras (Hofmeyr et al. 1973), and red deer (McAllum 1985). Kjellmer (1965) used perfused cat calf muscle to demonstrate that the K^+ concentration of the venous plasma increased by up to 100% during contractions and that this excess K^+ could account for as much as 65% of the vasodilation during exercise. Plasma osmolarity was significantly higher in chased zebras after 1 h than in a group of zebras that were immobilized without chasing (Hofmeyr et al. 1973). This situation parallels chasing prior to capture of wild red deer which is associated with postcapture myopathy (McAllum 1985). However, the stags in our study were well habituated to the regular yarding and handling, although considerable care was taken to ensure that the stags were yarded with the minimum of chasing and were handled such that there were no signs of visible distress. Although the possibility of tissue damage cannot be ruled out, both handling and fighting were consistent between years, and therefore tissue damage must have been minimal as plasma K^+ was not elevated in the second year of the experiment. Also examination of carcasses revealed that tissue damage associated with yarding and handling was minimal as virtually no bruising was apparent. Tissue damage associated with treatment may further be ruled out for two reasons. First, in a separate study, 2-year-old red stags that were severely distressed, and had been fighting, never had plasma K^+ concentrations in excess of 7.2 mmol/l (N.B. Jopson, unpublished data). Second, if elevated K^+ concentrations are caused by tissue damage, it is most unlikely that all the sampled animals would have the same amount of damage on the same day as would be indicated by the minimal variation about the mean K^+ concentrations (Fig. 99.1).

Separation of plasma or serum erythrocytes as quickly as possible is of vital importance to accuracy. Henry et al. (1974) reported an incrase of 3.8 and 7.8 mmol/l 18 h after venipuncture with storage at 4° C and 37° C, respectively. In a study of our own, plasma K^+ concentration increased by 1.3 mmol/l when stored at 7° C for 7.5 h before plasma was separated from erythrocytes compared with blood separated within 1 h of death (N.B. Jopson, unpublished data). Blood collected for this study was stored at 7° C and the plasma separated from erythrocytes within 4 h of sample collection. Consequently time between collection and separation cannot account for the change in plasma K^+.

It is difficult to draw firm conclusions from the data. However, our hypothesis is that the increase in plasma K^+ concentration is due to the fact that having insufficient energy stored as fat, the stags have been forced to catabolize protein stores in order to survive through the endogenously induced rut fast. It is important to note that plasma K^+ is elevated only during periods when ad libitum food intake is known to be very low such as the 4- to 8-week rut (Fennessy 1981). This period is physiologically different from an environmentally induced starvation over the winter. Fat is stored prior to the rut to enable male deer to gather, defend, and serve their harems without being distracted by the need to feed (Tyler 1987). Protein stores are largely spared because some protein-saving mechanism is operating. Fat, having largely been utilized over the rut (Drew 1985; Wallace and Davies 1985), is therefore not likely to be a major source of energy for adult stags in winter. Hence if stags are on a submaintenance diet in winter, protein catabolism is the normal response. This is in contrast to the rutting situation where a high level of protein catabolism is not the usual response to fasting. Protein catabolism is accompanied by release of K^+ and other intracellular ions into the blood (Tepperman 1980) and could account for the elevated K^+ concentrations we observed. However, the kidneys would normally remove excess K^+ from circulation. A chronically elevated K^+ concentration suggests there has been a modification of kidney function to conserve important muscle constituents, namely nitrogen and K^+, which would otherwise be lost in the urine. Similar systems are known to exist in other species. For example, the black bear during hibernation is able to use only fat to meet energy demands while actively conserving protein. This is achieved by incorporation of amino acids into plasma proteins, degradation of urea, and a shut-down of kidney function (Nelson et al. 1983). The rut is thus a period when a protein-conserving mechanism, similar in function (but not necessarily in mechanism) to that of the black bear, is likely to operate.

Acknowledgements

We are indebted to the Otago University Renal and Electrolyte Physiology Unit for use of the Flame Photometer and A.D.C. MacKnight and J.P. Leader for discussions on the data used in this paper. We also thank C.G. Mackintosh, A.J. Whaanga, G.D. Bruce, R.E. Labes, R.J. Young, J. Mawhinney, F.J. Laas, R.D. Dutton, G.J. Dunbar, and N.B. Grant for their willing assistance at various times through this study.

References

Chapman DI (1977) Haematology of the Deer. In Archer RK, Jeffcott LB (eds) Comparative Clinical Haematology. Blackwell Scientific Publications, Oxford

Cherel Y, Robin JP, Walsh O, Karmann H, Netchitailo P, Le Maho Y (1988) Fasting in king penguin. 1. Hormonal and metabolic changes during breeding. Am J Physiol 254:R170-R177

Drew KR (1985) Meat production from farmed deer. In Fennessy PF, Drew KR (eds) Biology of Deer Production. R Soc N Z Wellington Bull 22:285–290

Fennessy PF (1981) Nutrition of red deer. Proceedings of a Deer Seminar for Veterinarians. Deer Advisory Panel N Z Vet Assoc, pp 8–15

Fennessy PF, Thompson JM, Suttie JM (1991) Season and growth strategy in red deer: evolutionary implications and nutritional management. In LA Renecker (ed) Conservation and Sustainable

Development. Proc 2nd Int Wildlife Ranching Symp (in press)

Fisher MW, Fennessy PF, Milne JD (1988) Effects of melatonin on seasonal physiology of red deer. Proc N Z Soc Anim Prod 48:113–116

Ganong WF (1973) Review of Medical Physiology, 6th ed. Lange Medical Publications, Los Altos, California

Hays VW, Swenson MJ (1970) Minerals. In Swenson MJ (ed) Dukes' Physiology of Domestic Animals, 8th ed. Cornell University Press, Ithaca, New York, pp 663–690

Henry RJ, Cannon DC, Winkelman JW (1974) Clinical Chemistry: Principles and Techniques, 2nd ed. Harper and Row, Hagerstown, MD.

Hirche H, Schumacher E, Hagemann H (1980) Extracellular K^+ concentration and K^+ balance of the gastrocnemius muscle of the dog during exercise. Pflugers Archiv 387:231–237

Hofmeyr JM, Louw GN, du Preez JS (1973) Incipient capture myopathy as revealed by blood chemistry of chased zebras. Madoqua (Series 1) 7:45–50

Keenan DM (1979) Changes of blood metabolites in horses after racing, with particular reference to uric acid. Aust Vet J 55:54–57

Kjellmer I (1965) The potassium ion as a vasodilator during muscular exercise. Acta Physiol Scand 63:460–468

McAllum HJF (1985) Stress and post capture myopathy in red deer. In Fennessy PF, Drew KR (eds). Biology of Deer Production. R Soc N Z Wellington Bull 22:65–72

McAllum HJF (1981) Normal values: haematology and serum biochemistry. Proc. of a Deer Seminar for Veterinarians. Deer Advisory Panel N Z Vet Assoc, pp 50–53

Mitchell B, McCowan D, Nicholson IA (1976) Annual cycles in body weight and condition in Scottish red deer, *Cervus elaphus*. J Zool London 180:107–127

Nelson RA, Steiger DL, Beck TDI (1983) Neuroendocrine and metabolic interactions in the hibernating black bear. Acta Zoologica Fennica 174:137–141

Rose RJ, Arnold KS, Church S, Paris R (1980) Plasma and sweat electrolyte concentrations in the horse during long distance exercise. Equine Vet J 12:19–22

Rose LI, Carroll DR, Lowe SL, Peterson EW, Cooper KH (1970) Serum electrolyte changes after marathon running. J Appl Phys 29:449–451

Tasker JB (1980) Fluids, Electrolytes and acid-base balance. In Kaneko JJ (ed) Clinical Biochemistry of Domestic Animals, 3rd ed. Academic Press, New York, pp 401–446

Tepperman J (1980) Metabolic and Endocrine Physiology 4th ed. Year Book Medical Publishers, Chicago

Tyler NJC (1987) Body composition and energy balance of pregnant and non-pregnant Svalbard reindeer during winter. Symp Zool Soc London 57:203–229

Wallace V, Davies AS (1985) Pre- and post-rut body composition of red deer stags. In Fennessy PF, Drew KR (eds) Biology of Deer Production. R Soc N Z Wellington Bull 22:291–293

Wilson PR, Pauli JV (1983) Blood constituents of farmed red deer (*Cervus elaphus*). 2. Biochemical values. N Z Vet J 31:1–3

100
Indigestible Particulate Passage in White-Tailed Deer

Thomas G. Barnes, Larry W. Varner, Lytle H. Blankenship, and James F. Gallagher

Abstract

Relationships among intake, digestion, and passage rates for white-tailed deer (*Odocoileus virginianus*) eating guajillo (*Acacia berlandieri*) were examined in a completely randomized balanced design metabolism experiment. Dry matter intake was positively correlated ($P = 0.0005$) with rate of indigestible particle disappearance from the rumen. The more highly digested diets had longer ruminal and total tract mean retention times (MRT). There were seasonal differences in ruminal and total tract MRT. Differences in total tract MRT were a function of ruminal turnover rates, not of time spent in the lower tract. No chemical or physical criteria were related directly to ruminal or total tract MRT. Dietary NDF concentrations varied seasonally, but permanganate lignin concentration was similar across seasons. Passage rates were not a function of permanganate lignin concentration; rather, low cell wall content of the forage allowed for rapid excretion of indigestible particles.

Key words: Digestibility, digestion, mean compartment residence time, passage rate, white-tailed deer

Introduction

The value of forage in providing essential nutrients for ruminants is dependent upon animal nutrient requirements, intake, and digestibility, which in ruminants, are related to rate of passage through the alimentary tract (Allison 1985). Thus, interpretation of foraging strategy and an understanding of digestive function can be enhanced by measuring indigestible particulate passage.

Forage availability, animal energy requirements, gut capacity, body mass, and anatomical criteria dictate diet selection and digestion in ruminants (Kay et al. 1980; Hanley 1982). Concentrate selectors (browsers) select a diet high in cell solubles and low in cell wall content to meet their relatively high weight-specific metabolic requirements. For this strategy to be successful, browsers must be adapted for rapid fermentation and nutrient absorption (Kay et al. 1980), along with rapid removal of indigestible forage particles (Baker and Hobbs 1987). There is limited quantitative information on the passage of indigestible particulate matter in natural forages consumed by white-tailed deer. The present understanding of white-tailed deer digestive strategy has been based on speculation from dietary (Henke et al. 1988) and anatomical observations (Hoffman 1989). The objective of this study was to quantify passage rate and digestibility criteria in white-tailed deer for a single species of browse across seasons.

Methods

Six randomly selected 2- to 6-year-old male and female white-tailed deer (average initial weight, 40.3 ± 6.3 kg) were used in a completely randomized balanced design experiment with five seasonal (early summer, late summer, fall, winter, and spring) treatments. Individual balance trials (Harris 1970) were conducted from June 1986 to April 1987 at the Texas A&M Research and Extension Center in Uvalde. Experimental animals were born in captivity, hand-reared on evaporated milk and calf starter, and thereafter maintained on a nutritionally complete pelleted diet (Barnes 1988) supplemented with guajillo and alfalfa or native hay.

Guajillo was selected for study because it is a dominant forage item found in deer and cattle diets in the Northern Rio Grande Plains (Varner and Blankenship 1987), is a commonly occurring range plant, and has large, bipinnately compound leaves which facilitate collection. Guajillo stems, less than 20 mm in diameter, were removed with hand clippers from plants growing in shallow soils in Zavala and Uvalde counties, Texas. Only the current year's growth was collected to resemble deer browsing, which is important because deer usually select live tissue growing on stems less than 2.1 mm in diameter (Shafer 1963). Leaves were hand stripped and fed fresh to reduce chemical changes due to browning, respiration, and phenolics (Servello et al. 1987).

For each balance trial, deer were confined in individual 1.2 m^3 metabolism crates housed in a climate-controlled room. Room temperature closely approximated outdoor temperature during all seasons, except summer when daytime temperatures never exceeded 27° C. A mixture of adult male and female animals were used as Barnes (1988) found no differences in nutrient digestibilities among sexes for animals conditioned to the metabolism crates in three previous experiments with a pelleted diet. To ensure rapid acclimation to metabolism crates and to allow for maximal intakes, deer were not drugged prior to experimental periods.

Animals were weighed prior to each trial which consisted of a 5-day adjustment period followed by a 5-day total fecal collection period. Forage was offered ad libitum during the adjustment period and fed at maximal pretrial consumption levels during the collection period. A sample of forage was collected daily for each animal, weighed, oven-dried at 60° C for 72 h, and reweighed to determine DM. Water was available ad libitum during all experiments.

Before each metabolism study, 1400 g of fresh forage was subsampled and labeled (Teeter et al. 1984) with ytterbium (Yb). Coarsely chopped leaves were soaked for 24 h in a Yb acetate solution containing 15 mg Yb/ml solution. After soaking, excess fluid was poured off, and remaining forage was rinsed three times with deionized water to remove unbound Yb. The excess water was removed from leaves by air drying overnight. Leaves were maintained at 6.0 to 7.0 pH to minimize marker migration in the rumen. Two hundred grams of dried forage were used as a pulse dose marker. Deer were fed once in the morning throughout the adjustment and collection periods. At the beginning of each collection period, Yb marked forage was offered at 0800 h. Marked forage not consumed was removed at 0900 h, and unmarked forage was offered immediately thereafter.

Composite fecal samples were collected from fecal collection pans, weighed and frozen every 6 h postdose for 5 days. If no fecal samples were available, animals in a prone position were forced to stand. This invariably resulted in the animal defecating. Fecal samples were cleaned, removing hair and other extraneous materials. Feed, ort, and fecal samples were oven-dried to a constant weight at 60° C, ground (1 mm screen in a Wiley Mill), and analyzed for DM, ash, gross energy, CP (Laubner 1975), NDF (Goering and Van Soest 1970) without the addition of sodium sulfite (Mould and Robbins 1981), and permanganate lignin (Goering and Van Soest 1970).

Ytterbium in feed and fecal samples was extracted by chelating 200-mg samples with 20 ml diethylenetriaminepentaacetic acid and potassium chloride solution (Karimi et al. 1986). Samples were placed on a wrist-action shaker for 30 min and filtered through Whatman filter paper number 4 into glass vials where they were stored until analyzed. Ytterbium content was determined using an AOAC approved method utilizing atomic absorption spectrophotometry with an acetylene/nitrous oxide flame (McCollum and Galyean 1982).

Turnover rates, compartmental retention times, and total tract mean retention time of indigestible solid particles were estimated by fitting a series of two-compartmental age-dependent and age-independent models to fecal excretion data (Ellis et al. 1984; Pond et al. 1988). The nonlinear, least squares, iterative procedure of SAS (1979) was used to fit models to the data. The residual sums of squares (EMS) for each model were compared to determine model goodness of fit (Pond et al. 1988). Ruminal and total tract mean retention times were calculated using the algorithms described by Pond et al. (1988). A balanced one-way ANOVA was used to detect differences in DMI, DMD, and NDF digestibility. An unbalanced one-way ANOVA was used to detect treatment differences in compartment retention times, total tract mean retention time, Tau, and digestibility coefficients (Steel and Torrie 1980), because not all animals consumed the Yb marker. Season was the treatment effect. The least significance different procedure was used to compare means when a significant F-value was obtained from the ANOVA (Steel and Torrie 1980). Simple linear regression was used to determine relationships between DMI and DMD and relationships between chemical and digestibility characteristics and retention times (Steel and Torrie 1980).

Results and Discussion

One individual deer during each trial period went off feed and had to be removed from the metabolism crate; however, five animals remained in each trial so a balanced ANOVA could be used to detect seasonal differences in DMD, DMI, and NDF digestibility (Table 100.1). No seasonal comparison of forage quality could be made because daily samples were taken for each deer and composited for analysis. Not all deer accepted the Yb marked forage (Table 100.2), and an unbalanced ANOVA was used to analyze passage rate data.

The chemical composition of guajillo is presented in Table 100.1. As expected, concentrations of CP, digestible protein, DE, and NDF varied seasonally. The quality of guajillo followed plant growth patterns. The poorer, late summer forage quality was attributed to drought. As a result of mild temperatures and adequate rainfall, forage quality increased during the fall and winter. This is a common occurrence in south Texas (Varner et al. 1977). Gross energy was similar among balance trials and averaged $4.7 \pm .03$ kcal/g DM. Permanganate lignin of all diets was similar.

Dry matter intake varied seasonally (Table 100.1). Deer consumed less ($P < 0.05$) guajillo during the spring, early summer, and fall than in late summer and winter periods. Dry matter intake was correlated to ruminal turnover rate ($Y = 12.3 + 93.8X$, $R = 0.68$, $P = 0.0005$, $n = 22$), which may indicate that inherent physiological mechanisms contribute to seasonal intake differences, and passage rate is one of the primary mechanisms regulating DMI (Poppi et al. 1981; Holleman et al. 1984). Differences in seasonal intake of guajillo were expected and are

TABLE 100.1. Mean weight, dm intake, chemical composition, and digestibility of guajillo fed to white-tailed deer in metabolism studies

Item	Spring	Early summer	Season[a] Late summer	Fall	Winter	Mean	SEM
No.	5	5	5	5	5	40.3	1.86
DMI (g/kg$BW^{.75}$) per day	29.4^{A}	37.8^{AB}	50.4^{B}	35.8^{A}	49.2^{B}	1.89	0.96
DE (kcal/g)	2.27	1.91	1.60	1.75	1.93	18.0	0.17
CP (%)	20.0	20.2	15.6	16.8	17.6	5.2	1.70
Digestible protein (g/100 g feed)	9.1	7.3	2.1	3.5	4.0	54.4	0.26
NDF (%)	57.4	50.9	50.8	53.8	59.2	11.6	0.17
Permanganate lignin (%)	11.9	10.8	11.9	10.4	12.8	40.9	0.90
DMD (%)	48.1^{C}	41.2^{AB}	35.2^{A}	38.3^{AB}	41.5^{B}	11.0	1.38
Digestible NDF (%)	22.8^{B}	7.6^{A}	0^{A}	6.7^{A}	17.9^{B}		

[a]Means in the same row with different letters differ ($P < 0.05$)

TABLE 100.2. Mean time delay (τ) and compartmental residence times (cmrt, H) estimated using the G_2 model (Pond et al. 1988) when fitted to fecal excretion data from white-tailed deer used in metabolism studies

Item	Season[a] Spring	Early summer	Late summer	Fall	Winter	Mean	SEM
No.	5	3	4	5	5		
τ	15.4	11.6	13.6	14.1	11.6	13.3	0.56
CMRT1	15.3	18.1	6.8	9.1	6.4	11.1	1.47
CMRT2	22.8	29.9	16.6	19.6	24.4	22.7	1.45
CMRTS[b]	38.2[B,C,D]	48.1[D]	23.5[A]	28.7[A,B]	30.8[A,B,C]	33.9	1.41
GMRT	53.6[C]	59.5[C]	37.1[A]	42.8[A,B]	42.2[A,B]	47.0	1.42

[a]Means in the same row with differ letters differ ($P < 0.05$)
[b]CMRTS is the compartmental residence time in the two collective compartments (CMRT1 + CMRT2); GMRT is gastrointestinal residence time (τ + CMRTS) calculated using the algorithms of Pond et al. (1988)

determined by behavioral, morphological, and physiological mechanisms driven by photoperiod and associated endocrine changes (Chao and Brown 1984). Intake levels reported in our study are similar to levels reported in previous studies using mule deer (*Odocoileus hemonius*) or white-tailed deer (Mautz et al. 1976; Baker and Hobbs 1987).

Predictions from the two-compartmental age-independent and higher age-dependent models (G_3 to G_6) did not improve parameter estimates because further increases in age dependency beyond G_2 resulted in reduced fit by the model to observed values. This was due to a more accentuated and delayed peak in concentration coupled with a rapid subsequent decline in concentration of stained particles. The G_2 model was selected because it yielded the lowest EMS and is the most descriptive of compartmental flow of digesta, consistent with biological expectations, and results in reduction of SE which improves overall fit to the data (Pond et al. 1988). Figure 100.1 diagrammatically portrays the fit of the G_2 model to fecal Yb concentration.

There were no differences in time delay (Tau) among seasons; therefore, differences in total tract MRT appear to be due to ruminal turnover rates, not time spent in the lower tract (Table 100.2). Estimates of retention time in the lower tract of deer were similar to those of cattle (Ellis et al. 1984; Pond et al. 1988) and sheep (Krysl et al. 1985) when data were fitted to the same model. Thus, differences in total tract MRT between animal species appear to be a function of differences in ruminal MRT.

A considerable degree of variability in compartmental passage rates between individual deer was evident throughout the study, such that although mean CMRT 1 varied from 6.4 h to 18.1 h and CMRT 2 varied from 16.6 to 29.9 h (Tabe 100.2), significant differences among seasons were not detected. However, when combined into ruminal MRT (CMRTS), much of the variation was reduced, and seasonal differences were apparent (Table 100.2). The spring and early summer diets had longer ($P < 0.05$) total tract MRT than the other diets (Table 100.2). Periods with greater DM and NDF digestibility coefficients had longer ($P < 0.05$) ruminal and total tract MRT (GMRT) (Tables 100.1 and 100.2), although there was no statistically significant correlation or relationship between turnover rates or mean retention time and DMD or NDF digestibility

Percentage NDF (mean = 11.0%) digested by deer in our study was small. Fiber digestibility ultimately determines DMD (Huston et al. 1986). This could help explain the low DMD values obtained for deer in this study although a variety of factors including intake, passage rate, and secondary plant metabolites affect digestibility of forage. However, deer appeared to compensate for this low digestive efficiency by increasing

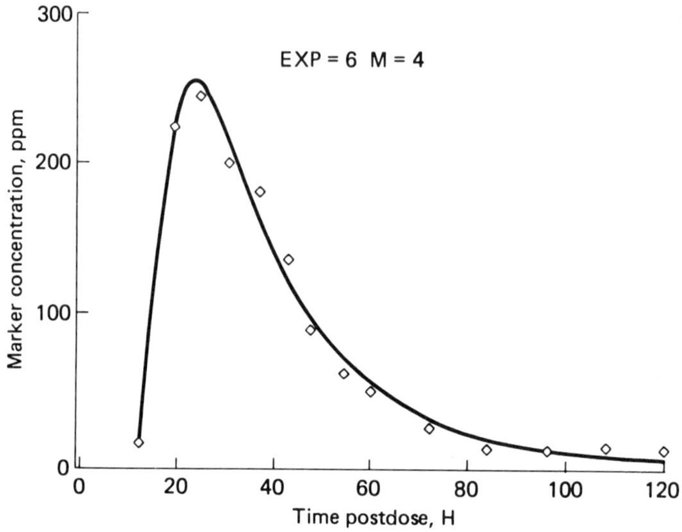

FIGURE 100.1. Observed results (◇) and predicted curves for a deer (no. 4) fed a late summer guajillo diet plotted arithmetically using the G_2 two-compartmental age-dependent model (Pond et al 1988).

ruminal and total tract MRT (37.1 to 59.5 h). Deer also may be able to compensate for low digestive efficiency by increasing intake and gut fill (Baker and Hobbs 1987). We did not measure gut fill; however, the relationship between ruminal turnover rate and intake suggests that removal of indigestible particles from the rumen is one of numerous factors in increasing digestive efficiency.

Total tract MRTs observed in this study closely approximate reported values for other concentrate selectors such as white-tailed deer (Huston et al. 1986) and mixed adaptable feeders consuming hay diets such as mule deer and muskoxen (*Ovibus moschatus*) (Holleman et al. 1984; Baker and Hobbs 1987). Total tract MRTs were much lower than reported values for roughage feeders (Warner 1981). Caution should be exercised when comparing retention time and turnover rates because various methods of parameter estimation and quantification, coupled with strong associative effects when feeding multispecies diets, lead to different data interpretation. We attempted to minimize total variation and associative effects by feeding a single species of browse with a similar lignin concentration.

Deer must be adapted for rapid excretion of indigestible particles when consuming browse (Baker and Hobbs 1987). Data from our study, while limited in scope, support the concept that digestibility of diets by deer is related to ruminal, not intestinal, rate factors. Furthermore, these rate factors are important in determining adaptability of deer to their habitat. Thus, digestive capabilities of white-tailed deer are compatible with natural diet selection of forages high in cell solubles and low in fiber.

There was no correlation between any chemical constituent, including lignin, and ruminal or total tract MRT. Two hypotheses have been presented in the literature to explain the effects of lignin and other chemical constituents on passage rates in small-bodied ruminants. Milchunas et al. (1978) indicated that lignin renders plant cell walls brittle. This allows for increased breakdown of forage particles during mastication and rumination, thus decreasing retention time by increasing the probability of ruminal escape. Spalinger et al. (1986) observed negative correlations between lignin concentration and particle breakdown rates in mule deer. They hypothesized that lignin makes particles more resistant to breakdown and thereby increases retention time. They argued that it is the low cell wall content of browse, not high lignin concentration, that allows for rapid removal from the gut and makes a particular forage suitable for deer. Baker and Hobbs (1987) fed diets with varying levels of browse to deer and concurred with Spalinger et al. (1986) that it is the low cell wall component that allows for rapid passage. Indirect evidence from our study, including static lignin concentrations (11.6%) accompanied by varying total tract MRT (37.1 to 59.5 h) and varying NDF concentrations (50.8 to 59.2%), indicates that it is the low cell wall content that allows for rapid excretion of indigestible particles in agreement with Spalinger et al. (1986). Supporting evidence (varying retention times and static lignin concentrations) indicates that passage rate is not a function of lignin concentration. The low cell wall hypothesis is supported by differences in passage rate as a function of changes in specific gravity (Ehle and Stern 1986) and buoyancy (Sutherland 1988). The latter may be a function of the amount of potentially digestible NDF available in the forage (Deswysen and Ellis 1988).

Acknowledgements

Research was supported by Caesar Kleberg Program in wildlife Ecology at Texas A&M University and the National Rifle Association of America. Approved as article 24130 of the Texas Agriculture Experimental Station. The authors gratefully acknowledge W.C. Ellis, Department of Animal. Science, Texas A&M University, for advice and assistance with model fitting and parameter estimation. D.E. Spalinger, M.J. Lacki, W.E. Grant, and J.O. Fritz reviewed earlier drafts of this manuscript.

References

Allison CD (1985) Factors affecting forage intake by range ruminants: a review. J Range Manage 38:305–311

Baker DL, Hobbs NT (1987) Strategies of digestion and digestive efficiency and retention time of forage diets in montane ungulates. Can J Zool 65:1978–1984

Barnes TG (1988) Digestion dynamics in white-tailed deer. PhD dissertation Texas A&M University, College Station

Chao CC, Brown RD (1984) Seasonal relationships of thyroid, sexual and adrenocortical hormones to nutritional parameters and climatic factors in white-tailed deer (*Odocoileus virginianus*) in South Texas. Comp Biochem Physiol 77A:299–306

Deswysen AG, Ellis WC (1988) Site and extent of neutral detergent fiber digestion, efficiency of ruminal digesta flux and fecal output as related to variations in voluntary intake and chewing behavior in heifers. J Anim Sci 66:2678–2686

Ehle FR, Stern MD (1986) Influence of particle size and density on particulate passage through alimentary tract of Holstein heifers. J Dairy Sci 69:564–568

Ellis WC, Matis JH, Pond KR, Lascano KR, Telford JP (1984) Dietary influences on flow rate and digestive capacity. In Gilchrist FMC, Mackie RI (eds) Herbivore Nutrition in the Tropics and Subtropics. Science Press, Pretoria, South Africa, pp 224–268

Goering HK, Van Soest PJ (1970) Forage Fiber Analyses (Apparatus, Reagents, Procedures and Some Applications). Agric Handbook 379. ARS, USDA, Washington, D.C.

Hanley TA (1982) The nutritional basis for food selection by ungulates. J Range Manage 35: 146–151

Harris LE (1970) Nutritional Research Techniques for Domestic and Wild Animals, vol 1. Anim Sci Dept Publ, Utah State University, Logan

Henke SE, Demarais S, Pfister JA (1988) Digestive capacity and diets of white-tailed deer and exotic ruminants. J Wildl Manage 52:595–598

Hoffman RR (1989) Evolutionary steps of ecophysiological adaptation and diversification of ruminants: a comparative view of their digestive system. Oecologia 78:443–457

Holleman DF, White RG, Frisby K, Jourdan M, Henrichsen P, Tallas PG (1984) Food passage rates in captive muskoxen as measured with non-absorbed radio labeled markers. Biol Pap Univ Alaska Spec Rep 4:188–192

Huston JE, Rector BS, Ellis WC, Allen ML (1986) Dynamics of digestion in cattle, sheep, goats, and deer. J Anim Sci 62:208–215

Karimi AR, Owens FN, Horn GW (1986) Simultaneous extraction of Yb, Dy, and Co from feces with EDTA, DCTA, or DTPA. J Anim Sci (Suppl 1) 63:447

Kay RNB, Englehardt WV, White RG (1980) The digestive physiology of wild ruminants. In Ruckebusch T, Thivend P (eds) Digestive Physiology and Metabolism in Ruminants. AVI, Westport, Conn. pp 743–761

Krysl LJ, McCollum FT, Galyean ML (1985) Estimation of fecal output and particulate passage rate with a pulse dose of ytterbium-labeled forage. J Range Manage 38:180–182

Laubner K (1975) Photometric determination of nitrogen. Wet incineration of indophenol blue with salicylate/hypochlorite. Clin Chim Acta 67:107–110

Mautz WW, Silver H, Holter JB, Hayes HH, Urban WE, Jr (1976) Digestibility and related nutritional data for seven northern deer browse species. J Wildl Manage 40:630-638

McCollum FT, Galyean ML (1982) Effects of forage type and prelabelling treatment on Yb binding efficiency. Yb concentration and in vitro organic matter digestibility of esophageal masticate samples. Proc West Sec Am Soc Anim Sci 33:33–36

Milchunas DG, Dyer MI, Wallmo OC, Johnson DE (1978) In vivo/in vitro relationships of Colorado mule deer forages. Colorado Div Wildl Spec Rep 43, Fort Collins

Mould ED, Robbins, CT (1981) Evaluation of detergent analysis in estimating nutritional value of browse. J Wildl Manage 45:937–947

Pond KR, Ellis WC, Matis JH, Ferreiro HH, Sutton JD (1988) Compartment models for estimating attributes of digesta flow in cattle. Br J Nutr 60:571–595

Poppi DP, Minson DJ, Ternouth JH (1981) Studies of cattle and sheep eating leaf and stem fractions of grasses. 1. The voluntary intake, digestability, and retention time in the reticulo-rumen. Aust J Agric Res 32:99–108

SAS (1979) SAS Users Guide, SAS Institute, Cary, North Carolina

Stervello FA, Kirkpatrick RL, Webb KE Jr (1987) Predicting metabolizable energy in the diet of ruffed grouse. J Wildl Manage 51:560–567

Shafer EL (1963) The twig-count method for measuring hardwood deer browse. J Wildl Manage 27:428–438

Spalinger DE, Robbins CT, Hanley TA (1986) The assessment of handling time in ruminants: the effect of plant chemical and physical structure on the rate of breakdown of plant particles in the rumen of mule deer and elk. Can J Zool 64:312–321

Steel RGD, Torrie JH (1980) Principles and Procedures of Statistics. A Biometrical Approach. McGraw-Hill, New York, NY

Sutherland TM (1988) Particle separation in the forestomach of sheep. In Dobson A, Dobson MH (eds) Aspects of Digestive Physiology in Ruminants. Cornell University Press, Ithaca, NY, pp 43–73

Teeter RG, Owens FN, Mader TL (1984) Ytterbium chloride as a marker for particulate matter in the rumen. J Anim Sci 58:465–473

Varner LW, Blankenship LH (1987) South Texas shrubs: nutritive value and utilization by herbivores. Proc. Symp Plant-Herbivore Interactions. USDA For Ser Gen Tech Rep INT-222, pp 108–112

Varner LW, Blankenship LH, Lynch GW (1977) Seasonal changes in nutritive value of deer food plants in south Texas. Proc Annu Conf SE Fish Wildl Agen 31:99–106

Warner ACI (1981) Rate of passage of digesta through the gut of mammals and birds. Nutrition Abs Rev Series B51:789–820

101

Effects of Thyroidectomy on Seasonal Patterns of Live Weight, Testicular Function, Antler Development, and Molting in Red Deer

Zhen D. Shi and Graham K. Barrell

Abstract

Yearling and deer (*Cervus elaphus*) stags were either thyroidectomized (THX; $n = 3$) or served as nonoperated controls ($n = 5$). During the following 18 months, live weight of the stags was recorded every month, and testicular diameter and neck girth were measured every 2 weeks. Blood samples were collected each week and assayed for testosterone. THX stags maintained hard antlers and did not undergo the same seasonal regression of testicular diameter and plasma testosterone concentration during spring observed in the control stags. Over the subsequent summer and autumn, THX stags exhibited seasonal changes of live weight, pelage, testicular diameter, neck girth, and plasma testosterone concentration, but these were of smaller magnitude than those of control animals. These results indicate that thyroid hormones are involved in regulation of seasonal physiological processes in red deer stags.

Key words: Antlers, deer, live weight, pelage, reproduction, seasonality, testis, testosterone, thyroid

Introduction

Red deer stags undergo an annual pattern of change in live weight, reproductive activity, and antler development (Lincoln 1971), and it has been established that the regular occurrence of these events is regulated by photoperiod (Simpson et al. 1984; Suttie et al. 1984). Increased secretion of testosterone in autumn has been associated with reduced food intake and live weight and the increase in mass of neck muscles. Cessation of testosterone secretion in spring leads to antler casting and growth of new antler tissue. However, not all of these events can be explained by changes in secretion of testosterone, because castrated deer also exhibit changes in food intake and live weight (Pollock 1975; Ryg and Jacobsen 1982). Increased food intake, and consequently live weight gain, in spring and summer may be regulated by changes in the secretion of hormones such as prolactin or the thyroid hormones in red deer (Loudon et al. 1989) and white-tailed deer (Bubenik and Leatherland 1984). These hormones exhibit seasonal patterns of secretion in red deer hinds with rises in plasma levels during spring and early summer, coincident with the onset of increases in live weight and the change of pelage (Loudon et al. 1989). One approach for investigating the role of thyroid hormones is to eliminate their secretion by thyroidectomy. This procedure has been shown to affect seasonal regulation of reproductive activity in mink (Jacquet et al. 1986) and sheep (Nicholls et al. 1988; Moenter et al. 1989). The present study was carried out to examine the effects of thyroidectomy on seasonal changes of live weight, reproductive activity, antler development, and pelage in adult red deer stags.

Materials and Methods

Ten 14-month-old red deer stags were allocated to thyroidectomy or nonoperated control groups ($n = 5$). In January 1988 thyroidectomy was performed following sedation with IM injection of 30 to 40 mg xylazine hydrochloride (Rompun, Bayer) and anesthesia with IV injection of 10 to 12 mg/kg live weight sodium thiopentone (Intraval sodium, May and Baker). A 4 to 5-cm medial ventral incision was made through the skin and fascia overlying the trachea. Thyroid glands were located on both sides of the trachea and removed following blunt dissection of the superficial muscles and fascia and tying off the thyroid blood vessels.

For the next 18 months all animals grazed outdoors on predominantly ryegrass/white clover pasture of the Lincoln University research farm. Water was provided ad libitum, and the diet was supplemented with lucerne hay during winter. Every 2 weeks, testicular diameter was measured by taking the craniocaudal distance between two lightly spring-loaded brass plates held on to the intact scrotum of standing stags, and neck girth, immediately cranial to the larynx, was measured with a plastic tape measure. Live weight was recorded every month, and changes of pelage were recorded in spring. A 10-ml blood sample was collected by venipuncture every week, and plasma was separated and stored at $-18°$ C until hormone assays were carried out. The stags were herded into a pen and held by manual restraint for these procedures.

Hormone Assays

Plasma testosterone concentration was measured with a direct radioimmunoassay described by Barrell et al. (1985). Duplicate 100-μl samples of plasma, diluted in phosphate buffer containing 0.1% gelatine, were heated to 65° C for 30 min. After cooling tritiated tracer ([1,2,6,7, $-{}^{3}$H] testosterone, Amersham, TRK402, 107 Ci/mmol) and antiserum (no. 250, rabbit anti-testosterone-11-BSA serum supplied by G.D. Niswender) were added. Free tracer was removed the next day by addition of dextran-coated charcoal and centrifugation. Sensitivity of the assay was 0.15 ng/ml, and inter- and intraassay coefficients of variation were 16.2% and 6.9%, respectively. Plasma thyroxine (T4) concentrations were determined at Christchurch Public Hospital by courtesy of W.A. Sadler.

Effects of season and treatment were examined by split-plot design for analysis of variance (Gill and Hafs 1971). Data on plasma testosterone concentrations were evaluated after log transformation.

Results

Thyroidectomy

Of the five stags which had surgery to remove the thyroid glands, only three proved to be completely thyroidectomized. In spring (October), plasma from these three animals had T4 concentrations ranging from 0 to 2 nmol/l, which was below the sensitivity of the assay (3 nmol/l). These animals were considered to be completely thyroidectomized (THX). Higher levels of thyroxine (11 and 38 nmol/l) were measured in plasma from the other two stags which had surgery. These two animals proved to have residual or regrown thyroid tissue when surgery was repeated at a later date. They were designated as being partially thyroidectomized and were excluded from the study. (Plasma T4 levels in the five control stags ranged from 51 to 84 nmol/l).

Live Weight

Live weight changed significantly ($P < 0.01$) throughout the year in both groups of animals (Fig. 101.1). Live weight of all stags was low in late winter and early spring of 1988. Thereafter, mean live weight of both groups rose. It increased by 43 ± 4.2 kg in control animals, reaching peak value in late summer–early autumn of 1989, whereas in THX stags it rose by 20 ± 1.5 kg and peaked 2 months earlier. From February onwards control animals had a significantly ($P < 0.05$) higher mean live weight than THX stags. The magnitude of weight loss following the peak live weight in autumn until midwinter was greater in controls: 17.1 ± 5.3 kg in 3 months compared with 9.5 ± 0.8 kg in 4 months in THX stags.

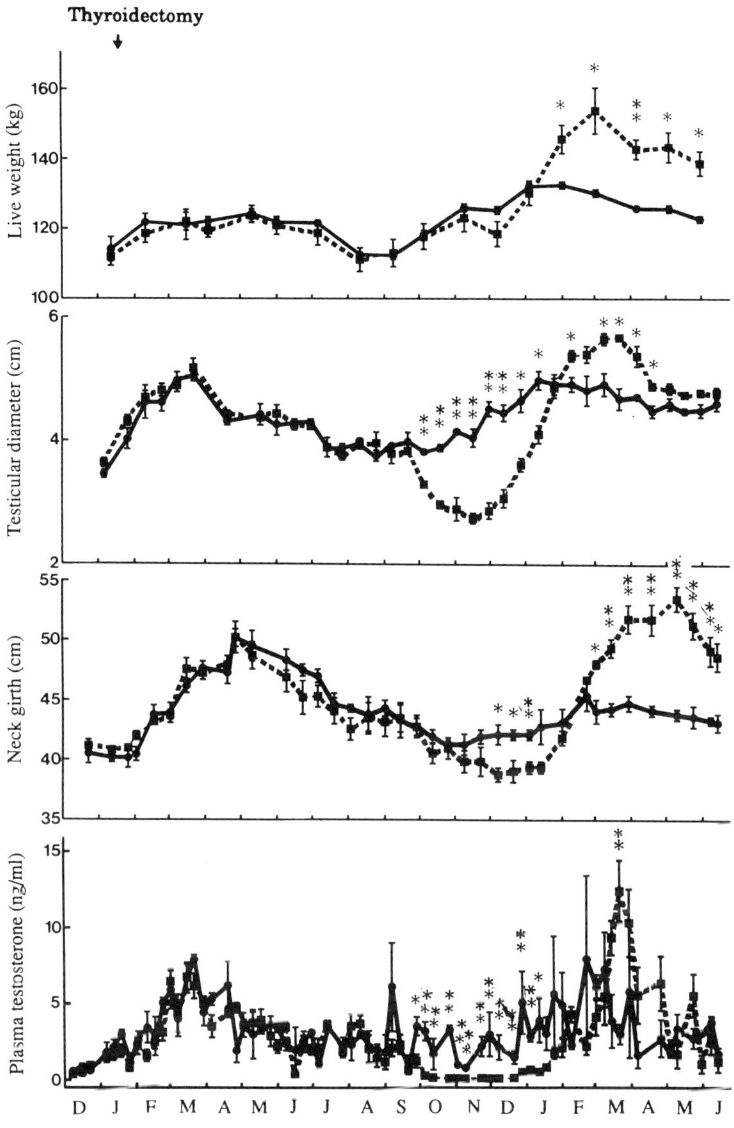

FIGURE 101.1 Mean live weight, testicular diameter, neck girth and plasma testosterone concentration of thyroidectomized (●——●, $n = 3$) and control (■----■, $n = 4$ for testicular diameter, otherwise $n = 5$) red deer stags recorded from December 1987 to June 1989. Vertical bars indicate SEM. Asterisks indicate times with significant treatment effects (*, $P < 0.05$; **, $P < 0.01$).

Testicular Diameter

One of the five control stags had abnormally small testes, so data from this animal have been excluded. From January until late September 1988 there was no effect of thyroidectomy on testis diameter (Fig. 101.1). Both groups of stags exhibited a similar significant ($P < 0.01$) prerut increase and postrut decrease of testis size. However, in the spring of 1988 mean testis diameter of control animals decreased markedly; from 3.85 ± 0.03 cm in September down to

2.75 ± 0.08 cm in mid-November. Thereafter, mean testis diameters of control stags rose, peaking at 5.70 ± 0.07 cm in autumn 1989 and decreasing ($P < 0.05$) again during the following rut. In THX stags changes in testis diameter during the latter half of the study had a smaller amplitude than in controls (Fig. 101.1). Mean testis diameter of THX stags rose from early October 1988 (3.9 ± 0.03 cm) to a peak of 5.0 ± 0.15 cm in mid-January 1989.

Plasma Testosterone Levels

All stags exhibited an increase in plasma testosterone concentration in the first autumn followed by a decrease in that winter (Fig. 101.1). However, from early spring to summer (October 1988 to January 1989) plasma testosterone concentrations of control stags were undetectable (i.e., < 0.15 ng/ml), whereas in THX stags the mean value never fell below 0.8 ± 0.13 ng/ml. In both groups a peak of plasma testosterone concentration associated with the rut was detected in the second year.

Neck Girth

Significant ($P < 0.01$) seasonal changes of neck girth were observed in both groups of animals (Fig. 101.1). Initially, both groups had a similar pattern of changes in neck girth (i.e., increase before the rut and decrease after it). However, the postrut decline in mean neck girth of THX stags ceased in November 1988 (at 41.3 ± 0.3 cm), whereas in control stags neck size continued to decrease until December (when mean girth was 38.8 ± 0.6 cm). From spring 1988 onwards, mean neck girth of THX stags increased moderately to a peak value of 44.7 ± 0.6 cm in February to March 1989, in contrast to the control stags in which neck musculature underwent massive changes and the mean girth reached 53.5 ± 1.0 cm in April to May (Fig. 101.1). Thus, similar to testicular diameter, the seasonal cycles of neck girth were attenuated in THX stags.

Development of Antlers

All five control stags cast their antlers in September and October 1988 (mean date 11 October) and grew immediately a new set of antlers; however, none of the three THX stags cast antlers during the study.

Pelage Changes

All stags molted their summer coats in the first autumn and control stags underwent the spring molt; however, the spring molt was severely disrupted in THX stags. These animals lost some hair at that time, but there was no complete progression of molting and the new hair grown was pale-colored and open in texture.

Discussion

This study has shown that thyroidectomy profoundly disrupted seasonal patterns of live weight, testicular function, antler development, and pelage in red deer stags. Therefore, it appears that thyroid hormones, triiodothyronine (T3), thyroxine, or both are involved in the regulation of the seasonal changes in deer. These events can be manifested as an annual cycle controlled by changes in daily photoperiod acting at the level of the hypothalamus (Karsch et al. 1984), but the processes which cause this rhythm to be expressed by other tissues are not well understood. The regulatory pathways probably involve various hormones operating in sequence, so it is possible that thyroid activity either causes or has a permissive role in the reduction of testosterone secretion in red deer stags during spring. This would explain the casting of antlers and hypotrophy of neck muscles that occurs at this time. However, changes in appetite, and hence, of live weight, have been reported in castrated deer (Pollock 1975; Ryg and Jacobsen 1982) and cannot be explained by changes in the secretion of testosterone alone.

Other hormones which might cause such changes in voluntary food intake and live weight are thyroid hormones and prolactin. Seasonal patterns of peripheral concentrations of these hormones in deer resemble the changes of appetite and live weight (Bubenik and Leatherland 1984; Barrell et al. 1985; Loudon et al. 1989). Red deer hinds treated with T4 from midwinter increased food intake and live weight earlier than controls (A.S.I. Loudon, personal communication) and a similar result occurred in reindeer

(*Rangifer tarandus*) following treatment with T3 (Ryg and Jacobsen 1982). Administration of prolactin to reindeer in winter caused an increase in food intake and live weight (Ryg and Jacobsen 1982); conversely suppression of prolactin secretion by bromocriptine diminished the gain of these parameters during spring in red deer hinds (Curlewis et al. 1988). However, the fact that prolactin and thyrotropin (TSH) are both readily released by thyrotropin-releasing hormone (TRH) (Fraser and McNeilly 1982) may explain why prolactin and the thyroid hormones appear to have associated roles in the mediation of some of these seasonal changes.

In this study thyroidectomy had similar consequences for seasonal reproductive activity as those already reported in birds, mink, and sheep. Administration of exogenous T4 to sexually active male starlings induced photorefractoriness and caused testicular regression (Goldsmith and Nicholls 1984a). Conversely thyroidectomy prevented the development of photorefractoriness in male starlings (*Sturnus vulgaris*) under long daily photoperiods, allowing them to maintain testicular activity (Wieselthier and van Tienhoven 1972; Goldsmith and Nicholls 1984b); this situation could be overcome by administration of T4 (Goldsmith et al. 1985). Thyroidectomy prolonged the duration of maximum testicular activity of mink (*Mustela vison*) (Jacquet et al. 1986), and in sheep, this operation enabled ewes to maintain estrous cyclicity into the normal anestrous period (Nicholls et al. 1988; Moenter et al. 1989). In addition, administration of T4 advanced the onset of anestrus by about 1 week in red deer hinds (A.S.I. Loudon, personal communication). These findings, together with the present results for red deer stags, suggest a role of thyroid function in termination of the breeding season in a variety of seasonally breeding animals.

Casting of hard antlers and growth of new antlers normally occur when plasma testosterone concentration is very low (Lincoln 1971; Barrell et al. 1985), as in the case of our control animals. Levels of testosterone recorded during spring and summer in the blood of THX stags were near to, or above, the immediate postrut values and so were sufficient to maintain hard antlers and prevent casting. During spring, the virtual absence of testosterone secretion in control stags, reflected by the reduction in testis diameter, would account for the regression of neck girth. Throughout the following summer, these parameters increased massively, in concert with the large gain in live weight of control stags (Fig. 101.1). Consequently, the larger testes and neck girths of control stags recorded during this latter period may be partly attributable to the greater size of the animals. It should be noted that the seasonal changes in live weight were superimposed on the general increase in size of all the stags in this study. Peak live weight of red deer stags is achieved after 3 years of age.

Seasonal changes of pelage result from changes in activity of hair follicles (Ryder 1971). Red deer exhibit two molts annually, each corresponding to a period of hair follicle inactivity (Ryder and Kay 1973). In Soay rams, molting of the coat occurred under artificial long daily photoperiods when plasma concentrations of T3 and T4 were high (Lincoln et al. 1980). Exogenous T4 stimulated growth of wool in hypophysectomized (Girard and Legault 1969) and normal sheep (Coop and Clark 1958; Lambourne 1964), presumably by stimulating follicular activity. Also, in mink, thyroidectomy prevented the spring molt, and this was overcome by feeding them T4 (Reineke et al. 1962). These results, as well those of the present study, indicate that molting in spring is achieved by the increased thyroid gland activity. However, there is a close temporal link between increasing circulating levels of prolactin and the spring molt in species such as roe deer (*Capreolus capreolus*), mink, blue fox (*Alopex lagopus*), and Soay sheep (Lincoln 1989). Also, administration of prolactin advanced the spring molt in mink (Martinet et al. 1981), and suppression of prolactin secretion by bromocriptine delayed molting in red deer (Curlewis et al. 1988) and white-tailed deer (*Odocoileus virginianus*) (Bubenik et al. 1985), and prevented it in the blue fox (Smith et al. 1987). This suggests that secretion of prolactin must be considered together with that of thyroid hormones in the regulation of pelage changes, as well as other seasonal changes discussed above.

In conclusion, the present findings have shown that thyroidectomy of red deer stags diminishes the amplitude of seasonal changes and prevents the suppression of testosterone secretion that normally occurs during spring.

Acknowledgements

The authors would like to thank A.S. Familton for performing thyroidectomy surgery, M.J. Keeley for taking care of animals and assistance with sample collection and recording of data, L.K. Lewis for her expertise with radioimmunoassays, and W.A. Sadler of Christchurch Public Hospital for the T4 measurements. Testosterone antibody was kindly provided by G.D. Niswender, Colorado State University.

References

Barrell GK, Muir PD, Sykes AR (1985) Seasonal profiles of plasma testosterone, prolactin, and growth hormone in red deer stags. In Fennessy PF, Drew KR (eds) Biology of Deer Production. R Soc N Z Wellington Bull 22:185–190

Bubenik GA, Leatherland JF (1984) Seasonal levels of cortisol and thyroid hormones in intact and castrated mature male white-tailed deer. Can J Zool 62:783–787

Bubenik GA, Schams D, Leatherland JF (1985) Seasonal rhythms of prolactin and its role in the antler cycle of white-tailed deer. In Fennessy PF, Drew KR (eds) Biology of Deer Production. R. Soc N Z Wellington Bull 22:257–262

Coop IE, Clark VR (1958) The effect of thyroxine on wool growth and lamb production. N Z J Agric Res 1:365–381

Curlewis JD, Loudon ASI, Milne JA, McNeilly AS (1988) Effect of chronic long-acting bromocriptine treatment on liveweight, voluntary food intake, coat growth and breeding season in non-pregnant red deer hinds. J Endocrinol 119:413–420

Fraser HM, McNeilly AS (1982) Effect of chronic immunoneutralization of thyrotropin-releasing hormone on the hypothalamic-pituitary-thyroid axis, prolactin and reproductive function in the ewe. Endocrinology 111:1964–1973

Gill JL, Hafs HD (1971) Analysis of repeated measurements of animals. J Anim Sci 33:331–336

Girard J, Legault L (1969) Resumption of wool growth induced by thyroxine injection in hypophysectomized Ile-de-France ewes. Annal Biol Anim Bioch et Biophy 9:497–502

Goldsmith AR, Nicholls TJ (1984a) Thyroxine induces photorefractoriness and stimulates prolactin secretion in European starlings (*Sturnus vulgaris*). J Endocrinol 101:R1–R3

Goldsmith AR, Nicholls TJ (1984b) Thyroidectomy prevents the development of photorefractoriness and the associated rise in plasma prolactin in starlings. Gen Comp Endocrinol 54:256–263

Goldsmith AR, Nicholls TJ, Plowman G (1985) Thyroxine treatment facilitates prolactin secretion and induces a state of photorefractoriness in thyroidectomized starlings. J Endocrinol 104:99–103

Jacquet J-M, Coutant C, Maurel D, Boissin-Agasse L, Boissin J (1986) Effects of thyroidectomy on spring and summer variation of testicular activity and plasma prolactin in the mink. Comptes Rendus Acad Sci Paris (Serie 111) 303:367–370

Karsch FJ, Bittman EL, Foster DL, Goodman RL, Legan SJ, Robinson JE (1984) Neuroendocrine basis of seasonal reproduction. Recent Prog Horm Res 40:185–232

Lambourne LJ (1964) Stimulation of wool growth by thyroxine implantation 1. Live weight changes and wool growth of fine-wool Merino wethers. Aust J Agric Res 15:657–675

Lincoln GA (1989) Significance of seasonal cycles in prolactin secretion in male mammals. In Serio M (ed) Perspectives in Andrology. Raven Press, New York, pp 299–306

Lincoln GA (1971) The seasonal reproductive changes in the red deer stag (*Cervus elaphus*). J Zool London 163:105–123

Lincoln GA, Klandorf H, Anderson N (1980) Photoperiodic control of thyroid function and wool and horn growth in rams and the effect of cranial sympathectomy. Endocrinology 107:1543–1548

Loudon ASI, Milne JA, Curlewis JD, McNeilly AS (1989) A comparison of the seasonal hormone changes and patterns of growth, voluntary food intake and reproduction in juvenile and adult red deer (*Cervus elaphus*) and Père David's deer (*Elaphurus davidianus*) hinds. J Endocrinol 122:733–745

Martinet L, Meunier M, Allain D (1981) Control of delayed implantation and onset of spring moult in the mink (*Mustela vison*) by daylight ratio, prolactin and melatonin. In Ortavant R, Pelletier J, Ravault JP (eds) Photoperiodism and Reproduction in Vertebrates. International Colloquium, INRA, Paris, pp 253–261

Moenter SM, Woodfill CJI, Karsch FJ (1989) Thyroidectomy blocks transition to anestrous season by sustaining LH-pulse generating activity in ewes. Soc Neurosci 15:950 (Abstract)

Nicholls TJ, Follett BK Goldsmith AR, Pearson H (1988) Possible homologies between photorefractoriness in sheep and birds: the effect of thyroidectomy on the length of the ewe's breeding season. Reprod Nutr Développment 28:375–385

Pollock A (1975) Seasonal changes in appetite and sexual condition in red deer stags maintained on a six-month photoperiod. J Physiol London 244:95P-96P

Reineke EP, Travis HF, Dolnick EH (1962) The effects of thyroid gland destruction and replacement therapy on fur growth in mink (*Mustela vison*) given a thyroxine-free diet. Am J Vet Res 23:121–127

Ryder ML (1971) Wool growth cycles in Soay sheep. J Agric Sci Cambridge 76:183–197

Ryder ML, Kay RNB (1973) Structure of and seasonal change in the coat of red deer (*Cervus elaphus*). J Zool London 170:69–77

Ryg M, Jacobsen E (1982) Effects of thyroid hormones and prolactin on food intake and weight changes in young male reindeer (*Rangifer tarandus*). Can J Zool 60:1562–1567

Simpson AM, Suttie JM, Kay RNB (1984) The influence of artificial photoperiod on the growth, appetite and reproductive status of male red deer and sheep. Anim Reprod Sci 6:291–299

Smith AJ, Mondain-Monval M, Simon P, Andersen Berg K, Clausen OPF, Hofmo PO, Scholler R (1987) Preliminary studies of the effects of bromocriptine on testicular regression and the spring moult in a seasonal breeder, the male blue fox (*Alopex lagopus*). J Reprod Fert 81:517–524

Suttie JM, Corson ID, Fennessy PF (1984) Voluntary intake, testis development and antler growth patterns of male red deer under a manipulated photoperiod. Proc N Z Soc Anim Prod 44:167–170

Weiselthier AS, van Tienhoven A (1972) The effect of thyroidectomy on testicular size and on the photorefractory period in the starling (*Sturnus vulgaris* L.). J Exp Zool 179:331–338

102

Volatile Fatty Acids, pH, Dry Matter, and Lactic Acid of the Rumen Contents of Sika Deer

Shangren Li and Rongqian Qin

Abstract

Volatile fatty acids, pH, dry matter, and lactic acid of rumen contents were determined on 20 female sika deer (*Cervus nippon*) collected in winter. The total content of VFAs, containing C_2, C_3, Iso-C_4, C_4, Iso-C_5, and C_5, was 114.6 mmol/1. In order of concentration, C_2 was first followed by C_3, C_4, Iso-C_4, C_5, and Iso-C_5. Iso-C_4 was found in only three samples. The pH averaged 6.1 ± 0.25; dry matter averaged $88.0\% \pm 7.3\%$. These data were similar to those of other ruminants on a high-roughage, low-concentrate diet, such as cattle and sheep. The lactic acid concentration, however, was low at 0.9 ± 0.68 mg/dl.

Key words: Lactic acid, rumen, sika deer, VFA

Introduction

Deer are characterized by physiological and biological peculiarities that differ from other ruminants. Digestibility of lignin by deer is generally better than by cattle or sheep. The digestion rate of crude fiber by deer has been shown to be double that by bulls (62.4% and 31.0%, respectively, Li and Qin, unpublished data). Deer often eat branches of oak and bushes containing great quantities of lignin and fiber in winter and early spring. In addition, deer are less susceptible to toxins when fed a variety of poisonous plants, such as fischer euphorbia (*Euphorbia fischeriana*), diversileaf jack-in-the-pulpit (*Arisaema heterophyllum*), and black falsehellebore (*Veratrum nigrum*) (Qin and Liang 1983). The extent to which the rumen function of deer supports these differences in lignin and fiber digestion and detoxification processes is not known. We determined the volatile fatty acids (VFAs), pH, dry matter (DM), and lactic acid of the rumen contents of sika deer to obtain basic information to allow future investigations on digestion, nutrition, metabolism, and feeding, which will be necessary to maximize production practices for the husbandry of sika deer.

Methods

Samples were taken from mixed rumen contents of 20 female sika deer immediately after death. The deer were on a diet of natural forage including corn straw and large oak leaves with a small

TABLE 102.1. Chemical composition (%) of the forage fed to sika deer

Ingredient	Crude protein	Crude fat	Crude fiber
Corn straw	6.1	1.8	47.8
Oak leaves	15.7	3.1	25.2
Corn	8.1	5.3	1.7
Bean cake	44.5	8.1	5.1

TABLE 102.2. Concentrations (mmol) and percentages (%) of VFAs in the rumen contents of sika deer ($n = 20$)

	C_2		C_3		C_4		Iso-C_5		C_5		Total	
	(mmol)	(%)	(mmol)	(%)	(mmol)	(%)	(mmol)	(%)	(mmol)	(%)	(mmol)	(%)
Mean	81.3	64.6	18.7	18.0	11.8	13.4	1.0	1.3	1.8	2.5	114.6	100
SD	13.6	3.8	5.0	2.0	4.5	2.5	0.4	0.5	0.4	0.3		
Range												
Low	58.6	54.0	10.8	15.0	7.1	9.6	0.4	0.5	1.1	1.9		
High	111.9	69.5	32.4	23.8	21.4	16.7	1.8	2.5	2.9	3.0		

C_2, acetic acid; C_3, propionic acid; C_4, butyric acid; Iso-C_5, iso-valeric acid; C_5, valeric acid

amount of concentrates composed of ground corn and a bean cake (0.75 kg each per day). The chemical composition of the forage is shown in Table 102.1. Volatile fatty acids were determined using gas chromatography according to the methods of Qin (1982). Lactic acid was analyzed by the use of an autobiochemical analyzer (MCA, IL Multistat III Plus System). Dry matter and pH were determined by routine methods.

TABLE 102.3. pH, dry matter, and lactic acid concentration in the rumen contents of sika deer

	pH	DM (%)	Lactic acid (mg/dl)
Number of deer	20	20	17
Mean	6.1	12.69	0.94
SD	0.25	7.44	0.68
Range			
Low	5.8	9.70	0.1
High	6.6	16.80	2.4

Results and Discussion

Concentrations of VFAs and the ratios among VFAs often vary greatly with the composition of the diet and the physiological condition of animals. Total ruminal VFA range is usually 60 to 120 mEq/l. However, levels below 100 mEq/l are found for ruminants fed hay. Levels may be as high as 200 mEq/l when animals graze on young summer grass or are fed starch-rich diets (Swenson 1984). In our study, the total VFA concentrations and the ratio of VFAs were approximately the same as those of cattle and sheep fed a high roughage, low concentrate diet with the exception that the C_4 proportion was higher than that of other ruminants (Table 102.2). The high ratio of C_4 was presumably due to the corn straw, which contained a large portion of soluble carbohydrate and comprised 50% of the roughage fed to the deer. Alternately, the high ratio of C_4 could be a distinguishing feature of deer digestion (see Table 102.4). In addition, Iso-C_4 accounted for 1% of the total VFAs, was found in only three samples, and was not considered for statistical analysis.

TABLE 102.4. Comparison of percentages of VFAs in the rumen contents among cattle, sheep, and sika deer

Animal	Diet	C_2	C_3	C_4	C_5	Reference
Cattle	High roughage, low concentrate	65–75	15–21	5–14	2.5	Balch and Rowland (1957)
	Low roughage, high concentrate	40–66	18–41	7–15	3–14	
Sheep	Hay (2.8% = *n*)	65–71	15–24	8–15	2–5	Johns (1955)
	Low hay, high maize	51–68	23–42	3–14	1–2	Phillipson (1952)
Sika deer	High roughage, low concentrate	51–70	15–24	10–17	2–5	Present study

The pH of the ruminoreticulum, which ranged from 5.8 to 7.0, decreased after feeding (Table 102.3). A fall in pH is common when the diet contains large amounts of fermentable carbohydrate (Ambo 1987). The pH of the rumen contents in this study averaged 6.1 ± 0.25. Considering the fact that the deer being studied were killed soon after feeding, a low value was expected. The DM of the rumen contents of sika deer were similar to those of other ruminants such as cattle and sheep (Table 102.3).

Lactate concentrations in the rumens of these deer were usually low (Table 102.3). Sudden consumption of large amounts of highly digestible cereals (corn, wheat) or foods rich in sugar (sugar beets) can cause accumulation of high concentration of lactic acid in the rumen (Swenson 1984). Infusion of a saccharose solution into the rumen of goats rapidly reduced the pH and caused a marked rise in lactic acid to as high as 410 mg/dl, compared with 10 mg/dl before infusion (Ambo 1987). In our study, the deer were fed a low nutrient diet, and lactate synthesis should have been minimal. In addition, absorption of lactic acid is faster from acidic rather than from neutral solutions, and a low level would be expected (pH 6.1), which agrees with that found (i.e., 0.98 ± 0.68 mg/dl).

Acknowledgements

We thank E.L. Yuan of Jilin Agricultural University for his design and for performing the VFA determinations, and H.F. Zhao and J.H. Yang for their help in this study.

References

Ambo K (1987) Nutrition of Ruminants. Yanbian Agricultural College Press, pp 26–27

Balch DA, Rowland SJ (1957) Volatile fatty acids and lactic acid in the rumen of dairy cows receiving a variety of diets. Br J Nutr 11:288–89

Johns AT (1955) Pasture quality and ruminant digestion. 2. Levels of volatile fatty acids and ammonia in the rumen of sheep on a high production pasture. N Z J Sci Tech (Series A) 37:323–331

Phillipson AT (1952) The fatty acids present in the rumen of lambs fed on a flaked maize ration. Br J Nutr 6:190–198

Qin WL (1982) Determination of rumen volatile fatty acids by means of gas chromatography. J Nanjing Agric Coll 4:110–116

Qin RQ, Liang FX (1983) Deer Feeding. Jilin Agricultural University Press, pp 23, 88–89

Swenson MJ (ed) (1984) Physiology of Domestic Animals, 10th ed, Comstock, Cornell University Press, 329 pp

103
Importance of Copper Supplementation in Farmed Red Deer

László Sugár, Zsuzsa Lassu-Merényi, and János Kelemen

Abstract

Very high liver copper concentrations were measured in the fetuses and neonates of different deer species: 2168 mg/kg DM in roe deer (*Capreolus capreolus*), 1570 mg/kg in red deer (*Cervus elaphus*), and 624.5 mg/kg in fallow deer (*Dama dama*). Farmed male red deer showed marginal liver copper content (9.7 mg/kg). Copper concentrations of neck hair in the same farmed herd decreased from 9.4 to 12.1 to 1.5 to 4.2 mg/kg in 1988–89, indicating copper deficiency. These findings suggested the necessity of copper supplementation in farmed red deer.

Key words: Copper, deer farming, hair analysis, liver, red deer, supplementation

Introduction

Information about the trace elements in deer is minimal. Copper is important for this ruminant group, and the liver is the primary storage site of copper, especially in fetuses and neonates (Reid et al. 1980; Hänichen et al. 1985; Sugár et al. 1985). While wild deer seldom show serious copper deficiency (Flynn et al. 1977), a lack of this mineral seems to be a problem in farmed deer herds (Reid et al. 1980; Mackintosh et al. 1986). Therefore, an evaluation of deer's copper status may be necessary. The most common method is to measure the copper concentration of the liver, but the biopsy technique is unsuitable for routine diagnostic purposes. Alternately, Kellaway et al. (1978) have proposed hair sampling to diagnose hypocuprosis, and Mackintosh et al. (1986) found a close correlation between serum and liver copper levels in farmed deer with marginal copper status.

The purpose of this investigation was to collect information on the normal copper concentrations of deer organs and tissues and to evaluate them for diagnosis of copper deficiency in farmed deer.

Materials and Methods

Samples were collected in southwestern Hungary, a rolling land with 50% to 70% forest cover. Liver samples were obtained from half-term red deer fetuses and from newborn neonates to 2-day-old red, fallow, and roe deer in the course of hunting or accidents during winter, and when newborn animals were captured for farming purposes in May–June. Hair and liver samples were taken from young and adult wild deer shot during the hunting season (winter). Hair samples were collected each year (1986–89) in February (late winter) from farmed red deer held on an experimental deer farm in Gálosfa. Sampling was done in conjunction with other management activities such as annual tuberculin (TB) testing and blood sampling. Liver samples were obtained when farmed deer were killed by feral dogs or were slaughtered after failing the TB test. The diet was a native pasture vegetation during the growing season and grass-silage with some supplementary food (mixed grain: maize and barley). The copper concentration of the diet was not examined.

Copper supplementation for the farmed red deer was started on 8 December 1989 with a salt-lick containing 2.5 g/kg copper for ad libitum consumption all year round.

Sampling methods

Hair samples were taken from the lateral part of the neck using electric clippers, and liver samples were taken from the ventral side of the central lobe. Liver samples were put into an electric kiln and ashed at 550° C for 10 h. The ash was dissolved in 5 *M* nitric acid, and the copper content of the solution was measured by atomic adsorption spectrophotometry in an air-acetylene flame (Pye Unicam SP 191). Hair samples were washed in benzine and distilled water and then subjected to the same procedure as the liver samples. Concentrations were expressed as mg/kg dry matter basis.

Results and Discussion

Highest liver copper concentrations were found in two newborn roe deer fawns, at 2350 and 1986 mg/kg. Very high and similar levels were recorded in seven red deer fetuses (1521 mg/kg) and five newborns (1570 mg/kg). Relatively high levels were found in six 1-month-old roe deer fawns (1040 mg/kg) and two newborn fallow deer (624 mg/kg). The mean concentration in two 2- to 3-month-old roe deer fawns was 104.4 mg/kg, but in 6-month-old and older roe and red deer the mean concentrations declined to below 50 mg/ kg. Similar trends were observed by Reid et al. (1980), who measured mean copper concentrations in fetus/neonate livers of 1175 mg/kg. This latter value was 35-fold higher than the mean concentrations in mature animals. Hänichen et al. (1985) also found extremely high copper concentrations in the livers of neonate Barasingha (*Cervus duvauceli*), moose (*Alces alces*), and red deer (Fig. 103.1). It is clear that the neonate liver is a very rich storage site of copper in all species of deer investigated. In red deer, the fetal liver copper concentration reaches its saturation point 4 months before birth, i.e., the half-term fetus. After birth the copper concentration averaged 20 to 50 mg/kg at 6 months and was 10 mg/kg, the lowest level, in mature farmed male red deer (Fig. 103.1). This lower value is probably indicative of hypocuprosis in these animals (Kellaway et al. 1978).

The mean copper concentrations of the hair of wild red deer were at or above 7 mg/kg, which is considered to be normal for copper in red deer hair (Anke et al. 1980). The copper levels in wild female red deer hair were similar to those found by Franzmann (1975) in Alaskan moose (*Alces alces*) hair (14.7 ± 4.8 mg/kg). In farmed deer, copper concentrations of the hair samples were

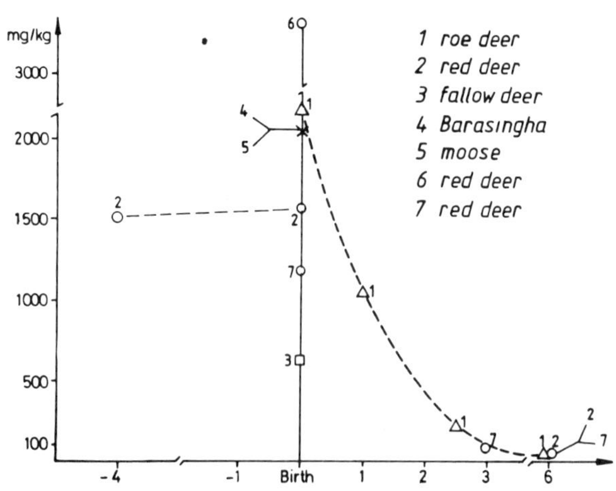

FIGURE 103.1. Postnatal monthly changes in liver copper concentrations in deer. Deer 4, 5, 6 are data from Hanichen et al. (1985); deer 7 data from Reid et al. (1980). Curve is fitted by eye.

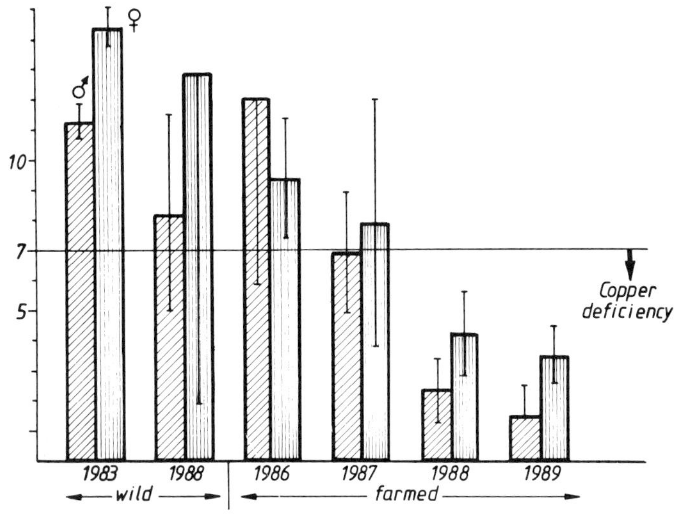

FIGURE 103.2. Hair copper concentrations in Hungarian red deer populations according to environment, year, and sex (mg/kg DM).

normal in 1986 and 1987; however, the levels in this farmed population markedly decreased in 1988, and remained low, 1 to 4 mg/kg, through 1989 (Fig. 103.2). Although enzootic ataxia (McAllum 1980) has not been observed in the herd thus far, other symptoms such as reduced keratinization of hair and hoof in some deer, lower weaner weights of the 1989 calves, and the relatively low fertility of yearling (41%) and older hinds (91.6%) could be attributable to copper deficiency (Kellaway et al. 1978). Based on this supposition of hypocuprosis in the farmed deer, a copper supplementation study was started in December 1989.

References

Anke M, Groppel B, Prien S, Briedermann L, Mehlitz S (1980) Die Mengen- und Spurenelementenversorgung der Wildwiederkäuer 4. Arch Tierernährung 30:707–721

Dunbar MS, Foreyt WJ (1985) Copper and manganese concentrations in the hair of mule deer. JAVMA 187:1180

Flynn A, Franzmann AW, Arneson PD, Olemeyer JL (1977) Indications of copper deficiency in a subpopulation of Alaskan moose. J Nutr 107:1182–1189

Franzmann AW (1975) Levels of some mineral elements in Alaskan moose hair. J Wildl Manage 39: 374–378

Hänichen T, Wiesner H, Günzler, Hegel G (1985) Abortion and still-birth among ovines due to excessive copper intake and comparative studies into other zoo-kept ruminants. Verhber Erkrg Zootiere 27:243–249

Kellaway RC, Sitorus P, Leibholz JML (1978) The use of copper levels in hair to diagnose hypocuprosis. Res Vet Sci 24:352–357

Mackintosh CG, Wilson PR, Beatson NS, Turner K, Johnstone P (1986) Preliminary report of the liver: serum copper relationship in deer. Proc Deer Course for Vets, NZ Deer Branch Course 3:156–163

McAllum HJF (1980) Enzootic ataxia in deer: copper deficiency. AgLink AST 94, NZ Min Agric and Fish, Wellington

Reid TC, McAllum HJF, Johnstone PD (1980) Liver copper concentrations in red deer (*Cervus elaphus*) and wapiti (*C. canadensis*) in New Zealand. Res Vet Sci 28:261–262

Sugár L, Lassu Zs, Horn A (1985) High copper or zinc levels in livers and kidneys of wild ruminants. Verhber Erkrg Zootiere 27:251–252

104

Effects of Dietary Protein and Energy Intake on Body Reserve Deposition and Milk Production in Caribou and Reindeer

Ann C. Allaye-Chan and Robert G. White*

An experiment was conducted in 1989 to determine the effects of energy and protein intake and the energy/protein ratio on (1) the deposition of body fat and body protein and (2) milk production in female caribou (*Rangifer tarandus granti*) and reindeer (*Rangifer tarandus tarandus*). Four diet treatments (high energy/high protein; high energy/low protein; low energy/high protein; low energy/low protein) were randomly assigned within each group of four lactating caribou, four non-lactating caribou, four lactating reindeer, and four nonlactating reindeer. Daily feed intakes were recorded for each individual, and body weights were measured every 2 to 4 days for both cows and calves. Body composition, determined using the tritiated water technique, and milk production, determined using the doubly labeled water method, were obtained every 3 weeks for 12 weeks. Body weight changes in both lactating and nonlactating females were primarily governed by energy intake. The efficiency of utilization above maintenance was the same for both lactation classes. The propensity to deposit fat was greater in lean than in fat animals and increased with time. Energy intake was significantly correlated with fat deposition in nonlactating females only. Energy intake was not significantly correlated with fat deposition in lactating females because surplus energy was directed towards increased milk production. However, females with low fat reserves allocated a lower portion of their resources to milk production than females with large fat reserves. The energy/protein ratio, protein intake, and subspecies were generally insignificant factors in determining patterns of tissue deposition or milk production.

*Institute of Arctic Biology, University of Alaska Fairbanks, Fairbanks, Alaska 99775, USA

105
A Bioenergetics Model for Evaluating the Quality of Moose Winter Habitats

Ken L. Risenhoover and John H. Roese*

A simulation model was developed to predict energy balance and survival of moose (*Alces alces*) feeding in different winter habitats. The energetic cost of foraging (including locomotion), rumination, thermoregulation, and maintenance were combined with estimates of nutrient ingestion and assimilation to predict changes in animal condition during winters of varying length and severity. The relative importance of forage quality and quantity, duration and severity of the winter period, and physiological state to model predictions were examined. The model was used to compare the relative quality of winter habitats near the northern and southern limits of moose distribution in North America.

*Department of Wildlife and Fisheries Science, Texas A&M University, College Station, Texas, USA

106
Effects of Supplemental Feeding on White-Tailed Deer Migration in Wisconsin

Timothy L. Lewis and Orrin J. Rongstad*

In 1986 we began a 3-year study to examine the effects of different food levels on migration dates and patterns of 169 free ranging white-taileld deer (*Odocoileus virginianus*). All study deer were trapped in or immediately adjacent to the Chequamegon National Forest in northwestern Wisconsin. Each deer was equipped with a radio transmitter and located weekly. Two groups of deer received supplemental feed (corn) ad libitum, one group on its summer range, the other on its winter range. A third group received no supplemental food. Summer and winter ranges of deer groups overlapped. Feeders were observed over 2000 h to determine frequency of feeder use. Feeding regime had no effect on the timing of fall migration, which ranged from 14 November to 21 January. Weather appeared to determine the last available day for migration, but had no detectable effect on early departures. Spring migration was delayed up to 1 week in deer receiving supplemental winter food, but weather was again the principle determinant of migration date. Thirty-five of 68 winter-fed deer were nonmigratory. This included 21 of 25 deer in a deer yard receiving food for 40 years, and 14 of 43 deer in a deer yard receiving food for 7 years. Before feeding commenced in that yard, all of 54 radio-tagged deer from a previous study were migratory. Only 2 of 101 deer not receiving winter food were nonmigratory.

*Department of Wildlife Ecology, University of Wisconsin, Madison, Wisconsin, 53706, USA

107
Voluntary Feed Intake and Digestion in Farmed Red Deer Stags: A Comparison with Goats and Sheep

B.M. Francoise Domingue, D.W. Dellow, P.R. Wilson, G.F. Wilson and T.N. Barry*

Castrate male red deer (*Cervus elaphus*), sheep, and goats, aged 2 to 3 years and fistulated in the rumen, were fed chopped lucerne (*Medicago sativa*) hay during experimental periods in winter (w) and summer (s) to determine effects of seasonality. All animals were fed at hourly intervals. Deer showed a marked (+34%) increase in voluntary food intake (VFI) and rumen pool size (+51%) during s, and this was accompanied by nonsignificant but consistent decreases in fractional outflow rate (FOR) of both rumen liquid and particulate matter, resulting in no change in apparent digestibility. Goats also showed an increase in VFI (+19%) and rumen pool size (27%) during s, accompanied by nonsignificant but consistent increases in FOR of rumen liquid and particulate matter, causing apparent DM digestibility to markedly decline. Sheep showed no differences between s and w in VFI, apparent dry matter (DM) digestibility, rumen pool size, and FOR of liquid or particulate matter. Rumen FOR was much greater for deer than for goats and sheep, both for liquid (0.16 vs 0.10/h) and for particulate matter (0.044) vs 0.035/h), and the ratio FOR liquid/FOR particulate was consistently higher for deer (3.6) than for sheep and goats (3.0). In a second experiment, implants of melatonin were given to the red deer for a 10-week period during spring. These resulted in a reversal of seasonal patterns during the ensuing 12 months, with VFI, rumen pool size, rumen capacity, and heart rate being depressed during summer and then rising at maximum values during winter. These data show seasonality in castrate male red deer in both VFI and a number of rumen criteria, and their strong entrainment by a period of high plasma melatonin concentration.

*Massey University, Palmerston North, New Zealand

108
Fatty Acid Composition in the Bone Marrow Fats of Cervidae

László Sugár and Istvan Nagy*

The depletion patterns of bone marrow fat as an indication of body condition are better documented than the fatty acid composition changes of bone marrow fats in Cervidae. The fatty acid composition in the marrow of the proximal bones (mandibula, humerus, femur) is dominated by saturated fatty acids. With stepwise changes, the composition turns into unsaturated dominance in the distal bone marrows (metacarpus, m tarsus, phalanges), lowering the melting point of the fat. The ratio of palmitic acid (C 16) decreases from 25% to 32% to 6% to 14% in red deer (*Cervus elaphus*), from 26% to 52% to 11% to 14% in fallow deer (*Dama, dama*), but does not change in roe deer (*Capreolus capreolus*). Stearic acid (C 18) decreases from 23% to 40% to 1% to 4% in red, from 26% to 39% to 2% to 2.5% in fallow and from 30% to 48% to 4% to 5% in roe deer. Palmitoleic acid (C 16:1) increases from 2% to 6% to 17% to 25% in red, from 1.5% to 3% to 15% to 21% in fallow and from 1% to 1.2% to 5.6% to 5.8% in roe. Oleic acid (C 18:1) increases from 18% to 27% to 46% to 56% in red, from 17% to 21% to 37% to 43% in fallow and from 20% to 43% to 66% to 67% in roe deer. In fallow deer fats the pentadecanoic acid (C 15:1) content is surprisingly high: 3.7% to 12.7%.

*Faculty of Animal Science, Pannonian Agricultural University, H-7401 Kaposvar, P.O. Box 16, Hungary

109
Urine Sampling Techniques for Semidomesticated Reindeer

Mauri Nieminen, Ulla Heiskari, and Timo Soveri*

Attempts to identify nutritional indices in reindeer (*Rangifer tarandus*) have been extensive. Recently, the search for indicators of nutriton in deer and other cervids has turned to urinalysis. Altogether, 52 semidomesticated reindeer (49 adult females and 3 castrated males) were used in the present study. Reindeer were freely grazing in Kaamanen Reindeer Research Station or fed with *Equisetum fluviatile* or concentrates during winter and summer 1989. The blood samples were taken from the jugular vein, and urine samples from the bladder of castrated males during slaughtering and from the living females by catheterization. After sampling, reindeer were weighed. Urine samples in snow were collected from different feeding groups. Serum parameters indicating nutritional condition of reindeer were determined, as was urea N(U) and creatinine (C). Urine collection from bladder by catheterization was successful. There was a significant positive correlation between serum U and bladder urine C ($r = 0.504$, $n = 28$, $P < 0.01$), between serum C and bladder urine C ($r = 0.729$, $n = 28$, $P < 0.001$) and between serum U:C and bladder U:C ratios ($r = 0.920$, $n = 28$, $P < 0.001$). No significant changes were measured in urine U:C ratio collected from the snow during the first 12 h. Urinalysis seems to be an alternative and perhaps more sensitive means of assessing nutritional status of semidomesticated reindeer.

*Finnish Game and Fisheries Research Institute, Reindeer Research, Rovaniemi and College of Veterinary Medicine, Helsinki, Finland

110
Annual Periodicity of Total Blood Thyroxine (T_4) in Male Fallow Deer: Problems and Experiences

Klaus Fischer*

The annual T_4 profile in adult fallow (*Dama dama*) bucks plateaus between January and August ($\bar{x} \approx 50$ ng/ml). T_4 has minimal values during the rut (October: $\bar{x} = 28.3$ ng/ml; November: 30.4 ng/ml) with differences of 44% ($P \leqq 0.001$). There were no differences between serum and plasma samples ($P \leqslant 0.001$; $r > 0.9$). Annual profiles with two commercial kits show $r = 0.98$, but differences of 33.1% to 34.6% ($P \leqslant 0.001$). T_4 measurcment was repeated later with the same blood samples and the same coded test kit. Results showed $r = 0.94$, but values differed at 10.14 ng/ml ($P \leqslant 0.001$). Other values were also lower in this test. One yearling fell sick and 4 weeks before death $T_4 = 14.9$ ng/ml. Another buck had a severe parasitic infection, but in conjunction with a social environment change, T_4 almost normalized after 1 year. T_4 measurements were made in animals that had lived together more than 1 year before sampling. *R* over 1.5 years were: 0.89, 0.79, and 0.85 between half-brothers, and 0.73, 0.79, and 0.86 between adult bucks. Annual T_4 curves of other bucks from other groups had faint positive correlations with the two groups mentioned, but all individuals demonstrated a strong T_4 depression around rut. The annual T_4 pattern in a fawn/yearling living in a small enclosure with several old animals showed extremely high T_4 due to the social stress. There was no difference in T_4 between fawns/yearlings and adults, but in fawns/yearlings T_4 oscillated.

*I. Zoological Institute, University, D-3400 Gottingen, Germany

111

Seasonal Differences in Pasture Species Preferences by Red and Fallow Deer in New Zealand

W.F. Hunt and R.J.M. Hay*

Alternative pasture species to the traditionally used endophyte-infected perennial ryegrass have been investigated for deer in New Zealand for two reasons. Productivity peaks of ryegrass-based pastures are out of phase with the summer lactation feed demand of deer, and endophyte infection risks the neuromuscular disorder "Ryegrass staggers." Alternative pasture species with high productivity in summer were compared with ryegrasses for acceptability by deer in all seasons of the year, using a photographic technique specifically developed for recording pasture species preferences and behavior by grazing animals. Research plots were established on a commercial deer farm at Wardville, Matamata, New Zealand, to determine pasture species preferences by both lactating red deer (*Cervus elaphus*) hinds and stags. Four replicates of each of 16 grasses, herbs, and legumes were sown in 7 × 7 m plots, and when established were used to test the preferences of stags and hinds in each season of the year. Both hinds and stags showed a preference for legumes and herbs compared with grasses, but hinds showed a stronger preference for red clover than did stags. Grasses were more acceptable in winter than in summer, but only "Matua" prairie grass and "Moata" Italian ryegrass were ever as palatable to deer as were the legumes. Red deer grazing behavior was dominated by alternative periods of grazing and lying down, with shorter times spent walking or standing.

*Grasslands Division, D.S.I.R., Palmerston North, New Zealand

112
In vitro Digestability of Fall Forages by Coexisting Deer Species in Texas

James J. Jackley[1], Stephen Demarais[1], Larry W. Varner[2], and Thomas G. Barnes[2]

Percent in vitro digestible dry matter (%DDM) of seven common fall forages was estimated using rumen inocula from three fallow deer (*Dama dama*), two sika deer (*Cervus nippon*), and two white-tailed deer (*Odocoileus virginianus*). Inocula were collected from animals sacrificed on 4 November 1987 on a ranch in Bandera County, in the Edwards Plateau Region of south-central Texas. Forage samples were collected during October 1987 in Bandera County and adjacent Kerr County. Estimates of %DDM were compared to determine if exotic falllow and sika deer are better able to digest common fall forages than native white-tailed deer. Sika deer inocula digested three of the seven forages better ($P < 0.05$) than inocula from fallow and white-tailed deer. Percent DDM for the warm-season grass common curly mesquite (*Hilaria belangeri*) and the cool-season grass Texas wintergrass (*Stipa leucotricha*) was higher ($P < 0.05$) for sika deer (30.6% and 35.3%, respectively) than for fallow deer (18.5% and 23.3%, respectively) and white-tailed deer (15.1% and 22.5%, respectively). Percent DDM for evergreen leaves of live oak (*Quercus virginiana*) also was higher ($P < 0.05$) for sika deer (25.2%) than for fallow deer (15.8%) and white-tailed deer (15.7%). Percent DDM for live oak acorns was higher ($P < 0.05$) for sika deer (24.6%) and white-tailed deer (25.0%) than for fallow deer (14.0%). The warm-season forbs yellow woodsorrel (*Oxalis dillenii*) and upright prairie-coneflower (*Ratibida columaris*) and the evergreen leaves of rock cedar (*Juniperus ashei*) did not differ ($P > 0.05$) in %DDM among deer species. These results support the hypothesis that sika deer are at a competitive advantage over white-tailed deer due to their ability to more completely digest common range forages.

[1]Range and Wildlife Management Department, Texas Tech University, Lubbock, 79409 and [2]Texas Agricultural Experiment Station, Texas A&M University, Uvalde, Texas 78801, USA

113
Biomechanics and Ecological Application of Escape Gaits by White-Tailed Deer, Mule Deer, and Their Hybrids

Susan Lingle*

White-tailed deer (*Odocoileus virginianus*) and mule deer (*O. hemionus*) traditionally prefer landscapes with distinct topographical features, but the basis for their preferences is not understood. Widespread changes in the distribution of white-tails and mule deer are occurring, and reports of hybridization have been received all along the north-south gradient of overlap. Two questions were posed to investigate these situations: (1) Do the escape gaits and behaviors of white-tails and mule deer account for the distinct habitat preferences of these species? (2) Are the locomotory and escape patterns of hybrids confused, possibly impairing their survival relative to purebred individuals? I used high-speed cinematography to film captive white-tails, mule deer, and hybrids as they travelled at fast speeds across flat ground and over obstacless. I adapted methods from the fields of human and equine biomechanics. Body points were digitized from film images of moving deer. A computer program was written to generate gait diagrams, calculate limb timing variables, generate stick-figures of each stride, calculate landing and takeoff angles of the limbs, and calculate angles of limb and vertebral flexion. F1 hybrids, when most alarmed, used a gait intermediate between the white-tail gallop and mule deer stott, but a gait that was unique to hybrids. Evaluation of the biomechanics and potential ecological application of hybrid gaits indicated the hybrid gaits are slow and mechanically inefficient. F1 hybrid also showed extreme "indecision" in their general behavioral response to the dog. It is predicted that hybrids would have difficulty in locations where they faced strong predator pressure or other challenging stimuli, such as tricky terrain. Gaits of individual three-quarter and seven-eights mule deer were more variable than those of individual F1s. Nevertheless, these backcrosses rarely reproduced the specialized mule deer gaits.

*Faculty of Environmental Design, University of Calgary, Calgary, Alberta T2N 1N4, Canada

114
Intake Digestion, and Passage Kinetics of Two Ruminants

Glenn W. Stout and Ken L. Risenhoover*

Four yearling white-tailed deer (*Odocoileus virginianus*) and four yearling angora goats were used in eight digestive balance trials to evaluate associative effects of forages in the rumen and the effects of plant tannin on protein assimilation. Comparisons of the utilization of plant chemical constituents were made across species to compare digestive strategies. Attributes of passage for the forages offered were estimated with the use of rare earth elements as markers. Different rare earths were applied to each diet to measure the kinetics of high and low quality feeds in the rumen relative to one another. Diets consisted of either a browse, a forb, or a browse-forb combination in six of the balance trials. In separate trials, low quality browse diets, highly lignified and containing high concentrations of tannins, were supplemented with protein or treated with a tannin-binding compound to determine whether protein complexing or tannin toxicity were limiting intake. Increased intake on browse diets as a result of supplementation provided more accurate estimation of ruminal turnover, rather than only passsage rate of undigested fractions of the diet as reported in previous studies. Analysis of comparative attributes of passage and gut fill are pending.

*Department of Wildlife and Fisheries Sciences, Texas A&M University, College Station, Texas, USA

115
Value of "Hobble Creek" Mountain Big Sagebrush as a Winter Forage for Mule Deer

Bruce L. Welch*

This study describes the development, naming, and release of a superior selection of big sagebrush (*Artemisia tridentata*) for use on mule deer (*Odocoileus hemionus*) winter ranges. Of the 186 big sagebrush selections tested, "Hobble Creek" was the most preferred by wintering mule deer. It was also preferred over an unknown selection of antelope bitterbrush (*Purshia tridentata*). The winter digestibility, crude protein, and phosphorus content of "Hobble Creek" forage ranks high compared with deciduous shrub species such as antelope bitterbrush, fourwing saltbush (*Atriplex canescens*), and true mountain-mahogany (*Cercocarpus montanus*). Also, "Hobble Creek" forage exceeds dormant grasses in winter digestibility, crude protein, and phosphorus content. "Hobble Creek" does not contain substances that lower grass cell wall digestion. "Hobble Creek" can be established and maintained on sites that have deep, well-drained soils with an annual precipitation of at least 35 cm (preferably more). Soil textures should not be any finer than a clay loam (40% clay or less). Soil pH may vary from 6.6 to 8.6. "Hobble Creek" can be established by direct seeding, by transplanting bareroot or containerized stock, and by a technique we term "mother plant." Hobble Creek big sagebrush is a highly preferred winter forage that can raise the nutrient level of mule deer diets.

*USDA-FS, Intermountain Research Station, Shrub Sciences Laboratory, 735 N. 500 E., Provo, Utah 84606 USA

Part 8
Antler Physiology

116
Recent Advances in the Physiological Control of Velvet Antler Growth

James M. Suttie and Peter F. Fennessy

Abstract

Antlers are organs of bone which develop by a process of modified endochondral ossification, that is, bone is formed as a replacement of a cartilagenous matrix. The growth of the velvet antler, which is mainly cartilage/calcified cartilage, provides a framework for the antler, and as this growth is from the tip, we have mainly directed our studies to the proximal 2 cm of the antler. Hormones/growth factors, known in other organs to stimulate cartilage growth, have been studied in this context. Plasma levels of insulin like growth factor 1 (IGF1) correlate strongly and positively with the rate of velvet antler growth, and the fact that they increase after the antler is surgically prevented from growing indicates that the antler may be a target organ for IGF. Studies indicate that in deer IGF1 secretion is under photoperiodic control, which can be manipulated by melatonin. Clearly many of the actions of growth factors are not classically endocrine but rather are autocrine or paracrine. That is, instead of a chemical being secreted by one organ and having an effect via the blood stream on a distant organ, chemical messages influence the cell which produces them or an adjacent cell. We have studied binding of growth hormone (GH) and IGF1 in the antler and shown no binding of GH, indicating no unoccupied receptors but substantial binding of IGF1. No messenger ribonucleic acid (RNA) for fibroblast growth factor has been detected in extracts of antler total RNA. Selenium concentration and levels of antioxidant enzymes in velvet antlers were studied.

Key words: Antler, growth factors, melatonin, photoperiod, physiology, selenium

Introduction

Antlers are cranial organs composed of bone which are fully regenerated each year; they are unique to the extant family, Cervidae (A. Bubenik 1966; Chapman 1975; Goss 1983). Antlers differ from the keratinous horns of the Bovidae in that horns are not cast and regrown annually, and the most active zone of cell division is at the base of the horn, whereas antler growth occurs at the tip (Goss 1983). While the antler is growing (regenerating), it is covered by a skin from which grows a dense mat of fine hair that leads to the term "velvet antler" being applied to the growing tissue. After growth is complete, mineralization has taken place, and antler bone has been formed and organized into compact and cancellous zones, the velvety skin peels off to reveal clean, hard, dead antler. Antler growth is seasonal in boreal and temperate species of deer. For example, in the seasonal red deer (*Cervus elaphus*) the velvet antler grows in spring, cleans in late summer or autumn, and is hard for the breeding season in midautumn. Previous antlers are cast (that is they are autotomized) in spring, and the annual cycle of regeneration ensues (Fig. 116.1). It is well known that the seasonal rhythm is strongly influenced

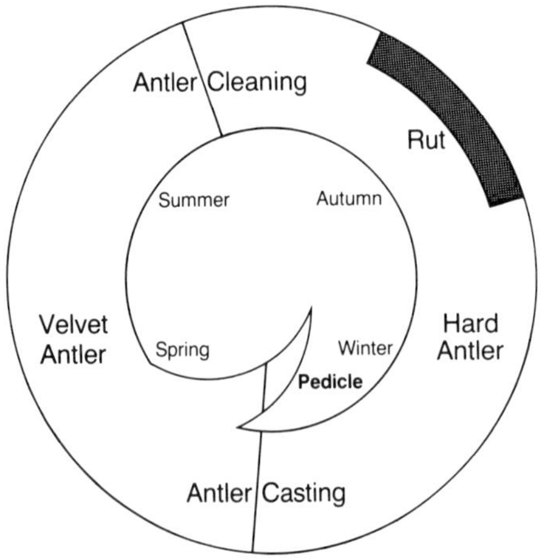

FIGURE 116.1. A diagrammatic representation of the antler growth cycle in red deer stags.

by changes in day length (Goss 1983), mediated by changes in the secretion of reproductive hormones (Suttie et al. 1989a).

The velvet antler grows by a process of modified endochondral ossification; that is, bone formation takes place on a preformed frame of cartilage rather than there being a growth plate

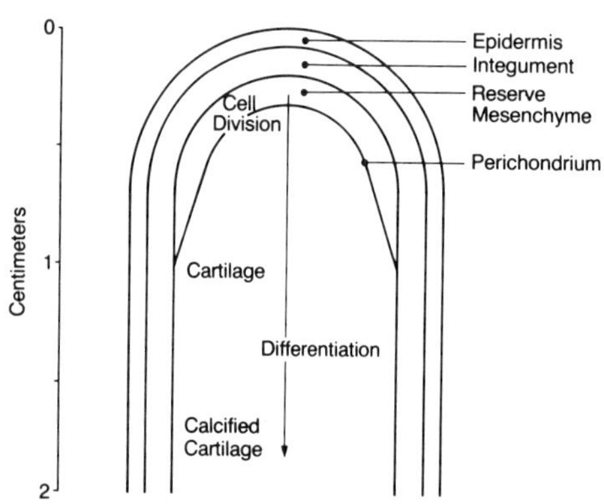

FIGURE 116.2. A schematic diagram ("apical meristem model") of the antler tip (top 2 cm of the growing velvet antler) showing tissue zones and distribution referred to in the text. We consider the growth pattern of the apical meristem of a plant to be a good analogy for the pattern of velvet antler growth. Progressive differentiation occurs from rudimentary fibroblasts at the very tip of the antler to cartilage and ultimately bone, distally. The diagram is based on computerized axial tomographs of antlers (Suttie and Fennessy 1990).

as is the case for the growth of limb bones (Banks 1974). At the very tip of the antler, therefore, there is a zone of fibroblasts (reserve mesenchyme) followed by successive layers of cartilage, calcified cartilage, and ultimately bone (Fig. 116.2). There is a gradient of progressive differentiation running from the zone of active cell division at the tip to organized bone some distance from the tip (Suttie and Fennessy 1990). In the red deer, the main beam of the antler may grow at up to 1 cm/day, and it follows that the zones of cell division and differentiation must turn over very rapidly indeed. This process must include the rapid growth of blood vessels, nerve fibers, hair (velvet) follicles, sweat glands, and epidermal and dermal (integument) tissue. There is a dense capillary bed at the growing tip of the antler (Fig. 116.3). Superficial arterioles supply the tip, and the venous drainage is through the core of the antler (Suittie et al. 1985a). This is an interesting finding in itself as veins are normally superficial in tissues and arteries are deep; the antler is thus opposite. The tip is the vibrant regulator of antler growth in the sense that its rate of cell division is the antler growth rate. It follows that angiogenesis and nerve proliferation must keep pace with the rapid enhancement of antler growth. The tip also contains the actual site where changes in branching and shape take place to form the well-recognized species-specific antler shape. Whether the genetic and positional information necessary to produce this shape, and also the potential antler size, is held in cells at the tip or whether it comes from nerve cells is at present controversial and under active investigation (Bubenik and Bubenik 1990; Suttie 1990). The tip must also be responsive to changes in the nutritional status of the animal as variations in antler size must be a consequence of the rate of cell division in the tip.

Our concept of the velvet antler tip is that it is the single most important tissue influencing antler size and shape. Therefore, we began to study the mechanisms and, in particular, the hormonal mechanisms that control the growth of tissues in the antler tip. As the tissues of the antler tip are primarily composed of fibroblasts (reserve mesenchyme) and chondroblasts/cytes (cartilage) we set out to measure hormones and growth factors that were known to be

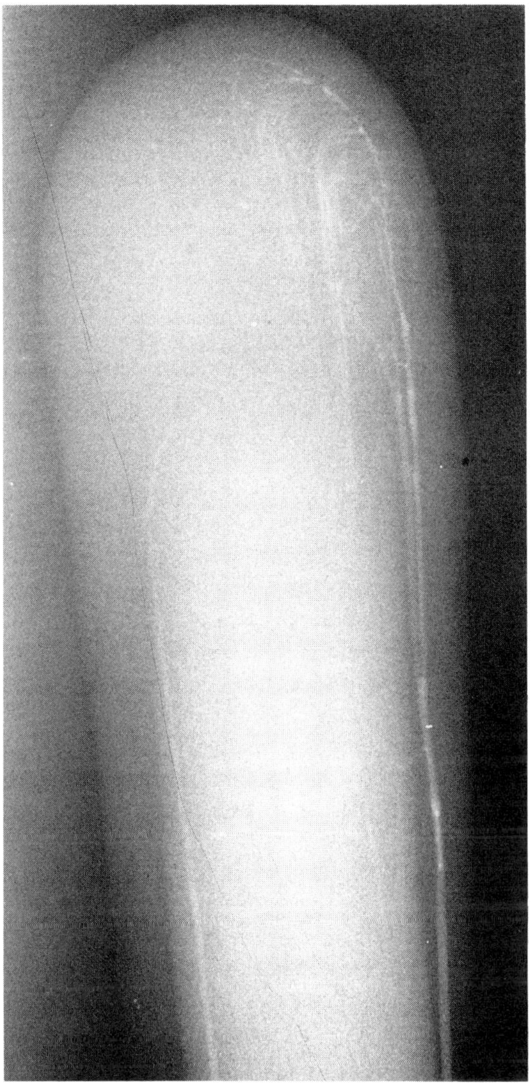

FIGURE 116.3. Radiograph of a velvet antler taken on Kodal Min-R film, 46pkV, 20mAS, 12 s after injection of 15 ml radio opaque contrast media. Note the blood vessels (arterioles) running superficially, the capillary bed 5 mm distal to the proximal tip of the antler and the venous drainage running distally through the core of the antler bone.

stimulatory for these cell types in other tissues. Our studies have taken us from the investigation of peripheral (plasma-borne) growth factors to the examination of their tissue receptors. They span the disciplines of physiology, biochemistry, and molecular biology.

In this chapter we wish to distinguish between two levels of antler regulatory hormones: that which controls the timing of the antler cycle and that which controls the trophic control of antler growth. The antler cycle is closely associated with the seasonal rhythm of reproduction such that antlers are clean and hard when they are required for offense and defense during the breeding season. There seems little doubt that antler casting and cleaning are dependent on seasonal rhythms of steroids, notably testosterone. Antler casting takes place when testosterone levels are very low, and cleaning occurs when testosterone levels are high (Lincoln 1971). Castration and hormonal treatment to block testosterone secretion or action result in antler casting at any time when stags are in hard antler (Wislocki et al. 1947; Fennessy and Suttie 1985). Testosterone treatment prevents antler casting and results in premature antler cleaning (if given to castrated male deer or intact stags during the velvet antler growing period). It is likely that testosterone is necessary for the organization of antler tissues into true Haversian bone, subsequent mineralization and the antler cleaning process possibly via the aromatase and 5α-reductase pathways, respectively (Morris and Bubenik 1982; Fennessy and Suttie 1985). However, despite the presence of testosterone (G. Bubenik et al. 1974) and estradiol receptors (Barrell et al. 1987) in growing antler, there is no evidence that steroids play a role in the stimulation of velvet antler growth. Indeed in vitro testosterone inhibits the growth of fibroblasts derived from the velvet antler tip (Gray et al., this volume). Plasma levels of testosterone are undetectable or very low throughout the velvet antler growth period (Fennessy et al. 1988), even in response to challenge from exogenous luteinizing hormone (J.M. Suttie and P.F. Fennessy, unpublished data). Thus, although the timing of the antler cycle is dependent on testosterone, it appears that the reproductive steroids do not play a primary role in the trophic control of the velvet antler and in particular the tip. In contrast, the trophic control is more likely to be vested in hormones/growth factors that stimulate growth in the relevant cell types in other organs. Evidence that the two levels of endocrine control, namely timing hormones and trophic hormones, are in synchrony and interdigitate with each other will be presented below.

Since 1983, research on the physiological control of velvet antler growth has centered on four main areas: (1) the role of calcium-regulating hormones and bone formation (Ramirez and Brown 1987; Baksi and Newbrey 1988, 1989; Sempéré et al. 1989); (2) the role of plasma steroids in vivo (G. Bubenik et al. 1987a, 1987b) and in vitro (Gray et al. Chapter 121, this volume); (3) the role of endorphins and thyroxine (G. Bubenik et al. 1987b, 1988); and (4) the role of growth factors (Suttie et al. 1985b, 1988, 1989b). It is not our intention to review all of these areas, but instead to address five topics which we have investigated recently; GH and IGF1 receptors in antlers, photoperiodic control of IGF1 and antler growth, melatonin effects on IGF1 and antler growth, fibroblast growth factor gene expression in the antler, and selenium and antioxidant enzymes of the antler tip.

GH and IGF1 Antler Receptors

GH is known to be responsible for anabolic and metabolic effects in ruminants (Davis et al. 1984). Growth hormone releasing hormone stimulates GH release and somatostatin inhibits it (Fig. 116.4). GH stimulates IGF1 release. IGF2 and insulin are both involved in growth process, but neither are believed to be under GH control. IGF1 and IGF2 are both carried in the plasma by binding proteins (Hodgkinson et al. 1987) in ruminants. IGF1, IGF2, and insulin each have specific tissue receptors (type 1, type 2, and insulin respectively), but each is capable of binding usually with lower affinity to the other's receptor (Centrella et al. 1990). In addition to its growth effects via IGF1, GH has direct effects on adipose tissue (Cameron et al. 1987) and possibly bone (Isgaard et al. 1986). The growth-promoting effect of GH on cartilage is believed to be through IGF which was formerly known as somatomedin; see Fig. 116.4 (Salmon and Daughaday 1957). However, whether or not IGF1 acts as classical hormone, that is, it is released from one organ, transported in the blood and influences the action of a second organ, is the subject of intense debate (Zapf et al.

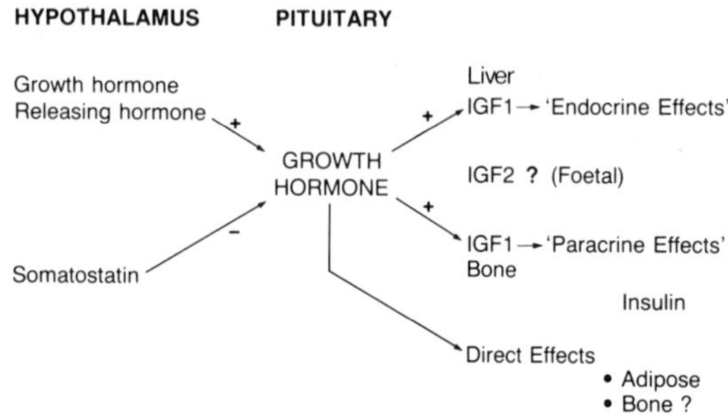

FIGURE 116.4. A simplified diagram of the major growth endocrine pathways linking the brain (hypothalamus and pituitary) with growth of peripheral tissues.

1989). There is some evidence that at least IGF1 is not always released into circulation but influences the actions of the cell in which it is produced (autocrine effect) or acts on an adjacent cell (paracrine effect). The fact that IGF1 might be capable of both functions is termed the "dual effector hypothesis." Both IGF1 and IGF2 stimulate collagen (cartilage) synthesis in vitro (Schmid et al. 1989; McCarthy et al. 1989a) and both are synthesized in bone (McCarthy et al. 1989b).

Because the tip of the antler is composed of cartilage, we have studied the relationships among GH, IGF, and antler growth in ad libitum fed tame red deer stags. At monthly intervals from 3 to 15 months of age, a cannula was positioned in the jugular vein of each stag under local anesthesia. Blood samples were withdrawn at 30-min intervals for 26 h from each stag, and the resultant plasma was analyzed for GH (all samples) and IGF1 and 2 (two samples each month for each stag) (Suttie et al. 1989b). We found that GH was released in a pulsatile manner in red deer stags (Fig. 116.5); that is, there were discrete outburts of GH secretion followed by periods of very low or undetectable concentrations of GH in the plasma. Furthermore, the pattern of pulsatility differed with season. Because of this finding, we used a pulse detection routine (PULSAR; Merriam and Wachter 1982) to analyze the GH plasma profiles. We found that the pulse frequency varied between two and five pulses/day and was highest in late autumn/spring and late summer

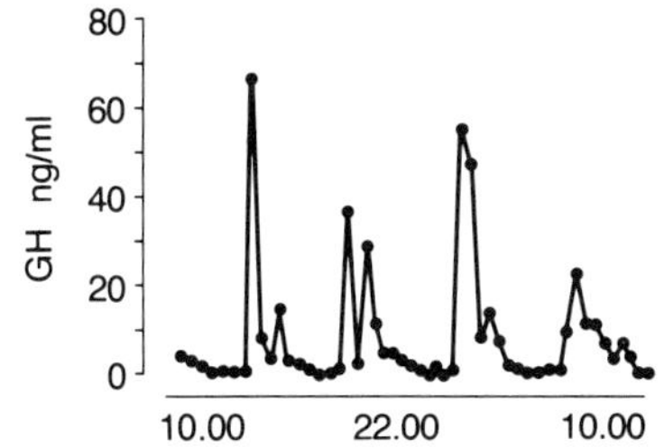

FIGURE 116.5. A typical GH (ng/ml) pulse profile for a 9-month-old red deer stag in spring. Notice the well-defined pulses with low inter-pulse concentrations of GH.

(Fig. 116.6). In contrast, pulse amplitude (the height of the pulses) was high only in spring. In addition only in spring a combination of high frequency, high amplitude pulses was detected.

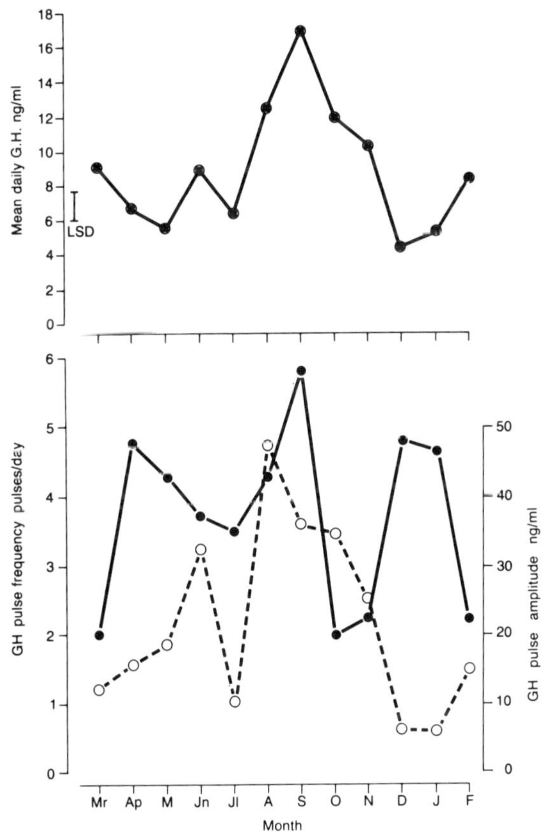

FIGURE 116.6. Daily mean (upper panel), pulse amplitude, and pulse frequency of GH (lower panel) GH (ng/ml) for each month of the study; each value is the mean for six stags.

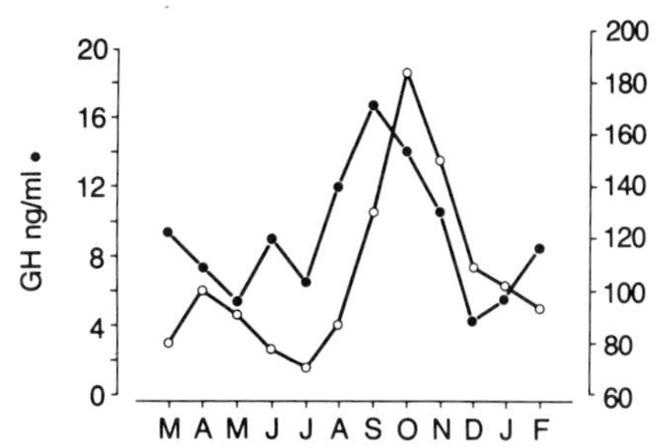

FIGURE 116.7. Monthly mean GH (ng/ml) plotted with mean monthly IGF1 (ng/ml). Notice how closely the curves are related with IGF1 out of phase with GH by 1 month.

This resulted (Fig. 116.6 upper panel) in a highly seasonal pattern of mean plasma concentrations of GH ("mean" is defined as the arithmetic mean of all samples within a profile), with high values being detected in spring and low values at all other times of the year. When the IGF1 levels were compared with the mean GH levels, it was found that IGF1 secretion was also strikingly seasonal in the young stags (Fig. 116.7). Peak IGF1 levels were detected in the plasma in midspring and early summer, but low values were detected at all other times of the year. As GH is known to stimulate IGF1 release, we concluded that the seasonal peak of GH was responsible for the seasonal peak of IGF1. Similarly, as IGF1 is known to feed back negatively on GH (i.e., to limit its release) (Zapf et al. 1989), we considered that the seasonal peak of IGF1 probably reduced plasma GH concentrations. When IGF1 concentrations were examined with the pattern of velvet antler growth, we found that the seasonal peak in IGF1

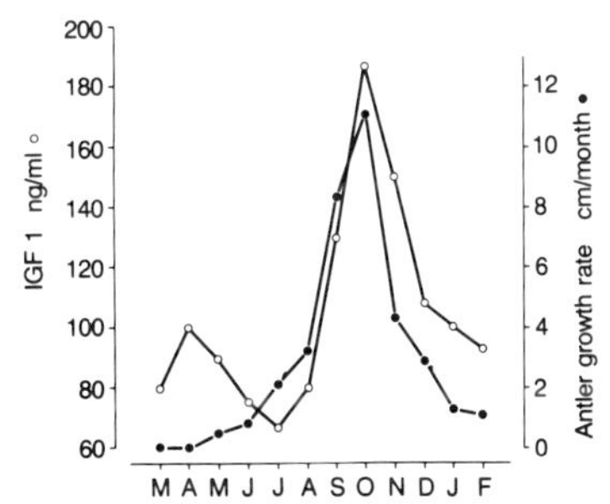

FIGURE 116.8. Monthly mean IGF1 (ng/ml) and antler growth rate (cm/month).

almost perfectly coincided with peak velvet antler growth rate (Fig. 116.8). Indeed the correlation coefficient between IGF1 and antler growth rate was highly significant ($P < 0.001$) (Suttie et al. 1985b). Thus, a growth factor which was known to stimulate cartilage growth in vivo was elevated in deer plasma during velvet antler growth. We concluded that this close association was likely to mean that IGF1 was stimulating antler growth. In contrast, IGF2 had no strong seasonal pattern of secretion and did not appear closely related either to velvet antler growth or GH level. However, we could not distinguish whether the antler was a net source or a target organ for IGF1.

As it was known that cultured fibroblasts released IGF1 (Atkinson et al. 1980), there was a possibility that the seasonal peak in IGF1 during antler growth was in fact caused by IGF1 released by the antler. In order to investigate this, we measured IGF1 in a group of stags whose antlers and pedicles were bilaterally and surgically removed to prevent any antler growth (Suttie et al. 1988). Surgery was performed when the previous hard antler was cast. Data shown in Fig. 116.9 showed that plasma concentrations of IGF1 were significantly elevated in the stags prevented from growing antlers. This indicates that the antler was not a source of IGF1, and it seemed more likely that it was a target organ for IGF1. As it was known that articular cartilage had IGF1 receptors (McQuillian et al.

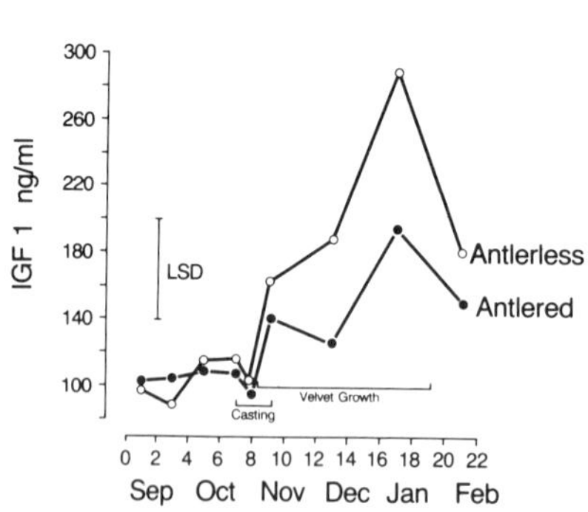

FIGURE 116.9. Plasma IGF1 (ng/ml) in stags in which antlers were surgically prevented from growing compared with normal stags. Differences in IGF1 levels between the groups were significantly different (ANOVA $P < 0.05$) in December and January.

1986), we speculated that specific IGF1 receptors would be found in cells from the antler. A further possibility which could not be discounted was that the antler secreted a binding protein for IGF1, the lack of which resulted in higher plasma IGF1 levels in antlerless stags; however, the data provided strong evidence of an endocrine mode of action of IGF1.

At this stage we decided to study whether there were available specific receptors for IGF1 and GH in the antler. The hypothesis postulated that if IGF1 was indeed stimulating antler growth by an endocrine mechanism, then IGF1 receptors would be expected in the antler. If IGF1 had combined endocrine and paracrine effects, then GH and IGF1 receptors would be present. A further reason for studying tissue/cellular binding of IGF1 was the fact that the whole animal studies could only provide a correlation/association between antlers and IGF1; if significant specific binding could be shown this would provide stronger evidence of a cause-effect relationship between antlers and IGF1.

Velvet antler was removed from four adult red deer stags under local anesthesia using standard New Zealand (NZ) husbandry conditions 60 days after casting of the previous hard antler. Within 1 min of removal, the top 2 cm of the antler tip was excised with a scalpel and placed into liquid nitrogen. The antler tissue was held at $-70°$ C until required for analysis. Subsequently, 8 μm cryosections, taken approximately 1 cm distal to the tip of the antler, were mounted on a histological slide, and air dried. After a 10-min preincubation in 0.17 *M* Tris HCl buffer, the sections were incubated for 2 h in 50 μl 0.17 *M* Tris HCl buffer with 1% BSA 5 m*M* MgCl, 1 m*M* phenyl methyl sulphonyl fluoride, and 400 000 cpm/ml of ^{125}I-IGF1 or ^{125}I-GH. Excess unlabeled GH or prolactin was added as a blocker to the ^{125}I-GH buffer, and excess IGF2, insulin (at high and low concentrations) or IGF1 was added to the ^{125}I-IGF1 buffer. The blockers permit specific binding to IGF and insulin receptors to be measured by difference from sections incubated only with ^{125}I-GH or ^{125}I-IGF1. After 2 h, the sections were washed twice for 5 min with 0.17 *M* Tris HCl, 0.25% BSA and 0.1% Triton X-100. The sections were fixed for 10 min in neutral buffered formalin, then washed in distilled water. The dried sections were placed on Kodak X-omat AR film and exposed for 1 to 5 days, then covered with Amersham LM1 emulsion and exposed for 14 days. The resultant autoradiographs were developed. The sections were stained with hematoxylin and eosin. Specific binding was measured by counting specific labeled binding sites in random areas of each tissue type defined histologically under 100 × magnification with an oil immersion objective lens. A schematic representation of the antler section X-ray autoradiographs incubated with ^{125}I-IGF1 and ^{125}I-GH is given in Fig. 116.10. There was no specific binding for ^{125}I-GH. In contrast, specific binding of IGF1 was present in all tissues, although binding was higher in the epidermis and the cartilage than in the other tissues. The use of the blockers to compete with ^{125}I-IGF1 for binding sites permits a quantification not only of total IGF1 tissue binding but also allows specific IGF1 binding to the insulin, type 1 and type 2 IGF receptors to be measured. The use of pharmacological concentrations of insulin did not distinguish between residual binding to type 2 IGF receptors or plasma binding proteins. Total binding was highest in the epidermis and cartilage, although significant binding was found in all tissues (Table 116.1). There appeared to be little specific binding to either the insulin or, surprisingly, the type 1 IGF receptor, but instead IGF1 bound strongly to the type 2 IGF receptor or the binding proteins.

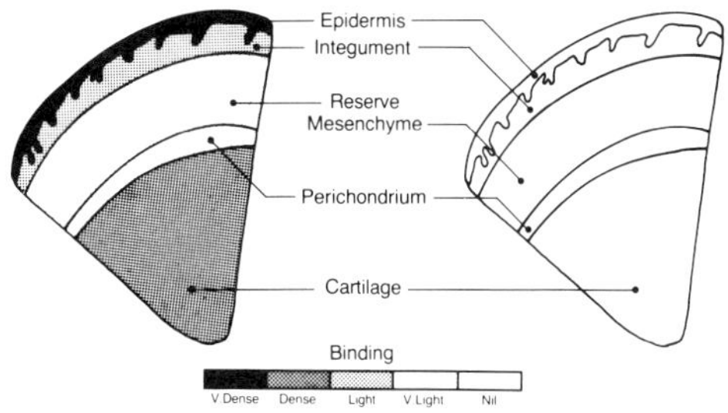

FIGURE 116.10. Schematic representation of antler autoradiographs. Diagram of a section incubated with ^{125}I-IGF1 is on the left and a section incubated with ^{125}I-GH is on the right.

The fact that there was no specific binding of ^{125}I-GH may indicate that all GH binding sites were occupied, none were at the cell surface, the specific activity of the label was too low to identify specific binding, or there were no GH binding sites. If the latter possibility is correct, then this would suggest that the antler tip does not produce IGF1, or if it does, this is not under GH control. The fact that there is significant specific binding of IGF1, albeit to the type 2 IGF receptor or an IGF binding protein, leads to the suggestion that IGF1 is likely to be transported to the antler, and the antler is responsive to IGF. However, the data provide good support for the hypothesis that IGF levels influence antler growth, although the possibility cannot be discounted that IGF2 and IGF1 each play a distinct role.

As plasma IGF1 levels are known to be highly dependent on plane of nutrition of the animal (Breier et al. 1986), there is a possible mechanism how poor nutrition could influence antler size by reducing plasma IGF1.

TABLE 116.1. Autoradiograph grain counts of specific and nonspecific binding of IGF1 to binding sites in each tissue of red deer antler

Binding site	Tissue type: Cartilage	Reserve mesenchyme	Epidermis	Integument	Perichondrium	SED
NSB	5.6	5.6	34.7	8.8	4.5	1.1
Insulin	0	4.3	11.7	1.3	0.7	2.2
IGF1 (type 1)	3.6	0	2.2	2.2	2.0	2.2
IGF2 (type 2) +Binding Proteins	23.0	13.7	38.0	15.4	14.9	2.2
Total	32.2	23.6	86.6	27.7	22.1	1.1

Photoperiodic Control of IGF and Antler Growth

It has been known for many years that the annual cycle of antler growth and replacement is under the control of the photoperiod (Goss 1983). We have previously shown that the growth of velvet antler in spring is associated with high plasma levels of IGF1 (Suttie et al. 1985b). However, it was not clear whether the increase in IGF1 was induced by changes in day length or was a consequence of velvet antler growth. To attempt to answer this question we exposed 5-month-old reindeer (*Rangifer tarandus*) calves ($n = 3$ per group) to either 16 h of light and 8 h dark each day (16L:8D) or 8L:16D at the Institute of Arctic Biology, Fairbanks. The Alaskan study began 2 weeks after the autumnal equinox and ended 2 weeks after the vernal equinox. The animals were fed ad libitum a high quality concentrate ration, and blood samples were withdrawn into pre-heparinized tubes from each animal every 2 weeks. The plasma was analyzed for IGF1.

For 6 weeks after the calves were exposed to the day-length treatments there were no significant differences in plasma IGF1 concentrations (Fig. 116.11). However, the calves exposed to the long day length (16L:8D) showed an increase in IGF1 plasma concentration, which was significant ($P < 0.05$) from week 8 to the end of the study. In contrast, IGF1 plasma levels in the calves exposed to 8L:16D remained virtually unchanged throughout the study.

All three of the calves on 16L:8D cast their antlers about 7 weeks after the start of the experiment and regrew them in velvet during the study. None of the calves exposed to 8L:16D cast their antlers, and thus they retained their previous hard antlers throughout the study. The rise in plasma IGF1 occurred about the time of antler casting in the 16L:8D group. We conclude that the long photoperiod (16L:8D) induced the increase in plasma concentration of IGF1. As we knew that antlerless red deer stags also show a spring increase in plasma IGF1, we can discount the possibility that the antler is itself the cause of this elevation. We believe that antler casting is a result of decreased plasma levels of testosterone (Fennessy and Suttie 1985), and that these levels in themselves are photoperiodically regulated (Suttie et al. 1989a), and therefore we can conclude that there must be synchrony in phase at a high level (i.e., the brain) between the areas controlling the growth and reproductive cycles to enable precise regulation of this phase of the antler cycle. The source of this synchrony is unknown, but conceivably it is due, in part, to the changing pattern/changing physiological response to melatonin. It appears that the short day length 8L:16D may have actively suppressed IGF1 or delayed the phase of the seasonal increase in secretion. This finding provides further evidence of an endocrine mechanism of action for IGF1 on antler growth.

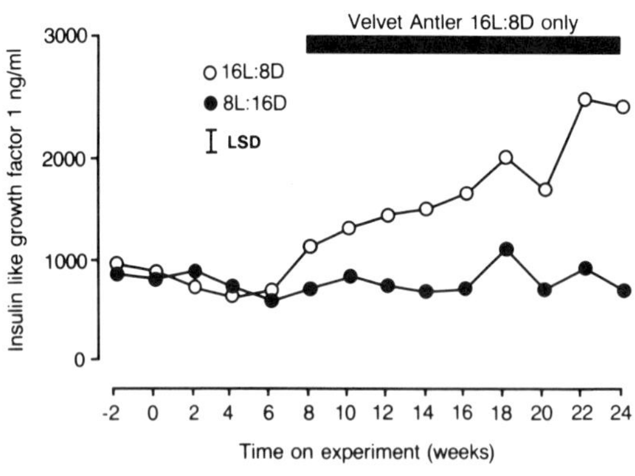

FIGURE 116.11. IGF1 (ng/ml) in two groups of reindeer calves maintained on either 16L:8D or 8L:16D from the autumnal to the vernal equinox. Period of antler growth of the 16L:8D group is shown in a block, diagrammatically.

Effects of Melatonin Treatment on Plasma Levels of IGF1 and Antler Growth

Melatonin implants (Regulin), which are used to advance the timing of the breeding season in farmed red deer (Fisher et al. 1988), also advance the annual cycle of body weight changes (Webster et al. 1991). If stags are treated in spring, then not only the peak weight gain phase of the seasonal growth rhythm is advanced, but these stags also clean their antlers of velvet early and lose weight (rut-associated weight loss) early. Regulin implants elevate the plasma levels of melatonin in excess of normal night-time levels continuously for up to 40 days in red deer (Webster et al 1991). Thus, the typical circadian rhythm of high melatonin levels during the night and low levels during the day (Bittman et al. 1983; Bubenik and Smith 1987) is altered, although the basic pattern remains, albeit at a higher plasma level of melatonin.

Given the fact that it is the circadian rhythm of melatonin level which is thought to provide day length information to the animal, the precise mechanism of action of the Regulin is not known. However, we believe that Regulin prevents the animal from obtaining photoperiodic information from the environment, and it either may simulate a "short day" or prevent the animal from perceiving a "long day."

Although Regulin treatment advances the phase of antler growth and body growth when given in late spring, the effect of melatonin on seasonal IGF1 secretion is of interest. We have therefore studied the effect of Regulin treatment on IGF1 secretion and the antler cycle in adult red deer stags. The experimental concept was to determine whether the stags required a period of long days in late winter/early spring in order to respond to Regulin with a phase advance of the breeding season in autumn.

Thirty adult red deer stags at pasture were allocated to one of six treatment groups as follows: (1) control, no Regulin (control), (2) 12 months of Regulin, beginning in June (year), (3) 6 months of Regulin, beginning in June (June), (4) 6 months of Regulin, beginning in August (August), (5) 6 months of Regulin, beginning in September (September), and (6) 6 months of Regulin, beginning in October (October). The Regulin implants were placed subcutaneously distal to the ear at monthly intervals in each treatment group. A blood sample was withdrawn from all stags monthly, and IGF1 concentration was determined. Antler casting date, antler length, and status (velvet or hard) were recorded monthly.

In the control animals (Fig. 116.12), there was a clear annual rhythm of plasma IGF1. IGF1 plasma concentration rose between August and September, coinciding with antler casting. There was a well-defined fall in plasma IGF1 from February to March. In those stags treated with Regulin for 1 year, there was a seasonal rhythm

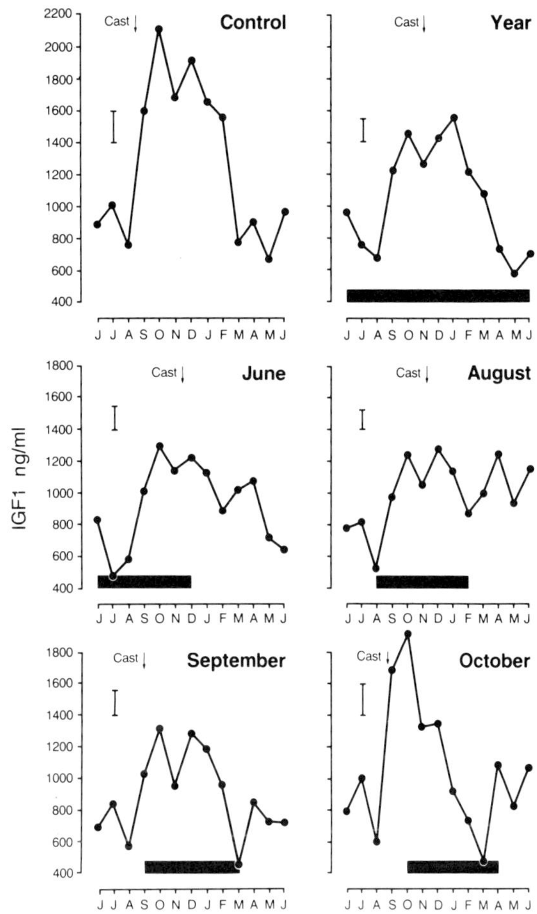

FIGURE 116.12. IGF1 (ng/ml) in stags treated with Regulin (shown as a solid block). Cast refers to antler casting. Bar is the LSD, 5%.

of IGF1 which, although it appeared in phase with that of control stags, was significantly reduced in amplitude (Table 116.2, $P < 0.05$). Casting was phase-delayed by 3 months compared with the control group. The groups in which Regulin treatment began in June and August for 6 months were similar. Casting was delayed for 3 months compared with controls. There was an attenuation of the seasonal rhythm of IGF1, which nevertheless appeared in phase with the control stags, because the plasma IGF1 increased markedly between August and September in both groups. The September group had cast their antlers before Regulin treatment began. Regulin significantly reduced the amplitude of the seasonal rhythm of IGF1. Likewise,

TABLE 116.2. Mean peak IGF1 concentrations ng/ml (mean of September, October, and November values) and antler length (cm) at 50 days after previous antler casting for red deer stags treated with Regulin for 12 months (Year) or for 6 months beginning from the month indicated. Means with different superscripts differ significantly within rows ($P < 0.05$).

	Regulin treatment group						
	Year	June	August	September	October	Control	SED
Peak IGF1 (ng/ml)	1381^{BC}	1206^{C}	1198^{C}	1168^{C}	1530^{B}	1899^{A}	104
Antler length (cm)	35.4^{CD}	30.1^{E}	31.6^{E}	34.6^{ED}	40.3^{B}	46.3^{A}	2.3

the group which began Regulin treatment in October had cast their antlers prior to the onset of treatment. IGF1 rhythm had already risen to a peak level of 1908 ng/ml in October, which was not significantly different from 2109 ng/ml (SE = 130) observed in the control stags at the same time. However, 2 months after Regulin treatment began, IGF1 levels had significantly fallen to 1336 ng/ml, compared with 1915 ng/ml in the control stags (SE = 238, $P < 0.05$). Overall peak (September, October, and December) levels of IGF1 were reduced by Regulin treatment. Antler growth was reduced by Regulin treatment (Table 116.2).

This, IGF1 plasma levels were reduced by Regulin treatment, although phase did not appear to be altered. Regulin delayed antler casting and reduced antler size even in the September group where casting date had not been altered. Thus, Regulin had an effect by phase-delaying antler casting, but an apparently separate effect on IGF1, namely a reduction in amplitude. The consequences of this are the seasonal rhythm of IGF1 must be endogenous, since the absence of photoperiodic information during Regulin treatment did not prevent a seasonal elevation, and also the timing of seasonal IGF1 is clearly not precisely linked to the rhythm of antler growth but rather exhibits a separate rhythm. The direct suppressive effect of Regulin on IGF1 argues that melatonin (photoperiod) acts on IGF1 secretion itself rather than influencing the phase of its secretion. If this is true, then the mechanism of action of Regulin on IGF1 is fundamentally different from its effect on reproductive hormones. The extrapolation that Regulin treatment is the same as the effect of photoperiod must be made with due caution because Regulin is a pharmacological agent which results in a nonphysiological circadian rhythm of melatonin.

The fact that plasma IGF1 is reduced by treatment, and antler growth is also reduced, provides further support for the hypothesis that IGF1 plasma level and antler size are causally related.

Fibroblast Growth Factor and Antler Growth

Acidic and basic fibroblast growth factors (aFGF and bFGF respectively) are similar molecules that share many biological properties (Gospodarowicz et al. 1987). bFGF is found in a wider range of tissues than aFGF, but they share a common receptor. Both are known to be mitogenic for vascular, nervous, and connective tissue acting both by endocrine and paracrine mechanisms (Gospodarowicz et al. 1987). In that bFGF is known to be mitogenic for cells of mesodermal origin (thus including fibroblasts and chondroblasts) and is known to be involved in limb regeneration in amphibians (Gospodarowicz and Mescher 1981), we have studied the possibility that FGF is involved in antler growth (Eagles 1989).

We have used an approach from molecular biology to attempt to localize bFGF gene expression using the following technique. Total ribonucleic acid (RNA) was purified from the antler using the CsCl and the acid guanidium phenol chloroform (AGPC) methods. As Southern blotting revealed homology between human and deer FGF sequences, human aFGF and bFGF cDNA (complementary deoxy ribonucleic acid) probes were used to detect FGF mRNA (messenger RNA) by dot blotting and northern analysis. No specific hybridization was detected (Eagles 1989). Subsequent to this finding a nonredundant oligonucleotide probe was used in an attempt to detect specific hybridization for FGF in antler mRNA. This also proved unsuccessful (Merriman 1989).

It is possible that the human probes did not hybridize with the deer antler mRNA because of specific differences in deer and human FGF. It is also possible that FGF is expressed at such low levels in the antler that it could not be detected with the current methodology. However, it seems unlikely that such a rapidly growing tissue would not be expressing an important mitogen at a high level. Conceivably, the antler may lack FGF control, or perhaps has a specific growth factor which fulfils the function of FGF without showing sufficient sequence homology to be detected by FGF probes. Our current approach is to form a cDNA library for deer, with the aim of making a deer bFGF probe. Signal amplification may also be required to magnify bFGF mRNA enough for detection.

Selenium and Antioxidant Enzymes of the Antler Tip

The selenium (Se) distribution in the velvet antler shows a remarkable pattern (Fig. 116.13). The concentration in the growing tips is around five- to ten-fold higher than any other part of the antler. Therefore, we have investigated the possibility that the high selenium concentration is due to the presence of the selenium-containing enzyme, glutathione peroxidase (GSHPx) whose function is to remove potentially harmful oxygen radicals from particularly rapidly growing tissues (Behne and Wolters 1983). As well as glutathione peroxidase we have also measured two other enzymes involved in the removal of oxygen radicals, superoxide dismutase and catalase. Velvet antler was taken from three red deer stags

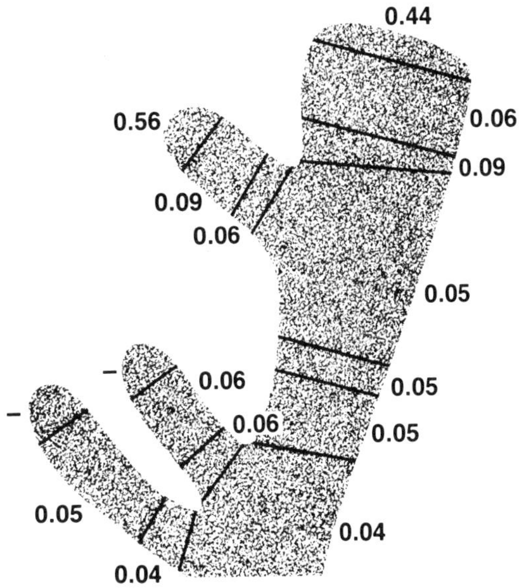

FIGURE 116.13. Selenium concentrations (mg/kg DM) of a velvet antler (cut 65 days after casting); -, means insufficient sample for analysis.

60 days after hard antler casting, and the tips of the trez tines analyzed; the antler tips were compared with rat liver as a well-characterized control tissue.

The selenium content was highest (0.90 mg/kg DM) in the most proximal section and lowest at the distal section (0.47 mg/kg DM) (Table 116.3). Thus, all sections were higher or as high as described in Fig. 116.13 in the preliminary study. Superoxide dismutase and catalase levels did not vary significantly with location in the trez tip, but the highest level of GSHPx ($P < 0.05$) was detected in the most proximal section. Se concentrations in the velvet antler were 13% to 25% of those of the rat liver, but GSHPx levels were around 1% of those found in the liver. Consequently, the top 1 cm of the growing tip of the deer antler had 0.59 mg Se/IU GSHPx compared with a concentration of 0.0082 mg Se/IU GSHPx in the rat liver.

The antler had very low levels of antioxidant enzyme activity compared with the rat liver, and although Se levels were lower in the antler than the rat liver, the amount of Se for each unit of glutathione peroxidase was higher in the antler.

TABLE 116.3. Mean enzyme activity/mg protein and selenium content of pieces of trez tine from three red deer stags; values for rat liver are provided for comparison

Distance from trez tip (cm)	Selenium (mg/kg DM)	Glutathione peroxidase (IU × 10^{-6})	Superoxide dismutase (μg)	Catalase (k)
0–0.5	0.90	20.3	3.03	0.82
0.5–1.0	0.87	15.3	2.80	0.89
1.0–1.5	0.60	13.7	2.43	0.78
1.5–2.0	0.56	13.7	2.70	0.88
2.0–2.5	0.47	14.7	2.87	0.94
SED	0.07	1.8	0.24	0.07
Rat liver	3.63	1289	20.6	9.67

That is, more than 98% of the Se in the antler cannot be associated with GSHPx. It is possible that it is associated with selenoprotein P, another Se-containing protein, which possibly also has antioxidant properties (Burk 1989), but whether the Se in the antler tip is organic or inorganic is not known at this stage.

The overall picture with regard to the Se status and antioxidant enzymes in the growing antler is one of a tissue which appears to be quite similar to some tumors which also have very low levels of antioxidant enzymes. Tumors have low levels of such enzymes possibly becase they largely rely for energy metabolism on glycolysis rather than oxidative phosphorylation (R. Munday, personal communication), and these former pathways produce fewer oxygen radicals which must be eliminated by antioxidants. Thus, there is the intriguing possibility that the antler is behaving like a tumor, reinforcing the speculation of numerous authors based largely on the rapid growth and microscopic examination of the growing tip (Steinwedel 1929; Goss 1983).

Speculation About Other Growth Factors and Antlers

Several other growth factors not yet or little studied in the antler could, from their known effects in other tissues similar to those found in the antler, have effects on antler growth and are briefly considered, with their possible role, below. Epidermal growth factor (EGF) has already been purified from velvet antler (Ko et al. 1986). EGF purified from the velvet antler epidermis was capable of inducing ^{3}H-thymidine incorporation in quiescent human fetal lung fibroblasts. Thus EGF could clearly be a candidate for growth promotion in the antler, and it may well be locally produced. Recent evidencc indicates that GH regulates the EGF receptor in rats (Ekberg et al. 1989). This implies a further possible mechanism where a hormone could influence an "antler" growth factor.

Platelet-derived growth actor (PDGF) is a known cell competence factor (Heldin et al. 1988), that is, it permits cells to begin the cell cycle. The cell cycle can only be completed if progression factors such as IGF1 are present (Russell et al. 1984). As PDGF can increase bone matrix protein synthesis in vitro (Centrella et al. 1989), we believe that it may be an important cell competence factor in the antler as well.

Tumor necrosis factor (TNF) is a macrophage-derived cytokine which is known to have immunoinflammatory responses as well as cytostatic and cytotoxic effects on normal or tumor cells (Sugarman et al. 1985). Sugarman et al. (1985) also reported that TNF was mitogenic for the growth of normal human fibroblasts in culture. TNF, as part of its inflammatory response, produces prostaglandins which negatively influences its growth-promoting activities (Hori et al. 1989). Therefore, TNF could have positive and negative effects on antler fibroblasts.

Two different types of transforming growth factor, α and β (TGF-α and TGF-β) have been described (Centrella et al. 1988). TGF-α is related to EGF and can bind to the EGF receptor. It is also known to be produced by transformed and tumor cells and can transform normal cells (Han et al. 1988). TGF-α stimulates bone resorption with a mechanism dependent on prostaglandins (Lorenzo et al. 1988). TGF-α might have an influence on antler bone remodeling.

In contrast, TGF-β does not bind to the EGF receptor, but it is known to regulate cell replication and differentiation of mesenchymal precursor cells and chondrocytes probably by a paracrine mechanism. Centrella et al. (1988) and Rosier et al. (1989) have studied the role of TGF-β in endochondral calcification (that is, similar to the form of antler growth). They found that TGF-β was produced by chondrocytes. The biological activity was correlated with binding to two receptors of high and low affinity. Maximal stimulation of cell division occurred in proliferating and early hypertrophic cells, because the smallest, least mature cells had only the high affinity receptor. Narayanan et al. (1989) have found that TGF-β independently regulates collagen synthesis by cultured human fibroblasts. On the basis of the above observations, we conclude that TGF-β is a very likely candidate for antler growth stimulation.

Discussion

The theme of this paper from the outset has been to develop the concept of the uniqueness of the antler tip with respect to tissue regeneration. A wide range of topics relating to the growth factor control of cartilage growth and the possible regulation of that growth factor control have been discussed. In addition, the Se status of the tip of the antler and its possible implications for growth have been briefly considered. The antler itself, is unique as it is the only organ to be fully regenerated by any mammal. It is also the only organ composed solely of skin and bone to be regenerated by any vertebrate. As an organ, it has intrigued generations of biologists and, likely, it will intrigue generations to come, but is the antler tip itself really unique?

We have found IGF receptors but not GH receptors in the cartilage of the growing antler tip. However, Hill et al. (1989) found human fetal fibroblasts that were capable of secreting IGF1, but did not do so under GH control. This probably means that if IGF1 were to be produced by the antler, it is not under GH control, hence the apparent lack of receptors (see below). We speculate that IGF has both endocrine and paracrine effects in the antler. However, we have only produced strong evidence that the antler is a target organ for IGF1 (Suttie et al. 1988). The IGF1 in the antler which binds to the type 2 IGF2 receptor may well be biologicaly active; the reason for the apparent low binding to IGF1 receptors in he antler of ^{125}I-IGF1 could be that there are quantitatively more type 2 IGF2 receptors, or the type 1 IGF1 receptor sites might have been fully occupied. Therefore, there is a necessity to design specific experiments where velvet antler is grown at different times of the year to examine the relationship between IGF1 and the rate of antler growth. The role of the high levels of Se in the antler is enigmatic; we speculate that the antler could be dependent on anaerobic biochemical pathways for energy. The possible interrelationships between Se and IGF merit further study.

At this stage, therefore, we are not fully able to answer the question as to whether the antler tip is unique insofar as aspects of its physiological control have parallels elsewhere in the body or in neoplastic cells. The parallels with neoplastic cells are exciting. Oncogenes are genes which cause the cancerous transformation of cells. They derive from normal genes which alter in such a way so that either an abnormal product is made or the control of the genes is disrupted and they are expressed in excessive amounts or at inappropriate times (Marx 1987). The interrelationships between growth factors and oncogenes are fascinating (Bishop 1987). Therefore, we ask the following question: Could the first antler have been the product of a transformed gene which promoted a tumor which, due to the forces of natural selection, became an advantage and controlled by the seasonal male cycle? Could the antler have been the product of more than one transformed gene and even possibly the loss of a suppressor gene? Indeed many mutations are needed for cancer development because cells control their normal growth very well (Marx

1989). The "oncogene/antler hypothesis" is based solely, at this time, on: (1) the similarity between tumors and antlers, (2) the likelihood that antlers and tumors share common growth factors, and (3) the fact that oncogenes provide a well-studied comparison which links genes to growth factors and to tissue growth. A great deal of study will be required to test this hypothesis.

This chapter has set out to consider the growing tip. It has covered physiological and biochemical control mechanisms for which there are parallels in other tissues, but many questions remain unanswered. Why does the antler's "apical meristem" grow upwards (i.e., against gravity)? Why do antlers curve? Is the information for this carried in cells in the antler tip or does this information come from elsewhere? What determines if and when the growing tip will divide to form a branch in the antler? How does the growing antler tip form a palmate antler? Other aspects of velvet antler regeneration which we have not considered include wound healing following casting, the inflammatory process preceding casting and its relationship with the mechanism of casting itself. These questions and many more will keep those of us in the antler field intrigued for years (lifetimes?) to come.

Acknowledgements

We are indebted to our colleagues without whose expertise the studies reported here could not have begun, let alone been completed. J. Webster, R. Littlejohn, M. Smith, P. Gluckman, B. Breier, J. Bass, R. Munday, B. White, I. Corson, B. Veenvliet, R. Labes, R. Eagles, J. Elliot, J. Oldham and D. Hartbaur. Studies were funded by the New Zealand Ministry of Agriculture and Fisheries and the Institute of Arctic Biology.

References

Atkinson PR, Weidman ER, Bhaumick B, Bala RM (1980) Release of somatomedin like activity by cultured W1/38 human fibroblasts. Endocrinology 106:2006–2012

Baksi SN, Newbrey JW (1989) Metabolism during antler growth in female reindeer. Calcif Tiss Int 45:314–317

Baksi SN, Newbrey JW (1988) Plasma calcemic hormones in mature female reindeer (*Rangifer tarandus*). Gen Comp Endocrinol 69:262–266

Banks WJ (1974) The ossification process of the developing antler in the white-tailed deer (*Odocoileus virginianus*). Calc Tiss Res 14:257–274

Barrell GK, Lengoc CM, Muir PD (1987) Receptors for oestradiol in growing antlers of red deer stags. 18th Cong Int Union Game Biologists Jagiellonian University, Karkow, Poland, pp 17–18

Behne D, Wolters W (1983) Distribution of selenium and glutathione peroxidase in the rat. J Nutr 113:456–461

Bishop MJ (1987) The molecular genetics of cancer. Science 235:305–311

Bittman EL, Dempsey RJ, Karsch FJ (1983) Pineal melatonin secretion drives the reproductive response to day length in the ewe. Endocrinology 113:2276–2283

Breier BH, Bass JJ, Butler JH, Gluckman PD (1986) The somatotrophic axis in young steers: influence of nutritional status on pulsatile release of growth hormone and circulating concentrations of insulin-like growth factor 1. J Endocrinol 111:209–215

Bubenik AB (1966) Das Geweih. Parey, Hamburg

Bubenik GA, Smith PS (1987) Circadian and circannual rhythms of melatonin in plasma of male white-tailed deer and the effect of oral administration of melatonin. J Exptl Zool 241:81–89

Bubenik GA, Smith JH, Flynn A (1988) Plasma levels of β-endorphin in white-tailed deer: seasonal variation and the effect of thyroxine. GnRH, dexamethasone, and ACTH administration. Comp Biochem Physiol 90A:309–313

Bubenik GA, Sempéré AJ, Hamr J (1987a) Developing antler, a model for endocrine regulation of bone growth. Concentration gradient of T_3, T_4 and alkaline phosphatase in the antler, jugular and the saphenous veins. Calc Tiss Int 41:38–43

Bubenik GA, Pomerantz DK, Schams D, Smith PS (1987b) The role of androstenedione and testosterone in the reproduction and antler growth of a male white-tailed deer. Acta Endocrinol 114: 147–152

Bubenik GA, Brown GM, Bubenik AB, Grota LJ (1974) Immunohistochemical localisation of testosterone in the growing antlers of white-tailed deer (*Odocoileus virginianus*). Calc Tiss Ress 14:121–130

Bubenik GA, Bubenik A (eds) (1990) Antlers, Horns, Cranial Appendages of Pecoran. Springer-Verlag, New York

Burk RF (1989) Recent developments in trace element metabolism and function: newer roles of selenium in nutrition. J Nutr 119:1051–1054

Cameron CM, Kostyo JL, Adamafio NA, Dunbar JC (1987) Metabolic basis for the diabetogenic action of growth hormone in the obese (ob/ob) mouse. Endocrinology 120:1568–1575

Chapman DI (1975) Antlers—bones of contention. Mammal Rev 5:121–172

Centrella M, McCarthy TL, Canalis E (1990) Receptors for insulin-like growth factors-1 and -11 in osteoblast-enriched cultures from fetal rat bone. Endocrinology 126:39–44

Centrella M, McCarthy TL, Canalis E (1989) Platelet-derived growth factor enhances deoxyribonucleic acid and collagen synthesis in osteoblast-enriched cultures from fetal rat parietal bone. Endocrinology 125:13–19

Centrella M, McCarthy TL, Canalis E (1988) Skeletal tissue and transforming growth factor β. FASEB J 2:3066–3073

Davis SL, Hossner KL, Ohlson DL (1984) Endocrine regulation of growth in ruminants. In Roche JF, O'Callaghan D (eds) Manipulation of Growth in Farm Animals. Martinus Nijhoff, Boston, pp 151–178

Eagles R (1989) Fibroblast growth factor involvement in deer antler development. MSc thesis, University of Otago, Dunedin, New Zealand

Ekberg S, Carlsson L, Carlsson B, Billig H, Jansson JO (1989) Plasma growth hormone pattern regulates epidermal growth factor (EGF) receptor messenger ribonucleic acid levels and EGF binding in the rat liver. Endocrinology 125:2158–2166

Fennessy PF, Suttie JM (1985) Antler growth: nutritional and endocrine factors. In Fennessy PF, Drew K (eds) Biology of Deer Production. R Soc N Z Wellington Bull 22:239–250

Fennessy PF, Suttie JM, Crosbie SF, Corson ID, Elgar HJ, Lapwood KR (1988) Plasma LH and testosterone responses to gonadotrophin-releasing hormone in adult red deer (*Cervus elaphus*) stags during the annual antler cycle. J Endocrinol 117:35–41

Fisher MW, Fennessy PF, Milne JD (1988) Effects of melatonin on seasonal physiology of red deer. Proc N Z Soc Anim Prod 48:113–116

Gospodarowicz D, Mescher AL (1981) Fibroblast growth factor and vertebrate regeneration. In Riccardi VM, Mulvill JJ (eds) Advances in Neurology: Neurofibromatosis. Raven Press, New York, 29:149–157

Gospodarowicz D, Ferrara N, Schweigerer L, Neufield G (1987) Structural characterisation and biological functions of fibroblast growth factor. Endocrinol Rev 8:95–115

Goss RJ (1983) Deer Antlers. Regeneration, Function and Evolution. Academic Press, New York

Han VKD, D'Ercole AJ, Lee DC (1988) Expression of transforming growth factor alpha during development. Can J Physiol Pharmacol 66:1113–1121

Heldin CH, Backstrong G, Ostamn A, Hammacher A, Ronnstvand L, Rubin K, Nister M, Westermark B (1988) Binding of different dimeric forms of PDGF to human fibroblasts: evidence for two separate receptor types. EMBO J 7:1387–1393

Hill DJ, Camacho-Hubner C, Rashid P, Strain AJ, Clemmons DR (1989) Insulin-like growth factor (IGF)-binding protein release by human fetal fibroblasts: dependency on cell density and IGF peptides. J Endocrinol 122:87–98

Hori T, Kashiyama S, Hayakawa M, Shibamoto S, Tsujimoto M, Oku N, Ito F (1989) Possible role of prostaglandins as negative regulators in growth stimulation by tumor necrosis factor and epidermal growth factor in human fibroblasts. J Cell Phys 141:275–280

Hodgkinson SC, Davis SR, Burleigh BD, Henderson HV, Gluckman PD (1987) Metabolic clearance rate of insulin-like growth factor-1 in fed and starved sheep. J Endocrinol 115:233–240

Isgaard J, Nilsson A, Lindahl A, Jansson JO, Isaksson GP (1986) Effects of local administration of GH and IGF-1 on longitudinal bone growth in rats. Am J Physiol 250:E367–E372

Ko KM, Yip TT, Tsao SW, Kong YC, Fennessy PF, Belew MC, Porath J (1986) Epidermal growth factor from deer (*Cervus elaphus*) submaxillary gland and velvet antler. Gen Comp Endocrinol 63:431–440

Lincoln GA (1971) The seasonal reproductive changes in the red deer stag (*Cervus elaphus*). J Zool 163:105–123

Lorenzo JA, Sousa SL, Centrella M (1988) Interleukin-1 in combination with transforming growth factor α produces enhanced bone resorption in vitro. Endocrinology 123:2194–2200

Marx JL (1989) Many gene changes found in cancer. Science 246:1386–1388

Marx JL (1987) Oncogene action probed. Science 237:602–603

Merriam GR, Wachter KW (1982) Algorithms for the study of episodic hormone secretion Am J Physiol 243:E310–E318

Merriman T (1989) Oligonucleotide probes for fibroblast growth factor expression in deer antler. Honors Thesis, University of Otago, Biochemistry Department, Dunedin, New Zealand

Morris JM, Bubenik GA (1982) The effects of androgens on the development of antler bone. In Brown RD (ed) Antler Development in Cervidae. Caesar Kleberg Wildlife Research Institute, Kingsville, Texas, pp 123–139

McCarthy TL, Centrella M, Canalis E (1989a) Regulatory effects of insulin-like growth factors 1 and 11 on bone collagen synthesis in rat calvarial cultures. Endocrinology 124:301–309

McCarthy TL, Centrella M, Canalis E (1989b) Insulin-like growth factor (IGF) and bone. Conn Tiss Res 20:277–282

McQuillan DJ, Handley CJ, Campbell MA, Bolis S, Milway VE, Herington AC (1986) Stimulation of proteoglycan biosynthesis by serum and insulin-like growth factor-1 in cultured bovine articular cartilage. Biochem J 240:423–430

Narayanan AS, Page RC, Swanson J (1989) Collagen synthesis by human fibroblasts. Biochem J 260: 463–469

Ramirez V, Brown R (1987) Effect of biopsy site and stage of growth on in vitro antler tissue. 18th Cong Int Union Game Biologists. Jagiellonian University, Krakow, Poland, pp 161–162

Rosier RN, O'Keefe RJ, Crabb ID, Puzas JE (1989) Transforming growth factor beta: an autocrine regulator of chondrocytes. Conn Tiss Res 20: 295–301

Russell WE, van Wyk JJ, Pledger WJ (1984) Inhibition of the mitogenic effects of plasma by a monoclonal antibody to somatomedin-C. Proc Natl Acad Sci USA 81:2389–2392

Salmon WD, Daughaday WH (1957) A hormonally controlled serum factor which stimulates sulphate incorporation by cartilage in vitro. J Lab Clin Med 49:825–836

Schmid C, Guler HP, Rowe D, Froesch RE (1989) Insulin-like growth factor 1 regulates type 1 procollagen messenger ribonucleic acid steady state levels in bone of rats. Endocrinology 125:1575–1580

Sempéré AJ, Grimberg R, Silve C, Tau C, Garabedian M (1989) Evidence for extrarenal production of 1,25-dihydroxyvitamin during physiological bone growth: in vivo and in vitro production by deer antler cells. Endocrinology 125:2312–2319

Steinwedel M (1929) Histologische und histogentische Studien über das Geweih von *Cervus elaphus*, von *Cervus capreolus und Rangifer tarandus*; zugleich ein Beitrag zur Kenntnis des Verknocherungsvorganges. Inaugural dissertation, Doctor medicinae veterinariae der Tierarztlichen Hochschule, Berlin

Sugarman BJ, Aggarwal BB, Figari IS, Palladino Jr MA, Shephard HM (1985) Recombinant human tumor necrosis factor-α: Effects on proliferation of normal and transformed cells in vitro. Science 230:943–945

Suttie JM (1990) Experimental manipulation of the neural control of antler growth. In Bubenik G, Bubenik A (eds) Antlers, Horns, Cranial Appendages of Pecoran. Springer-Verlag, New York, pp 359–371

Suttie JM, Fennessy PF (1990) Antler regeneration–studies with antler removal, axial tomography and angiography. In Bubenik G, Bubenik A (eds) Antlers, Horns, Cranial Appendages of Pecoran. Springer-Verlag, New York, pp 313–338

Suttie JM, Fennessy PF, Corson ID, Laas FJ, Elgar HJ, Lapwood KR (1989a) LH and testosterone responses to GnRH in red deer (*Cervus elaphus*) stags kept in a manipulated photoperiod. J Reprod Fert 85:213–219

Suttie JM, Fennessy PF, Corson ID, Laas FJ, Crosbie SF, Butler JH, Gluckman PD (1989b) Pulsatile growth hormone, insulin-like growth factor and antler development in red deer (*Cervus elaphus scoticus*) stags. J Endocrinol 121:351–360

Suttie JM, Fennessy PF, Gluckman PD, Corson ID (1988) Elevated plasma IGF1 levels in stags prevented from growing antlers. Endocrinology 122:3005–3007

Suttie JM, Fennessy PF, Mackintosh CG, Corson ID, Christie R, Heap SW (1985a) Sequential cranial angiography of young red deer stags. In Fennessy PF, Drew K (eds) Biology of Deer Production. R Soc N Z Wellington Bull 22:263–268

Suttie JM, Gluckman PD, Butler JH, Fennessy PF, Corson ID, Laas FJ (1985b) Insulin-like growth factor 1: antler stimulating hormone? Endocrinology 116:846–848

Webster JR, Suttie JM, Corson ID (1991) Effects of melatonin implants on reproductive seasonality of male red deer (*Cervus elaphus*). J Reprod Fert 92: 1–11

Wislocki GB, Aub JC, Waldo CM (1947) The effects of gonadectomy and the administration of testosterone propionate on the growth of antlers in male and female deer. Endocrinology 40:220–224

Zapf J, Guler HP, Schmid C, Kurtz A, Froesch ER (1989) in vivo actions of insulin-like growth factor-1. In Eugenio E, Cocchi D, Locatelli V (eds) Advances in Growth Hormone and Growth Factor Research. Springer-Verlag, Berlin, pp 145–162

117
Antler Growth Patterns in Young Red Deer Stags

Peter F. Fennessy, Ian D. Corson, James M. Suttie, and Roger P. Littlejohn

Abstract

Stages of the antler cycle were recorded for up to 16 red deer stags (*Cervus elaphus*) as 2-, 3- and 4-year-olds. Antler dimensions were recorded at weekly intervals for five stags over 2 successive years; the pattern of growth in antler length and estimated volume were analyzed for these stags using the generalized logistic growth curve. The results show marked effects of age on the date of casting and the period in velvet, and indicate the appropriateness of the generalized logistic growth curve in analysing antler growth patterns. At the completion of growth, the main beam made up approximately 71% of the total volume, the brow tine 7%, the trez tine 8%, and the royal tines 14%.

Key words: Antlers, growth curves, growth patterns, red deer

Introduction

There is a dearth of published information on the pattern of growth of individual antlers recorded through the period of antler growth. Goss (1983) describes the elongation of the antler as a typical S-shaped growth curve citing studies by Rorig (1906, 1908), which show that the more proximal tines complete their development before the more distal ones, while successive stages of growth may be taking place simultaneously in different branches of a growing antler. Recently, growth curves for antler length measured at frequent intervals in individual white-tailed (*Odocoileus virginianus*) and red deer have been published (Suttie 1981; Fennessy 1982; Jacobson and Griffin 1982; Suttie et al. 1984). Using another approach, Muir et al. (1987) cut one antler from several red stags at various stages of growth and expressed the length of the cut antler as a proportion of the contralateral antler which was removed after velvet cleaning. In general these data support Goss' (1983) statement that the curve of antler growth is S-shaped (i.e., sigmoidal or Gompertz). Therefore, the purpose of the present study was to obtain detailed information on the timing and pattern and the rate of antler growth in red stags.

Methods

Animals and Recording

Sixteen red stags were housed in individual pens and fed high quality (11 MJ ME/kg DM) pelleted diets from about 12 months of age in December 1979. For all but a short period as 4-year-olds, all stags were fed ad libitum. The dates of hard antler casting and the first appearance of discernible antler tines were recorded for all 16 stags as 2- and 3-year-olds and for 11 of the stags as 4-year-olds. One antler from each stag was cut at around 55 to 65 days of growth as velvet antler

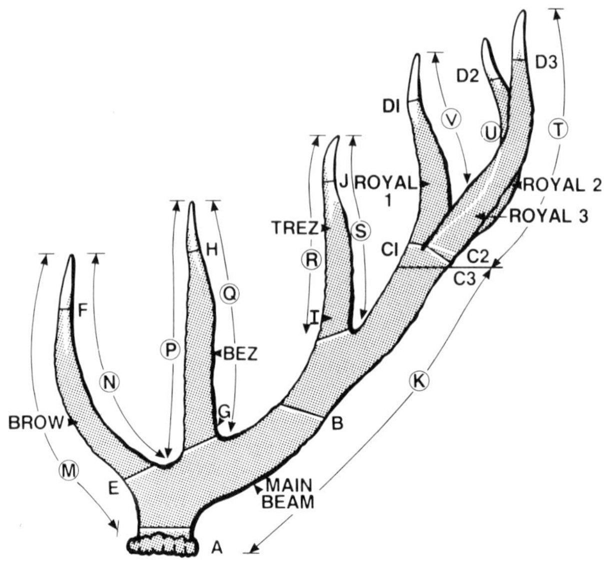

FIGURE 117.1. Diagram of an antler showing the various measurement sites.

while the stags were under sedation. The date for the start of velvet cleaning was recorded for the remaining antler in each of the 3 years. The weight of the hard antler, sawn off approximately 7 days after velvet cleaning, was recorded. As 2- and 3-year-olds in 1980–81 and 1981–82, five stags were run into a crush at weekly intervals and the length and circumference of the main beam and all tines recorded (Fig. 117.1).

Data Analysis

The differences in time of casting, tine initiation, and velvet cleaning between consecutive years were analyzed using t-tests for the 11 stags for which data were available throughout; where there were two values for each animal (e.g., casting), the mean value was used. The antler volumes for the five stags at both 2 and 3 years were estimated by assuming that the antler was made up of a number of tines, each being a tapered cylinder supplanted with a cone. The tapered cylinder was assumed to be 0.9 of the tine length, with the cone being the remaining 0.1 of the tine length (this proportion was estimated from examination of the antlers during growth). The data for one antler per stag were analyzed for volume and length, fitting the generalized logistic curve,

$$y = a + c/(1 + t\exp(-b(\text{time} - m)))^{1/t},$$

where y is the volume or antler length, a is the lower asymptote, $(a + c)$ is the upper asymptote, m is the point of inflexion (days), b is a slope parameter, and t is a power law parameter, with high values of b indicating a fast overall rate of growth and high values of t indicating a sudden conclusion to growth.

Results

Antler Growth

The date of casting was significantly advanced by 21 ± 2.8 (SED) days as the stags aged from 2 to 3 years and by 21 ± 3.6 days as they aged from 3 to 4 years (Table 117.1). The appearance of the antler tines (days after casting) was similar at all three ages, although the trez tines were significantly later at 2 years than 3 years of age. The dates of velvet antler cleaning (across all stags over the 3 years the range was 24 January to 21 February) were similar for all three ages so that there were significant differences in the period of time in velvet between the three ages,

TABLE 117.1. Mean values for casting date, the time of appearance of antler tines, the date of velvet cleaning, and the period of antler growth (casting to velvet cleaning) by age in red stags

Age (years):	2	3	4
Number of stags	16	16	11
Antler casting[a]			
Mean date $\pm$ SD	Oct 26 $\pm$ 9.6	Oct 8 $\pm$ 14.1	Sept 15 $\pm$ 17.8
Appearance of antler tines (days after casting $\pm$ SD; number of stags)			
Brow	13 $\pm$ 5.4 (16)	12 $\pm$ 5.1 (16)	14 $\pm$ 3.6 (11)
Bez	25 $\pm$ 6.2 (8)	20 $\pm$ 6.9 (13)	22 $\pm$ 2.7 (8)
Trez[b]	43 $\pm$ 7.7 (16)	38 $\pm$ 6.5 (16)	38 $\pm$ 5.0 (11)
Royals[c]	69 $\pm$ 5.2 (4)	66 $\pm$ 6.4 (16)	68 $\pm$ 5.6 (10)[d]
Antler velvet cleaning (date $\pm$ SD)	Feb 10 $\pm$ 4.1	Feb 9 $\pm$ 6.1	Feb 3 $\pm$ 6.9
Period in velvet (casting to velvet cleaning $\pm$ SD)[a]	106 $\pm$ 7.2	124 $\pm$ 11.2	141 $\pm$ 15.3
Hard antler weight (kg $\pm$ SD)[e]	0.62 $\pm$ 0.233	1.05 $\pm$ 0.400	1.61 $\pm$ 0.510

[a] All differences between age groups are significant ($P < 0.01$)
[b] 2-year-olds are significantly later than 3-year-olds
[c] Data for only one antler per stag, the other having been removed as velvet antler; only 4 of 16 stags grew royal tines as 2-year-olds
[d] One antler broken
[e] Weight of one antler removed approximately 7 days after velvet cleaning (cut about 1 cm above coronet)

with 3-year-olds being in velvet for 19 $\pm$ 2.6 days longer than 2-year-olds and 4-year-olds for 17 $\pm$ 3.0 days longer than 3-year-olds. Hard antler weight increased markedly with age, with the weight for 3- and 4-year-olds being 1.69 $\pm$ 0.093 (SE) and 2.72 $\pm$ 0.23 times that for the individual stags as 2-year-olds.

Growth Patterns

The data for stag 884 (as a 3-year-old), the stag which grew the largest antlers of the five recorded, indicate the apparent S-shape curve with a period of about 40 days of approximately linear increase in volume (from about day 30 to about day 70) (Fig. 117.2). For this stag the increase in volume over the 42-day period was about 32 cm^3/day for the one antler. Over this same period, the growth in length of the main beam was 0.93 cm/day, compared with an overall average growth rate over the 100-day antler growth period of 0.67 cm/day. If the growth of all tines is included, the growth rate in length over the 42-day period was about 1.8 cm/day.

The estimated mean volume of the fully grown mature hard antler (Table 117.2) was about 46% greater in 3-year-olds than 2-year-olds, but the proportional contributions of the different parts of the antler (i.e., main beam and the individual tines) were very similar with the main beam making up an average 71% of the total volume. On an individual stag basis, the relative volume of 3-year-olds compared with 2-year-olds was 1.46 $\pm$ 0.038 (SE), the relative antler length was 1.11 $\pm$ 0.026 (extended main beam), and the relative hard antler weight was 1.60 $\pm$ 0.13.

Table 117.3 presents a summary of the parameter estimates for the growth in volume and length of the antler (i.e., extended main beam being the main beam plus the main royal tine) analyzed using the generalized logistic curve (Ross 1980) for the five stags as 2- and 3-year-olds. In all cases, more than 98% of the variance was accounted for. The only significant difference between the age groups was in the upper asymptotic value where 3-year-olds had both a greater antler volume and length. There was a tendency for the point of inflexion of the

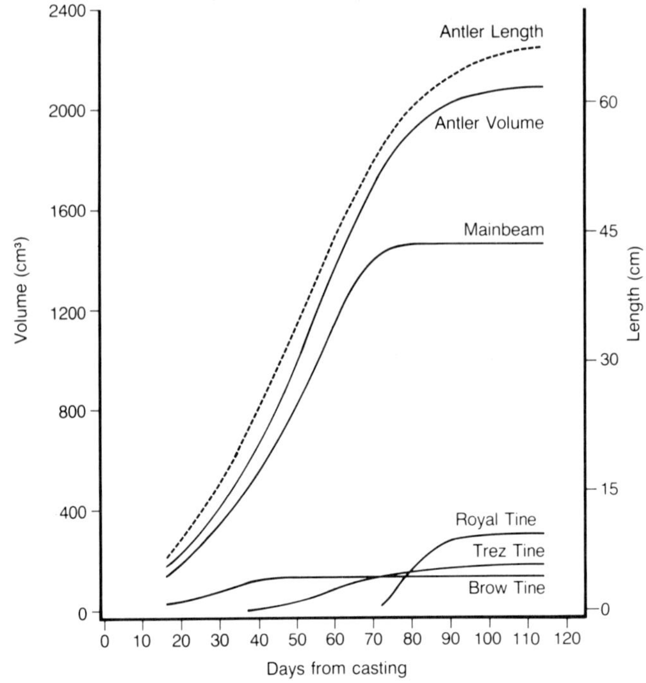

FIGURE 117.2. Patterns of growth in estimated volume of the individual tines and total antler and antler length (extended main beam) for stag 884 as a 3-year-old.

TABLE 117.2. Mean values for five red deer stags as 2- and 3-year-olds for total volume of one antler, the proportional volumes for the different parts of the fully grown antler (i.e., antler at maturity), length of the antler[a] and weight of the one hard antler[b]

Age years	Total volume (cm^3)	Proportional volume of individual tines at antler maturity					Antler length		Hard antler weight (kg)
		Main beam	Brow	Bez	Trez	Royals	Main beam	Extended main beam	
2	1005	0.713	0.072	0.004	0.077	0.136	38.6	54.3	0.58
3	1466	0.705	0.059	0.016	0.083	0.137	45.7	60.3	0.95

[a] Extended main beam = main beam + main royal tine
[b] Weight of one antler removed approximately 7 days after velvet cleaning (cut about 1 cm above coronet)

TABLE 117.3. Mean values for the parameters of the generalized logistic growth curves for antler volume and antler length for the groups of red deer stags as 2- and 3-year-olds

	Volume (cm^3)			Antler length (cm)[a]		
Age (years):	2	3	SED	2	3	SED
b	0.13	0.094	0.035	0.12	0.088	0.015
m	54	40	4.8	61	51	3.6
t	4.7	2.5	2.4	5.5	3.4	2.0
$(a + c)$	1011	1485	76**	55.0	61.4	1.56*

[a] Antler length = length of the extended main beam = main beam + main royal tine

growth curve to be earlier in the 3-year-olds (e.g., for length, $0.05 < P < 0.10$). There was also a tendency for the growth pattern of royal tines to follow the pattern of a Mitscherlich curve (see Fig. 117.2) rather than a sigmoidal curve, due to royal tines in effect being an extension of the main beam. This was the source of a minor anomaly in fitting the generalized logistic curve.

Interpolation of the logistic growth curves indicated that the antlers had attained 17%, 56%, 91% and 99% of their final volume at maturity by 25, 50, 75, and 100 days of growth, respectively. The corresponding values for antler length were 17%, 50%, 85%, and 98% Analysis of the data for the five stags over the 2 years indicates that the growth in length of the individual tines was complete by 56 ± 10.7 days after casting for the brow, 69 ± 5.5 for the main beam (i.e., the point of royal tine initiation) 83 ± 13.3 for the trez, and 96 ± 7.8 for the royal tines (i.e., the whole antler).

Discussion

Antler Growth

The effect of age on casting date of hard antler was considerable (21 days advancement between 2 and 3 and between 3 and 4 years) and greater than that recorded by Muir and Sykes (1988) of 8 days between 4- and 5-year-olds and 6 days between 3- and 4-year-olds. In the latter study, the 3- and 4-year olds had mean casting dates of 25 and 19 September compared with 8 October and 15 September in the present study. The relatively late antler casting in 2-year-olds is a well-known phenomenon in farmed red deer in New Zealand.

In contrast to the situation with antler casting, velvet cleaning occurred over a very narrow time span as indicated by the relative standard deviations and the narrow range in the date of velvet cleaning. Interestingly, the mean date of velvet cleaning was several days earlier than that reported by Muir et al. (1987) of 20 to 27 February for groups of older stags also kept indoors.

An increase in antler weight with age has been reported by several workers, notably Huxley (1926, 1931) and Moore et al. (1988). Within an age group there is also a strong relationship with live weight (Muir and Sykes 1988; Moore et al. 1988; Fennessy and Suttie 1985).

Growth Patterns

The contributions of the individual tines to the total antler volume were very similar in stags at both ages, despite a considerable difference in antler volume and length. The analysis of the growth curve confirmed the appropriateness of the generalized logistic equation for modeling the pattern of antler growth. The hard antler weights for the five stags measured were slightly below the average for the whole age group in both years (0.58 compared with 0.62 for the whole group as 2-year-olds and 0.95 compared with 1.05 as 3-year-olds).

The data on growth of individual tines confirms the observations of Rorig (1906, 1908, cited by Goss 1983) that the more proximal tines complete their development before the distal ones. This pattern is also similar to that reported by Muir et al. (1987), who found that in stags in velvet antler for an average of 164 days, the brow tine was almost fully formed by 56 days after casting when it had reached 84% of its final length, whereas the main shaft was still growing rapidly, reaching 95% of its final length by 112 days.

References

Fennessy PF (1982) Growth and nutrition. In Yerex D (ed) The Farming of Deer: World Trends and Modern Techniques. Agricultural Promotion Associates, Wellington, pp 105–114

Fennessy PF, Suttie JM (1985) Antler growth: nutritional and endocrine factors. In Fennessy PF, Drew KR (eds) Biology of Deer Production. R Soc N Z Wellington Bull 22:239–250

Goss RJ (1983) Deer Antlers: Regeneration, Function and Evolution. Academic Press, New York

Huxley JS (1931) The relative size of antlers in deer. Proc Zool Soc London pp 819–864

Huxley JS (1926) The annual increments of the antlers of red deer (*Cervus elaphus*). Proc Zool Soc London, pp 1021–1035

Jacobson HA, Griffin RN (1982) Antler cycles of white-tailed deer in Mississippi. In Brown RD (ed) Antler Development in Cervidae. Caesar Kleberg

Wildlife Research Institute, Kingsville, Texas, pp 15–22

Moore GH, Littlejohn RP, Cowie GM (1988) Liveweights, growth rates, and antler measurements of farmed red deer stags and their usefulness as predictors of performance. N Z J Agric Res 31:285–291

Muir PD, Sykes AR (1988) Effect of winter nutrition on antler development in red deer (*Cervus elaphus*): a field study. N Z J Agric Res 31:145–150

Muir PD, Sykes AR, Barrell GK (1987) Growth and mineralisation of antlers in red deer (*Cervus elaphus*) N Z J Agric Res 30:305–315

Rorig A (1908) Das Wachstum des Geweihes von *Capreolus vulgaris*. Arch Engwicklungs mech Org 25:423–430; cited by Goss (1983)

Rorig A (1906) Das Wachstum des Geweihes von *Cervus elaphus*, *Cervus barbarus* und *Cervus canadensis*. Arch Entwicklungs mech Org 20:507–536; cited by Goss (1983)

Ross GJS (1980) Maximum Likelihood Programme. Rothamsted Experimental Station, Lawes Agricultural Trust

Suttie JM (1981) The influence of nutrition and photoperiod on the growth, development and endocrine status of captive red deer and Soay rams. PhD thesis, University of Aberdeen, Scotland

Suttie JM, Corson ID, Fennessy PF (1984) Voluntary intake, testis development and antler growth patterns of male red deer under a manipulated photoperiod. Proc N Z Soc Anim Prod 44:167–170

118
Antler Growth in Male and Female Reindeer Calves Occurs in the Absence of the Gonads

Gerald A. Lincoln and Nicholas J.C. Tyler

Abstract

Reindeer/caribou (*Rangifer tarandus*) differ from all other species of deer in their antler physiology in three notable ways: (1) antlers are developed in both sexes, (2) antlers begin to grow in young calves well in advance of puberty, and (3) antlers are cast and regrown annually in females, and both intact and castrated males. These observations suggest that, in contrast to the situation in other species, the initiation of antler growth in the young animal and the seasonal cycle of antler regeneration in the adult animal are temporal events not controlled by the secretion of gonadal hormones. To investigate the control of the initiation of antler growth, we monitored antler growth in two castrated male and two ovariectomized (OVX) female reindeer which had their gonads surgically removed at 5 to 15 days of age when there were no palpable antler pedicles (births 16 to 28 May). The animals were reared outdoors by their mothers in Tromsø, Norway (70° N). All experimental animals initiated antler growth. The two castrated males and one of the OVX females developed normal spike antlers 2 to 36 cm long in the first year of life, while one of the OVX females developed only antler buds. In the animals that developed antlers, these were cleaned of velvet in the winter, cast in the spring, and new antler growth was initiated. There were small differences in the pattern of cleaning and casting compared with gonadally intact animals. These observations provide experimental evidence that the development, growth, and casting of the first antler can occur after postnatal gonadectomy. Nevertheless, gonadal hormones are thought to play an important role in dictating the timing of the antler cycle in reindeer as in other cervids, synchronizing the hard antler phase with other reproductive events.

Keywords: Antler growth, secondary sexual characteristics, ovary, reindeer, testis, testosterone

Introduction

In most species of cervids, growth of the antlers is a male secondary sexual characteristic dependent on the secretion of testosterone from the testes. Only males develop antlers, and if a male is castrated before puberty no antlers develop. The single exception is reindeer/caribou (*Rangifer tarandus*) in which both sexes produce antlers. In this species the antlers begin to develop soon after birth, well in advance of puberty, and removal of the gonads in adults apparently does not block growth and annual replacement of antlers. These observations have been known for over 100 years and were recorded by Charles Darwin (1891): "stags after (castration) never renew their horns. The male reindeer, however, must be excepted as after castration he does renew them. This fact, as well as possession of horns by both sexes, seems at first to prove that the horns in this species do not constitute a sexual character; but as they are developed at a very early age, before the sexes differ in constitution, it is not surprising that they should be unaffected by castration; even if they were aboriginally acquired by the male."

It appears that gonadal hormones are not required for development and annual replacement of antlers in either male or female reindeer. If this is true, surgical removal of the gonads in newborn animals will not prevent development of antlers. To our knowledge there is no published account to establish this simple fact. The current study was carried out to test whether growth and development of antlers would occur normally in reindeer calves following castration or ovariectomy performed soon after birth.

Materials and Methods

Observations were made on five Norwegian semidomesticated reindeer calves which included two castrated males, two ovariectomized females, and one intact male. The animals were maintained at the Department of Arctic Biology, University of Tromsø, Norway (70° N, 19° E). They were born at the Department 16 to 28 May 1989 and were reared normally by their mothers. The calves and their mothers were kept in small groups in two adjacent outdoor enclosures where concentrate rations, RF-71 (Jakobsen and Skjenneberg 1979) or RF-80, and water or snow were available ad libitum.

The male calves were castrated at 5 and 8 days of age, and the female calves were ovariectomized at 13 and 15 days of age. Surgical gonadalectomy was performed under sterile conditions while the animals were deeply sedated using a single intramuscular injection of 54 to 62 μg/kg medetomidine (Farmos Group, Turku, Finland) (Tyler et al. 1991). Lidocain (Xylocain, 10 mg/ml; Astra, Södertälje, Sweden) was administered locally around the ovarian pedicle. In the males, each spermatic cord was ligated using absorbable sutures, and testes and epididymides were removed bilaterally (weight of left testes for the two calves was 420 and 251 mg). In the females, each ovarian pedicle was ligated using two absorbable sutures passed through the midline, and the ovaries, together with a small portion of each pedicle, were removed bilaterally (weight of left ovaries was 274 and 106 mg). Sedation was reversed using a single intramuscular injection of 276 to 320 μg/kg atipamezole (Farmos Group, Turku, Finland) (Tyler et al. 1991). Each calf was returned to its mother within 30 min of arousal. All the calves were recognized and accepted by their mothers at once, and all made an uneventful recovery from the operation.

Growth and development of antlers were observed weekly in all calves from May 1989 until May 1990. The length of the antlers was measured at approximately monthly intervals. For comparison, information on the timing of the antler cycle in reindeer calves living in the locality was obtained from records kept at the Department of Arctic Biology and at the Reindeer Research Station, Lødingen.

Results

All the experimental animals initiated antler growth. Antler buds were palpable at 2 to 4 weeks of age in the castrated and ovariectomized calves, at the same age as in the gonadally intact male kept in the same group. This was regarded as the normal age for the initiation of antler development based on previous observations for both intact male and female calves living under similar conditions. The subsequent development of the antlers in the experimental animals over the first year of life is summarized in Fig. 118.1.

Both castrated males grew antlers from June onwards. These were 20 to 36 cm long in October, and one of the animals developed large antlers with a brow tine on one side (Fig. 118.2, upper panel). The antlers became dried in appearance and the velvet was cleaned slowly to expose the polished antler bone by January, 2 months later than in the accompanying intact male. The two castrated calves cast their antlers in late April, again 2 months later than the intact male, while new antler growth occurred synchronously in all animals in April (Fig. 118.1). The antler cycle in the castrated males was unusual both because the left and right antlers were cast up to 10 days apart, and because regrowth of new antlers began slightly before casting of the old antlers was completed. In addition, in the only antler recovered from the castrated males, the erosion surface at the base was slightly concave and unusually irregular.

The antler cycle of one ovariectomized female was similar to that of the castrated males except that one antler was small (2 cm) and was incompletely cleaned of velvet in the winter

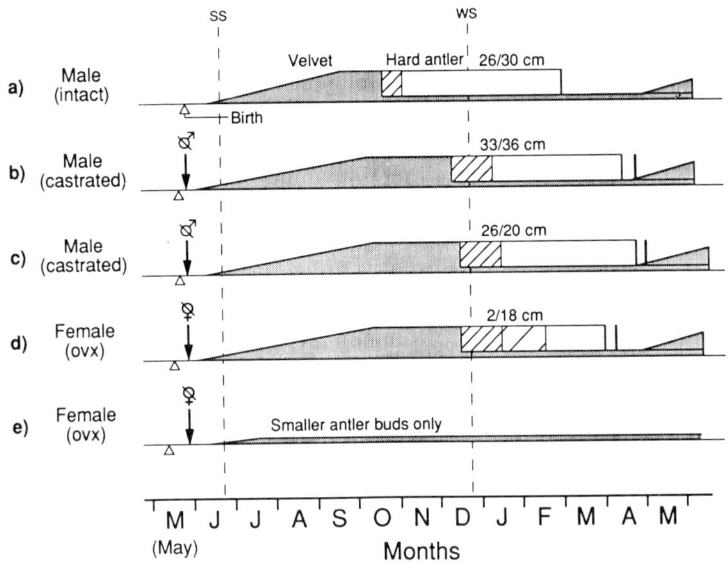

FIGURE 118.1. Summary of the antler cycle during the first year of life in five reindeer calves (a) intact male, (b and c) castrated males, and (d and e) ovariectomized females. Stages of the antler cycle are shown as periods of growth (velvet, closed bar), cleaning (hatched bar), and hard antler (open bar) terminating with casting of the old antlers (vertical line). The time of the summer solstice (SS) and winter solstice (WS) is shown by vertical dotted lines.

(Figs. 118.1, 118.2). Again, casting of left and right antlers was slightly asynchronous, but new antler growth began in April at the normal time (Fig. 118.1). The other ovariectomized female failed to continue to develop antlers, and the antler buds remained small, <2 cm long, throughout the winter and had not changed by May 1990 at 1 year of age.

Discussion

These results demonstrate that juvenile reindeer will initiate antler growth following surgical removal of the gonads soon after birth. Furthermore, both sexes can grow, clean, and cast their first antlers and will grow a second set of antlers in the absence of gonads. This contrasts to the situation in other cervids. In these, antlers are a male secondary sexual characteristic, and removal of the testes in the newborn totally blocks growth of antlers (Wislocki et al. 1947; Chapman 1975; Goss 1983).

The castrated reindeer calves in our study completed an antler cycle in their first year of life. However, the temporal pattern of events differed from the intact male kept with the experimental group, and from the normal pattern observed previously in intact male calves living in captivity. Cleaning of the velvet was more protracted in the castrates, and both cleaning and casting of the polished antlers were delayed by 2 months (Fig. 118.1). In these animals casting occurred at the time of, or even slightly after, initiation of new antler growth in April. The asynchronous casting of the antlers in the castrates and the irregular erosion surface at the base of the cast antler was also unusual. However, our observation of protracted cleaning and delayed casting in the castrated calves corresponds closely with the pattern reported for castrated adult reindeer (Jacobi 1931; Skjenneberg and Slagsvold 1968; Wika 1980). Abnormal erosion of the base of the antler also occurs in other species of deer if antler casting is induced prematurely by castration (Goss 1983). This may be due to the incomplete death of antler

FIGURE 118.2. Antler development in reindeer calves. Upper panels: castrated male calf (16/89) on the day of surgical removal of the testes in May at 8 days old (left) and the same animal in October at 5 months old showing the complete development of antlers (right). Middle panel: intact male calf (18/89) in May (left) and October (right) for comparison with the castrated male. Lower panel: ovariectomized female calf (20/89) on the day of surgical removal of the ovaries in May at 13 days old (left) and the same animal in October at 5 months old showing the development of antlers (right).

tissue (also indicated by protracted cleaning in castrated reindeer) which results in a poorly defined junction between the dead antler and the living pedicle where erosion of the bone occurs to permit casting.

In most species of deer it is the changes in the secretion of testosterone from the testes which induces the growth of the first antlers at puberty and also the seasonal changes in the cleaning and casting of the antlers in the adults (Wislocki et al. 1947; Goss 1968; Lincoln 1984). The first antlers grow out from clearly defined antler pedicles which are totally dependent for their growth on stimulation by testosterone. Thus, if a male is castrated before puberty the antler pedicles never develop, while administration of testosterone to a prepubertally castrated male or a normal female causes the growth of pedicles. Reindeer differ from other cervids in several ways. Their first antlers covered by fine hair characteristic of the antler velvet, develop from the skull before differentiation of distinct antler pedicles covered by the normal scalp hair (Leader-Williams 1988). The initiation of growth in males occurs before the pubertal increase in testosterone (Whitehead and McEwan 1973; Leader-Williams 1979) and occurs in females with no associated increase in the circulating concentration of testosterone (Leader-Williams 1988). Furthermore, it will proceed in either sex even if the gonads are removed from calves before the pedicle buds are palpable (as shown in the current study).

Nevertheless, the role of testosterone can be invoked to explain some of the events during the antler cycle of intact reindeer. In particular, calcification and cleaning of the antlers in the animals' first autumn occur following elevated blood levels of testosterone (Whitehead and McEwan 1973; Leader-Williams 1979), and the administration of testosterone to animals in velvet has been shown to induce premature cleaning of the velvet in yearling reindeer (Ryg 1983). Castration of adult reindeer in hard antler appears to cause premature casting of the antlers (Skjenneberg and Slagsvold 1968), and this event normally occurs following the seasonal decline in testosterone secretion (Whitehead and West 1977; Leader-Williams 1979; Mossing and Damber 1981). Thus, it is most probable that in male reindeer, testosterone secreted by the testes, though not essential for the growth and development of antlers, has an overriding influence on the timing of the antler cycle. The role of androgens is to ensure that the antlers are fully developed as functional weapons shortly before the rut. At this time the overt sexual behavior is itself induced by high levels of testosterone secretion (Lincoln et al. 1972).

One of the OVX reindeer calves completed an antler cycle in its first year of life while the second animal unaccountably failed to develop full antlers. The temporal pattern of the antler cycle in the first animal was similar to that of the castrated male calves; cleaning of the velvet was protracted and casting was asynchronous (Fig. 118.1). Similar results have been reported for a single ovariectomized adult female reindeer (Tandler and Grosz 1913). As argued for the males, it still seems likely that gonadal hormones also contribute to the control of the normal antler cycles in females. This is based on several observations. Cleaning of antlers normally occurs close to the onset of ovarian activity in the autumn (Leader-Williams 1988), pregnancy delays casting of the antlers in the spring (Espmark 1971), and ovariectomy of nonpregnant adult females in hard antler induces premature casting of the antlers (G.A. Lincoln and N.J.C. Tyler, unpublished data). Furthermore, in other species of cervids, administration of the ovarian hormone estrogen acts like testosterone. It causes cleaning of the velvet and development of hard antlers, and on its withdrawal, results in casting of the hard antlers (Goss 1968; Fletcher 1978). Such a role for reproductive hormones in the antler cycle in females would also ensure synchrony between antler development and the annual cycle in reproduction. Possession of hard antlers enhances social dominance, and females use their antlers in winter (when they are normally pregnant) when competing for food.

In conclusion, the current study provides clear evidence that the growth, development, and casting of antlers proceed in reindeer calves even after surgical removal of the gonads at an early age, and is not totally dependent on secretion of gonadal hormones. However, our results do not preclude an important role of gonadal hormones in modifying the timing of the antler cycle in reindeer to ensure synchrony with other reproductive events.

Acknowledgements

This study was supported in part by the Norwegian Research Council for Science and the Humanities and the Norwegian Reindeer Administration.

References

Chapman DI (1975) Antlers—bones of contention. Mammal Rev, 5:121–172

Darwin C (1891) The Descent of Man and Selection in Relation to Sex. John Murray, London, pp 226

Espmark Y (1971) Antler shedding in relation to parturition in female reindeer. J Wildl Manage 35:175–177

Fletcher TJ (1978) The induction of male sexual behaviour in red deer (*Cervus elaphus* L.) by administration of testosterone to hinds and estradiol-17β to stags. Hormones Behav 11:74–88

Goss RJ (1983) Deer Antlers. Regeneration, Function and Evolution. Academic Press, New York

Goss RJ (1968) Inhibition of growth and shedding of antlers by sex hormones. Nature, London 220:83–85

Jacobson E, Skjenneberg S (1979) Experiments with different diets to reindeer. Feeding value of reindeer feed (RF-71). Sci Rep Agric Coll, Norway 58:1–11

Jacobi A (1931) Das Rentier, eine Zoologische Monographie der Gattung Rangifer. Zoologicher Anzeiger, Ergänzungsband zu Band 96

Leader-Williams N (1988) Reindeer on South Georgia. Cambridge University Press, Cambridge

Leader-Williams N (1979) Age-related changes in the testicular and antler cycles of reindeer, *Rangifer tarandus*. J Reprod Fert 57:117—126

Lincoln GA (1984) Antlers and their regeneration—a study using hummels, hinds and haviers. Proc R Soc Edinburgh 82B:243–259

Lincoln GA, Guinness F, Short RV (1972) The way in which testosterone controls the social and sexual behaviour of the red deer stag *Cervus elaphus*. Hormones Behav 3:375–396

Mossing T, Damber J-E (1981) Rutting behaviour and androgen variation in reindeer (*Ranifer tarandus* L.) J Chem Ecol 7:377–389

Ryg M (1983) Effect of testosterone on antler growth in yearling male reindeer. Rangifer 3:6–9

Skjenneberg S, Slagsvold L (1968) Reindeer husbandry. Universitersforlaget, Oslo, pp 22–27

Tandler J, Grosz S (1913) Die biologischen Grundlagen der sekundären Geschlechtscharaktere. Springer-Verlag, Berlin, pp 25–27

Tyler NJC, Hotvedt R, Blix AS, Sørensen D (1991) Immobilization of Norwegian reindeer (*Rangifer tarandus tarandus*) and Svalbard reindeer (*Rangifer tarandus platyrhynchus*) with medetomidine and medetomidine-ketamine and reversal of immobilization with atipamezole. Acta Vet Scan 31:479–488

Whitehead PE, McEwan EH (1973) Seasonal variation in the plasma testosterone concentration of reindeer and caribou. Can J Zool 51:651–658

Wika M (1980) On growth of reindeer antlers. In Reimers E, Gaare E, Skjenneberg S (eds) Proc 2nd Int Reindeer/Caribou Symp. Direktoratet for vilt og ferskvannsfisk, Trondheim, pp 416–421

Wislocki GB, Aub JC, Waldo CM (1947) The effects of gonadectomy in male and female deer. Endocrinology 40:202–224

119
Seasonal Regulation of the Sexual Cycle and Antler Growth in Roe Deer: Evidence for an Endogenous Rhythm

Antoine J. Sempéré, Robert Mauget, and André Lacroix

Abstract

Seasonal variations of luteinizing hormone (LH), prolactin (PRL), and testosterone were analyzed in relation to the annual cycle of antlers in four adult roe deer (*Capreolus capreolus*) living under the natural photoperiod and adult males exposed to manipulated photoperiods. In the second experiment, one male was exposed for 15 months to an artificial photoperiod providing two complete cycles per year. In the third experiment, two males were exposed for 9 months to an artificial photoperiod providing 8 h of light and 16 h of dark (8L:16D), while two control males were kept in a light-proof room subject to the natural photoperiod. At monthly intervals blood samples were taken every 15 min 2 h after an injection of gonadotrophin-releasing hormone (GnRH). High LH activities were correlated with increasing day length, (in January, 2.69 ± 0.25 ng/ml, and in May, 3.27 ± 0.56 ng/ml), while high LH levels were again recorded in the fall when day length was decreasing (in October, 1.89 ± 0.66 ng/ml) in the animals living outside in natural photoperiod. In deer kept under manipulated photoperiods, sexual cycles and antler cycles were modified. For the second experiment, a biphasic pattern of LH was observed during the first 6-month photoperiod, whereas during each following photoperiod cycle a single peak was observed. The bucks exposed to the 8L:16D photoperiod exhibited two periods of high LH levels (January–February and August). Moreover, two antler cycles recorded during the 9-month experiment. These results provide evidence for an endogenous rhythm of LH occurring after the winter solstice. The pituitary activity could be later entrained by the long photoperiod (> 12L) after the spring equinox.

Key words: LH, testosterone, PRL, photoperiod, roe deer

Introduction

In most cervids considered to be short-day breeders, decreasing day length induces gonadotrophin secretions, whereas increasing day length inhibits this process. In deer species, peak levels of prolactin (PRL) occur in spring and summer (Lincoln 1985), whereas the timing of the rut is different. The rutting period of most cervids is in the fall, whereas the European roe deer mates in summer (Bramley 1970; Hoffman et al. 1978; Sempéré 1982, 1990). However, the roe deer is the only temperate cervid with high gonadotrophin levels from January to September (Sempéré and Lacroix 1982; Sempéré and Boissin 1983). Their peak testosterone levels were recorded in July during the breeding season. In temperate cervids the rutting season is very short (Bramley 1970; Plotka et al. 1980; Sempéré 1982; Lincoln 1985). However, although female roe deer are monoestrous (Hoffman et al. 1978), does mate in summer (Bramley 1970; Hoffman et al. 1978; Sempéré 1982; Sempéré et al. 1989). It has been shown that antlers grow in relation to the photoperiod (Jaczewski 1956; Goss 1969a, 1969b; Goss 1983), but there is also evidence that casting and regrowth of antlers occurred when testosterone levels are indetectable (Sempéré

1990). Since testosterone directly affects the antler cycle, antlers can be considered an index of reproductive status (Mirarchi et al. 1978; Sempéré et al. 1980). The aim of this study was to analyze the influence of the photoperiod on pituitary and sexual hormones in male roe deer and their effect on the antler cycle. We first investigated the seasonal changes in luteinizing hormone (LH) and testosterone responses to gonadotrophin-releasing hormone (GnRH) administration in roe bucks living outside. We also tested the influence of artificial photoperiods on the hormone responses to GnRH administration.

Material and Methods

The study was carried out at the Laboratory located in Chizé National Reserve, midwest France (46°10' N and 0°30' E). All adult males (3 to 6 years old) utilized in this study were previously living outside in a 2-ha park for 2 years. In the first experiment, one group of four males was kept outside in the park. The other males were kept in light-proof rooms. The artificial photoperiod was obtained using six 40 W fluorescent tubes that produced an intensity of 350 lux at 1 m from the floor. In the second experiment, one male was exposed for 15 months to an artificial photoperiod providing two complete cycles per year. In the third experiment, two males were exposed to constant short days (8L:16D) during 9 months, and two males were kept for control in identical pens subject to an artificial light providing the natural photoperiod variations.

For monthly blood sampling, all animals were anesthetized with Rompun (25 mg of xylazine hydrochloride) and 50 mg of ketamine. A catheter was inserted in the jugular vein. For the analysis of the basal LH levels, blood samples were taken after catheterization and before GnRH injection, then blood samples were also collected every 15 min for 2 h after the injection of 100 μg of GnRH. Such high doses of GnRH induced high LH responses throughout the year, with no interaction with the antler cycle in red deer (*Cervus elaphus*) (Fennessy et al. 1988). Blood samples were centrifuged and stored at $-20°$ C until assayed.

Plasma testosterone was determined using a nonchromatographic assay similar to that described by Sempéré and Boissin (1983). Plasma LH was analyzed in duplicate using the heterologous radioimmunoassay described by Sempéré and Lacroix (1982). Plasma PRL was analyzed in duplicated tubes using the technique of Kann (1971). Statistical comparisons were examined by Student's *t*-tests for differences among paired variates. The comparisons between hormones for the different phases of the sexual cycle were carried out by analysis of variance (ANOVA).

Results

High LH levels were observed in January–February, May–June, and September–October in animals living outside (Fig. 119.1). Significantly low levels were recorded in March–April ($P < 0.05$), July–August ($P < 0.01$), and November–December ($P < 0.05$). The lowest plasma testosterone levels were observed from October to January. Testosterone rose in February and reached a peak in July, prior to the rut. All the animals cast their antlers in November. New antlers in velvet started to grow immediately. The antlers were fully developed by the end of January and were polished by the end of February. PRL concentrations followed the annual photoperiod variations, and the peaks were obseved in June–July. The greatest responses of LH to GnRH administration were found in January–February, and in June. A slight but nonsignificant increase was also recorded in October. LH responses decreased significantly in November (Fig. 119.2).

In the 6-month photoperiod cycle experiment, LH and PRL followed the artificial photoperiod (Fig. 119.3). The peaks of PRL rose progressively during the successive photoperiod cycles. The second and the third antler cycles were shortened (Fig. 119.3).

In the 8L:16D experiment, a first LH peak was observed in January–February and a second peak was recorded in July–August (Fig. 119.4). Testosterone levels were high from February to May, then dropped in June. The first antler cycle was from November to March. The antlers were cast in July, and a new set started to grow immediately. In experimental bucks, plasma

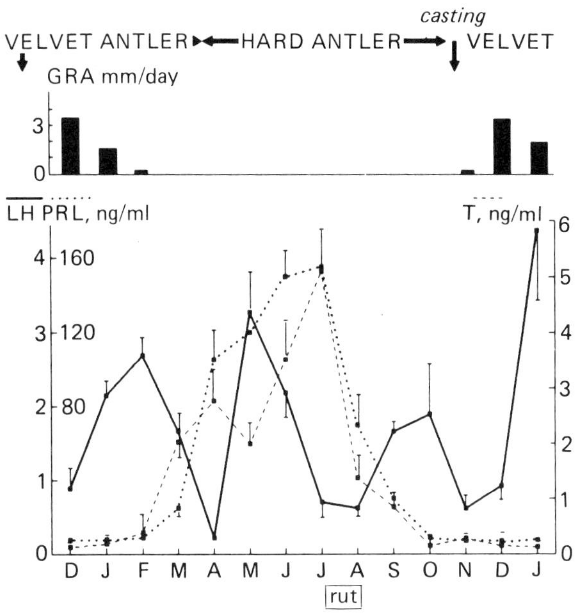

FIGURE 119.1. Annual variations of plasma LH, testosterone (T) and prolactin (PRL) concentrations in four adult bucks living in natural conditions. Antler cycle represented in the upper graph; growth rate of antlers (GRA) indicated by vertical bars.

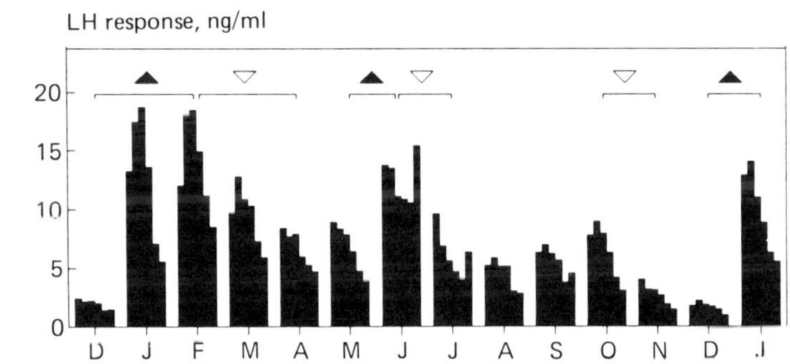

	D	J	F	M	A	M	J	J	A	S	O	N	D	J
mean	11.7	75.7	83.0	57.0	39.8	40.2	75.3	37.1	27.2	33.5	38.3	16.5	10.0	58.7
SE	2.2	29.0	12.0	5.8	9.0	7.7	7.7	4.3	3.5	4.0	6.4	2.4	1.1	14.2

FIGURE 119.2. LH response (area under the curve) to 100 μg of GnRH in four adult bucks living outside. Pooled SE were 1.46 for LH responses; ▲, significant increase and, ▽, significant decrease —$P < 0.05$.

PRL exhibited high levels from February to April. A drastic decrease was observed in June (Fig. 119.4).

The comparative LH responses between the controls, the 6-month photoperiod experiment, and the 8L:16D experiment are shown in Fig. 119.5. These data indicate that in all conditions, the first maximum in the LH response was always recorded in January–February. However, considering the following cycles, high LH levels were recorded 3 or 4 months later in controls, whereas an interval of 6 months was observed between two peaks of LH response in both experimental conditions.

Discussion

Plasma PRL in control males followed the photoperiodic variations as demonstrated previously (Van Mourik et al. 1985). In this study, LH fluctuations did not precisely follow the annual photoperiod cycles. The rise of LH first occurred when the photoperiod started to increase in January, and secondly by the end of

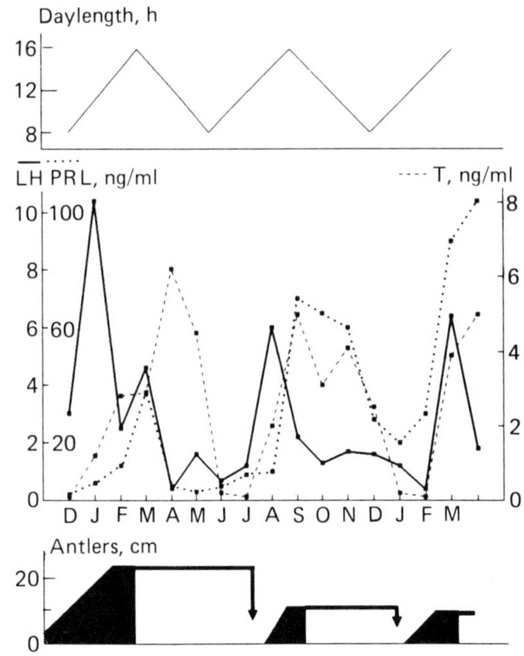

FIGURE 119.3. Variations of plasma LH, T, and PRL concentrations in an adult male exposed during 15 months to an artificial 6-month photoperiod cycle. Antler cycle is shown on the lower part of the graph: antler velvets (black areas) and casting (arrows).

the increasing day length in May–June, whereas in most temperate cervids, the reactivation of the pituitary activity occurs when day lengths are decreasing (Lincoln and Kay 1979; Bubenik et al. 1982; Brown et al. 1983; Fennessy et al. 1988; Asher et al. 1989). In roe deer, the decline of LH observed in August was followed by an increase in LH concentrations by the fall equinox.

In our animals living outside, the pattern of the LH responses and the basal LH levels obtained before GnRH injections differed only in May–June. LH response was greatest in May–June when the basal LH plasma concentrations started to decrease. This indicates that the decrease in basal LH was related to a lower secretion of GnRH from the hypothalamus, as has been shown in red deer (Suttie et al. 1984). The seasonal cycle of testosterone reflected the change in LH activity. Meanwhile, prior to the rut, testosterone levels were high when LH levels were low. At the same time, maximum testosterone response to GnRH injection in July would indicate a high sensitivity of the testis to gonadotrophins (A.J. Sempéré and R. Mauget, unpublished data).

These annual patterns of LH and testosterone raise the question of the influence of the photoperiod in the sexual cycle in roe deer. LH levels in roe deer fluctuate largely during increasing day length, from January to June, but also during the decreasing day length, in September–October (Sempéré and Lacroix 1982). In our second experiment (6-month photoperiod), except for the first LH peak observed in January, the remaining LH peaks were recorded around the artificial summer solstice. PRL also peaked during each summer solstice, but PRL progressively increased to the maximum concentrations. In this experiment, the second and third set of antlers were affected by testosterone which had probably caused the premature ossification (Suttie and Simpson 1985).

In the third experiment (8L:16D), the first LH peak was in January and the second was located 6 months later in July. Considering the antler

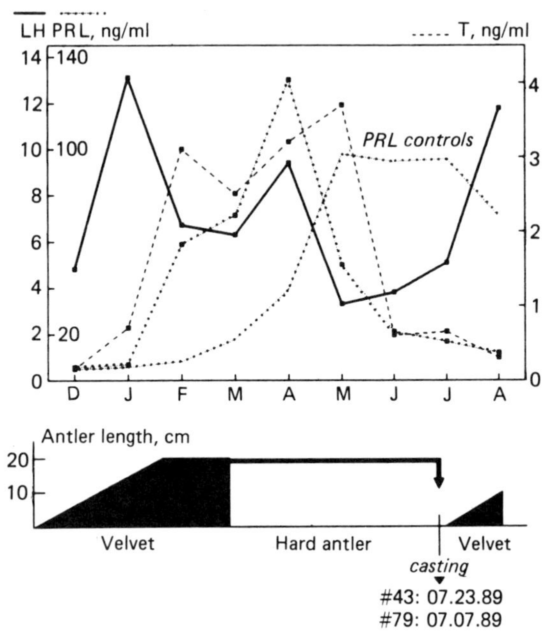

FIGURE 119.4. Variations of plasma LH, T, and PRL concentrations in two adult males (mean values) exposed to a continuous short-day photoperiod (8L:16D). The individual casting dates of antlers are mentioned.

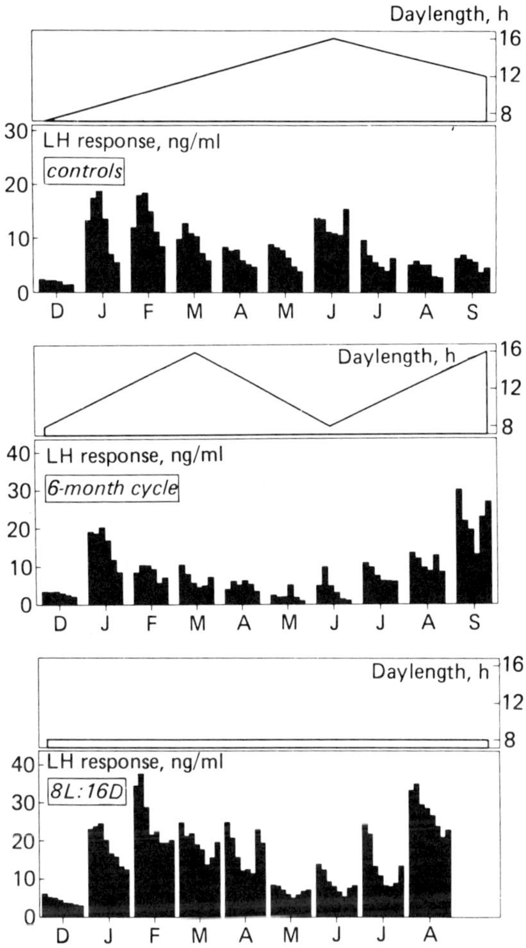

FIGURE 119.5. Comparative LH responses to 100 μg of GnRH of roe bucks living under different photoperiods.

cycle as an indicator of the reproductive cycle, it is evident that the change in the frequency of the antler cycle suggests that the reproductive cycle exibited an endogenous rthythm of 6 months in the roe deer. Numerous studies have shown that mammals maintained under continuous photoperiod exhibit the endogenous rhythm of the reproductive function (Lincoln and Short 1980; Howles et al. 1982). The PRL cycle could also be the reflection of an endogenous rhythm, as such a rhythm has been shown in rams (Howles et al. 1982; Almeida and Lincoln 1984). In male roe deer, the reproductive cycle starts endogenously after the winter solstice. In spring, the reactivation of the sexual function could, this time, be induced by the long photoperiod.

Acknowledgements

The authors wish to thank Z. Marillet and C. Trouvé for skilled technical assistance. We also thank C. Mauget for performing computer programs and preparation of the manuscript, and N. Guillon for blood sampling and animal care.

References

Almeida OFX, Lincoln GA (1984) Reproductive photorefractoriness in rams and accompanying changes in the patterns of melatonin and prolactin secretion. Biol Reprod 30:143–158

Asher GW, Peterson AJ, Bass JJ (1989) Seasonal pattern of LH and testosterone secretion in adult male fallow deer, (*Dama dama*). J Reprod Fert 85: 657–665

Barth D, Gimenez T, Hoffman B, Karg H (1976) Testosteronkonzentrationen im peripheren Blut beim Rehbock (*Capreolus capreolus*). Z Jagdwiss 22:134–148

Bramley PS (1970) Territoriality and reproductive behaviour of roe deer. J Reprod Fert Suppl 11: 43–70

Brown RD, Chao CC, Faulkner LW (1983) The endocrine control of the initiation and growth of antlers in while-tailed deer. Acta Endocrinol 103:138–144

Bubenik GA, Morris JM, Schams D, Claus A (1982) Photoperiodicity and circannual levels of LH, FSH, and testosterone in normal and castrated male, white-tailed deer. Can J Physiol Pharmacol 60: 788–793

Fennessy PF, Suttie JM, Crosbie SF, Corson ID, Elgar HJ, Lapwood KR (1988) Plasma LH and testosterone responses to gonadotrophin releasing hormone in adult red deer (*Cervus elaphus*) stags during the annual antler cycle. J Endocrinol 117:35–41

Goss RJ (1969a) Photoperiodic control of antler cycles in deer. I—phase shift and frequency changes. J Exp Zool 170:311–324

Goss RJ (1969b) Photoperiodic control of antler cycles in deer. II—alteration in amplitude. J Exp Zool 171:223–234

Goss RJ (1983) Control of deer antler cycles by the photoperiod in Antler Development in Cervidae. RD Brown (ed) Caesar Kleberg Wildl Res Inst, Kingsville, Texas, 1–13

Hoffman B, Barth T, Karg H (1978) Progesterone and

estrogen levels in peripheral plasma of the pregnant roe deer (*Capreolus capreolus*). Biol Reprod 19:931–935

Howles CM, Craigon J, Haynes NB (1982) Long-term rhythms of testicular volume and plasma prolactin concentrations in rams reared for 3 years on constant photoperiod. J Reprod Fert 65:439–446

Jaczewski Z (1954) The effect of changes in length of daylight on the growth of antlers in the deer (*Cervus elaphus* L). Folia Biol 2:133–143

Kann G (1971) Dosage radioimmunologique de la prolactine plasmatique chez les ovins. CR Acad Sci Paris 272:2808–2811

Lincoln GA (1985) Seasonal breeding in deer. In Fennessy P, Drew K (eds). The Biology of Deer Production. R Soc N Z Wellington Bull 22:165–179

Lincoln GA, Kay RN (1979) Effects of season on the secretion of LH and testosterone in intact and castrated red deer stags (*Cervus elaphus*). J Reprod Fert 55:75–80

Lincoln GA, Short RV (1980) Seasonal breeding: nature's contraceptive. Rec Prog Horm Res 36:1–52

Mirarchi RE, Howland BE, Scanlon PF, Kirkpatrick RL, Sanford LM (1978) Seasonal variation in plasma LH, FSH, prolactin and testosterone concentrations in adult male white-tailed deer. Can J Zool 56:121–128

Plotka ED, Seal US, Verme LJ, Ozoga JJ (1980) Reproductive steroids in deer. 3. Luteinizing hormone, estradiol and progesterone around estrus. Biol Reprod 26:258–262

Schams D, Barth D (1982) Annual profiles of reproductive hormones in peripheral plasma of the male roe deer (*Capreolus capreolus*). J Reprod Fert 66: 463–468

Sempéré A (1990) The annual antler cycle of the European roe deer (*Capreolus capreolus*) in relationship to the reproductive cycle. In Bubenik AB, Bubenik GA (eds) Horns and Antlers. Springer-Verlag, New York, pp 396–415

Sempéré A (1982) Fonction de reproduction et caractères sexuels secondaires chez le chevreuil (*Capreolus capreolus* L): Variations saisonnières et incidences sur l'utilisation du budget temps-espace. Thèse Doct. ès Sci, Université de Tours

Sempéré AJ, Boissin J (1983) Neuroendocrine and endocrine control of testicular activity and the antler cycle from birth to adulthood in the male roe deer (*Capreolus capreolus* L). In Brown RD (ed) Antler Development in Cervidae. Caesar Kleberg Wildlife Research Institute, Kingsville, Texas, pp 396–415

Sempéré AJ, Lacroix A (1982) Temporal and seasonal relationship between LH, testosterone and antlers in fawn and adult male roe deer. Acta Endocrinol Copenhagen 99:295–301

Sempéré AJ, Renaud G, Bariteau F (1989) Embryonic development measured by ultrasonography and plasma progesterone concentrations in roe deer (*Capreolus capreolus* L). Anim Reprod Sci 20:155–164

Sempéré A, Garreau JJ, Boissin J (1980) Variations saisonnières de l'activité de marquage territorial et de la testostéronémie chez le Chevreuil mâle adulte (*Capreolus capreolus*). CR Acad Sci Paris 290:803–806

Suttie JM, Simpson AM (1985) Photoperiodic control of appetite, growth, antlers and endocrine status of red deer. In Fennessy P, Drew K (eds) The Biology of Deer Production. R Soc N Z Wellington Bull 22:429–432

Suttie JM, Lincoln GA, Kay RNB (1984) Endocrine control of antler growth in red deer stags. J Reprod Fert 71:7–15

Van Mourik S, Stelmalsiak T, Outch KH (1985) Seasonal variation in plasma testosterone, luteinizing hormone concentrations and LH-RH responsiveness in nature, male rusa deer (*Cervus rusa timorensis*). Comp Biochem Physiol 83A:347–351

120

Antler Growth in Roe Deer: Effects of Casting and Correlations with Growth Hormone and Insulin Like Growth Factor I

Dieter Schams, Dietrich Barth, Eva-Maria Heinze-Mutz, Gabriele Pflaum, and Heinrich Karg

Abstract

We investigated the influence of casting, castration, growth hormone (GH), and insulin-like growth factor I (IGF-I) on antler growth in adult male roe deer (*Capreolus capreolus*). Casting was induced at different times of the year by antiandrogen (cyproterone acetate, 150 mg). Castration was performed at the end of August. A slow-release preparation of recombinant bovine GH (somedibove, 320 mg or 160 mg) was given during the normal antler growth period (January–March) or after induced casting during September–October. Casting was correlated with low steroid hormone levels, and antler growth was blocked by high endogenous steroid concentrations. Casting (induced any time of the year) seemed in itself to be a stimulus for new growth. If casting was blocked by temporarily elevated endogenous steroid levels, growth was delayed until the clear decrease of testosterone (December). Antler growth ceased at the end of March regardless of manipulated hormone levels. Normal antler growth was positively correlated with IGF-I levels. Growth hormone treatment stimulated IGF-I values but not antler growth. Local changes of hormones and receptors within the velvet may be more important for control of antler cycles than systemic changes in roe deer.

Key words: Antler, casting, castration, growth hormone, IGF-I, roe buck, season

Introduction

The roe deer is a seasonal breeder with a rutting season from mid-July to mid-August. Marked changes in testicular size, hormone secretion, behavior, and antler cycle during the year have been reported (Stieve 1950; Short and Mann 1966; Barth et al. 1976; Sempéré 1978; Sempéré and Boissin 1981; Schams and Barth 1982; Sempéré and Lacroix 1982; Schams et al. 1986). Antlers begin to develop at the end of December. Shedding of the velvet occurs soon after the testicular testosterone concentrations have risen in the spring, and casting of the antlers in late fall coincides with very low testosterone levels. In the roe buck, the function of the different hormones for antler growth is not well defined, but involvement of testosterone in mineralization of antlers, shedding the velvet, and maintenance and casting of antlers are evident (Schams et al. 1986).

Recently (Schams et al. 1986), it was shown that inhibition of testosterone and LH by an antiandrogen treatment (cyproterone acetate) during the antler growth period (late December to April) did not influence antler growth but clearly delayed shedding of velvet. Application of an estrogen implant 1 week after casting for 50 days blocked antler growth. It was concluded that concentrations of androgens and estrogens must be low for antler growth in the roe buck. However, for mineralization and shedding of the velvet, levels of testosterone must be high. More recently plasma insulin like growth factor I (IGF-I) concentrations correlate strongly and positively with rate of velvet antler growth and it has been suggested that this hormone is stimulating for antler growth (Suttie et al. 1985a). Evidence was given that the antler is a target for IGF-I (Suttie et al. 1988). Growth hormone (GH) is one of the most important factors stimulating IGF-I production, which occurs predominantly in the liver (Clemmons and Van Wyk 1984). Therefore a stimulating effect of GH on antler growth could be expected. The present study was conducted to further elucidate stimulatory effects on antler growth, especially possible influences of growth hormone and insulin-like growth factor I (IGF-I).

Materials and Methods

The bucks were born in captivity and housed individually or paired with a doe in enclosures. They were fed a standard pelleted diet containing 20% crude protein and hay ad libitum. Blood was withdrawn by venipuncture from the jugular vein of nonanesthesized animals between 1300 and 1400 h at weekly intervals. Blood was collected in heparinized tubes, immediately cooled, and centrifuged. The plasma was kept frozen until assayed.

Growth hormone, insulin like growth factor I, and insulin like growth factor II (IGF-II) levels were determined radioimmunologically as described by Schams et al. (1988). A homologous system against bovine GH was applied for determination of GH. Pituitary GH (USDA-bGH-I-1) was used for iodination by the iodogen method, and USDA-bGH-B-1 served as the reference preparation. Cross-reaction and parallelism with the bovine standards were checked by inhibition curves of an extract from roe deer pituitaries and plasma samples of roe deer, containing higher concentrations of endogenous hormone. These were parallel as shown for example for LH and FSH in pituitary extracts by Schams et al. (1980). The intraassay coefficient of variation (CV) estimated from control samples was less than 9%, and the interassay CV averaged 14%.

IGF-I was determined by RIA after acid-ethanol extraction of 50 μl blood plasma. Recombinant human IGF-I (Monsanto, St. Louis, Mo.) was used for iodination according to the lactoperoxidase method. Synthetic IGF-I (C H Li, San Francisco, Calif.) was used for calibration of a bovine pool standard reference preparation. The intra-assay CV and interassay CV were less than 8% and 15%, respectively.

IGF-II was determined by RIA after acid-ethanol extraction of 100 μl blood plasma. A monoclonal antibody against rat insulin like growth factor II (Serva-lab., Germany) was used. Cross reaction against human IGF-I, bovine insulin was less than 1%. Recombinant human IGF-II (Monsanto, St. Louis was used for iodination according to the lactoperoxidase method and as a reference preparation. The intrassay CV and interassay CV were less than 9% and 14%, respectively.

Testosterone was evaluated by RIA (Karg et al. 1976). Specificity of the original method was improved by using a highly specific antiserum raised against testosterone-11-hemisuccinate-BSA. There was a negligible cross-reaction with androstenolone and androstendione. Testosterone was determined directly using 20 μl plasma. Sensitivity of the assay was about 50 pg/ml. The interassay CV was about 7.5% to 12%.

Recombinant bovine somatotropin (somedibove), in a slow-release formulation (Eli Lilly, Elanco, Indianapolis), was injected at 320 mg s.c. at 2-week intervals or at 160 mg s.c. at 1-week intervals. Statistical analysis was conducted by ANOVA using the Statistical Analysis System (SAS, 1985).

In order to study the induction of antler casting by antiandrogen treatment, three adult bucks were injected i.m. with 150 mg Androcur (cyproterone acetate) during the hard antler period (high steroid and gonadotropin levels) beginning in May 1987 at 1-week intervals until casting. In addition, the seasonal changes of GH, IGF-I, and IGF-II were examined in blood collected from four adult bucks at weekly intervals from September 1985 until August 1986. To study the effect of exogenous growth hormone on growth of antlers, eight adult bucks were used (four controls and four for GH treatment). Recombinant bovine GH (320 mg s.c.) in a slow-release formulation was injected during the growth period (January–March 1988). Five injections were given at 2-week intervals.

Antler growth was measured with and without treatment of growth hormone at an aseasonal time after induced casting by antiandrogen treatment. Eight adult bucks were injected with 150 mg Androcur i.m. beginning at the end of August 1988 and thereafter at weekly intervals. All eight bucks cast antlers after three injections at the middle to end of September. One week after casting, four bucks were injected with 160 mg recombinant-derived bovine GH at weekly intervals (five injections). The other four bucks were treated with vehicle only and served as controls. Two bucks of the treatment group died during the experiment.

Finally, an experiment examined the effect of castration on antler growth, GH and IGF-I levels. Four bucks were castrated at the end of August and four intact bucks served as controls. One buck died in each group during the experiment.

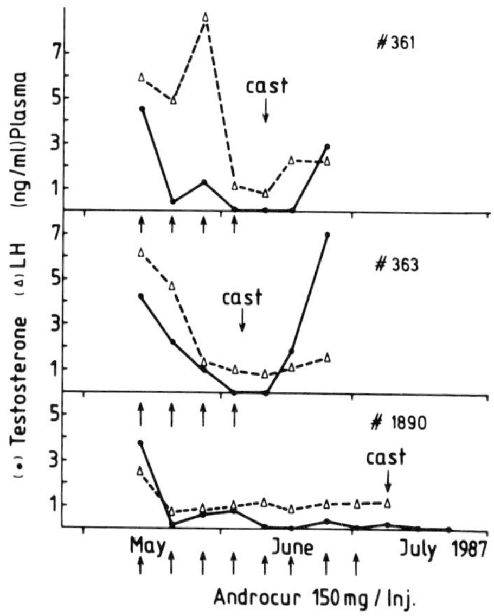

FIGURE 120.1. Circulating LH and testosterone in three roe bucks during and after treatment with an antiandrogen (Androcur). Arrows indicate Androcur injections.

Results

In the first experiment, casting occurred after four injections in two bucks and after eight injections in one buck. Concentrations of testosterone and LH decreased during treatment (Fig. 120.1). Casting occurred when testosterone levels were below 0.1 ng/ml. Testosterone values increased again in two bucks 2 to 3 weeks after the last injections, although concentrations were still low at 3 weeks after the end of treatment in one buck. New antler growth started at the regular time during December.

In the second experiment, there was high variability for GH between and within animals (Fig. 120.2) and no clear seasonal profile. There was a tendency for values to be lower from December until February, the period of antler growth. In contrast, IGF-I showed a distinct seasonal pattern with an increase in January and elevated levels until May followed by a decline. Concentrations of IGF-II did not change.

Antler growth of animals treated with GH started later than that of control bucks, whose growth began earlier in December (Fig. 120.3). Consequently, in January the antler length was longer in the controls; however, the growth rate was similar for both groups, and there was no difference in the ultimate antler size.

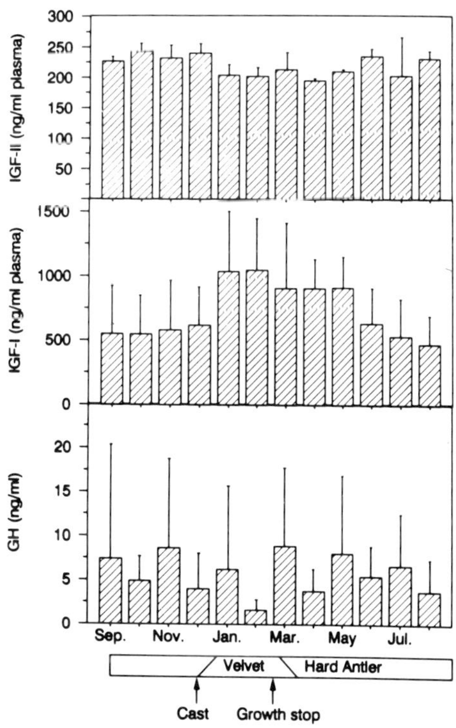

FIGURE 120.2. Antler growth and seasonal changes of growth hormone, IGF-I and IGF-II in the peripheral blood of four roe bucks (mean + SD).

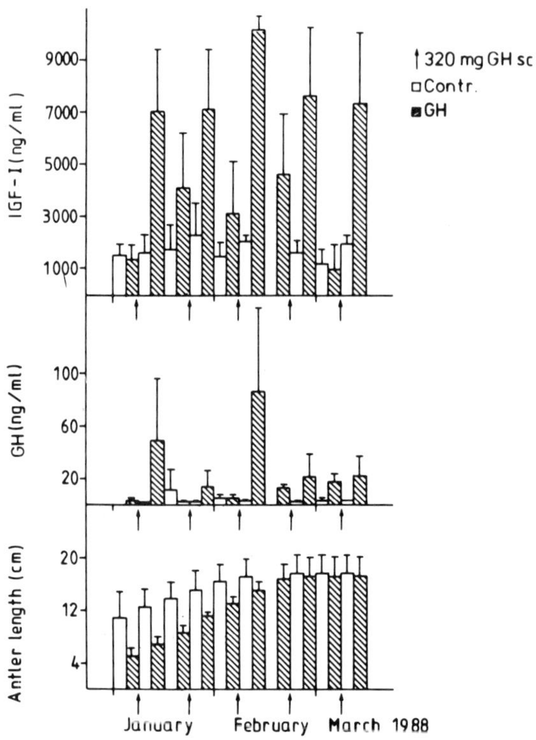

FIGURE 120.3. Antler growth and IGF-I and GH (mean + SD) in blood of four control roe bucks and four bucks injected with recombinant-derived bGH during the antler growth period. Arrows indicate injections (320 mg GH s.c.); □ controls; ▧ GH-injected bucks.

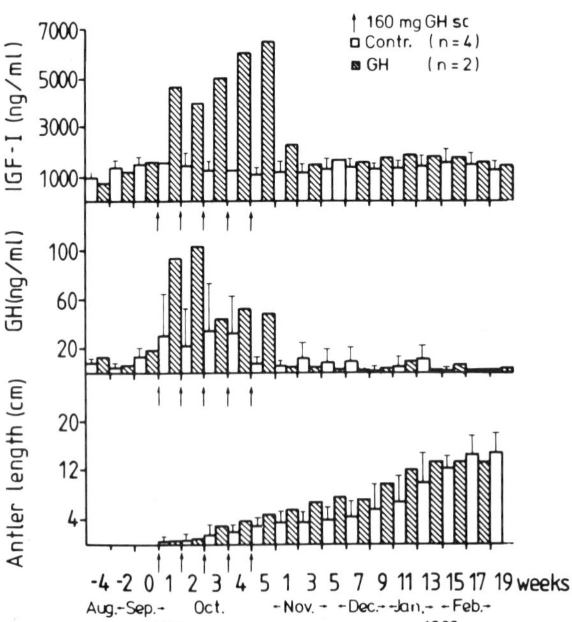

FIGURE 120.4. Antler growth and IGF-I and GH in blood of roe bucks after antler casting induced by antiandrogen treatment with and without injection of recombinant-derived GH at weakly intervals. Week 0 indicates time of casting; arrows indicate injections (160 mg GH s.c.); □, controls ($n = 4$); ▧ GH-injected bucks ($n = 2$).

In the deer first treated with Androcur, antler growth started slowly in both groups immediately after casting and was more pronounced during the normal growing season in December–February (Fig. 120.4). There was a tendency for faster growth in GH-treated bucks in November–January, and growth stopped earlier. GH and especially IGF-I were clearly elevated in the treated group. Growth hormone increased 7 days after each injection and decreased to control values 7 days later. There was a three- to nine-fold increase in IGF-I concentrations. It was concluded that ultimate antler size was not affected by GH treatment.

Antler growth of the castrated deer started at the end of September, 1 week after casting. In contrast, controls cast at the beginning of December. Antler growth terminated in castrated animals in January (apart from some limited growth on antler basis) and at the beginning of March in controls. However, there was no difference in final antler length between the two groups. There was a tendency for lower GH values in castrated bucks throughout the experiment (Fig. 120.5). Concentrations of IGF-I were relatively low in both groups and increased from January in controls and clearly during February–March in the experimental group.

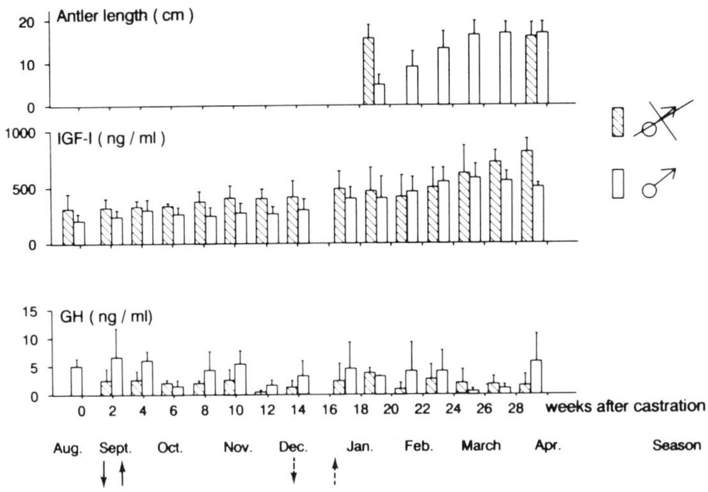

FIGURE 120.5. Antler growth and IGF-I and GH (mean + SD) in blood of three intact □ and three castrated roe bucks ▨. Solid arrow indicates casting and start of antler growth for castrated bucks and dashed arrow for control bucks.

Discussion

Minimal levels of testosterone are a stimulus for casting in roe deer and may be the stimulus for new antler growth. This hypothesis is supported by other authors (Sempéré and Boissin 1981, 1983; Sempéré and Lacroix 1982; Bubenik 1983; Goss 1983). The increase of testosterone after the end of antiandrogen treatment prevented new growth of antlers and confirms observations made by Goss (1969), Sempéré and Boissin (1983), and Sempéré et al. (1983). The seasonal cycle of intact bucks started as usual in December, when testosterone levels were again low. Further support that low testosterone levels are a stimulus for antler growth was shown after depression of testosterone with an antiandrogen at the end of testis activity (August–September) and after castration.

Only few data are available for GH levels in roe bucks. Sempéré and Boissin (1983) found low values from November to March (growing period of antlers) followed by an increase during April, a decrease thereafter, and then a further increase. No clear correlation of season and period of antler growth can be seen in our data. The high variation of GH levels obtained from single samples may reflect pulsatile release pattern as demonstrated in ruminants (Gluckman et al. 1987) and red deer (*Cervus elaphus*) (Suttie et al. 1985b). In contrast, levels of IGF-I increased clearly during the antler growth period. A similar elevation of IGF-I during the velvet antler growing phase was observed in red deer stags by Suttie et al. (1985a). In contrast, stimulation of IGF-I levels in blood by administration of exogenous GH did not stimulate the ultimate size of antlers in roe bucks. We conclude that there may be local changes of hormones and receptors within the velvet which are more important than systemic changes. A season-dependent and as yet unknown factor cannot be excluded since antler growth was more pronounced during January-March in our deer.

Acknowledgements

We are most grateful to the National Institute of Health, Bethesda and USDA Beltsville, USA, for gifts of bovine pituitary GH preparations (USDA-bGH-I-1 and USDA-bGH-B-1), P. Gluckman, Auckland, New Zealand for the supply of IGF-I antiserum, the late C.H. Li, San Francisco, and Monsanto Corp., St. Louis, for the supply of IGF-I.

References

Barth D, Gimenez T, Hoffmann B, Karg H (1976) Testosteron-konzentrationen im peripheren Blut beim Rehbock (*Capreolus capreolus*). Zeit Jagdwiss 22:134–148

Bubenik GA (1983) Shift of seasonal cycle in white-tailed deer by oral administration of melatonin. J Expt Zool 225:155–156

Clemmons DR, Van Wyk JJ (1984) Factors controlling blood concentration of somatomedin C. J Clin Endocrin Metabolism 13:113–119

Gluckman PD, Breier BH and Davis SR (1987) Physiology of the somatotropic axis with particular reference to the ruminant. J Dairy Sci 70:442–466

Goss R (1983) Deer Antlers. Regeneration, Function and Evolution. Academic Press, New York

Goss R (1969) Photoperiodic control of antler cycles in deer. 2. Alterations in amplitude. J Expt Zool 171:223–234

Karg H, Gimenez T, Hartl H, Hoffmann B, Schallenberger E, Schams D (1976) Testosterone, luteinizing hormone (LH) and follicle stimulating hormone (FSH) in peripheral plasma of bulls: levels from birth through puberty and short term variations. Ztb Veterinärmed A23:793–803

SAS (1985) SAS User's Guide: Statistics, SAS Institute, Cary, North Carolina

Schams D, Barth D (1982) Annual profiles of reproductive hormones in peripheral plasma of the male roe deer (*Capreolus capreolus*). J Reprod Fert 66:463–468

Schams D, Winkler U, Schallenberger E, Karg H (1988) Wachstumshormon- und insulin-likc growth factor I (Somatomedin C)-Blutspiegel bei Rindern von der Geburt bis nach der Pubertät. Deutsche tierärztliche Wochenschrift 95:360–362

Schams D, Barth D, Karg H (1986) Is antler growth in adult roe deer (*Capreolus capreolus*) influenced by treatment with estradiol-17β, an antiandrogen, or a prolactin inhibitor? In Assenmacher I, Boissin J (eds) Endocrine Regulations as Adaptive Mechanisms to the Environment. Editions du Centre National de la Recherche Scientifique, 15, quai Anatole-France-75700 Paris, pp 475–481

Schams D, Barth D, Karg H (1980) LH, FSH and progesterone concentrations in peripheral plasma of the female roe deer (*Capreolus capreolus*) during the rutting season. J Reprod Fert 60:109–114

Sempéré AJ (1978) The annual cycle of plasma testosterone and territorial behavior in the roe deer. In Assenmacher I, Farner D (eds) Environmental Endocrinology. Springer-Verlag, Berlin, pp 73–74

Sempéré AJ, Boissin J (1983) Neuroendocrine and endocrine control of the antler cycle in roe deer. In Brown RD (ed) Antler Development in Cervidae. Caesar Kleberg Wildlife Research Institute, Kingsville, Texas, pp 109–122

Sempéré AJ, Boissin J (1981) Relationship between antler development and plasma androgen concentrations in adult roe deer (*Capreolus capreolus*). J Reprod Fert 62:49–53

Sempéré AJ, Lacroix A (1982) Temporal and seasonal relationships between LH, testosterone and antlers in fawn and adult male roe deer (*Capreolus capreolus*). Acta Endocrinol 99:295–301

Sempéré AJ, Boissin J, Dutourne B, Lacroix A, Blanc MR, Ravault JP (1983) Variations de la concentration plasmatique en prolactine, LH, et FSH et de l'activité testiculaire au cours de la première année de vie chez le chevreuil (*C. capreolus*). Gen Comp Endocrinol 52:247–254

Short RV, Mann T (1966) The sexual cycle of a seasonally breeding mammal, the roe buck (*Caprcolus caprcolus*). J Reprod Fert 12:337 351

Stieve H (1950) Anatomisch-biologische Untersuchungen über die Fortpflanzungstätigkeit des europäischen Rehes. Z mikroskopisch-anatomische Forsch 55:427–530

Suttie JM, Fennessy PF, Gluckman PD, Corson ID (1988) Elevated plasma IGF 1 levels in stags prevented from growing antlers. Endocrinology 122:3005–3007

Suttie JM, Gluckman PD, Butler JH (1985a) Insulin-like growth factor I (IGF-I), antler-stimulating hormone? Endocrinology 116:846–848

Suttie JM, Fennessy PF, Crosbie SF, Butler JH, Gluckman PD, Corson ID, Laas FJ (1985b) Growth hormone release in relation to IGF 1 level in young red deer stags. Proc Aust Soc Endocrinol Suppl 2:37

121
Studies on Cells Derived from Growing Deer Antler

Colin Gray, M. Louise Taylor, Michael A. Horton, Andrew S.I. Loudon, and Timothy R. Arnett

Abstract

The aim of this study was to characterize cells cultured from growing deer antler tissue. Positive immunocytochemical staining of multinucleate cells in isolated cell preparations or tissue imprints from a 20-cm fallow deer (*Dama dama*) antler was seen with the osteoclast-recognizing monoclonal antibodies 23C6 and 13C2. A quantitative in vitro resorption assay was established using cells disaggregated from small fragments of midportion axis deer (*Axis axis*) mineralizing antler tissue incubated on slices of devitalized sperm whale dentin. Antler osteoclasts were highly active in this system, leaving extensive resorption trails after 24-h culture. Resorption, assessed morphometrically, was strongly inhibited by synthetic salmon calcitonin. Cultures of proliferating cells were derived from outgrowths of tissue fragments explanted from the mid- and tip regions of growing antler from Père David's deer (*Elaphorus davidianus*) and fallow deer. These cells exhibited low alkaline phosphatase activity but positive immunocytochemical staining with an antibody recognizing human Von Willebrand Factor (factor VIII-related antigen), indicating that cultures were possibly endothelial rather than osteoblastic in origin. Cell proliferation was strongly inhibited by 1,25 dihydroxyvitamin D_3 and retinoic acid and, to a lesser extent, by testosterone and 17β-estradiol.

Key words: Antler, calcitonin, cell culture, endothelial, retinoic acid, testosterone, 1,25 dihydroxyvitamin D

Introduction

Relatively little is known about the cell biology of deer antler and the factors regulating the remarkable rates of growth seen in this mineralized tissue. Investigation of the endocrine control of antler function has been largely confined to in vivo studies. Among the factors which may be implicated in the regulation of antler growth are androgens (Bubenik 1983; Morris and Bubenik 1983; Brown et al. 1983; Suttie et al. 1984), parathyroid hormone and calcitonin (Chao et al. 1984), 1,25 dihydroxyvitamin D_3 (Eiben et al. 1984; Van der Eems et al. 1988), thyroid hormones (Brown et al. 1983; Bubenik et al. 1987), and insulin-like growth factors (Suttie et al. 1989).

Cell culture methods have become widely used in bone cell biology but, surprisingly, antler has been somewhat neglected in this regard, with the exception of recent work by Ramirez and Brown (1987) and Sempéré and colleagues (1989). In this study, we examined the hormonal regulation of cells cultured from growing antler using simple functional approaches.

Materials and Methods

Materials

Fresh antler tissue was obtained from fallow deer, Père David's deer, and axis deer culled at Whipsnade Zoo, Bedfordshire, UK. Whole antlers were briefly immersed in 70% ethanol before removal of the epidermal and dermal velvet tissue and excision of tissues for culture. Tissue culture media were obtained from Gibco (Paisley, UK) and other reagents came from Sigma (Poole, UK). Tissue culture ware was purchased from Falcon (Oxford, UK). Rabbit anti-human Von Willebrand factor (factor VIII-related antigen) serum was purchased from Dakopatts (Glostrup, Denmark). Sperm whale dentin, cut into polished slices $10\,mm^2 \times 0.2\,mm$ thick using a low-speed diamond saw (Buehler, Coventry, UK) was provided by A. Boyde. Synthetic salmon calcitonin (sCT, 4673 IU/mg) was a gift of D. Cockayne of Rorer Health Care, Eastbourne, UK, and 1,25 dihydroxyvitamin D_3 ($1{,}25(OH)_2D_3$) was provided by K.W. Colston (St. George's Hospital Medical School, London).

Osteoclast Culture

For osteoclast disaggregation, a few small pieces of tissue (approximately 2-mm cubes) were excised from the annular region of forming bone 10 to 15 cm below the growing antler tip and rapidly chopped with a scalpel blade in 2 ml of phosphate buffered saline (PBS). The tissue fragments were triturated a few times using a wide-bore polyethylene transfer pipette, and cells were allowed to sediment from the resultant suspension for 60 min onto dentin (Boyde et al. 1985) or plastic surfaces. Adherent cells were washed vigorously with PBS, leaving a mixed population, enriched for osteoclasts. In resorption experiments, cells were incubated for 24 h in medium 199 containing 0.7 g/l (rather than 2.2 g/l) sodium bicarbonate, supplemented with 10% fetal calf serum (FCS), penicillin (100 IU/ml), streptomycin (100 μg/ml), amphotericin (0.25 μg/ml), and L-glutamine (2 m*M*) in a humidified 5% CO_2 atmosphere. The reduced bicarbonate concentration generated a medium pH of 6.9 to 7.0, previously shown to promote resorptive function of osteoclasts in vitro (Arnett and Dempster 1986, 1987). Following fixation for 5 min in 3% glutaraldehyde/PBS, tartrate-resistant acid phosphatase-positive multinucleate osteoclasts were counted on each dentin slice, and the cells were then stripped off. Encoded slices were stained with toluidine blue, and the plane surface area of each discrete resorption pit was assessed "blind" by point counting on a stereological grid using video output from an Olympus IMT-2 inverted microscope with bright field illumination. Data were evaluated using one-way analysis of variance. Time lapse, phase contrast observations were made on cells cultured in bicarbonate-free medium 199/10% FCS in plastic dishes, using the same microscope equipped with a 37° C incubation chamber.

Proliferating Cell Culture

Cultures of proliferating cells were derived from explanted fragments of PBS-washed tissue excised from various regions of developing antler. Cells were grown initially in Dulbecco's minimum essential medium, without the estrogenic indicator dye phenol red (Berthois et al. 1986), supplemented with L-glutamine, penicillin/streptomycin/amphotericin (DMEM), and 10% FCS, in a humidified 5% CO_2 atmosphere. For quantitative studies, cells at passages two to three were plated into 24 multiwells at an initial density of 10^4/well and cultured for 24 h in DMEM with 5% charcoal-stripped FCS. Cells were then cultured for a further 3 to 6 days in the presence of hormone or ethanol vehicle; in the case of 6-day cultures, media were changed after 3 days. Cell proliferation was estimated as ^3H-thymidine incorporation into DNA over the final 6 h of culture according to the method of Beresford et al. (1986) Data were evaluated by one-way analysis of variance.

Cell Staining

Tissue imprints and isolated cell preparations from developing antler tissue were stained on glass slides using the osteoclast-recognizing monoclonal antibodies 13C2 and 23C6 (Horton and Davies 1989; Davies et al. 1990) using the peroxidase-antiperoxidase method, with diaminobenzidine visualization. Proliferating cultures were stained using rabbit anti-human Von Willebrand factor (factor VIII-related antigen) serum (peroxidase-antiperoxidase/diaminobenzidine method). Alkaline phosphatase (ALP) and tartrate-resistant acid phosphatase (TRAP) staining were performed using standard azo dye-based kits (Sigma).

Results

Osteoclast Culture

Routine histology (Fig. 121.1) and immunocytochemical staining of tissue imprints (Fig. 121.2) revealed that osteoclasts were abundant in the regions of forming trabeculae below the soft tip of growing antler. Tissue from this region proved to be a rich source of functional disaggregated osteoclasts. Antler osteoclasts cultured on dentin slices produces extensive resorption tracks, suggesting that many cells were excavating continuously during the experiments. Motility (Fig. 121.3) and dentin resorption (Fig. 121.4) were both inhibited by salmon calcitonin; however, not all TRAP-positive multinucleate cells showed the characteristic cytoplasmic retraction response to sCT.

Proliferating Cell Culture

Routine histology (Fig. 121.1) and cytochemical staining of frozen sections showed active, alkaline phosphatase-positive osteoblasts lining forming bone surfaces in developing antler. Cells proliferated rapidly from explanted antler tissue fragments but growth generally ceased effectively after about 4 to 6 weeks in culture. Alkaline phosphatase activity was very low, as evidenced by quantitative assay of cell layers and the sparse numbers of cells exhibiting positive cytochemical staining for this enzyme; testosterone exerted a small stimulatory effect on ALP activity (data not shown). However, proliferating cells stained positively using an antibody recognizing human Von Willebrand factor, a marker for the

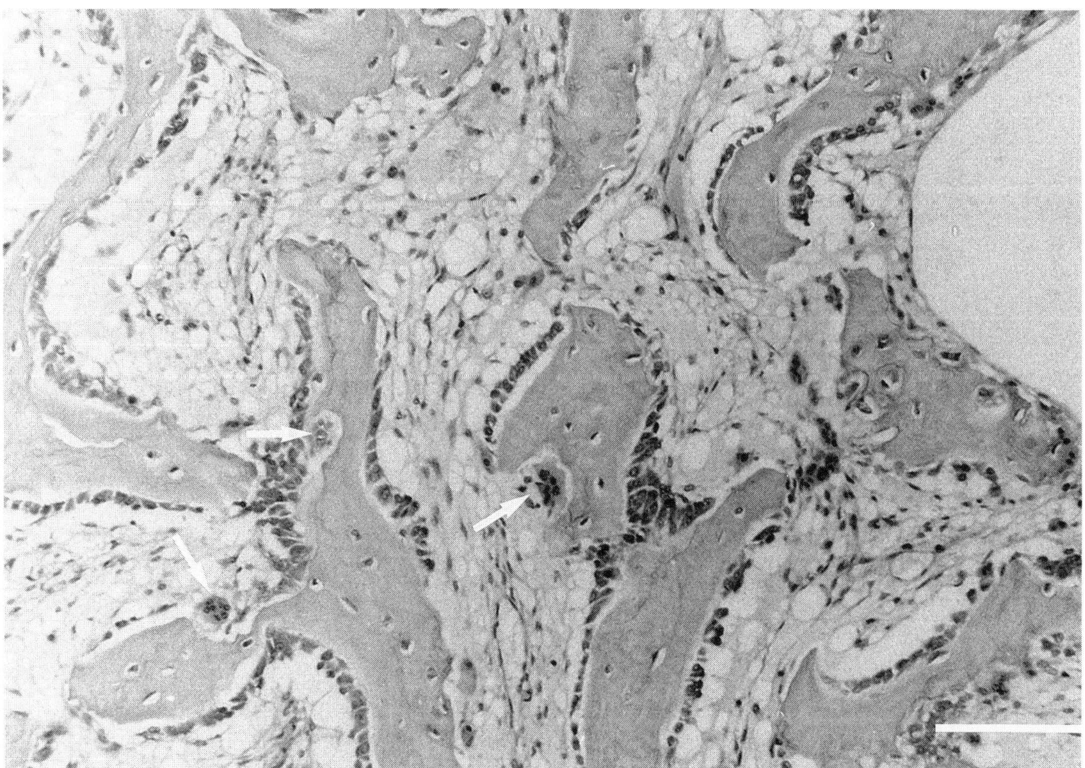

FIGURE 121.1. Transverse section (hematoxylin and eosin) across the midregion of a 20-cm developing fallow deer antler in the region of forming trabeculae, showing active osteoclasts (arrowed), osteoblasts and part of a large vascular channel; scale bar, 100 μm.

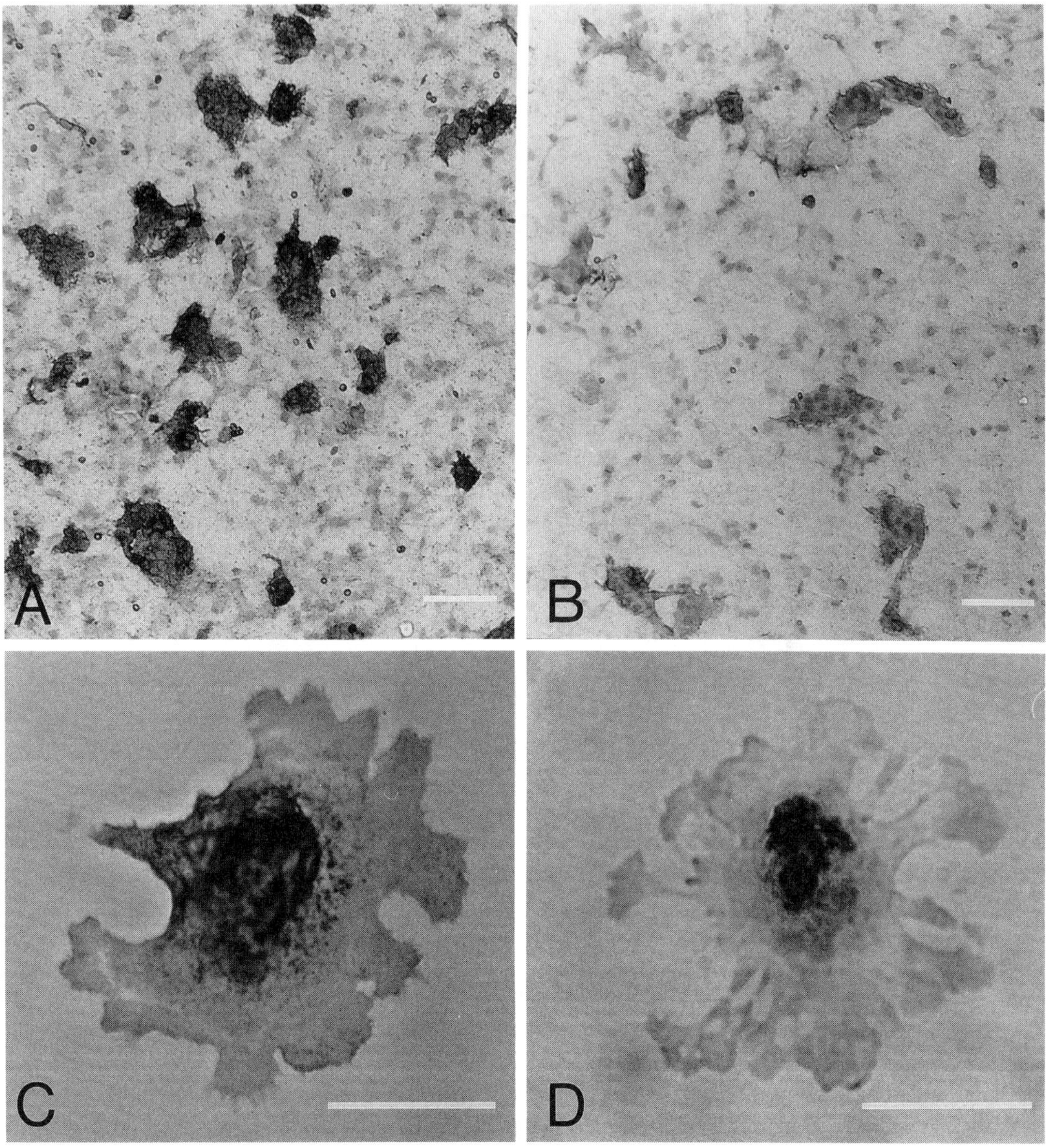

FIGURE 121.2A–D. Positive immunocytochemical staining of multinucleate cells in tissue imprints (A, B) and isolated cell preparations (C, D) from the midregion of a 20-cm developing fallow deer antler using the osteoclast-recognizing monoclonal antibodies 23C6 (A, C) and 13C2 (B, D); scale bar 50 μm.

endothelial phenotype, in contrast with trabecular bone cells (Fig. 121.5). The proliferation of antler-derived cells was generally strongly inhibited by $1,25(OH)_2D_3$ and, to a lesser extent, by testosterone (Fig. 121.6). Significant growth inhibition was additionally observed in cultures treated with 1 n*M* 17β-estradiol; the magnitude of the response was similar to the inhibition produced by 10 n*M* testosterone (data not shown). Retinoic acid also produced striking, dose-dependent inhibitions of cell proliferation (Fig. 121.7); the decline in basal ^{3}H-thymidine

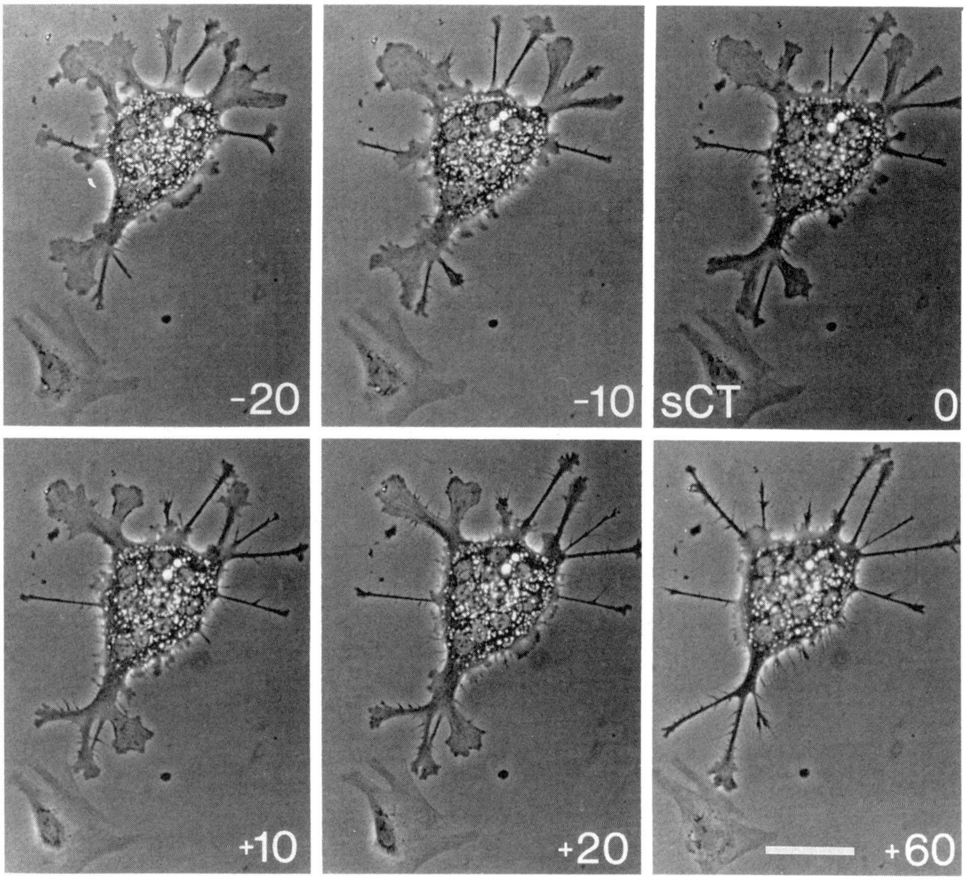

FIGURE 121.3. Time lapse sequence illustrating inhibition of membrane ruffling activity in an axis deer antler osteoclast (note adjacent endothelial/fibroblast-like cell) cultured on plastic, following addition of synthetic salmon calcitonin (sCT) to a final concentration of 10 ng/ml. Elapsed time is indicated in minutes; scale bar, 50 μm.

incorporation between days 3 and 6 in this experiment reflects the limited proliferative lifespan of antler-derived cells (see above).

Discussion

The studies described here represent the first investigation of the regulation of antler cell function in vitro. Developing antler proved to be an abundant source of active osteoclasts for in vitro study. In common with other mammalian osteoclasts studied to date (Chambers et al. 1985; Arnett and Dempster 1986, 1987; Nicholson et al. 1986), antler osteoclasts were strongly inhibited by calcitonin, suggesting a possible physiologic role for this hormone in antler development. Elevated serum calcitonin levels have been reported in white-tailed deer (*Odocoileus virginianus*) during the rapid antler growth period (Chao et al. 1984).

Proliferating cultures derived from explanted fragments of antler tissue invariably exhibited low alkaline phosphatase activity, in contrast with the situation for human trabecular bone explants (Beresford et al. 1986; Colston et al. 1989). The positive staining observed for Von Willebrand factor in cultured antler cells is thus suggestive of a predominantly endothelial, rather than osteoblastic phenotype for these cells. This finding may reflect a greater proliferation rate

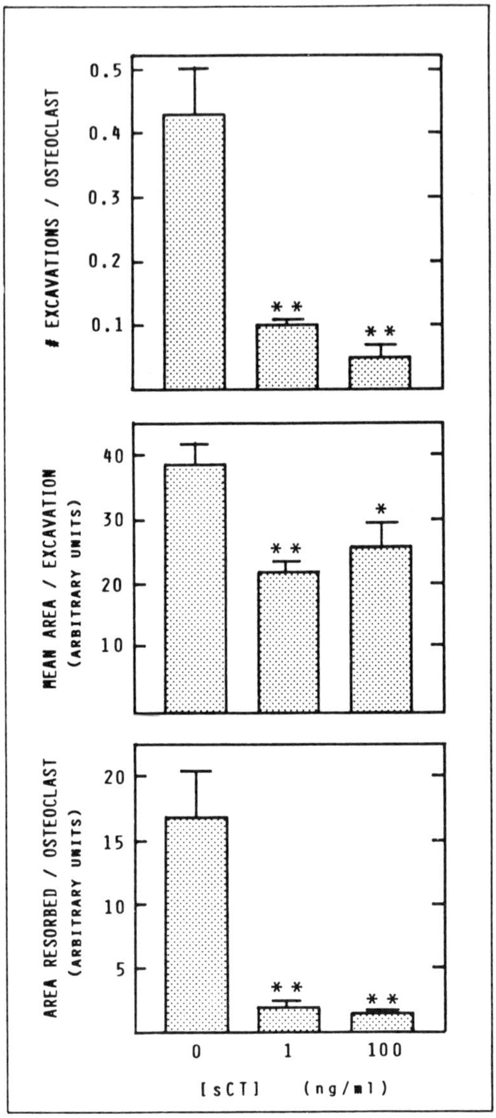

FIGURE 121.4. Inhibition of resorptive activity of fallow deer antler osteoclasts cultured for 24 h on sperm whale dentine slices by synthetic salmon calcitonin (sCT). One area unit = 103 μm^2. Values are means $\pm$ SEM ($n = 4$ slices/treatment; mean number of pits/slice for controls = 90 $\pm$ 18); $*P < 0.05$, $**P < 0.01$.

or, possibly, migratory capability for endothelial versus osteoblastic populations in developing antler. Deer antler offers an interesting model for investigation of the vascular control of bone growth.

Cells cultured from growing antler showed varying responses to steroid hormones. Cell proliferation was inhibited to a modest extent by testosterone or 17β-estradiol. This finding accords with the cessation of antler growth in vivo associated with the rise in circulating testosterone levels at the onset of the rutting season (Bubenik 1983; Brown et al. 1983; Suttie et al. 1984), but contrasts with the proliferative, anabolic actions reported for gonadal steroids on skeletal bone-derived cells (Vaishnav et al. 1988; Ernst et al. 1989; Kasperk et al. 1989). With regard to the testosterone effects observed here, the possibility that aromatization to estrogen occurred in culture cannot be ruled out, and further studies with non-metabolizable androgen analogs will be required.

The proliferation of most cultures was generally strongly inhibited by 1,25 dihydroxyvitamin D_3 and retinoic acid, findings which are in accordance with the well-recognized antiproliferative, differentiation-inducing actions of these compounds in other cell systems. Two recent studies are of particular interest with regard to the $1,25(OH)_2D_3$ effect. First, the presence of the $1,25(OH)_2D_3$ receptor, together with $1,25(OH)_2D_3$ biosynthesis has been demonstrated in bovine aortic endothelial cells (Merke et al. 1989). Second, cultured antler velvet tissue and cells have also been shown to synthesize $1,25(OH)_2D_3$ (Sempéré et al. 1989). Circulating $1,25(OH)_2D_3$ levels in deer peak markedly during the later phase of antler growth (Eiben et al. 1984; Van der Eems et al. 1988; Sempéré et al. 1989). Taken together, these data suggest that $1,25(OH)_2D_3$ may be acting as both a systemic and local regulator of antler development, inhibiting vascular growth and promoting mineralization in the final stages of antler growth. Future studies will also need to address the roles of locally acting growth factors and cytokines in antler development.

Acknowledgements

The authors would like to thank S. Jones for original suggestions which led to this work. This work was supported in part by grants from the Nuffield Foundation and Research into Ageing.

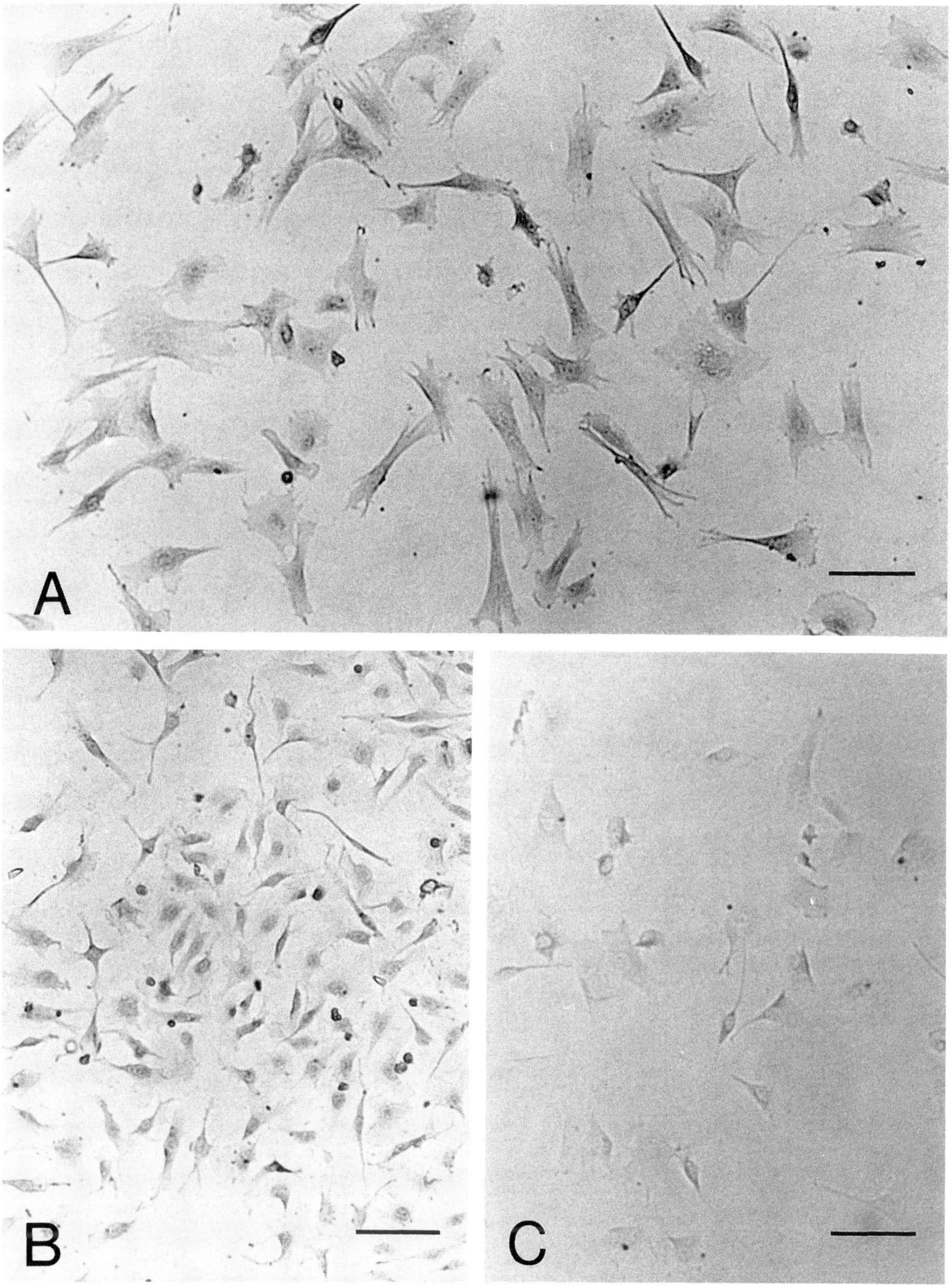

FIGURE 121.5A–C. Positive immunocytochemical staining using anti-Von Willebrand factor serum in fallow deer antler midregion-derived cells (A) and bovine aortic endothelial cells (B); negative staining in cells cultured from autologous vertebral trabecular bone explants (C). Scale bar, 100 μm.

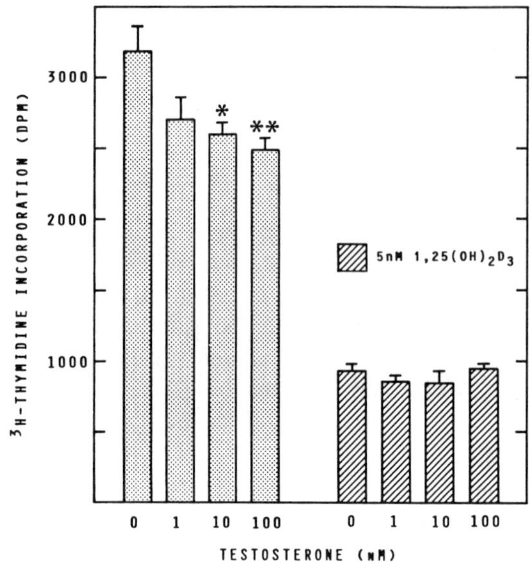

FIGURE 121.6. Effect of $1{,}25(OH)_2D_3$ and testosterone on proliferation of cells derived from fallow deer antler midregion (15 cm below growing tip) after 6 days in culture. Values are means $\pm$ SEM ($n = 6$); $*P < 0.05$, $**P < 0.01$. (For clarity, significance levels associated with the strongly inhibitory effect of $1{,}25(OH)_2D_3$ are omitted.)

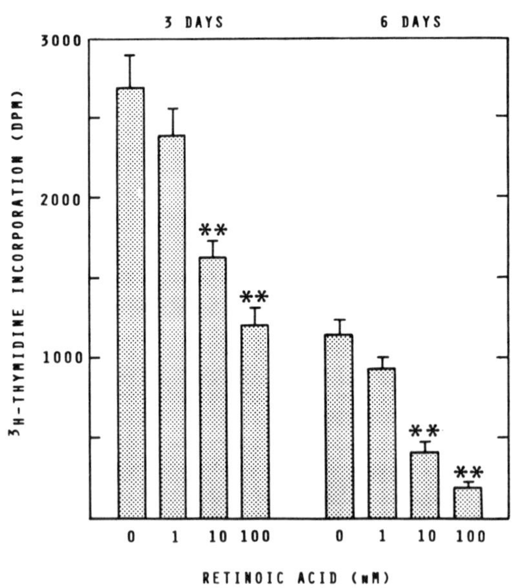

FIGURE 121.7. Effect of retinoic acid on proliferation of cells derived from fallow deer antler mid-region (15 cm below growing tip) after 3 and 6 days in culture. Values are means $\pm$ SEM ($n = 6$); $**P < 0.01$.

References

Arnett TR, Dempster DW (1987) A comparative study of disaggregated chick and rat osteoclasts in vitro: effect of calcitonin and prostaglandins. Endocrinology 120:602–608

Arnett TR, Dempster DW (1986) Effect of pH on bone resorption by rat osteoclasts in vitro. Endocrinology 119:119–124

Beresford JN, Gallagher JA, Russell RGG (1986) 1,25-dihydroxyvitamin D_3 and human bone-derived cells in vitro: effects on alkaline phosphatase, type 1 collagen and proliferation. Endocrinology 119:1776–1785

Berthois Y, Katzenellenbogen JA, Katzenellenbogen BS (1986) Phenol red in tissue culture media is a weak estrogen: implications concerning the study of estrogen-responsive cells in culture. Proc Natl Acad Sci USA 83:2496–2500

Brown RD, Chao CC, Faulkner LW (1983) Hormone levels and antler development in white tailed and sika fawns. Comp Biochem Physiol 75:385–390

Boyde A, Ali NN, Jones SJ (1984) Resorption of dentin by isolated osteoclasts in vitro. Br Dent J 156:216–220

Bubenik GA (1983) The endocrine regulation of the antler cycle. In Brown RD (ed) Antler Development in Cervidae. Caesar Kleberg Wildlife Research Institute, Kingsville, Texas, pp 73–107

Bubenik GA, Sempere AJ, Hamr J (1987) Developing antler, a model for endocrine regulation of bone growth. Concentration and gradient of T_3, T_4 and alkaline phosphatase in the antler, jugular and the saphenous veins. Calcif Tissue Int 41:38–43

Chambers TJ, McSheehy PMJ, Thomson BM, Fuller K (1985) The effect of calcium regulating hormones and prostaglandins on bone resorption by osteoclasts disaggregated from neonatal rabbit bones. Endocrinology 116:234–239

Chao CC, Brown RD, Deftos LJ (1984) Seasonal levels of serum parathyroid hormone, calcitonin and alkaline phosphatase in relation to antler cycles in white-tailed deer. Acta Endocrinol 106:234–240

Colston KW, King RJB, Hayward J, Fraser DI, Horton MA, Stevenson JC, Arnett TR (1989) Estrogen receptors and human bone cells: immunocytochemical studies. J Bone Min Res 4:625–631

Davies J, Warwick J, Totty, N, Philp R, Helfrich M, Horton MA (1990) The osteoclast functional antigen, implicated in the regulation of bone resorption, is biochemically related to the vitronectin receptor. J Cell Biol 109:1817–1826

Eiben B, Scharla S, Fischer K, Schmidt-Gayk H (1984) Seasonal variations of serum 1,25-dihydroxyvitamin D_3 and alkaline phosphatase in relation to the antler formation in the fallow deer (*Dama dama* L.). Acta Endocrinol 107:141–144

Ernst M, Heath JK, Rodan GA (1989) Estradiol effects on proliferation, messenger ribonucleic acid for collagen and insulin-like growth factor, and parathyroid hormone-stimulated adenylate cyclase activity in osteoblastic cells from calvariae and long bones. Endocrinology 125:825–833

Horton MA, Davies J (1989) Adhesion receptors in bone. J Bone Min Res 4:803–807

Kasperk CH, Wergedal JE, Farley JR, Linkhart TA, Turner, RT, Baylink DJ (1989) Androgens directly stimulate proliferation of bone cells in vitro. Endrocrinology 124:1576–1478

Merke J, Milde P, Lewicka S, Hugel U, Klaus G, Mangelsdorf DJ, Haussler MR, Rauterberg EW, Ritz E (1989) Identification and regulation of 1,25-dihydroxyvitamin D_3 receptor activity and biosynthesis of 1,25-dihydroxyvitamin D_3: studies in cultured bovine aortic endothelial cells and human dermal capillaries. J Clin Invest 83:1903–1915

Morris JM, Bubenik GA (1983) The effect of androgens on the development of antler bone. In Brown RD (ed) Antler Development in Cervidae. Caesar Kleberg Wildlife Research Institute, Kingsville, Texas, pp 123–141

Nicholson GC, Moseley JM, Sexton PM, Mendelsohn FAO, Martin TJ (1986) Abundant calcitonin receptors in isolated rat osteoclasts: biochemical and autoradiographic characterization. J Clin Invest 78:355–360

Ramirez V, Brown RD (1987) A technique for the in vitro incubation of deer antler tissue. Comp Biochem Physiol 89A:279–281

Sempéré AJ, Grimberg R, Silve C, Tau C, Garabedian M (1989) Evidence for extrarenal production of 1,25-dihydroxyvitamin D during physiological bone growth: in vivo and in vitro production by deer antler cells. Endocrinology 125:2312–2319

Suttie JM, Fennessy, PF, Corson ID, Laas FJ, Crosbie SF, Butler JH, Gluckman PD (1989) Pulsatile growth hormone, insulin-like growth factors and antler development in red deer (*Cervus elaphus scoticus*) stags. J Endocrinol 121:351–360

Suttie JM, Lincoln GA, Kay, RNB (1984) Endocrine control of antler growth in red deer stags. J Reprod Fert 71:7–15

Vaishnav R, Beresford JN, Gallagher JA, Russell RGG (1988) Effects of the anabolic steroid stanozolol on cells derived from human bone. Clin Sci 74:455–460

Van der Eems KL, Brown RD, Gundberg CM (1988) Circulating levels of 1,25 dihydroxyvitamin D_3, alkaline phosphatase, hydroxyproline and osteocalcin associated with antler growth in white-tailed deer. Acta Endocrinol 118:407–414

122
Antler Cycles of a White-Tailed Deer with Congenital Anophthalmia

Harry A. Jacobson and Steven J. Waldhalm

Abstract

A male white-tailed deer (*Odocoileus virginianus*) with anophthalmia was captured as a fawn on 3 November 1981 at approximately 6 weeks of age and raised in the research facilities at Mississippi State University; events of his antler cycle were recorded. Between 1982 and 1986, he had an average of 373 (range, 370 to 380) days between velvet shedding dates. Antler casting dates averaged 378 (range, 368 to 391) days apart. Treatment with a melatonin-blocking agent (atenolol) in March and April 1987 reduced testosterone levels and shortened his antler casting cycle to 336 days and velvet shedding cycle to 352 days apart for the 1987 events. Antler casting in 1988 returned to a 380-day cycle. A melatonin implant immediately following molt to summer coat on 8 July 1989 had no apparent effect on subsequent molting and velvet shedding. Results support presence of an endogenous circannual rhythm of antler regulation and that ophthalmic innervation is required for the pineal to adjust the endogenous rhythm to match photoperiod cues.

Key words: Anophthalmia, white-tailed deer, antler cycle, endogenous rhythm

Introduction

On 3 November 1981, a male white-tailed deer fawn with congenital anophthalmia was captured in Copiah County, Mississippi (Fig. 122.1). The fawn had survived for some time in the wild, as it was estimated to be 6 weeks or more of age and in excellent health. The blind deer presented a unique opportunity to examine the role of retinal light reception on regulation of seasonal cycles.

Materials and Methods

The deer was maintained in a 7 m × 7 m outdoor pen on natural light. The pen was in close proximity to other captive deer pens, and during summer and fall months, fawns were housed in an adjoining pen. The deer was bottle-reared to about 4 months of age and then weaned to a 16% protein commercial dairy cattle pellet provided ad libitum.

Velvet shedding dates and antler casting dates were recorded annually. From capture until 27 January 1987, no treatment was imposed. From 27 January 1987 to 6 April 1987, weekly 10 cc blood samples were collected by jugular venipuncture, and serum was stored frozen. Testosterone levels were later determined by radioimmunoassay (Coker and Davidson 1978). From 19 to 21 March 1987 and 1 to 5 April 1987, 50 mg of atenolol, a melatonin blocking agent (Cowen et al. 1983, 1985), was daily administered between 1500–1700 h. Blood samples were collected at each dosing session and the day following the last dosage. At the start of treatment, the buck was in hard antlers and in physical condition corresponding to that of peak rut for normal intact males.

On 6 April 1987, the buck was returned to his

FIGURE 122.1. Male white-tailed deer fawn with congenital anophthalmia, soon after capture.

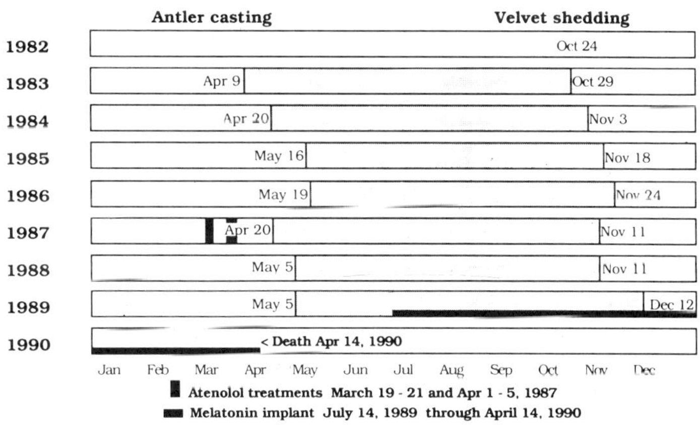

FIGURE 122.2. Antler cycles of an anopthalmic white-tailed buck, 1982–90.

pen, and no further treatments were administered until 7 July 1989. At that time, he was sedated with 100 mg xylazine hydrochloride, and a melatonin implant was administered (Jacobson and Waldhalm, this volume). At the time of the implant, the deer had unbranched antlers of less than 9 cm length and had just completed his vernal molt. On 14 April 1990, the experiment was prematurely terminated when feral dogs killed the deer.

Results

Velvet shedding dates and antler casting dates of the blind deer prior to 1987 indicated photoperiod or other environmental cues were not synchronizing these events (Fig. 122.2). An average period of 373 days elapsed between velvet shedding dates and 378 days between antler casting. A shift in velvet shedding occurred from 24 October 1982 to 24 November 1986,

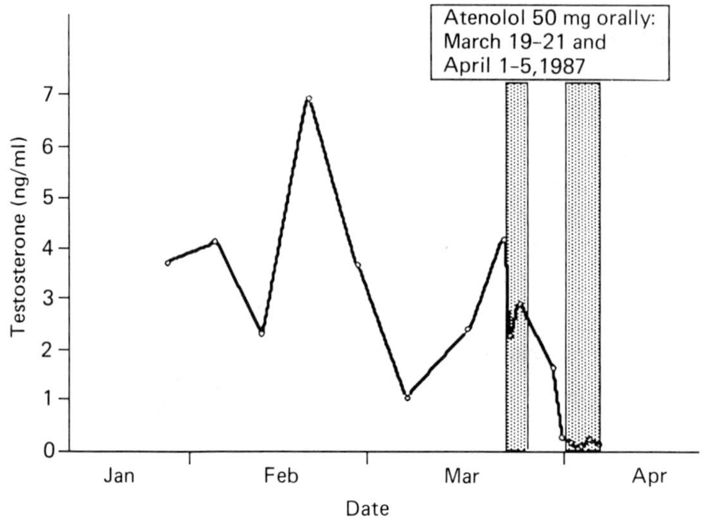

FIGURE 122.3. Blood testosterone levels in an anophthalmic white-tailed buck 27 January to 6 April 1987.

and antler casting shifted from 6 April 1983 to 19 May 1986. General observations on molting indicated vernal and autumn molts also shifted to later occurrences in the same general way that antler cycles changed.

Atenolol treatment effectively changed the antler casting dates in 1987. Testosterone levels declined upon atenolol treatment (Fig. 122.3), and this was associated with antler casting on 20 April, 30 days earlier than the previous year. Testosterone withdrawal induces antler casting (Wislocki et al. 1947; Bubenick 1982). After casting antlers in 1987, the blind buck had an extended antler growth period, and velvet was shed on 11 November 1987. For 2 years after atenolol treatment, antler casting and velvet shedding dates were identical. However, antler velvet shedding in 1989 demonstrated the blind buck was apparently regaining the same antler pattern he had prior to atenolol treatment.

Unlike atenolol administration, melatonin treatment had little apparent effect on the timing of antler events. Velvet shedding was consistent with the first five antler cycles for this buck showing longer than a 1-year period between the same events, in this case, 392 days. Likewise, molting from vernal to autumn coat showed this same delay with autumn molt not being completed until the end of November.

The death of the animal prevented the observation of normal antler casting. At the time of his death, 14 April 1990, the antlers had demineralized at their base, indicating casting would have occurred in the near future.

Antler characteristics of this buck were asymmetric and not typical of those seen in normal male white-tailed deer. The most striking characteristic was the backward projecting antler tines which originated near the tip of the main beams (Fig. 122.4). In other respects, the antlers were of similar size and weights to normal intact males.

Discussion

Photoperiod regulates the antler cycle of male cervids (Goss 1982, 1983; Jaczewski 1954; Brown et al. 1978; Bubenik 1982; Goss 1983) mediated through the pineal. Melatonin release by the pineal is thought to be the mechanism that allows the animal to interpret photoperiod and entrain seasonal events. Presence of endogenous circannual rhythms of antler growth and other associated reproductive phenomena in the absence of photoperiod or other environmental cues have also been speculated (Bubenik 1982; Goss 1983). The evidence for endogenous circannual rhythms is inconclusive in deer but has been

FIGURE 122.4. Antlers with backward projecting tines of an 8-year-old anopthalmic white-tailed buck, November 1989.

established for another seasonal breeder, the sheep (Legan and Karsch 1983; Robinson and Karsh 1984; Karsh et al. 1988).

We believe the observations of antler events for the blind buck offers new evidence of an endogenous circannual cycle of antler growth for the white-tailed deer. Antler events for the blind deer were considerably later than normal. Normal males within our research pens had mean velvet rubout dates of 24 September and antler casting dates of 9 April (Jacobson and Griffin 1982). The blind buck's antler events reoccurring consistently with an approximate 375-day cycle for an extended 5-year period indicates seasonal environmental cues were inoperative for this animal. The use of a melatonin-blocking agent appeared to successfully alter his antler cycle, whereas a melatonin implant had no apparent effect. Since this buck was blind from birth and essentially in a state of constant darkness, it might be expected that melatonin would have no effect. Melatonin is secreted in response to darkness, and it is the duration of the melatonin signal that sets the timing of seasonal physiological events (Karsch et al. 1988). Retinal light reception is required for regulation of the melatonin signal (Legan and Karsh 1983). The response of an intact male, at a similar stage of molt and antler development to that of the blind buck, when given an identical melatonin implant was to accelerate antler mineralization and rubout (Jacobson and Waldhalm, this volume). Rubout occurred within 63 days after implant in the intact male but did not occur until 153 days after implant in the blind buck. These events are consistent with the findings for sheep. Change in melatonin signals can induce reproductive events, but sheep on constant photoperiods become refractory to melatonin and then cycle by an endogenous clock (Robinson and Karsh 1984; Robinson et al. 1985; Malpaux et al. 1987; Foster et al. 1988; Karsh et al. 1986, 1988). Although it is apparent the pineal can regulate the seasonal clock for white-tailed deer and other nontropical cervids, it also seems apparent that an endogenous clock can regulate seasonal events in the absence of the pineal. This is further substantiated by the fact that the pinealectomized deer can still demonstrate seasonal molting and antler cycles, but that these are often irregular or out of phase from normal (Brown et al. 1978). In conclusion, new evidence has been presented that an endogenous circannual clock exists in the white-tailed deer. The pineal and, more specifically, melatonin appears to be the mechanism by which deer entrain endogenous rhythms of antler growth and molting by responding to retinal reception of photoperiod.

References

Brown RD, Cowan RL, Kavanaugh JF (1978) Effects of pinealectomy on seasonal androgen titers, antler growth, and feed intake in white-tailed deer. J Anim Sci 47:435–440

Bubenik GA (1982) The endocrine regulation of the antler cycle. In Brown RD (ed) Antler Development of Cervidae. Caesar Kleberg Wildlife Research Institute, Kingsville, Texas, pp 72–108

Coker CS, Davidson D (1978) The radioimmunoassay of testosterone in various biological fluids without chromatography. J Steroid Biochem 9:373–374

Cowen PJ, Bevan JS, Gosden B, Elliott SA (1985) Treatment with beta-blockers reduces plasma melatonin concentration. Br J Clin Pharmacol 19:258–260

Cowen PJ, Frasier S, Sammons R, Green AR (1983) Atenolol reduces plasma melatonin concentrations in man. Br J Clin Pharmacol 15:579–581

Foster DL, Ebling FP, Claypoal LE, Woodfill CJI (1988) Endocrinology 123:1636–1641

Goss RJ (1983) Deer Antlers: Regeneration, Function and Evolution. Academic Press, New York

Jacobson HA, Griffin RN (1982) Antler cycles of white-tailed deer in Mississippi. In Brown RD (ed) Antler Development in Cervidae. Caesar Kleberg, Wildlife Research Institute, Kingsville, Texas, pp 15–22

Jaczewski Z (1954) The effect of changes in length of daylight on the growth of antlers in the deer (*Cervus elaphus* L.). Folia Biol 2:133–143

Legan SJ, Karsch FJ (1983) Importance of retinal photoreceptors to the photoperiodic control of seasonal breeding in the ewe. Biol Reprod 29:316–325

Karsch FJ, Malpaux B, Wayne NL, Robinson JR (1988) Characteristics of the melatonin signal that provide the photoperiodic code for timing seasonal reproduction in the ewe. Reprod Nutr Develop 28:459–472

Karsch FJ, Bittman EC, Robinson JE, Yellon SM, Wayne NL, Olster DH Kaynard AH (1986) Melatonin and photorefractoriness: loss of response to the melatonin signal leads to seasonal reproductive transitions in the ewe. Biol Reprod 34:265–274

Malpaux B, Robinson JE, Brown MB, Karsch FJ (1987) Reproductive refractoriness of the ewe to inductive photoperiod is not caused by inappropriate secretion of melatonin. Biol Reprod 36:1333–1341

Robinson J, Karsch FJ (1984) Refractoriness to inductive day length terminates the breeding season of the suffolk ewe. Biol Reprod 31:656–663

Robinson JE, Wayne NL, Karsch FJ (1985) Refractoriness to inhibitory day lengths initiates the breeding season of the suffolk ewe. Biol Reprod 32:1024–1030

Wislocki GB, Aub JC, Walde CM (1947) The effect of gonadectomy and the administration of testosterone propionate on the growth of antlers in male and female deer. Endocrinology 40:202–244

123
State of Determination of the Antlerogenic Tissues with Special Reference to Double-head Formation

Horst Kierdorf and Uwe Kierdorf

Abstract

The developmental conditions for the growth of primary cranial appendages and secondary or subsequent antlers are discussed with special regard to ortho- and heterotopic double-head formation. We hypothesize that the cells of the pedicle periosteum, being derivatives of the initial or antlerogenic periosteum which is determined for the formation of primary cranial appendages, are equally endowed with special antlerogenic tendencies. The hormonally controlled activation of these tendencies is possible even if previous casting of the old antlers has not taken place, and no casting wound arises at the top of the pedicle. The fact that antler regeneration occurs regardless of previous detachment of the old antlers indicates that the stimulus for regeneration is not antler casting but the death of the velvet antler. Antler regeneration, like sequestration of the old antler, is thus hormonally suppressed for some months after velvet shedding.

Key words: Antler, double-head formation, histogenesis, periosteum

Introduction

The question of the origin and state of determination of the antlerogenic tissues arises twice in the course of antler ontogeny. Primary cranial appendages (i.e., pedicles and primary antlers) result from neogenesis of bony structures, whereas the formation of subsequent or secondary antlers is a regenerative event unparalleled in other mammals. The purpose of this chapter is to review the developmental conditions of primary and secondary antler growth.

Formation of Primary Cranial Appendages

According to the studies of Hartwig and Schrudde (1974) in roe deer (*Capreolus capreolus*) and of Goss and Powel (1985) in fallow deer (*Dama dama*), primary cranial appendages are formed by the action of a specialized antlerogenic or "initial" periosteum located in the distal parts of the cristae externae of the frontals. The autonomous antlerogenic tendencies of this group of determined periosteal cells are activated by rising androgen levels in the fawn's blood. By autologous transplantation of antlerogenic periosteum, growth of antler structures can be provoked in heterotopic locations such as the front leg (Hartwig and Schrudde 1974; Goss and Powel 1985). It was further demonstrated (Goss et al. 1964; Hartwig 1967) that skin plays a passive role in antler development, reacting in a specific way (velvet differentiation) to subdermal formation of antler bone.

With the exception of the reindeer (*Rangifer tarandus*), female deer normally do not grow antlers. Yet they also possess an antlerogenic periosteum which experimentally can be activated by hormonal (Wislocki et al. 1947; Jaczewski 1983) as well as by unspecific stimuli (Hartwig et al. 1989). Normally the antlerogenic periosteum of female deer remains unstimulated because the hormonal requirements necessary for this process are not present (Kierdorf 1985).

Formation of Secondary Antlers

At present, the question of origin of the cells forming the regenerative blastema after antler casting is unsolved. The same applies to the question, whether, in correspondence to the situation in primary antler growth, specifically determined cells are also present for secondary antler formation.

Based on histological studies of very young antler buds of white-tailed deer (*Odocoileus virginianus*), Wislocki (1942) and Waldo and Wislocki (1951) concluded that the proliferation zone giving rise to new antler growth is made up from cells of the mesodermal component of pedicle skin, not of the pedicle's periosteum, while derivatives of the pedicle epidermis first seal off the casting wound and later cover the young antler bud. According to Goss (1972, 1984), cells derived from the dermal component of the pedicle skin constitute the bulk of the regenerative tissue giving rise to the new antler. In roe deer, Gruber (1937) discovered, that the proliferative zone of the antler bud is built by cambial cells migrating onto the casting surface and by cells originating from the marrow cavities of the pedicle.

The question if determined cell material exists for subsequent antler growth has been addressed by different authors. Olt (1927) supposed that in the skin of the upper part of the pedicle, the so-called *Bildungssaum* (formative rim), cells are present that are determined for secondary antler formation. In a number of extirpation experiments it could, however, be shown that neither the existence of pedicle skin nor of pedicle periosteum or bone is a necessary prerequisite for antler regeneration. It was not possible to inhibit this process by removing the skin rim of the *Bildungssaum* (Bubenik and Pavlansky 1965), by amputating the distal part of the pedicle (Goss 1961), or by amputating the whole pedicle including its periosteal and integumental covering (Jaczewski 1955). Even when Bubenik and Pavlansky (1956) resected the whole pedicle along with the adjacent parts of the tabula externa and the diploë of the frontal, wound healing was followed by the regeneration of an antler beam. In comparing antler regeneration with epimorphic regeneration in other vertebrates, Goss (1972, 1984, 1985) supposed that a necessary precondition for secondary antler formation is the existence of a healing epidermal wound occurring normally on the casting surface of the pedicle or experimentally, e.g., after pedicle extirpation.

In conclusion, at present the developmental conditions governing secondary antler formation are not as clear as in the case of primary cranial appendages. In this situation the occurrence of a rare antler malformation, the so-called double-head, allows some insight into the process of antler regeneration. Double-head antlers develop as a result of new antler growth without previous casting of the old antlers (Fig. 123.1), due to lack of or incomplete osteoclastic resorption within the distal part of the pedicle. The undetached antlers are mostly primary antlers or small antlers of very old stags (Nitsche 1898). Because there is no casting surface onto which cells can migrate in order to form a regeneration blastema, the proliferating velvet antler tissue forms a ring-like bulge beneath the base of the undetached beam (Figs. 123.2 and 123.3). Sometimes beam- or tine-like outgrowths emerge from this basal bulge (Figs. 123.1 and 123.2).

Double-head formation proves that casting of the old antlers is not the cause for the onset of antler regeneration. In the case of double-heads, there is no healing epidermal wound on the casting surface; therefore, the existence of such a casting wound is not a necessary prerequisite for antler regeneration. This statement does not deny the importance of an interaction between the epidermis of the antler bud and underlying tissues for normal development of secondary antlers. In amphibians an interaction of this kind between epidermis and adjacent undifferentiated mesodermal cells is essential for the formation and growth of regenerative blastemata (Stocum 1985).

In the case of secondary antler formation, it

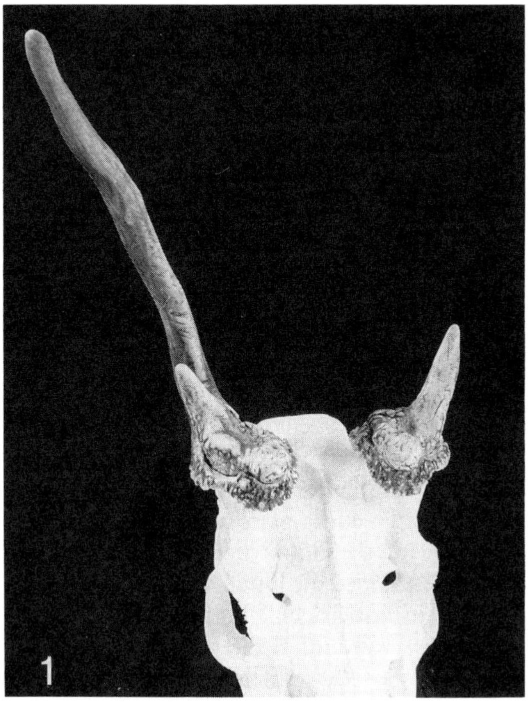

FIGURE 123.1. Double-head antler of a 3-year-old fallow buck. On the right side a beam-like outgrowth has developed from the basal bulge of the double-head's second antler generation.

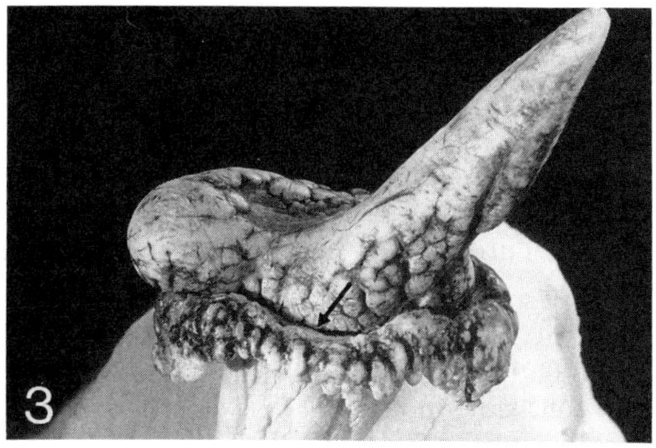

FIGURE 123.3. Higher magnification of left double-head structure of Fig. 123.1 (lateral aspect). First and second antler generation are separated by a ringlike groove (arrow) previously occupied by velvet.

seems reasonable to assume that the stimulus for regeneration is not antler casting but the death of the velvet antler which occurs several months earlier. According to our hypothesis, antler regeneration, like sequestration of the dead bony antlers, is hormonally inhibited for a certain timespan, mainly because of high plasma androgen levels. The fact that antler casting and the start of new antler growth are not causally related, although these two processes occur in some (not all) cervid species more or less simultaneously, is further proven by the observation that in these species (e.g., fallow deer) a first sign of antler regeneration (swelling of the distal part of the pedicle covering) is already seen before detachment of the old antlers (personal observation).

Double-head formation is also of interest in respect to the question of origin of the cells forming the young antler bud. Based on morphological studies Nitsche (1898) concluded that it is the pedicle's periosteum which builds up the second antler generation of the double-head. He deduced this from the fact that the bony structures of the pedicle periphery are continuous with those of the basal part of the double-head's second antler generation. If bone formation would occur extraperiostally (i.e., in the pedicle integument) there should be a distinct boundary between the two components (Nitsche 1898). He therefore designates the second antler

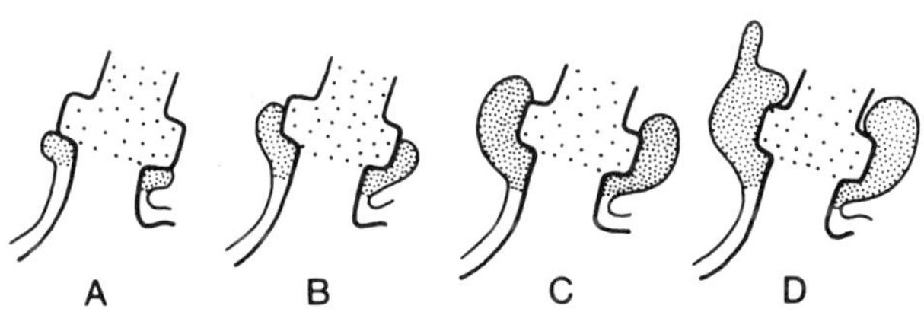

FIGURE 123.2. Schematic illustration of subsequent stages of double-head formation. Lightly stipped, first antler generation; heavily stippled, second antler generation of double-head.

generation of the double-head as a periosteal exostosis. If Nitsche's statement is correct that in double-heads antler tissue is formed exclusively by periosteal activity, this would suggest that in the case of antler bud formation on a casting surface cells of periosteal origin are responsible for chondrogenesis and osteogenesis. The question is, whether the zone of reserve mesenchyme (Banks and Newbrey 1983) forming the outermost portion of the antler's proliferating zone(s) is populated exclusively by cells derived from the pedicle's periosteum or if mesenchymal cells of skin origin (possibly under the influence of periosteal derivatives) participate in the formation of the reserve mesenchyme.

The state of determination of the pedicle's periosteal cells cannot be elucidated by means of morphological investigations alone. From the deletion experiments of Jaczewski (1955) and Bubenik and Pavlansky (1956), it can only be concluded that pedicle periosteum, like pedicle skin and bone, is not indispensable for antler regeneration. These studies further demonstrate that the antler-forming territory (i.e., the area of the skull capable of antler formation) expands beyond the normal site of pedicle location. The question of the range of this territory can be addressed by looking at the phenomenon of ectopic antlers. As already stated by Nitsche (1898), Brandt (1901), and Rörig (1901), ectopic antler beams can be formed as a result of local irritation of the skull's periosteum. They are mostly found on the frontal bones, especially in the region of the crista externa. However, ectopic antler beams have been reported from the parietal (Nitsche 1898; Brandt 1901) and nasal bones (Dixon 1934; Wislocki 1952) and in one case even from the zygomatic arch of the jugal (Nellis 1965). Wislocki (1952) assumes that the periosteum of all membrane bones of the cervid skull is potentially capable of producing antler structures, the ability to do so being highest in the frontals.

Hartwig and Schrudde (1974) and Goss and Powel (1985) have shown that only the antlerogenic or initial periosteum of the pedicle anlage area, but not the periosteum of other skull regions, has autonomous antlerogenic tendencies. Nevertheless, if adequate hormonal conditions are prevailing, the latter can by stimuli of unphysiologic nature and intensity be induced to form antler structures. Presumably the responsiveness of the periosteum to such stimuli decreases from the center of the antler-forming territory, the site of pedicle location, towards the periphery. In normogenesis the autonomous antlerogenic tendencies of the initial periosteum and its high responsiveness to the physiological androgen stimulus ensure that primary cranial appendages develop in an orthotopic location.

From the fact that the periosteum of the whole antler-forming territory has the potential to give rise to antler structures, it can not definitely be concluded that no cells exist which are specially determined for secondary antler formation. A likely candidate for such a specialized tissue would be the pedicle periosteum, because it is a derivative of the initial periosteum and, thus, may have retained the latters' state of determination (i.e., its antlerogenic tendencies).

The observation that the growth of regenerating antlers starts from the upper (free) border of the pedicle's integumental and periosteal covering (Wislocki 1942; Gruber 1952; authors' own observation), regardless of previous detachment of the old antlers, is hypothetically attributed to the fact that after velvet shedding the periosteum of this region is part of a free wound lip. The wound postulated by Goss (1972, 1984, 1985) as a prerequisite for antler regeneration would, thus, develop not because of antler casting but because of velvet shedding, a process in which the antler's periosteum is removed.

Experimental results exist that are of importance when discussing the state of determination of the cells forming the antler bud on the casting surface. Hartwig and Schrudde (1974) transplanted initial periosteum autologously between autochthonous periosteum and the metacarpal bone of the left front leg of a roe buck. Subsequently, a small bony protuberance covered by normal skin developed at the transplantation site. After a standstill of growth for 16 months, a small velvet covered antler beam developed from this protuberance. The antler structure was in direct contact with the metacarpal bone (i.e., a pedicle had not been formed). In the following 4 years velvet shedding, antler casting, and renewed antler growth occurred at regular times. In the fifth year, the antler structure was not cast, and renewed antler growth led to the formation of an ectopic double-head, which was cast in the following year (Figs. 123.4, 123.5). This process happened again during the next 2 years. Subsequently, only a thin, bony plate was formed that was cast in due time. This was again followed by the development of a double-head structure.

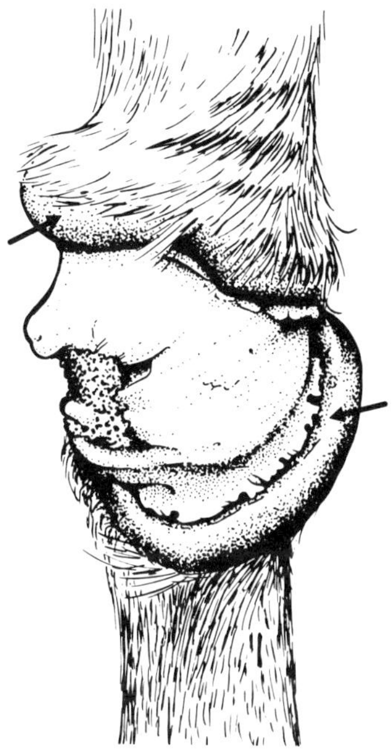

FIGURE 123.4. Ectopic double-head formation on the metacarpus of a roe-buck. Arrows, second antler generation.

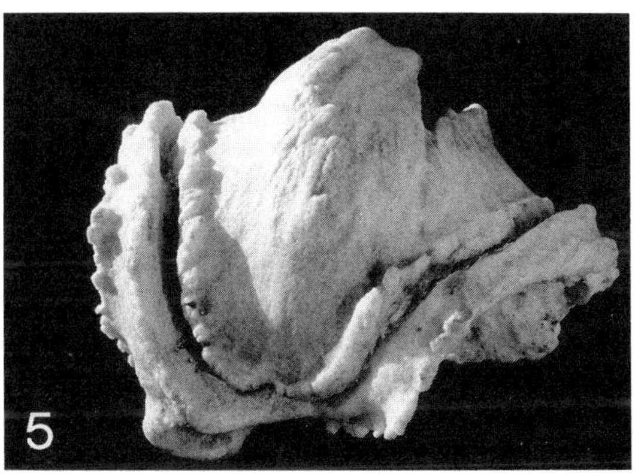

FIGURE 123.5. Double-head structure shown in Fig. 123.4 after casting. Note ring-like groove between the two antler generations.

The repeated formation of ectopic double-head structures in this experiment is attributed to the lack of a pedicle on the animal's metacarpus. Thus, with increasing circumference of the antler tissue, its connecting area with the metacarpal bone increased steadily. Presumably the limited time-span of osteoclastic activity during the "precasting period" was too short for total resorption of the bony trabeculae connecting antler structure and metacarpal bone when their connecting face became increasingly larger. Only in the following year could the remaining connection, as well as the one newly formed between second antler generation of the double-head and the metacarpus, be loosened.

The question of which cells were responsible for the formation of the ectopic secondary antlers and the double-head structures and what was the state of determination of these cells cannot be answered definitely. We assume that periosteal cells located at the base of the first formed antler beam were responsible for renewed growth of antler bone, and, that these cells were derived from the transplanted initial periosteum and thus possessed autonomous antlerogenic tendencies.

Observations by Goss and Powel (1985) support this hypothesis. They autologously transplanted initial periosteum into the subcutis of the front leg (metacarpal region) of fallow bucks. This led to the formation of pedicles, and subsequently antlers, unconnected to the metacarpal bone. Antlers exceeding several centimeters in length bounced around when the animals ran. Antlers and pedicles were lost from some of the deer, either because they were torn out of the surrounding skin or because of rupture of the blood vessels at the base of the pedicle and subsequent detachment. Antler regeneration did not occur in any of these cases. The nature of the skin healing over the lesions caused by antler and pedicle loss suggested a velvet-like character (Goss and Powel 1985), which was attributed to residual antlerogenic cells in the mesoderm. According to our hypothesis, the lack of antler regeneration was caused by the fact that no, or only a few, periosteal cells endowed with antlerogenic tendencies remained in the wound area after loss of the ectopic pedicles and antlers.

If the assumption that ectopic antler and double-head formation was initiated by cells derived from the transplanted initial periosteum and that these cells were endowed with a morphogenetic tendency for antler growth is correct, this would constitute a certain parallel to heterotopic limb formation in amphibians. In this group the formation of supernumerary limbs outside the limb field can only be induced when previously a part of limb skin or skin of the shoulder region has been transplanted to this site (Wallace 1981). Experimental studies have shown that the morphogenetic information for such ectopic limb growth is contained in the mesodermal part of the transplanted pieces of skin (Stocum 1985).

References

Banks WJ, Newbrey JW (1983) Light microscopic studies of the ossification process in developing antlers. In Brown RD (ed) Antler Development in Cervidae. Caesar Kleberg, Wildlife Research Institute, Kingsville, Texas, pp 231–260

Brandt K (1901) Das Gehörn und die Entstehung monströser Formen. Parey, Berlin

Bubenik AB, Pavlansky R (1965) Trophic responses to trauma in growing antlers. J Exp Zool 159:289–302

Bubenik A, Pavlansky R (1956) Von welchem Gewebe geht der eigentliche Reiz zur Geweihbildung aus? 2. Mitteilung: Operative Eingriffe auf den Rosenstöcken der Rehböcke, *Capreolus c. capreolus* (L., 1758). Säugetierkdl Mitt 4:97–103

Dixon JS (1934) A study of the life history and food habits of mule deer in California. Calif Fish Game 20:1–146

Goss RJ (1985) Tissue differentiation in regenerating antlers. In Fennessy PF, Drew KR (eds) Biology of Deer Production. R Soc N Z Wellington Bull 22:229–238

Goss RJ (1984) Epimorphic regeneration in mammals. In Hunt TK, Heppenstall RB, Pines E, Rovee D (eds) Soft and Hard Tissue Repair. Praeger, New York, pp 554–573

Goss RJ (1972) Wound healing and antler regeneration. In Maibach HI, Rovee DT (eds) Epidermal Wound Healing. Year Book Medical Publishers, Chicago, pp 219–228

Goss RJ (1961) Experimental investigations of morphogenesis in the growing antler. J Embryol Exp Morphol 9:342–354

Goss RJ, Powel RS (1985) Induction of antlers by transplanted periosteum. 1. Graft size and shape. J Exp Zool 235:359–373

Goss RJ, Severinghaus CW, Free S (1964) Tissue relationships in the development of pedicles and antlers in the virginia deer. J Mammal 45: 61–68

Gruber GB (1952) Studienergebnisse am Geweih des *Cervus capreolus*. Zbl allg Pathol pathol Anat 88:336–345

Gruber GB (1937) Morphobiologische Untersuchungen am Cerviden-Geweih. Nachr Ges Wiss Göttingen, Math-phys Klasse NF 3:9–63

Hartwig H (1967) Experimentelle Untersuchungen zur Entwicklungsphysiologie der Stangenbildung beim Reh (*Capreolus c. capreolus* L. 1758). Roux's Arch Entw mech 158:358–384

Hartwig H, Schrudde J (1974) Experimentelle Untersuchungen zur Bildung der primären Stirnauswüchse beim Reh (*Capreolus capreolus* L.). Z Jagdwiss 20:1–13

Hartwig H, Schrudde J, Kierdorf H, Kierdorf U (1989) Durch unspezifische lokale Periostaktivierung provozierte Bildung eines knöchernen Stirnfortsatzes bei einem weiblichen Reh (*Capreolus capreolus* L.). Z Jagdwiss 35:130–136

Jaczewski Z (1983) The artificial induction of antler growth in deer. In Brown RD (ed) Antler Development in Cervidae. Caesar Kleberg, Wildlife Research Institute, Kingsville, Texas, pp 143–162

Jaczewski Z (1955) Regeneration of antlers in red deer, *Cervus elaphus* L. Bull Acad Polon Sci (Ser 2) 3:273–278

Kierdorf U (1985) Gehörnte Ricken. WJSC-Blätter 61:4–14

Nellis CH (1965) Antler from right zygomatic arch of white-tailed deer. J Mammal 46:108–109

Nitsche H (1898) Studien über Hirsche. Engelmann, Leipzig

Olt A (1927) Die Perücke der Cerviden und das Karzinom. Dtsch Tierärztl Wochenschr 35:131–133

Rörig A (1901) Ueber Geweihentwickelung und Geweihbildung IV Abnorme Geweihbildungen und ihre Ursachen. Arch Entw-mech Org 21:225–301

Stocum DL (1985) The role of the skin in urodele limb regeneration. In Sicard RE (ed) Regulation of Vertebrate Limb Regeneration, Oxford University Press, New York, Oxford pp 32–53

Waldo CM, Wislocki GB (1951) Observations of the shedding of antlers in Virginia deer (*Odocoileus virginianus borealis*). Am J Anat 88:351–396

Wallace H (1981) Vertebrate Limb Regeneration. Wiley, Chichester

Wislocki GB (1952) A possible antler rudiment on the nasal bones of a whitetail deer (*Odocoileus virginianus borealis*). J Mammal 33:73–76

Wislocki GB (1942) Studies of the growth of deer antlers. 1. On the structure and histogenesis of the antlers of the Virginia deer (*Odocoileus virginianus borealis*). Am J Anat 71:371–415

Wislocki GB, Aub JC, Waldo CM (1947) The effects of gonadectomy and the administration of testosterone propionate on the growth of antlers in male and female deer. Endocrinology 40:202–224

124
Investigation of Endocrine Parameters in Chital Deer in Relation to Their Antler Cycle

George Bubenik[1], Robert D. Brown[2], and Dieter Schams[3]

Endocrine investigations of chital deer (*Axis axis*) were performed in the captive group of six adult males maintained in the deer facility of Caesar Kleberg Wildlife Research Institute in Kingsville, Texas (latitude 27° N). Average seasonal levels of LH measured between November and June did not exhibit a seasonal trend. Concentrations varied between 1.17 and 3.6 ng/ml. Similarly, testosterone (T) levels fluctuated between 0.71 and 2.08 ng/ml. Levels of both hormones did not appear to correlate with the antler cycle. FSH concentrations were lowest in November (27 ng/ml) and highest in April (59 ng/ml). Prolactin (PRL) exhibited a typical seasonal variation with lowest level observed in December (21 ng/ml) and the highest level in May (103 ng/ml). All three adult bucks were challenged in March with 100 ng of GnRH and exhibited increased concentrations of LH. Maximal values (achieved 40 to 60 min after GnRH administration) varied from 7.7 to 11.2 ng/ml. T increased above baseline (0.2 to 0.5 ng/ml) in only two bucks, reaching 1.3 and 1.6 ng/ml. Three adult bucks injected in March with 20 IU of ACTH responded with an immediate elevation of cortisol. Pretreatment values, 4 to 8 μg/dl, at 20-min mark, 12 to 16 μg/dl, maximum 16 to 21 μg/dl detected 140 min after ACTH administration. The antlers were cast between December and March, and the new growth started almost immediately. The mineralization and polishing of antlers occurred from March to June. The velvet was polished first in the strongest buck. Compared with boreal cervids, in chital only FSH and PRL exhibited seasonal variations related to antler cycle. In contrast to recent reports, the antler cycle was relatively well synchronized within the herd.

[1] Department of Zoology, University of Guelph, Guelph, Ontario, Canada N1G 2W1, [2] Department of Wildlife and Fisheries, P.O. Drawer LW, Mississippi State, MS 39762, USA, and [3] Department of Physiology, Technical University München-Weihenstephan, D-805 Freising, Germany

125
Antler Growth in Red Deer Stags Implanted with an Antiandrogen

Ludek Bartos*

The red deer is a species with a seasonal antler cycle. The development and mineralization of antlers is closely related to the sexual cycle. Variations in levels of sexual hormones play a major role in the timing of individual phases of antler growth. Testosterone (T) appears to be the hormone which most directly affects the antler cycle. The casting of antlers in cervid species in temperate regions is generally believed to be caused by a sudden drop in T levels. It has been already suggested, however, that this fits only to one group of species such as white-tailed deer (*Odocoileus virginianus*), mule-deer (*O. hemionus*), moose (*Alces*), reindeer, and caribou (*Rangifer*), where an interval exists between antler casting and new antler growth. Species where casting of old antlers is followed immediately by the growth of new antlers, such as wapiti (*Cervus canadensis*), red (*C. elaphus*), roe (*Capreolus*), fallow (*Dama dama*), Père David's (*Elaphurus davidianus*) and sika deer (*C.e. nippon*), may cast their antlers after a short-termed restoration of T levels following the seasonal T drop. Twelve male red deer were used. Before the antler casting the deer were either implanted with cyproterone acetate in silastic tubes or sham-implanted. In the present study we take into consideration the following factors: the time of the antiandrogen application before antler casting; antler casting, growth, and cleaning; and hair moulting.

*Group of Ethology, Research Institute of Animal Production, CS-104 00 Praha 10-Uhrineves, Czechoslovakia

126
Artificial Production of Three-Antlered Red Deer Stags

Z. Jaczewski and R. Bartecki*

When pedicles of stags (*Cervus elaphus*) were about 1 cm long, a compound-flap-graft method was used to move the left pedicle more to the middle line of the skull. In two stags, a narrow flap of skin was crossed with the compound flap graft to cover the medial part of the wound after pedicle amputation. Stag no. 1, born 27 May 1986, was operated on 14 April 1987. Stag no. 2, born 13 July 1987, was operated on 19 May 1988. In two other stags, the skin was excised completely, and the pedicle amputation wound remained open. Stag no. 3, born 4 June 1987, and stag no. 4, born 16 June 1987, were both operated on 22 April 1988. Each stag produced three antlers as follows: a regenerated antler at the place of pedicle amputation (R), a grafted antler (G), and a control antler (C). Stag no. 1 cast R on 19 April 1988 and G and C on 20 April (C weighed 54.4 g and was 25.1 cm long; G, 36.3 g and 15.2 cm; R, 70.8 g and 33.4 cm, respectively). Antlers on the injured side (G + R) produced 196% more antler mass than the control side (C). The next year, stag no. 1 cast G on 13 March, but R and C on 17 March 1989. These antlers were branched, and their weights were as follows: R, 486.4 g; G, 324 g; C, 424 g. Again, the injured side produced 191% more antler tissue than the control side. Stag no. 2 cast G on 22 April 1989 and C and R on 23 April. C weighed 25.6 g and was 8.6 cm long (a small part of C was broken on 25 January 89); G, 8.2 g and 3.8 cm; R, 73.7 g and 37.5 cm, respectively. The injured side (G + R) produced nearly 320% more antler tissue than the control side (c). Stag no. 3 cast C on 23 April 1989. It weighed 52.6 g and was 18 cm. G and R were cast on 24 April 1989. G and R were joined with a narrow bridge of antler tissue. This bridge was probably caused by an accidental injury of R on 4 June 1988. The basal surface of that antler (G + R) was 8-shaped. G + R weighed 151.95 g; G was 30.5 cm and R 9.9 cm long. The injured side produced 288.9% more antler tissue than the control side. Stag no. 4 cast all three antlers on 22 April 1989. C weighed 52 g and was 21.6 cm long; G, 48.3 g and 16.3 cm; R, 88.6 g and 35 cm, respectively. The injured side produced 263% more antler tissue than the control side. It is interesting that during the first antler cycle after the operation, R grew quicker and mostly larger than G or C in spite of the total removal of the periosteum.

*Institute of Genetics and Animal Breeding, Polish Academy of Sciences, Popielno, 12-222 Wejsuny, Poland

127
Influence of Ecological Factors on the Development of Sika Deer Antlers

A. Kravchenko, A. Velizhanin, and E. Nivin*

Older sika deer's (*Cervus nippon*) antlers begin to grow earlier (March–April) than younger deer's antlers (April–May). Accordingly from June to August antlers of similar form can be seen in a deer herd of varying ages. We observed 2483 deer on a farm from 1976 to 1979. The dimensions of the antlers were not the same in summer; they were maximum in the middle of June and minimum at the end of August. Year to year, the dynamics of antler dimensions were not identical. In our summers, ambient temperature and solar radiation are minimum in the middle of June and maximum at the beginning of August; the humidity is relatively constant. Correlative dependence of antlers' dimensions on temperature and relative humidity, summary and direct solar radiation were not important. We decreased the quantity of solar radiation getting to the antlers by use of fabric covers. These experimental deer developed anomalous antlers. The quantity of erythrocytes in blood increased, and their volume decreased. Speed subsidence of erythrocytes were reduced by one-third, and the mineral content of blood serum changed. Behavioral activities of the deer were also observed to be reduced.

*Primorsky Agricultural Institute, Ussuriisk, USSR

128
Production of 1,25(OH)2D by Deer Antler Cells

A.J. Sempéré, C. Silve, and M. Garabedian*

The development of deer antler follows a pattern similar to that described for mammalian endochondral ossification and has been proposed as a suitable model for studies of bone growth. Seasonal changes in the plasma concentration of 25(OH)D, 1,25(OH)2D, calcium, phosphorus, and the activity of alkaline phosphatase in relation to the antler cycle have been investigated over 1 year in captive roe deer (*Capreolus capreolus*) and in wild roe deer during their antler cycle. A significant increase of 1,25(OH)2D in peripheral plasma with no parallel variation in the concentration of its precursor 25 hydroxyvitamin D, was observed during the rapid growth phase of the antler cycle in captive and wild deer. During the same phase, there was a gradient in levels of 1,25(OH)2D in antler vs jugular blood ($P < 0.01$). In addition, velvet cells in culture proved to have the ability to convert 25 hydroxyvitamin D in 1,25(OH)2D. These in vivo and in vitro results strongly suggest the evidence of a local production of 1,25(OH)2D by the antler cells and may contribute to the increase in plasma 1,25(OH)2D during the bone growth.

*C.E.B.A.S.—C.N.R.S.—Villers-en-Bois, 79360 Beauvoir-sur-NIORT, France

Special Section
The Commercialization of Deer

129
Commercialization of Wildlife: A Value-Added Incentive for Conservation

Delwin E. Benson

Abstract

The debate about commercialization of deer and other wildlife by hunting or captive rearing is confused with the unrelated history of market hunting during the 1800s and undermined by false presumptions and fears about wildlife and public lands being taken over by private interests. Opponents underestimate the private role in landscape and wildlife management and the control that landowners have on wildlife populations and recreational hunting. Assertions are incorrect that harvested animals and sport hunting are not integral to the success of wildlife conservation. Fallacious arguments about hunting may have entered the commercialization issue into an antihunting debate. The issue should not be whether the public or private sectors should benefit from wildlife, but rather how providing wildlife values to landowners can increase their stewardship. I argue that it is right and proper that private individuals should be entitled to manage and benefit from wildlife and recreation on their properties. It is pragmatic for governments to enable landowners to do so in conjunction with rules established and enforced on behalf of all the people. Public wildlife ownership and management have been imposed upon the private sector even though land is controlled privately on which wildlife are maintained and where recreationists seek access. Private lands (which encompass 66% of the United States) will continually grow in value to sustain biotic diversity, wildlife populations, and to provide recreation. To deny landowners a role in the management of their own lands is a mistake. Pressures are mounting for private landowners to account for costs and benefits and to operate effectively. Wildlife must be treated as a benefit, because as a cost, wildlife will be treated negatively either directly or indirectly. Lands not economically competitive for wildlife enterprises will likely be put to agricultural, industrial, or residential uses that diminish natural environmental values.

Key words: Commercialization, fee hunting, game farming, game ranching, privatization, wildlife enterprise, wildlife utilization

Introduction

I agreed to debate with Dr Geist not as a crusader, but because his published arguments were not in the best interest of preserving wildlife and their habitats on private land. He overlooks the private sector as a positive or negative force in conservation. His assertions deemphasize the role of sport hunting and hunters in conservation, and he overlooks the role of hunting as a way to provide incentives for landowners to encourage wildlife and to allow access for persons who are willing to pay for their activities. His arguments support the success story of public land management for wildlife in North America, but they fail to deal adequately with wildlife on private land. The reality is that many landowners are rebelling against governments asking for fewer big game on their properties, and landowners are seeking compensation from government for wildlife damages. For example, landowners in Arizona (Anonymous 1990) and

New Mexico (Latham 1989) are seeking elk herd reductions of 61% and 60% respectively. In some cases, landowners are taking a positive approach to their problems and opportunities by charging for wildlife-based recreation. I argue that it is pragmatic and right for wildlife agencies to enable landowners to manage wildlife on their properties. Society benefits from wildlife if landowners and recreationists benefit; society loses if landowners are alienated or even disinterested.

Considerations for using wildlife commercially are based upon the following thoughts: (1) Values of wildlife on private land are underutilized, and the role of landowners needs redefining. (2) Historical commercialization problems are irrelevant in modern times, yet the lessons are valuable. (3). Land and wildlife resources are declining caused by the private sector (but for societal good), and wildlife without values cannot compete with other land used. (4) Demands for wildlife-based recreation are increasing, and private supplies are decreasing. (5) Current wildlife management philosophy for private lands incorrectly works to perpetuate resources rather than to create them. (6) The real issue for debate and for conservation is how to conserve wildlife without evicting the civilization upon which we depend. Once the overview of issues is complete, points made against commercial use of wildlife will be reviewed specifically and counter-points will be asserted.

Private Wildlife Management: A New Frontier

Commercialization of wildlife on private land in partnership with states will open a new resource base and constituency for wildlife conservation if landowners have a reason to want wildlife and recreationists. The unique role of private lands for conservation has been asserted since the formal wildlife profession began (Leopold 1930). Meaningful, long-term incentives by governmental agencies to bring the private sector into the wildlife conservation field have been slow to develop (Burger and Teer 1981; Teer et al. 1983), although wildlife administrators in 48 states considered that their programs were successful (Wigley and Melchiors 1987). Even agencies of the Department of Agriculture such as the Cooperative Extension System, Soil Conservation Service, and Agricultural Stabilization and Conservation Service have relatively few staff and programs that benefit wildlife and recreation on private land compared with other uses of the land. Negative relations between landowners and recreations have persisted in spite of studies and programs for solutions dating back to the early 1900s (Jackson and Anderson 1982).

Fee-hunting (Benson 1989a) and other commercial uses represent market-based incentives for landowners to manage wildlife based upon a willingness of various publics to pay. I contend that government should enable reasonable commercialization of wildlife for recreational opportunities, live sale, and production of venison on private lands under managed conditions. Establishing cooperative partnerships between public and private sectors for integrated wildlife and recreation management should be a goal in this era of wildlife management (Benson 1991). The consequence of no action by government is for disenfranchised landowners to exert their will on resources and to become the de facto wildlife and recreation managers.

Evolution from Market to Sport Hunting

Unfortunately, exploitative activities by the private sector to acquire and manage lands during the 1800s formed indelible images of destruction that are slow to change among wildlife managers and within society. Uncontrolled commercialization of wildlife for food, subsistence hunting and trapping, or to control wildlife that were causing agricultural and forest losses were recognized as problems and brought under control near the turn of the twentieth century.

The new commercialization, sport hunting, emerged as the driving force behind the conservation movement (Reiger 1975; Trefethen 1975). Sportsmen were the most consistent force for passage of game legislation. Game laws were written largely by sportsmen with little organized opposition and were enforced entirely by sportsmen until they prevailed upon the states to provide licensing and wardens (Tober 1981).

Private clubs were formed by sportsmen to acquire habitats and hunting grounds, and legislation was suggested to preserve large tracts of land under governmental control. The blending of public and private conservation initiatives began.

Role of Public and Private Lands

The doctrine of President Theodore Roosevelt forecasted a combination of public and private contributions to wildlife conservation. He wrote in 1909 that the most democratic and intellectual form of conservation was by creating public reserves, and the other form was through intelligently managed private reserves (Leopold 1933). Roosevelt was instrumental, along with the sporting press such as *Forest and Stream*, and sportsman organizations, most notably the Boone and Crocket Club, to press for legislation to protect large tracts of land (Reiger 1975; Trefethan 1975; Tober 1981). His wisdom and leadership set the stage for acquiring state and federal parks, wildlife refuges, forests, rangelands, and recreation areas.

Leopold (1933) contended that the dominate idea until about 1905 was to perpetuate hunting rather than to improve or create it. Through restrictions on hunting, the supply of wildlife could be divided among the users to make it last longer. This policy reflected the democratic ideas of our political systems, a game system based on an equally distributed citizenship, rather than, as in Europe, on an unequally distributed landownership. This novel system, however admirable in theory, failed to halt the decline in game supply. Public-spirited sportsmen groped for solutions and developed further the idea of restrictions. Better enforcement and prohibition of market hunting were their answer. Roosevelt's "wise use" programs began to preserve habitats for public benefits and to conserve their uses. Leopold contended that the issue was still cloudy about what kinds of game could best be renewed under public initiative, and what kinds could be recovered by public encouragement and regulation of private initiatives. Unfortunately, supplies of wild public and private land resources were dwindling while demand for wild-based recreation was increasing. The Leopold Committee (Leopold 1930) set the stage for including the private sector in wildlife and recreation management.

Recent Loss of Wild Resources

The rate of accumulating public interest land and water has decreased since Roosevelt's time. Yet, the consumption of public and private lands and waters for uses that are incompatible with landscape integrity and wildlife has increased. Wild landscapes and animals suffer if other land uses have commercial value greater than those that can be derived from wildlife and recreation.

The loss of wild lands and production of crops using chemicals continued to increase. Wetlands were drained at a rate of 45 000 acres each year in the United States leaving one-half of the nation's original base (Chandler 1988). During the drought of 1988, an estimated 50% of the existing wetlands dried up on the Northern Great Plains. Rangelands in the Great Plains were converted to crops at a level of 4.5 million acres in the 1970s and 1980s (Laycock 1987). Cropland conversions to urban areas, transportation corridors, and water bodies between 1967 and 1975 were at the rate of 1.87 million acres per year (Batie and Taylor 1990). However, planted area rose from 303 to 356 million acres during the 1970 to 1980 period representing "fence row to fence row" production (Batie and Taylor 1990). Pesticide use increased from under 200 million pounds applied in 1964 to 479 million pounds applied in 1982. Herbicides were applied to 80 million acres in 1964, and by 1982 over 200 million acres received treatment (Batie and Taylor 1990). Fertilizers increased from 25 million tons applied in 1960 to 45 million tons applied in 1987 (Batie and Taylor 1990). Timber removal increased by 27% between 1970 and 1986 (Hagenstein 1990).

Recent Private Supply and Demand for Wildlife-Based Recreation

Land and water have decreased as a supply of suitable habitats for wildlife in the US, yet demands have increased for wildlife-based recreation. The private sector will be called upon for solutions. Hunter numbers increased from 11.8 million in 1955 to 16.3 million in 1985, although the total is down by less than 1 million since the high of 1975 (Anonymous 1988). About 140 million Americans, 77% of the population 16 years of age or older, enjoyed some form of wildlife-related recreation (Anonymous 1988).

Visits to national recreation areas and forests averaged 4% annual growth since 1977 (Cordell 1990). In 1972, most overnight trips were over 100 miles, and day outings were more than 50 miles from home (Cordell 1990). In the 1980s the majority of overnight trips was within 100 miles of home, and day trips were within 25 miles (Cordell 1990). Cordell concluded that both public and private opportunities need expansion through cooperation and partnerships. Hunters will find increasingly poor opportunities as wild lands are converted to other uses and as other recreation competes with hunting uses.

Private lands account for 1.28 billion acres in the United States of which all but 5% are non industrial and most are not available for access (Cordell 1990). Eighty-two percent of hunters (13.7 million persons) hunted on private lands, and 51% hunted only on private lands (Anonymous 1988). Access to 48% of private lands was estimated to be restricted to persons acquainted with the owner. Only 23% of land was open to the public. About 20% of the open lands were under recreational leases, and 80% were open for free or for a daily fee (Cordell et al. 1988a, and unpublished data). Cordell's study was based on a survey of 16000 private landowners in 48 states. Brown et al. (1984) reported a steady increase of lands being posted against hunting in New York between 1963 and 1980 and predicted an eventual decline of hunters in the eastern US (Brown et al. 1987). By 1980, approximately 50% of private land in upstate New York was posted. If the trend continues, all private land in New York will be posted by 1993. In the west, Guynn and Schmidt (1984) reported that 79% of private land was closed to hunting in Colorado during 1977, up from 68% in 1969 (Rounds 1975). Access was considered a major problem by wildlife administrators in 26 of 50 states, especially where public lands were most abundant (Wright and Kaiser 1986).

Decisions for providing access by landowners are affected by their socioeconomic characteristics, ownership objectives, residency, posting policy, concerns about liability, and attitudes regarding economic incentives (Brown 1974; Holecek and Westfall 1977; Thowardson 1979; Dumke et al. 1981:17–161; Kaiser and Wright 1985; Wright et al. 1988). Unsafe and destructive behavior by hunters and other recreation users has also been cited as a major problem that contributes to prohibition of access (Larson 1958; Brown 1974; Rounds 1975). Landowners (95%) considered income from recreational access to their lands as an important reason for owning their lands and waters (Cordell et al. 1988b and unpublished data).

Fee-hunting is not a widespread practice by hunters, yet a willingness to pay exists. Langner (1987) reported from the 1985 National Survey of Fishing, Hunting, and Associated Recreation that 8% of American hunters paid to use private lands. Benson (1988) reported from a random sample of 355 hunters in five states that 6% paid to hunt on private land. A willingness to pay was expressed by 51% of the sample. In the same study, 55% of National Wildlife Federation affiliate leaders representing 34 states expressed a willingness to pay for hunting. Sixty-five percent of pronghorn hunters in Colorado were willing to pay private landowners for hunting access (Conquist 1990).

Hunters report a loss of quantity and quality in their sport. A 1986 survey of hunters by National Family Opinion Research Inc. reported that hunters had become discouraged because of the following five factors: (1) poor access to hunting land, (2) crowded hunting areas, (3) finding time to go hunting, (4) less cooperation from landowners, and (5) less game in general. Hunters want more from their sport than the US system has offered, yet managers could be confused by apparent inconsistencies from reported motives for hunting. Applegate (1989) reviewed studies about the meaning of success for hunter satisfaction and concluded that killing of wildlife for recreation was not the image that hunters wished to project as their primary motivation in taking to the field. However, higher rates of enjoyable hunts were reported in locations with higher kills. Potter et al. (1973) suggested that the kill is only one part of a person's reasons for hunting. It is not surprising that hunters rate companionship, esthetic values, family ties, sport, and meat as reasons for hunting, especially since "killing for recreation" is not pleasant sounding or interpreted correctly by nonhunters. Geist (1986) wrote that "in the eyes of the public the killing of creatures for sport is, often enough, a contemptible activity, a frivolity for such who have nothing better to do with their time. There is, therefore, real danger in emphasizing the 'sporting' aspect of wildlife harvest." Ortega y Gasset (1972) justified the dichotomy by writing that, one kills to have hunted, one does not hunt to kill. That distinction is lost in the minds of many. In a study of first-year hunters in New Jersey (Applegate and Otto 1982), 72% of respondents hunted for "sport" or "to be close to nature" rather than "for the meat," although success at bagging game was important to their persistance. Twenty years earlier, Gilbert (1977) reviewed various authors whose studies were establishing multidimensional reasons why people hunt. The traditional "game-bagged" or "days afield" theories of satisfaction used by managers were criticized by Hendee (1974) as being too limited a way to express hunting values. Gilbert (1977) also found many dimensions to the hunt, but the probability of success was a major consideration. Meat had lesser importance. Applegate (1989) concluded that survery research has probably underestimated the significance of the kill to the rewards of hunting.

The sporting fraternity that formed the conservation movement in North America has become discouraged about the status quo. Wildlife conservation cannot afford to lose historical allies, political champions of conservation, and an economic force that generated $10.1 billion in the US during 1985 (Anonymous 1988)? Resources on private land can support the existing demand and open new opportunities for the future if landowners have appropriate incentives and if governments enable them to have a proactive role.

Resisted Change for Modern Problems

Wildlife agency professionals fear loss of control and resist the social and economic necessities that dictate management in the private sector. Berryman (1987) contended that socioeconomic evaluations were used as academic disciplines in the profession without commensurate application to the real world. He said that wildlife researchers and managers find it somewhat repugnant to apply the obvious economic values to the uses of wildlife resources. He said that we have documented the values but resisted commercialism of the resource and application of the known economic tools. We have resisted virtually every form of paid hunting while quoting the economic impact of expenditures made for the uses of fish and wildlife. If wildlife resources are to remain a part of the fabric of our total landscape and environment, if this is really our "conservation ethic," then we must undertake research and find ways to make the maintenance of habitat compatible and competitive with other land uses. Berryman said earlier (1981) that "the plain average sportsman has been ahead of the professionals and the administrators, and has been willing for many years to accept the kind of system that will assure the future of habitat and the opportunity to harvest game surpluses."

Perhaps our system of allocating wildlife resources among hunters with crowded conditions or imposing hunting on private lands with limited and decreasing access has contributed most significantly to the problems of present-day hunting management and wildlife conservation. Perhaps it is true today as Leopold (1933) contended in the 1930s, that we were still merely allocating scarce resources that are caused by human changes to the land and water base. We use limited licenses, bag limits, season structures, laws, and enforcement to share the joys and burdens of wildlife among society rather than making more wildlife and access available to those who use it in disproportional amounts or to enable those willing to produce it to reap commensurate rewards. Governments, wildlife professionals, landowners, hunters, and all of society must now answer Leopold's (1933: 19) question of, "How shall we conserve wildlife without evicting ourselves?"

Conserving Wildlife Without Evicting Ourselves

The concerns, opportunities, and perhaps answers to how wildlife can be conserved on private land without evicting ourselves is the true issue for debate! The issue is not whether deer and other wildlife should have commercial value. Rather, will commercial value help to preserve wildlife on landscapes threatened continuously by reductions of habitat? Will commercial costs from wildlife and negative attitudes of land owners who receive inadequate support and encouragement from wildlife professionals, lead to less cooperation, more defiance, and eventually fewer wildlife? Finally, will anti-hunting sentiments destroy the wildlife conservation system in North America?

The assumption is made that society will do the best management job possible within the existing acres of public lands and waters, and they will possess the foresight to acquire more public interest properties when possible. There are numerous support groups, laws, governmental policies, and professional employees in which we will entrust our good will.

We cannot continue to overlook private landowners who produce wildlife and access at personal costs on two-thirds of the nation's acres and who have few outside supporters. To deny their role in wildlife and recreation management is to deny a resource base, a latent constituency and an important future in modern wildlife management. They are the de facto managers of wildlife and recreation on their own lands and affect wildlife and recreation on adjoining public lands if wildlife numbers are diminished by their action or inactions. For governments to enable landowners to take a more proactive role in management of their own properties is *not* to give them de jure control of the resources at the expense of society and traditional managers. Rather, providing a positive role for the private sector will create partners in conservation that governments cannot afford to lose.

The Geist Debate and Counter-Points

Points addressed thus far have been indirect to the debate but direct to the issue of commercializing deer and wildlife in general. Dr. Geist attributes the success of North American wildlife conservation to four nondocumented "policies" (Geist 1989) and uses those and related arguments (Geist 1985, 1986, 1987, 1988) against commercialization of wildlife. The derivation, definitions, and relevance of these "policies" are questionable and will be debated.

Point 1: Public ownership of wildlife is de jure and (usually) de facto (Geist 1989).

There is no controversy about "de jure" wildlife ownership, but "de facto" ownership is controlled most by the private sector. Private landowners control access, habitats, and ultimately populations because of property control. Pronghorn (*Antilocapra americana*) are below ecological carrying capacity and bison (*Bison bison*) have no place to roam because landowners will not support them. Property in wildlife resources has become increasingly well defined with scarcity (Tober 1981). Early unregulated and uncontrolled exploitation of public property in wildlife environments created scarcity which was countered by protection, land and water acquisition, and conservation through the allocation of property rights in public resources for private gain. A scarcity today is with access to property that provides hunting as one, but a major demand. Current trends are to use property rights (de jure) to close private lands for public access (only 23% are currently open for hunting, Cordell 1990). The percentage of persons who charge (20% of lands open to public access, Cordell 1988b) or who pay (< 10%, Langner 1987; Benson 1988) is small. However, desires for income by landowners are high (95%, Cordell 1988b) and willingness-to-pay by hunters exists (> 50%, Benson 1988).

Public users also preempt property rights from other users. Hunters with bow in early seasons may take property rights from hunters using firearms in later seasons, or snowmobile users may frighten wildlife away from cross-country skiers. A national park may provide refuge for deer or wapiti (*Cervus elaphus*) facilitating the opportunity for property acquisition by photographers which in turn prevents persons from access to the same property for hunting. The same park may create herds of wildlife that cause damage to private property.

Why is legal ownership of wildlife by the public considered to be essential? Ownership is actually a moot point. The critical issue is how "owners" treat wildlife. There are no quarrels with the present de jure system of wildlife ownership. Public ownership is quite workable because privileges for managing and using wildlife can be granted to the private sector through the present system. Society does not give up its property in wildlife and governments do not give up their custodial role to manage wildlife and to allocate property to various users. Hunting is still regulated by government, based upon wildlife surpluses and societal needs. Landowners are not told to sell access privileges, and recreationists are not told to pay. However, willing buyers and sellers can be helped to become respected partners in conservation with each expected to maintain and enhance wildlife and habitat values in the process. Governmental regulations and policy can be used to help landowners by removing the uncertainty (Swank 1981; Sheldon 1987) that currently exists caused by season structure and timing, hunting permits allocated by lottery, and bag limits based upon allocation of limited resources rather than for creating more wildlife.

Point 2: Elimination of markets in meat, parts, and products of game animals, shore and songbirds, that is of vulnerable wildlife, [is good] while retaining markets in furs of the less vulnerable furbearers [is also good] (Geist 1989).

There have always been markets for wildlife in the US Markets shifted from food and animal products in the 1800s to markets for services, manufacturing and recreation in modern times. Economic value of the new markets is acknowledged and cited positively by Geist (1989). Markets were not the problem. Uncontrolled exploitation without knowledge, fear, or care about animal losses was the problem. It is false to assert that furbearers are "less vulnerable." Uncontrolled trapping in North America caused problems 50 to 100 years ago (Novak et al. 1987). However, when and where fur harvest was regulated and managed, harvest has not lead to unabated population problems. In fact, harvest provides incentives to manage, obtain knowledge, and to maintain a human support base.

Point 3: The allocation of the material benefits of wildlife is appropriate by law, not by the marketplace, one's birthright, land ownership or social position (Geist 1989).

This is a philosophy that theoretically sounds just; however, it is not possible because allocation of resources is never uniform even in true socialistic systems. There are externalities associated with de facto rights (or access) to property in wildlife resources. These externalities have two properties: (1) interdependency—one person's behavior creates a cost or benefit to other persons; and (2) lack of compensation—the one who creates cost is not made to pay for it, nor is the one who creates the benefit completely rewarded for it (Barkley and Seckler 1972). For example, a poor person living near wildlife has less expensive access to property in wildlife. Conversely, a wealthy member of a royal family may be far from wildlife but can afford the costs associated with obtaining property in wildlife.

Landowners who control land, control property in wildlife and can deny access. Some are demanding a decrease in wildlife populations where depredation problems exist. Perry Olson, Colorado Division of Wildlife Director, identified two major concerns in a letter to grazing interests: (1) damage payments for forage eaten by big game, and (2) the numbers and distribution of big game animals (Jordan 1989). He said, "It is clear to me that some landowners who support significant numbers of big game animals are doing so at considerable financial cost, while others gain financial benefits from those same animals." The goal of the Division is to develop a method to resolve landowner concerns regarding big game on private land, a goal that will increase benefits for the landowner, sportsman, and the wildlife resource. Colorado, other states, and provinces have the ability, by law, to achieve this goal. As program coordinator Jim Lipscomb stated, "Without a new approach, the Division could eventually be required to lower population objectives for big game" (Jordan 1989). Lowering wildlife populations affects the property of society and benefits the property held by relatively few persons. This represents a "frivolous" loss of wildlife which leads to Geist's final presumed "policy" of North American conservation.

Point 4: Frivolous killing of wildlife is prohibited. It prohibits waste of wildlife once killed, recognizes subsistence hunting as a priority and permits the destruction of noxious wildlife (Geist 1989). [Hunting for "fun," "sport," and "recreation" is considered frivolous.]

This point contradicts itself and does not represent policy or practice in North America. The contradiction equates recreational hunting with frivolity, then suggests that frivolous killing was prohibited. Subsistence hunting is not a priority; sport hunting is the foundation. If anything, persons with greater wealth and free time have greater opportunity for wildlife property than do persons who are poor and need food for subsistence. Subsistence hunting, we must remember, was also part of the market hunting era, a way of life of homesteaders in the west, and continued as a problem during the mining years at the turn of the century and during the depression of the 1930s. It led to the near demise of wildlife. Sport hunting led to conservation. Finally, "destruction of noxious wildlife" was and is practiced, but could hardly be considered a universally accepted policy. Destruction of noxious wildlife, whether they are large ungulates or songbirds, is a practice to be stopped or regulated, not a policy to be heralded as the foundation of wildlife conservation.

The argument about frivolous killing precipitates an antihunting debate. I will not argue the philosophy of hunting here. Comments about the contributions of sport hunters to conservation and to the recovery of wildlife populations (to abundant, huntable numbers) through the leadership of hunters were made previously in this paper, supported by Kallman et al. (1987) and acknowledged by Geist (1986, 1987, 1988, 1989).

A real form of frivolous killing is illegal killing of wildlife for food, damage control, or "personal" reasons. Native persons in remote parts of North America "subsist" on legally and illegally taken wildlife. The use of modern firearms, vehicles, and illegal markets threaten wildlife in some areas (Klein 1989). Illegal poaching is speculated to exceed the legal harvest, and apprehension rates may be as low as 1% (Sigler 1972). The current system in North America does not stop illegal wildlife killing, but Bubenick (1989) contended that losses due to poaching were minimal in Europe where the private sector, including landowners and hunters, have a much greater role in management. Laws and policy did not stop frivolity nor did they develop around the concept. Rather, laws and policies allocated legal kill to account for it.

Point 5: The "tragedy of the commons" was misused by The Economist *in reference to wildlife management in North America (Geist 1989).*

Geist explained correctly how the "tragedy of the commons" (Hardin 1968) existed during the days of exploitation. The essence of the commons is not public ownership, but "the *absence of controls over exploitation*, an absence of responsibility for care, planning, control, and restitution." I argue that private lands and landowners are now facing the tragedy because the public's wildlife affects them, and they are not enabled to have responsibility for care, planning, and restitution. Commercialization of wildlife enabled and controlled by governments will provide positive opportunities for restitution: the making good of or giving an equivalent for some injury (as a loss of or damage to property).

Point 6: Legal trafficking in wildlife removes the most important protection from wildlife, the absence of a market (Geist 1989).

No, the problem is with absence of controls to supply the market: a question of sustainable or nonsustainable use. The market hunting "days of ignorance" do not prevail. We are in a modern era where philosophies are positive and esthetic ideologies are strong. The conservation leadership is concerned about biotic diversity and species richness, conservation biology, resource integrity, and animal welfare. However, persons who interact with wildlife—ranging from American landowners and sport hunters through African subsistence-based tribes—are faced with utilitarian perspectives in addition to their esthetic philosophies. Pragmatically, utilitarian use of land and its influences on wildlife must be made compatible with wildlife resources.

Production and harvest of wildlife for commercial recreation, meat, and products are high in European countries where wildlife management by the citizens is more common than in North America (Bubenik 1989). Production of biomass is similar to land areas in the US. Yet, human populations in Europe are considerably higher than those in the US, suggesting that Europeans have a system that promotes conservation.

In the case of wild meats, the US imports most of its venison. Markets are present and growing (Hughes 1984; White 1987). Venison is one of the few agricultural products where demand exceeds supply (von Kerckerinck 1987). Illegal harvest has not filled the demand.

Selling live, big game to persons in the private sector from public lands such as national parks or urban areas with an overpopulation is also in the best interest of modern management. Once public populations of wildlife have reached saturation, the private sector can help to solve public sector problems while also benefiting. Models of successful management programs exist for legal markets and systems of production and removal for meat and animal products, live sale or hunting (Eltringham 1984; Benson 1989b; Hudson et al. 1989). They work through enabling legislation, policy, and goodwill.

Point 7: Wildlife management is "food production first and foremost" (Geist 1986).

Reasons why persons hunt in the US have little to do with meat as the primary motive (most references were cited earlier), yet food from wildlife is important to hunters (Applegate 1989 and authors cited therein). It is uneconomical to hunt for meat except for those having unequal opportunities for property in wildlife (primarily rural persons and their friends or family). Hunters in the United States spent an average of $603 dollars and 20 days in the field during 1985 (Anonymous 1988). Considering a liberal success rate of 50% on deer (Yorks 1989) and most big game, plus the opportunity cost of time, motives for hunting must be greater than for obtaining inexpensive protein. Contrary to Geist, European publics and governments do not consider game management primary as food production either (Bubenik 1989).

If food production is the first and foremost reason for wildlife management, then pen-raised production systems should undoubtedly be encouraged. Geist criticizes the wildlife management system in Germany yet acknowledges that it is "an excellent production system but a poor conservation system" (Geist 1987). If food production is the goal, then the German system should be favored by Geist.

Point 8: Remove the consumer of wildlife and the land becomes available to various forms of agricultural and forestry uses (Geist 1986).

This is correct and an argument in support of commercial wildlife values. In Texas, wildlife was valued at over $4 US billion in terms of land values and contributed over $50 per acre to the average value of deer range (Pope et al. 1984).

Point 9: The cost of privatizing wildlife will be to shift onto the public purse the cost of protecting not only the public stocks of wildlife but also the private ones (Geist 1986).

Public agencies in Mississippi, California, and Texas have not been destroyed or even threatened by existing programs (Morrill 1987). The wildlife agency budget in Texas increased over 100% from 1977 to 1986 ($49.8 million compared with $108 million). Employee numbers rose from 1754 to 2436. There are many professionals working independently who provide habitat management and habitat plans. Texas Parks and Wildlife almost doubled the public hunting areas under its control and opened state parks for hunting. The number of licensed hunters increased from

850 000 to 1 001 000 between 1977 and 1986. I interpret these changes as positive for the economy, the wildlife profession, employment, wildlife, and conservation. Surely, as Morrill contended, "hunters not hunting will decreases the number of licenses bought, thereby reducing support for fish and wildlife departments."

Bubenik (1989) reported that in Europe, where commercial use of wildlife is most talked about, game management and law enforcement are provided almost entirely by private reserve operators and private gamekeepers. The revier system offers inexpensive game and environmental protection, it produces employment and meat, and is self-sustaining. It is considered more democratic than centralized licensing systems, because it does not drain taxpayers for game protection and supplemental feeding. Venison can be sold, and markets are accessible by citizens. Losses due to poaching are minimal.

Point 10: The pressure to privatize wildlife is on, and it will grow (Geist 1986)!

Fee-hunting and other commercial uses of wildlife will likely grow owing to the ability of landowners to control their land and to charge access fees. Hunting lease income to landowners in Texas rose from an estimated $13 US million in 1965 (Klussmann 1966) to a range of $100 to $300 million in 1987 (Steinbach et al. 1987). White (1987) reviewed studies where benefits to landowners from fee-hunting were evaluated, and he predicted increased demand for using private lands and paying fees.

South Africa is the number one big game hunting destination in Africa according to *Finance Week* (1990). Provincial wildlife agencies allowed control of game to landowners who fenced animals within private properties. This caused the game industry to "skyrocket." Although I do not recommend fencing large tracts because of environmental and esthetic reasons, the opportunity in South Africa did prompt farmers to see value in wildlife and recreation. Consequently today, private enterprise owns more land specified as having wildlife use than the government (8.6 million ha compared with 5.4 million ha) (*Finance Week* 1990). In South Africa 46% of landowners planned to increase their wildlife enterprise business and 45% planned the same output (Benson 1989b). Bothma and Rabie (1983) asserted that "All possible attempts should thus be made to conserve the habitat for so-called farm game, and to reconcile the needs of humans and wild animals. It has been pointed out that unavoidable conflicts of interest where it is either the survival of man or the survival of wildlife that is at stake are rare, and that in most cases all it takes it a little thought, ingenuity and goodwill for man and wildlife to find a way of coexisting harmoniously. If it is borne in mind that wildlife is a national asset conserved by the state for the benefit of the whole nation, the landowner's responsibility should be clear. Basically, conservation by landowners will have to be accomplished through education, which will hopefully occur in time to save our wildlife. There are also other stimulants, mainly of an economic nature, such as surrendering a part of the hunting license fee to the landowner, and encouraging the establishment of private game reserves and game farming. The fact remains that the maintenance of wild animals on private land often depends on an incentive for the landowner to keep them there." Luxmoore (1985) provided further evidence that conservation was improved in South Africa through profit incentives to the landowner. Commercial growth in the wildlife management field can be valuable for conservation.

"Privatize" is a loaded word designed to generate negative attitudes. It means "to move ownership from the public to private sector." Privatization is not what the commercialization debate is about. Enabling landowners to manage their land and wildlife resources to reap some of the allotted wildlife surpluses through equitable fee-hunting and production systems is all that is asked.

Point 11: Game ranching and market hunting [to be confused with Market Hunting of old?] are not compatible with predators (Geist 1988).

Uncontrolled predator removal is not appropriate in modern times.

Point 12: Introduction of exotics causes disease

problems, competition and, genetic pollution (Geist 1988).

Do not allow exotics to be introduced except where their populations are controlled and where accidental release will not be an unsolvable problem. Allowing use of native species with wild blood lines for meat production in enclosed environments would avoid all but the disease problem. Modern husbandry practices and veterinary care are well known.

There are numerous supplies of overpopulated and unwanted ungulates that need reduction in urban areas, military reservations, and parks. If wildlife are removed under the present system, the costs are borne by the public and animals usually are killed. It they could be trapped and used privately, costs could be borne by the person enabled by government to acquire and use them productively.

Placing wildlife in small enclosures removes them from the "wild." They are no longer wildlife and must be treated as domesticated animals. Wildlife agencies are correct to be concerned about problems caused if these animals escape.

Point 13: Paid hunting will form an "elite" who can pay and is a deterrent that excludes those with low incomes, the young and early-married, and poor ethnic minorities (Geist 1988).

Results from the 1985 National Survey of Fishing, Hunting, and Wildlife Associated Recreation (Anonymous 1988) places a perspective on actual hunting participation rates and ways that expenditures were made. Of all hunters, 96% were white, 24% were 16 to 24 years old (the total population in that category was 25%), and all income levels in the US were represented. The majority (54%) made $25000 or more, and 43% made less. Only a small part of hunters' expenditures supports the basic production of wildlife. Nine percent of hunters' expenditures were to lease or own private land. Only 4% of expenditures went for licenses, stamps, tags, and permits (nondiscretionary costs). Hunters spent $10.1 US billion for hunting with 86% going for discretionary purchases to service industries for equipment (49%) and trip-related activities (37%). Discretionary income was not spent in any great amount to improve the resource base. Fortunately excise taxes on firearms and ammunition and federal tax suppport can be added indirectly to the total financial picture or the discretionary contributions by hunters for wildlife management would be small.

Fee-hunting could cause some hunters to stop hunting, but that is not currently a major factor (Applegate 1989). Conversely, Fitzhugh (1989) suggested that if ranches with smaller properties in California are discouraged from an organized program of "wildlife ranching," access would be lost and the program would be driven from market control toward an elitist system of pricing. Demographics of hunters in Texas suggest that incomes are higher than the averages in Texas or nationally but that 62% hunted without purchasing a lease (Adams and Thomas 1986).

Point 14: Game ranching would remove public land from recreational purposes because game ranching and unfetttered access by the public to such [public] land are not compatible (Geist 1985).

Private landowners can legally exclude persons from access to public land only by land juxtaposition. Access should be provided where possible through governmental acquisitions, easements, agreements, goodwill, or condemnation when situations are totally against the good of society. Laws and good sense provide for each. Game ranching, if managed in cooperation with governments, should never deny access of the public unless in so doing the public will ultimately benefit. Conversely, if governments refuse to cooperate, access may be denied across private land to public areas to an even greater extent.

Public lands tend to be managed to allow "unfettered" access in a negative sense causing too many users to be spread over too small areas in too little time. Managed private land should enable proportioned use and consumption of wildlife in relation to harvest quotas (according to public and private objectives), behavior and social considerations of wildlife, and time needed to have quality recreational experiences.

Conclusion

Commercialization of wildlife may be the only way to bring professional wildlife management and the private sector together. The current system of wildlife management in North America has tended to focus on acquisition of public lands, managed by public servants, for the public good with little recognition of the problems, rights, contributions, and influences from the private sector of production. Private landowners are part of the public, and they represent a significant source of hunting recreation (82%) and production of wildlife.

Leopold's game policy (1930) and the World Conservation Strategy of 1980 (Luxmoore 1989: 413–414) recognized the need to balance environmental needs with sustainable utilization of ecosystems and species. To address that need we cannot be constrained by only allowing government to solve problems on their own. The historical exploitation of wildlife represented public responses to public policy of the times. Later, policy of governments prompted by enlightened citizens played a vital role in reversing the trends. Times are changing again and the enlightened private sector can now aid governments.

Governments that exclude private landowners from thoughtfully integrating wildlife and recreation management into their plans are now the problem. Agriculture, range, forestry, and water management practices harmed wildlife habitats. Incentives for landowners to do otherwise on behalf of the public's wildlife were not strong enough. The old benefits from complacency and altruism by landowners are being replaced with economic analyses that examine costs and benefits and the margins of production and profit. Landowners are asking whether wildlife and public access are in their best interests and many are saying "no!"

Governmental supported commercialization of wildlife is recommended as a management need and solution. Not all lands are capable of providing viable wildlife habitats, populations, or enterprises and not all landowners want to be involved. But, if persons on the land who are closest to wildlife influences perceive that governments do not have concern for their needs, they may reach the point of rebellion and thus take a proactive role to oust wildlife and recreationists. Or, they may take the enterprise issue into their own hands exercising de jure rights to land access and de facto property authority over the animals that reside there. If landowners are made partners in wildlife management they will seek education to do the job properly just as the South African landowners are doing (Benson 1989b). The wildlife profession must be ready to help when needed. The goals will be the same: to preserve environmental integrity, wildlife populations, and human-related values on a sustained yield basis.

This debate with Dr. Geist may cast us in the light of crusaders. Please do not give that distinction to me. Criticize me only as a thinker who perceives the present system as grand, but flawed, and who seeks to find other useful and sustaining tools. Leopold (1933) warned us of the crusaders: "The Crusader wrote many volumes, but these told us why rather than how wildlife and civilization should be adjusted to each other. These men were mostly biologists, but strangely enough their technique was not biological. It was, rather, an intensification of the pre-existing idea of protective legislation, which experience has now shown does not alone suffice, even when enforced. If retards, but does not reverse, the forces of destruction."

References

Adams CE, Thomas JK (1986) Characteristics and opinions of Texas hunters. International Ranchers Roundup. Kerrville, Texas, Texas A & M University, Agricultural Research and Extension Center, Uvalde, TX, pp 255–261

Anonymous (1990) Arizona public land policy. Hunting Ranch Bus 5:4

Anonymous (1988) 1985 National Survey of Fishing, Hunting, and Wildlife Associated Recreation. USDI Fish and Wildlife Service, Washington, D.C.

Applegate JE (1989) Patterns of early desertion among New Jersey hunters. Wildl Soc Bull 17:476–481

Applegate JE, Otto RA (1982) Characteristics of first-year hunters in New Jersey. NJ Agric Exp Stn Bull R-12381-(1)-82

Barkley PW, Seckler DW (1972) Economic Growth and Environmental Decay—the Solution Becomes the Problem. Harcourt Brace Javanovich, New York

Batie SS, Taylor DB (1990) Cropland and soil sustainability. In Sampson RN, Hain D (eds) Natural Resources for the 21st Century. American Forestry Assoc, Island Press, Washington, D.C., pp 56–77

Benson DE (1991) Integrated wildlife management partnership among agriculture, natural resources professions and business. Proc 2nd Int Wildl Ranching Symp, Alberta, Canada (in press)

Benson DE (1989a) Changes from free to fee hunting. Rangelands 11:176–180

Benson DE (1989b) Private values and management of wildlife and recreation in South Africa with comparisons to the USA. PhD dissertation, Colorado State University, Fort Collins, CO

Benson DE (1988) What fee hunting means to sportsmen in the USA: a preliminary analysis. In Valdez R (ed) Proc 1st Int Wildl Ranching Symp, las Cruces, New Mexico, Wild Sheep and Goat International Box 244 Mesilla, NM, pp 296–302

Berryman JH (1987) Socio-economic values of wildlife resource: Are we really serious? In Decker DJ, Goff GR (eds) Valuing Wildlife—Economic and Social Perspectives. Westview Press, Boulder, Colorado, pp 5–11

Berryman JH (1981) Needed now: an action program to maintain and manage wildlife habitat on private lands. In Dumke RT, Burger GV, March JR (eds) Wildlife Management on Private Lands. La Crosse Printing, La Crosse, Wisconsin, pp 6–10

Bothma JDu P, Rabie MA (1983) Wild Animals. In Fuggle RF, Rabbie MA (eds) Environmental Concerns in South Africa. Technical and Legal Perspectives. Juta, Cape Town, Republic of South Africa, pp 190–236

Brown TL (1974) New York landowners' attitudes toward recreation activities. Trans North Am Wildl Nat Res Conf 39: 173–179

Brown TL, Decker DJ, Purdy KG, Mattfeld GF (1987) The future of hunting in New York. Trans North Am Wildl Nat Res Conf 53:553–566

Brown TL, Decker DG, Kelley JW (1984) Access to private lands for hunting in New York: 1963–1980. Wildl Soc Bull 12:344–349

Bubenik AB (1989) Sport hunting in continental Europe. In Hudson RJ, Drew KR, Baskin LM (eds) Wildlife Production Systems. Cambridge University Press, Cambridge, pp 115–133

Burger GV, Teer JG (1981) Economic and Socioeconomic issues influencing wildlife management on private land. In Dumke RT, Burger GV, March JR (eds) Wildlife Management on Private Lands. Dept Nat Resources, Madison, Wisconsin, pp 252–278

Chandler WJ (ed) (1988) Audubon wildlife Report 1988/1989. National Audubon Soc, New York

Cordell HK (1990) Outdoor recreation and wilderness. In Sampson RN, Hain D (eds) Natural Resources for the 21st Century. American Forestry Assoc, Island Press, Washington, D.C., pp 242–268

Cordell HK et al. (1988a) An Analysis of the Outdoor Recreation and wilderness Situation in the United States: 1989–2040. A Technical Document Supporting the 1989 RPA Assessment. USDA Outdoor Recreation and Wilderness Assessment Res Unit, Southeast Forest Exp Station,

Cordell HK, Wright B, Powell B, Guldin R, Kaiser F (1988b) Leasing values and management of wildlife and fish resources on private lands in the United States. Trans North Am Wildl Nat Res Conf, Louisville, Kentucky, 53: (Abstract)

Cronquist MV (1990) Attitudes or pronghorn hunters for paying to hunt on private lands in Colorado. MSc thesis, Colorado State University, Fort Collins, Co

Dumke RT, Burger GV, March JR (1981) Wildlife Management and Private Lands. La Crosse Printing, La Crosse, Wisconsin

Eltringham SK (1984) Wildlife Resources and Economic Development. Wiley, Chichester, New York

Finance Week (1990) Buck in the bag SA is a number one big game hunting destination. Focus On South Africa. March, p 15

Fitzhugh EL (1989) Innovation of the private land wildlife management program: a history of fee hunting in California. Trans West Sect Wildl Soc 25:49–59

Geist V (1989) Legal trafficking and paid hunting threaten conservation. Trans North Am Wildl Nat Res Conf 54:171–178

Geist V (1988) How markets in wildlife meat and parts, and the sale of hunting privileges, jeopardize wildlife conservation. Conserv Biol 2:1–12

Geist V (1987) Three threats to wildlife: game markets, pay hunting and hunting for "fun." Proc Privatization of Wildl Public Land Access Symp, Wyoming Game and Fish Dept, Cheyenne, pp 46–58

Geist V (1986) Wildlife a public trust. Western Sportsman. July/August, pp 52–55

Geist V (1985) Game ranching: threat to wildlife conservation in North America. Wildl Soc Bull 13:594–598

Gilbert AH (1977) Influence of hunter attitudes and characteristics on wildlife management. Trans North Am Wildl Nat Res Conf 42:227–235

Guynn DE, Schmidt JL (1984) Managing deer hunters on private lands in Colorado. Wildl Soc Bull 12:12–19

Hagenstein P (1990) Forests. In Sampson RN, Hain D

(eds) American Forestry Assoc, Island Press, Washington, D.C., pp 78–100
Hardin G (1968) The tragedy of the commons. Science 162:1243–1248.
Hendee JC (1974) A multiple-satisfaction approach to game management. Wildl Soc Bull 2:104–113
Holecek DF, Westfall RD (1977) Public recreation on private lands—the landowners' perspective. Mich Agric Exp Stn Res Rep 335
Hudson RJ, Drew KR, Baskin ML (1989) Wildlife Production Systems—Economic Utilization of Wild Ungulates. Cambridge University Press, Cambridge
Hughes DM (1984) Meat processors view of game ranching for meat production. International Ranchers Roundup. San Angelo, Texas, Texas A & M University, Agricultural Research and Extension Center, Uvalde, TX, pp 336–343
Jackson RM, Anderson RK (1982) Hunter-landowner relationship: a management and educational perspective. Trans North Am Wildl Nat Res Conf 47:693–703
Jordan G (assoc ed) (1989) New landowner compensation sought by DOW. News for Landowners 7:1–2
Kaiser RA, Wright BA (1985) Recreational use of private lands: beyond the liability hurdle. J Soil Water Conserv 40:478–481
Kallman, H, Agee CD, Goforth WR, Linduska JP (eds) (1987) Restoring America's Wildlife 1937–1987. USDI Fish and Wildlife Service, Washington, D.C.
Klein DR (1989) Northern subsistence hunting economics. In Hudson RJ, Drew KR, Baskin LM (eds) Wildlife Production Systems. Cambridge University Press, Cambridge, pp 96–111
Klussmann WG (1966) Deer and commercialized hunting systems in Texas. Proc the White-Tailed Deer: Its Problems and Potentials. Texas A&M University, College Station, pp 18–21
Langner LL (1987) Hunter participation in fee access hunting. Trans North Am Wildl Nat Res Conf 52:475–481
Larson JS (1958) Straight answers about posted land. Trans North Am Wildl Nat Res Conf 24:480–487
Latham A (1989) The cosmic elk factor: Who owns New Mexico's wildlife? N M Res Coll Agric Home Econ 2:5–9, 12
Laycock WA (1987) History of grassland plowing and grass planting on the Great Plains. In Mitchell JE (ed) Impacts of the Conservation Reserve Program in the Great Plains. USDA Forest Service Gen Tech Rept RM-158, Rocky Mountain Forest and Range Experiment Station, Fort Collins, CO, pp 3–8
Leopold A (1933) Game Management. Scribner's, New York
Leopold A (1930) The American game policy. Trans Am Game Conf 17:284–307
Luxmoore RA (1989) Impact on conservation. In Hudson RJ, Drew KA, Baskin LM (eds) Wildlife Production Systems. Cambridge University Press, Cambridge, pp 413–423
Luxmoore RA (1985) Game farming in South Africa as a force in conservation. Oryx 19:225–231
Morrill WI (1987) Fee access views of a private wildlife management consultant. Trans North Am Wildl Nat Res Conf 52:530–543
Novak M, Baker JA, Obbard ME, Mallech B (eds) (1987) Wild furbearer management and conservation in North America. Ontario Ministry of Natural Resources, Toronto, Canada
Ortega Y Gasset J (1972) Meditations on Hunting. Scribner's, New York
Pope CA III, Adams CE, Thomas JK (1984) The recreational and aesthetic value of wildlife in Texas. J Leisure Res 16:51–60
Potter DR, Hendee JC, Clarke RN (1973) Hunting satisfaction: game, guns, or nature? Trans North Am Wildl Nat Resour Conf 38:220–229
Reiger JF (1975) American sportsmen and the origins of conservation. Winchester Press, New York
Rounds RC (1975) Public access to private land for hunting. Colorado Div Wildl Report 2
Sheldon R (1987) Fee hunting systems and important factors in wildlife commercialization on private lands. In Decker DJ, Goff GR (eds) Valuing Wildlife—Economic and Social Perspectives. Westview Press, Boulder, Colorado, pp 109–116
Sigler WF (1972) Wildlife law Enforcement, 3rd ed. Wm C Brown Publishers, Dubuque, Iowa
Steinbach DW, Conner JR, Glover MK, Inglis JM (1987) Economic and operational characteristics of recreational leasing in the Edwards Plateau and Rio Grande Plains in Texas. Trans North Am Wildl Nat Res Conf 52:496–515
Swank WG (1981) Wildlife laws and their effect on deer management. International Ranchers Roundup. Texas A & M University, Agricultural Research and Extension Center, Uvalde, TX, pp 361–366
Teer JG, Burger GV, Deknatel CY (1983) State-supported habitat management and commercial hunting on private lands in the United States. Trans North Am Wildl Nat Res Conf 48:445–456
Thowardson N (1979) Landowner constraints on Oklahoma hunting opportunities. MSc thesis, Okahoma State University, Stillwater
Tober JA (1981) Who Owns the Wildlife? Greenwood Press, Westport, Conn.
Trefethen JB (1975) An American Crusade for Wildlife. Winchester Press, New York

Von Kerckerinck J (1987) Deer Farming in North America. Phanter Press, Rhinebeck, New York

White RJ (1987) Big Game Ranching in the United States. Wild Sheep and Goat International. Box 244, Mesilla, New Mexico

Wright BA, Kaiser RA (1986) Wildlife administrator's perceptions of hunter access problems: a national overview. Wildl Soc Bull 14:30–35

Wright BA, Kaiser RA, Fletcher JE (1988) Hunter access decisions by rural landowners: an East Texas example. Wildl Soc Bull 16:152–158

Wigley TB, Melchiors MA (1987) State wildlife management programs for private lands. Wildl Soc Bull 15:580–584

Yorks TP (1989) Ranching native and exotic ungulates in the United States. In Hudson RJ, Drew KR, Baskin LM (eds) Wildlife Production Systems. Cambridge University Press, Cambridge, pp 268–285

130

Deer Ranching for Products and Paid Hunting: Threat to Conservation and Biodiversity by Luxury Markets

Valerius Geist

Abstract

Farming deer for antlers, venison, salable parts, or for paid hunting poses unsurmountable hazards to conservation of native American deer. Despite warnings, these activities continue with support from agriculture. Some realized consequences of this commercialization are as follows: (1) transmission of foreign disease to susceptible deer species, their commensals, and predators; (2) inadvertant and deliberate genetic pollution of native deer; (3) competition from feral populations of nonnative deer; (4) uncontrollable, illegal killings, and capture of deer supplying meat, parts, and trophy markets, and living animals to game ranches; (5) corruption of conservation legislation by eliminating proven conservation policies; (6) loss of power by law enforcement and regulatory agencies due to attacks by powerful interest groups; (7) loss of public accountability and control over conservation by making native wildlife de facto private property; (8) loss of support for wildlife management through disenfranchising the public and de facto allocating benefits from public wildlife along racial, social, and income lines; (9) loss of jobs, income, and influence for the wildlife profession as an elite of wealthy landowners take control. For the sake of a $5 US billion world market in *dead* wildlife, a North American market worth $60 US billion annually, but centered on *living* wildlife, is put at risk, as is North America's successful system of wildlife conservation.

Key words: Wildlife policy, deer ranching/farming, paid hunting

Introduction

The debate about "fee hunting" in North America is part of a larger, complex conservation issue. At risk is the future of wildlife and our continental system of wildlife conservation. It is a successful system, and an illuminating example of how to manage a renewable resource on a sustained basis with great economic benefits, in a democratic, industrial society, with widespread ownership of firearms. Under the banner of "privatization" it faces determined attempts by the wealthy to abbrogate public wildlife into private ownership. This grab for wildlife has historic precedence. The rich and powerful have been drawn irresistably to wildlife throughout occidental history and have used it for their pleasure and profit as they saw fit. The "privatization" of public wildlife has many ramifications, including the destruction of North America's effective conservation policies, the threat to native large mammals from diseases and competition, but also the shrinkage of civic liberties, the loss of public lands, and the loss of treaty rights by native people (Swenson 1983; Kahdren 1983; Geist 1985a, 1988, 1989). Wildlife policies have had nontrivial implications for the ownership of firearms (Lund 1980), freedom from categories of wildlife crimes, freedom to use public lands (Caughley 1983), and freedom from personal dangers arising from irresponsible trespass and wildlife protection (Gesit 1988). Paid hunting advocates argued for the conversion of public land into privately controlled wildlife ranches as happened in New Zealand (Massey 1986a). The conversion of public land to private control was an expressed goal of the Hageman Bill defeated in Wyoming by public opposition. I shall emphasize nine points relevant to recent developments.

Transmissions of Foreign Diseases to Susceptible Deer Species, Their Commensals and Predators

Parasitologists warn of disease and parasite introudctions (Holmes 1982; Samuel 1987). Examples are: (1) the spread of *Brucella suis* from reindeer (*Rangifer tarandus*) introduced from Siberia to Alaska and the spread of this disease to native caribou, carnivores, and humans (Meyer 1966; Neiland et al. 1968; Broughton et al. 1970; Neiland 1975; Tessaro and Forbes 1986); (2) the poor performance of white-tailed deer (*Odocoileus virginianus*) infected with parasites from sika deer (*Cervus nippon*; Davidson and Crow 1983; Davidson et al. 1985); (3) the spread of *Elaphostrongylus cervi* from European reindeer introduced into Newfoundland to native caribou and possibly to moose (*Alces alces*; Lankester 1987). However, there is for today's North American large mammals a much larger problem.

North American large mammals, due to a quirk of history, appear neither resistant to competition from Eurasian relatives, nor to their diseases (Geist 1985a, 1985b, 1988). Postglacial megafaunal extinctions in North America (14 000 to 8 000 BC) removed an ecologically highly specialized fauna (Guthrie 1984) and led to the formation of a new, species-poor megafauna based on Siberian immigrants and Rancholabrean survivors. This generates unique conservation problems. The Siberian wapiti (*Cervus elaphus canadensis*, mose (*Alces alces*), bison (*Bison occidentalis*), grizzly bear (*Ursus arctos*), timber wolf (*Canais lupus*), man (*Homo sapiens*), plus two species that came from Siberia earlier, the bighorn sheep (*Ovis canadensis*) and mountain goat (*Oreamnos americanus*), appear to be susceptible to diseases of southern relatives, and as ecological generalists, are expected to be poor competitors. They are vulnerable due to long evolutionary histories in the dry, cold climates of Siberia that are inhospitable to pathogens and parasites (Geist 1985a). The best data supporting this hypothesis comes from the effects of the diseases of Eurasian immigrants on American Indians (Baruzzi et al. 1979; Joralemon 1982; Dobyns 1983). The susceptibility to diseases may have overshadowed the poorer competitive abilities of northerners wherever they had to compete against Eurasian species. American and Siberian wapiti, released in Europe, readily became ill and fared poorly against red deer (Beninde 1937), as they did in New Zealand (see below). American moose brought into Europe died inexplicably soon thereafter. Siberian roe deer, introduced repatedly into Europe, also failed to thrive (Stubbe and Passarge 1979). Foreyt (personal communication) placed six clinically healthy domestic sheep together with six bighorn sheep. The first bighon was dead in 4 days, the sixth in 71 days. The bighorns were victims of *Pasteurella hemolytica*, carried by four of the six domestic sheep. Similar experiments were conducted in Alberta on bighorns by infecting them with pathogens from domestic sheep and cattle (Onderka et al. 1988; Onderka and Wishart 1988). This confirmed earlier conclusions about the direction of disease transfer between bighorns and livestock (Goodson 1982). Domestic sheep and cattle originate from wild forms adapted to warmer latitudes and altitudes than the ancestors of bighorn sheep (Zeuner 1967; Geist 1985b).

These experiments illustrate what is likely a general problem for North American large mammals of Siberian origin: weak immune systems due to a long evolutionary history in the dry, cold climates of Pleistocene Siberia and Beringia. Consequently, Siberian mammals are likely to "sponge-up" diseases and parasites of American relatives as illustrated by those of *Odocoileus* (and introduced Eurasian livestock) on moose, wapiti, and caribou (Anderson 1972; Anderson and Lankeaster 1974; Samuel 1979; Hibbler 1981; Lankester 1987). Contrary to the above, White (1987) states that bison are more, not less, resistant to diseases than cattle.

America's postglacial fauna is of theoretical

interest, as it appears poorly adapted to North America and to one another. Inadequacy of adaptation is seen in the overlap in antler growth and rutting in California elk (McCullough 1969; Geist 1982: 248, 272–274), in the frozen "cauliflower" ears of the mule deer (*Odocoileus hemionus*) in the Yukon Territory (Geist 1981), in the intolerance to the diseases and parasites of *Odocoileus* by the Siberian cervids, in the poor fit of bighorn sheep to American deserts in contrast to the success of reintroduced "old Americans" such as the subgenus *Asinus* and *Equus* in the same badlands (Bailey 1980, 1984; McCutchen 1981; Weyhausen 1984; Geist 1985b), or the success of desert-adapted caprids introduced from Eurasia and Africa (Decker 1978). Also, the Siberians never entered Middle and South America.

Inadvertant and Deliberate Genetic Pollution of Native Deer

Game ranching, regardless of type, poses a threat to the genetic integrity of North American wildlife; it threatens extinction by degree, loss of genetic and cultural information, and is a threat to the hybrids during severe conditions. While hybrids can readily survive in captivity where food and security are assured, such need not be the case for free-living hybrids.

The white-tailed deer, at the subspecies level, is already genetically polluted due to attempts to breed trophy bucks by introducing large northern whitetails to populations of small-bodied deer (Etling 1985). However, since white-tailed deer in their recovery from near extinction in the 1880s were often transferred into vacant habitat by human hands, one may ask whether additional mixing matters. In Europe escaped sika deer have hybridized so extensively with red deer that the red deer may be lost as a species (Harrington 1973; Lowe and Gardiner 1975; Bartos et al. 1981).

While studies of how hybridizations affects the adaptations of American deer are not available, excepting Lingle (1989), there is some information from other species. Genetic pollution essentially disrupts the integrity of a time-tested, coadapted gene complex as Turcek (1951) illustrated for crosses of ibex (*Capra ibex*) with *C. hircus* and *C. nubiana*. The timing of breeding seasons was disrupted and resulted in births during the cold months, followed by the extinction of the hybrid population through high neonatal mortality. Crosses of Siberian roe bucks (*Capreolus pygargus*) and European roe deer (*C. capreolus*) females led to birth complications due to larger than normal fetuses (Turcek 1951; Stubbe and Passarge 1979). European wild boar (*Sus scrofa*) appear to have various admixtures of domestic pig, recognizable by the presence of spotted individuals. Spotting is associated with decreased survival, lower oxygen binding capacity in the blood, decreased thermoregulatory ability, and poorer development of the underwool (Briedermann 1986: 94). Our work with the behavior of mule deer × white-tailed deer hybrids showed that the antipredator strategies of the hybrids are totally disrupted. In addition, the ability of the hybrids to run is impaired by a variety of inefficiencies not found in the comparable gait of the parent species; stotting, a specialized bound by the mule deer, was not witnessed even in seven-eighths mule deer by one-eighth white-tailed deer hybrids (Lingle 1989).

The claim by trophy ranchers and various schemes to farm and market dead wildlife, that they act in the interest of conservation, is unsubstantiated (Ehrenfeld 1974, 1981; Geist 1985a, 1988).

Competition from Feral Populations of Nonnative Deer

The surviving Rancholabrean species in North America's extant large mammal fauna appear to be specialists in opportunistic resource exploitation. They are primitive, old species with a preference for concentrated foods, and are characterized by high reproductive potentials and great ability at dispersal. As opportunists, that is, as forms that deliberately avoid competition in favor of dispersal and searching for opportunities to exploit locally abundant resources, they are poor competitors for material resources. They include the white-tailed deer, the black-tailed deer, the pronghorn (*Antilocapra americana*), the collared peccary (*Tayassu tajacu*), the black bear (*Ursus americanus*) and the coyote (*Canis latrans*).

Circumstantial evidence also suggests that these are indeed poor competitors:

1. These species are rare among Rancholabrean fossils, but are common after megafaunal extinction while the peccary greatly extended its range (Kurten and Anderson 1980).
2. The best evidence for competitive inferiority comes from the introductions of American and Eurasian deer species into New Zealand and Europe. In New Zealand, white-tailed deer are reduced to two populations, one of doubtful viability; black-tailed deer and American moose died out. The wapiti established itself, but spread much less than the red deer. Wapiti, white-tailed deer, and moose in New Zealand shrunk in body and antler size; wapiti have thin, translucent skulls (Batcheler and McLennan 1977). Red deer, sika deer, fallow deer (*Dama dama*), rusa deer (*Rusa timorensis*), and sambar (*Rusa unicolor*), Eurasians all, became established, grew to fairly large sizes, and are still thriving (Wodzicki 1950; Caughley 1983; Harris 1984). In Europe, white-tailed deer have done poorly when pitted against roe deer (Bojovic and Halls 1984). In Finland, on agricultural land in the absence of roe deer, they are a qualified success (Nygren 1984). Elsewhere in Europe introduced white-tails live in small, closely managed populations (Bojovic and Halls 1984).

Fallow deer are now established in Montana (near Sidney) and on some Gulf Island in British Columbia (Cowan and Guiguet 1965), despite native white-tailed, black-tailed, and mule deer.

On the eastern shores of Maryland, sika deer, introduced in 1916 are outcompeting and displacing white-tailed deer (Flyger 1960; Feldhamer et al. 1978; Keiper et al. 1984; Keiper 1985); sika deer are better foragers than white-tailed deer, as shown in field and experimental studies (Armstrong and Harmel 1981; Feldhamer et al. 1978). In a long-term experimental study, sika deer displaced white-tailed deer within an enclosure (Armstrong and Harmel 1981; White 1987: 63). That exotics can be permanently kept behind fences is not borne out by experience here and abroad (Massey 1986b; Rennie 1986).

In Texas extensive introductions of foreign species of ungulates have been carried out. In a careful review, White (1987: 56–68) points out that these are proving very difficult to control; on some ranches they have virtually displaced white-tailed deer. White (1987), a proponent of wildlife ranching, cautions about the introduction of exotics, pointing out that the results are unpredictable, that long-term consequences are unknown, and that the introduction and spread of parasites and pathogens are a painful reality. A growing threat to North America's unique desert bighorn sheep and to mule deer are introductions of barbary sheep (*Ammotragus lervia*) in the southern United States and Mexico. This species disperses rapidly, has a high reproductive rate, broad food habits, and is a hardy carrier of diseases (Simpson et al. 1978, 1980; Seegmiller and Simpson 1979). Game ranches in Alberta are vying to breed pedigreed elk with exceptionally large antlers for the pantin trade, or have attempted to breed "super elk" for sale to wealthy trophy hunters.

Uncontrollable Illegal Killings and Capture of Deer Supplying Meat, Parts and Trophy Markets, and Living Animals to Game Ranches

The surpise in recent years is not that common criminals have enterd into the profitable illegal trade in wildlife, but that large ranch companies have been charged with such crimes, or have been successfully prosecuted. Here I point to the investigative reporting by the *Albuquerque Tribune* (16–24 October *l.c.* 1989); I participated as an expert witness in Alberta in one successful prosecution of such a case in 1989 and am aware of conviction in a second.

Corruption of Conservation Legislation by Eliminating Proven Conservation Policies

Geist (1988) pointed to four policies basic to the success of wildlife conservation in North America: (1) public ownership and de facto control of wildlife; (2) prohibition of markets in dead, vulnerable wildlife; (3) allocation for consumption of surplus wildlife by law; and (4) prohibition on nonfrivolous killing of wildlife.

Legalization of game ranching, in whatever jurisdiction, violates the first policy. It transfers affected wildlife into the private domain to be manipulated for the sake of the market place by whatever means possible. It is an abdication of public responsibility for wildlife.

Legalization of game ranching aims to create legal markets in venison and wildlife parts, and thus violates the second policy. Laundering of illegally killed wildlife is beyond control in a large infrastructure of producers and retail outlets.

Paid hunting on private land is a violation of the third policy. It allocates wildlife for consumption by ability to pay, and de facto kills recruitment into the hunting force by discriminating against the less than affluent young hunters and the less than affluent minorities. Thus, it generates de facto racial discrimination in wildlife consumption.

The policy against frivolous killing is violated not legally, but morally, where hunting is done for entertainment. When hunting no longer is the necessary annual harvest of the land participated in by the many, hunting for "fun, sport and trophies" becomes a reprehensible activity by the few. It is hardly what Leopold (1933) stood for. Current tests of public opinion bear this out (MacDonald 1987).

The game ranching regulations passed in 1988 by the province of Sakatchewan show how sloppy wording can endanger wildlife and public health: the species allowed for ranching are so poorly defined that virtually anything may be imported; the wild animals are subject to checking for diseases of domestic diseases only; stock for game ranching must be imported from outside the province. The legislation is written as if to spread wildlife diseases deliberately.

Loss of Power by Law Enforcement and Regulatory Agencies due to Attacks by Powerful Interest Groups

Babington (1989) in *The Texas Lawyer* and Marshall (1990) in *Science* reported on the fate of law and science when in conflict with trophy collecting. The case, unresolved in its ramifications, illustrates the manner in which influence could be exerted.

Loss of Public Accountability and Control over Conservation by Making Native Wildlife De Facto Private Property

Paid hunting entails the raising of native or exotic big game for sale to affluent hunters for sport hunting or trophies (White 1987; Roseborough 1986). For obvious political reasons we know little about management practices on these private ranches. However, the German system of wildlife management is de facto game ranching of this type; its consequences are well recorded. It favors game over nongame species, exterminates large predators, and severely culls small ones. It promotes agricultural-type habitat manipulation, introductions of nonnative species and strains, and genetic manipulation of wildlife (Beninde 1937; Draskovich 1951; Stahl 1979; Nuesslein 1983; Eggeling 1983). While delivering only fair wildlife production and economic return (see Geist 1988), the German wildlife management system is a poor conservation system. It reflects a history of autocratic rule (Stahl 1979), has little support among the public (*Der Spiegel* 1983, 37[4]–100, 37[4]102–105), and is struggling to adjust as hunting becomes less popular (Schroeder 1986). Because of its close association with elitism and the perception of hunting as a frivolous pasttime of the affluent, it does not protect wildlife against local interests. In a recent case in southern Germany, in a choice between removing from state forests domestic sheep or wildlife, the vote was cast to eliminate wildlife (Hesper 1988). Nevertheless, the German system compared with those in Italy or France, retains self-sustaining game populations, a significant success, though it falls short of the American success in conservation.

Loss of Support for Wildlife Conservation Through Disenfranchising the Public and De Facto Allocating Benefits from Public Wildlife Along Racial, Social, and Income Lines

I pointed out this danger in Geist (1988), but it appears to be reality already, judging from investigative reporting by the Albuquerque

Tribune (25 October 1989, p. 25): Latin Americans barred from wildlife and condemned as poachers, white (sometimes absentee) ranchers who sell local wildlife to affluent guests. One rancher resorted to a bullet-proof vest and a bullet-proof truck to protect himself against local wrath.

Loss of Jobs, Income, and Influence for the Wildlife Profession as Agriculture and an Elite of Wealthy Landowners Take Control

With wildlife private property, be it de jure or de facto, wildlife managers become gamekeepers, aides in fulfilling private wishes. In my home province of Alberta, with the growth of overt and covert support of the game ranching industry by the government, the Wildlife and Fisheries Branch has lost its research capabilities, provincial biologists search for private funds to do elementry wildlife work, and the department is due for downsizing as personnel are to be transferred to Agriculture with game ranching. Big game populations are one-third of those found in Wyoming or less, though Alberta is three times larger than Wyoming. This, despite wildlife's economic competitiveness (Johnson and Lindner 1986; Filion et al. 1983), despite some $60 US billion spent annually by North Americans on wildlife viewing, hunting and fishing, despite some 50 000 jobs created per billion dollars spent on such activities, and despite some 160 million from 270 million North Americans participating in viewing, hunting and fishing. The $5 US billion world market in dead wildlife is seen as more promising.

References

Albuquerque Tribune (1989) Series, 16 Oct to 3 Nov: A price on their heads. Special Reprint. Albuquerque, New Mexico

Anderson RC (1972) The ecological relationships of meningeal worm and native cervids in North America. J Wildl Dis 8:304–310

Anderson RC, Lankester WM (1974) Infectious and parasitic diseases and arthropod pests of moose in North America. Natural Can 101:25–50

Armstrong WE, Harmel DE (1981) Exotic mammals competing with the natives. Texas Parks Wildl 39:6–7

Bailey JA (1984) Bighorn zoogeography: response to McCutchen, Hansen and Weyhausen. Wildl Soc Bull 12:86–89

Bailey JA (1980) Desert bighorn, forage competition and zoogeography. Wildl Soc Bull 8:208–216

Babington C (1989) Big guns flex political muscle in wildlife case. Texas Lawyer, 18 Dec, 10

Bartos L, Hyanek J, Zirovnicky J (1981) Hybridization between red and sika deer. 1. Craniological analysis. Zool Anz Jena 207:260–270

Baruzzi RC, Marcopito LF, Serra MLC, Souza FAA, Stabile C (1979) The Kren-Akorore: a recent contacted indigenous tribe. In Health and Disease in Tribal Society. Ciba Foundation Symp 49. Elsevier, Amsterdam, pp 179–200

Batcheler LC, McLennan MJ (1977) Craniometric study of allometry, adaptation and hybridization of red deer (*Cervus elaphus* L.) and wapiti (*C.e. nelsoni*, Bailey) in Fjordland, New Zealand. Proc N Z Ecol Soc 24:57–75

Beninde J (1937) Zur Naturgeschichte des Rothirsches. Monographie der Sáugetiere, vol 4. P Schoeps, Leipzig

Bojovic D, Halls LK (1984) Central Europe. In Halls LK (ed) The White-tailed Deer. Stackpole Books, Harrisburg, Pa. pp 557–560

Briedermann L (1986) Schwarzwild. Verlag Neumann-Neudamm, Melsungen

Broughton E, Choquette LPE, Cousineau JG, Miller FL (1970) Brucellosis in reindeer (*R. tarandus*) and the migratory barren-ground caribou, *R.t. groenlandicus* (L.), in Canada. Can J Zool 48:1023–1027

Caughley G (1983) The Deer Wars. Heinemann, Auckland, New Zealand

Cowan IMcT, Guiguet CJ (1965) The Mammals of British Columbia, 3rd ed. Provincial Museum, Victoria, B.C.

Dasmann RF (1964) African Game Ranching. Pergamon Press, Oxford

Davidson WR, Crow CB (1983) Parasites, diseases and health status of sympatric populations of sika deer and white-tailed deer in Maryland and Virginia. J Wildl Dis 19:345–348

Davidson WR, Crum JM, Blue JL, Sharp DW, Phillips JH (1985) Parasites, diseases and health status of sympatric populations of fallow deer and white-tailed deer in Kentucky. J Wildl Dis 21:153–159

Decker E (1978) Exotics. In Schmidt JL, Gilbert DC (eds) Big Game of North America. Stackpole Books, Harrisburg, Pa. pp 249–256

Dobyns HF (1983) Their Numbers Become Thinned: Native American Population Dynamics in Eastern North America. University of Tennessee Press

Draskovich I (1951) Rotwildhege. Roehrer-Verlag, Innsbruck

Eggeling FK von (1983) Diezels Niederjagd, 23rd ed. Paul Parey, Hamburg

Ehrenfeld DE (1981) Options and limitations in the conservation of sea turtles. In Bjorndal K (ed) Biology and Conservation of Sea Turtles. Smithsonian Institution Press, Washington, D.C., pp 457–463

Ehrenfeld DE (1974) Conserving the edible sea turtle: can mariculture help? Am Sci 62:23–31

Etling K (1985) Can science produce a race of super bucks? Outdoor Life, Jan issue, p 21

Feldhamer GA, Chapman JA, Miller RL (1978) Sika deer and white-tailed deer on Maryland eastern shores. Wildl Soc Bull 6:155–157

Filion FL, James SW, Ducharm J, Pepper W, Reid R, Boxall P, Teillet D (1983) The importance of wildlife to Canadians. Environment Canada. Supply and Services, Cat. No. CW66-62/1983E, Ottawa

Flyger V (1960) Sika deer on islands in Maryland and Virginia. J Mammal 41:140

Geist V (1989) Legal trafficking and paid hunting threaten conservation. Trans 54th North Am Wildl Nat Res Conf, pp 171–178

Geist V (1988) How markets in wildlife meat and parts, and the sale of hunting privileges, jeopardize wildlife conservation. Conserv Biol 2:1–12

Geist V (1985a) Game ranching: threat to wildlife conservation in North America. Wildl Soc Bull 13:594–598

Geist V (1985b) On Pleistocene bighorn sheep: some problems of adaptation, and relevance to today's American megafauna. Wildl Soc Bull 13:351–359

Geist V (1982) Adaptive behavioral strategies. In Thomas JW, Toweill DE (eds) Elk of North America. Stackpole Books, Harrisburg, Pa., pp 219–277

Geist V (1981) Behavior: adaptive strategies in mule deer. In Wallmo OC (ed) Mule and Black-Tailed Deer of North America University of Nebraska Press, Lincoln, pp 157–223

Goodson N (1982) Effects of domestic sheep grazing on bighorn sheep populations: review. Bienn Symp North Am Wild Sheep Goat Counc 3:287–313

Guthrie RD (1984) Mosaics, allochemics and nutrients. An ecological theory of Late Pleistocene megafaunal extinctions. In Martin PS, Klein RG (eds) Quaternary Extinctions. University of Arizona Press, pp 259–298

Harrington R (1973) Hybridization among deer and its implication to conservation. Inst Forestry 30:64–78

Harris LH (1984) New Zealand. In Halls LK (ed) The White-Tailed Deer. Stackpole Books, Harrisburg, Pa., pp 547–556

Hesper B (1988) Gams oder Schaf—wer frisst den Bergwald? Wild und Hund 91 (August 12):4–8

Hibbler CP (1981) Diseases. In Wallmo OC (ed) The Mule and Black-Tailed Deer of North America. University of Nebraska Press, Lincoln, pp 129–155

Holmes JC (1982) Impact of infectious disease agents on the population growth and geographical distribution of animals. In Anderson RM, May RM (eds) Population Biology of Infectious Diseases. Springer-Verlag, New York, pp 37–51

Joralemon D (1982) New World depopulation and the case of disease. J Anthropol Res 38:108–127

Johnson CW, Lindner RL (1986) An economic valuation of South Dakota wetlands as a recreation resource for resident hunters. Landscape J 5:33–38

Kahdren P (1983) Dying wilderness. Western Can Outdoors 7:2

Keiper RR (1985) Are sika deer responsible for the decline of white-tailed deer on Assateague Island, Maryland? Wildl Soc Bull 13:144–146

Keiper RR, Stephens J, Baldwin D (1984) Sex, age, and dressed weights of hunter-killed sika and white-tailed deer from Assateague Island, Maryland. Proc Penn Acad Sci 58:101–102

Kurten B, Anderson E (1980) Pleistocene mammals of North America. Columbia University Press, New York

Lankester MW (1987) Pests, parasites and diseases of moose (*Alces alces*) in North America. Swedish Wildl Res Suppl 1:461–489

Leopold A (1933) Game Management. Scribner's, New York

Lingle S (1989) Escape gates of white-tailed deer, mule deer and their hybrids: implication for hybrid success, and for management of white-tailed and mule deer. MSc thesis, University of Calgary, Canada

Lowe VPW, Gardiner AS (1975) Hybridization between red deer (Cervus elaphus) and sika deer (*C. nippon*) with particular reference to stocks in N.W. England. J Zool London 177:553–566

Lund TA (1980) American Wildlife Law. University of California Press, Berkeley

McCullough DR (1969) The tule elk: its history, behaviour and ecology. University of California Publications in Zool 88, Berkeley and Los Angeles

McCutchen HE (1981) Desert bighorn zoogeography and adaptation in relation to historic land use. Wildl Soc Bull 9:171–179

MacDonald D (1987) Hunting—an exercise in pluralistic democracy. Wildl Soc Bull 15:463–465

Marshall E (1990) Mountain sheep experts draw hunter's fire. Science 248:437–38

Massey W (1986a) The Lilybank safari. Deer Farmer, Jan issue, pp 12–13

Massey W (1986b) Escape! The crisis faced by Robbie and Barbara Oldeman. Deer Farmer, Sept issue, pp 6–10

Meyer ME (1966) The epizoology of brucellosis and its relationship to the identification of Brucella organisms. Am J Vet Res 25:553–557

Neiland KA (1975) Further observations on rangiferine brucellosis in Alska carnivores. J Wildl Dis 11:45–53

Neiland KA, King AJ, Huntley BE, Skoog R (1968) The diseases and parasites of Alaskan populations. Part I. Some observations on brucellosis in caribou. Bull Wildl Dis Assoc 4:27–37

Nuesslein F (1983) Jagdkunde, 11th ed. BLV Verlagsgesellschaft, Munich

Nygren KFA (1984) Finland. In Halls LK (ed) The White-Tailed Deer. Stackpole Books, Harrisburg, Pa., pp 561–568

Onderka DK, Wishart WD (1988) Experimental contact transmission of *pasteurella haemolytica* from clinically normal domestic sheep causing pneumonia in rocky Mountain bighorn sheep. J Wildl Dis 24:663–667

Onderka DK, Rawluk SA, Wishart WD (1988) Susceptibility of Rocky Mountain bighorn sheep and domestic sheep to pneumonia induced by bighorn and domestic livestock strains of *Pasteurella haemolytica*. Can J Vet Res 52:439–444

Rennie N (1986) Good insurance deals are available. Deer Farmer, Sep issue, pp 11–12

Roseborough DJ (1986) Wildlife as a cash crop? Globe and Mail, July 17, p 7A

Samuel WM (1987) Moving the zoo—or—the potential introduction of a dangerous parasite into Alberta with its translocated host. In Renecker LA (ed) Focus on a New Industry. Proc Alberta Game Growers Assoc Conf 1:85–92

Samuel WM (1979) The winter tick *Dermacentor albipictus* (Packard, 1869) on moose *Alces alces* (L.) of central Alberta. Proc North Am Moose Conf Workshop 15:303–348

Seegmiller RF, Simpson CD (1979) The Barbary sheep: some conceptual implication of competition with desert bighorns. Trans Desert Bighorn Counc 23:47–49

Schroeder W (1986) Jagd 2000. Die Hirsch 38:773–777

Simpson CD, Krysl JB, Dickinson TG (1980) Food habits of Barbary sheep in the Guadalupe Mountains, New Mexico. In Simpson CD (ed) Proceedings of a Symposium on Ecology and Management of Barbary Sheep. Texas Tech University, Lubbock, pp 87–91

Simpson CD, Krysl JB, Hampy DB, Gray GG (1978) The Barbary sheep: a threat to desert bighorn survival. Trans Desert Bighorn Counc 22:26–31

Stahl D (1979) Wild, Lebendige Umwelt. K Alber, Freiburg/Munich

Stubbe C, Passarge H (1979) Rehwild. Verlag Neumann–Neudamm, Berlin

Swenson JE (1983) Free public hunting and the conservation of public wildlife resources. Wildl Soc Bull 8:75–87

Tessaro SV, Forbes LB (1986) *Brucella suis* biotype 4: a case of granulomatous nephritis in a barren ground caribou (*Rangifer tarandus groenlandicus* L.) with a review of the distribution of rangiferine brucellosis in Canada. J Wildl Dis 22:479–483

Turcek FJ (1951) Effect of introductions on two game populations in Czechoslovakia. J Wildl Manage 15:113–114

Weyhausen JD (1984) Comment on desert bighorn as relict: further considerations. Wildl Soc Bull 12:82–85

White RJ (1987) Big game ranching in the United States. Wild Sheep Goat Int, Mesilia, New Mexico

Wodzicki KA (1950) Introduced mammals in New Zealand. N Z Dept Sci Indust Res Bull 98, Wellington

Zeuner FE (1967) Geschichte der Haustiere. In Bosseneck J, Halthenorth T (eds) Translated from English by Ross-Rath R. Bayerischer Landwirtschaftsverlag, Munich

Closing Address

131
Biology of Seasonal Breeding in Deer

Gerald A. Lincoln

Abstract

Deer are long-lived herbivores, and most species are seasonal breeders. In females, the most energy-demanding phase of the reproductive cycle (late pregnancy and lactation) is timed to coincide with the period of maximum food abundance and quality. In males, the period of fertility is timed to the availability of estrous females. The annual cycle in gonadal activity is generated endogenously and is entrained to a particular time of year by responses to environmental cues; these are principally changes in day length in the cold/temperate zones, and possibly nutrition in the tropics.

Key words: Annual rhythms, environmental cues, latitude, ovary, pineal gland, pituitary gland, seasonal reproductive cycles, testis

Introduction

The purpose of this chapter is to illustrate general points about seasonal reproduction in deer. For this it will be convenient to divide the species into two categories according to their natural distribution by latitude (Lincoln 1985). Of the existing 40 species, some 15 species live primarily in the temperate and cold zones (30°–80° latitude), and 25 live in the tropics (0° to 30° latitude; Fig. 131.1). The species adapted to temperate and cold climates are markedly seasonal in their reproductive physiology and behavior in contrast to the tropical species. The great majority of studies on the biology of deer deal with these seasonal species which are important as game animals and/or kept on deer farms (e.g., red deer (*Cervus elaphus*); wapiti (*Cervus canadensis*); fallow deer (*Dama dama*) and white-tailed deer (*Odocoileus virginianas*)).

Female Strategy for Seasonal Breeding

For reproduction to occur, energy and nutrients are required above those normally needed for body maintenance. Since it is the female that produces the fetus and produces milk to rear the offspring, it is logical that reproduction in the female will be timed so that the most energy-demanding phase of the reproductive cycle (e.g., late pregnancy, birth, and lactation) is synchronized with the season of greatest food abundance and quality (Bronson 1989; Fig. 131.2) This will be most critical in highly seasonal environments, and as predicted, all species of deer living in cold and temperate climates have a restricted birth season. Deer are large animals with a long gestation period (Lincoln 1985), and the timing of conception occurs under quite different environmental conditions to those favoring

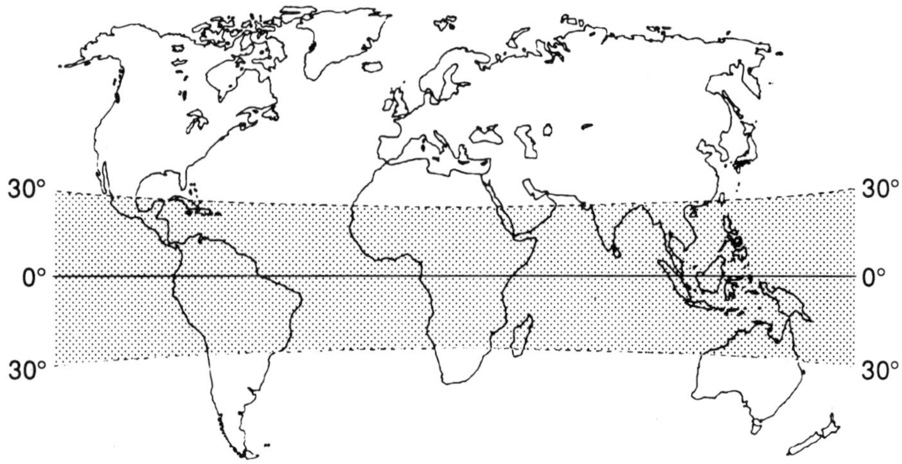

Temperate / Cold 30° - 80° lat. 15 species e.g.	Subtropical / Tropical 0° - 30° lat. 25 species e.g.
Temperate species	**Tropical species**
Reindeer / Caribou	Sambar
Moose	Swamp
Red / Wapiti	Elds
Père David's	Axis
Sika	Rusa
White-tailed	Hog
Mule	Muntjac
Fallow	White-tailed
Roe	Marsh
Water	Pampus
Musk	Brocket
S Huemul	N Huemul
S Pudu	N Pudu

FIGURE 131.1. Deer species divided according to their natural distribution by latitude: (a) species adapted to temperate and cold climates, and (b) species adapted to subtropical and tropical climates (modified from Lincoln 1985).

survival of the offspring. In this situation, it is possible to distinguish between *ultimate* environmental factors that affect the survival of the mother and offspring (directly influencing reproductive success), and *proximate* environmental factors to which the animals time conception (Baker 1938).

The different components influencing the birth season can be illustrated using the timing of ovulation and birth in seven species of deer living in the British Isles (Fig 131.3). All the species, with the exception of the introduced muntjac (*Muntiacus reevesi*), have a restricted birth season. The offspring are produced in early summer which appears to be the most favorable time when food of high nutritional quality is most abundant. Studies in red deer living in the wild have shown that calves born in the middle of the birth season survive better than those born early or late (Clutton-Brock et al. 1982). This indicates that it is the climatic and feeding conditions particularly in spring and early summer which favor the survival and rearing of the young; these constitute the ultimate environmental factors dictating the timing of the birth season.

To achieve births at the optimum time, conception must occur in the late summer and

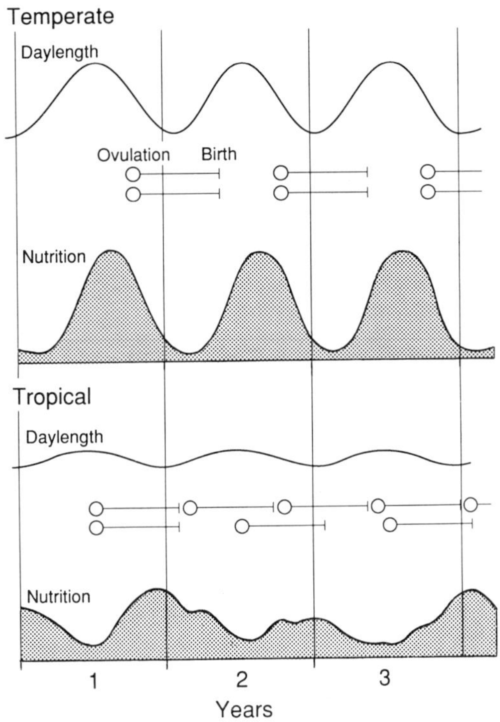

FIGURE 131.2. Seasonal breeding strategy in female deer living in temperate and tropical climates. In temperate climates, seasonal changes in food quality and quantity are very marked, and animals maximize reproductive success by synchronizing ovulation using *predictive* environmental cues, particularly the annual cycle in day length. In tropical species, seasonal changes in climate and food supply are usually less marked, and species may breed continuously or more commonly synchronize ovulations in response to *local* cues. This includes the effect of changes in nutrition and the secondary effect of social stimuli from other reproductively active individuals in the population (see Bronson 1989).

autumn 6 to 10 months earlier (Fig. 131.3). This involves a response to proximate environmental cues which influence the timing of ovulation. The principle environmental cue in these temperate species is provided by the annual cycle in day length. This is evidenced by the way the reproductive cycle is readily manipulated under artificial lighting [sika deer (*Cervus nippon*); Goss 1983; fallow deer: Schnare and Fischer 1987; red deer: Suttie et al. 1989; roe deer (*Capreolus capreolus*): A.J. Sempéré, personal communication]. Differences between species in the physiological mechanisms involved in transducing the photoperiodic signal are presumed to account for the differences in the mating season.

In the species adapted to subtropical and tropical regions, where the seasonal constraints on reproduction are less defined and predictable, births may occur in all months of the year; however, studies on these species usually reveal changes in the frequency of births with season. For example, muntjac, sambar (*Cervus unicolor*), hog deer (*Axis porcinus*), and axis deer (*Axis axis*) living in the Chitawan National Park in Nepal have clearly defined birth seasons, although these occur at different times for the various species (Mishra 1982). In these situations, it is difficult to distiguish the ultimate and proximate factors governing the timing of the birth season. These tropical species do not apparently respond to changes in day length. This has not been studied directly, but is evident when tropical species of deer are transferred to temperate climates. Under these seasonal conditions they continue to produce offspring throughout the year and the winter-born animals suffer high mortality (Lincoln 1985: A.S.I. Loudon, personal communication). In the natural environment, changes in nutrition related to the rainy season may be the principle factor influencing the number of females ovulating and the seasonal pattern of births (Sadlier 1987).

Male Strategy for Seasonal Breeding

Since male reproductive success is determined by the number of offspring produced (and the number which survive to sexual maturity), it is logical that the breeding cycle in males will be geared to the availability of females in estrus. This may be either a restricted period, as in the highly seasonal species, or extended throughout the year according to the female reproductive strategy. The variation between species can be illustrated by comparing the seasonality in reindeer (*Rangifer tarandus*) and red deer (cold and temperate species), and muntjac and axis deer (subtropical and tropical species; Fig. 131.4). In reindeer and red reer, mature males show highly synchronized testicular cycles with the period of maximum activity corresponding to the time when the females ovulate (Lincoln 1971;

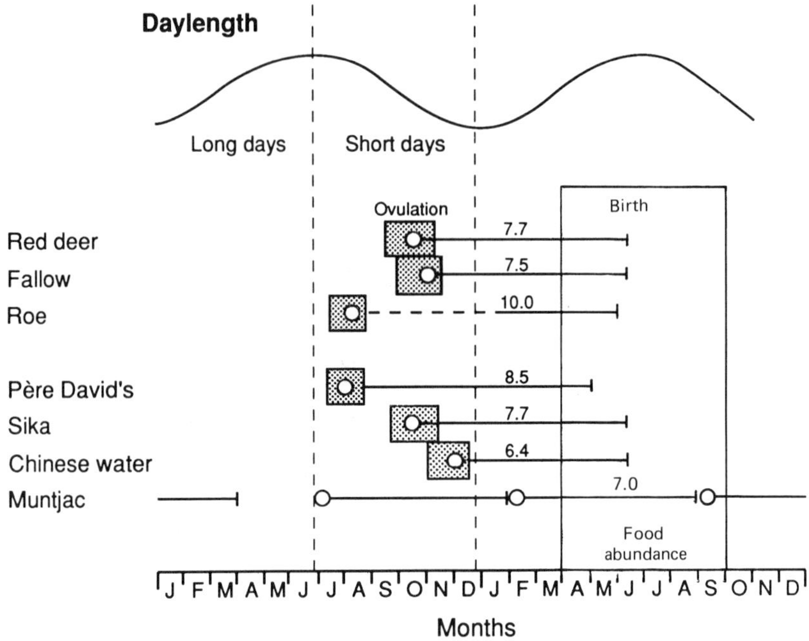

FIGURE 131.3. The timing of ovulation for three native (above) and four introduced species of deer living in England (British Isles, 50–52° latitude). Six of the species synchronize births to spring and early summer when climatic conditions are favorable and food is abundant. To achieve this timing the seasonal cycle controlling ovulation is influenced by environmental cues that accurately predict the time of year. The annual cycle in day length is used as the principle predictive cue but differences in the photoperiodic response allow each species to have a specifically timed mating season (stippled box indicates the timing of the rut and numbers indicate gestation period in months). The muntjac is a subtropical species and does not show a restricted birth season even in the markedly seasonal climate (rutting and birth dates from Asdell 1964; Corbet and Southern 1977).

Mitchell and Lincoln 1973; Leader-Williams 1988). The seasonal increase in the secretion of testosterone from the testes induces the hard antler phase, and also induces the overt sexual and aggressive behavior characteristic of the rut. Male red deer remain fertile and retain their antlers throughout the winter and females may show estrus over this period. In contrast, male reindeer cast their antlers soon after the rut and are infertile in winter; there are no females in estrus in winter in this highly seasonal species.

In the axis deer, the temporal events during the male reproductive cycle are similar to those described in temperate species, but the individuals are less synchronized and some males may be in hard antler and fertile in all months of the year (Loudon and Curlewis 1988). In this species, females are potentially in estrus in all months of the year, and different males take part in mating according to the stage of their reproductive cycle. A similar situation occurs in white-tailed deer living in the tropics (Fig. 131.8). The muntjac is the other species in which there is an unrestricted mating season. In this case, the males are synchronized in their testicular cycles, but seasonal regression of the testes in spring and summer is relatively minor compared with the more seasonal species, and the production of spermatizoa continues throughout the year (Chapman et al. 1984, N.G. Chapman and S. Harris personal communication; Fig. 131.5).

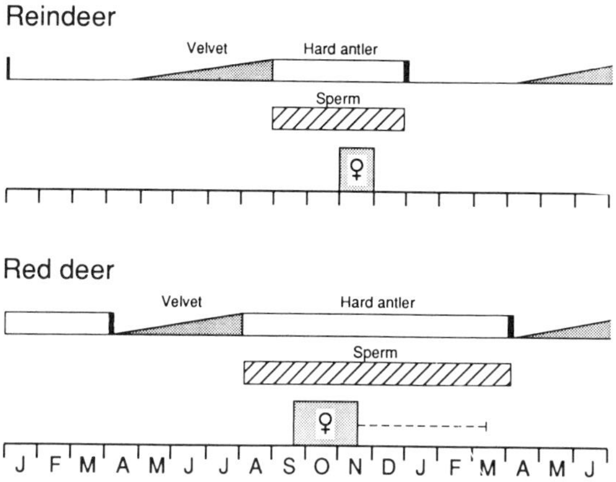

FIGURE 131.4. Seasonal reproductive cycle in two species of deer from cold and temperate climates with a restricted mating season, showing the period of sperm production and hard antler, relative to the time when females are likely to be in estrus. Data for reindeer in South Georgia with dates converted to northern hemisphere, from Leader-Williams (1988), and for red deer in Scotland, from Lincoln (1971) and Mitchell and Lincoln (1973).

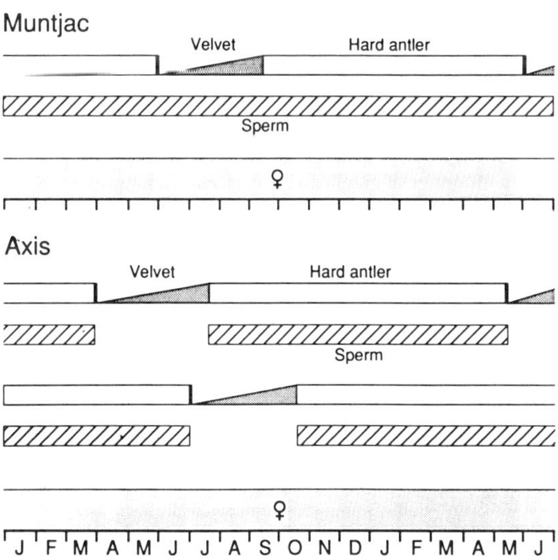

FIGURE 131.5. Seasonal reproductive cycle in two species of deer from subtropical and tropical climates with a protracted mating season, showing the period of sperm production and hard antler. Data for muntjac in England, from Chapman et al. (1984) and for axis deer in Whipsnade Zoo, England, from Loudon and Curlewis (1988).

Endogenous Annual Rhythms

Detailed studies of sika deer have shown that cycles in antler growth and casting can persist in deer kept indoors under constant light conditions (Goss 1983; Fig. 131.6). This indicates that the reproductive cycle which governs the antler cycle is an endogenously generated annual rhythm. Since this species readily adjusts its antler cycle to changes in day length, it is evident that timing of the seasonal reproductive cycle depends on the way the annual cycle in day length entrains the endogenous annual rhythm governing reproduction. In white-tailed deer and red deer, it has been shown that annual cycles in reproduction persist in animals rendered unresponsive to changes in day length following removal or denervation of the pineal gland (Brown et al. 1978; Plotka et al. 1979; Lincoln 1985; Figs. 131.7, 131.8). In this situation the cycles do not occur at the normal time, however they can persist throughout life. The same seasonal pattern occurs in congenitally blind deer (H.A. Jacobson and S.J. Waldhalm, Chapter 122 this volume).

With the long-term cycle being generated endogenously, even very subtle cues from the environment, occurring on an annual basis, may be sufficient to synchronize the breeding cycle to a particular time of year. This could account for the timing of the breeding season in tropical species of deer where changes in temperature or nutrition entrains (to a greater or lesser extent) the endogenous annual cycle in the individual animals. These effects may be accentuated by social cues (e.g., behavior and pheromones) acting to synchronize sexual activity which is a feature noted in tropical species (K.B. Woodford and A. Dunning, personal communication).

Melatonin and Photoperiodic Effects on Reproduction

Melatonin, secreted by the pineal gland, relays the effects of photoperiod to influence the timing of the seasonal reproductive cycle in the photoperiodic species (Arendt 1986; Fig. 131.9). Melatonin is secreted at night and the duration of the peak varies with night length providing

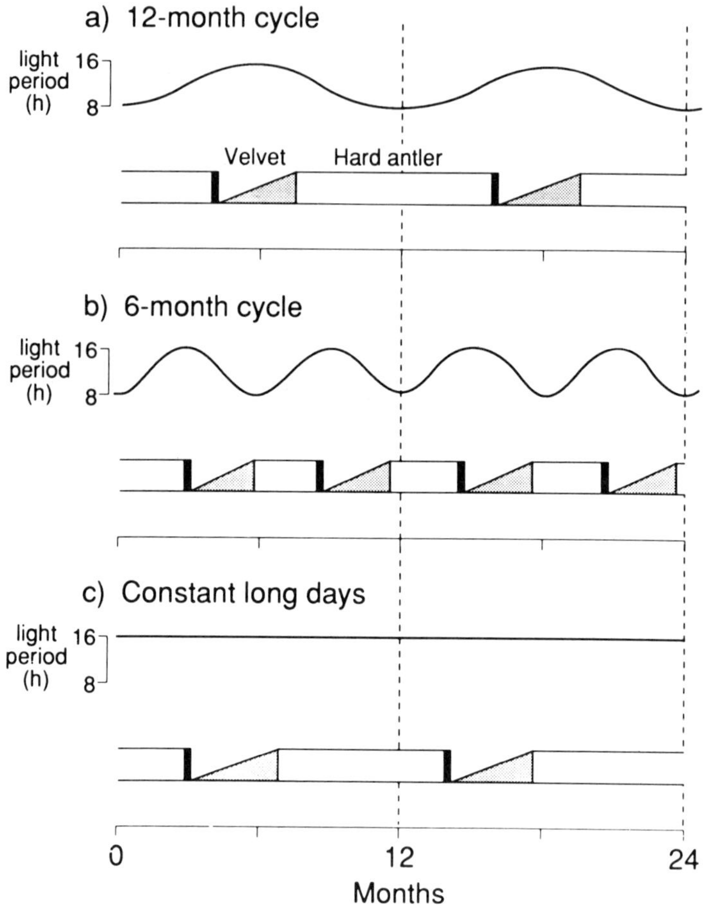

FIGURE 131.6. Antler cycle in axis deer housed under different photoperiod regimens: (a) natural 12-month cycle, (b) 6-month cycle, (c) constant long days. The results illustrate that the antler cycle (and the controlling reproductive cycle) constitutes a circannual rhythm generated endogenously but the period can be precisely regulated in response to changes in photoperiod (modified from Goss 1983).

an endocrine index of the prevailing photoperiod. How melatonin acts within the brain to influence reproduction is still unresolved, but it is assumed to influence the neural systems within the hypothalamus which regulate pulsatile secretion of LH releasing hormone (LHRH, also known as GnRH). This decapeptide controls the reproductive axis by stimulating the secretion of LH and FSH from the pituitary gland (Lincoln et al. 1985).

The current hypothesis for species with a mating season in the autumn, including red, fallow and sika deer, is that the change from long days to short days following the summer solstice causes an increase in the duration of the nocturnal peak in melatonin secretion. This, in turn induces an increase in pulsatile LHRH secretion which activates the reproductive axis (Fig. 131.9). The pattern of melatonin secretion has been measured in fallow deer, along with a detailed study of the changes in pulsatile LH and testosterone secretion, which supports this model (Asher et al. 1987, 1988, 1989). It is also well established for several species of deer that administration of

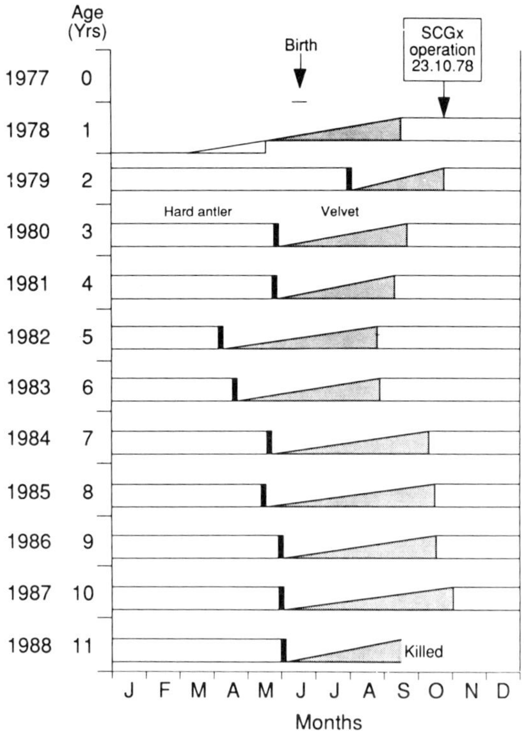

FIGURE 131.7. Antler cycle in a SCGX red deer stag living out doors at Reedie Hill Deer Farm, Fife, Scotland. The SCGX operation (removal of the superior cervical ganglia to denervate the pineal gland) was carried out in October 1978 when the animal was 1 year old. This caused a delay in the casting and cleaning of the antlers by about 8 weeks in the following year. The antlers were then replaced annually (usually 4 to 8 weeks later than normal) throughout life. This animal showed correspondingly late rutting behavior compared with normal stags (Lincoln 1985; G.A. Lincoln and J.T. Fletcher, unpublished data; Fig. 131.8, bottom panel).

melatonin to animals under long days (thereby increasing the daily exposure to melatonin and thus mimicking the effect of short days) causes premature gonadal activity in both sexes (e.g. white tailed deer: Bubenik 1983; red deer: Adams and Aitkinson 1984; fallow deer: Asher et al. 1988). This has been shown to be dependent on the secretion of LHRH (Lincoln et al. 1984). This effect can be induced by the continuous administration of melatonin from a subcutaneous implant, which provides a potentially convenient way to manipulate the onset of the breeding season in commercially important species of deer (see papers in this volume).

The effect of day length acts by entraining the endogenous annual rhythm which governs the long-term cycle in reproduction. This means that the melatonin signal which relays the effect of day length must influence the timing of this endogenous rhythm by actions within the brain which ultimately influence the secretion of LHRH. In this context, melatonin is not viewed as strictly a stimulator or inhibitor of the reproductive system but as a coordinator, influencing the timing of a sequence of events occurring within the brain which constitutes the endogenous annual rhythm (Tarmarkin et al. 1985). It is probable that intermittent signals from the pattern of melatonin secretion, present during only part of the year, would be sufficient to entrain the annual rhythm. This is supported by a recent study in reindeer living at high latitude in Finland which demonstrates that the daily melatonin rhythm is absent in the long days of summer (E. Eloranta, personal communication). In this situation the change from no melatonin signal in summer to a long duration signal in winter, is apparently effective in entraining the annual reproductive cycle.

While the pineal gland and its secretion of melatonin plays an important role in the timing of reproduction in cold and temperate species of deer, it is unlikely to be of such significance in tropical species. This is based on the assumption that tropical species of deer do not respond to day length to time their breeding cycles. In one study, melatonin was administered to male axis deer, and this was ineffective at altering the timing of the antler cycle (Loudon and Curlewis 1988).

Conclusion

In considering the biology of seasonal breeding in deer it has been convenient to divide the species into two groups according to the latitude of origin (north and south of 30°lat). The species adapted to cold and temperate climates have a rigidly programmed seasonal physiology with an endogenous circannual rhythm generating the long-term cycle, and a neuroendocrine mechanism which relays effects of the annual cycle in

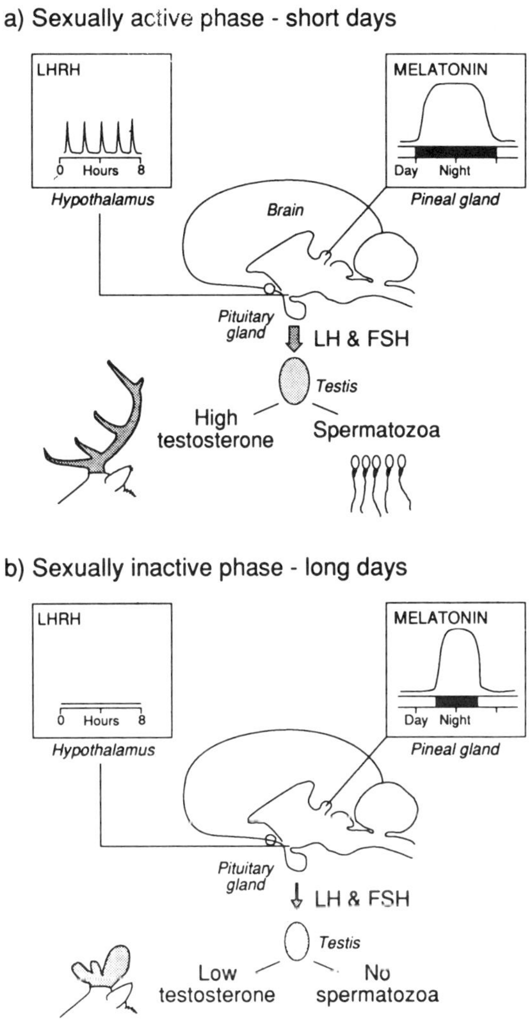

FIGURE 131.9. Diagram of the endocrine control of seasonal reproduction, e.g., red deer: (a) sexually active phase (short days) induced by high frequency LHRH secretion fron the neurosecretory cells in the hypothalamus that stimulates LH and FSH secretion from the pituitary gland and thus gonadal activity. Short days cause a long duration daily period of melatonin secretion from the pineal gland which acts within the brain to affect LHRH secretion, (b) sexually inactive phase (long day is induced by a "switch off" of LHRH secretion (modified from Lincoln 1989).

day length to entrain the cycle. Each species has a short mating season timed to produce the offspring during the season of food abundance. The species adapted to subtropical and tropical climates have a less marked seasonal physiology but still show many examples of seasonal breeding. In this situation, factors such as nutrition, temperature, and social cues may act to synchronize conception at a particular time of year. Detailed studies of the physiology of some tropical species of deer are required to reveal the factors which regulate seasonal reproduction in these species. The white-tailed deer provides an especially interesting case for study since there are genetically distinct races adapted to both temperate and tropical climates.

References

Adam CL, Atkinson T (1984) Effect of feeding melatonin to red deer (*Cervus elaphus*) on the onset of the breeding season. J Reprod Fert 72:463–466

Arendt J (1986) Role of the pineal gland and melatonin in seasonal reproductive function in mammals. Oxford Reviews Reprod Biol 8:266–320

Asdell SA (1964) Patterns of Mammalian Reproduction, 2nd ed. Comstock, Ithaca, New York

Asher GW, Peterson AJ, Bass JJ (1989) Seasonal pattern of LH and testosterone secretion in adult male fallow deer *Dama dama*. J Reprod Fert 85:657–665

Asher GW, Barrell GH, Adam JL, Staples L (1988) Effects of subcutaneous melatonin implants on reproductive seasonality of farmed fallow deer (*Dama dama*). J Reprod Fert 84:679–691

Asher GW, Day AM, Barrel GK (1987) Annual cycle of liveweight and reproductive changes of farmed male fallow deer (*Dama dama*) and the effect of daily oral administration of melatonin in summer on the attainment of seasonal fertility. J Reprod Fert 79:353–362

Baker JR (1938) The evolution of breeding seasons. In de Beer GR (ed) Evolution: Essays on Aspects of Evolutionary Biology, Oxford University Press, Oxford, pp 161–177

FIGURE 131.8. Top panels: two male white-tailed deer photographed on the same day in March 1987 in Parque de las Leyendas, Lima, Peru at lat. 12° S at different stages of the reproductive cycle illustrating the way the cycles of individuals can be asynchronous in tropical species. Bottom panels: SCGX red deer stag living out doors showing a delayed reproductive cycle, (left) velvet cleaning in late September 1985, and (right) hard antler (sawn off) and swollen neck in April 1986. The SCGX operation (removal of the superior cervical ganglia to denervate the pineal gland) was performed at 1 year old (1978) and the animal replaced its antlers annually throughout life but up to 8 weeks later than normal (Fig. 131.6).

Bronson FH (1989) Mammalian Reproductive Biology. University of Chicago Press, Chicago

Brown RD, Cowan RL, Kavanaugh JF (1978) Effect of pinealectomy on seasonal androgen titres, antler growth and feed intake in white-tailed deer. J Anim Sci 47:435–440

Bubenik GA (1983) Shift of seasonal cycle in white-tailed deer by oral administration of melatonin. J Exp Zool 225:155–156

Chapman DI, Chapman NG, Dansie O (1984) The periods of conception and parturition in feral Reeves muntjac (*Muntiacus reevesi*) in Southern England based upon age of juvenile animals. J Zool London 204:575–578

Clutton-Brock TH, Guinness FE, Albon SD (1982) Red Deer: Behaviour and Ecology of Two Sexes. University of Chicago Press, Chicago

Corbet GB, Southern HN (1977) The Handbook of British Mammals. Blackwell Scientific Publications, Oxford.

Goss RJ (1983) Deer Antlers: Regeneration, Function and Evolution. Academic Press, New York

Leader-Williams N (1988) Reindeer on South Georgia. Cambridge University Press, Cambridge

Lincoln GA (1989) Seasonal aspects of testicular function. In Burger H, de Kretser D (eds) The Testis, 2nd ed. Raven Press, New York, pp 329–386

Lincoln GA (1985) Seasonal breeding in deer. In Fennessy PF, Drew KR (eds) Biology of Deer Production. R Soc N Z Wellington Bull 22: 165–180

Lincoln GA (1971) The seasonal reprodutive changes in the red deer stag (*Cervus elaphus*). J Zool London 163:105–123

Lincoln DW, Fraser HM, Lincoln GA, Martin GB, McNeilly AS (1985) Hypothalamic pulse generators. Recent Prog Hormone Res 41:369–419

Lincoln GA, Fraser HM, Fletcher TJ (1984) Induction of early rutting in male red deer (*Cervus elaphus*) by melatonin and its dependence on LHRH. J Reprod Fert 72:339–343

Loudon ASI, Curlewis JD (1988) Cycles of antler and testicular growth in an aseasonal tropical deer (*Axis axis*). J Reprod Fert 83:729–738

Mishra HR (1982) The ecology and behaviour of chital (*Axis axis*) in the Royal Chitawan National Park, Nepal. PhD thesis, University of Edinburgh, Scotland

Mitchell B, Lincoln GA (1973) Conception dates in relation to age and condition in two populations of red deer in Scotland. J Zool London 171:141–152

Plotka ED, Seal US, Letellier MA, Verme LJ, Ozoga JJ (1979) Endocrine and morphological effects of pinealectomy in white-tailed deer. In Alexander J (ed) Animal Models for Research on Contraception and Fertility Research. Harper and Row, Hagerstown, pp 452–466

Schnare H, Fischer K (1987) Secondary sex characteristics and connected physiological values in male fallow deer (*Dama dama* L) and their relationship to changes of the annual photoperiod: doubling the frequency. J Zool 244:463–471

Suttie JM, Fennessy PF, Corson ID, Laas FJ, Elgar JH, Lapwood KR (1989) LH and testosterone response to GnRH in red deer (*Cervus elaphus*) stags kept in a manipulated photoperiod. J Reprod Fert 85:213–219

Tamarkin L, Baird CJ, Almeida OFX (1985) Melatonin: a co-ordinating signal for mammalian reproduction. Science 227:714–720

Sadleir RMFS (1987) Reproduction of female cervids. In Wemmer CM (ed) Biology and Management of the Cervidae. Smithsonian Institution Press, Washington, D.C., pp 123–144

Workshops

Diseases and Immune Systems

Ed Couvillion and Richard Randall

There was discussion of herd health programs that would maximize control of common infections and parasitic diseases of the various deer species. In general, health programs and vaccines were similar to those described for cattle and other domestic ruminants and included diseases such as *Clostridium* spp. and *Leptospirosis*. The need for attention to other factors such as nutrition and forage management was reiterated.

Most of the discussion focused on two of the most important diseases, brucellosis and tuberculosis (TB). In particular it was noted that tests for detection of these diseases in cattle are not sensitive nor specific in deer. Considerable research effort is needed to develop tuberculin tests specific for the various deer species. In the absence of suitable tests there is potential for introduction of tuberculosis into clean herds when movement of infected animals occurs. In addition to the basic microbiological problem of developing suitable tests in deer, regulatory agencies in states and federal governments must develop appropriate regulatory controls. The problem of introduction of the diseases into clean deer herds is exacerbated by apparent lack of uniform state regulation in the United States.

Another fruitful area of research involves parasitologic problems in deer herds. In many areas of the world, especially the United States, the epidemiology of gastrointestinal parasites has received little, if any, study. In fact, the effect of many of these parasites on the productivity of deer herds is unknown. Considerable research effort should be expended to determine the patterns of transmission, disease states, and methods of control of gastrointestinal worms of deer.

Bioenergetics and Estimation of Carrying Capacity

Ken L. Risenhoover and Robert G. White

There is renewed interest among managers in assessing habitat quality and carrying capacity. This interest is perpetuated by the desire of ecologists to improve their understanding of the link between animal and population characteristics and habitat quality. This workshop provided a forum for the discussion of several key issues relating to ungulate bioenergetics and the estimation of carrying capacity.

Animals as Indicators of Carrying Capacity

Management concerns for populations regulated primarily by hunter harvest led to a discussion of the potential utility of using animal characteristics as gauges of the "ecological density" of a population in relation to environmental carrying capacity. There was general agreement that both short-term (e.g., body composition, weight dynamics) and long-term (adult size, femur and mandible length, etc.) indicators of animal physiological condition and growth are useful indicators of animal nutritional status in relation to population density. These indicators also are assumed to be indicators of habitat carrying capacity.

However, the question was raised "Are animal indicators adequate for management purposes?" It was argued that this approach lacked the predictive value which often is needed for effective decision-making by managers. Several limitations of this approach were identified and debated. First, animal indicators may only be detectable over a narrow range of habitat conditions, and thus, may not provide the resolution necessary for intensive management. Second, there will be time delays between animal indicators and changes in population density and/or habitat quality. Consequently, indices may not reflect the current relationship between animals and their habitats. There was concern that management decisions based on this approach would be less effective. Finally, caution was urged when applying indices between areas and between populations. The phenotypic and genetic plasticity of widespread species may require that indices be developed for each population before they can be meaningful. This problem is especially troublesome for species in arid ecosystems because of reduced animal densities or population isolation.

Vegetative Indicators of Carrying Capacity

It was argued that vegetative indicators of ecological density may provide a more direct means of assessing carrying capacity. The problems of quantifying food supply (both quantity and quality) were discussed, as was their value for predicting patterns of habitat utilization by ungulates. Predation and disturbance were mentioned as other important factors influencing animal distributions. It was advocated that, given the limits of our present understanding, a conservative approach might be to manage to maintain habitat diversity where this is a feasible option. It was suggested that gaps in our understanding of factors influencing diet selection and patterns of habitat utilization were limiting our ability to use the carrying capacity concept as a management tool.

Despite the complexity of this problem, the consensus was that research should focus more on understanding the mechanisms causing changes in animal indicators. Foraging behavior and diet selection were mentioned as an approach for examining animal and vegetative indicators simultaneously.

Habitat Evaluation Procedures

A discussion of the application of the carrying capacity concept in management led to a debate over habitat evaluation procedures. Many felt that the value of the "Habitat Suitability Index" (HSI) technology, currently employed by most federal and state agencies, is flawed and of limited value. It was argued that current HSI models ignore mechanisms explaining animal-habitat relationships, and, therefore, they are inconsistent with the concept of nutritional carrying capacity. They are impossible to validate, difficult to test, and potentially "dangerous" in the hands of unqualified individuals. Others argued that, despite these shortcomings, the HSI idea is useful because it provides a starting point and a framework to build upon. Overall, the consensus was that there continues to be an urgent need for meaningful habitat evaluation procedures, and that new approaches should be validated and tested before broad applications are attempted.

Endangered Species

Harry Jacobson and Anthony English

Maintenance of Gene Pools

There was general agreement that high priority was needed to insure existing gene pools were preserved. Of the 39 endangered deer species, many have inadequate captive populations. South American species may be the most vulnerable. Most zoos participate in species survival plans, and progress is being made to maintain viable breeding populations with adequate genetic variability.

What constitutes a genetically viable population is dependent on social structure and breeding biology. Species that are highly social and have harem breeding are likely to require more individuals than species that are composed of solitary individuals that do not form social groupings. A minimum of 50 to 500 individuals would be required to maintain genetically viable populations depending on the social makeup of a species.

Artificial Propagation

Much discussion centered around artificial propagation of endangered species. It was apparent that the group placed a high priority and value on the development of artificial propagation procedures for cervids. Some progress has been made in this area, particularly in semen processing and artificial insemination procedures. However, only limited progress has been made in the embryo transfer area. Technology that exists for embryo transfer of domestic livestock has not been similarly developed for deer. Embryo collection, superovulation, cryopreservation of embryos, and estrous synchronization are not well developed for even the highly studied red deer (*Cervus elaphus*). For many of the excitable, easily stressed, small species of endangered deer we do not even know the rudiments of their reproductive biology. Thus, another priority for study of these species should be their reproductive biology and physiology.

Embryo collection and cryopreservation of endangered deer should have a very high priority placed on it for obvious reasons. The potential exists to safeguard existing gene pools and to disperse these to several different global locations. This could possibly prevent a catastrophic event from causing extinction. Development of procedures for embryo transfer from one species of deer to surrogate mothers of another deer species should similarly have high priority. This would facilitate propagation of endangered deer and also provide opportunity for reintroduction to the wild.

Reintroduction and Management

The outlook for management of endangered deer is not encouraging. The pressures that cause extinction, principally habitat loss, poaching, and other human disturbances are not likely to change in the near future. This provides a strong argument for establishment of captive populations. There was some concern that captivity might alter the species ability to later adapt back to the wild. However, this should not pose a problem as long as existing gene pools are maintained and the general consensus was that cervids readily revert back to a wild state.

Another topic was whether endangered species like the Mesopotamian fallow deer (*Dama dama*) or Pére David's deer (*Elapharus davidianus*) should be allowed to hybridize with closely related species to reduce inbreeding depression or to increase the chance of survival. No conclusion was reached, but the comment was made that the dusky seaside sparrow was allowed to go extinct because the US Fish and Wildlife Service would not allow the last live captive males to breed closely related species in captivity.

Commercial use of endangered deer was also

discussed. Although there were some minority views, it appeared most participants felt commercial use of these animals was justified, if as a result species propogation and/or habitat protection and restoration resulted.

Commercial utilization, of course, can take many forms, from photographic safaris, to commercial production of meat, hides, antlers, musk, and other by-products. It should be recognized that social and political conditions are going to govern what commercial uses are acceptable. However, it should also be recognized that commercial exploitation, if wisely regulated, may offer the most expedient means of achieving protection and restoration of some species. Similarly, illegal commercial exploitation and subsistence hunting are contributing factors to the endangerment of some deer species.

Venison Marketing and International Trade

K.R. Drew and T.C. Williams

Definition of Product

The workshop spent some considerable time discussing the product as game, meat, or venison. Venison in the United States is largely perceived as a product from white-tailed deer while meat from wapiti is sold as elk, not venison. In Norway the meat from reindeer is commonly regarded as a low quality product with a poor image and a new top quality lean product should be classified as game for a luxury market. Game products can come from a large range of animals including nilgai and pronghorn antelope, bison, and moose, as well as the deer species. Venison should be specified as meet from deer.

Market Image

There was general agreement that while venison volumes are small, the target market should be at the top restaurant level. There was some disagreement about the tactic of aiming at a segment of the beef market. There is rising interest in lean red meat in the United States and some people felt that minimal promotion might achieve some of the beef market share for venison. It could be seen as a healthy alternative to beef. Other people felt that venison is an elite food with dietary advantages and a clean, green chemical-free image. As such it should avoid competing with beef but aim at a niche luxury, gourmet food market. As venison volume increases, it is probably inevitable that it will compete to some extent with beef.

Product Pricing

New Zealand is perceived to be the price setter because of its trading volume, and there was a view expressed that in North America local producers had their product price driven down by New Zealand prices. In some situations, international pricing can mean the marketing of the best quality cuts of venison in a high value market and selling the rest at cut rates in other markets. There is often a real lack of discipline. The question was asked about the practicality of forming a venison cartel. The clear problem is that a cartel is only as strong as the weakest link. Sooner or later (often sooner!) a member will want a break on the field and do so by price cutting. Producer control may help marketing discipline. In Europe, beef and lamb sales are controlled by producer boards. To what extent can venison producers do the same? Mr. Tom Williams (Chairman of the New Zealand Game Industry Board) briefly outlined the function of this Board which is to provide international marketing of New Zealand's deer products. This had nothing to do with price setting which was determined by the competing commercial trading companies. If other countries had similar organizations there would be a good forum for communication. The point was made that any national venison organization would likely clash with heavily subsidized traditional meat industry organizations in several parts of the world.

Product Quality and Standards

The public perception of venison in the United Kingdom was stated to be something shot in the wild, dragged through the mud, and eventually chopped into meat pieces. A great deal of education will be required to establish ranched venison as a top quality product. There may be advantages to be gained in North America by maketing venison from "exotic" deer (not native to North America) rather than wild game because the latter is often associated with poor quality rough shot feral white-tailed deer.

Chemical Immobilization

Eric C. Hellgren

Discussion during this workshop focused on three main questions: (1) What do we have (e.g., immobilizing drugs, drug antagonists, and drug delivery systems)? (2) What do we need?, and (3) What pressures are being felt and will be felt in the field of chemical immobilization?

Drug Availability

A wide variety of chemical immobilants are available today to restrain wild and farmed deer species for veterinary care, reproductive control, and captive and wild animal research. Drugs used for animal restraint by workshop participants included ketamine, xylazine, carfentanil, M-99, immobilon, and succinyl choline chloride (SCC). Although SCC has a narrow safety margin, its use was discussed because of its potential in free-ranging capture due to low volume requirements and rapid metabolism. In recent years, there also has been a proliferation of drug antagonists, including yohimbine, naloxone, and diprenorphine (M50-50) that reverse the effects of the above immobilizing drugs.

Virtually every workshop participant had experience with ketamine and/or xylazine hydrochloride, indicative of the ready availability and general effectiveness of these drugs. Participants from outside North America and New Zealand indicated that ketamine and xylazine were the only options available to them. Regulation of chemical immobilants is more rigorous in Canada and New Zealand, where drug access is limited to veterinarians, and costs are higher than in the United States. Certain immobilizing drugs, such as the opioids, are controlled by state and federal law in the United States, but others, such as ketamine and xylazine, are available, in some cases in local feed stores.

Delivery Systems

Drug delivery systems were briefly discussed. Scientists were generally pleased with available systems, particularly blowgun dart syringe and power-projected systems. Scientists and game ranchers interested in remote capture of free-ranging wildlife are always searching for greater accuracy and range on drug delivery systems without sacrificing safety and animal welfare. New developments in power-projected systems, such as adjustable power settings and disposable, preloaded darts, are beneficial in this area.

Drug Needs

Needs identified in the area of chemical immobilization include tighter regulation of drug availability, standardized training for researchers and producers using immobilizing drugs, and a computerized clearinghouse summarizing data on cervid immobilizations. Stricter regulations and enforcement of existent regulations were deemed important for controlling abuses and improper usages of immobilizing drugs. A short course in wildlife restraint techniques, such as that offered by David Jessup, Wildlife Investigations Laboratory, California Fish and Game Department, would provide essential information to wildlife and veterinary researchers and interested deer producers. Standardized training on emergency and safety procedures, anatomy and physiology of wild deer, and correct drug dosages and handling techniques are extremely necessary today, particularly with mounting pressure from animal rights groups for proper animal welfare. An idea for a wildlife immobilization data bank was presented by Keith D. Amass of Safe Capture International, Inc. He is developing, in conjuction with Leon Nielsen, a standard data collection form and a five-section compendium of data covering technology and equipment, techniques, pharmacology, emergency treatment, and individual species information.

Future pressures from animal rights groups are expected in the area of animal handling and immobilization. The above-mentioned training course and the use of institutional animal research in the United States were considered strategic responses to these pressures.

Genetics

Peter A. Dratch and Olin E. Rhodes

The primary concerns of the participants focused on three major issues: hybridization of cervid species, conservation of genetic variation, and wildlife management with regard to genetic resources. The discussion of each topic is summarized and followed by a compilation of recommendations.

Hybridization

The session began with a videotape by Susan Lingle showing the irregular escape gaits of white-tailed/mule deer hybrids. It raised the issue that when predation has been reduced by human intervention, perhaps the selection pressures against hybridization have been reduced as well. The taxonomic status of several cervid species and subspecies was challenged. However, legal and management decisions are made on the basis of the current taxonomic status, usually determined from morphology. Such decisions are often flawed, given the different circumstances under which hybridization occurs. The creation of new "species" by means of hybridization now occurs commonly in the context of deer farming. Unintentional releases of hybrids into competition with wild populations could result in competitive exclusion of native species. Currently, there are few laws which address the problem of hybridization. Failure to properly regulate the creation and handling of hybrid deer could result in conflicts between the farming and wildlife management communities.

Conservation of Genetic Variation

The question was raised, "Why and how should we be concerned about conserving deer species?" Examples of species such as white-tailed deer (*Odocoileus virginianus*) in North American and red deer (*Cervus elaphus*) in Europe show that natural systems are effective in preserving genetic variation. The absence of variation in other species such as the Père David's deer (*Elaphurus davidianus*) shows that it is necessary to conserve genetic variation, particularly where population numbers and habitat are being reduced by human activity. The means of maintaining this variation have been developed and in some cases applied during the past decade. Plans are now under way to ascertain whether there are effective population sizes of 500 or more in wild populations and to maintain distinct subpopulations or lineages in zoo animals. New techniques, such as mitochondrial DNA analysis and DNA fingerprinting have an important role in these cases. The application of conservation genetic principles becomes most critical in situations involving endangered species and translocation programs. If wild populations are founded from small numbers of breeding individuals, variation loss due to drift or inbreeding can be substantial, and can be followed by concomitant loss of fitness.

Management

For many wildlife managers at the workshop, the issue has shifted from should we manage for genetic variations to how can we do so, routinely and economically. It was pointed out that heterozygosity is a measure of individual genetic variability, which has been correlated with many valued characteristics such as body weight, antler development, and reproductive rates in some deer species. Heterozygosity can be measured from tissues obtained from hunter-killed animals. The same sample can be used to determine if there are significant genetic differences in adjacent populations, perhaps requiring different management plans. It also can contribute baseline forensic data to aid in the prosecution of poachers. Many management decisions directly impact the genetic structure of wild populations. Harvest strategies determine herd sex ratios which in turn determine effective population size and the quality of available mates. Selective prebreeding season harvest of branch-antlered males, for example, will prevent the transfer of their genes to the next generation.

Recommendations

1. Sampling of legally hunted populations as well as blood-sampling of threatened and endangered species can be used to complete genetic surveys of deer in their natural ranges.
2. F-statistics can be applied to determine genetic discordance while also accounting for within-species variation.
3. The status of hybrids and nonnative deer must be addressed in deer-farming regulations.
4. Genetic survey data and knowledge of species breeding structure should be used to develop management plans that do not reduce gene flow and maximize breeding opportunities among individuals.
5. Prolonged breeding bottlenecks should be prevented and large effective population sizes maintained whenever possible in natural and reintroduced populations.
6. Natural variation and genetic drift should be left to maintain variation within and between breeding units by maintaining habitat of sufficient size and preserving dispersal corridors between locally distinct breeding groups.
7. Harvest regulations should be implemented with consideration from the time of breeding and the concurrent transfer of desirable genes into the next generation.
8. Whenever deer species are hunted or tranquilized, efforts should be made to obtain blood and/or tissue samples for subsequent genetic analysis. Such samples can provide a genetic baseline that will be useful in management.

Artificial Breeding

Jerry C. Haigh and Steve J. Waldhalm

Estrus Control

Best results were obtained during the breeding season, with progestin devices having been most studied. The CIDR had been most often used, and with proper insertion was associated with a very high retention rate. Prostaglandins have also been used, but can only be expected to be effective during the breeding season. They also have potential to be useful if given either just before or at the time of removal of the progestogen device. PMSG at the time of removal of intravaginal devices is contraindicated in fallow deer (*Dama dama*), but is used in red deer and wapiti (*Cervus elaphus*), although good evidence as to the efficacy of this practice is lacking. When it is used, synchrony is probably tightened, and the onset of estrus may be reduced to about 36 h. Without its use, estrus is expected in about 48 h.

Adjusting the Breeding Season

Forage availability in some deer-farming areas makes it desirable to bring the breeding season forward. This technique can advance the season as much as 3 or 4 weeks, but some question remains as to its true usefulness because semen quality improves during and into the rut. None-the-less it is probably better used in natural mating systems than in artificial insemination dependent systems.

Estrus Detection

All agreed that detection of estrus in female deer could only be effectively achieved if a male was present. Marking (tupping) harnesses are considered effective for small deer, of doubtful use in red deer, and of no use in wapiti. When used, the crayon should be changed regularly. A good teaser is essential. Subordinate males can be suppressed by dominant males in adjacent pens. Yearlings are easy to handle, but may miss some females, and testosterone-treated castrates will work, but can be difficult to handle.

Insemination

Artificial insemination should be seen as a genetic tool, and not as a seasonality tool, especially with expensive semen. Results have been encouraging, especially in white-tailed deer (*Odocoileus virginianus*), with conceptions as high as 85% after intravaginal or intracervical insemination. Fallow and red deer conceptions have been as high as 65% and 60% respectively.

Currently, the numbers of sperm being inseminated are higher than in the cattle industry. For fallow deer, at least 30×10^6 are deemed necessary, and for white-tailed deer, from 30 to 100×10^6 are satisfactory. For reds (wapiti), 50% calving resulted from 12×10^6 spermatozoa.

Laparoscopic intrauterine insemination is considered to be more effective than any other method in fallow and red deer. One reason may be that the insemination occurs after the preovulatory luteinizing hormone (LH) surge, so that the stress of handling is unlikely to have an adverse effect.

The timing of insemination is important, especially in relation to site of insemination. Best results in white-tailed deeer are obtained with intravaginal insemination. It is estimated that estrus starts about 30 h after removal of progestagens, and single inseminations 6 to 30 h after this gave uniform results (67% conception). For fallow deer, best results have been obtained with insemination 65 to 70 h after CIDR removal. For red deer given PMSG, best results are obtained with inseminations 48 to 60 h after CIDR removal. If no PMSG is used, insemination is best in the 65 to 70 h window.

Superovulation

Of the two standard treatment systems, PMSG has frequently proven frustrating and inconsistent, although recent use of specific PMSG antibody has improved results. Follicular stimulating hormone (FSH) has been used with considerable success in commercial red deer programs. Doses have ranged from 20 to 36 mg total in red deer. In 1990, a regime of twice daily injections of purified FSH (4 mg b.i.d. for 4 days) produced 900 embryos and an average of four transferable embryos per red deer donor (range, 0 to 22). These purified FSH products were ineffective in wapiti, and some encouraging results have been achieved using less purified FSH products in conjunction with low dose PMSG at the time of CIDR removal. Osmotic minipumps have also been used with some success. They have the advantage that the animal requires much less handling, but the disadvantage that minor surgery is required for their use and postoperative complications have been seen.

Embryo Collection

It was recognized that stress on donors reduces success. Nonsurgical methods were considered too slow for commercial use except in wapiti. Collection on days 6 to 7 after breeding is standard. There was a consensus that the red deer embryo is hardier than that of cattle. Successes as high as 75% have routinely been obtained with thawed embryos after surgical transfer, although genetic line differences in conception rates are being seen. Oocyte recovery and in vitro fertilization is just beginning.

Referencing

The session ended with a discussion of selection of breeding stock. Both sire and group referencing schemes exist in New Zealand, and selection upon genetic merit is practiced. Artificial insemination should improve and enlarge the data base.

Evolution of the Cervidae

Robert Brown

This group discussed a variety of hypotheses, theories, and speculations concerning evolution. The group decried the lack of definitive information about the evolution of New World and Old World species. Little true genetic work has been done, although some morphological information is available from Dr. Hoffman's work on digestive physiology.

Odocoileus

There seems to be recent information that there is as much similarity between mule deer (*O. hemonius*) and white-tailed deer (*O. virginianus*) as variation within these species. Mule deer and white-tailed deer have similar mitochondrial DNA, which is only passed down the maternal line. Hybridization does occur, but it is not clear how the black-tailed deer (*O. h. columbianus*) fits in. The theory that the mule deer is a hybrid of black-tailed and white-tailed deer was not accepted by this group. It was generally believed that black-tailed deer are probably a subspecies of mule deer, although the discussion group would like more definitive genetic information, and perhaps even a better definition of subspecies.

Roe Deer

There are five or six groups of roe deer (*Capreolus capreolus*), which are all remarkably similar.

Perhaps this is another instance where the definition of subspecies should be challenged. It was agreed that biochemical analyses are more powerful tools in evolutionary genetics than morphological characteristics or even geographic information. The Western European roe population seems to be fairly homogeneous. In Eastern Europe, the Siberian and Chinese roe have microchromosomes which are not found in European roe deer, but which are found in muntjac. Behaviorally, these Eastern European roe deer are much different than their Western European counterparts, often migrating hundreds of miles. This may be due to habitat, however, and may not be an evolutionary characteristic. There is not enough fossil information to determine if Western roe descended from Eastern roe.

Primitive Deer

Likewise, we do not know whether muntjac (*Muntiacus muntjac*) are primitive or where the musk deer and Chinese water deer (*Hydropotes inermis*) fit in. Regression of antlers could be due to habitat rather than evolution, but those species with pronounced canines are generally believed to be more primitive. Similarly the theories of evolution based on "r" and "K" strategies were largely dismissed by this discussion group. Such strategies are believed to be due to more recent changes in habitat, rather than earlier evolution.

Immunology as a Genetic Tool

A blood parasite, babesia, has been found in white-tailed deer in Texas, roe deer in Germany, red deer (*Cervus elaphus*) in Edinburgh, roe deer in Russia, and possibly mule deer in the Western Rocky Mountains of the United States. Ticks are the vectors, and the ticks transmitting the disease are morphologically similar. The blood parasites are serologically similar, but experimental efforts to intertransmit the disease have not been very successful. The question is, did the tick and parasite exist around the world and perhaps adapt as new hosts arrived, or did deer spread the parasite as cervids evolved and migrated around the world? This could offer a secondary method of looking at genetic differences between species of deer.

Another example was a brainworm which is tolerated by the white-tailed deer but which is pathogenic to the mule deer. Liver flukes are tolerated well by the white-tailed deer, but are not tolerated by caribou (*Rangifer*), reindeer (*Rangifer*), mule deer, or axis deer (*Axis axis*) in North America. One must be cautious in cross-reaction studies, however, as many deer, especially sika deer (*Cervus nippon*), may be hybrids. The differing susceptibility of fallow deer (*Dama dama*) and red deer to tuberculosis was discussed, with conflicting evidence.

Parasite loads on deer farms in Texas and in the United Kingdom have often shown differing loads between species pastured together. In Texas, white-tails were found with much higher parasite loads than axis, sika, and fallow in the same pastures. In the UK, red and sika deer seemed less resistant than fallow. Possible differences in resistance to parasites between white-tailed deer and moose (*Alces alces*) in Maine were discussed. Again caution was suggested in interpreting this information, as age of the animals, nutritional conditions, and different feeding strategies could lead to differences in parasite loads.

Other Relationships

The similarities between roe deer and white-tailed deer were discussed. It is believed that roe were the progenators of New World deer, but it could have been the other way around. However, there are no fossil records of roe deer in North America. Literature on roe deer fossils in Eastern Europe is largely not available. Roe were not present in the Pleistocene era, and no one in this group was aware of evidence of the paleontological information on white-tailed deer. We assume South American deer, such as marsh (*Blastocerus dichotomous*) and pampas deer (*Ozotoceros bezoarticus*), branched off from white-tails, but we do not know when.

The giant Irish elk (*Megalocerus giganteus*) was discussed briefly. This species had widespread distribution, and therefore was neither Irish nor an elk. It may not have been the only large deer around at the time.

Dr. Harrington's theories were discussed, especially those of the relationships between red deer and sika deer. Dr. Harrington's paper in the previous symposium (in 1982) was used for discussion. Speakers in this discussion group believed that the Irish sika and red deer could be hybrids. The group questioned the use of behavior, feeding strategies, and "r" and "K" strategies as evidence of evolutionary change. In addition, Norma Chapman has shown large variations in metacarpi in fallow deer, bringing the use of this morphological characteristic into question. The use of immunological information was considered more substantial evidence, although little such information is available.

Antler Development

James M. Suttie and George A. Bubenik

Role of Steroidal Hormones

High androgen levels are required for initiation of pedicle growth and antler mineralization; lowering of androgen levels is responsible for antler casting. Testosterone (the most abundant androgen in peripheral circulation) can be converted to estradiol (by the aromatase pathway) or to dihydrotestosterone (DHT) (by the 5-α reductase pathway). Experimental data indicate that estradiol may be responsible for the antler ossification process and DHT for the desiccation and the death of the velvet. A possible antagonistic action between DHT and estrogens was suggested. Problems in investigating the role of estrogens may occur because some aromatase blockers (used to prevent conversion of testosterone to estrogens) have estrogenic activity. A different ratio between androgens and estrogens in females may be the cause of their diminished capacity to produce antlers or horns. The initiation of pedicle growth by androgens may be not dose-dependent; dose dependency to testosterone is not a necessary prerequisite for the growth of some other secondary sexual characteristics such as human beard or bovine horns. Female deer that were induced to grow antlers never reached the potential of males; although female antlers were grown and cast each year, they never were cleaned of velvet.

The pedicle and the antler must have different sensitivities to androgens, as the pedicle grows under conditions of increasing and the antler under conditions of decreasing androgens. It was speculated that the longterm attachment of dead antler to the pedicle during high androgen levels in the circulation was probably due to suppression of osteoclasts.

Role of "Inductive Periosteum"

The term of "inductive periosteum" was criticized as a poor choice. The periosteum is not an inducer itself, but it may induce "dermal periosteal ossification" leading to the pedicle development. The initiation of pedicle transformation may be under growth factor control, possibly testosterone-primed. Transplanted periosteum induced antler growth when transplanted to heterotopical locations in deer or produced chondroidal tissue under the scalp of a nude mouse. Periosteal tissue from the presumptive pedicle regions exhibited a higher number of receptors than from control regions.

Role of Vitamins

Retinoic acid (RA), a derivative of vitamin A, serves as a signal which provides a positional memory along a proximodistal axis. A deer was injected in the lateral periosteum of growing antler buds with RA in oil about 2 weeks after commencement of regrowth. The antler grew abnormally (laterally), before turning upwards. Discussion took place on the fate of a retinoic acid gradient in tissues with a high blood flow and the dangers of extrapolation on the role of RA from studies of embryonic tissues. It was hypothesized that a second brow tine may develop if a growing antler beam was injected with RA. However, Bubenik was unsuccessful in inducing changes in shape of antlers by RA administration in later developmental phases.

Data were provided indicating that 1,25 dihydroxyvitamin D may be produced in the growing antler tissues, and the role of this factor in growth of mesenchymal tissues and its action on osteoblasts was discussed.

Role of Nerves in Antler Growth

Parasympatethic nerves appear to be involved in the regulation of antler shape and size but not in the process of initiation of antler regeneration. Unilateral denervation of antlers resulted in smaller and less branched antlers than on the control side. Bubenik described the response of a white-tailed buck to a severe antler trauma. This caused a massive hypertrophy not only to the previously slightly damaged antler but also to the contralateral antler as well. The hypertrophy persisted till the death of this deer 6 years later. The pertinent question remains: where is this information stored to "code" for the changes leading to a massive abnormality in antler growth year after year? The most likely answer is the nerves. Nerve stimulation by a cardiac pacemaker led to a distorted antler growth which was most pronounced only 1 year after the stimulation ceased.

Pheromones

Karl V. Miller

Because of the expertise of the majority of workshop participants, much of the discussion on the role(s) of pheromones in cervid behavioral and physiological ecology focused on white-tailed deer (*Odocoileus virginianus*). Additional input was provided on fallow deer (*Dama dama*) chemical ecology. Focal points of discussion during this workshop concentrated around three areas: (1) priming pheromones and the role of flehmen, (2) estrus detection in cervids, and (3) the role of the tarsal gland in white-tailed deer behavior.

Priming Pheromones and the Role of Flehmen

Considerable evidence has been presented recently in the literature on the stag's role in estrus induction and synchronization in several deer species. Participants agreed that the priming mechanism may be chemical in nature, and possibly of urinary origin. Male cervids employ some mechanism to anoint themselves with their own urinary products (wallowing, rub-urination, urine-splashing, etc.). The resulting characteristic rutting odor may serve a dual purpose: an advertisement of dominance status and a mechanism of estrus induction.

Flehmen by male cervids was suggested as a means of assaying a priming pheromone of female origin. The neurologic pathway from the vomeronasal organ (VNO) suggests that its role is physiological and not behavioral. In addition, flehmen is not necessary for estrus detection, may occur before or after copulation, and occurs year-round. Therefore, the suggestion was offered that flehmen and VNO stimulation served as a priming mechanism to enhance the male's reproductive physiology to ensure that both sexes were at reproductive peaks simultaneously.

Estrus Detection in Cervids

Traditional beliefs indicate that urine is the source of a pheromone signaling estrus status in cervids. This belief was based partly on the assumption that flehmen was a mechanism for detection of this pheromone. Recent studies raise doubt on the role of flehmen in estrus detection and the role of urine in signaling estrous status. These studies have implicated the female reproductive tract itself as the primary source of reproductive pheromones and not urine. However, it must be acknowledged that urine may become contaminated with vaginal washing during urination. The role of voided urine in signaling estrus was identified as a particularly important area for future research.

Fallow deer produce copious amounts of cervical mucus during estrus, and fallow bucks show considerable interest in this secretory product. White-tails also produce this mucus during estrus, albeit to a lesser extent. Recent research has determined that odors arising from this material can elicit courtship behavior in males. The synthesis was offered that estrus detection can be accomplished through normal olfactory mechanisms from odors arising from the reproductive tract. Normal olfaction and VNO stimulation during flehmen may act in concert to enhance or intensify the estrus signal.

Tarsal Gland in White-Tailed Deer Behavior

The role of the tarsal gland and rub-urination in white-tailed deer behavior was discussed. The staining on the tarsal gland appears related to dominance and aggression among males. Dominant males urinate frequently on the tarsal hairs during the rut. Young males and subordinate males rub-urinate less frequently, and often lick excess urine from the tarsal hairs. Castrate males rub-urinate infrequently regardless of season. It is unsure whethr the frequency of rub-urination by females is related to dominance status or reproductive state. Ovariectomized females were noted to have lighter-stained tarsal glands than intact females.

The tarsal hairs are underlain with an area of enlarged sebaceous glands. The activity of these sebaceous glands does not change with changes in the reproductive state of the animal. It was suggested that these glands secrete a lipid material that adheres to the tarsal hairs. During rub-urination, this material selectively retains certain fat-soluble material from the urine. These materials, or decompositional products of these materials, perhaps are the origin of the individual odor of the deer.

Differences in the degree of staining of the tarsal among individuals likely result from differing frequencies of rub-urination and/or differences in concentration of urinary constituents, not from differences in secretions from the tarsal gland itself.

Conclusion

Workshop participants agreed that very little is known about the chemical ecology of deer in comparison with other aspects of deer behavioral and physiological ecology. Considerable research is necessary to understand the roles of chemical communication in deer social behavior and reproductive physiology. Many aspects of scent communication will likely have important applications in the management of wild and farmed deer.

Index

Acid rain, 1, 10
Activity patterns, 30, 56–61
Adrenocorticotropin (ACTH), 121, 211, 532
Adrenal gland, 120
Afghanistan, 4, 6
Africa, 95, 114, 556
Age
  determination, 18, 23, 295
  effect on reproduction, 233–236, 293–298
  effect on fat deposits, 233–236, 253–257
  effect on body weight, 343–349
Agriculture, 5–6, 22–28, 37, 541
Alabama, 246–250
Alaska, 5, 7–10, 87, 241–251, 376, 407, 411, 425, 478
Alces alces (see Moose)
Alkaline phosphatase, 536
Androstenedione, 357
Anopthalmia, 520–524
Antiandrogen (cyproterone acetate), 505–509, 533
Antelope, Pronghorn (*Antilocapra americana*), 43, 136, 544, 556
Antioxidant enzymes, 481–482
Antler
  cells, 471–484, 525–530
  cell culture, 511–518, 536
  cycle, 191, 199, 493–498, 499–503, 520–524, 532, 568–571
  growth, 302, 370, 443–446, 471–474, 487–491, 493–498, 499–503, 505–509, 525–530, 533, 535, 587–588
  production, 11, 159–170, 173–178, 222, 225–231, 401
  regeneration, 525–530, 534
  removal, 192
  size, 10, 346, 403
  weight, 201, 221, 222
Appetite cycle, 164–165, 261–274, 282
Argentina, 4, 5, 32, 160
Arizona, 43–46, 69–75, 246–250, 539
Arkansas, 246–250
Arthritis, 101–105
Artificial insemenation, 166, 267, 334–337, 374, 401, 584
Asia, 4, 7, 373, 377
Australia, 4, 94, 113, 164, 168, 189–195, 197–202, 242, 297, 334, 352, 374, 414
Axis axis (see Axis deer)
  A. calamianensis (see Calamian deer)
  A. kuhlii (see Bawean deer)
  A. porcinus (see Hog deer)
Axis deer (*Axis axis*), 4, 6, 7, 93, 115, 154–155, 162, 163, 166, 168, 189–195, 230, 244, 253–257, 261–274, 334–337, 352, 374, 511–518, 532, 565–573, 586

Barasingha (see Swamp deer)
Barking deer (see Muntjac)
Bawean deer (*Axis kuhlii*), 374
Behavior, 82, 339, 189–195
  agonistic, 352, 568
  defensive, 6, 56–61
  dispersal, 24, 30–36, 39
  dominance, 161
  feeding, 411, 463
  hiding, 61–75
  homosexual, 352
  sexual (see also Breeding, Reproduction, Rut), 54, 263, 287, 313–319, 320–325, 331, 352, 353, 360, 375, 568
  territorial, 353
Bioenergetic Models (see also Metabolism), 457
Birth (see also Parturition)
  dates, 271, 293–298, 300–305
  rate, 15–20, 175, 293–298
  weights, 104, 293–298
Bison (*Bison bison*), 107, 385, 544, 555
Black bear (*Ursus americanus*), 556
Black-tailed deer (*Odocoielus hemionus columbianus*), 4, 8, 61, 89, 104, 169, 266, 361, 369, 411, 556–557
Blackbuck antelope (*Antilope cervicapra*), 244
Blastocerus dichotomous (see Marsh deer)
Bluetongue, 154, 158
Bobcat (*Lynx rufus*), 57
Body, live (see also Carcass)
  condition, 11, 48–55, 63, 67, 82, 84, 85, 283, 293–298, 411–413
  composition, 92, 233–236, 423–424, 456, 460
  weight, 192, 199–200, 203–209, 217, 221, 261–274, 283, 293–298, 301, 306–311, 338–342, 343–349, 362, 370, 403, 425, 431–432, 443–446, 478
  fat, 10, 233–236, 262, 410, 453–457, 460
Bolivia, 5
Bone marrow fat, 460
Bovidae (see Cattle)
Breeding season (see also Reproduction, Rut), 198–199, 261–262, 300–305, 313–319, 320–325, 352
Brocket deer (*Mazama*), 566
  Brown (*M. gouazoubira*), 5, 371, 377
  Dwarf (*M. chunyi*), 5, 371
  Little red (*M. rufina*), 5, 371
  Red (*M. americana*), 5, 6, 371, 377
Bromocriptine, 269, 447
Burma, 4, 6

Calamian deer (*Axis calamianensis*), 4, 374
Calcitonin, 511

California, 78, 190, 246–250, 547, 548, 556, 582
Calving (see Birth, Parturition)
Canada, 4, 5–8, 10, 101–105, 117, 160–162, 174, 369, 370, 376
Capreolus capreolus (see Roe deer)
Capture
 myopathy, 101–105, 190
 techniques, 23
Carcass
 compostion, 18, 175, 192, 194, 200–201, 221, 225–231, 238–243, 253–257, 432
 weight, 48–55, 222, 225–231, 432
Caribou (*Rangifer*) (see also Reindeer), 5, 8–11, 456, 533, 555, 566
Carrying capacity, 84, 86, 577–579
Castration, 169, 193, 238–243, 266–267, 446, 493–498, 505–509
Cattle (*Bovidae*), 162, 263, 324, 327, 332, 456, 584
 and deer densities, 6, 80
 and deer disease, 94, 107, 127
Central America, 4, 5, 165
Cervus
 C. albirostris (scc Thorold's dccr)
 C. canadensis (see Wapiti)
 C. duvauceli (see Swamp deer)
 C. elaphus (see Red deer)
 C. e. bactrianus, 4
 C. e. nannodes (see Elk, Tule)
 C. e. nelsoni (see Elk, Rocky Mountain)
 C. e. roosevelti (see Elk, Roosevelt)
 C. eldi (see Eld's deer)
 C. nippon (see Sika deer)
 C. timorensis (see Rusa deer)
 C. unicolor (see Sambar deer)
Chemical immobilization, 105, 136–140, 153, 321, 334, 581–582
Chile, 5, 32–36
China, 4–6, 160, 161, 164–166, 267, 359, 373, 377, 414, 450–452, 586
Chinese water deer (*Hydropotes inermis*), 4, 567, 568
Chital (see Axis deer)
Chronic wasting disease, 56–61
Circadian activity patterns, 104
Circannual cycles (see also Antler, Reproduction Cycles), 261–274, 306
Climate, 15–20
Coat (see Pelage)
Cobalt, 178
Collared peccary (*Tayassu tajacu*), 556
Colorado, 104, 246–250, 255, 542, 545
Commercialization, 11, 539–550, 554–559, 579
Conception, 67, 85, 361, 370
Condition indices (see Body, live)
Copper, 113–117, 177–178, 450–452
Cortisol, 211–215, 220, 532
Coues white-tailed deer (*O. v. couesi*), 80
Coyote (*Canis latrans*), 556
Cyproterone acetate, 505–509, 553
Czechoslovakia, 7

Dama
 D. dama (see Fallow deer)
 D. mesopotamica (see Mesopotamian fallow deer)
Death rate (see Mortality)
Dehydration, 101–105
Deleware, 246–250
Desert bighorn sheep (*Ovis canadensis mexicana*), 43
Desert mule deer (*Odocoileus hemionus crooki*), 43–46
Diarrhea, 178
Diet (see also Food habits)
 composition, 10, 424–428
 protein and energy intake, 456
Digestion and Digestability, 412–413, 424–428, 436–441, 459, 464, 466
Dihydrotestosterone, 587
Disease (see also specific diseases), 34, 93–98, 101–105, 130–135, 149–156, 177, 201–202, 450–455, 577, 586
 prevention, 107–111, 128, 151, 166–167
 suseptibility/resistance, 113–117, 120–125, 126–128, 141–148, 150
Dispersal, 22–28, 79
DNA, 367–378, 384–389, 582
Drownings, 43–46
Duhl's deer (see Bawean deer)

Ecuador, 5
Elaphurus davidiensis (see Pere David's deer)
 E. cephalophus (see Tufted deer)
Eld's deer (*Cervus eldi*), 4, 6, 94, 155, 374, 566
Electroejaculation, 334, 355, 401
Electrophoresis, 367–378, 390–396, 398, 399, 402, 409
Elk (*Cervus elaphus canadensis*) (see Wapiti)
 Rocky Mountain (*C. e. nelsoni*), 107–111, 116
 Roosevelt (*C. e. roosevelti*), 77
 Tule (*C. e. nannodes*), 78, 556
Embryo
 counts, 38, 65, 390–396
 recovery, 356–357, 579, 585
Endangered deer, 4–5, 579
Endorphins, 474
Energy (see also Metabolism, Nutrition)
 intake, 407–415, 438, 456
 requirements, 163–165, 262, 266, 418–422, 425, 577
England, 4, 93, 162, 180–187, 373, 401
Enteritis, 101–105
Environment, 3–11, 15–20, 22–23, 183, 262–271, 565–573
Enzootic ataxia, 113–117, 177–178, 455
Epidermal growth factor (EGF), 482
Estrogens, 263–264, 289, 320, 357, 473, 505, 511–518
Estrous cycle, 55, 193, 221, 263–274, 279–283, 285–290, 300–305, 313–319, 320–325, 354
Estrus, 221, 263, 320–325, 263, 327–332, 390–396, 584–585, 588
 synchronization, 314, 334–337, 351, 355–357, 577, 584
Europe, 4, 5, 10, 160, 222, 374, 376–377, 548, 554–559, 582
Evolution, 585–586

Facial exema, 113–117
Fading elk syndrome, 113–117
Fallow deer (*Dama dama*), 4, 88, 113–117, 152, 162–169, 180–187, 189–195, 211–215, 217, 228, 238–243, 245, 251, 253–257, 263–264, 270–271, 286, 293–298, 319, 336, 351, 354, 357, 360, 368, 369, 374, 385–387, 400, 453–455, 460, 462, 463–464, 511–518, 525, 533, 557, 566–571, 579, 584, 586, 588
Farming of deer (see also Management), 159–170, 173–178, 180–187, 189–195, 197–202, 203–209, 217, 219, 222, 225–231, 233–236, 238–243, 244–252, 253–257, 274, 300–305, 338–342, 357, 459, 539–550, 554–559
Fat (see also Body and Carcass composition)
 content, 225–231, 233–236
 reserves, 10, 233–236, 262, 456
 deposition, 253–257, 460
Fatty acids, 460
Fecundity (see also Birth rate), 15–20, 76, 83, 343
Feral dogs, 6, 57
Fertility (see also Conception, Reproduction), 52, 233–236, 390–396
Fetus, 390

age estimation, 63–68, 313–319, 358
counts, 49, 85
growth rate, 63–68
Fibroblast growth factor (aFGF/bFGF), 480–481
Fiji, 4
Finland, 4, 7, 364, 461, 557, 571
Florida, 76, 246–250, 332, 346, 375
Florida panther (*Felis concolor*), 76
Flushing, 395, 412
Follicle Stimulating Hormone (FSH), 261–274, 356, 357, 532, 570–573
Food
habits, 30, 77, 423–428
intake, 15–20, 262, 279–283, 410–415, 423–428, 436–441, 443, 459, 466
Foot and mouth disease, 154, 155
Foot rot, 101–105
Forage quality and quantity, 87, 411, 450–452, 464, 467
Forestry, 7–9, 28, 83
Fox (*Vulpes vulpes*), 193, 447
France, 127, 558

Gaits, 465
Game ranching, 67, 244–252, 253–257, 539–550, 554–559
Genetics, 7, 16–17, 176, 343–349, 367–378, 384–389, 390–396, 398–403, 465, 582–583
Georgia, 246–250, 327
Germany, 32–33, 374, 400, 547, 586
Gestation (see also Pregnancy), 65, 198, 221, 267, 358
Global warming, 15, 20
Goats (*Capra*), 86, 127, 320, 459, 466, 555, 558
Goiter, 102
Gonadotrophin Releasing Hormone (GnRH), 263–266, 285–290, 320–325, 474–477, 500, 532, 570
Gray fox (*Pseudalopex griseus*), 34
Greenhouse effect, 10
Grizzly bear (*Ursus arctos*), 555
Growth
compensatory, 18, 413–414
factors, 282, 471–474
hormone (GH), 410, 474–477, 505–509
patterns, 203–209, 261–274, 343–349, 407–415, 487–491
rates, 175, 221, 99–200, 238–243, 255, 302, 338–342
Guemal (see Huemul)

Habitat
preferences, 22–28, 30, 69–75, 77, 84, 343–348, 360, 465
loss, 5–6, 9–11, 541
conditions, 87, 89, 395, 403, 457, 577
Hartebeest (*Alcelaphus buselaphus*), 95
Harvest, of wild deer (see also Hunting), 37–42, 48–55, 253–257
Hawaii, 4, 190, 246–250
Health (see Disease)
Heterozygosity (see also Genetics), 343–348, 390–396, 398, 582
Hippocamelus
H. antisensis (see Huemul, Peruvian)
H. bisculcus (see Huemul, Chilean)
Hiding
behavior, 56–61, 61–75
cover, 22–28
habitat, 61–75
Histology, 130–135, 136–146, 141–148, 156, 338–342, 471–484, 511–518, 525–530
Hog deer (*Axis porcinus*), 154, 155, 189, 273, 374, 566–567
Home range, 24, 57
Hormones (see specific hormones)
Huemul (*Hippocamelus*)
Chilean (*H. bisulcus*), 5, 377, 566
Peruvian (*H. antisensis*), 5, 6, 377, 566
Hungary, 32, 37–42, 222, 453–455, 460
Hunting
fee, 78, 250, 539–550, 554–559
pressures, 6, 11, 24–28, 34
ranches, 67, 244–252, 253–257, 539–550, 554–559
Hybridization (see also Genetics), 7, 174, 367–378, 384–389, 400–402, 465, 556, 582–583
Hydropotes inermis (see Chinese water deer)

Ibex (*Capra ibex*), 556
Iceland, 5
Idaho, 8, 108, 246–250
Illinois, 22–28, 246–250
Immunity (see also Disease), 120–125, 141–148, 150, 577
India, 4–6, 190, 261, 334, 347, 377
Indiana, 246–250
Indonesia, 4, 374
Insulin, 410, 474
Insulin-like growth factors (IGF1/IGF2), 474–480, 505–509, 511
Intake (see Food)
Introductions, 30–36
Iowa, 24, 60, 246–250, 375
Irac, 374
Iran, 4, 374
Ireland, 7, 369
Irish elk (*Megaloceros*), 586
Isozyme, 367–378
Israel, 374
Italy, 558

Japan, 4
Jimela topi (*Damaliscus lunatus*), 96
Jordan, 374

K-selected species, 76, 162, 166, 168, 368, 586
Kansas, 246–250
Kentucky, 246–250
Ketamine hydrochloride, 153, 334
Kidney Fat Index (KFI), 253–257, 390–396
Korea, 5

Labrador, 10, 11
Lactate dehydrogenese (LDH), 211–215, 220
Lactation, 233–236, 262, 263, 267, 271, 279–283, 301–305, 319, 407–415, 456
Lactic acid, 220, 450–452
Laparoscopy/Laparotomy, 351, 357, 401, 584
Leather, 225–231
Legal issues/Legislation, 7, 160, 244–252, 557–559
Leuteinizing Hormone (LH), 261–274, 285–290, 320–325, 351, 357, 499–503, 505–509, 532, 570–573, 584
Leutinizing Hormone Releasing Hormone (LHRH), 570–573
Live weight (see Body weight)
Louisiana, 48, 246–250
Lungworm, 103, 114, 178

Maine, 246–250, 586
Malignant catarrhal fever, 93–98, 113–117, 120
Management (see also Farming)
farmed deer, 120–125, 159–170, 173–178, 180–187, 189–195, 197–202, 203–209, 217–222, 453–455
wild deer, 20–28, 30–36, 37–42, 43–46, 48–55, 67–68, 74, 78, 79, 80, 81, 86, 107–111, 253–257, 458
Manchuria, 4, 5
Marsh deer (*Blastocerus dichotomous*), 5, 377, 566, 586
Maryland, 7, 246–250, 557
Massachusetts, 246–250
Mating (see Behavior, Reproduction)

Maximum sustained yield (MSY), 48–49
Mazama (see Brocket deer)
M. americana (see Brocket deer, Red)
M. chunyi (see Brocket deer, Dwarf)
M. gouazoubira (see Brocket deer, Brown)
M. rufina (see Brocket deer, Little red)
Meat (see Venison)
Melatonin, 203–209, 261–274, 279–283, 300–305, 306–311, 313–319, 354, 355, 364, 407, 430, 478–480, 523, 569–573
Mesopotamian fallow deer (*Dama mesopotamica*), 369, 374, 400
Metabolism (see also Energy, Nutrition), 266, 398, 418–422, 457
Mexico, 5, 80
Michigan, 246–250
Migration, 22–28, 30–36, 458
Milk (see also Lactation), 373, 456
Minnesota, 17, 245, 246–250, 346, 375
Mississippi, 48–55, 246–250, 541
Missouri, 246–251
Molt (see Pelage)
Molybdenum, 178
Monoclonal antibodies, 141–148
Mongolia, 4, 5
Montana, 108, 111, 246–250, 375, 557
Moose (*Alces alces*), 4, 6, 7, 10, 15–20, 83, 94, 162–163, 368, 369, 376–377, 387, 418, 423, 425, 427, 450–455, 457, 533, 555, 566
Mortality, 15–20, 22–28, 38–42, 43–46, 60, 76, 79, 101–105, 111, 217, 271, 343, 390–396
Moschus (see Musk deer)
M. berezovskii (see Musk deer, Berezouski)
M. sibiricus (see Musk deer, Siberian)
M. moschiferus (see Musk deer, Southern)
Movements, 22–28, 56–61, 69–75, 85
Mule deer (*Odocoileus hemonius*), 4, 10, 43–46, 67, 84, 85, 94, 104, 162–164, 169, 186, 251, 255, 369, 375–376, 385–387, 411, 415, 424, 439, 465, 467, 533, 556, 566, 582, 585
Muntiacus (see Muntjac)
M. muntjak (see Muntjac, Indian)
M. reevesi (see Muntjac, Reeve's)
M. crinifrons (see Muntjac, Black)
M. feae (see Muntjac, Fea's)
M. rooseveltorum (see Muntjac, Roosevelt's)
Muntjac (*Muntiacus*), 6, 272, 566–569, 586
Black (*M. crinifrons*), 4, 377
Fea's (*M. feae*), 4, 6, 377
Indian (*M. muntjak*), 4, 377, 387
Reeve's (*M. reevesi*), 4, 154, 155, 359
Roosevelt's (*M. rooseveltorum*), 4, 377
Musk, 160
Musk deer (*Moschas*), 10, 11, 160, 163, 566
Berezovski (*M. berezoskii*), 5
siberian (*M. sibiricus*), 5
southern (*M. moschiferus*), 5
Muskoxen (*Ovibus moschatus*), 412, 440
Myocarditis, 101–105
Myopathy (see also Mortality)
post-capture, 101–105, 211, 430

Natality (see Fecundity)
Nebraska, 246–250
Nepal, 4, 82, 190, 261, 567
Nephritis, 101–105
Nerves, 588
Nevada, 246–250
New Caledonia, 113
Newfoundland, 555
New Guinea, 4, 81
New Hampshire, 246–250
New Jersey, 246–250, 543
New Mexico, 6, 246–250, 540
New York, 246–250, 346, 399, 542
New Zealand, 4, 5, 7, 30, 94, 113–117, 120–125, 126–128, 136–140, 160–162, 164–166, 173–178, 203–209, 217–219, 225–231, 242, 244, 255, 297, 338, 354, 358, 369–374, 400, 402, 414, 430–434, 463, 555, 557, 580, 585
North Carolina, 246–250
North Dakota, 246–250
North America, 4, 5, 6, 10, 11, 113–117, 160, 371, 373, 390, 407–415, 554–559, 582
Norway, 6, 7, 15–20, 186, 233–236, 371, 407, 423–428, 580
Nutrition, 15–20, 63, 163–165, 218, 270, 282, 338, 407–415, 418–422, 423–428, 429–434, 436–441, 450–452, 453–455, 456–467, 567

Odocoileus
O. hemionus (see Mule deer)
O. h. columbianus (see Black-tailed deer)
O. h. crooki (see Desert mule deer)
O. h. cerroseusis, 4
O. h. sithensis (see Sitka black-tailed deer)
O. virginianus (see White-tailed deer)
O. v. couesi (see Coues white-tailed deer)
O. v. leucurus, 4
Ohio, 246–250
Oklahoma, 246–250
Oregon, 8, 84, 246–250
Ovariectomy, 290, 328, 493–498
Ozotoceros bezoarticus (see Pampas deer)
Ovulation, 263, 320–325, 351, 565–573

Pakistan, 5
Pampas deer (*Ozotoceros bezoarticus*), 3, 5, 566, 586
Paraguay, 5
Parasites, 30, 103, 113–117, 149, 178, 201, 555, 577, 586
Parathyroid hormone, 511
Paratuberculosis, 151
Parturition (see also Birth), 65, 293–298
Pastures, 203–209, 218, 283, 463
Pastuerella, 152
Pelage, 191, 267, 269, 273, 306–311, 355, 443–448, 446
Pennsylvania, 246–250
Pere David's deer (*Elaphurus davidiensis*), 4, 93, 94, 97, 113–117, 141, 166, 168, 263–269, 285–290, 320–325, 368, 373, 511, 518, 566, 568, 579, 582
X red deer hybrids, 369, 374, 401, 402
Peru, 4, 5, 573
Pheromone, 55, 202, 318, 327–332, 588
Philippines, 5
Photoperiod, 261–274, 279, 300–305, 478, 567–573
Pineal gland (see also Melatonin), 270, 300, 306, 523, 571–573
Pinealectomy, 303
Pituitary, 324, 565
Placentation, 362
Platelet-Derived Growth Factor (PDGF), 482
Poaching, 30–36, 79, 81, 546
Poland, 10, 534
PMSG, 334–337, 356, 357, 584
Pollution, 9–11
Pneumonia, 101–105
Polymorphism, 367–378, 403
Population
structure, 87

growth, 77
density, 83, 360
characteristics, 181
status, 3–11
demography, 15–20, 22–28, 37–42
models, 18–19, 23, 27, 37–42, 48
dynamics, 22–28, 48–55, 77, 79, 89, 348
Potassium, 429–434
Predation, 3, 6, 19, 34, 56–61, 74, 76, 87, 193, 201
Pregnancy (see also Gestation), 266, 358, 391, 565–573
rates, 38, 76, 316–319
diagnosis, 363
Products, 11, 225–231, 233–236, 238–243, 244–252, 253–257
Progesterone, 193, 263–265, 280, 286–290, 301, 306–311, 320–325, 334–337, 351, 355, 357, 303, 584
Prolactin, 208, 268–274, 306-311, 319, 447, 476, 499–503, 532
Prostaglandin, 286, 351, 355, 356, 483
Protein intake, 407–415, 456
Puberty, 293, 300–305, 306–311, 338–342, 493
Pudu (*Pudu*)
northern (*P. mephistophiles*), 5, 377, 566
southern (*P. pudu*), 5, 377, 566
Pudu
P. mephistophiles (see Pudu, northern)
P. pudu (see Pudu, southern)
Puma (*Felis concolor*), 34

Quebec, 10, 11

R-selected species, 162, 166, 168, 368, 586
Radiotelemetry, 23, 56–61, 69–75, 76, 79, 399, 458
Rangifer (see Caribou and Reindeer)
Recruitment, 22–28, 38, 48, 77, 403
Red deer (*Cervus elaphus*), 4, 6, 7, 15–20, 30–36, 37–42, 61, 94, 97, 104, 113–117, 120–125, 126–128, 130–135, 136–140, 141, 149, 161–165, 167–168, 180–187, 189–195, 197, 203–209, 217–220, 222, 225–231, 245, 255, 261–274, 279–283, 293, 297, 300–305, 306–311, 313–319, 336, 338–342, 357, 358, 363, 368, 369–373, 402, 407, 423, 427, 429–434, 443–446, 450–455, 459, 460, 463, 481–484, 487–491, 509, 533, 534, 557, 566–673, 579, 582, 584, 586
X Pere David hybrids, 267, 369, 374, 401
X sika hybrids, 7, 369
X wapiti hybrids, 165, 173–178, 369
Reindeer (*Rangifer*) (see also Caribou), 5, 10, 94, 167, 176, 209, 233–236, 262, 267, 293, 297, 336, 359, 364, 368, 369, 370, 376, 407–415, 427, 446, 461, 478, 493–498, 525, 567, 586
Reproduction, 15–20, 67, 76, 85, 233–236, 261–274, 279–283, 293–298, 306–311, 320–325, 327–332, 338–342, 351–364, 370, 390–396, 429–434, 443–448, 524
cycles, 165–166, 191–192, 198–199, 217, 285–290, 300–305, 313–319, 411–413, 478, 499–503, 565–573
Rhode Island, 246–250
Roan antelope (*Hippotragus equinus*), 96
Roe deer (*Capreolus capreolus*), 4, 6, 7, 10, 30, 94, 263, 267, 269, 286, 353, 368, 369, 377, 418, 423–428, 447, 450–455, 459, 460, 495–503, 505–509, 525, 533, 536, 556, 566–568, 585–586
Rumen
capacity, 424–428
retention time, 424–428, 436–441, 466
Rusa deer (*Cervus timorensis*), 4, 81, 91, 113–117, 189, 197–202, 369, 374, 557, 566
Russia, 160, 162, 167
Rut, 54, 266, 331, 338
Ryegrass staggers, 113–117, 177

Sambar deer (*Cervus unicolor*), 4, 113–117, 154, 155, 189, 273, 369, 374, 557, 566, 567
Scandinavia, 4, 5, 7, 376
Scotland, 5, 15–20, 164, 220, 271, 279, 338, 568, 586
Scours, 113
Seasonality (see Antler and Reproduction Cycles)
Selenium, 481–482
Semen, 192, 334–337
Sexual dimorphism, 343–349, 367–378, 403
Sheep, 95, 114, 263, 266, 271, 282, 300, 311, 320, 447, 450–452, 459, 555
Siberia, 4, 10, 586
Sika deer (*Cervus nippon*), 4, 7, 94, 113–117, 149, 163, 167, 221, 245, 269, 273, 368, 373, 385–387, 450–452, 464, 533, 535, 555–557, 566–576, 586
X red deer hybrids, 7, 165, 369, 373, 402
Sitka black-tailed deer (*Odocoileus hemionus sithensis*), 8, 87
Social structure, 24, 149, 399, 430–431
Sodium, 430–431
Somatostatin, 474–477
South Africa, 548
South Carolina, 246–250, 343–349, 375, 390, 398
South Dakota, 246–250
South America, 4, 5, 165, 374, 377, 526
Southeast Asia, 7, 374, 377
Soviet Union, 4, 5, 8, 377, 535, 586
Spermatogenesis, 334–337, 338–342, 568
Spotted deer (see Sika deer)
Sri Lanka, 4, 113, 334
Stag effect, 318
Starvation, 434
Stress, 211–215, 220, 433
and disease, 117, 120–125, 126–128, 178, 201
Sulphur, 178
Superovulation, 356, 357, 400, 579, 585
Supplementary feeding, 163–165, 178, 450–455, 458
Survival (see also Mortality, Natality), 15–20, 22–28, 38, 88, 300, 423–428
Swamp deer (*Cervus duvaucell*), 4, 94, 374, 454, 566
Sweden, 6, 7, 83, 353, 423
Svalbard reindeer (*Rangifer tarandus platyrhynchus*), 153, 376
Syria, 374

Taiwan, 4
Tennessee, 246–250
Testes, 191, 261, 338, 370, 443, 565
Testosterone, 191, 263, 273, 443, 473, 478, 497, 499–501, 505–509, 511–518, 532, 533, 573, 587
Texas, 6, 7, 85, 86, 90, 190, 245–250, 253–257, 346, 369, 375, 464, 547, 557, 586
Thailand, 4, 6
Thamin (see Eld's deer)
Thorold's deer (*Cervus albirostris*), 4, 6, 374
Threatened species, 3–5
Thyroid, 443–446
Thyroidectomy, 443
Thyrotropin (TSH), 447

Thyrotropin Releasing Hormone (TRH), 447
Thyroxine (T4), 462, 474, 511
Tibet, 4, 5, 6
Ticks, 586
Timor deer (see Rusa deer)
Transforming growth factors (TGF a/ TGF b), 482–483
Translocation, 78
Trauma, effects of, 101–105, 190, 201
Triiodothyronine (T3), 268, 443–446
Tuberculosis, 113, 120–125, 130–135, 154–156, 577, 586
Tufted deer (*Elaphodus cephalophus*), 4, 374
Tumor necrosis factor (TNF), 482

Ultrasonography, 307, 313–319, 334–337, 358
United Kingdom, 4, 7, 113, 160–161, 164, 191, 244, 262, 371, 374 377, 586
United States, 4, 7, 10, 15–20, 22–28, 160, 163, 165–168, 244–252, 369, 373, 374, 539–540, 577, 586
Urea Nitrogen, 461
Urine, 461, 589
Utah, 246–250, 467

Vaccination, 107–111, 128, 157
Venison (see also Carcass)
  nutritive value, 229
  regulations, 250
  production (see also Farming), 34–35, 159–170, 173–178, 189–195, 203–209, 222, 225–231
  quality, 228, 253–257
  wild-shot, 253–257, 547
Venzuela, 5
Vermont, 245–250
Vietnam, 4, 6
Virginia, 56–61, 246–250
Vitamin A (Retinoic acid), 511–587
Vitamin D (1, 25 dihydroxy), 511–518, 588
Vocalization, 82, 318, 369
Volatile fatty acids (VFA), 221, 450–452
Voluntary food intake (VFI) (see also Food), 163–165, 266, 269, 272

Wapiti (*Cervus elaphus canadensis*), 4, 69–75, 101–105, 113–117, 160, 163, 173–178, 229, 251, 336, 368, 369, 373, 385–387, 411, 418, 540, 555, 557, 565–573, 586, 580, 584
  X red deer hybrids, 7, 113–117, 165, 173–178, 230, 369, 373, 402
  X Sika deer hybrids, 387
Washington, 8, 77, 89, 246–250, 361
West Virginia, 246–250
White-tailed deer (*Odocoileus virginianus*), 4, 6, 7, 10, 17, 22–28, 48–55, 56–61, 63–68, 76, 79, 86, 94, 104, 113, 136, 161–163, 165–166, 168, 186, 251, 255, 262, 263, 269, 270, 286, 293, 297, 327–332, 336, 343–349, 355, 356, 362, 369, 371, 374–375, 385–389, 390–398, 399, 415, 418–421, 436–441, 447, 458, 464, 465, 466, 487, 520–524, 526, 533, 555–557, 565–573, 580–581, 584
  X mule deer hybrids, 369, 374–375, 385–387, 465, 556, 582–583, 585, 588
Wildebeest (*Connochaetes*), 95
Winter Death Syndrome, 165
Wisconsin, 246–250, 403, 458
Wolf (*Canis lupus*), 19, 87
Wyoming, 101–105, 107–111, 246–251, 555, 559

Xylazine hydrochloride, 136–140, 321, 334

Yersinosis, 113–117, 120, 126–128